PROGRESS IN BRAIN RESEARCH

VOLUME 169

ESSENCE OF MEMORY

Other volumes in PROGRESS IN BRAIN RESEARCH

Volume 131: Concepts and Challenges in Retinal Biology, by H. Kolb, H. Ripps and S. Wu (Eds.) – 2001, ISBN 0-444-50677-2.
Volume 132: Glial Cell Function, by B. Castellano López and M. Nieto-Sampedro (Eds.) – 2001, ISBN 0-444-50508-3.
Volume 133: The Maternal Brain. Neurobiological and Neuroendocrine Adaptation and Disorders in Pregnancy and Post Partum, by J.A. Russell, A.J. Douglas, R.J. Windle and C.D. Ingram (Eds.) – 2001, ISBN 0-444-50548-2.
Volume 134: Vision: From Neurons to Cognition, by C. Casanova and M. Ptito (Eds.) – 2001, ISBN 0-444-50586-5.
Volume 135: Do Seizures Damage the Brain, by A. Pitkänen and T. Sutula (Eds.) – 2002, ISBN 0-444-50814-7.
Volume 136: Changing Views of Cajal's Neuron, by E.C. Azmitia, J. DeFelipe, E.G. Jones, P. Rakic and C.E. Ribak (Eds.) – 2002, ISBN 0-444-50815-5.
Volume 137: Spinal Cord Trauma: Regeneration, Neural Repair and Functional Recovery, by L. McKerracher, G. Doucet and S. Rossignol (Eds.) – 2002, ISBN 0-444-50817-1.
Volume 138: Plasticity in the Adult Brain: From Genes to Neurotherapy, by M.A. Hofman, G.J. Boer, A.J.G.D. Holtmaat, E.J.W. Van Someren, J. Verhaagen and D.F. Swaab (Eds.) – 2002, ISBN 0-444-50981-X.
Volume 139: Vasopressin and Oxytocin: From Genes to Clinical Applications, by D. Poulain, S. Oliet and D. Theodosis (Eds.) – 2002, ISBN 0-444-50982-8.
Volume 140: The Brain's Eye, by J. Hyönä, D.P. Munoz, W. Heide and R. Radach (Eds.) – 2002, ISBN 0-444-51097-4.
Volume 141: Gonadotropin-Releasing Hormone: Molecules and Receptors, by I.S. Parhar (Ed.) – 2002, ISBN 0-444-50979-8.
Volume 142: Neural Control of Space Coding, and Action Production, by C. Prablanc, D. Pélisson and Y. Rossetti (Eds.) – 2003, ISBN 0-444-509771.
Volume 143: Brain Mechanisms for the Integration of Posture and Movement, by S. Mori, D.G. Stuart and M. Wiesendanger (Eds.) – 2004, ISBN 0-444-513892.
Volume 144: The Roots of Visual Awareness, by C.A. Heywood, A.D. Milner and C. Blakemore (Eds.) – 2004, ISBN 0-444-50978-X.
Volume 145: Acetylcholine in the Cerebral Cortex, by L. Descarries, K. Krnjević and M. Steriade (Eds.) – 2004, ISBN 0-444-51125-3.
Volume 146: NGF and Related Molecules in Health and Disease, by L. Aloe and L. Calzà (Eds.) – 2004, ISBN 0-444-51472-4.
Volume 147: Development, Dynamics and Pathology of Neuronal Networks: From Molecules to Functional Circuits, by J. Van Pelt, M. Kamermans, C.N. Levelt, A. Van Ooyen, G.J.A. Ramakers and P.R. Roelfsema (Eds.) – 2005, ISBN 0-444-51663-8.
Volume 148: Creating Coordination in the Cerebellum, by C.I. De Zeeuw and F. Cicirata (Eds.) – 2005, ISBN 0-444-51754-5.
Volume 149: Cortical Function: A View from the Thalamus, by V.A. Casagrande, R.W. Guillery and S.M. Sherman (Eds.) – 2005, ISBN 0-444-51679-4.
Volume 150: The Boundaries of Consciousness: Neurobiology and Neuropathology, by Steven Laureys (Ed.) – 2005, ISBN 0-444-51851-7.
Volume 151: Neuroanatomy of the Oculomotor System, by J.A. Büttner-Ennever (Ed.) – 2006, ISBN 0-444-51696-4.
Volume 152: Autonomic Dysfunction after Spinal Cord Injury, by L.C. Weaver and C. Polosa (Eds.) – 2006, ISBN 0-444-51925-4.
Volume 153: Hypothalamic Integration of Energy Metabolism, by A. Kalsbeek, E. Fliers, M.A. Hofman, D.F. Swaab, E.J.W. Van Someren and R.M. Buijs (Eds.) – 2006, ISBN 978-0-444-52261-0.
Volume 154: Visual Perception, Part 1, Fundamentals of Vision: Low and Mid-Level Processes in Perception, by S. Martinez-Conde, S.L. Macknik, L.M. Martinez, J.M. Alonso and P.U. Tse (Eds.) – 2006, ISBN 978-0-444-52966-4.
Volume 155: Visual Perception, Part 2, Fundamentals of Awareness, Multi-Sensory Integration and High-Order Perception, by S. Martinez-Conde, S.L. Macknik, L.M. Martinez, J.M. Alonso and P.U. Tse (Eds.) – 2006, ISBN 978-0-444-51927-6.
Volume 156: Understanding Emotions, by S. Anders, G. Ende, M. Junghofer, J. Kissler and D. Wildgruber (Eds.) – 2006, ISBN 978-0-444-52182-8.
Volume 157: Reprogramming of the Brain, by A.R. Møller (Ed.) – 2006, ISBN 978-0-444-51602-2.
Volume 158: Functional Genomics and Proteomics in the Clinical Neurosciences, by S.E. Hemby and S. Bahn (Eds.) – 2006, ISBN 978-0-444-51853-8.
Volume 159: Event-Related Dynamics of Brain Oscillations, by C. Neuper and W. Klimesch (Eds.) – 2006, ISBN 978-0-444-52183-5.
Volume 160: GABA and the Basal Ganglia: From Molecules to Systems, by J.M. Tepper, E.D. Abercrombie and J.P. Bolam (Eds.) – 2007, ISBN 978-0-444-52184-2.
Volume 161: Neurotrauma: New Insights into Pathology and Treatment, by J.T. Weber and A.I.R. Maas (Eds.) – 2007, ISBN 978-0-444-53017-2.
Volume 162: Neurobiology of Hyperthermia, by H.S. Sharma (Ed.) – 2007, ISBN 978-0-444-51926-9.
Volume 163: The Dentate Gyrus: A Comprehensive Guide to Structure, Function, and Clinical Implications, by H.E. Scharfman (Ed.) – 2007, ISBN 978-0-444-53015-8.
Volume 164: From Action to Cognition, by C. von Hofsten and K. Rosander (Eds.) – 2007, ISBN 978-0-444-53016-5.
Volume 165: Computational Neuroscience: Theoretical Insights into Brain Function, by P. Cisek, T. Drew and J.F. Kalaska (Eds.) – 2007, ISBN 978-0-444-52823-0.
Volume 166: Tinnitus: Pathophysiology and Treatment, by B. Langguth, G. Hajak, T. Kleinjung, A. Cacace and A.R. Møller (Eds.) – 2007, ISBN 978-0-444-53167-4.
Volume 167: Stress Hormones and Post Traumatic Stress Disorder: Basic Studies and Clinical Perspectives, by E.R. de Kloet, M.S. Oitzl and E. Vermetten (Eds.) – 2008, ISBN 978-0-444-53140-7.
Volume 168: Models of Brain and Mind: Physical, Computational and Psychological Approaches, by R. Banerjee and B.K. Chakrabarti (Eds.) – 2008, ISBN 978-0-444-53050-9.

PROGRESS IN BRAIN RESEARCH

VOLUME 169

ESSENCE OF MEMORY

EDITED BY

WAYNE S. SOSSIN
Montreal Neurological Institute, McGill University, Montreal, QC, Canada

JEAN-CLAUDE LACAILLE
Département de Physiologie, Université de Montréal, Montreal, QC, Canada

VINCENT F. CASTELLUCCI
Département de Physiologie, Université de Montréal, Montreal, QC, Canada

SYLVIE BELLEVILLE
Centre de Recherche, Institut Universitaire de Gériatrie de Montréal, Montreal, QC, Canada

ELSEVIER

AMSTERDAM – BOSTON – HEIDELBERG – LONDON – NEW YORK – OXFORD
PARIS – SAN DIEGO – SAN FRANCISCO – SINGAPORE – SYDNEY – TOKYO

Elsevier
Radarweg 29, PO Box 211, 1000 AE Amsterdam, The Netherlands
Linacre House, Jordan Hill, Oxford OX2 8DP, UK

First edition 2008

Copyright © 2008 Elsevier B.V. All rights reserved

No part of this publication may be reproduced, stored in a retrieval system
or transmitted in any form or by any means electronic, mechanical, photocopying,
recording or otherwise without the prior written permission of the publisher

Permissions may be sought directly from Elsevier's Science & Technology Rights
Department in Oxford, UK: phone (+44) (0) 1865 843830; fax (+44) (0) 1865 853333;
email: permissions@elsevier.com. Alternatively you can submit your request online by
visiting the Elsevier web site at http://www.elsevier.com/locate/permissions, and selecting
Obtaining permission to use Elsevier material

Notice
No responsibility is assumed by the publisher for any injury and/or damage to persons
or property as a matter of products liability, negligence or otherwise, or from any use
or operation of any methods, products, instructions or ideas contained in the material
herein. Because of rapid advances in the medical sciences, in particular, independent
verification of diagnoses and drug dosages should be made

Library of Congress Cataloging-in-Publication Data
A catalog record for this book is available from the Library of Congress

British Library Cataloguing in Publication Data
A catalogue record for this book is available from the British Library

ISBN: 978-0-444-53164-3 (this volume)
ISSN: 0079-6123 (Series)

For information on all Elsevier publications
visit our website at books.elsevier.com

Printed and bound in The Netherlands

08 09 10 11 12   10 9 8 7 6 5 4 3 2 1

Working together to grow
libraries in developing countries

www.elsevier.com | www.bookaid.org | www.sabre.org

ELSEVIER   BOOK AID International   Sabre Foundation

List of contributors

T. Abel, University of Pennsylvania, Department of Biology, Biological Basis of Behavior Program, Philadelphia, PA 19104, USA

C.H. Bailey, Department of Neuroscience, College of Physicians and Surgeons of Columbia University, New York State Psychiatric Institute, New York, NY 10032, USA; Kavli Institute for Brain Sciences, New York, NY 10032, USA

J.L. Banko, Department of Molecular Medicine, University of South Florida — Health, 12901 Bruce B. Downs Boulevard, MDC 61, Tampa, FL 33612, USA

D.M. Bannerman, Department of Experimental Psychology, University of Oxford, South Parks Road, Oxford, OX1 3UD, UK

S. Belleville, Department of Psychology, Research Center Institut Universitaire de Gériatrie de Montréal, Montreal, QC H3W 1W5, Canada

J. Berry, Program in Developmental Biology, Baylor College of Medicine, Houston, TX 77030, USA

G. Blaise, CHUM, Hôpital Notre-Dame, 1560 Sherbrooke Est, Pavillon Deschamps-FS-1136, Montréal, QC H2L 4M1, Canada

K.L. Campbell, Department of Psychology, University of Toronto, Toronto, ON M5S 3G3, Canada; The Rotman Research Institute, Baycrest, 3560 Bathurst Street, Toronto, ON M6A 2E1, Canada

V.F. Castellucci, Department of Physiology, Université de Montréal, Pavillon Paul-G. Desmarais, 2960, Chemin de la Tour, CP 6128, Succ. Centre-ville, Montreal, QC H3C 3J7, Canada

N. Caza, Centre de Recherche, Institut Universitaire de Gériatrie de Montréal, 4565 Chemin Queen-Mary, Montréal, QC H3W 1W5, Canada; Psychology Department, Université de Montréal, C.P. 6128, Succursale Centre-Ville, Montréal, QC H3C 3J7, Canada

H. Chertkow, Bloomfield Centre for Research in Aging, Sir Mortimer B. Davis–Jewish General Hospital, McGill University, 3999 Chemin de la Côte Ste–Catherine, Montréal, QC, H3T 1E2, Canada; Department of Neurology and Neurosurgery, McGill University, Montréal, QC, Canada; Division of Geriatric Medicine, Department of Medicine, Sir Mortimer B. Davis–Jewish General Hospital, McGill University, Montréal, QC, Canada; Centre de Recherche, Institut Universitaire de Gériatrie de Montréal, Université de Montréal, Montréal, QC, Canada

M. Costa-Mattioli, Department of Biochemistry and McGill Cancer Center, 3655 Promenade Sir William Osler, McGill University, Montréal, QC H3G 1Y6, Canada

N. Cowan, Department of Psychological Sciences, University of Missouri, 18 McAlester Hall, Columbia, MO 65211, USA

R.L. Davis, Department of Molecular and Cellular Biology, Baylor College of Medicine, Houston, TX 77030, USA; Menninger Department of Psychiatry and Behavioral Sciences, Baylor College of Medicine, Houston, TX 77030, USA

C. de Boysson, Department of Psychology, Research Center Institut Universitaire de Gériatrie de Montréal, Montreal, QC H3W 1W5, Canada

M. De Roo, Department of Neuroscience, Centre Médical Universitaire, 1211 Geneva 4, Switzerland

L. DesGroseillers, Département de Biochimie et Physiologie, Université de Montréal, Montréal, QC H3C 3J7, Canada; Groupe de Recherche sur le Système Nerveux Central (GRSNC), Université de Montréal, Montréal, QC H3C 3J7, Canada

A. Duong, Bloomfield Centre for Research in Aging, Sir Mortimer B. Davis–Jewish General Hospital, McGill University, Montréal, QC, Canada; Neurochem Inc., Laval, QC H7V 4A7, Canada

J.U. Frey, Department of Neurophysiology, Leibniz Institute for Neurobiology, Brenneckestrasse 6, D-39118 Magdeburg, Germany

S. Frey, Department of Neurophysiology, Leibniz Institute for Neurobiology, Brenneckestrasse 6, D-39118 Magdeburg, Germany

E.M. Galloway, Section on Neural Development and Plasticity, NICHD, NIH, Bethesda, MD 20892-3714, USA; Department of Physiology, Anatomy and Genetics, University of Oxford, Oxford, UK

P.M. Garcia, Department of Neuroscience, Centre Médical Universitaire, 1211 Geneva 4, Switzerland

D.L. Glanzman, Department of Physiological Science, UCLA College, Los Angeles, CA 90095-1606, USA; Department of Neurobiology, David Geffen School of Medicine, UCLA, Los Angeles, CA 90095-1761, USA; Brain Research Institute, David Geffen School of Medicine, UCLA, Los Angeles, CA 90095-1761, USA

M.A. Good, School of Psychology, Cardiff University, Tower Building, Park Place, P.O. Box 901, Cardiff, CF10 3YG, UK

L. Hasher, Department of Psychology, University of Toronto, Toronto, ON M5S 3G3, Canada; The Rotman Research Institute, Baycrest, 3560 Bathurst Street, Toronto, ON M6A 2E1, Canada

M.K. Healey, Department of Psychology, University of Toronto, Toronto, ON M5S 3G3, Canada

J.G. Howland, Neural Systems and Plasticity Research Group, Department of Psychology, University of Saskatchewan, Saskatoon, SK, S7N 5A5, Canada

M. Isingrini, University François-Rabelais de Tours, UMR CNRS 6215 LMDC, 3 rue des Tanneurs, BP 4103, 37041 Tours Cedex 1, France

J.D. Johnson, Center for the Neurobiology of Learning and Memory, and Department of Neurobiology and Behavior, Bonney Research Laboratory, University of California at Irvine, Irvine, CA 92697-3800, USA

E.R. Kandel, Howard Hughes Medical Institute, New York, NY 10032, USA; Department of Neuroscience, College of Physicians and Surgeons of Columbia University New York State Psychiatric Institute, New York, NY 10032, USA; Kavli Institute for Brain Sciences, New York, NY 10032, USA

E. Klann, Center for Neural Science, New York University, New York, NY 10003, USA

P. Klauser, Department of Neuroscience, Centre Médical Universitaire, 1211 Geneva 4, Switzerland

W.C. Krause, Department of Molecular and Cellular Biology, Baylor College of Medicine, Houston, TX 77030, USA

J.-C. Lacaille, Département de Physiologie, Université de Montréal, GRSNC, C.P. 6128, Succ. Centre-ville, Montréal, QC H3C 3J7, Canada

B. Lu, Section on Neural Development and Plasticity, NICHD, NIH, Bethesda, MD 20892-3714, USA; Genes, Cognition and Psychosis Program, NIMH, Bethesda, MD, USA

A. Maffei, Department of Biology, Brandeis University, Waltham, MA 02544, USA; National Center for Behavioral Genomics, Brandeis University, Waltham, MA 02544, USA

C.J. McBain, Laboratory of Cellular and Synaptic Neurophysiology, Program in Developmental Neurobiology, NICHD, NIH, Porter Neuroscience Building 35, Bethesda, MD 20892, USA

M.-C. Ménard, Department of Psychology, Research Center Institut Universitaire de Gériatrie de Montréal, Montreal, QC H3W 1W5, Canada

D. Muller, Department of Neuroscience, Centre Médical Universitaire, 1211 Geneva 4, Switzerland

P.V. Nguyen, Departments of Physiology and Psychiatry, University of Alberta School of Medicine, Centre for Neuroscience, Edmonton, AB T6G 2H7, Canada

H. Park, Center for the Neurobiology of Learning and Memory, and Department of Neurobiology and Behavior, Bonney Research Laboratory, University of California at Irvine, Irvine, CA 92697-3800, USA

J.G. Pelletier, Département de Physiologie, Université de Montréal, GRSNC, C.P. 6128, Succ. Centre-ville, Montréal, QC H3C 3J7, Canada

A. Perrotin, University François-Rabelais de Tours, UMR CNRS 6215 LMDC, 3 rue des Tanneurs, PB 4103, 37041 Tours Cedex 1, France

L. Poglia, Department of Neuroscience, Centre Médical Universitaire, 1211 Geneva 4, Switzerland

Y. Qi, CHUM, Hôpital Notre-Dame, 1560 Sherbrooke Est, Pavillon Deschamps-FS-1136, Montréal, QC H2L 4M1, Canada

J.N.P. Rawlins, Department of Experimental Psychology, University of Oxford, South Parks Road, Oxford, OX1 3UD, UK

M.D. Rugg, Center for the Neurobiology of Learning and Memory, and Department of Neurobiology and Behavior, Bonney Research Laboratory, University of California at Irvine, Irvine, CA 92697-3800, USA

T.C. Sacktor, The Robert F. Furchgott Center for Neural and Behavioral Science, Departments of Physiology, Pharmacology, and Neurology, SUNY Downstate Medical Center, 450 Clarkson Avenue, New York, NY 11203, USA

M. R. Sánchez-Carbente, Département de Biochimie et Physiologie, Université de Montréal, Montréal, QC H3C 3J7, Canada

D.J. Sanderson, Department of Experimental Psychology, University of Oxford, South Parks Road, Oxford, OX1 3UD, UK

D. Saumier, Department of Neurology and Neurosurgery, McGill University, Montréal, QC, Canada; Centre de Recherche, Institut Universitaire de Gériatrie de Montréal, Université de Montréal, Montréal, QC, Canada; Neurochem Inc., Laval, QC H7V 4A7, Canada

P.H. Seeburg, Max-Planck Institute of Medical Research, Department of Molecular Neurobiology, Jahnstrasse 29, D-69120 Heidelberg, Germany

N. Sonenberg, Department of Biochemistry and McGill Cancer Center, 3655 Promenade Sir William Osler, McGill University, Montréal, QC H3G 1Y6, Canada

W.S. Sossin, Department of Neurology and Neurosurgery, McGill University, Montreal Neurological Institute, BT 110, 3801 University Street, Montreal, QC H3A 2B4, Canada

C. Souchay, Institute of Psychological Sciences, University of Leeds, Leeds LS2 9JT, UK

R. Sprengel, Max-Planck Institute of Medical Research, Department of Molecular Neurobiology, Jahnstrasse 29, D-69120 Heidelberg, Germany

W.A. Suzuki, Center for Neural Science, New York University, 4 Washington Place Room 809, New York, NY 10003, USA

S. Sylvain-Roy, Department of Psychology, Research Center Institut Universitaire de Gériatrie de Montréal, Montreal, QC H3W 1W5, Canada

R. Taha, CHUM, Hôpital Notre-Dame, 1560 Sherbrooke Est, Pavillon Deschamps-FS-1136, Montréal, QC H2L 4M1, Canada

G. Turrigiano, Department of Biology, Brandeis University, Waltham, MA 02454, USA; National Center for Behavioral Genomics, Brandeis University, Waltham, MA 02544, USA

M.R. Uncapher, Center for the Neurobiology of Learning and Memory, and Department of Neurobiology and Behavior, Bonney Research Laboratory, University of California at Irvine, Irvine, CA 92697-3800, USA

Y.T. Wang, Department of Medicine and the Brain Research Centre, University of British Columbia, Vancouver, BC V6T 2B5, Canada

C. Whatmough, Bloomfield Centre for Research in Aging, Sir Mortimer B. Davis–Jewish General Hospital, McGill University, Montréal, QC, Canada; Department of Neurology and Neurosurgery, McGill University, Montréal, QC, Canada; Centre de Recherche, Institut Universitaire de Gériatrie de Montréal, Université de Montréal, Montréal, QC, Canada

N.H. Woo, Section on Neural Development and Plasticity, NICHD, NIH, Bethesda, MD 20892-3714, USA

Preface

The following volume stems from a meeting of the same name "The Essence of Memory" held in Montreal from May 12 to 14, 2007 organized by the editors of this volume. The editors would like to acknowledge the support of the Groupe de Recherche sur le Système Nerveux Central, Université de Montréal, for the organization of the meeting. This meeting brought together an international group of investigators studying memory from the molecular, cellular, physiological and behavioral levels. The goal of the meeting was to see how these levels could talk to one another and to inform researchers in this field. In addition, for this book, we have recruited a number of additional leading investigators to fill in gaps necessitated by the limited time for presentations at the meeting. Several of these were from other attendees of the meeting, and several were by invitation. By trying to discuss the 'essence' of memory, we mean to distill the fundamental issues in the memory field that should transcend boundaries and hopefully eventually lead to an understanding of memory at the molecular level that will be able to be translated into pharmacological and behavioral treatments of memory disorders.

The volume begins with cellular and molecular approaches to understanding memory. There is an overview chapter by Dr. Sossin on molecular memory traces, discussing how our growing understanding of the molecular and cellular basis of the memory trace opens up opportunities and challenges for cellular and systems understanding of memory. In particular, an argument is made for multiple independent memory traces that are formed after an experience. This is followed by a series of chapters further elucidating these distinct molecular traces. Dr. Sacktor describes his research on PKMζ, which fulfills many of the requirements for a molecular memory trace. Another memory trace requires gene expression and local translation. Dr. DesGroseillers describes how mRNAs are transported in neurons and Dr. Klann then describes how the translation of these mRNAs is regulated, and how dysregulation of this pathway can disrupt memory. Dr. Sonenberg describes his recent findings that protein translation also regulates an important transcriptional switch that determines when long-term memory is induced. Drs. Abel and Nguyen describe the role of another critical signal transduction pathway, production of cyclic AMP, in multiple kinds of memory traces. Together these investigators highlight the recent advances in molecular cognition. After identifying important molecules involved in memory formation, mice are generated either lacking these molecules, or expressing blockers of this pathway, and then behavioral tests of memory are used to determine the importance of the molecular pathway.

The discussion of the molecular and cellular memory trace continues with a chapter from Dr. Frey, the discoverer of synaptic tagging. The memory trace is thought to be synapse-specific; however cellular processes such as translation and transcription may not be this local. Dr. Frey describes the mechanism of 'tagging' to allow for plasticity-related proteins to be restricted to activated synapses. Dr. Wang introduces another well-understood memory trace, the trafficking of ionotropic receptors, notably the glutamate sensing AMPA receptors, and describes how other systems like stress impact on the ability of this plasticity to occur. Dr. Bannerman examines memory in the absence of this pathway, suggesting interesting dissociations between different kinds of memory based on the trace that they use. Another type of memory trace is structural; both structural modification of synapses already present, and the generation of new synapses. Dr. Bailey and Kandel describe the evidence that structural modifications occur during memory in invertebrates and then the question of whether similar changes occur in the rodent hippocampus are discussed by Dr. Muller.

An understanding of the 'essence' of memory must also elucidate memory at the systems level; understanding the anatomical constraints of the circuits underlying memory and identifying at which synapses the molecular and cellular events occur. Dr. Turrigiano discusses her work on how intrinsic properties are important during development to set up the system for efficient memory formation. Dr. McBain and Dr. Lacaille both discuss the importance of plasticity in inhibitory neurons both in controlling the memory systems, and perhaps in forming memories themselves. Dr. Lu describes his work on how molecules such as BDNF can impact the persistent activation of neurons seen during working memory.

The volume now moves into investigations of the 'essence' of memory at the whole animal level, where memory traces are examined during behavior. This problem is introduced by Dr. Castellucci in an introductory chapter. Dr. Glanzman describes the growing importance of post-synaptic changes in regulating the defensive reflex in *Aplysia* during associative conditioning; Dr. Davis describes how new molecular techniques in the fly allow monitoring of cellular memory traces during learning and the growing knowledge of the cellular organization of memory in the fruit fly. Dr. Suzuki describes her work on following memory formation using recordings in the monkey limbic area as they learn a new task.

The 'essence' of memory must reflect an understanding of our memory, human memory. In particular, a major justification for elucidating the molecular and cellular basis of memory traces is to help solve problems and pathologies of memory formation in humans. Over the last decade, studies of human memory have provided cognitive and neurobiological models of memory to explain, characterize and organize the act of memory within a coherent framework. This effort should lead to a better understanding of how events are typically encoded and retrieved by memory. It should also shed light on the memory changes that accompany human development and aging, the impact of memory-related disorders such as Alzheimer's disease or mild cognitive impairment, and the effect of stressful biological events such as anesthesia and surgery. This part of the volume starts with a chapter by Dr. Cowan who proposes a theoretical view regarding the distinction between important components of memory: short-term memory, working memory and long-term memory. He presents defining principles to dissociate those forms of memory and discusses why working memory tasks are such good predictors of high-level performance. In the following chapter, Dr. Rugg describes imaging studies to support a neurocognitive model of episodic memory. The proposed model addresses the correspondence between encoding and retrieval cues and the role of binding processes in episodic memory. Dr. Hasher describes her work on the role of distraction on memory processes and on complex cognitive tasks. She discusses that the age-related changes in attentional regulation can result in both costs and potential benefits on older persons' ability to complete memory and complex cognitive activities. Dr. Belleville introduces the notion of mild cognitive impairment, a potential precursor state of Alzheimer's disease. She discusses its impact on episodic memory and working memory and proposes that both components are impaired during the earliest phase of the disease. Dr. Isingrini presents a chapter where metamemory is conceived as a crucial component of memory performance. He discusses findings showing that this is a component critically modified by healthy aging and by Alzheimer's disease. Dr. Chertkow discusses the impact of Alzheimer's disease on memory by focusing on the notion of semantic memory, that is memory for words and concepts. In this chapter, Dr. Chertkow proposes a cognitive and a neurobiological model to account for the semantic memory breakdown characterizing Alzheimer's. Finally, Dr. Caza discusses a new area of research investigating the impact of anesthesia and surgery on memory processes. She introduces hypotheses regarding the underlying biological models for such an impact on memory and highlights areas for future investigation.

Wayne S. Sossin
Jean-Claude Lacaille
Vincent F. Castellucci
Sylvie Belleville
Montreal, Canada

Contents

List of contributors v

Preface ix

I. Cellular and Molecular Approaches to the Essence of Memory

1. Molecular memory traces
   W.S. Sossin (Montreal, QC, Canada). 3

2. PKMζ, LTP maintenance, and the dynamic molecular biology of memory storage
   T.C. Sacktor (New York, NY, USA). 27

3. Understanding the importance of mRNA transport in memory
   M.R. Sánchez-Carbente and L. DesGroseillers (Montréal, QC, Canada). 41

4. Cap-dependent translation initiation and memory
   J.L. Banko and E. Klann (Tampa, FL and New York, NY, USA). 59

5. Translational control of gene expression: a molecular switch for memory storage
   M. Costa-Mattioli and N. Sonenberg (Montréal, QC, Canada). 81

6. Regulation of hippocampus-dependent memory by cyclic AMP-dependent protein kinase
   T. Abel and P.V. Nguyen (Philadelphia, PA, USA and Edmonton, AB, Canada). 97

7. 'Synaptic tagging' and 'cross-tagging' and related associative reinforcement processes of functional plasticity as the cellular basis for memory formation
   S. Frey and J.U. Frey (Magdeburg, Germany) 117

8. Synaptic plasticity in learning and memory: stress effects in the hippocampus
   J.G. Howland and Y.T. Wang (Saskatoon, SK and Vancouver, BC, Canada) 145

9. The role of the GluR-A (GluR1) AMPA receptor subunit in learning and memory
   D.J. Sanderson, M.A. Good, P.H. Seeburg, R. Sprengel, J.N.P. Rawlins and
   D.M. Bannerman (Oxford and Cardiff, UK and Heidelberg, Germany) 159

10. Synaptic remodeling, synaptic growth and the storage of long-term memory in *Aplysia*
    C.H. Bailey and E.R. Kandel (New York, NY, USA) 179

11. Spine dynamics and synapse remodeling during LTP and memory processes
    M. De Roo, P. Klauser, P.M. Garcia, L. Poglia and D. Muller (Geneva, Switzerland) ... 199

II. Systems Approaches to the Essence of Memory

12. The age of plasticity: developmental regulation of synaptic plasticity in neocortical microcircuits
    A. Maffei and G. Turrigiano (Waltham, MA, USA) 211

13. Differential mechanisms of transmission and plasticity at mossy fiber synapses
    C.J. McBain (Bethesda, MD, USA) 225

14. Long-term synaptic plasticity in hippocampal feedback inhibitory networks
    J.G. Pelletier and J.-C. Lacaille (Montréal, QC, Canada) 241

15. Persistent neural activity in the prefrontal cortex: a mechanism by which BDNF regulates working memory?
    E.M. Galloway, N.H. Woo and B. Lu (Bethesda, MD, USA and Oxford, UK) 251

III. Animal Approaches to the Essence of Memory

16. Animal models and behaviour: their importance for the study of memory
    V.F. Castellucci (Montreal, QC, Canada) 269

17. New tricks for an old slug: the critical role of postsynaptic mechanisms in learning and memory in *Aplysia*
    D.L. Glanzman (Los Angeles, CA, USA) 277

18. Olfactory memory traces in *Drosophila*
    J. Berry, W.C. Krause and R.L. Davis (Houston, TX, USA) 293

19. Associative learning signals in the brain
    W.A. Suzuki (New York, NY, USA) 305

IV. Human Approaches to the Essence of Memory

20. What are the differences between long-term, short-term, and working memory?
    N. Cowan (Columbia, MO, USA) 323

21. Encoding-retrieval overlap in human episodic memory: a functional neuroimaging perspective
    M.D. Rugg, J.D. Johnson, H. Park and M.R. Uncapher (Irvine, CA, USA) 339

22. Cognitive aging and increased distractibility: costs and potential benefits
    M.K. Healey, K.L. Campbell and L. Hasher (Toronto, ON, Canada) 353

23. Characterizing the memory changes in persons with mild cognitive impairment
    S. Belleville, S. Sylvain-Roy, C. De Boysson and M.-C. Ménard (Montreal, QC, Canada)   365

24. Aging, metamemory regulation and executive functioning
    M. Isingrini, A. Perrotin and C. Souchay (Tours, France and Leeds, UK) 377

25. Cognitive neuroscience studies of semantic memory in Alzheimer's disease
    H. Chertkow, C. Whatmough, D. Saumier and A. Duong
    (Montréal and Laval, QC, Canada) 393

26. The effects of surgery and anesthesia on memory and cognition
    N. Caza, R. Taha, Y. Qi and G. Blaise (Montréal, QC, Canada) 409

Subject Index 423

See Color Plate Section at the end of this book

SECTION I

Cellular and Molecular Approaches to the Essence of Memory

CHAPTER 1

Molecular memory traces

Wayne S. Sossin*

*Department of Neurology and Neurosurgery, McGill University, Montreal Neurological Institute, BT 110,
3801 University Street, Montreal, QC H3A 2B4, Canada*

Abstract: To understand the essence of memory, one must examine the working of the brain on many levels. It is important to find the appropriate level to study the particular aspect of memory under investigation. In this review, I will focus on insights gained from examining memory at the molecular level. I will illustrate these insights with specific examples from examining the molecular and cellular mechanisms underlying long-term facilitation in the marine mollusk *Aplysia* and long-term potentiation, studied mainly in rodents. In particular, I will discuss how molecular memory traces are formed and focus in detail on what role increasing the level of proteins through protein synthesis and gene expression plays in memory formation. I will point out three important constraints from molecular work that should impact on cognitive modeling of the nervous system: (i) the induction of plasticity depends on the 'state' of the synapse; (ii) there are multiple independent molecular traces formed after experience with different half-lives; and (iii) the requirement for the conjunction of synaptic activation and new protein synthesis implies that new conjunctions are required to induce long-term memory formation.

Keywords: *Aplysia*; long-term facilitation; long-term potentiation; protein synthesis; gene expression

Introduction

Brains are not computers

Since our brain is made up of neurons, not silicon chips; and because the brain evolved, instead of being designed, there are molecular constraints on brain function. For an alternative example, let us examine the storage of memory on a computer. The fundamental mechanism of memory storage has changed multiple times (from electrostatic memories, to magnetic memories, to semiconductors, perhaps eventually to nanotubes), without the need for changes in higher-level computer languages or computer design. This is due to layers of engineering software between access and writing to memory and the actual mechanism involved in storing the memory. Being able to treat each module as a 'black box', without requiring knowledge of the underlying mechanism of action, is a strong engineering principle. The brain does not work this way, since it was created by evolution, not design. Thus, how memories are stored in the brain is constrained by the limitations of biology and could not be fundamentally changed as brains became more complex.

Most neuroscientists believe that memories are stored in the strengths of the synaptic connections between neurons. Consequently, the properties of neurons and the synapses that interconnect them are pivotal for understanding how our memory works. Nevertheless, just because molecules are

*Corresponding author. Tel.: +1-514-398-1486;
Fax: +1-514-398-8106; E-mail: wayne.sossin@mcgill.ca

important does not mean that one can understand complex mnemonic processes solely by understanding molecules or synapses; valuable insights are also needed from systems and cognitive level analysis. The key from molecular studies is to extract the rules from which these systems can be built up. For example, the most ambitious extraction from neurons to cognitive systems was the parallel distributed processing concept (PDP), where cognitive scientists attempted to extract the 'essence' of how neuronal systems work to generate interesting models for learning (McClelland and Rumelhart, 1985). A major rule that was extracted was that neurons connect to each other with a specific strength, this connection can be modified by some learning rule, and that each connection can be modified independently. Interestingly, to make these models work, they also required a new rule that the overall strength of connections to a neuron had to remain approximately the same so that as the strength of connections increased, other connections had to decrease. This kind of homeostatic rule is also present in neurons and our understanding of its importance for how the brain works is just beginning (Turrigiano and Nelson, 2000). This is an example of an engineering principle that is invariant, a system that works a certain way needs to encode it whether biological or silicon based. However, as discussed below, there are probably additional important constraints not captured by the PDP model that are required to understand how memories are encoded by the brain.

Model systems

Aplysia, a cellular model for studying memory at the molecular level

The reductionist approach is a powerful way to determine molecular mechanisms of memory. This approach assumes the mechanisms that underlie how memory works in simple systems will be conserved in higher systems due to conservation of molecular mechanisms affecting how synapses work. *Aplysia* is a marine mollusk with a simple nervous system. By a comprehensive examination of this system at the molecular, cellular and behavioral level, a number of important insights have been gained into the molecular mechanisms of memory formation (Kandel, 2001). The major memory examined in this system is changes in the defensive reflex of the animal, exemplified by the withdrawal of the gill or siphon, based on experience. At the simplest level, this reflex is mediated by a direct connection between sensory neurons that detect a touch and motor neurons that withdraw the gill or siphon. This reflex can be modulated by experience in a number of ways exemplifying many of the basic non-associative and associative learning paradigms (Hawkins et al., 2006). Habituation is the decrease in the reflex seen after multiple occurrences of a harmless touch. This can be explained at the cellular level by a decrease in the strength of the sensory–motor neuron connection, termed depression (Castellucci et al., 1970). Sensitization is the increase in the reflex after experiencing a noxious stimulus. This is a non-associative memory since the noxious stimulus does not have to be associated with the activation of the reflex. The noxious stimulus causes the release of serotonin (5-HT) (Marinesco and Carew, 2002), and 5-HT acts at multiple levels of the circuit to increase the defensive reflex (Frost et al., 1988), including increasing the strength of the sensory–motor neuron connection, a process known as facilitation (Castellucci and Kandel, 1976). Finally, if the animal is touched while it receives the noxious stimulus, an additional increase in the reflex occurs that reflects the associative change (Hawkins et al., 1983; Walters and Byrne, 1983). Here the coupling of 5-HT and firing of the sensory neuron leads to additional increases in the connection between the sensory and motor neurons.

One of the great strengths of this system is that a behavioral memory has been strongly associated with a specific change in synaptic strength between two identified neurons. Thus, one can study molecular mechanisms of memory at this specific synapse. Moreover, the sensory and motor neuron can be cultured from the animal and in culture they re-grow their synaptic connection and one can study depression and facilitation in a reduced system that mimics many of the mechanisms observed in the animal (Montarolo et al., 1986).

The CA3–CA1 synapse in vertebrates as a cellular model for associative learning

The hippocampus is the anatomical region required for many forms of associative learning in vertebrates. Also, due to its organized fiber tracts, it is a wonderful model for examining cellular plasticity, as it is simple to activate presynaptic fibers onto postsynaptic neurons and record the synaptic strength in a field recording. While long-term potentiation (LTP) was discovered in the fibers entering the hippocampus (the perforant pathway) (Bliss and Lomo, 1973), it has been studied most extensively in the connection between the CA3 neurons and the CA1 neurons. Unlike the non-associative facilitation induced by 5-HT in *Aplysia*, the increase in strength between CA3 neurons and CA1 neurons is associative, requiring both presynaptic firing in the CA3 neuron and postsynaptic depolarization in the CA1 neuron. This conjunction of inputs is required to open the *N*-methyl-D-aspartic acid (NMDA) receptor and calcium entry through this receptor induces LTP (Bliss and Collingridge, 1993). One major weakness of this system is the lack of knowledge of the specific function of the CA3–CA1 connection and how changes in the strength of these connections is related to the memories encoded in the hippocampus. Nevertheless, there is abundant evidence that this plasticity is important for memory and many of the fundamental molecular insights into memory have come from this system (Morris, 2003).

What types of cellular changes underlie memory formation?

Excitability changes

The major advantage of changing memories at synapses is the ability to strengthen specific inputs to a neuron based on those inputs coupling with the firing of the output neuron. This can lead to the forming of neuronal associations underlying associative memory as was famously outlined by Hebb (Cooper, 2005). However, while this review will focus on synaptic plasticity, not all memories need be encoded only by changes in synaptic strength. A neuron can also change the rate at which it fires action potentials to the same inputs, a change in excitability (Daoudal and Debanne, 2003; Schulz, 2006). In *Aplysia*, 5-HT increases the excitability of the sensory neuron through activation of a kinase, protein kinase A (PKA), and phosphorylation of potassium channels, S channels, that are open at rest (Siegelbaum et al., 1982). Closing these channels causes the neurons to depolarize and thus less input is required for them to reach the threshold for firing an action potential. It also closes channels that normally prevent the firing of multiple action potentials, a process known as anti-accommodation (Klein et al., 1986). In combination, these two molecular changes allow stimulation of the sensory neuron (a touch) to fire more action potentials, and thus lead to an increase in the defensive reflex. Excitability changes in *Aplysia* are both seen immediately after sensitization, or 5-HT addition in culture, and can last at least 24 h after training (Scholz and Byrne, 1987). The prolonged increase in sensitization is probably due to the persistent activation of PKA (see below).

Excitability changes have also been detected in vertebrate systems. During LTP there is an increase in excitability of the CA1 neurons as well as increases in synaptic strength (Xu and Kang, 2005). Moreover there are multiple examples of changes is ionic channels in dendrites affecting the propagation and integration of inputs and these changes can be local, either synapse or branch specific (Frick et al., 2004; Frick and Johnston, 2005). It will be interesting in the future to determine how important these changes are for memory. Clearly genetic changes that affect excitability do affect memory (Nolan et al., 2004; Hammond et al., 2006) but it is hard to determine if their effect is to change the state of the neuron to undergo plasticity, or is intrinsically involved in the plasticity itself. For example, a genetic change that reduces the ability of a dendritic region to depolarize will inhibit the ability of coincident inputs to activate NMDA channels and thus inhibit the ability to induce plasticity. In contrast, NMDA channel opening leading to calcium entry, activation of kinases and phosphorylation of an ion channel that leads to a long-term increase in the ability of local excitatory postsynaptic potentials (EPSPs) to propagate down the dendritic branch would be an example where excitability changes are intrinsically involved in the memory trace.

Changes in synaptic strength

Synaptic strength is regulated by three parameters, the probability of release (p), the size of the response to a single vesicular release (q) and number of functional synapses (n). All are probably changed in distinct memory traces. There are also many different mechanisms to regulate each of these parameters. For example, the value of q is determined by: (i) the number and efficacy of postsynaptic receptors; (ii) the amount of transmitter/vesicle (which can be regulated by the number of vesicular transporters); (iii) the efficacy of neurotransmitter uptake; and (iv) the ionic concentrations that determine the driving force for the ionotropic receptors. All of these can be modulated by plasticity (Levenson et al., 2002; Malinow and Malenka, 2002; Woodin et al., 2003; Shen and Linden, 2005; Takamori, 2006).

Facilitation in Aplysia sensory–motor neuron synapses is mainly determined by changes in the probability of release

In *Aplysia*, depression, the cellular model for habituation, is due to a decrease in p and facilitation (Castellucci and Kandel, 1974), the cellular model for sensitization, is due to increases in p (Castellucci and Kandel, 1976). The decrease in p is not linked to a decrease in the frequency of spontaneous release (Eliot et al., 1994) and is mainly due to a decrease in calcium-secretion coupling (Zhao and Klein, 2002), the ability of calcium to induce release of vesicles. Facilitation is coupled with an increase in spontaneous release (Dale et al., 1988), however this increase is mediated by 5-HT activation of protein kinase C (PKC), but inhibitors of PKC do not block short-term facilitation (STF); thus this increase may not be related to facilitation (Ghirardi et al., 1992). Instead, facilitation is caused both by increased calcium entry due to action potential broadening, and additional PKA-dependent changes in the probability of release that have not yet been identified (Klein, 1994; Byrne and Kandel, 1996). At later points, 5-HT can increase n both due to the activation of silent synapses and the formation of new synapses (see Chapter 10 by Bailey and Kandel, this volume). Finally, 5-HT can also increase the number of alpha-amino-3-hydroxy-5-methyl-4-isoxazolepropionic acid (AMPA) receptors as measured by the response to extracellular glutamate (Trudeau and Castellucci, 1995; Chitwood et al., 2001), however it has not yet been shown that these receptors are synaptic (and thus alter q) and the contribution of the increase in these receptors to sensitization is still under debate (see Chapter 17 by Glanzman, this volume).

Potentiation in CA3–CA1 synapses is mainly determined by changes in the reception of neurotransmitter

In CA3–CA1 neurons early LTP is due to increased levels of AMPA receptor activity. Thus, there is an increase in the reception of neurotransmitter or q, the size of the response to the release of a single vesicle. This is due to increased activity of AMPA receptors already present in the synaptic membrane, the insertion of more AMPA receptors into previously active synapses and the insertion of AMPA receptors into silent synapses (Liao et al., 1995). Changes in the probability of release have been inferred at later times after plasticity, due to a measured increase in the probability of release (Bolshakov et al., 1997; Zakharenko et al., 2001) and the requirement of presynaptic changes for prolonged potentiation after some stimulation paradigms (Huang et al., 2005). It is not clear if later stages of LTP at CA3–CA1 synapses are due to increased numbers of synaptic connections (see De Roo et al., Chapter 11).

The difference in the locus of the change in *Aplysia* sensory–motor neuron synapses and CA3–CA1 synapses does not reflect a fundamental difference between invertebrates and vertebrates, but is simply a function of the choice of model synapses. *Aplysia* still retains the mechanisms required for insertion of AMPA receptors and this may play a role in associative sensitization at sensory–motor neuron synapses and probably at other synapses that have not been examined in detail (Roberts and Glanzman, 2003). Moreover, there are numerous examples of LTP in vertebrates that are mainly mediated by changes in p, notably the mossy fiber synapses connecting the dentate gyrus to CA3 (Zalutsky and Nicoll, 1990).

Molecular memory can be dissociated into temporal phases

Memory has been roughly divided into three temporal domains. Working memory refers to retention of events that can be represented by the continued firing of neurons that represent the stored information. Short-term memory refers to memories that are vulnerable to interference until they are consolidated; at the cellular and molecular level this is often associated with the lack of a requirement for protein synthesis. Finally long-term or consolidated memory represents a long-term store that is less vulnerable to interference and at the cellular and molecular level requires protein synthesis. This classification serves some purposes, but does not fully capture the cellular changes that underlie the memory trace. I would argue instead that there are multiple molecular traces that are induced by experience that last for various times and do or do not require new protein synthesis. These traces can be independent of each other.

If memories are stored in the strength of synapses, how are these changes stored. Is it in the state or level of a single molecule, or is it a more complex interaction between many molecules? One of the key problems to overcome is the long lifetime of a memory compared to the lifetime of an individual molecule. Moreover, unlike computer memories in neurons, there are distinct types of biological synaptic changes that last for different time periods. While memory in biological systems is vastly different than memory in computers, there is a useful analogy. In computers, memory storage devices are divided into volatile (such as dynamic random access memory where constant energy is required to maintain the memory) or static (such as memory encoded on a magnetic disc). In biological memory, this difference is not a dichotomy but a continuum where different types of memory storage require different amounts of energy to maintain. For example, if memory is encoded by changes in phosphorylation (i.e. due to production of a persistently active kinase), the memory is volatile and inhibiting the production of the kinase should eliminate the memory. In contrast, if memory is encoded by constructing new synapses, inhibiting production of new proteins should not immediately remove the synapse, and thus, the memory is more static.

Below, we will examine the molecular memory traces at the two model synapses under discussion and examine what traces underlie the different temporal domains in detail, concentrating on short- and long-term traces. For discussions of working memory, see Chapter 19 by Suzuki, this volume; Chapter 15 by Galloway et al., this volume.

Molecular traces of different volatility are induced by protein phosphorylation

Most short-term changes in synaptic strength are due to phosphorylation, either kinases adding phosphates, or phosphatases removing them. While the phosphorylation event itself is usually short lasting, depending on the consequence of phosphorylation the trace is graded in its volatility (Fig. 1). The half-life of messengers (calcium or second messengers (cyclic AMP, diacylglycerol, etc.), that activate the kinases/phosphatases ranges from seconds to minutes. Furthermore, the change can be easily reversed, i.e. a phosphate added by a kinase can be removed by a phosphatase. Thus in *Aplysia*, a single 5 min application of 5-HT leads to activation of PKA, phosphorylation of target molecules including ion channels and proteins involved in vesicular release and an increase in synaptic strength that lasts for about 20–30 min. Thus, this is a highly volatile memory trace. However, transient changes in phosphorylation sites can also lead to a longer-lasting change by inducing other events that may not reverse when the site is removed. For example, insertion and removal of AMPA receptors at CA3/CA1

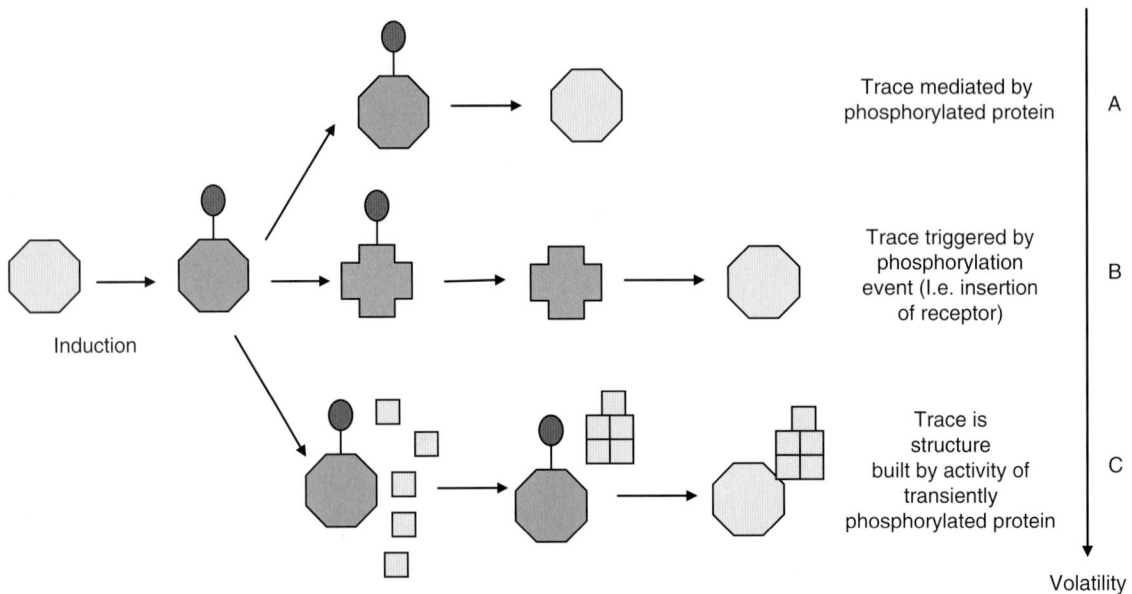

Fig. 1. Different molecular traces induced by phosphorylation have different volatility. (A) The simplest memory trace involves activation of kinase that phosphorylates a substrate. Phosphorylation of the substrate itself leads to an increase in synaptic strength through either increasing transmitter release (e.g. SNAP-25) or transmitter reception (e.g. CAMKII phosphorylation of AMPA receptors). Once the kinase is no longer active (degradation of second messengers), phosphatases remove the phosphate and the trace is erased. (B) Phosphorylation can induce a change in the protein (shift from hexagon to cross) that lasts longer than the phosphorylation itself. For example, phosphorylation of AMPA receptors may lead to insertion of the receptor that can persist after the phosphorylation has been removed. Some additional event then has to occur to erase the memory trace. (C) Phosphorylation can activate a protein (i.e. a regulator of actin polymerization, cofilin) that leads to a structural change (assembly of blocks into a house-like structure) that persists after the phosphorylation is removed and requires an active process to disassemble. The three changes are graded in their volatility. (See Color Plate 1.1 in color plate section.)

synapses requires several phosphorylation events, but the insertion or removal of the receptor may last longer than the change in phosphorylation. This would represent a less volatile trace, however, insertion of AMPA receptors can still be reversed by endocytosis of the receptors, and does not persist in the absence of additional events (i.e. the early form of LTP lasts only for a few hours). Thus this trace is less volatile than a simple phosphorylation event, but still is easily reversed (Fig. 1). Morphological changes that are induced by kinase activation are probably not reversed simply by removing the phosphate off the cytoskeletal regulators involved in inducing the change, and may persist in the absence of an active process that would lead to a reversal of the cytoskeletal change. Thus, morphological changes could represent a more 'static' change still induced by phosphorylation (Fig. 1). Thus, a common finding for most kinases involved in longer-lasting changes is that they are required to induce the change, but in many cases not for the maintenance of synaptic changes. An exception to this is when the molecular trace itself is a persistently active kinase (see below).

State-dependence of cellular memory traces

An important aspect of the memory trace is its induction; what is required to induce the memory trace. The state of the synapse is a critical regulator of the induction step, and thus depending on the state of the synapse, the identical stimulus can have different effects and induce different traces. The synapse regulates its own ability to be plastic, or metaplasticity (Abraham and Bear, 1996), at the molecular level by altering the rate-limiting steps in the induction of plasticity.

The actions of 5-HT depend on the past history of the synapse

In *Aplysia*, adult animals respond to a shock similarly whether or not they were previously habituated, showing an increase in the reflex. However, in juvenile animals, response to the shock occurred only after initial habituation (Rankin and Carew, 1988). Thus, dishabituation appeared to be distinguishable from sensitization. At the cellular level, both the reversal of depression and facilitation are due to increases in transmitter release, but at the molecular level the mechanisms are distinct. For facilitation, 5-HT requires PKA activation and for the reversal of depression, 5-HT requires PKC activation (Ghirardi et al., 1992). This could be explained in two ways: (i) the mode of actions was the same, but based on the differences in the state of the synapse, 5-HT either activated PKA or PKC (Fig. 2A); (ii) the mode of actions was different, PKA phosphorylated proteins important for increasing the strength of resting synapses, but a different phosphorylation event was necessary for increasing the strength of depressed synapses (Fig. 2B). Indeed, this second model (Fig. 2B) is correct; studies of PKC activation showed that a specific isoform of PKC, the novel PKC Apl II, was responsible for the

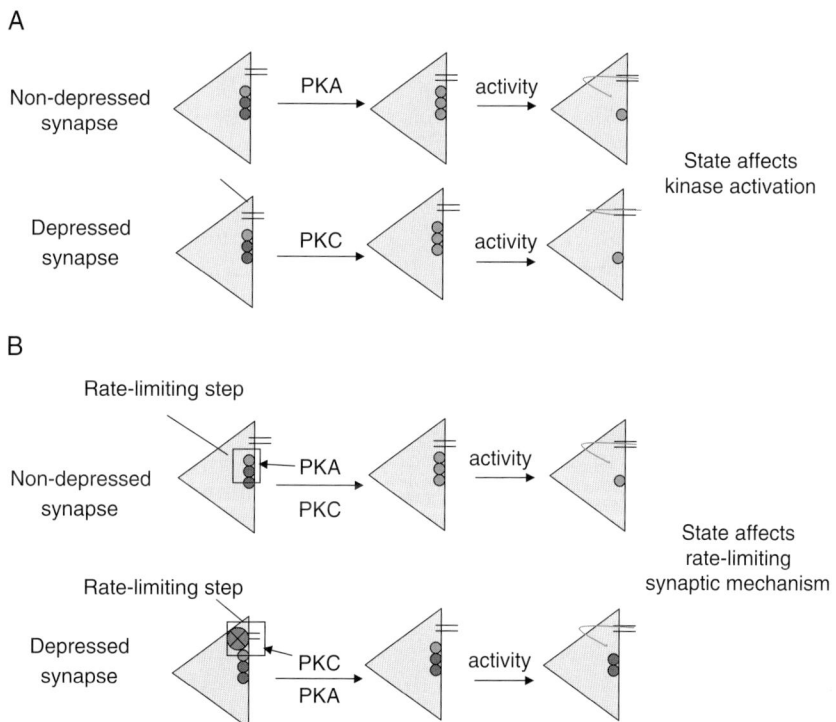

Fig. 2. The state of the synapse can determine the requirements for increases in synaptic strength. (A) This figure describes the possibility that the kinase activated can be determined by the state of the synapse. In the non-depressed synapse PKA is activated, increases the readiness of the synaptic vesicles to be released (shift from red-green), and thus more transmitters are released after an action potential (activity). At the depressed synapse, PKC is activated instead of PKA, but the change at the synapse is the same. (B) This figure describes the possibility that the state of the synapse determines which kinase is required. At the non-depressed synapse, the rate-limiting step is the readiness of the synaptic vesicles to be released. While both PKA and PKC are activated by 5-HT, only PKA phosphorylations that increase the readiness of the synaptic vesicles to be released are affected. At the depressed synapse, calcium-secretion coupling becomes the rate-limiting step and activation of PKC, but not PKA can remove this inhibition. The bars represent calcium channels and the arrow represents the influx of calcium leading to the release of synaptic vesicles. (See Color Plate 1.2 in color plate section.)

response of 5-HT at depressed synapses, and this isoform was equally activated whether the synapses was depressed or not (Manseau et al., 2001; Zhao et al., 2006; Sossin, 2007). Thus, the difference between the two states lay in the rate-limiting step for synaptic activation. In depressed synapses, since PKA could not remove the depression, it is not sufficient to increase synaptic strength. In contrast, PKC could reverse depression, but could not phosphorylate the protein that was rate limiting in non-depressed synapses (Fig. 2B).

The rate-limiting step for GluR1 regulation depends on the past history of the synapse

There is an interesting parallel in state-dependence of phosphorylation in the examination of AMPA receptor regulation during LTP. The regulation of glutamate receptor (GluR)1 receptors depends on three phosphorylation events, one mediated by PKC, one mediated by PKA and one mediated by calcium calmodulin-dependent kinase II (CAMKII) (Mammen et al., 1997; Esteban et al., 2003; Boehm et al., 2006); however the important site is determined by the previous history of the synapse. The PKA site is normally phosphorylated at rest, activation of phosphatases removes this site, and is thought to be important for the removal of this receptor during long-term depression (LTD) (Malinow and Malenka, 2002). However, PKA activation is not normally required to insert the receptor, because at rest this site is already phosphorylated (Lee et al., 2000). However, following LTD, there is now a requirement for PKA to phosphorylate this site to 'reverse' the LTD (Lee et al., 2000). The CAMKII site is normally not phosphorylated, but is phosphorylated during LTP (Lee et al., 2000). While this site is not thought to be important for insertion (Hayashi et al., 2000), it does lead to increased efficacy of receptors already at the synapse (Derkach et al., 1999). Again, after this stimulation, activation of phosphatases now removes this site, causing a 'reversal' of LTP. Thus the kinases/phosphatases that are important depend on the previous history of the synapse.

The state of the synapse is also regulated by development

The state of the synapse is not only determined by the past history of the synapse, but also by the developmental state. This has been examined in detail at CA3–CA1 synapses in vertebrates where a number of developmental switches are important in determining the stimulation that is required for plasticity. In particular, developmental switches in the ionotropic receptors are important. Both NMDA receptors and AMPA receptors are tetramers. For NMDA receptor, the tetramer consists of an NMDA receptor (NR)1, which is always the same and NR2 that consists of four isoforms A–D, of which the most prominent are NR2A and NR2B. There is a developmental switch from NR2B to NR2A (Sheng et al., 1994; Cull-Candy et al., 2001). NR2B is thought to induce plasticity more easily; this switching limits plasticity in adult animals (Dumas, 2005). Indeed, in animals where NR2B is expressed in the adult using transgenic technology, animals have enhanced learning ability (Tang et al., 1999). Similarly there are four major subunits of AMPA receptors GluR1-4, although each isoform also contains a number of splice forms as well. GluR4 and a spliced version of GluR2 are abundant early in development and are inserted into synapse more easily than GluR1 (Zhu et al., 2000; Kolleker et al., 2003). Thus, in GluR1 KOs, LTP exists in juveniles, but not in adults (Zamanillo et al., 1999; Lee et al., 2003). These are a few examples, but many more exist, demonstrating that the molecular constituents of the synapse regulate the ability of that synapse to be modified, and thus not all synapses are equal. This is an important concept that has not yet been used in computational models extracted from the nervous system.

Persistent activation of protein kinases

One of the best examples of a molecular memory trace is the persistent activation of a protein kinase. Since activation of the kinase was sufficient to increase synaptic strength in the short-term, prolonged activation of the kinase can lead to a prolonged memory trace. Kinases usually consist of two domains, a catalytic domain that is a phosphotransferase, and a regulatory domain containing a pseudosubstrate sequence that inhibits the enzyme in the absence of activators. Persistent kinase activity is generated by the removal of the regulatory domain, leaving a persistently active catalytic domain. In this case, inhibition of the kinase can temporarily erase the memory, although the memory could return if the inhibitor is reversible and the mechanism of persistent activation does not depend on continued activation of the kinase.

Persistent kinases are important for intermediate forms of facilitation in Aplysia

While one pulse of 5-HT increases synaptic strength for 20 min, five spaced pulses of 5-HT can increase synaptic strength for much longer times, at least 72 h in cultured neurons (Casadio et al., 1999). For the first 12 h, inhibitors of PKA can block the increase in synaptic strength, suggesting that at this point, the memory trace is due to persistent activation of PKA (Greenberg et al., 1987; Hegde et al., 1997; Chain et al., 1999). Initially (for 1–3 h), the persistent activation of PKA and the increase in synaptic strength depend on protein synthesis, but not gene expression. At later times both persistent activation of PKA and the increase in synaptic strength require gene expression (Muller and Carew, 1998). PKA consists of two separate proteins, a regulatory subunit and a catalytic subunit. In *Aplysia*, two regulatory subunits have been described, RI a type I and RII, a type II subunit (Bergold et al., 1992; Liu et al., 2004). The RI subunit is degraded after 5-HT treatment by the proteosome. This requires up-regulation of the messenger ribonucleic acid (mRNA) encoding ubiquitin hydrolase, whose activity increases levels of free ubiquitin by liberating ubiquitin from poly-ubiquitin chains. This suggests levels of ubiquitin are rate limiting for this degradation of RI (Hegde et al., 1997). The down-regulation of the RI subunit will lead to the persistent activation of PKA, and this is necessary for the persistent activation of gene expression (Chain et al., 1999). However, the increase in synaptic strength requires phosphorylation of proteins at the synapse. At synapses, the majority of PKA consists of the RII subunit, which is not degraded (Liu et al., 2004; Kurosu et al., 2007). RII subunits are distinguished from RI subunits due to their specific binding to an A-kinase anchoring protein (AKAP). Disruption of this binding blocks both short and long-term facilitation (LTF) suggesting that this binding is important for localizing PKA close to its substrates at the synapse (Liu et al., 2004). It is not clear if the down-regulation of the RI subunit is necessary for the persistent activation of PKA at the synapse that serves as the memory trace. It is possible that catalytic subunits freed from RI serve as the active kinases during the intermediate phase of facilitation, but this has not been directly demonstrated. It is also not clear what protein is synthesized that is required for the persistent activation of PKA at the synapse. There is a gene expression dependent up-regulation of the RII subunit that is required for LTF (Liu et al., 2004). While up-regulation of the RII subunit could shift the balance between RI and RII, it cannot generate a persistently active kinase.

A different persistent kinase is activated when one pulse of 5-HT is coupled to sensory neuron action potentials, or when a touch is coupled to a shock (Sutton and Carew, 2000; Sutton et al., 2004). In this case an intermediate facilitation/sensitization is induced that is independent of protein synthesis. This form of intermediate-term facilitation (ITF) is blocked by inhibitors of PKC, and not PKA (Sutton and Carew, 2000; Sutton et al., 2004). PKCs can be cleaved in the hinge domain separating the regulatory and catalytic domain to form a constitutively active catalytic domain called protein kinase M (PKM). The coupling of 5-HT and activity is necessary both to activate the calcium-dependent isoform of PKC, Apl I (Zhao et al., 2006), and to activate a protease (probably calpain) that may cleave PKC Apl I into a PKM form (Sutton et al., 2004). The substrate phosphorylated by PKC Apl I may be SNAP-25, since activators of PKC, such as phorbol esters increase synaptic strength through phosphorylation of SNAP-25 (Houeland et al., 2007). The coupling of activity and 5-HT also can lead to a longer-term facilitation (Schacher et al., 1997). It is not clear if this later form still requires persistent activation of PKC for its maintenance.

A persistently active PKC is also important for the maintenance of LTP

The first kinase to be identified as a persistently active kinase after LTP was CAMKII that becomes persistently activated through autophosphorylation. Initial excitement came from the fact that CAMKII is an oligomer of 12 subunits, and if activated by intrasubunit phosphorylation (Miller and Kennedy, 1986), one could envision a kinase that could remain active for long periods of time in the absence of additional stimulation (Lisman, 1985). However, further mechanistic studies showed that continued phosphorylation was not possible in the absence of calcium–calmodulin, and thus the persistent activity was time-limited (Hanson et al., 1994). Moreover, inhibitors of CAMKII activity do not block ongoing LTP, although they do block the induction of LTP (Chen et al., 2001). In contrast, inhibitors of PKC activity can block the maintenance of LTP (Malinow et al., 1988) suggesting a role for PKM formation. Only one form of PKC appears to have a PKM form in the hippocampus, the atypical zeta form (Sacktor et al., 1993). Moreover, specific inhibitors of this PKC isoform can block the maintenance of LTP (Serrano et al., 2005). This isoform is not generated by proteolysis, but by translation of an alternative transcript that does not contain the regulatory domain (Hernandez et al., 2003). Production of this kinase is induced at the translational level after LTP (Osten et al., 1996) and one model is that the kinase activity of PKMζ activates the protein synthesis of its own transcript. This positive feedback system could explain both the long persistence of the memory trace encoded by PKMζ (weeks) and the ability to erase this memory trace with inhibitors of PKMζ (Pastalkova et al., 2006). For a more complete description of the role of PKM in memory (see Chapter 2 by Sacktor, this volume).

Morphological changes at pre-existing synapses

Morphological changes can occur either on the presynaptic or postsynaptic side of the synapse. On the presynaptic side, there can be more docked vesicles and an increased size of the active zone where vesicles are released. On the postsynaptic side there can be an increased spine head, or increased size of the postsynaptic density (PSD) where the ionotropic receptors are clustered (see Chapter 11 by De Roo et al., this volume). While one can envision initial changes occurring on one side only, there is a strong correlation between presynaptic morphology and postsynaptic morphology (Harris and Stevens, 1989) suggesting that overtime a change in one side will cause a corresponding change on the other. For example, recent studies show that overexpression of PSD-95, a postsynaptic density protein can increase presynaptic release (Futai et al., 2007).

In *Aplysia* behavioral sensitization was shown to increase the number of docked vesicles and the size of active zones (Bailey and Chen, 1989). In cultures, increases in staining for synaptic vesicle proteins are seen in 'silent' synapses within hours after stimulation (Kim et al., 2003). Interestingly, the initial increase in synaptic vesicles occurred in the absence of protein synthesis, but required protein synthesis to be retained (Kim et al., 2003) (see Chapter 10 by Bailey and Kandel, in this volume, for more details). It is still not clear when these morphological changes contribute to the memory trace, but it is likely to be important for 24 h LTF when PKA inhibitors can no longer inhibit the facilitated EPSP (Hegde et al., 1997).

There have been a number of studies that observe morphological changes after LTP (Yuste and Bonhoeffer, 2001), also see Chapter 11 by De Roo et al., in this volume. In some paradigms, there is an increase in FM1-43 labeled release sites and in puncta of presynaptic proteins, similar to what was observed at *Aplysia* presynaptic terminal (Zakharenko et al., 2001). Increases in spine size have been correlated with the same synapses having undergone LTP, suggesting synapse specificity of these changes (Toni et al., 2001; Fedulov et al., 2007). The increase in spine size is associated with increases in the size of the PSD and increases in the number of AMPA receptors, although the increase in spine size can be dissociated from increases in AMPA receptors (Kopec et al., 2006). Changes in spine size are due to regulation of the actin cytoskeleton. Indeed LTP induces changes in the actin cytoskeleton (Colicos et al., 2001) and inhibitor of actin reorganization can block LTP (Kim and Lisman, 1999) suggesting a requirement for these changes. It should be noted that actin reorganization is required for recruitment of AMPA receptors (Korkotian and Segal, 2007) and formation of PKMζ (Kelly et al., 2007), and thus whether the requirement of actin dynamics for establishing the memory trace is due to its requirement for a change in spine size or due to other downstream actions is still an open question.

New synaptic connections

The formation of new synaptic connections requires coordination between the pre- and post-synaptic sides. Synapse formation require a number of steps: (i) new growth or sprouting from a neurite, or often from a pre-existing synapse; (ii) this growth has to connect to its partner and this connection has to initiate biochemical cascades leading to synapse formation; (iii) the nascent synapse has to become functional (i.e. release and receive transmitter; (iv) the synapse needs to be stabilized. This process has been studied extensively at the sensory–motor neuron synapse in *Aplysia* (see Chapter 10 by Bailey and Kandel, this volume). New growth can be seen after a single pulse of 5-HT; thus the requirement for multiple pulses is not to induce the synaptic growth (Udo et al., 2005). It requires activation of PI-3 kinase (Udo et al., 2005) and often initiates from presynaptic varicosities that already contain pre-existing synapses (Hatada et al., 2000). While the induction of growth does not require protein synthesis (Udo et al., 2005), the stabilization of the growth and the formation of synapses does require protein synthesis and eventually gene expression (Grabham et al., 2005; Udo et al., 2005). Stabilization of the new synapses requires an increase in the translational capacity of the synapse that is initiated at the same time as the synaptic growth (Casadio et al., 1999). New synaptic connections are seen both in cultures neurons and in the animal after sensitization training. These connections are new connections between the sensory and motor neuron and there is no evidence that new connections between neurons that were not previously synaptically connected occurs.

Training in enriched environments leads to rodents with a significantly higher number of synapses than caged animals (Volkmar and Greenough, 1972); however no specific increase in the density of synapses after a specific learning event has been demonstrated. At CA3–CA1 synapses there is little evidence for new synaptic connections forming soon after LTP (Sorra and Harris, 1998), however LTP induces new filipodia (Engert and Bonhoeffer, 1999), and these may turn into synapse at later times (Nagerl et al., 2007). In other areas of the brain, long-term imaging studies in living mice do observe new synapse formation, largely between already connected neurons; however stabilization of these new synapses is rare (Holtmaat et al., 2005). Clearly new techniques may be required to visualize if new synapses are formed after learning in the rodent model.

Protein-synthesis/gene-expression-dependent memory

Studies using various methodologies at both the cellular and behavioral level suggest that memories that persist require protein synthesis and gene expression. These do not only involve using inhibitors of protein synthesis and gene expression, but also biochemical and genetic manipulations that specifically regulate translation and transcription factors. Most convincingly, stimulation that would normally lead to a short-term change leads to long-term changes when manipulations are conducted such that the stimulation now activates gene expression. In *Aplysia*, activation of the cyclic-AMP response element binding protein (CREB) is required for gene-expression-dependent memory (Bartsch et al., 1998). After adding of an antibody to a repressor of the transcription factor CREB, making CREB easier to activate, a single pulse of 5-HT was sufficient to give LTF (Bartsch et al., 1995). In vertebrates, the CREB repressor is translationally regulated downstream of eukaryotic initiation factor 2 (eIF2)-alpha phosphorylation (Costa-Mattioli et al., 2005). Decreasing eIF2-alpha phosphorylation decreased levels of the CREB repressor and allowed lowering of not only the stimulation required for gene-expression-dependent late LTP (L-LTP), but also the stimulations required to generate a long-term memory in a number of different paradigms (Costa-Mattioli et al., 2007), see Chapter 5 by Costa-Mattioli and Sonenberg, this volume. The fact that gene-expression is the rate-limiting step for the formation of long-term memory has many implications for how memories are stored and raises a number of important questions that I will address in turn. First, is protein synthesis and gene expression required to 'consolidate' or maintain changes in synaptic strength, or do they represent a parallel mechanism to make longer-lasting memories. Second, is the role of protein synthesis and/or gene expression to make new 'memory' molecules or to increase the levels of proteins that already exist, but where the number of these molecules is rate limiting for synaptic strength. Finally, we will discuss the challenge that a requirement for gene expression means for synapse specificity.

Serial vs. parallel mechanisms for protein-synthesis-dependent memory

A major issue in the molecular mechanism of memory is whether long-term memory is simply a 'consolidation' of short-term memory or whether long-term memory represents a parallel molecular pathway. This can be made clear by an analogy where long-term memory is represented by a brick house. The serial model has postulates that the frame of the house is the short-term memory that is then consolidated (with plaster and brick) to solidify the structure into a long-term memory. In contrast, memory could work similarly to the houses of the three little pigs; houses of different stability (straw, stick and brick) are built in parallel, but only the most stable house is present at the end. A major difference between these models is that the parallel model predicts that short-term memories are not required for long-term memories. Thus while normally both short- and long-term memory traces are induced by the same stimulus, by looking downstream of the stimulus one should be able to discover manipulations to dissociate the two memory stores.

In *Aplysia*, the evidence is strong for the parallel mechanism of memory formation. First, one can generate long-term increases in synaptic strength in the absence of short-term changes as activation of gene expression in the sensory neurons is sufficient to cause increases in synaptic strength that last for 24 h in the absence of STF (Casadio et al., 1999). However, since longer-lasting changes do require synaptic activation (Casadio et al., 1999), does this suggest a serial pathway? To one extent it does, the longest lasting memory in *Aplysia* is probably stored in new synaptic connections between the sensory neuron and the motor neuron. This requires initial activation at the synapse to induce synaptic growth, which in a protein-synthesis and gene-expression-dependent manner is eventually consolidated into a new synaptic connection (see above). However, even at 24 h, this new synaptic connection is not required for either synaptic facilitation or memory. Thus, in cultures, while new synapses are observed at 24 h, there is no difference in the level of facilitation even when the formation of new synapses is blocked by not applying 5-HT to the synapse (Sun and Schacher, 1998). Moreover, behavioral sensitization leads to normal 24 h behavioral memory in the absence of new synapse formation (Wainwright et al., 2004). Thus, there is another mechanism, involving increases in synaptic strength at previously existing synapses, that is responsible for maintaining memory at earlier times, while in parallel new synapses are being built. It is important to note, that not only is gene expression required for the stabilization and consolidation of new synapses, but also for the long-term changes at pre-existing synapses. So in this case, the parallel pathways both involve gene-expression-dependent steps.

In CA3–CA1 synapses there are a number of indications that parallel processes occur during LTP, although the evidence is not as strong as in *Aplysia*. LTP can be induced as either early LTP (E-LTP) that is protein-synthesis-independent, or L-LTP that is blocked by inhibitors of protein synthesis and gene expression (Nguyen et al., 1994). Insertion of AMPA receptors is critical for E-LTP (Malinow and Malenka, 2002), but is L-LTP simply anchoring of these receptors, or is it due to a parallel pathway? In genetic models with loss of the GluR1 subunit of the AMPA receptor, E-LTP is abolished, but later forms of LTP may be spared (Hoffman et al., 2002). Indeed, in these mice short-term memories show severe impairment, while some long-term memories (within the same task) are spared (Schmitt et al., 2003) and see Chapter 9 by Sanderson et al., this volume. As mentioned earlier, retention of L-LTP depends on translation of PKMζ (Serrano et al., 2005). However, overexpression of PKMζ is sufficient to increase the number of AMPA receptors independently of E-LTP (Ling et al., 2002, 2006). Thus, it is possible that this phase of L-LTP does not depend on the previous insertion of AMPA receptors during E-LTP. This would suggest a parallel pathway. A major product of gene expression downstream of L-LTP is brain derived neurotrophic factor (BDNF) (Barco et al., 2005). However BDNF acts mainly in paradigms where presynaptic changes are critical for LTP (Zakharenko et al., 2003). This could act as a third parallel pathway underlying L-LTP.

What is produced by protein synthesis and/or gene expression?

The requirement for new proteins can be interpreted in three ways. First, protein synthesis inhibitors block basal translation, reducing levels of critical proteins required to maintain the memory trace. The second possibility is that protein synthesis inhibitors block an increase in translation induced by plasticity, and these new proteins act to maintain the memory trace by stabilizing the initial protein-synthesis-independent trace. The third possibility is that the new proteins induced by plasticity are sufficient for the memory trace in the presence of an additional latent trace (a tag), but the tag is independent of the initial memory trace. We will use the theoretical production of GluR1 in CA1 neurons to illustrate this concept (Fig. 3), although the same logic will be true for almost any memory trace. In the first possibility, protein-synthesis-independent insertion of GluR1 receptors into synapses leads to increases in synaptic strength, but perhaps the pool of receptors to be inserted has a short half-life and must be continually generated through protein synthesis. Blocking protein synthesis will remove this pool of GluR1 and thus block the maintenance of plasticity; however the stimulation that generates the plasticity does not need to regulate GluR1 synthesis (Fig. 3A). In this model, no regulation of translation is required for memory. The second possibility is quite similar, but in this case increases in translation of GluR1 stimulated by plasticity are required to generate the replacement pool required to maintain the memory trace (Fig. 3B). Indeed, LTP stimulation is known to increase local protein synthesis and genetic disruption of these pathways can lead to deficits in L-LTP and learning (see Chapter 4 by Banko and Klann, this volume). Since the production of the new protein is only required to maintain insertion in the face of turnover, the new proteins will have no effect unless the non-translation dependent insertion occurred first (Fig. 3B). Finally, the third explanation is that increases in translation are sufficient to generate a change in synaptic strength at synapses that have been tagged, but the tag is a parallel pathway to the increase in synaptic strength (Fig. 3C). In this case a signal is set in the synapse to allow newly synthesized GluR1 to be inserted into the synapse, but this signal is independent of the protein-synthesis-independent insertion of GluR1 in E-LTP. The conjunction of the signal and the newly synthesized protein allows for a parallel path of insertion that lasts longer than the initial phosphorylation-mediated pathway, but does not depend on E-LTP.

Depending on the synapse and the plasticity, it is probable that all of these explanations may be appropriate. An example where translation is required but not sufficient is metabotropic glutamate receptor (mGLUR)-mediated LTD.

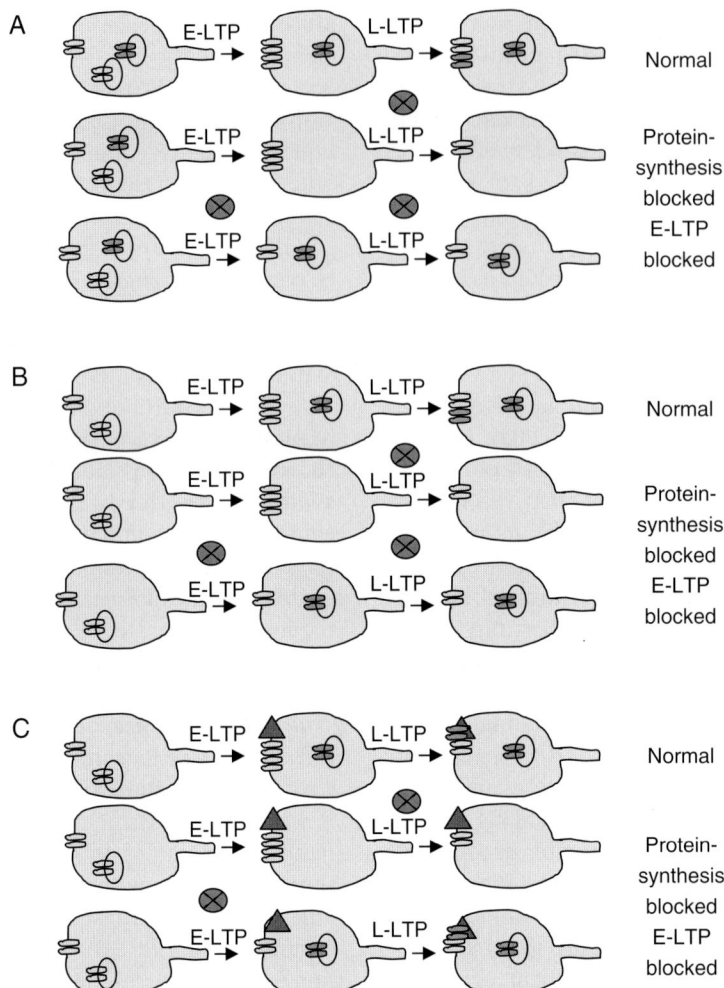

Fig. 3. Three models for the role of new protein synthesis in memory. (A) A model where only basal protein synthesis is required. Old AMPA receptors 'yellow' are inserted into the membrane to give E-LTP. These receptors are replaced with more newly synthesized AMPA receptors (green) to give L-LTP. In the presence of a protein synthesis inhibitor, the newer AMPA receptors were never synthesized and thus there are no AMPA receptors to replace the old ones and L-LTP is blocked. Blocking the initial insertion of AMPA receptors blocks L-LTP as the new AMPA receptors are only serving to replace the old ones. (B) A model where regulated protein synthesis is required to consolidate the synaptic change. The same as in (A) but the newly synthesized AMPA receptors (green) are newly synthesized by the E-LTP stimulation. (C) A parallel model. In this case the stimulus for E-LTP both inserts AMPA receptors and sets a tag (red diamond) such that newly synthesized receptors can insert into the membrane. Protein synthesis blocks the production of these new receptors and thus blocks L-LTP. In contrast if E-LTP is blocked but not protein synthesis, or tag production, L-LTP will exist in the absence of E-LTP. (See Color Plate 1.3 in color plate section.)

Retention of mGLUR-mediated LTD requires local protein synthesis (Huber et al., 2000). mGLUR LTD is larger in the absence of fragile X mental retardation protein (FMRP), the protein lost in patients suffering from fragile X mental retardation (Huber et al., 2002). Remarkably, this larger LTD no longer requires protein synthesis (Hou et al., 2006; Nosyreva and Huber, 2006). Thus, normally the proteins required for mGLUR LTD are encoded locally by mRNAs repressed by FMRP. In the absence of FMRP, these proteins are translated already and thus de-novo translation is not required. Importantly, this demonstrates that the protein synthesis required for LTD does not act in parallel with the initial induction of LTD, since in this case the long-lasting LTD would be occluded. Instead it must act in a serial fashion to maintain the removal of GluR1 receptors, without being sufficient to remove the receptors themselves.

Earlier we discussed persistent protein kinase activation as a mechanism for persistent traces. If the product of translation is a protein kinase (i.e. PKMζ), then this protein itself may be sufficient to activate the molecular trace in the absence of the initial mechanism underlying the early phase. Indeed PKMζ has been identified as a plasticity related protein at some synapses (i.e. a protein that interacts with a tag to lead to L-LTP (Sajikumar et al., 2005) (see Chapter 7 by Frey and Frey, this volume). However, it is still not clear in this case if the tag is the E-LTP (Fig. 3B) or the tag is a latent molecular trace (Fig. 3C) that is independent of E-LTP.

An example of a manipulation that seems to represent the example in Fig. 3A comes from the Bonhoeffer lab. First, they showed that one could induce protein-synthesis-dependence, even to what is normally called E-LTP by increasing the frequency of stimulation (Fonseca et al., 2006a). Indeed, stimulation of L-LTP is sensitive to inhibition of protein synthesis inhibitors, even at short times, when normally E-LTP is not sensitive to these inhibitors (Costa-Mattioli et al., 2005). This suggests that increasing the frequency of stimulation leads to the removal or degradation of proteins that now require protein synthesis for replacement, but under normal conditions basal synthesis is all that is required. Direct evidence for this came from experiments where the same group showed that inhibiting ongoing proteolysis (with proteosome inhibitors) made LTP protein-synthesis-independent, consistent in this case with the argument that the requirement for protein synthesis is mainly to replace proteins that are normally quickly degraded, but if this degradation is inhibited, protein synthesis is not necessary (Fonseca et al., 2006b).

One can extend this argument to the requirement for gene expression, which might be required only to again, replace mRNAs whose rate of degradation is high. Indeed, in *Aplysia*, there are several interesting experiments suggesting that the requirement for gene expression for 24 h LTF is not mandatory. After removing the cell body, 5-HT addition leads to 24 h LTF that requires protein synthesis, but not gene expression (Liu et al., 2003). Similarly, inhibition of the proteosome system can lead to 24 h changes in synaptic strength that again, require protein synthesis, but not gene expression (Zhao et al., 2003). Thus, in these cases, it is unlikely that gene expression is required to make some mRNA that did not previously exist, but rather to increase the levels of certain proteins by elevating the level of their mRNA, since other changes that can affect the stability or level of the proteins encoded by these mRNAs can overcome the requirement for gene expression, at least for 24 h LTF for these changes. However, while these manipulations can make gene-expression irrelevant, not only is gene expression normally required, but as discussed above, it is the rate-limiting step.

An interesting analogy for the relationship between gene expression and the synapse is a socket and a light bulb, where the increase in synaptic strength is represented by an increase in light. One imagines that the change at the synapse is the insertion of a socket for the light bulb (this would be the 'tag'). This socket is non-functional until a light bulb is placed in it, but there are a limited number of these bulbs. Thus, new light bulbs either made in the soma, or synthesized locally from mRNAs are required for the increase in light. Under normal conditions, the activation of gene expression is the rate-limiting step (i.e. light bulbs are rate limiting), since as discussed above, activating gene expression allows long-term changes from stimuli that normally led to short-term changes. Thus, normally the stronger or multiple stimulations required to generate a long-term memory are due to the need to make more light bulbs, while the socket can be set in place by stimuli that would normally only generate short-term memories. However, under certain conditions, such as axotomy or proteosome inhibition, there may be an increase in the levels of available light bulbs in the absence of gene expression and thus the need for protein synthesis can be alleviated, and the socket or tag becomes the rate-limiting step.

What are the light bulbs?

There have been a large number of studies attempting to determine the proteins that are newly synthesized after a learning experience. One confounding problem is that neuronal homeostasis also occurs during this time, thus firing a neuron at high frequency will not only cause LTP, but also will stimulate a homeostatic process that will lead to a decrease in synaptic strength to compensate for a perceived overexcitability (Turrigiano and Nelson, 2000). Thus, many of the proteins identified in these screens appear to have a negative regulatory function on synaptic strength (Pak and Sheng, 2003; Sala et al., 2003; Shepherd et al., 2006). It is also suggested that the proteins made by stimulation must serve both increases and decreases in synaptic strength, with the determining factor being the tag as first suggested by cross-tagging experiments, where stimulation of a synapse with an LTD stimulus could be extended into L-LTD with proteins generated from an L-LTP stimulus (Sajikumar and Frey, 2004) (see Chapter 7 by Frey and Frey, this volume). There are two good candidates for proteins that are synthesized after learning that may be important for increases in synaptic strength seen during L-LTP and have been discussed above. One is BDNF, a growth factor, whose presence may be sufficient to make L-LTP protein-synthesis-independent (Pang et al., 2004; Barco et al., 2005). The second is PKMζ, that is sufficient to increase synaptic strength through increasing the number of synaptic AMPA receptors (Ling et al., 2006).

In *Aplysia*, several mRNAs are up-regulated after learning. Some of them are involved in a positive feedback loop (either transcription factors, or proteins involved in up-regulating additional transcription) (Alberini et al., 1994; Hegde et al., 1997). Others may relate to stabilization of synaptic growth and the up-regulation of the translational machinery that is necessary for this (Kennedy et al., 1992; Kuhl et al., 1992; Zwartjes et al., 1998; Giustetto et al., 2003; Moccia et al., 2003). In vertebrates as well, components of the translational machinery are sent out processes and translation of these components is up-regulated after LTP (Tsokas et al., 2005, 2007; Poon et al., 2006). One critical factor induced at the mRNA and protein level in *Aplysia* is the neuropeptide sensorin (Sun et al., 2001; Hu et al., 2006). Sensorin is important both for activation of gene expression through ERK activation, and for the maintaining new synapses (Hu et al., 2004a, b), and in this sense may be analogous to BDNF in vertebrates. Thus, in *Aplysia* where morphological changes are more definitively associated with memory, much of the synthesis appears to be involved in the stabilization of morphological changes. It will be interesting to see in vertebrates if the protein synthesis/gene expression requirements will eventually be tied to these mechanisms.

The synaptic tag and synapse specificity in the face of new protein synthesis

The products of protein synthesis and gene expression need to be restricted to activated synapses to retain the synapse specificity of the memory trace. A mechanism must therefore exist to restrict the new proteins synthesized to act only at synapses that were activated; in other words the activated synapses must somehow be tagged in order to capture the products on new synthesis (Sossin, 1996; Frey and Morris, 1998). However, while data supports the existence of such a tag (Frey and Morris, 1997; Martin et al., 1997), the physical nature of the 'tag' has been elusive. There are three main models for the tag that are linked to the roles that protein synthesis plays in the memory trace (see Fig. 3). In the first model, the tag acts by allowing for specific protein synthesis at the tagged synapse and the proteins themselves lead to the increase in synaptic strength. An example of this mechanism would be that the tag would allow anchoring of PKMζ mRNA to a synapse tagged for potentiation, the protein would then be translated at that site, and restricted to act in this locale. The second model is that the tag is the short-term synaptic change, and the protein products act to consolidate this change. For example, removal of AMPA receptors leading to LTD are stabilized by a new protein product. This protein would have no action at synapses where no internalization had occurred. In this model the protein product consolidates the change. Finally, the third model assumes that the tag is a latent memory trace; and the protein process is necessary for it to lead to changes in synaptic strength. For example, new filipodia grow out and require protein products for stabilization, synapse formation and synapse stabilization. In this case, the tag and protein synthesis become a parallel process independent of the initial event that occur changing synaptic strength in the short-term. In all of these models, there can be a time differential between the tag and the production of proteins on the order of minutes to hours.

Conclusions

So, what are the new insights discussed above that may be important to extract for cognitive models. The first major point is that the synapse in not always the same. The induction of a memory trace depends on the state of the synapse. The state of the synapse itself changes based on its past experience, and this again will affect the type of molecular trace that will be made. The second major point is that not all memory traces are the same. The change in synaptic strength is not a simple binary decision after an experience. There are multiple independent traces that last for various periods of time and differ in their volatility. Cognitive models could use this fact to more completely model how weights are altered after experience. The third and last major point is that the requirement for protein synthesis and its possible interaction with a tag at activated synapses challenges a simple associative Hebbian model, since the associations between the generation of proteins and the half-life of a tag lasts many minutes, not milliseconds as in the association between presynaptic and postsynaptic input. In this review, we have also identified several important questions concerning the role for new proteins, the identity of these proteins and the composition of the tag. Answers to these questions will allow for more sophisticated understanding of the molecular constraints on how memories are formed and stabilized.

Abbreviations

5-HT	5-hydoxytryptamine (serotonin)
AMPA	alpha-amino-3-hydroxy-5-methyl-4-isoxazolepropionic acid
BDNF	brain derived neurotrophic factor
cAMP	cyclic adenosine mono-phosphate
CAMKII	calcium calmodulin-dependent kinase II
CREB	cyclic-AMP response element binding protein
eIF2	eukaryotic initiation factor 2
FM1-43	Fei Mao
LTD	long-term depression
LTP	long-term potentiation
E-LTP	early long-term potentiation
EPSP	excitatory postsynaptic potential
FMRP	fragile X mental retardation protein
GluR	glutamate receptor
ITF	intermediate-term facilitation
L-LTP	late long-term potentiation
LTF	long-term facilitation
mGLUR	metabotropic glutamate receptor
mRNA	messenger ribonucleic acid
n	number of synapses
NMDA	N-methyl-D-aspartic acid
NR	NMDA receptor
p	probability of release
PKA	protein kinase A (cAMP-dependent protein kinase)
PKC	protein kinase C
PKM	protein kinase M (magnesium-dependent protein kinase)
PSD	postsynaptic density
q	size of response to a single vesicular release
STF	short-term facilitation

Acknowledgments

This work was supported by Canadian Institute of Health Research grants MT-12046, MT1512 and group grant MGC-57079. WSS is a William Dawson scholar and a Fonds de la Recherche santé du Québec (FRSQ) Chercheur nationaux. Thanks to Dr. Jean-Claude Lacaille for helpful comments.

References

Abraham, W.C. and Bear, M.F. (1996) Metaplasticity: the plasticity of synaptic plasticity. Trends Neurosci., 19: 126–130.

Alberini, C.M., Ghirardi, M., Metz, R. and Kandel, E.R. (1994) C/EBP is an immediate-early gene required for the consolidation of long-term facilitation in Aplysia. Cell, 76: 1099–1114.

Bailey, C.H. and Chen, M. (1989) Structural plasticity at identified synapses during long-term memory in Aplysia. J. Neurobiol., 20: 356–372.

Barco, A., Patterson, S., Alarcon, J.M., Gromova, P., Mata-Roig, M., Morozov, A. and Kandel, E.R. (2005) Gene expression profiling of facilitated L-LTP in VP16-CREB mice reveals that BDNF is critical for the maintenance of LTP and its synaptic capture. Neuron, 48: 123–137.

Bartsch, D., Casadio, A., Karl, K.A., Serodio, P. and Kandel, E.R. (1998) CREB1 encodes a nuclear activator, a repressor, and a cytoplasmic modulator that form a regulatory unit critical for long-term facilitation. Cell, 95: 211–223.

Bartsch, D., Ghirardi, M., Skehel, P.A., Karl, K.A., Herder, S.P., Chen, M., Bailey, C.H. and Kandel, E.R. (1995) Aplysia CREB2 represses long-term facilitation: relief of repression converts transient facilitation into long-term functional and structural change. Cell, 83: 979–992.

Bergold, P.J., Beushausen, S.A., Sacktor, T.C., Cheley, S., Bayley, H. and Schwartz, J.H. (1992) A regulatory subunit of the cAMP-dependent protein kinase down-regulated in Aplysia sensory neurons during long-term sensitization. Neuron, 8: 387–397.

Bliss, T.V. and Collingridge, G.L. (1993) A synaptic model of memory: long-term potentiation in the hippocampus. Nature, 361: 31–39.

Bliss, T.V. and Lomo, T. (1973) Long-lasting potentiation of synaptic transmission in the dentate area of the anaesthetized rabbit following stimulation of the perforant path. J. Physiol., 232: 331–356.

Boehm, J., Kang, M.G., Johnson, R.C., Esteban, J., Huganir, R.L. and Malinow, R. (2006) Synaptic incorporation of AMPA receptors during LTP is controlled by a PKC phosphorylation site on GluR1. Neuron, 51: 213–225.

Bolshakov, V.Y., Golan, H., Kandel, E.R. and Siegelbaum, S.A. (1997) Recruitment of new sites of synaptic transmission during the cAMP-dependent late phase of LTP at CA3–CA1 synapses in the hippocampus. Neuron, 19: 635–651.

Byrne, J.H. and Kandel, E.R. (1996) Presynaptic facilitation revisited: state and time dependence. J. Neurosci., 16: 425–435.

Casadio, A., Martin, K.C., Giustetto, M., Zhu, H., Chen, M., Bartsch, D., Bailey, C.H. and Kandel, E.R. (1999) A transient, neuron-wide form of CREB-mediated long-term facilitation can be stabilized at specific synapses by local protein synthesis. Cell, 99: 221–237.

Castellucci, V. and Kandel, E.R. (1976) Presynaptic facilitation as a mechanism for behavioral sensitization in Aplysia. Science, 194: 1176–1178.

Castellucci, V., Pinsker, H., Kupfermann, I. and Kandel, E.R. (1970) Neuronal mechanisms of habituation and dishabituation of the gill-withdrawal reflex in Aplysia. Science, 167: 1745–1748.

Castellucci, V.F. and Kandel, E.R. (1974) A quantal analysis of the synaptic depression underlying habituation of the gill-withdrawal reflex in Aplysia. Proc. Natl. Acad. Sci. U.S.A., 71: 5004–5008.

Chain, D.G., Casadio, A., Schacher, S., Hegde, A.N., Valbrun, M., Yamamoto, N., Goldberg, A.L., Bartsch, D., Kandel, E.R. and Schwartz, J.H. (1999) Mechanisms for generating the autonomous cAMP-dependent protein kinase required for long-term facilitation in Aplysia. Neuron, 22: 147–156.

Chen, H.X., Otmakhov, N., Strack, S., Colbran, R.J. and Lisman, J.E. (2001) Is persistent activity of calcium/calmodulin-dependent kinase required for the maintenance of LTP? J. Neurophysiol., 85: 1368–1376.

Chitwood, R.A., Li, Q. and Glanzman, D.L. (2001) Serotonin facilitates AMPA-type responses in isolated siphon motor neurons of Aplysia in culture. J. Physiol., 534: 501–510.

Colicos, M.A., Collins, B.E., Sailor, M.J. and Goda, Y. (2001) Remodeling of synaptic actin induced by photoconductive stimulation. Cell, 107: 605–616.

Cooper, S.J. (2005) Donald O. Hebb's synapse and learning rule: a history and commentary. Neurosci. Biobehav. Rev., 28: 851–874.

Costa-Mattioli, M., Gobert, D., Harding, H., Herdy, B., Azzi, M., Bruno, M., Bidinosti, M., Ben Mamou, C., Marcinkiewicz, E., Yoshida, M., Imataka, H., Cuello, A.C., Seidah, N., Sossin, W., Lacaille, J.C., Ron, D., Nader, K. and Sonenberg, N. (2005) Translational control of hippocampal synaptic plasticity and memory by the eIF2alpha kinase GCN2. Nature, 436: 1166–1173.

Costa-Mattioli, M., Gobert, D., Stern, E., Gamache, K., Colina, R., Cuello, C., Sossin, W., Kaufman, R., Pelletier, J., Rosenblum, K., Krnjevic, K., Lacaille, J.C., Nader, K. and Sonenberg, N. (2007) eIF2alpha phosphorylation bidirectionally regulates the switch from short- to long-term synaptic plasticity and memory. Cell, 129: 195–206.

Cull-Candy, S., Brickley, S. and Farrant, M. (2001) NMDA receptor subunits: diversity, development and disease. Curr. Opin. Neurobiol., 11: 327–335.

Dale, N., Schacher, S. and Kandel, E.R. (1988) Long-term facilitation in Aplysia involves increase in transmitter release. Science, 239: 282–285.

Daoudal, G. and Debanne, D. (2003) Long-term plasticity of intrinsic excitability: learning rules and mechanisms. Learn. Mem., 10: 456–465.

Derkach, V., Barria, A. and Soderling, T.R. (1999) Ca^{2+}/calmodulin-kinase II enhances channel conductance of alpha-amino-3-hydroxy-5-methyl-4-isoxazolepropionate type glutamate receptors. Proc. Natl. Acad. Sci. U.S.A., 96: 3269–3274.

Dumas, T.C. (2005) Developmental regulation of cognitive abilities: modified composition of a molecular switch turns on associative learning. Prog. Neurobiol., 76: 189–211.

Eliot, L.S., Kandel, E.R. and Hawkins, R.D. (1994) Modulation of spontaneous transmitter release during depression and posttetanic potentiation of *Aplysia* sensory–motor neuron synapses isolated in culture. J. Neurosci., 14: 3280–3292.

Engert, F. and Bonhoeffer, T. (1999) Dendritic spine changes associated with hippocampal long-term synaptic plasticity. Nature, 399: 66–70.

Esteban, J.A., Shi, S.H., Wilson, C., Nuriya, M., Huganir, R.L. and Malinow, R. (2003) PKA phosphorylation of AMPA receptor subunits controls synaptic trafficking underlying plasticity. Nat. Neurosci., 6: 136–143.

Fedulov, V., Rex, C.S., Simmons, D.A., Palmer, L., Gall, C.M. and Lynch, G. (2007) Evidence that long-term potentiation occurs within individual hippocampal synapses during learning. J. Neurosci., 27: 8031–8039.

Fonseca, R., Nagerl, U.V. and Bonhoeffer, T. (2006a) Neuronal activity determines the protein synthesis dependence of long-term potentiation. Nat. Neurosci., 9: 478–480.

Fonseca, R., Vabulas, R.M., Hartl, F.U., Bonhoeffer, T. and Nagerl, U.V. (2006b) A balance of protein synthesis and proteasome-dependent degradation determines the maintenance of LTP. Neuron, 52: 239–245.

Frey, U. and Morris, R.G. (1997) Synaptic tagging and long-term potentiation. Nature, 385: 533–536.

Frey, U. and Morris, R.G. (1998) Synaptic tagging: implications for late maintenance of hippocampal long-term potentiation. Trends Neurosci., 21: 181–188.

Frick, A. and Johnston, D. (2005) Plasticity of dendritic excitability. J. Neurobiol., 64: 100–115.

Frick, A., Magee, J. and Johnston, D. (2004) LTP is accompanied by an enhanced local excitability of pyramidal neuron dendrites. Nat. Neurosci., 7: 126–135.

Frost, W.N., Clark, G.A. and Kandel, E.R. (1988) Parallel processing of short-term memory for sensitization in *Aplysia*. J. Neurobiol., 19: 297–334.

Futai, K., Kim, M.J., Hashikawa, T., Scheiffele, P., Sheng, M. and Hayashi, Y. (2007) Retrograde modulation of presynaptic release probability through signaling mediated by PSD-95-neuroligin. Nat. Neurosci., 10: 186–195.

Ghirardi, M., Braha, O., Hochner, B., Montarolo, P.G., Kandel, E.R. and Dale, N. (1992) Roles of PKA and PKC in facilitation of evoked and spontaneous transmitter release at depressed and nondepressed synapses in *Aplysia* sensory neurons. Neuron, 9: 479–489.

Giustetto, M., Hegde, A.N., Si, K., Casadio, A., Inokuchi, K., Pei, W., Kandel, E.R. and Schwartz, J.H. (2003) Axonal transport of eukaryotic translation elongation factor 1alpha mRNA couples transcription in the nucleus to long-term facilitation at the synapse. Proc. Natl. Acad. Sci. U.S.A., 100: 13680–13685.

Grabham, P.W., Wu, F., Schacher, S. and Goldberg, D.J. (2005) Initiating morphological changes associated with long-term facilitation in *Aplysia* is independent of transcription or translation in the cell body. J. Neurobiol., 64: 202–212.

Greenberg, S.M., Castellucci, V.F., Bayley, H. and Schwartz, J.H. (1987) A molecular mechanism for long-term sensitization in *Aplysia*. Nature, 329: 62–65.

Hammond, R.S., Bond, C.T., Strassmaier, T., Ngo-Anh, T.J., Adelman, J.P., Maylie, J. and Stackman, R.W. (2006) Small-conductance Ca^{2+}-activated $K+$ channel type 2 (SK2) modulates hippocampal learning, memory, and synaptic plasticity. J. Neurosci., 26: 1844–1853.

Hanson, P.I., Meyer, T., Stryer, L. and Schulman, H. (1994) Dual role of calmodulin in autophosphorylation of multifunctional CaM kinase may underlie decoding of calcium signals. Neuron, 12: 943–956.

Harris, K.M. and Stevens, J.K. (1989) Dendritic spines of CA 1 pyramidal cells in the rat hippocampus: serial electron microscopy with reference to their biophysical characteristics. J. Neurosci., 9: 2982–2997.

Hatada, Y., Wu, F., Sun, Z.Y., Schacher, S. and Goldberg, D.J. (2000) Presynaptic morphological changes associated with long-term synaptic facilitation are triggered by actin polymerization at preexisting varicosities. J. Neurosci., 20: p. RC82.

Hawkins, R.D., Abrams, T.W., Carew, T.J. and Kandel, E.R. (1983) A cellular mechanism of classical conditioning in *Aplysia*: activity-dependent amplification of presynaptic facilitation. Science, 219: 400–405.

Hawkins, R.D., Kandel, E.R. and Bailey, C.H. (2006) Molecular mechanisms of memory storage in *Aplysia*. Biol. Bull., 210: 174–191.

Hayashi, Y., Shi, S.H., Esteban, J.A., Piccini, A., Poncer, J.C. and Malinow, R. (2000) Driving AMPA receptors into synapses by LTP and CaMKII: requirement for GluR1 and PDZ domain interaction. Science, 287: 2262–2267.

Hegde, A.N., Inokuchi, K., Pei, W., Casadio, A., Ghirardi, M., Chain, D.G., Martin, K.C., Kandel, E.R. and Schwartz, J.H. (1997) Ubiquitin C-terminal hydrolase is an immediate-early gene essential for long-term facilitation in *Aplysia*. Cell, 89: 115–126.

Hernandez, A.I., Blace, N., Crary, J.F., Serrano, P.A., Leitges, M., Libien, J.M., Weinstein, G., Tcherapanov, A. and Sacktor, T.C. (2003) Protein kinase M zeta synthesis from a brain mRNA encoding an independent protein kinase C zeta catalytic domain. Implications for the molecular mechanism of memory. J. Biol. Chem., 278: 40305–40316.

Hoffman, D.A., Sprengel, R. and Sakmann, B. (2002) Molecular dissection of hippocampal theta-burst pairing potentiation. Proc. Natl. Acad. Sci. U.S.A., 99: 7740–7745.

Holtmaat, A.J., Trachtenberg, J.T., Wilbrecht, L., Shepherd, G.M., Zhang, X., Knott, G.W. and Svoboda, K. (2005) Transient and persistent dendritic spines in the neocortex in vivo. Neuron, 45: 279–291.

Hou, L., Antion, M.D., Hu, D., Spencer, C.M., Paylor, R. and Klann, E. (2006) Dynamic translational and proteasomal regulation of fragile X mental retardation protein controls mGluR-dependent long-term depression. Neuron, 51: 441–454.

Houeland, G., Nakhost, A., Sossin, W.S. and Castellucci, V.F. (2007) PKC modulation of transmitter release by SNAP-25 at sensory-to-motor synapses in Aplysia. J. Neurophysiol., 97: 134–143.

Hu, J.Y., Glickman, L., Wu, F. and Schacher, S. (2004a) Serotonin regulates the secretion and autocrine action of a neuropeptide to activate MAPK required for long-term facilitation in Aplysia. Neuron, 43: 373–385.

Hu, J.Y., Goldman, J., Wu, F. and Schacher, S. (2004b) Target-dependent release of a presynaptic neuropeptide regulates the formation and maturation of specific synapses in Aplysia. J. Neurosci., 24: 9933–9943.

Hu, J.Y., Wu, F. and Schacher, S. (2006) Two signaling pathways regulate the expression and secretion of a neuropeptide required for long-term facilitation in Aplysia. J. Neurosci., 26: 1026–1035.

Huang, Y.Y., Zakharenko, S.S., Schoch, S., Kaeser, P.S., Janz, R., Sudhof, T.C., Siegelbaum, S.A. and Kandel, E.R. (2005) Genetic evidence for a protein-kinase-A-mediated presynaptic component in NMDA-receptor-dependent forms of long-term synaptic potentiation. Proc. Natl. Acad. Sci. U.S.A., 102: 9365–9370.

Huber, K.M., Gallagher, S.M., Warren, S.T. and Bear, M.F. (2002) Altered synaptic plasticity in a mouse model of fragile X mental retardation. Proc. Natl. Acad. Sci. U.S.A., 99: 7746–7750.

Huber, K.M., Kayser, M.S. and Bear, M.F. (2000) Role for rapid dendritic protein synthesis in hippocampal mGluR-dependent long-term depression. Science, 288: 1254–1257.

Kandel, E.R. (2001) The molecular biology of memory storage: a dialogue between genes and synapses. Science, 294: 1030–1038.

Kelly, M.T., Yao, Y., Sondhi, R. and Sacktor, T.C. (2007) Actin polymerization regulates the synthesis of PKMzeta in LTP. Neuropharmacology, 52: 41–45.

Kennedy, T.E., Kuhl, D., Barzilai, A., Sweatt, J.D. and Kandel, E.R. (1992) Long-term sensitization training in Aplysia leads to an increase in calreticulin, a major presynaptic calcium-binding protein. Neuron, 9: 1013–1024.

Kim, C.H. and Lisman, J.E. (1999) A role of actin filament in synaptic transmission and long-term potentiation. J. Neurosci., 19: 4314–4324.

Kim, J.H., Udo, H., Li, H.L., Youn, T.Y., Chen, M., Kandel, E.R. and Bailey, C.H. (2003) Presynaptic activation of silent synapses and growth of new synapses contribute to intermediate and long-term facilitation in Aplysia. Neuron, 40: 151–165.

Klein, M. (1994) Synaptic augmentation by 5-HT at rested Aplysia sensorimotor synapses: independence of action potential prolongation. Neuron, 13: 159–166.

Klein, M., Hochner, B. and Kandel, E.R. (1986) Facilitatory transmitters and cAMP can modulate accommodation as well as transmitter release in Aplysia sensory neurons: evidence for parallel processing in a single cell. Proc. Natl. Acad. Sci. U.S.A., 83: 7994–7998.

Kolleker, A., Zhu, J.J., Schupp, B.J., Qin, Y., Mack, V., Borchardt, T., Kohr, G., Malinow, R., Seeburg, P.H. and Osten, P. (2003) Glutamatergic plasticity by synaptic delivery of GluR-B(long)-containing AMPA receptors. Neuron, 40: 1199–1212.

Kopec, C.D., Li, B., Wei, W., Boehm, J. and Malinow, R. (2006) Glutamate receptor exocytosis and spine enlargement during chemically induced long-term potentiation. J. Neurosci., 26: 2000–2009.

Korkotian, E. and Segal, M. (2007) Morphological constraints on calcium dependent glutamate receptor trafficking into individual dendritic spine. Cell Calcium, 42: 41–57.

Kuhl, D., Kennedy, T.E., Barzilai, A. and Kandel, E.R. (1992) Long-term sensitization training in Aplysia leads to an increase in the expression of BiP, the major protein chaperon of the ER. J. Cell. Biol., 119: 1069–1076.

Kurosu, T., Hernandez, A.I. and Schwartz, J.H. (2007) Serotonin induces selective cleavage of the PKA RI subunit but not RII subunit in Aplysia neurons. Biochem. Biophys. Res. Commun., 359: 563–567.

Lee, H.K., Barbarosie, M., Kameyama, K., Bear, M.F. and Huganir, R.L. (2000) Regulation of distinct AMPA receptor phosphorylation sites during bidirectional synaptic plasticity. Nature, 405: 955–959.

Lee, H.K., Takamiya, K., Han, J.S., Man, H., Kim, C.H., Rumbaugh, G., Yu, S., Ding, L., He, C., Petralia, R.S., Wenthold, R.J., Gallagher, M. and Huganir, R.L. (2003) Phosphorylation of the AMPA receptor GluR1 subunit is required for synaptic plasticity and retention of spatial memory. Cell, 112: 631–643.

Levenson, J., Weeber, E., Selcher, J.C., Kategaya, L.S., Sweatt, J.D. and Eskin, A. (2002) Long-term potentiation and contextual fear conditioning increase neuronal glutamate uptake. Nat. Neurosci., 5: 155–161.

Liao, D., Hessler, N.A. and Malinow, R. (1995) Activation of postsynaptically silent synapses during pairing-induced LTP in CA1 region of hippocampal slice. Nature, 375: 400–404.

Ling, D.S., Benardo, L.S. and Sacktor, T.C. (2006) Protein kinase Mzeta enhances excitatory synaptic transmission by increasing the number of active postsynaptic AMPA receptors. Hippocampus, 16: 443–452.

Ling, D.S., Benardo, L.S., Serrano, P.A., Blace, N., Kelly, M.T., Crary, J.F. and Sacktor, T.C. (2002) Protein kinase Mzeta is necessary and sufficient for LTP maintenance. Nat. Neurosci., 5: 295–296.

Lisman, J.E. (1985) A mechanism for memory storage insensitive to molecular turnover: a bistable autophosphorylating kinase. Proc. Natl. Acad. Sci. U.S.A., 82: 3055–3057.

Liu, J., Hu, J.Y., Schacher, S. and Schwartz, J.H. (2004) The two regulatory subunits of Aplysia cAMP-dependent protein kinase mediate distinct functions in producing synaptic plasticity. J. Neurosci., 24: 2465–2474.

Liu, K., Hu, J.Y., Wang, D. and Schacher, S. (2003) Protein synthesis at synapse versus cell body: enhanced but transient expression of long-term facilitation at isolated synapses. J. Neurobiol., 56: 275–286.

Malinow, R., Madison, D.V. and Tsien, R.W. (1988) Persistent protein kinase activity underlying long-term potentiation. Nature, 335: 820–824.

Malinow, R. and Malenka, R.C. (2002) AMPA receptor trafficking and synaptic plasticity. Annu. Rev. Neurosci., 25: 103–126.

Mammen, A.L., Kameyama, K., Roche, K.W. and Huganir, R.L. (1997) Phosphorylation of the alpha-amino-3-hydroxy-5-methylisoxazole4-propionic acid receptor GluR1 subunit by calcium/calmodulin-dependent kinase II. J. Biol. Chem., 272: 32528–32533.

Manseau, F., Fan, X., Hueftlein, T., Sossin, W. and Castellucci, V.F. (2001) Ca^{2+}-independent protein kinase C Apl II mediates the serotonin-induced facilitation at depressed *Aplysia* sensorimotor synapses. J. Neurosci., 21: 1247–1256.

Marinesco, S. and Carew, T.J. (2002) Serotonin release evoked by tail nerve stimulation in the CNS of *Aplysia*: characterization and relationship to heterosynaptic plasticity. J. Neurosci., 22: 2299–2312.

Martin, K.C., Casadio, A., Zhu, H., Yaping, E., Rose, J.C., Chen, M., Bailey, C.H. and Kandel, E.R. (1997) Synapse-specific, long-term facilitation of aplysia sensory to motor synapses: a function for local protein synthesis in memory storage. Cell, 91: 927–938.

McClelland, J.L. and Rumelhart, D.E. (1985) Distributed memory and the representation of general and specific information. J. Exp. Psychol. Gen., 114: 159–197.

Miller, S.G. and Kennedy, M.B. (1986) Regulation of brain type II Ca^{2+}/calmodulin-dependent protein kinase by autophosphorylation: a Ca^{2+}-triggered molecular switch. Cell, 44: 861–870.

Moccia, R., Chen, D., Lyles, V., Kapuya, E., E, Y., Kalachikov, S., Spahn, C.M., Frank, J., Kandel, E.R., Barad, M. and Martin, K.C. (2003) An unbiased cDNA library prepared from isolated *Aplysia* sensory neuron processes is enriched for cytoskeletal and translational mRNAs. J. Neurosci., 23: 9409–9417.

Montarolo, P.G., Goelet, P., Castellucci, V.F., Morgan, J., Kandel, E.R. and Schacher, S. (1986) A critical period for macromolecular synthesis in long-term heterosynaptic facilitation in *Aplysia*. Science, 234: 1249–1254.

Morris, R.G. (2003) Long-term potentiation and memory. Philos. Trans. R. Soc. Lond. B Biol. Sci., 358: 643–647.

Muller, U. and Carew, T.J. (1998) Serotonin induces temporally and mechanistically distinct phases of persistent PKA activity in *Aplysia* sensory neurons. Neuron, 21: 1423–1434.

Nagerl, U.V., Kostinger, G., Anderson, J.C., Martin, K.A. and Bonhoeffer, T. (2007) Protracted synaptogenesis after activity-dependent spinogenesis in hippocampal neurons. J. Neurosci., 27: 8149–8156.

Nguyen, P.V., Abel, T. and Kandel, E.R. (1994) Requirement of a critical period of transcription for induction of a late phase of LTP. Science, 265: 1104–1107.

Nolan, M.F., Malleret, G., Dudman, J.T., Buhl, D.L., Santoro, B., Gibbs, E., Vronskaya, S., Buzsaki, G., Siegelbaum, S.A., Kandel, A. and Morozov, A. (2004) A behavioral role for dendritic integration: HCN1 channels constrain spatial memory and plasticity at inputs to distal dendrites of CA1 pyramidal neurons. Cell, 119: 719–732.

Nosyreva, E.D. and Huber, K.M. (2006) Metabotropic receptor-dependent long-term depression persists in the absence of protein synthesis in the mouse model of fragile X syndrome. J. Neurophysiol., 95: 3291–3295.

Osten, P., Valsamis, L., Harris, A. and Sacktor, T.C. (1996) Protein synthesis-dependent formation of protein kinase Mzeta in long-term potentiation. J. Neurosci., 16: 2444–2451.

Pak, D.T. and Sheng, M. (2003) Targeted protein degradation and synapse remodeling by an inducible protein kinase. Science, 302: 1368–1373.

Pang, P.T., Teng, H.K., Zaitsev, E., Woo, N.T., Sakata, K., Zhen, S., Teng, K.K., Yung, W.H., Hempstead, B.L. and Lu, B. (2004) Cleavage of proBDNF by tPA/plasmin is essential for long-term hippocampal plasticity. Science, 306: 487–491.

Pastalkova, E., Serrano, P., Pinkhasova, D., Wallace, E., Fenton, A.A. and Sacktor, T.C. (2006) Storage of spatial information by the maintenance mechanism of LTP. Science, 313: 1141–1144.

Poon, M.M., Choi, S.H., Jamieson, C.A., Geschwind, D.H. and Martin, K.C. (2006) Identification of process-localized mRNAs from cultured rodent hippocampal neurons. J. Neurosci., 26: 13390–13399.

Rankin, C.H. and Carew, T.J. (1988) Dishabituation and sensitization emerge as separate processes during development in *Aplysia*. J. Neurosci., 8: 197–211.

Roberts, A.C. and Glanzman, D.L. (2003) Learning in *Aplysia*: looking at synaptic plasticity from both sides. Trends Neurosci., 26: 662–670.

Sacktor, T.C., Osten, P., Valsamis, H., Jiang, X., Naik, M.U. and Sublette, E. (1993) Persistent activation of the zeta isoform of protein kinase C in the maintenance of long-term potentiation. Proc. Natl. Acad. Sci. U.S.A., 90: 8342–8346.

Sajikumar, S. and Frey, J.U. (2004) Late-associativity, synaptic tagging, and the role of dopamine during LTP and LTD. Neurobiol. Learn. Mem., 82: 12–25.

Sajikumar, S., Navakkode, S., Sacktor, T.C. and Frey, J.U. (2005) Synaptic tagging and cross-tagging: the role of protein kinase Mzeta in maintaining long-term potentiation but not long-term depression. J. Neurosci., 25: 5750–5756.

Sala, C., Futai, K., Yamamoto, K., Worley, P.F., Hayashi, Y. and Sheng, M. (2003) Inhibition of dendritic spine morphogenesis and synaptic transmission by activity-inducible protein Homer1a. J. Neurosci., 23: 6327–6337.

Schacher, S., Wu, F. and Sun, Z.Y. (1997) Pathway-specific synaptic plasticity: activity-dependent enhancement and suppression of long-term heterosynaptic facilitation at converging inputs on a single target. J. Neurosci., 17: 597–606.

Schmitt, W.B., Deacon, R.M., Seeburg, P.H., Rawlins, J.N. and Bannerman, D.M. (2003) A within-subjects, within-task demonstration of intact spatial reference memory and impaired spatial working memory in glutamate receptor-A-deficient mice. J. Neurosci., 23: 3953–3959.

Scholz, K.P. and Byrne, J.H. (1987) Long-term sensitization in *Aplysia*: biophysical correlates in tail sensory neurons. Science, 235: 685–687.

Schulz, D.J. (2006) Plasticity and stability in neuronal output via changes in intrinsic excitability: it's what's inside that counts. J. Exp. Biol., 209: 4821–4827.

Serrano, P., Yao, Y. and Sacktor, T.C. (2005) Persistent phosphorylation by protein kinase Mzeta maintains late-phase long-term potentiation. J. Neurosci., 25: 1979–1984.

Shen, Y. and Linden, D.J. (2005) Long-term potentiation of neuronal glutamate transporters. Neuron, 46: 715–722.

Sheng, M., Cummings, J., Roldan, L.A., Jan, Y.N. and Jan, L.Y. (1994) Changing subunit composition of heteromeric NMDA receptors during development of rat cortex. Nature, 368: 144–147.

Shepherd, J.D., Rumbaugh, G., Wu, J., Chowdhury, S., Plath, N., Kuhl, D., Huganir, R.L. and Worley, P.F. (2006) Arc/Arg3.1 mediates homeostatic synaptic scaling of AMPA receptors. Neuron, 52: 475–484.

Siegelbaum, S.A., Camardo, J.S. and Kandel, E.R. (1982) Serotonin and cyclic AMP close single K+ channels in *Aplysia* sensory neurones. Nature, 299: 413–417.

Sorra, K.E. and Harris, K.M. (1998) Stability in synapse number and size at 2h after long-term potentiation in hippocampal area CA1. J. Neurosci., 18: 658–671.

Sossin, W.S. (1996) Mechanisms for the generation of synapse specificity in long-term memory: the implications of a requirement for transcription. Trends Neurosci., 19: 215–218.

Sossin, W.S. (2007) Isoform specificity of protein kinase Cs in synaptic plasticity. Learn. Mem., 14: 236–246.

Sun, Z.Y. and Schacher, S. (1998) Binding of serotonin to receptors at multiple sites is required for structural plasticity accompanying long-term facilitation of *Aplysia* sensorimotor synapses. J. Neurosci., 18: 3991–4000.

Sun, Z.Y., Wu, F. and Schacher, S. (2001) Rapid bidirectional modulation of mRNA expression and export accompany long-term facilitation and depression of *Aplysia* synapses. J. Neurobiol., 46: 41–47.

Sutton, M.A., Bagnall, M.W., Sharma, S.K., Shobe, J. and Carew, T.J. (2004) Intermediate-term memory for site-specific sensitization in *Aplysia* is maintained by persistent activation of protein kinase C. J. Neurosci., 24: 3600–3609.

Sutton, M.A. and Carew, T.J. (2000) Parallel molecular pathways mediate expression of distinct forms of intermediate-term facilitation at tail sensory–motor synapses in *Aplysia*. Neuron, 26: 219–231.

Takamori, S. (2006) VGLUTs: 'exciting' times for glutamatergic research? Neurosci. Res., 55: 343–351.

Tang, Y.P., Shimizu, E., Dube, G.R., Rampon, C., Kerchner, G.A., Zhuo, M., Liu, G. and Tsien, J.Z. (1999) Genetic enhancement of learning and memory in mice. Nature, 401: 63–69.

Toni, N., Buchs, P.A., Nikonenko, I., Povilaitite, P., Parisi, L. and Muller, D. (2001) Remodeling of synaptic membranes after induction of long-term potentiation. J. Neurosci., 21: 6245–6251.

Trudeau, L.E. and Castellucci, V.F. (1995) Postsynaptic modifications in long-term facilitation in *Aplysia*: upregulation of excitatory amino acid receptors. J. Neurosci., 15: 1275–1284.

Tsokas, P., Grace, E.A., Chan, P., Ma, T., Sealfon, S.C., Iyengar, R., Landau, E.M. and Blitzer, R.D. (2005) Local protein synthesis mediates a rapid increase in dendritic elongation factor 1A after induction of late long-term potentiation. J. Neurosci., 25: 5833–5843.

Tsokas, P., Ma, T., Iyengar, R., Landau, E.M. and Blitzer, R.D. (2007) Mitogen-activated protein kinase upregulates the dendritic translation machinery in long-term potentiation by controlling the mammalian target of rapamycin pathway. J. Neurosci., 27: 5885–5894.

Turrigiano, G.G. and Nelson, S.B. (2000) Hebb and homeostasis in neuronal plasticity. Curr. Opin. Neurobiol., 10: 358–364.

Udo, H., Jin, I., Kim, J.H., Li, H.L., Youn, T., Hawkins, R.D., Kandel, E.R. and Bailey, C.H. (2005) Serotonin-induced regulation of the actin network for learning-related synaptic growth requires Cdc42, N-WASP, and PAK in *Aplysia* sensory neurons. Neuron, 45: 887–901.

Volkmar, F.R. and Greenough, W.T. (1972) Rearing complexity affects branching of dendrites in the visual cortex of the rat. Science, 176: 1445–1447.

Wainwright, M.L., Byrne, J.H. and Cleary, L.J. (2004) Dissociation of morphological and physiological changes associated with long-term memory in *Aplysia*. J. Neurophysiol., 92: 2628–2632.

Walters, E.T. and Byrne, J.H. (1983) Associative conditioning of single sensory neurons suggests a cellular mechanism for learning. Science, 219: 405–408.

Woodin, M.A., Ganguly, K. and Poo, M.M. (2003) Coincident pre- and postsynaptic activity modifies GABAergic synapses by postsynaptic changes in Cl-transporter activity. Neuron, 39: 807–820.

Xu, J. and Kang, J. (2005) The mechanisms and functions of activity-dependent long-term potentiation of intrinsic excitability. Rev. Neurosci., 16: 311–323.

Yuste, R. and Bonhoeffer, T. (2001) Morphological changes in dendritic spines associated with long-term synaptic plasticity. Annu. Rev. Neurosci., 24: 1071–1089.

Zakharenko, S.S., Patterson, S.L., Dragatsis, I., Zeitlin, S.O., Siegelbaum, S.A., Kandel, E.R. and Morozov, A. (2003) Presynaptic BDNF required for a presynaptic but not postsynaptic component of LTP at hippocampal CA1–CA3 synapses. Neuron, 39: 975–990.

Zakharenko, S.S., Zablow, L. and Siegelbaum, S.A. (2001) Visualization of changes in presynaptic function during long-term synaptic plasticity. Nat. Neurosci., 4: 711–717.

Zalutsky, R.A. and Nicoll, R.A. (1990) Comparison of two forms of long-term potentiation in single hippocampal neurons. Science, 248: 1619–1624.

Zamanillo, D., Sprengel, R., Hvalby, O., Jensen, V., Burnashev, N., Rozov, A., Kaiser, K.M., Koster, H.J., Borchardt, T., Worley, P., Lubke, J., Frotscher, M., Kelly, P.H., Sommer, B., Andersen, P., Seeburg, P.H. and Sakmann, B. (1999)

Importance of AMPA receptors for hippocampal synaptic plasticity but not for spatial learning. Science, 284: 1805–1811.

Zhao, Y., Hegde, A.N. and Martin, K.C. (2003) The ubiquitin proteasome system functions as an inhibitory constraint on synaptic strengthening. Curr. Biol., 13: 887–898.

Zhao, Y. and Klein, M. (2002) Modulation of the readily releasable pool of transmitter and of excitation-secretion coupling by activity and by serotonin at *Aplysia* sensorimotor synapses in culture. J. Neurosci., 22: 10671–10679.

Zhao, Y., Leal, K., Abi-Farah, C., Martin, K.C., Sossin, W.S. and Klein, M. (2006) Isoform specificity of PKC translocation in living *Aplysia* sensory neurons and a role for Ca^{2+}-dependent PKC APL I in the induction of intermediate-term facilitation. J. Neurosci., 26: 8847–8856.

Zhu, J.J., Esteban, J.A., Hayashi, Y. and Malinow, R. (2000) Postnatal synaptic potentiation: delivery of GluR4-containing AMPA receptors by spontaneous activity. Nat. Neurosci., 3: 1098–1106.

Zwartjes, R.E., West, H., Hattar, S., Ren, X., Noel, F., Nunez-Regueiro, M., MacPhee, K., Homayouni, R., Crow, M.T., Byrne, J.H. and Eskin, A. (1998) Identification of specific mRNAs affected by treatments producing long-term facilitation in *Aplysia*. Learn. Mem., 4: 478–495.

CHAPTER 2

PKMζ, LTP maintenance, and the dynamic molecular biology of memory storage

Todd Charlton Sacktor*

The Robert F. Furchgott Center for Neural and Behavioral Science, Departments of Physiology, Pharmacology, and Neurology, SUNY Downstate Medical Center, 450 Clarkson Ave., Brooklyn, NY 11203, USA

Abstract: How memories persist is a fundamental neurobiological question. The most commonly studied physiological model of memory is long-term potentiation (LTP). The molecular mechanisms of LTP can be divided into two phases: *induction*, triggering the potentiation; and *maintenance*, sustaining the potentiation over time. Although many molecules participate in induction, very few have been implicated in the mechanism of maintenance. Understanding maintenance, however, is critical for testing the hypothesis that LTP sustains memory storage in the brain. Only a single molecule has been found both necessary and sufficient for maintaining LTP — the brain-specific, atypical PKC isoform, protein kinase Mzeta (PKMζ). Although full-length PKC isoforms respond to transient second messengers, and are involved in LTP induction, PKMζ is a second messenger-independent kinase, consisting of the independent catalytic domain of PKCζ, and is persistently active to sustain LTP maintenance. PKMζ is produced by a unique PKMζ mRNA, which is generated by an internal promoter within the PKCζ gene and transported to the dendrites of neurons. LTP induction increases new PKMζ synthesis, and the increased level of PKMζ then enhances synaptic transmission by doubling the number of postsynaptic AMPA receptors (AMPAR) through GluR2 subunit-mediated trafficking of the receptors to the synapse. PKMζ mediates synaptic potentiation specifically during the late-phase of LTP, as PKMζ inhibitors can reverse established LTP when applied several hours after tetanization in hippocampal slices or 1 day after tetanization in vivo. These studies set the stage for testing the hypothesis that the mechanism of LTP maintenance sustains memory storage. PKMζ inhibition in the hippocampus after learning eliminates the retention of spatial memory. Once the PKMζ inhibitor has been eliminated, the memory is still erased, but new spatial memories can be learned and stored. Similar results are found for conditioned taste aversion when the inhibitor is injected in the insular neocortex. Thus PKMζ is the first molecule found to be a component of the long-term memory trace.

Keywords: PKMζ; PKCζ; atypical PKC; long-term potentiation (LTP); memory

How memories persist over time is a central question in neuroscience. Despite extensive work on the signaling pathways that initiate the formation of memory, almost nothing has previously been known about the nature of the "engram," or physical memory trace, itself (Dudai, 2002). The formation of long-term memory requires new protein synthesis for a brief period of an hour or

*Corresponding author. Tel.: +1 718 270 3933;
Fax: +1 718 270 8974; E-mail: tsacktor@downstate.edu

so after learning, i.e., during the time window for cellular consolidation, but how do newly synthesized proteins store memory for days to weeks or longer? Which proteins are functionally important for memory storage, as opposed to those that may have other activity-dependent functions, such as regulating synaptic homeostasis, or increasing the neuronal connectivity that could enhance subsequent learning?

A general assumption in the field has been that these key memory storage proteins in the mammalian brain will directly mediate the persistence of changes in synaptic strength. This assumption is based upon a large body of work from model systems of memory in invertebrates that have demonstrated that changes in synaptic strength largely explain modifications in simple forms of behavior (Kandel and Schwartz, 1982). A further assumption has been that whereas short-term memory is due to posttranslational modifications of existing proteins, long-term memory is due to the synthesis of new proteins that increase the number of synapses between specific networks of neurons. Whereas de novo protein synthesis is required for long-term memory formation, the evidence that long-term memory storage is essentially due to an anatomical change, and is thus similar to a late stage of development, is correlative in invertebrates (Kandel and O'Dell, 1992), and such new synaptic growth, when it occurs, has not yet been shown to be functionally important for memory storage. Therefore, how long-term memories persist over time has remained an open question.

One approach to identify the mechanism of memory storage in the mammalian brain has been to examine long-term potentiation (LTP), a widely studied activity-dependent form of long-term synaptic enhancement (Bliss and Collingridge, 1993). Analogous to the consolidation and storage phases of behavioral memory, LTP can be divided into induction and maintenance phases. And like the initial phase of memory consolidation, the initial induction of the late-phase of LTP depends upon new protein synthesis. Over a hundred molecules have been implicated in LTP induction, and many of them have also been shown to be important in learning or the initial consolidation of memory (Sanes and Lichtman, 1999). Until recently, however, none has been found to function in LTP maintenance, meaning that the inhibition of such a molecule not just prevents LTP from forming, but *reverses* LTP maintenance after it has been established. If LTP does indeed employ for its maintenance a core mechanism of memory storage, then reversing this maintenance mechanism might erase previously encoded long-term memories.

In this chapter, we will present evidence that the first such molecule to maintain LTP has now been identified as an unusual, constitutively active form of protein kinase C (PKC), PKMζ, and that, confirming the functional relationship between LTP maintenance and behavioral memory storage, inhibiting PKMζ erases previously encoded associative long-term memories. Thus PKMζ is the first identified molecular component of the storage mechanism of associative long-term memory in the mammalian brain.

To begin, we will first discuss the unique properties of PKMζ that lead to its discovery as a memory storage molecule. Second, we will review the evidence for PKMζ's central role in the maintenance of the late, protein synthesis-dependent phase of LTP. Lastly, we will examine its role in the storage of long-term memory for several different learned behaviors.

The discovery of PKMζ

The notion that enzymes might maintain long-term memory was articulated by Crick (1984). Protein kinases have well-established roles in the transient enhancement of synaptic transmission thought to mediate short-term memory (Kandel and Schwartz, 1982). Therefore, one simple mechanism for long-term memory storage is the formation of a persistently active form of a kinase that could then sustain this enhancement over time. This notion was actively explored in the 1980s by the late Dr. James Schwartz in his research program on the mechanisms of learning and memory in the mollusk, *Aplysia californica* (Schwartz and Greenberg, 1987; Schwartz, 1993). Schwartz and his colleagues demonstrated the persistent activation of several kinases: the cAMP-dependent protein kinase A (PKA), the Ca^{2+}-calmodulin-dependent protein kinase II (CaMKII), and PKC, after synaptic stimulation that produces learning in *Aplysia*, but their persistent activity appeared functionally important in specific intermediate forms of memory (Sutton and Carew, 2000).

This early work in *Aplysia* and other model systems nonetheless suggested that protein kinases could have persistent effects that might explain at the molecular level long-term synaptic plasticity, and then ultimately, the long-term memory that this plasticity might mediate. Discovering the specific persistently active kinase that might underlie long-term memory in the mammalian brain would best be investigated by studying LTP, the synaptic correlate of mammalian associative memory. Indeed, PKC had a very simple mechanism for persistence.

PKC is a monomeric polypeptide consisting of an amino-terminal regulatory domain and a carboxy-terminal catalytic kinase domain (Nishizuka, 1988). Under basal conditions, PKC is held in an inactive state by interactions between the regulatory domain, which contains an autoinhibitory pseudosubstrate, and the catalytic domain. Second messengers, such as Ca^{2+} and diacylglycerol (depending on the isoform), activate the enzyme by binding to the regulatory domain, causing a translocation to membrane and a conformational change that releases this autoinhibition. This activation by translocation is usually transient because the second messengers are rapidly metabolized, and the enzyme returns to its inactive state, awaiting the next round of activation.

But in the very first papers on PKC, Nishizuka and colleagues reported another, more permanent mechanism for activating the kinase (Inoue et al., 1977; Takai et al., 1977). When PKC is in its open active conformation it is susceptible to limited proteolysis at its hinge, forming an independent catalytic domain, known as PKM. Because PKM lacks the inhibition of a PKC regulatory domain, it is an autonomously active form of the kinase, i.e., it is active even in the absence of second messengers and thus could have persistent effects.

PKM, however, had only been created by proteolysis in vitro and had never been observed in normal tissue.

By the early 1990s, however, it had become clear that PKC was not a single molecule, but a gene family of isoforms, each with distinct features and functions (Dekker et al., 1995; Newton, 2001). In mammals, PKC is gene family of ~15 protein kinases derived from 9 separate genes. Therefore, it was possible that some forms of PKC might be important for short-term synaptic plasticity, and others for long-term plasticity. Because it was not clear which isoform might be which, we generated antisera to all known PKC isoforms, using epitopes in the C-termini of the various PKCs to detect both a translocation to membrane and formation of PKM (Sacktor et al., 1993). We found that in the hippocampus, only a single isoform of PKC was expressed as a PKM form, the atypical PKC, PKCζ.

In this initial study, we tetanized hippocampal slices and then, at varying times later, froze the tissue and dissected out the CA1 regions. Then by Western blot we compared PKC levels in the cytosol and membrane from the slices that underwent LTP to control slices from the same hippocampus that had received only test stimulation for equivalent periods of time.

We found that there was a translocation of most isoforms of PKC from cytsosol to membrane during LTP — but it occurred within 15 s of the tetanus, and was gone by 1 min. The only persistent effect that we observed after tetanization was an increase in PKMζ. The increase required the activation of NMDA receptors (NMDARs) and was first observed 10 min after the tetanization and lasted for at least 2 h (Osten et al., 1996). The level of the persistent increase in PKMζ correlated with the extent of synaptic potentiation during LTP maintenance.

PKMζ is synthesized from a PKMζ mRNA

Curiously, however, while an increase in PKMζ was seen during LTP maintenance, a decrease in PKCζ was not, i.e., there was no precursor–product relationship, as might be expected if

PKMζ were produced by proteolytic cleavage. At that time, we reasoned that there might be new synthesis of PKCζ that could have replaced that cleaved to PKMζ. Therefore, to reveal the putative precursor–product relationship, we added protein synthesis inhibitors during the LTP experiments. To our surprise, however, in addition to blocking the late-phase of LTP, as previously reported, the protein synthesis inhibitors also blocked the persistent increase of PKMζ (Osten et al., 1996). Subsequent metabolic labeling of hippocampal slices with ^{3}H-methionine/cysteine followed by immunoprecipitation of PKMζ confirmed an increase in the rate of PKMζ synthesis between 10–45 min after tetanization (Hernandez et al., 2003). Thus the increase of PKMζ was due to de novo protein synthesis.

The mechanism for the increase in PKMζ during LTP turned out not to be due to proteolysis of PKCζ, but instead to transcription and translation of a unique PKMζ mRNA from a dedicated internal promoter within the PKCζ gene. The ζ gene produces two sets of RNAs from two distinct promoters (Marshall et al., 2000): a full-length PKCζ mRNA (Ono et al., 1989) and a PKMζ mRNA (Ono et al., 1988). The 5′ end of the PKMζ mRNA is a unique sequence not present in the PKCζ mRNA, whereas its 3′ end is identical to that of PKCζ mRNA. The open reading frame (ORF) for the kinase domain of PKMζ mRNA begins in the hinge and extends to the end of the catalytic domain, thus encoding the PKMζ protein (Hernandez et al., 2003). Both PKMζ mRNA and protein are expressed specifically in brain (except for a low abundant PKMζ mRNA in kidney that does not express detectable levels of protein) (Hernandez et al., 2003). The only other brain-specific PKC isozymes are PKCγ (Kikkawa et al., 1987) and a form of PKCη (Sublette et al., 1993).

During the initial induction phase of LTP, translational upregulation is believed to be the primary mechanism for the increase in PKMζ (Hernandez et al., 2003; Kelly et al., 2007). The PKMζ mRNA has a long 5′UTR and multiple short ORFs prior to the kinase ORF, consistent with an inhibitory constraint of the 5′UTR on the translation of the message (Hernandez et al., 2003). This inhibitory mechanism was confirmed by sequential truncation of the 5′UTR, which greatly increased PKMζ translation in vitro. Since the rate of DNA transcription is ~1–2 kb/min (Shilatifard, 1998) and the length of DNA encoding the PKMζ mRNA is ~100 kb (Hernandez et al., 2003), new transcription of PKMζ RNA could begin only as early as ~1 h after tetanization, whereas increases in PKMζ protein are observed within 10 min (Osten et al., 1996). In comparison, the genomic length of activity-dependent immediate early genes whose RNA expression increases rapidly after synaptic stimulation is typically short. The immediate early gene, Arc, for example, is only 3.5 kb.

Because the PKMζ message is dendritic (Muslimov et al., 2004), local translational regulation may provide a mechanism for the synapse-specificity that is a characteristic feature of LTP and would be important for its putative ability to store associative memory (Bliss and Collingridge, 1993). In situ hybridization with a specific probe to the 5′UTR of PKMζ mRNA showed PKMζ mRNA both in the pyramidal cell body layer and in dendritic layers of the hippocampus (Muslimov et al., 2004). The dendritic location of PKMζ mRNA was confirmed in primary cultures of hippocampal neurons where individual neuritic processes could be clearly identified. Two dendritic localization signals, one in the kinase ORF and a second in the 3′UTR, were identified (Muslimov et al., 2004). In addition to local dendritic synthesis, however, other mechanisms may also contribute to synapse-specific effects of PKMζ in LTP, including synaptic tagging and capture (Sajikumar et al., 2005) (see below).

Although translational regulation is important for the initial increase of PKMζ in LTP, transcriptional upregulation may also contribute to more persistent increases in PKMζ. Indeed, an early in vivo LTP study suggested a trend upward in the level of ζ mRNA at 2 and 24 h posttetanization (Thomas et al., 1994). The promoter region of the PKMζ mRNA in the mouse, rat, and human PKCζ genes contains a canonical cAMP response element (CRE), as well as putative sites for nuclear factor-κB (NF-κB) and CCAAT/enhancer binding protein (C/EBP) (Hernandez et al., 2003). Each of these transcription factors has been implicated in memory formation (Dash et al., 1990; Alberini et al., 1994; Yin and Tully, 1996; Silva et al., 1998; Merlo et al., 2002; Yeh et al., 2002; Meffert et al., 2003).

PKMζ levels can also be decreased by moderate frequency stimulation that induces long-term depression (LTD) (Hrabetova and Sacktor, 1996). This decrease requires protease activity and persists for at least several hours in hippocampal slices (Hrabetova and Sacktor, 2001).

PKMζ synthesis is regulated by other protein kinases in LTP induction

As mentioned in the introduction, many molecules are critical for inducing LTP, in particular the protein kinases CaMKII, mitogen-activated protein kinase (MAPK), and PKA (Roberson et al., 1996; Sweatt, 1999). Other kinases important for translational regulation, the lipid kinase, phosphatidyl 3-kinase (PI3-kinase) (Kelly and Lynch, 2000; Sanna et al., 2002; Opazo et al., 2003), and the mammalian target of rapamycin (mTOR) (Tang et al., 2002; Cammalleri et al., 2003; Cracco et al., 2005) are also important. We found that selective inhibitors of each of these kinases block the synthesis of PKMζ in LTP (Kelly et al., 2007); thus, several key kinases converge to regulate PKMζ synthesis in LTP induction.

One protein kinase, however, functions at a later step to regulate the activity of PKMζ to initiate LTP maintenance. Many kinases, including PKC isoforms, are inactive (or with relatively little activity) immediately after their synthesis because their kinase domains are initially in an inactive conformation that inefficiently binds protein substrates. These kinases require phosphorylation by phosphatidylinositol-dependent protein kinase 1 (PDK1) on the "activation loop" of their catalytic domain for conversion to a maximally active conformation (Newton, 2003; Biondi, 2004). In brain, basal PKMζ binds constitutively to PDK1, which maximally phosphorylates the PKC on its activation loop (Kelly et al., 2007). Within 1 h of LTP, the newly synthesized PKMζ is also maximally phosphorylated. Thus during LTP maintenance, PKMζ is constitutively phosphorylated by PDK1, ensuring maximal autonomous activity of the kinase for sustaining responses in neurons.

Another important effector mechanism for LTP is the actin-based cytoskeleton. Although usually thought of as part of structural changes assumed to contribute to the expression of potentiation (Fukazawa et al., 2003; Lang et al., 2004; Okamoto et al., 2004; Ouyang et al., 2005; Kramar et al., 2006), the function of actin in LTP is unknown. We found that the inhibition of actin filament formation with latrunculin blocked PKMζ synthesis (Kelly et al., 2006). This is consistent with the notion that actin filaments serve as a scaffolding for assembling proteins important for translation (Stapulionis et al., 1997), or perhaps as a local trafficking mechanism for targeting dendritic mRNAs into spines.

Finally, PKMζ activity, itself, is also necessary for the de novo synthesis of PKMζ in LTP, because both chelerythrine, a PKC inhibitor that appears selective for PKM forms of PKC at low doses (Ling et al., 2002), and the PKMζ-selective inhibitor ζ-inhibitory peptide (ZIP) block PKMζ synthesis after tetanization and prevent LTP (Kelly et al., 2007). These results suggest the possibility of a positive feedback loop at the level of translation that might maintain the persistent increased levels of PKMζ. Further characterization of this and other potential positive feedback mechanisms will be critical for future studies on the persistence of long-term synaptic plasticity and memory.

PKMζ potentiates postsynaptic AMPA receptor responses

The persistent increase in PKMζ correlates with the extent and duration of synaptic potentiation during LTP, but does the persistent kinase action of PKMζ cause synaptic potentiation? To address this question, we recombinantly expressed and purified PKMζ from the baculovirus/Sf9-overexpression system (in which endogenous PDK1 of the Sf9 insect cells phosphorylates the activation loop of the overexpressed PKMζ) (Ling et al., 2002). We then patched CA1 pyramidal cells in hippocampal slices with low concentrations (1–3 nM) of the purified kinase in the whole-cell recording solution, thus bypassing the synthesis of the kinase that occurs during LTP induction (Ling et al., 2002). Diffusion of PKMζ into cells within a few minutes doubled the amplitude of the evoked AMPA receptor (AMPAR)-mediated EPSC, which then stabilized for the length of the recordings. Bath application of the PKC catalytic domain inhibitor chelerythrine reversed the potentiation to baseline, indicating that PKMζ activity was continually required for the synaptic potentiation. Bath applications of the PKMζ-selective inhibitor ZIP also prevented the potentiation (Serrano et al., 2005).

In comparison with other studies using whole-cell recording pipettes to introduce proteins into hippocampal neurons, PKMζ is by far the most potent potentiating substance known. Postsynaptic perfusion of 100–1000 times more of autoactivated CaMKII, for example, produces approximately half the potentiation of AMPAR responses (Lledo et al., 1995; Shirke and Malinow, 1997). Furthermore, PKMζ-mediated increases in EPSCs completely occluded LTP induced by tetanization paired with postsynaptic depolarization, indicating that PKMζ and LTP enhance transmission through the same mechanism (Ling et al., 2002).

We next examined the site of synaptic transmission on which PKMζ acts to enhance synaptic strength (Ling et al., 2006). Analysis of the coefficient of variance of evoked responses recorded from CA1 pyramidal cells before and after PKMζ-mediated potentiation showed no evidence of retrograde presynaptic enhancement. Confirming a postsynaptic site, postsynaptic PKMζ perfusion doubled the amplitude of AMPAR-mediated, miniature excitatory postsynaptic currents (mEPSCs), without affecting their frequency. These results also indicate that PKMζ enhances without unsilencing NMDAR-only, "silent" synapses, which would have also lead to changes in the coefficient of variance of evoked responses and an increase in mEPSC frequency. The kinetics of the mEPSCs during PKMζ-mediated potentiation, dendritic filtering, and input resistance of the cell also did not change with PKMζ-mediated synaptic enhancement.

This left two possible sites of PKMζ action: an increase in AMPAR channel unit conductance or an increase in the number of active channels (i.e., the product of the number of physical channels and their open probability). To distinguish between these two possibilities, we used a statistical technique, peaked-scale, nonstationary fluctuation analysis (NSFA, also referred to as "noise analysis") (Traynelis et al., 1993) to examine the effect of PKMζ on mEPSCs. NSFA indicated that PKMζ did not affect unit conductance, but doubled the number of active AMPAR channels. The doubling was rapidly reversed by applying chelerythrine to PKMζ-potentiated mEPSCs (Ling et al., 2006), indicating that persistent activity of the kinase was necessary for maintaining the increased number of active channels.

These results indicated that PKMζ enhances synaptic transmission exclusively by increasing the number of functional postsynaptic AMPARs. This suggested that PKMζ may work up altering the regulation of AMPAR trafficking that normally maintains a constant number of receptors at synapses.

One of these pathways involves the trafficking protein N-ethylmaleimide-sensitive factor (NSF) that interacts directly with the AMPAR subunit GluR2 to maintain the receptors at postsynaptic sites (Nishimune et al., 1998; Osten et al., 1998; Song et al., 1998). Inhibiting NSF/GluR2 interactions by a peptide, called pep2m (Nishimune et al., 1998), that mimics the NSF binding site on GluR2 and prevents their interaction, blocked PKMζ-mediated potentiation of AMPARs (Yao et al., 2008). Furthermore, bath applications of a cell-permeable myristoylated version of the peptide both completely blocked PKMζ-mediated potentiation and reversed the late-phase of LTP (as observed for the PKMζ kinase inhibitor ZIP, see next section). The effect of PKMζ is specific to this trafficking pathway because inhibiting a second AMPAR trafficking pathway that involves VAMP/synaptobrevin-mediated, constitutive exocytosis of receptors from internal stores to the plasma membrane (Luscher et al., 1999) had no effect on PKMζ-potentiation of AMPARs.

The NSF/GluR2-dependent trafficking pathway mediated by PKMζ in LTP may involve a lateral movement of receptors along the plasma membrane from extrasynaptic sites into the synapse

(Gardner et al., 2005). During LTP maintenance, the AMPAR subunits GluR2 and 3 were found to increase in synaptosomes through the action of PKMζ and NSF/GluR2-dependent trafficking, but the total level of surface GluR2 measured by biotin-labeling (and its total level overall) did not change (Yao et al., 2008). This action appears to be mediated by the release of AMPARs from protein interacting with C-kinase (PICK1), because intracellular perfusion of a peptide that blocks GluR2-PICK1 interactions, mimics and occludes PKMζ-mediated AMPAR potentiation (Yao et al., 2008). Thus PKMζ enhances synaptic transmission through a highly specific molecular mechanism — persistently increasing NSF/GluR2-mediated trafficking of AMPARs from extrasynaptic to postsynaptic sites. The next key questions are if and when this mechanism of potentiation is important for LTP.

PKMζ maintains late-LTP

If PKMζ is sufficient for synaptic potentiation, is it also necessary for the potentiation that maintains LTP? As mentioned in the introduction, although many inhibitors block LTP induction when applied around the time of tetanization, none had reversed established late-phase LTP prior to our work on PKMζ (Ling et al., 2002). In our initial investigation, we examined LTP maintenance 1 h after tetanization by comparing the effects of applications of two PKMζ inhibitors, chelerythrine, an inhibitor of the catalytic domain of PKC, which selectively inhibits PKM forms such as PKMζ at low concentrations relative to full-length PKC isoforms and CaMKII (Ling et al., 2002), and the myristoylated ζ-pseudosubstrate peptide (ZIP), a cell-permeable selective PKMζ inhibitor (Bandyopadhyay et al., 1997; Ling et al., 2002; Braun and Mochly-Rosen, 2003). As a control, we examined staurosporine, a general kinase inhibitor that blocks conventional and novel PKCs (c/nPKCs), CaMKII, and PKA, but not atypical PKCs such as PKMζ (McGlynn et al., 1992; Kochs et al., 1993; Ling et al., 2002). Consistent with several earlier studies (Denny et al., 1990; Muller et al., 1992), staurosporine, while blocking LTP induction when applied prior to tetanization, had no effect on maintenance when bath applied 1–5 h after tetanization (Ling et al., 2002). Therefore, CaMKII, c/nPKCs, PKA, and the many other protein kinases that are inhibited by staurosporine appear not to be required for LTP maintenance. In contrast, both chelerythrine and ZIP reversed maintenance when applied 1 h after tetanization. The reversal was specific to potentiated synapses because synaptic strength in a nontetanized pathway, simultaneously recorded in each experiment, was not affected.

Because the ability to reverse established LTP is very unusual, we explored the effect of PKMζ inhibition by ZIP on LTP maintenance in more detail (Serrano et al., 2005). To avoid nonspecific effects of the drug, we determined the lowest concentration of ZIP that effectively blocks PKMζ's function in neurons, i.e., we determined the lowest concentration of ZIP that, when applied to the bath of hippocampal slices, blocked the ability of intracellular perfusion PKMζ to potentiate synaptic transmission. We found that 1 μM attenuated and 5 μM ZIP completely blocked PKMζ-mediated potentiation of synaptic transmission.

We then applied 5 μM ZIP to the bath continuously throughout an LTP experiment to determine the overall time window of synaptic potentiation that is dependent on PKMζ activity. We found that when PKMζ was blocked there was still a residual potentiation that was indistinguishable from the early, protein synthesis-independent phase of LTP (Serrano et al., 2005). Therefore, PKMζ did not contribute to early LTP, but is specifically required for late-LTP.

We then examined the ability of 5 μM ZIP to reverse established LTP when applied at 1, 3, and 5 h after tetanization (Serrano et al., 2005). The agent completely reversed LTP at all time points, but had no effects on the control, nontetanized pathways, simultaneously recorded within each slice. The IC50 of ZIP for reversal of established late-LTP at 3 h was nearly identical to its IC50 for blocking the potentiation of AMPAR responses by intracellular perfusion of the kinase. These results strongly support the hypothesis that the persistent activity of PKMζ maintains late-LTP.

PKMζ maintains potentiation after synaptic tagging

PKMζ is synthesized by strong, but not weak afferent stimulation of synapses (Osten et al., 1996; Hernandez et al., 2003), and maintains late, but not early LTP. How does newly synthesized PKMζ affect only specific, activated synapses? One possibility is that the kinase might be involved in the process of synaptic tagging and capture (Sossin, 1996; Frey and Morris, 1997; Sajikumar et al., 2005). Synaptic tagging is the hypothesis, proposed by Sossin and by Frey and Morris, to explain how newly synthesized "plasticity-related proteins" (PRPs), which are presumably synthesized away from synapses in the soma or dendrites, can maintain potentiation specifically at recently activated synapses, but not at neighboring synapses that had not been activated. They proposed that a "tag" could be set at activated synapses to which the PRPs might bind or be "captured." To test their hypothesis, they showed that a weak tetanization, which normally would produce only early, decremental LTP, could nonetheless result in persistent late-LTP, if late-LTP had been produced elsewhere in the neuron by an independent synaptic pathway (Frey and Morris, 1997). In their model, the weak tetanization, although not strong enough to stimulate protein synthesis, could nonetheless set synaptic tags that could capture the PRPs that had been synthesized by the strongly tetanized input.

Sajikumar and Frey (Leibniz Institute for Neurobiology, Magdeburg, Germany) tested the hypothesis that PKMζ was a PRP that could be captured to maintain potentiation at tagged synapses (Sajikumar et al., 2005). They first showed that ZIP reversed late-LTP, as expected, but did not affect the maintenance of transient early-LTP. Then during a tagging experiment, when the early-LTP pathway would normally have been converted to late-LTP, they found that ZIP reversed potentiation at both the strongly and the weakly tetanized synapses. These results indicated that persistent PKMζ activity not only maintains potentiation at strongly tetanized synapses, but also is critical for the conversion of early- into late-LTP at tagged synapses.

Sajikumar and Frey had also discovered another tagging phenomenon known as "cross-tagging," which revealed a surprising relationship between LTP and LTD (Sajikumar and Frey, 2004). They found that strong LTD, just like strong LTP, could allow a subsequent weak tetanus to produce late-LTP. Conversely, strong LTP could allow a subsequent weak LTD stimulus to produce a late, long-lasting depression. Is PKMζ's role to produce the persistence of any change in synaptic strength regardless of its sign, or is it specific to persistent increases? Sajikumar and Frey first found that ZIP had no effect on reversing LTD maintenance (although it did block LTD induction). They then showed ZIP could nonetheless reverse the persistent synaptic potentiation produced by cross-tagging after strong LTD. On the other hand, ZIP did not affect the persistent depression produced by cross-tagging after strong LTP, even though the agent reversed the persistent potentiation produced by the strong tetanization. These findings demonstrated that PKMζ specifically maintains synaptic potentiation, and not depression; thus, they are consistent with the results showing that the postsynaptic perfusion of PKMζ caused synaptic potentiation (Ling et al., 2002).

How is PKMζ's role in synaptic tagging commensurate with the possibility that the kinase is produced from a dendritic mRNA during LTP? One notion is that activity-dependent synthesis of PKMζ is local in dendrites, but that tagging is required for finer compartmentalization of the synthesized PKMζ to only activated synapses. Alternatively, the phenomenon of tagging may be due to an increase in the efficiency of local dendritic synthesis (Tsokas et al., 2006). Interestingly, whole-cell perfusion of PKMζ causes potentiation of presumably all synapses of the patched neuron (Ling et al., 2002, 2006), and thus this PKMζ-mediated synaptic potentiation appears to overcome the need for a synaptic tag. This may support the notion that the synaptic tag is actually increased local synthesis (Tsokas et al., 2006), since the intracellular infusion of PKMζ would simply substitute for the endogenous local synthesis and provide the kinase throughout the neuron. Alternatively, the activity-dependent synthesis of PKMζ during LTP may produce only limited quantities of the kinase, and a synaptic tag might be required for localizing and competing for the limited amounts of the enzyme. In contrast, during whole-cell perfusion the availability of the kinase is not restricted.

PKMζ maintains long-term spatial memory storage in the hippocampus

The first evidence that PKMζ plays a role in memory came from work by Jerry Yin, Eric Drier, and colleagues (University of Wisconsin, Madison, WI) on odor avoidance conditioning in *Drosophila* (Drier et al., 2002). They found that overexpression of PKMζ during a narrow time-window after learning enhanced the persistence of memory. Conversely, blocking atypical PKC activity with a dominant negative or chelerythrine prevented persistent memory formation, without affecting short-term memory formation.

Our work on PKMζ inhibition with ZIP showed that LTP could be reversed even hours after tetanization, without affecting basal synaptic transmission. These findings suggested that we could also test the hypothesis that PKMζ stores long-term memory. In order to translate our findings from hippocampal slices to behavior, we began by focusing on the role of PKMζ in spatial memory that is initially encoded and stored in the hippocampus (Pastalkova et al., 2006). Because the synapses that are thought to encode spatial information are widely distributed in the hippocampus (Moser and Moser, 1998; Kubik and Fenton, 2005), to test the hypothesis that PKMζ maintained this information it was important to inhibit PKMζ widely in the hippocampus. We therefore examined in vivo LTP at perforant path-dentate gyrus synapses, using an array of recording electrodes spaced up to 2 mm away from the injection site in CA3. We determined a dose of ZIP that rapidly reversed LTP when injected ~1 day after its induction by tetanization. Labeling the ZIP peptide with biotin showed that the agent diffused ~3–4 mm within the hippocampus but not to other brain regions.

We then were in a position to examine the role of PKMζ in maintaining hippocampus-dependent spatial memory. We chose active place avoidance (Cimadevilla et al., 2000a, b, 2001; Wesierska et al., 2005), a behavior with several experimental advantages for the initial study of PKMζ function in spatial memory. Hippocampus inactivation by anesthetics blocks both initial learning and short-term memory, as well as long-term memory retention of active place avoidance (Cimadevilla et al., 2000a, 2001). This would allow us to distinguish between an effect of PKMζ inhibition on short-term and long-term retention of the conditioned responses. Learning is rapid (optimal performance observed within ~2.5 h of training) and remains hippocampus-dependent for at least one month, allowing us also to examine the role of PKMζ in remote spatial memory.

The apparatus for active place avoidance consists of a slowly rotating platform, open to the room environment, within which a nonrotating 60° sector is a shock zone (Cimadevilla et al., 2000b). The rotation brings the animal into the shock zone, and the animal rapidly learns to avoid the shock by actively moving to the nonshock areas of the platform. Then ~1 day after the last training session, we injected either ZIP or saline into both hippocampi. Two hours later, retention testing on the apparatus without shock (i.e., extinction testing) showed that the saline-injected animals exhibited long-term memory by avoiding entry into the shock zone. In contrast, the ZIP-injected animals failed to avoid the shock zone and actively explored the entire apparatus as if naïve. Whereas the ZIP-injected animals did not recall the long-term memory, spending time in the shock zone at the level of chance, immediate reconditioning and then retesting without the shock showed that the animals could nonetheless recall the short-term memory of the conditioned response. These results indicate that ZIP produces retrograde amnesia specifically for long-term memory, not short-term memory of place avoidance. Staurosporine, the nonspecific kinase inhibitor that blocks conventional and novel PKCs, CaMKII, and PKA, and which blocks LTP induction, but not maintenance (Ling et al., 2002), had no effect on long-term memory retention at a dose that completely prevented learning when injected prior to training.

A critical question concerning the effect of PKMζ inhibition of memory retention is whether ZIP disrupts information retrieval or information storage. One way to distinguish between these possibilities is to inject ZIP after long-term memory has been established, wait for the drug to be eliminated, and then test whether the memory returns. If the conditioned behavior returns, this would indicate an effect on information retrieval; if the memory were persistently eliminated, this would suggest an effect on storage. We therefore trained animals and then after a day later injected ZIP without testing. One week later (when the ZIP had been eliminated), the previously ZIP-injected animals had no recall of the long-term memory, whereas the saline-injected animals avoided the shock zone. Then to test whether ZIP caused permanent damage to memory function, we immediately retrained the previously ZIP-injected animals and found that they learned and could remember place avoidance as a new long-term memory. Thus ZIP erases the storage of previously acquired memory, but does not affect the ability to subsequently store new long-term memory.

We also examined whether memories older than 1 day were also sustained by PKMζ and found that 1-month-old place avoidance memory could be eliminated by ZIP.

Subsequent to this initial study, other forms of spatial long-term memory have also been shown to be maintained by PKMζ. Retention of information required for spatial accuracy in the water maze and for conditioned behavior in the 8-arm radial maze are affected by intrahippocampal injections of ZIP ~1 day after training (Serrano et al., 2008). In the latter, the deficit was due to increases in reference memory errors, but not working memory errors, indicating a specific loss of spatial information. Interestingly, information required for contextual fear conditioning does not appear to be eliminated by injections of ZIP into the dorsal hippocampus. Because lesions of the dorsal hippocampus have been reported to disrupt retention of contextual information, this suggests that either the dorsal hippocampus does not store the contextual information, but is important in retrieval or processing of the information stored elsewhere, or, alternatively, contextual information is stored in the dorsal hippocampus by a mechanism that does not require PKMζ, such as LTD (Sajikumar et al., 2005). Fear-motivated memories, per se, however, may still be encoded by PKMζ because ZIP injection in the basolateral amygdala complex ~1 day after training blocked memory retention for auditory fear conditioning and inhibitory avoidance (Serrano et al., 2008).

PKMζ maintains long-term associative memory storage in the neocortex

Although the hippocampus stores spatial memories and the amygdala fear memories, the repository for most memories in the mammalian brain is thought to be neocortex. To test the role of PKMζ in storage in neocortex, Reut Shema and Yadin Dudai (Weizmann Institute, Rehovot, Israel) examined conditioned taste aversion (CTA), which is stored in the taste cortex located in the insular cortex (Shema et al., 2007). Rats were first presented with a novel taste, the conditioned stimulus (CS, such as saccharin), followed shortly by intraperitoneal lithium, the unconditioned stimulus (US), that makes the animals sick. This single training trial produced a long-term memory to avoid the taste that lasts for at least several weeks. Injection of ZIP into the insular neocortex erased memory for CTA when the agent was injected from 3 to 25 days after training. The memory for CTA did not return even a month after the agent's injection, even with repeated presentations of the CS or attempted reinstatement with the US alone. Multiple distinct CTA taste memories were erased, as well as CTA memory made stronger by multiple CS-US pairings. Remarkably, however, injections of ZIP had no anterograde effect on the formation of long-term memory when injected 2 h prior to training. Furthermore, memories for taste familiarity, such as attenuation of neophobia, which also require the insular cortex, were not erased by ZIP. This suggests that PKMζ specifically stores associative, but perhaps not nonassociative memories in the neocortex.

Conclusions

These studies show that PKMζ is the first known component of the storage mechanism for long-term associative memory. The unique structure and function of the PKMζ gene, mRNA, and protein provide a relatively simple, essential molecular mechanism for information storage through synthesis of a constitutively active protein kinase (Sacktor et al., 1993; Hernandez et al., 2003), which persistently maintains enhanced synaptic transmission at synapses. Although this mechanism is quite different from the standard notion of memory storage as a passive change in the large-scale structure of synaptic connections between neurons, as we have seen, it is consistent with a large range of experimental data on LTP and memory.

Why is this molecular mechanism of associative memory storage so simple, compared to the enormous complexity of the signaling mechanisms of initial memory formation? Although answers to such teleological questions are by necessity speculative, perhaps within complex systems, the mechanism of information storage is, in general, relatively simple, whereas the nature of the information stored and the regulation of the encoding of that information can be exceedingly complex. For example, the mechanism for storing 0's and 1's in a computer hard drive is relatively simple compared to the patterns of stored information or the software for encoding that information. Similarly, the structural basis for storing genetic information in DNA is relatively simple, compared to the actual sequence of genes and the highly complex biochemical networks that regulate transcription and translation. Furthermore, in addition to its persistent increase in LTP, PKMζ can be persistently decreased by LTD (Hrabetova and Sacktor, 1996). Thus storage of information by PKMζ levels at synapses may be flexible and have the additional benefit that it might allow the incorporation of new information into old memories (Tse et al., 2007). Thus, once a stable yet flexible mechanism of memory storage was arrived at through evolution, perhaps it was advantageous to stick with this single mechanism for rapid and efficient encoding of learned associations.

Several important avenues are now open for future investigation. First, what are the substrates of PKMζ that mediate its function at synapses? AMPAR trafficking through NSF/GluR2-mediated mechanisms is downstream of PKMζ, but the specific substrates mediating this trafficking are unknown.

Second, because the effects of PKMζ are essential for long-term memory maintenance yet rapidly reversible, memory storage can now be viewed as a dynamic process amenable to experimental, even therapeutic, manipulation. Can other dynamic features of long-term memory, such as reconsolidation, now be explained by the degradation and resynthesis of PKMζ? Can we visualize the "engram" by monitoring changes in PKMζ after conditioned behavior? Can human disorders thought to be mediated by pathophysiologically enhanced LTP, such as neuropathic pain, post-traumatic stress, even addictive behavior, now be reversed by local injections of ZIP?

Finally, the persistent activity of PKMζ maintains memories for at least 1 month (Pastalkova et al., 2006), far longer than the likely half-life of the PKMζ protein. Which transcriptional, translational, or posttranslational mechanisms form positive feedback loops that maintain persistently increased PKMζ at specific synapses? How these mechanisms of persistent PKMζ regulation work is the next essential question for understanding memory storage.

References

Alberini, C.M., Ghirardi, M., Metz, R. and Kandel, E.R. (1994) C/EBP is an immediate-early gene required for the consolidation of long-term facilitation in *Aplysia*. Cell, 76(6): 1099–1114.

Bandyopadhyay, G., Standaert, M.L., Galloway, L., Moscat, J. and Farese, R.V. (1997) Evidence for involvement of protein kinase C (PKC)-zeta and noninvolvement of diacylglycerol-sensitive PKCs in insulin-stimulated glucose transport in L6 myotubes. Endocrinology, 138(11): 4721–4731.

Biondi, R.M. (2004) Phosphoinositide-dependent protein kinase 1, a sensor of protein conformation. Trends Biochem. Sci., 29(3): 136–142.

Bliss, T.V.P. and Collingridge, G.L. (1993) A synaptic model of memory: long-term potentiation in the hippocampus. Nature, 361(6407): 31–39.

Braun, M.U. and Mochly-Rosen, D. (2003) Opposing effects of delta- and zeta-protein kinase C isozymes on cardiac fibroblast proliferation: use of isozyme-selective inhibitors. J. Mol. Cell Cardiol., 35(8): 895–903.

Cammalleri, M., Lutjens, R., Berton, F., King, A.R., Simpson, C., Francesconi, W. and Sanna, P.P. (2003) Time-restricted role for dendritic activation of the mTOR-p70S6K pathway in the induction of late-phase long-term potentiation in the CA1. Proc. Natl. Acad. Sci. U.S.A., 100(24): 14368–14373.

Cimadevilla, J.M., Fenton, A.A. and Bures, J. (2000a) Functional inactivation of dorsal hippocampus impairs active place avoidance in rats. Neurosci. Lett., 285(1): 53–56.

Cimadevilla, J.M., Kaminsky, Y., Fenton, A. and Bures, J. (2000b) Passive and active place avoidance as a tool of spatial memory research in rats. J. Neurosci. Methods, 102(2): 155–164.

Cimadevilla, J.M., Wesierska, M., Fenton, A.A. and Bures, J. (2001) Inactivating one hippocampus impairs avoidance of a stable room-defined place during dissociation of arena cues from room cues by rotation of the arena. Proc. Natl. Acad. Sci. U.S.A., 98(6): 3531–3536.

Cracco, J.B., Serrano, P., Moskowitz, S.I., Bergold, P.J. and Sacktor, T.C. (2005) Protein synthesis-dependent LTP in isolated dendrites of CA1 pyramidal cells. Hippocampus, 15(5): 551–556.

Crick, F. (1984) Memory and molecular turnover. Nature, 312(5990): 101.

Dash, P.K., Hochner, B. and Kandel, E.R. (1990) Injection of the cAMP-responsive element into the nucleus of Aplysia sensory neurons blocks long-term facilitation. Nature, 345(6277): 718–721.

Dekker, L.V., Palmer, R.H. and Parker, P.J. (1995) The protein kinase C and protein kinase C related gene families. Curr. Opin. Struct. Biol., 5(3): 396–402.

Denny, J.B., Polan-Curtain, J., Rodriguez, S., Wayner, M.J. and Armstrong, D.L. (1990) Evidence that protein kinase M does not maintain long-term potentiation. Brain Res., 534(1–2): 201–208.

Drier, E.A., Tello, M.K., Cowan, M., Wu, P., Blace, N., Sacktor, T.C. and Yin, J.C. (2002) Memory enhancement and formation by atypical PKM activity in *Drosophila melanogaster*. Nat. Neurosci., 5(4): 316–324.

Dudai, Y. (2002) Memory from A to Z: Keywords, Concepts, and Beyond. Oxford University Press, Oxford.

Frey, U. and Morris, R.G. (1997) Synaptic tagging and long-term potentiation. Nature, 385(6616): 533–536.

Fukazawa, Y., Saitoh, Y., Ozawa, F., Ohta, Y., Mizuno, K. and Inokuchi, K. (2003) Hippocampal LTP is accompanied by enhanced F-actin content within the dendritic spine that is essential for late LTP maintenance in vivo. Neuron, 38(3): 447–460.

Gardner, S.M., Takamiya, K., Xia, J., Suh, J.G., Johnson, R., Yu, S. and Huganir, R.L. (2005) Calcium-permeable AMPA receptor plasticity is mediated by subunit-specific interactions with PICK1 and NSF. Neuron, 45(6): 903–915.

Hernandez, A.I., Blace, N., Crary, J.F., Serrano, P.A., Leitges, M., Libien, J.M., Weinstein, G., Tcherapanov, A. and Sacktor, T.C. (2003) Protein kinase Mζ synthesis from a brain mRNA encoding an independent protein kinase Cζ catalytic domain. Implications for the molecular mechanism of memory. J. Biol. Chem., 278(41): 40305–40316.

Hrabetova, S. and Sacktor, T.C. (1996) Bidirectional regulation of protein kinase Mζ in the maintenance of long-term potentiation and long-term depression. J. Neurosci., 16(17): 5324–5333.

Hrabetova, S. and Sacktor, T.C. (2001) Transient translocation of conventional protein kinase C isoforms and persistent downregulation of atypical protein kinase Mζ in long-term depression. Brain Res. Mol. Brain Res., 95(1–2): 146–152.

Inoue, M., Kishimoto, A., Takai, Y. and Nishizuka, Y. (1977) Studies on a cyclic nucleotide-independent protein kinase and its proenzyme in mammalian tissues. II. Proenzyme and its activation by calcium-dependent protease from rat brain. J. Biol. Chem., 252(21): 7610–7616.

Kandel, E.R. and O'Dell, T.J. (1992) Are adult learning mechanisms also used for development? Science, 258(5080): 243–245.

Kandel, E.R. and Schwartz, J.H. (1982) Molecular biology of learning: modulation of transmitter release. Science, 218(4571): 433–443.

Kelly, A. and Lynch, M.A. (2000) Long-term potentiation in dentate gyrus of the rat is inhibited by the phosphoinositide 3-kinase inhibitor, wortmannin. Neuropharmacology, 39(4): 643–651.

Kelly, M.T., Crary, J.F. and Sacktor, T.C. (2007) Regulation of protein kinase Mζ synthesis by multiple kinases in long-term potentiation. J. Neurosci., 27(13): 3439–3444.

Kelly, M.T., Yao, Y., Sondhi, R. and Sacktor, T.C. (2006) Actin polymerization regulates the synthesis of PKMζ in LTP. Neuropharmacology, 52(1): 41–45.

Kikkawa, U., Ogita, K., Ono, Y., Asaoka, Y., Shearman, M.S., Fujii, T., Ase, K., Sekiguchi, K., Igarashi, K. and Nishizuka, Y. (1987) The common structure and activities of four subspecies of rat brain protein kinase C family. FEBS Lett., 223(2): 212–216.

Kochs, G., Hummel, R., Meyer, D., Hug, H., Marme, D. and Sarre, T.F. (1993) Activation and substrate specificity of the human protein kinase C alpha and zeta isoenzymes. Eur. J. Biochem., 216(2): 597–606.

Kramar, E.A., Lin, B., Rex, C.S., Gall, C.M. and Lynch, G. (2006) Integrin-driven actin polymerization consolidates long-term potentiation. Proc. Natl. Acad. Sci. U.S.A., 103(14): 5579–5584.

Kubik, S. and Fenton, A.A. (2005) Behavioral evidence that segregation and representation are dissociable hippocampal functions. J. Neurosci., 25(40): 9205–9212.

Lang, C., Barco, A., Zablow, L., Kandel, E.R., Siegelbaum, S.A. and Zakharenko, S.S. (2004) Transient expansion of synaptically connected dendritic spines upon induction of hippocampal long-term potentiation. Proc. Natl. Acad. Sci. U.S.A., 101(47): 16665–16670.

Ling, D.S., Benardo, L.S. and Sacktor, T.C. (2006) Protein kinase Mζ enhances excitatory synaptic transmission by increasing the number of active postsynaptic AMPA receptors. Hippocampus, 16(5): 443–452.

Ling, D.S., Benardo, L.S., Serrano, P.A., Blace, N., Kelly, M.T., Crary, J.F. and Sacktor, T.C. (2002) Protein kinase Mζ is necessary and sufficient for LTP maintenance. Nat. Neurosci., 5(4): 295–296.

Lledo, P.M., Hjelmstad, G.O., Mukherji, S., Soderling, T.R., Malenka, R.C. and Nicoll, R.A. (1995) Calcium/calmodulin-dependent kinase II and long-term potentiation enhance synaptic transmission by the same mechanism. Proc. Natl. Acad. Sci. U.S.A., 92(24): 11175–11179.

Luscher, C., Xia, H., Beattie, E.C., Carroll, R.C., von Zastrow, M., Malenka, R.C. and Nicoll, R.A. (1999) Role of AMPA receptor cycling in synaptic transmission and plasticity. Neuron, 24(3): 649–658.

Marshall, B.S., Price, G. and Powell, C.T. (2000) Rat protein kinase c zeta gene contains alternative promoters for generation of dual transcripts with 5′-end heterogeneity. DNA Cell Biol., 19(12): 707–719.

McGlynn, E., Liebetanz, J., Reutener, S., Wood, J., Lydon, N.B., Hofstetter, H., Vanek, M., Meyer, T. and Fabbro, D. (1992) Expression and partial characterization of rat protein kinase C-delta and protein kinase C-zeta in insect cells using recombinant baculovirus. J. Cell Biochem., 49(3): 239–250.

Meffert, M.K., Chang, J.M., Wiltgen, B.J., Fanselow, M.S. and Baltimore, D. (2003) NF-kappa B functions in synaptic signaling and behavior. Nat. Neurosci., 6(10): 1072–1078.

Merlo, E., Freudenthal, R. and Romano, A. (2002) The IkappaB kinase inhibitor sulfasalazine impairs long-term memory in the crab *Chasmagnathus*. Neuroscience, 112(1): 161–172.

Moser, M.B. and Moser, E.I. (1998) Distributed encoding and retrieval of spatial memory in the hippocampus. J. Neurosci., 18(18): 7535–7542.

Muller, D., Bittar, P. and Boddeke, H. (1992) Induction of stable long-term potentiation in the presence of the protein kinase C antagonist staurosporine. Neurosci. Lett., 135(1): 18–22.

Muslimov, I.A., Nimmrich, V., Hernandez, A.I., Tcherepanov, A., Sacktor, T.C. and Tiedge, H. (2004) Dendritic transport and localization of protein kinase Mζ mRNA: implications for molecular memory consolidation. J. Biol. Chem., 279(50): 52613–52622.

Newton, A.C. (2001) Protein kinase C: structural and spatial regulation by phosphorylation, cofactors, and macromolecular interactions. Chem. Rev., 101(8): 2353–2364.

Newton, A.C. (2003) Regulation of the ABC kinases by phosphorylation: protein kinase C as a paradigm. Biochem. J., 370(Pt 2): 361–371.

Nishimune, A., Isaac, J.T., Molnar, E., Noel, J., Nash, S.R., Tagaya, M., Collingridge, G.L., Nakanishi, S. and Henley, J.M. (1998) NSF binding to GluR2 regulates synaptic transmission. Neuron, 21(1): 87–97.

Nishizuka, Y. (1988) The molecular heterogeneity of protein kinase C and its implication for cellular recognition. Nature, 334(6184): 661–665.

Okamoto, K., Nagai, T., Miyawaki, A. and Hayashi, Y. (2004) Rapid and persistent modulation of actin dynamics regulates postsynaptic reorganization underlying bidirectional plasticity. Nat. Neurosci., 7(10): 1104–1112.

Ono, Y., Fujii, T., Ogita, K., Kikkawa, U., Igarashi, K. and Nishizuka, Y. (1988) The structure, expression, and properties of additional members of the protein kinase C family. J. Biol. Chem., 263(14): 6927–6932.

Ono, Y., Fujii, T., Ogita, K., Kikkawa, U., Igarashi, K. and Nishizuka, Y. (1989) Protein kinase C zeta subspecies from rat brain: its structure, expression, and properties. Proc. Natl. Acad. Sci. U.S.A., 86(9): 3099–3103.

Opazo, P., Watabe, A.M., Grant, S.G. and O'Dell, T.J. (2003) Phosphatidylinositol 3-kinase regulates the induction of long-term potentiation through extracellular signal-related kinase-independent mechanisms. J. Neurosci., 23(9): 3679–3688.

Osten, P., Srivastava, S., Inman, G.J., Vilim, F.S., Khatri, L., Lee, L.M., States, B.A., Einheber, S., Milner, T.A., Hanson, P.I. and Ziff, E.B. (1998) The AMPA receptor GluR2 C terminus can mediate a reversible, ATP-dependent interaction with NSF and alpha- and beta-SNAPs. Neuron, 21(1): 99–110.

Osten, P., Valsamis, L., Harris, A. and Sacktor, T.C. (1996) Protein synthesis-dependent formation of protein kinase Mζ in LTP. J. Neurosci., 16(8): 2444–2451.

Ouyang, Y., Wong, M., Capani, F., Rensing, N., Lee, C.S., Liu, Q., Neusch, C., Martone, M.E., Wu, J.Y., Yamada, K., Ellisman, M.H. and Choi, D.W. (2005) Transient decrease in F-actin may be necessary for translocation of proteins into dendritic spines. Eur. J. Neurosci., 22(12): 2995–3005.

Pastalkova, E., Serrano, P., Pinkhasova, D., Wallace, E., Fenton, A.A. and Sacktor, T.C. (2006) Storage of spatial information by the maintenance mechanism of LTP. Science, 313(5790): 1141–1144.

Roberson, E.D., English, J.D. and Sweatt, J.D. (1996) A biochemist's view of long-term potentiation. Learn. Mem., 3(1): 1–24.

Sacktor, T.C., Osten, P., Valsamis, H., Jiang, X., Naik, M.U. and Sublette, E. (1993) Persistent activation of the ζ isoform of protein kinase C in the maintenance of long-term potentiation. Proc. Natl. Acad. Sci. U.S.A., 90(18): 8342–8346.

Sajikumar, S. and Frey, J.U. (2004) Late-associativity, synaptic tagging, and the role of dopamine during LTP and LTD. Neurobiol. Learn. Mem., 82(1): 12–25.

Sajikumar, S., Navakkode, S., Sacktor, T.C. and Frey, J.U. (2005) Synaptic tagging and cross-tagging: the role of protein kinase Mζ in maintaining long-term potentiation but not long-term depression. J. Neurosci., 25(24): 5750–5756.

Sanes, J.R. and Lichtman, J.W. (1999) Can molecules explain long-term potentiation? Nat. Neurosci., 2(7): 597–604.

Sanna, P.P., Cammalleri, M., Berton, F., Simpson, C., Lutjens, R., Bloom, F.E. and Francesconi, W. (2002) Phosphatidylinositol 3-kinase is required for the expression but not for the induction or the maintenance of long-term potentiation in the hippocampal CA1 region. J. Neurosci., 22(9): 3359–3365.

Schwartz, J.H. (1993) Cognitive kinases. Proc. Natl. Acad. Sci. U.S.A., 90(18): 8310–8313.

Schwartz, J.H. and Greenberg, S.M. (1987) Molecular mechanisms for memory: second-messenger induced modifications of protein kinases in nerve cells. Annu. Rev. Neurosci., 10: 459–476.

Serrano, P., Friedman, E.L., Kenney, J., Taubenfeld, S.M., Zimmerman, J.M., Alberini, C., Kelley, A.E., Maren, S., Yin, J.C.P., Sacktor, T.C. and Fenton, A.A. (2008) PKMζ maintains spatial, instrumental, and classically-conditioned long-term memories (submitted).

Serrano, P., Yao, Y. and Sacktor, T.C. (2005) Persistent phosphorylation by protein kinase Mζ maintains late-phase long-term potentiation. J. Neurosci., 25(8): 1979–1984.

Shema, R., Sacktor, T.C. and Dudai, Y. (2007) Rapid erasure of long-term memory associations in cortex by an inhibitor of PKMζ. Science, 317(5840): 951–953.

Shilatifard, A. (1998) Factors regulating the transcriptional elongation activity of RNA polymerase II. FASEB J., 12(14): 1437–1446.

Shirke, A.M. and Malinow, R. (1997) Mechanisms of potentiation by calcium-calmodulin kinase II of postsynaptic sensitivity in rat hippocampal CA1 neurons. J. Neurophysiol., 78(5): 2682–2692.

Silva, A.J., Kogan, J.H., Frankland, P.W. and Kida, S. (1998) CREB and memory. Annu. Rev. Neurosci., 21: 127–148.

Song, I., Kamboj, S., Xia, J., Dong, H., Liao, D. and Huganir, R.L. (1998) Interaction of the N-ethylmaleimide-sensitive factor with AMPA receptors. Neuron, 21(2): 393–400.

Sossin, W.S. (1996) Mechanisms for the generation of synapse specificity in long-term memory: the implications of a requirement for transcription. Trends Neurosci., 19(6): 215–218.

Stapulionis, R., Kolli, S. and Deutscher, M.P. (1997) Efficient mammalian protein synthesis requires an intact F-actin system. J. Biol. Chem., 272(40): 24980–24986.

Sublette, E., Naik, M., Jiang, X. and Sacktor, T.C. (1993) Evidence for a new isoform of protein kinase C in rat hippocampus. Neurosci. Lett., 159(1-2): 175–178.

Sutton, M.A. and Carew, T.J. (2000) Parallel molecular pathways mediate expression of distinct forms of intermediate-term facilitation at tail sensory–motor synapses in Aplysia. Neuron, 26(1): 219–231.

Sweatt, J.D. (1999) Toward a molecular explanation for long-term potentiation. Learn. Mem., 6(5): 399–416.

Takai, Y., Kishimoto, A., Inoue, M. and Nishizuka, Y. (1977) Studies on a cyclic nucleotide-independent protein kinase and its proenzyme in mammalian tissues. I. Purification and characterization of an active enzyme from bovine cerebellum. J. Biol. Chem., 252(21): 7603–7609.

Tang, S.J., Reis, G., Kang, H., Gingras, A.C., Sonenberg, N. and Schuman, E.M. (2002) A rapamycin-sensitive signaling pathway contributes to long-term synaptic plasticity in the hippocampus. Proc. Natl. Acad. Sci. U.S.A., 99(1): 467–472.

Thomas, K.L., Laroche, S., Errington, M.L., Bliss, T.V.P. and Hunt, S.P. (1994) Spatial and temporal changes in signal transduction pathways during LTP. Neuron, 13(3): 737–745.

Traynelis, S.F., Silver, R.A. and Cull-Candy, S.G. (1993) Estimated conductance of glutamate receptor channels activated during EPSCs at the cerebellar mossy fiber-granule cell synapse. Neuron, 11(2): 279–289.

Tse, D., Langston, R.F., Kakeyama, M., Bethus, I., Spooner, P.A., Wood, E.R., Witter, M.P. and Morris, R.G. (2007) Schemas and memory consolidation. Science, 316(5821): 76–82.

Tsokas, P., Ma, T., Landau, E.M. and Blitzer, R.D. (2006) Synaptic capture requires protein synthesis. Soc. Neurosci. Abstr.

Wesierska, M., Dockery, C. and Fenton, A.A. (2005) Beyond memory, navigation, and inhibition: behavioral evidence for hippocampus-dependent cognitive coordination in the rat. J. Neurosci., 25(9): 2413–2419.

Yao, Y., Kelly, M.T., Sajikumar, S., Serrano, P., Tian, D., Bergold, P.J., Frey, J.U., and Sacktor, T.C. (2008). PKMζ maintains late-LTP by enhancing NSF/GluR2-mediated trafficking of postsynaptic AMPARs (submitted).

Yeh, S.H., Lin, C.H., Lee, C.F. and Gean, P.W. (2002) A requirement of nuclear factor-κB activation in fear-potentiated startle. J. Biol. Chem., 277(48): 46720–46729.

Yin, J.C. and Tully, T. (1996) CREB and the formation of long-term memory. Curr. Opin. Neurobiol., 6(2): 264–268.

CHAPTER 3

Understanding the importance of mRNA transport in memory

María del Rayo Sánchez-Carbente[1] and Luc DesGroseillers[1,2,*]

[1]*Département de Biochimie et Physiologie, Université de Montréal, Montréal, QC H3C 3J7, Canada*
[2]*Groupe de Recherche sur le Système Nerveux Central (GRSNC), Université de Montréal, Montréal, QC H3C 3J7, Canada*

Abstract: RNA localization is an important mechanism to sort proteins to specific subcellular domains. In neurons, several mRNAs are localized in dendrites and their presence allows autonomous control of local translation in response to stimulation of specific synapses. Active constitutive and activity-induced mechanisms of mRNA transport have been described that represent critical steps in the establishment and maintenance of synaptic plasticity. In recent years, the molecular composition of different transporting units has been reported and the identification of proteins and mRNAs in these RNA granules contributes to our understanding of the key steps that regulate mRNA transport and translation. Although RNA granules are heterogeneous, several proteins are common to different RNA granule populations, suggesting that they play important roles in the formation of the granules and/or their regulation during transport and translation. About 1–4% of the neuron transcriptome is found in RNA granules and the characterization of bound mRNAs reveal that they encode proteins of the cytoskeleton, the translation machinery, vesicle trafficking, and/or proteins involved in synaptic plasticity. Non-coding RNAs and microRNAs are also found in dendrites and likely regulate RNA translation. These mechanisms of mRNA transport and local translation are critical for synaptic plasticity mediated by activity or experience and memory.

Keywords: neurons; long-term potentiation; RNA granule; memory; ribonucleoprotein; translation; RNA transport

Introduction

Memory is the capacity of an organism to store, retain, and recall information. Recent studies suggest that activity-dependent modulation of synaptic strength, also known as synaptic plasticity, is a fundamental mechanism by which some forms of memory are encoded and stored (Bailey et al., 2004). Understanding the neuro-anatomical, physiological, biochemical, and molecular bases of memory is a major task in the field of neurosciences. Almost 60 years ago, it was proposed that the physical base of memory could be dependent on the synthesis of specific proteins (Monné, 1948; Sutton and Schuman, 2006). Since then, a growing number of biochemical and genetic approaches have supported the role of new protein synthesis in memory formation/consolidation (Banko et al., 2006; Sutton and Schuman, 2006).

*Corresponding author. Tel.: +1 514-343-5802;
Fax: +1 514-343-2210; E-mail: luc.desgroseillers@umontreal.ca

In response to different behavioral experiments, memory can be classified as short-term memory if it lasts few minutes/hours or long-term memory if it is maintained for days. In a time-dependent period, short-term memory can also be converted into long-term memory. It is now accepted that memory formation/consolidation relies on activity-dependent synaptic strengthening. Long-term potentiation (LTP) is a well-established model to study the cellular and molecular aspects of memory formation/consolidation. Early LTP (E-LTP) can be induced by a single episode of high-frequency stimulation and lasts about 1 h. In contrast, late LTP (L-LTP) that lasts for hours can be induced by repeated and spaced stimulations. At the molecular level, several changes at the synapses occur during LTP, ranging from modification of pre-existing proteins to synthesis of new proteins and activation of gene expression. While modification of pre-existing proteins is sufficient to induce E-LTP, new protein synthesis and gene expression are required for the formation and maintenance of L-LTP (Lynch, 2004; Kelleher et al., 2004b). Protein modification and local protein synthesis are required to mark activated synapses that in turn become targets for transcription-dependent morphological modifications (Kelleher et al., 2004b).

For many years, it was thought that protein synthesis only occurred in the neuronal cell body. However, it is now clear that local translation in dendrites, particularly at or near synapses, is a major determinant for neuronal plasticity, mediating the enduring morphological and functional synaptic changes that underlie long-term memory (Kelleher et al., 2004b; Sutton and Schuman, 2006). Local protein synthesis at synapses following stimulation and localization of newly synthesized mRNAs at tagged synapses depends on efficient mechanisms of mRNA transport within dendrites. Therefore, it was not surprising to observe a subset of specific mRNAs in dendrites and axons and the modulation of their transport following neuronal activity. By spatially restricting gene expression within neurons, localized mRNAs endow synapses with the capacity to change their synaptic strength, to modify their structure, and consequently to regulate the efficacy by which they respond to specific stimuli (Eberwine et al., 2001; Steward and Schuman, 2003; Martin and Zukin, 2006). Several questions emerge from these studies: How are mRNAs selected for transport? How are they transported? How do they find their final destination? How translation is repressed during transport and de-repressed following cell stimulation? What is the molecular nature of the transporting units? What is their role in memory formation/consolidation?

In theory, mRNA localization can be achieved by several mechanisms including active directional transport on the cytoskeleton, directional export from the nucleus, localized RNA stability, and random cytoplasmic diffusion with local anchoring (Palacios and St Johnston, 2001; St Johnston, 2005). In neurons, active mRNA transport on elements of the cytoskeleton is now well documented (Mohr, 1999; Kiebler and DesGroseillers, 2000; Kanai et al., 2004). Imaging studies and biochemical experiments reveal that mRNAs to be localized are recognized by several RNA-binding proteins that form large messenger ribonucleoprotein (mRNP) complexes. These proteins regulate the transport and translation of bound mRNAs (Yoshimura et al., 2006). These mRNPs may be viewed as post-transcriptional operons which, by binding differential subpopulations of mRNAs, allow concerted expression of genes required for coordinated functions (Keene and Tenenbaum, 2002). In this chapter, we will review the mechanisms that allow active mRNA transport in dendrites. First, we will briefly summarize critical findings that support mRNA transport and localization in neurons. Second, we will present a model that highlights major steps in the mechanism of mRNA transport. Then, we will describe the molecular components of mRNPs and their role in the regulation of transport and/or translation. Finally, we will describe critical experiments that link mRNA transport to memory.

mRNA transport in neurons is a dynamic and active mechanism

The first experimental evidence that supports the hypothesis that mRNAs are transported, localized, and locally translated in dendrites is the presence of polyribosomes and associated membranous cisternae in the neck of spine heads of granular neurons by electron microscopy (Steward and Levy, 1982). This hypothesis was strengthened by the observation that the number of polysomes in spine heads increases during periods of synaptic growth (Steward, 1983; Steward and Falk, 1986). These results lead to a more direct search for mRNAs in dendrites using radioactive labeling strategies and in situ hybridization. First, cultured hippocampal neurons were incubated with radioactive uridine to label newly synthesized mRNAs and the localization of radioactive material was traced by autoradiography after chase periods of increasing length (Davis et al., 1987). Whereas radioactive material was only found in the cell body at early time points, it was also observed in neurites following longer incubation times, suggesting that newly synthesized RNAs were transported into dendrites. Then, the next step was to identify these transcripts. In situ hybridization with known probes allowed the identification of the first mRNAs localized in the somato-dendritic compartments of neurons both on sections of cerebral cortex and hippocampus and in cultured neurons (Garner et al., 1988; Bruckenstein et al., 1990; Burgin et al., 1990; Kleiman et al., 1990). While most mRNAs were restricted to the cell body, additional mRNAs such as those coding for the high-molecular-weight microtubule-associated protein 2 (MAP2), the α-subunit of the calcium/calmodulin-dependent protein kinase II (αCaM-KII), and the growth-associated protein 43 (GAP43) were also found in the proximal and/or distal parts of dendrites. With the development of the polymerase chain reaction (PCR) and microarray screening approaches, several mRNAs have been identified in dendrites (see below), their number being nevertheless relatively small as compared to that for mRNAs present in the soma (Miyashiro et al., 1994; Eberwine et al., 2002). Therefore, an active and specific mechanism to transport a restricted population of mRNAs that is critical for synaptic functions must have been elaborated in neurons. These studies, although significant, give a steady-state picture of the mRNAs in neuronal processes. A more dynamic view of mRNA transport was provided by the microinjection of in vitro synthesized labeled RNAs coding for the myelin basic protein (MBP) into oligodendrocytes (Ainger et al., 1993) and of brain-specific cytoplasmic RNA 1 (BC1) into sympathetic neurons (Muslimov et al., 1997). These experiments clearly showed the translocation of specific RNAs into neurites, providing the first evidences that mRNA transport in neural cells is a dynamic process. Dynamic translocation of endogenous mRNAs in neuronal processes was then described in live cells, using the membrane-permeable nucleic acid stain SYTO 14 (Knowles et al., 1996). Labeled RNAs were distributed in discrete granules containing poly(A)$^+$ mRNA, ribosomes, and the eukaryotic elongation factor (eEF)-1a, suggesting that these motile structures are translational units. Finally, an elegant demonstration that a specific mRNA is translocated to neuronal processes was provided in cultured hippocampal neurons by tethering a green fluorescent protein (GFP) to the 3′-untranslated region (3′UTR) of the αCaMKII mRNA with the MS2 bacteriophage tagging system (Rook et al., 2000). Lately, different approaches have shown that RNA granules are mobile structures with anterograde, retrograde, and oscillatory movements (Bannai et al., 2004; Dynes and Steward, 2007) that can be stimulated by neuronal activity or neurotrophin signaling (Rook et al., 2000; Zhang et al., 2001; Tiruchinapalli et al., 2003; Elvira et al., 2006b).

The mechanisms of mRNA transport

A first model on how mRNAs are transported and localized came from work in oligodendrocytes (Carson et al., 1998). Since then, a growing number of experimental data in many cell types of many species, including mammalian neurons, provides clues on different steps of mRNA transport and allows to elaborate a working model (Fig. 1) that includes the following steps: (1) recognition of RNA cis-acting elements by RNA-binding proteins (trans-acting factors) in the nucleus; (2) export of mRNA/mRNPs from the nucleus to the cytoplasm; (3) association of

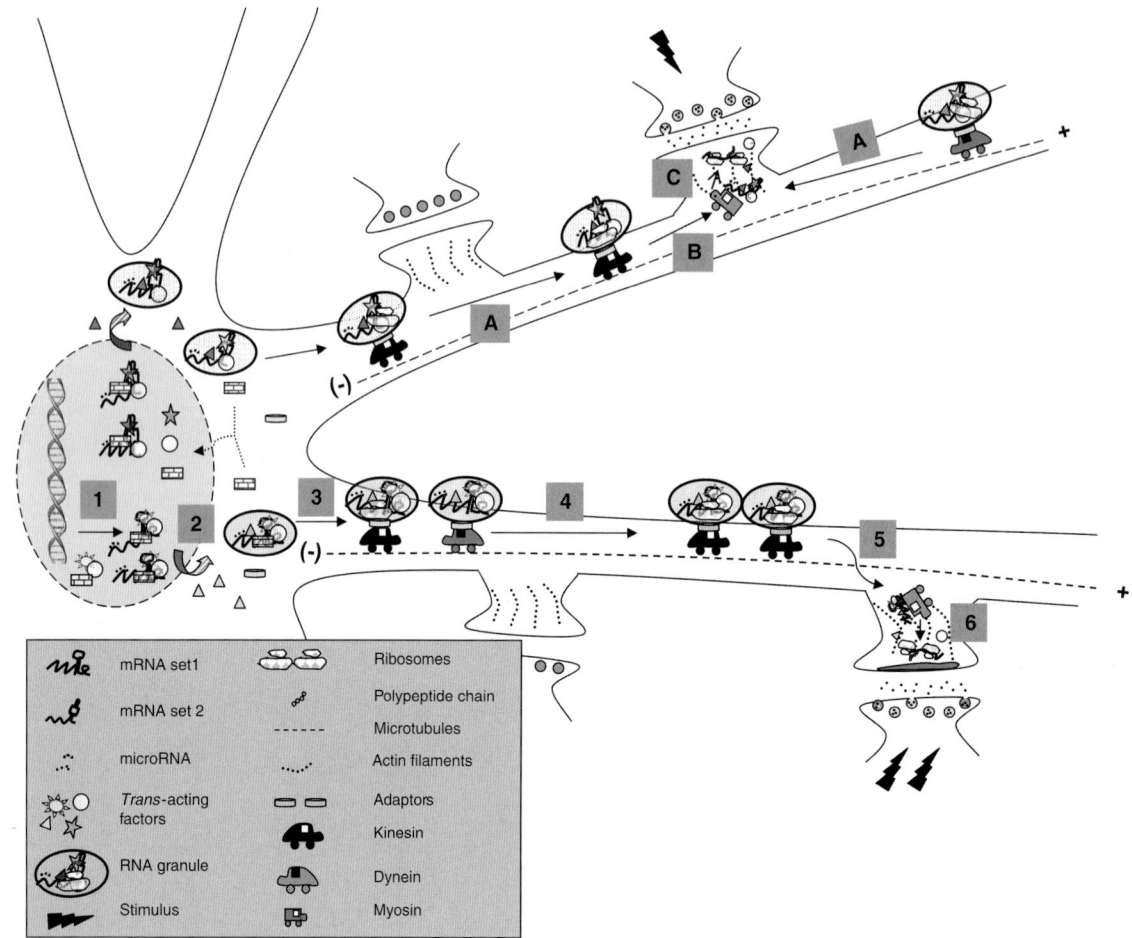

Fig. 1. Model of mRNA transport in neurons. In neurons, the presence of certain mRNAs in dendrites is constitutive (top neurite), whereas it is induced by activity for other mRNAs (bottom neurite). Top neurite: Transport of mRNAs constitutively present in the dendritic domain is increased in response to stimuli at specific synapses (A). Then, mRNAs are believed to be localized and/or anchored at the activated synapses (B) and locally translated at the activated sites (C). Bottom neurite: Transcription of other mRNAs is induced by synaptic activity. Transcribed mRNAs are recognized in the nucleus by RNA-binding proteins (1) and exported to the cytoplasm (2). RNPs are associated with additional factors including molecular motors to form functional mRNPs (3) that move on the cytoskeleton (4) and reach tagged synapses (5). mRNAs are then locally translated (6). The model also illustrates the heterogeneity of RNA granules in dendrites where different subpopulations of mRNAs are recognized by different although overlapping RNA-binding proteins.

additional factors including molecular motors to form functional mRNPs; (4) transport of mRNPs on the cytoskeleton; (5) localization and/or anchoring of mRNPs at their destination; and (6) translational derepression of the localized mRNAs by specific signals.

Several recent publications thoroughly review the mechanisms of mRNA transport and therefore the readers will be redirected to these reviews for more information (Kiebler and DesGroseillers, 2000; Kindler et al., 2005; St Johnston, 2005; Kiebler and Bassell, 2006; Dahm et al., 2007). Briefly, initial steps of RNA recognition and/or packaging of mRNP complexes to be transported in the cytoplasm likely occur in the nucleus (Hoek et al., 1998; Larocque et al., 2002; Oleynikov and Singer, 2003; Kindler et al., 2005). On one hand, several nuclear and/or nucleo-cytoplasmic shuttling proteins such as heterogeneous nuclear ribonucleoproteins (hnRNPs) and DEAD-box proteins are found in cytoplasmic mRNPs (Siomi and Dreyfuss, 1995; Weighardt et al., 1995; Michael et al., 1997; Weston and Sommerville, 2006). On the other hand, cytoplasmic proteins that are markers of dendritic RNA granules such as Staufen1 and Staufen2 transit in the nucleus/nucleolus (Macchi et al., 2004; Monshausen et al., 2004; Miki et al., 2005; Martel et al., 2006). The zinc-finger protein (ZFR) is likely a co-factor for the nucleo-cytoplasmic transport of Staufen2-containing mRNPs (Elvira et al., 2006a). This nuclear protein is nevertheless present in Staufen2-containing granules in the neuronal processes. In hippocampal neurons, down-regulation of ZFR induces the nuclear retention of Staufen2^{62} isoform, suggesting that the ZFR/Staufen2^{62} association first occurs in the nucleus and is required for the nuclear exit of the complex (Elvira et al., 2006a). Finally, direct evidences that the initial packaging of transported mRNAs may occur in the nucleus were provided when the β-actin mRNA-binding protein zipcode-binding protein 1 (ZBP1) was observed at the transcription sites of β-actin mRNA in serum-stimulated fibroblasts (Oleynikov and Singer, 2003). ZBP1 is required for β-actin mRNA transport to the growth cones of developing neurons (Zhang et al., 2001). In neurons, ZBP2 is also involved in β-actin mRNA transport (Gu et al., 2002). ZBP2 is a nuclear protein homologous to the KH-domain containing splicing regulatory protein (KSRP), suggesting that transport may be linked to splicing and/or to other nuclear mechanisms of RNA processing (Le Hir et al., 2001; Palacios, 2002).

Classically targeting of mRNAs to dendrites relies on the presence of dendritic targeting elements (DTEs) or "zipcodes". These RNA sequences are usually found in the 3′UTR of mRNAs (Blichenberg et al., 1999; Mori et al., 2000; Huang et al., 2003; Kremerskothen et al., 2006). These *cis*-acting elements are recognized by RNA-binding proteins and their association results in the formation of RNP complexes competent for their transport on the cytoskeleton (Mohr, 1999; Kindler et al., 2005). Deletion and expression studies have contributed to the identification of functional DTEs. These RNA elements are necessary and sufficient to cause the dendritic transport of reporter transcripts. DTEs from different transcripts are quite different in length, sequence, number, and position ranging from 54 nucleotides in the case of β-actin mRNA (Kislauskis et al., 1994; Ross et al., 1997) to around 1200 nucleotides for the αCaMKII transcript (Blichenberg et al., 1999). This diversity in the zipcode sequences/structures may parallel the complexity of the mechanisms of mRNA transport and translation (Kremerskothen et al., 2006). So far, no clear consensus "zipcode" sequence or structure has been identified. Giving the complexity of DTEs, it is predicted that DTEs are recognized by RNA-binding proteins through complex structural elements rather than via linear sequences. The growing list of mRNAs that are dendritically localized should allow the identification of more DTEs and by comparison the deciphering of a functional code.

Then, mRNPs are transported, docked, and stabilized at the target sites (Palacios and St Johnston, 2001; Kloc et al., 2002). In neurons, long-range transport is mediated by microtubules, whereas localization at destination sites is supported by actin filaments (Muslimov et al., 2002; Kindler et al., 2005). The presence of colchicine, a molecule that depolarizes microtubules, diminishes the amounts of poly(A)$^+$ mRNAs in dendrites. In contrast, cytochalasin D, a drug that disrupts actin-based microfilaments, has no effect (Bassell et al., 1994). Different motor proteins, members of the kinesin, dynein, or myosin families, are used to transport mRNPs along the cytoskeleton. Members of the kinesin (KIFs) family move toward the plus end of microtubules, whereas dynein moves in the opposite direction. These motor proteins participate in the anterograde and retrograde transport of mRNPs between the cell body and the axonal and dendritic terminals (Hirokawa and Takemura, 2005). Myosin members use actin filaments as rails to move their cargos (Langford, 2002). Therefore, the movement of RNA granules is bidirectional depending on the molecular motor with which they are associated. Transport of endogenous RNA granules (Knowles et al., 1996) and of RNA granule-associated proteins fused to

GFP (Kohrmann et al., 1999; Rook et al., 2000; Zhang et al., 2001; Bannai et al., 2004; Takano et al., 2007) was observed in live cells. The rapid anterograde and retrograde velocity of RNA granules (around 0.1–1.5 µm/s) supports the model of active transport of mRNAs in dendrites on the cytoskeleton. Consistently, down-regulation of members of the kinesin family impairs the ability of RNAs to translocate in neurites (Carson et al., 1998; Kanai et al., 2004). It is likely that mRNPs control their sorting by regulating the relative activities of opposing motors in microtubules (Bullock et al., 2006).

Final transport of at least some mRNAs into spines is mediated by the RNA-binding protein, translocation liposarcoma protein (TLS), using the microfilament network (Yoshimura et al., 2006). The actin-based molecular motor myosin-Va interacts with TLS to transport certain mRNAs from dendritic shafts into spines. Down-regulation of myosin-Va impairs the translocation of TLS and, in consequence, of its associated mRNAs into spines (Yoshimura et al., 2006). Interestingly, the translocation of TLS from microtubules to actin filaments is stimulated by metabotropic glutamate receptor (mGluR) agonists (Fujii and Takumi, 2005; Fujii et al., 2005; Yoshimura et al., 2006).

It is generally accepted that mRNAs are translationally repressed during transport and that repression is relieved after localization or in response to cell needs. It is believed that mRNAs in RNA granules that travel along with ribosomes (see below) are blocked at a translation elongation step (Ainger et al., 1993; Knowles et al., 1996; Kiebler and DesGroseillers, 2000; Sossin and DesGroseillers, 2006). The mechanism by which this may occur is completely unknown. It is possible that the absence of essential translation factor(s) in RNA granules prevents the initiation and/or elongation of proteins (Knowles et al., 1996). Interestingly, transcript coding for the translation elongation factor eEF-1a is present in dendrites and its translation is locally induced by protocols that trigger protein synthesis-dependent LTP in hippocampus, increasing dendritic translational capacity that contributes to LTP maintenance (Huang et al., 2005; Tsokas et al., 2005). Alternatively, RNA-binding proteins such as the fragile X mental retardation protein (FMRP) (Zalfa et al., 2006) or chaperons that recognized newly synthesized peptides as they emerge from translating ribosomes (Elvira et al., 2006b) may contribute to the inhibition of elongation steps. FMRP is a protein associated with polysomes that is known to repress translation both in vitro and in vivo. It is believed that FMRP in its heavily phosphorylated form represses translation elongation (Ceman et al., 2003). However, FMRP was also shown to associate with the non-coding BC1 RNA, a molecule that blocks translation initiation (Wang et al., 2005), and with the RNA-induced silencing complex (RISC) complex (Zalfa et al., 2006), suggesting that it may play other roles during translation. Similarly, the RNA-binding protein RNA granule 105 (RNG105) is found as puncta in dendrites of neurons and co-localizes with ribosomes (Shiina et al., 2005). Its overexpression inhibits translation in rabbit reticulocyte lysates and cell cultures. This protein interacts with the receptor for activated C kinase (RACK) (Angenstein et al., 2005), a protein that associates with the large ribosomal subunit and may regulate subunit joining through eukaryotic translation initiation factor (eIF) 6 (Ceci et al., 2003). On the other hand, translation of mRNAs in RNA particles devoid of ribosomes is likely repressed at the level of initiation (Anderson and Kedersha, 2006; Sossin and DesGroseillers, 2006). For example, ZBP1, a protein that binds β-actin mRNA and prevents its premature translation in the cytoplasm, was shown to impair the formation of 80S ribosomal complexes (Huttelmaier et al., 2005). It is not clear whether transient association of RNA granules/particles with stress granules, processing bodies (P bodies), or the RISC complex may contribute to the translational repression of associated mRNAs. Several granule-associated proteins such as Staufen and FMRP were indeed identified in stress granules, P bodies, and/or the RISC complex (Caudy et al., 2002; Mazroui et al., 2002; Lugli et al., 2005; Thomas et al., 2005; Barbee et al., 2006; Kim and Kim, 2006).

Interestingly, local translation seems to be regulated by neuronal activity. Blocking spontaneous neurotransmission in cultured hippocampal neurons results in a rapid increase of dendritic protein synthesis (Sutton et al., 2004). The mechanisms by which mRNAs are translationally derepressed in response to cell needs are still unclear. It was known that neuronal activity and brain-derived neurotrophic factor (BDNF) signaling increase mRNA transport in dendrites (Krichevsky and Kosik, 2001; Kanai et al., 2004; Elvira et al., 2006b), showing that mRNA transport is regulated. Neuronal activity was also shown to relocalize mRNAs from Staufen1-containing RNA granules to actively translating polyribosomes (Krichevsky and Kosik, 2001). However, the mechanism of translocation to the polysomes is not known. Similarly, RNG105 protein dissociates from RNA granules in dendrites following treatment of the cells with BDNF, and this is correlated with the translation of a reporter transcript (Shiina et al., 2005). Phosphorylation of repressor proteins may be a good way to regulate translation. Indeed, the protein kinase Src was shown to phosphorylate ZBP1, a protein modification that interferes with RNA binding and relieves the translational repression of β-actin mRNA (Huttelmaier et al., 2005). Similarly, in yeast, casein kinase II was shown to phosphorylate the translational repressor of ash1 mRNA and as a consequence to allow its active translation in the yeast bud (Paquin et al., 2007).

In contrast, dephosphorylation of FMRP was shown to allow ribosomes to translate mRNA (Zalfa et al., 2006). Phosphorylation of translation factors by the mammalian target of rapamycin (mTOR) pathway also regulates local protein synthesis and the rapamycin-dependent steps of LTP (Schratt et al., 2004; Kelleher et al., 2004b; Banko et al., 2006).

The molecular components of RNA granules

Protein components

One constant observation at the cellular level is that mRNAs are transported as molecular structures that form puncta in dendrites (Fig. 2). These structures are called RNA granules. They are distinguishable at the molecular levels from other RNA structures in the cells, such as translating polysomes, P bodies, stress granules, and microRNA particles (miRNPs) (Sossin and DesGroseillers, 2006). Nevertheless, these structures are dynamic and it is predicted that mRNA can transit between these structures. In addition to be transported, mRNA has to be translationally repressed during transport and eventually degraded to terminate the response.

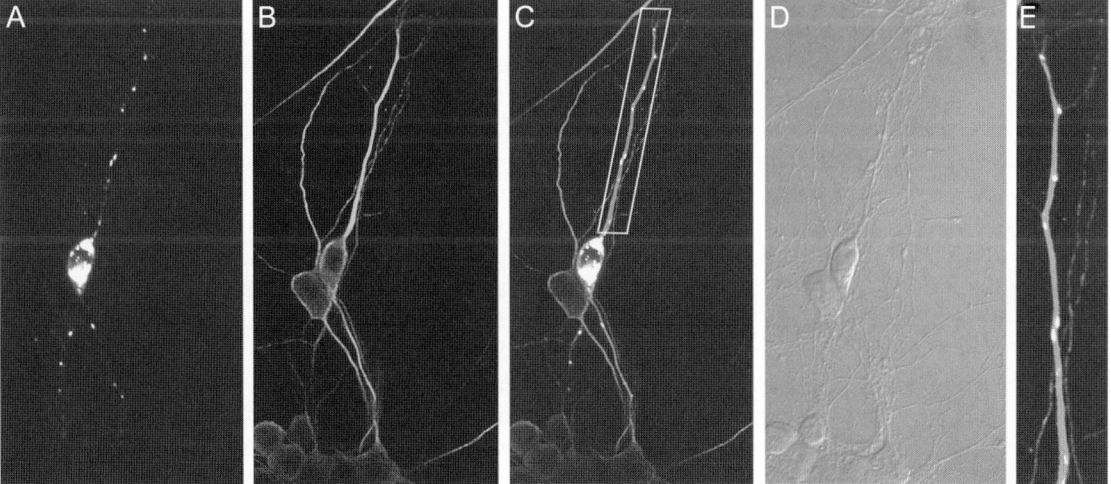

Fig. 2. DDX3 granules are present in the dendritic domain. (A) Enhanced yellow fluorescent protein (eYFP)-DDX3 was transfected in 6 DIV hippocampal neurons in culture and fixed after 16 h post-transfection. Note the granular distribution of DDX3 in neurites. (B) Immunofluorescence against MAP2 (dendritic marker). (C) Merge of A and B. (D) Phase contrast of the transfected neuron. (E) Magnification of the inset box in C.

Association with or transit through miRNPs, P bodies, active polysomes, or other structures may serve to fulfill complementary functions to fine-tune the whole process of mRNA localization.

Localized mRNAs travel as large mRNP complexes that may be associated with ribosomes. RNA granules contain both the large and small subunits of ribosomes, whereas RNA transporting particles are devoid of ribosomes (Sossin and DesGroseillers, 2006). In neurons, both RNA granules and RNA particles are believed to be components of the RNA transport machinery. Several studies focused on the molecular characterization of these structures as a mean to understand their roles in RNA transport and the nature of the molecules that regulate transport, translation, stability, etc. Studies using biochemical approaches first revealed that large macromolecular complexes can be purified by sucrose gradients (Krichevsky and Kosik, 2001) or gel filtration (Mallardo et al., 2003). The larger complexes co-fractionated with ribosomes and specific mRNA. Smaller particles do not co-fractionate with ribosomes but co-enrich with the molecular motor kinesin and with BC1 RNA and αCaMKII mRNA (Mallardo et al., 2003).

More recently, proteomic approaches were used to characterize RNA granules in developing (Elvira et al., 2006b) and newborn rat brains (Kanai et al., 2004). Kanai et al. (2004) identified proteins that co-purified with the kinesin KIF5. It was known that, in dendrites, RNA granules/particles used microtubules to be transported and that kinesin served as a molecular motor. Therefore, it was hypothesized that RNA granule components may be associated with kinesin. Several proteins were identified that included Purα, Purβ, hnRNP-U, PSF, DEAD-box1 (DDX1), DEAD-box3 (DDX3), SYNCRIP/hnRNP Q1, TLS, NonO, ALY, Staufen1, FMRP proteins, and eEF-1a. Interestingly, many of these proteins are RNA-binding proteins and some of them have previously been shown to be involved in RNA transport. Moreover, several of the identified proteins were known to co-localize in dendrites and/or to co-purify with Purα, a well-known protein involved in RNA transport (Ohashi et al., 2000). Over-expression of KIF5 enhanced the transport of Purα-containing granules toward the distal direction, whereas expression of a KIF5 dominant-negative protein reduced their transport (Kanai et al., 2004), confirming the link between KIF5, proteomically-identified proteins, and RNA transport. This conclusion was more directly tested using a protein knockdown approach. Transport in dendrites of a reporter transcript fused to the 3′UTR of αCaMKII mRNA was impaired when neurons were transfected with siRNAs that knocked down the expression of Purα, PSF, hnRNP-U, or Staufen1. The same approach that knocked down the expression of DDX3 and SYNCRIP/hnRNP Q1 failed to modify the localization of the transcript. These results show that many proteins are required for the transport of αCaMKII mRNA and suggest that other proteins may be involved in regulating other aspects of RNA transport such as translation or RNA stability. Alternatively, the RNA granule population is likely heterogeneous and the proteins that did not impair the αCaMKII mRNA localization may define other subpopulations of RNA granules that transport other RNAs (see below).

A partial response to these possibilities was provided in a second proteomic study (Elvira et al., 2006b). In this study, RNA granules were biochemically purified according to their physicochemical properties. Embryonic rat brains were homogenized and RNA granules purified through differential centrifugation steps and sucrose gradients. In the proteomics, several proteins were identified and many of them were shown to form granules in dendrites when expressed in neurons, confirming the role of these proteins. Many of these proteins identified in RNA granules from embryonic brains (Elvira et al., 2006b) were also identified as components of the KIF5-associated RNA granules in postnatal brains (Kanai et al., 2004) (Table 1). These include FMRP, Purα, Purβ, SYNCRIP/hnRNP Q1, DDX1, DDX3, and the uncharacterized protein CGI-99. Ribosomes were also found in these RNA granule preparations, a hallmark of RNA granules. Interestingly, the embryonic granules contain several proteins of the DEAD-box family. These proteins are RNA helicases that play diverse roles in the organization of RNA granules such as Vasa in *Drosophila*

Table 1. Proteomic comparison of different mRNPs from neurons and other cell types

	RNA granule[a]	RNA granule[b]	RNA particle[c]	Staufen1 RNA granule[d]	SYNCRIP/ hnRNPLQ1 RNA granule[e]	IMP1/ZBP1 granule[f]
Cell types						
Proteomics	Postnatal brain	Embryonic brain	Postnatal brain	HEK293	293EBNA	HEK293
Confocal	Neurons	Neurons	Neurons	Neuroblasts	Neurons	
Identified mRNAs	αCaMKII; Arc	β-actin	Poly(A)$^+$ RNA	RNA	IP3R1	IMP1; others
Molecular motors					No	No
Kinesin	KIF5			Kinesin		
Dynein		Heavy chain; dynactin	Heavy chain	Intermediate chain; Heavy chain		
Myosin						
Ribosomal proteins	Yes	Yes	No	Yes	Yes	40S
Translation factors						
eIF4A	No	Yes	No	No	No	No
eIF4E	No	No	No	No	No	No
eIF2α	Yes	Yes	No	No	No	No
eEF1α	Yes	Yes	No	No	Yes	No
RNA-binding proteins						
hnRNPs	Yes	Yes	Yes	Yes	Yes	Yes
Staufens	Staufen1	Staufen2	No	Staufen1	Staufen1	No
FMRPs	FMRP; FXR1; FXR2	FMRP	No	FMRP	No	No
DEAD-box	1; 3; 5	1; 3; 5; 6; 9; others	5	9	5; 9; 17; 21	9
PABPs	No	Yes	Yes	Yes	Yes	Yes
Pur proteins	α; β	α; β	α	No	No	No
Nucleolin	Yes	Yes	Yes	Yes	Yes	Yes
YB1	No	Yes	No	No	Yes	Yes

Note: mRNP-associated proteins were detected by mass spectrometry (underlined), immunoprecipitation, and/or confocal microscopy.
[a] Kanai et al. (2004).
[b] Elvira et al. (2006b).
[c] Angenstein et al. (2005).
[d] Brendel et al. (2004); Villace et al. (2004).
[e] Bannai et al. (2004).
[f] Jonson et al. (2007).

(Breitwieser et al., 1996) and Dhh1p in yeast (Coller and Parker, 2005), in the nuclear export of mRNA (Dbp5 and DDX3) (Snay-Hodge et al., 1998; Tseng et al., 1998; Yedavalli et al., 2004), and in the translation initiation (eIF4A) (Rocak and Linder, 2004) and translation repression (Dhh1p) (Coller and Parker, 2005). The expression of a tagged form of DDX3 in neurons lead to the formation of puncta in dendrites that contain ribosomes, poly(A)$^+$ RNA, and other proteomically identified proteins (Elvira et al., 2006b). These DDX3-containing granules move in dendrites on the cytoskeleton and their displacement increases in response to the neurotrophin BDNF, confirming their link with RNA transport.

One of the conclusions of this study is that RNA granules are heterogeneous (Elvira et al., 2006b). Indeed, two proteomically identified proteins, DDX3 and CGI-99, only poorly co-localized in dendrites, identifying different populations of RNA granules. In addition, each RNA granule population only partly co-localized with Staufen2, an identified protein of RNA granules. Similarly, expression of different members of the DEAD-box family tagged to fluorescent proteins reveals that they only partly co-localize in RNA granules (unpublished data). Altogether, these results indicate that the RNA granule population is heterogeneous and that specific proteins define subpopulation of granules that may transport different RNA molecules and/or differentially localize in subcellular domains. Despite many common components between the embryonic and postnatal RNA granule populations, associated mRNAs differ in these preparations. Whereas αCaMKII and activity-regulated cytoskeletal protein (Arc/Arg3.1) mRNA were found in the postnatal granules, they were absent in the embryonic preparations. In contrast, β-actin mRNA was enriched in the embryonic RNA granules. Interestingly, this mRNA co-localizes with DDX3-containing granules (unpublished data), suggesting that these RNA granules may regulate the transport of β-actin mRNA but not that of αCaMKII mRNA (see above). Whether these differences represent developmental differences or RNA granule diversity is not clear. However, it is possible to speculate that the RNA granules isolated from developing brain are involved in neurite outgrowth, whereas the RNA granules from adult brain are involved in the delivery of mRNAs important for synapse maintenance and/or synaptic plasticity.

Recently, proteomic analyses of RNA granules isolated from non-neuronal mammalian cells have been reported (Table 1). These analyses were done following immunoprecipitation of materials using antibodies directed against RNA granule-associated proteins such as insulin-like growth factor II mRNA-binding protein 1 (IMP1), Staufen1, and SYNCRIP/hnRNP Q1 (Bannai et al., 2004; Brendel et al., 2004; Villace et al., 2004; Jonson et al., 2007). Comparison of identified proteins shows that several components are shared between RNA granules such as hnRNP proteins, ribosomal proteins, Y-box-binding protein, nucleolin and zipcode-binding proteins IMP1 and IMP2. However, the composition of these RNPs is not identical and, for example, IMP1-containing granules do not contain components of the large ribosomal subunit.

RNA components

At least two strategies have been used to identify dendritically localized mRNAs: (1) immunopurification of selected RNA granules/particles-associated proteins and (2) mRNA purification from soma-free neurite preparations. mRNA identification in these preparations was performed by PCR-based approaches or microarray hybridization, and the validation was done by RT-PCR or in situ hybridization. Consistently, less than 4% of the neuron transcriptome is isolated from neurites (Miyashiro et al., 1994; Eberwine et al., 2001, 2002; Moccia et al., 2004; Angenstein et al., 2005; Willis et al., 2005; Poon et al., 2006; Zhong et al., 2006). One consistent finding is the presence of mRNAs coding for cytoskeletal elements and components of the translational machinery. Local translation of several of these mRNAs has been observed in neurite preparations (Wu et al., 1998; Wells et al., 2001). Local translation of cytoskeletal elements suggests that the cytoskeleton is quite dynamic and that protein synthesis serves to promote the growth and/or maintenance of new synaptic connections associated with synaptic strengthening (Yuste and Bonhoeffer, 2001; Ethell and Pasquale, 2005). Other functions such as growth cone guidance and motility (Leung et al., 2006; Yao et al., 2006) mediated by local synthesis of β-actin mRNA or regulation of AMPA receptors trafficking by Arc/Arg3.1 mRNA (Mokin et al., 2006; Rial Verde et al., 2006; Shepherd et al., 2006) have been described. Similarly, the presence of diverse mRNAs coding for components of the translational machinery suggests the presence of a positive feedback mechanism that increases the translational capacity of induced synapses. Moreover, a spatial and/or temporal regulation of specific components of the translation machinery may be used to repress protein synthesis in dendrites and allow their derepression according to cell needs. A good candidate for such a regulation is the rapamycin-sensitive pathway that is activated during LTP (Huang et al., 2005; Tsokas et al., 2007). Other mRNAs found in neurites code for receptors and channels, signaling molecules, and factors involved in membrane trafficking, post-translational protein modifications, protein degradation, RNA binding, etc. (Miyashiro et al., 1994; Eberwine et al., 2001, 2002; Moccia et al., 2004; Angenstein et al., 2005; Willis et al., 2005; Poon et al., 2006; Zhong et al., 2006).

Candidate-based approaches to identify mRNA associated with specific RNA granules result in the recovery of about 4% of the mRNAs expressed in mouse brain as putative components of FMRP-containing granules (Brown et al., 2001; Chen et al., 2003; Miyashiro et al., 2003; Jin et al., 2004; Darnell et al., 2005). Identified transcripts from FRMP–mRNP complexes code for a wide range of proteins involved in neuronal functions. Several of the mRNAs encode proteins involved in the Golgi vesicle maturation and vesicle transport in neurons, whereas others are associated with the postsynaptic components and are involved in maintaining the postsynaptic density (PSD) structure and neuronal cell signaling (Brown et al., 2001). Considering the heterogeneity of RNA granules, these candidate-based approaches are interesting to decipher the precise pathways regulated by specific RNA-binding proteins and dissect the mRNA transport machineries involved in the post-transcriptional regulation of gene expression.

Untranslated RNAs such as BC1 RNA and miRNAs are other components of RNA granules (Cao et al., 2006; Dahm et al., 2007). They are involved in the transport and/or translational regulation of mRNAs in dendritic domains. However, as they do not code for proteins, non-coding RNAs likely have structural, catalytic, or regulatory functions. The BC1 RNA (Brosius and Tiedge, 2001) is present in the postsynaptic dendritic domain and it is actively transported into dendrites on the cytoskeleton (Chicurel et al., 1993; Muslimov et al., 1997; Cristofanilli et al., 2006). BC1 was shown to repress translation by binding to poly(A)-binding protein (PABP) and eIF4A and targeting assembly of 48S initiation complexes (Wang et al., 2005). BC1 knockout mice show reduced exploration and increased anxiety (Lewejohann et al., 2004). Recently, more attention was placed on the presence and role of miRNA in dendrites. miRNAs are small RNA molecules involved in the translational repression of bound mRNAs or in their degradation (Bushati and Cohen, 2007). Several miRNAs have been isolated from neurons (Lagos-Quintana et al., 2002; Dostie et al., 2003; Krichevsky et al., 2003; Kim and Kim, 2006; Kosik, 2006; Wheeler et al., 2006). At least one of them was described in dendritic RNA granules and shown to inhibit the translation of Limk1 resulting in altered dendritic spines (Schratt et al., 2006). It is interesting to hypothesize that miRNAs may regulate the translational inhibition of specific mRNAs during transport and that this inhibition may be abolished by cell stimulation. Indeed, Limk1 translational inhibition is relieved following BDNF signaling. Local degradation of protein components of the RISC complex during formation of long-term memory may be an efficient mechanism to induce local protein synthesis from miRNA-repressed transcripts (Ashraf et al., 2006).

The relevance of mRNA transport in memory

Altogether, these results indicate that mRNA transport and their local translation in dendrites are important mechanisms for neuron functions. These mechanisms are regulated by neuron activity and neurotrophin signaling, suggesting that they modulate specific pathways related to synaptic functions. Nevertheless, the significance of these processes is not completely understood. One emerging field in which transport and translation of mRNAs seem essential is synaptic plasticity related to LTP. Although only few reports address the relevance of mRNA transport, several convincing cellular and behavioral results show that local translation plays a key role in synaptic plasticity (Pfeiffer and Huber, 2006; Sutton and Schuman, 2006).

Indirect evidences have accumulated for a role of mRNA transport during memory formation/consolidation. Several mRNAs such as those coding for αCamKII (Thomas et al., 1994), BDNF, tyrosine kinase receptor B (TrkB) (Tongiorgi et al., 1997), and protein kinase Mζ (PKMζ) (Muslimov et al., 2004), as well as the non-coding

BC1 RNA (Muslimov et al., 1998), are transported in dendrites. Similarly, in *Aplysia* neurons, syntaxin mRNA normally localized at one pole of the cell is re-localized to the axon hillock in response to long-term facilitation (Hu et al., 2003). These proteins are known to play critical roles during induction and/or maintenance of LTP following their local translation. The observation that their transport is stimulated by synaptic activity and LTP induction supports the notion that mRNA transport is crucial for these functions. The transport of Arc/Arg3.1 mRNA is also informative since it occurs in response to LTP-producing stimulation and only targets stimulated synapses (Steward et al., 1998). The molecular path that links Arc/Arg3.1 expression to the late phases of LTP is not clear. Recently, a correlation was made between Arc/Arg3.1 expression and an increase rate of AMPA receptor endocytosis through interaction of Arc/Arg3.1 with components of the endocytic machinery (Chowdhury et al., 2006). However, this observation is not consistent with a putative role of Arc/Arg3.1 during LTP, which rather requires membrane insertion of AMPA receptors. Similarly, it is not clear why LTP is lost in *Arc/Arg3.1* knockout mice that are shown to express more AMPA receptors at the membranes (Tzingounis and Nicoll, 2006). It is likely that other molecular mechanisms contribute to the Arc/Arg3.1-dependent late phases of LTP. Interestingly, Arc/Arg3.1 was shown to interact with αCaMKII and potentiate its action for neurite extension in neuroblastoma cells (Donai et al., 2003).

More direct evidences are provided by genetic manipulation of mRNA transport. First, a mutant mouse was generated in which the *αCaMKII* gene expression was restricted to the cell body as a consequence of the deletion of the DTE in its 3′UTR. Maintenance of stable L-LTP and spatial memory are impaired in these mice (Miller et al., 2002). However, it is not clear whether it is the impaired mRNA transport or the observed decrease in the translation of αCaMKII mRNA in the cell body that is most detrimental for LTP (Steward and Worley, 2002). Other studies elegantly addressed this problem with antisense oligonucleotides that bind the RNA *cis*-acting targeting element and therefore prevent mRNA recognition by *trans*-acting RNA-binding proteins involved in mRNA transport. Antisense oligonucleotides directed against the targeting element of the β-actin mRNA prevent its association with ZBP1, its transport into growth cones, and abolish growth cone motility (Zhang et al., 2001; Yao et al., 2006). Similarly, blocking the interaction between syntaxin mRNA and RNA-binding proteins prevents the translocation of syntaxin mRNA to the axon hillock in *Aplysia* neurons and abolishes long-term facilitation (Liu et al., 2006).

Whereas E-LTP is mediated by post-translational modifications of proteins, it is now evident that specific stages of L-LTP are dependent on local translation of transported mRNA in dendrites independent of transcription (Kang and Schuman, 1996; Kelleher et al., 2004b; Cracco et al., 2005). Increased synthesis of specific proteins in dendrites following LTP induction has been reported (Ouyang et al., 1999; Steward and Halpain, 1999; Aakalu et al., 2001; Schratt et al., 2004; Tsokas et al., 2005). However, there is so far no clear evidence of an overall increase in protein synthesis in dendrites, suggesting that LTP-inducing protocols, rather than increasing global protein synthesis in dendrites, differentially regulate translation of particular mRNAs (Schuman et al., 2006). The mechanisms that link LTP-inducing protocols to increase local translation and/or differential protein synthesis in dendrites are not clear. Post-translational modifications by the extracellular signal-regulated kinase (ERK)/mitogen-activated protein kinase (MAPK) and/or the phosphatidylinositol-3 kinase (PI3K) pathways of translation initiation factors and/or of RNA-binding proteins involved in the translational regulation of specific mRNAs have been shown to enhance translation initiation in response to LTP-inducing protocols (Wells et al., 2001; Tang et al., 2002; Banko et al., 2004; Kelleher et al., 2004a; Pfeiffer and Huber, 2006; Wells, 2006; Tsokas et al., 2007). Consistently, genetic ablation of suppressors of translation initiation results in a deficit of some forms of protein synthesis-dependent LTP (Alarcon et al., 2004; Banko et al., 2005; Costa-Mattioli et al., 2005). Whether these mechanisms occur in dendrites to regulate local protein synthesis is still unknown.

Conclusions

Altogether these results indicate that mRNA transport and local protein synthesis are regulated mechanisms that are required for the maintenance of synaptic plasticity. Recent studies contribute to our understanding of the molecular composition of RNA granules in neurons that in turn will concur to the elucidation of the role(s) of specific molecules in the phenomenon of learning and memory. Many questions are still unanswered. Is there a functional link between RNA granules and RNA particles? What is the mechanism of recognition and assembly of these RNA granules/particles and in which subcellular domains does it occur? What is the role(s) of each component of the RNA granules/particles? Which cellular pathways do regulate transport, localization, and translation of mRNAs in response to cell needs? What is the time frame for efficient regulation of protein synthesis during memory formation/consolidation? Obviously, the coming years will provide exciting new concepts.

Abbreviations

Arc/Arg3.1	activity-regulated cytoskeletal protein
BC1	brain-specific cytoplasmic RNA 1
BDNF	brain-derived neurotrophic factor
αCaMKII	α-subunit of the calcium/calmodulin-dependent protein kinase II
DDX	DEAD-box protein
DTE	dendritic target element
eIF	eukaryotic initiation factor
eEF	eukaryotic elongation factor
E-LTP	early LTP
FMRP	fragile X mental retardation protein
GFP	green fluorescent protein
hnRNP	heterogeneous nuclear ribonucleoprotein
IMP1	insulin-like growth factor II mRNA-binding protein 1
L-LTP	late LTP
LTP	long-term potentiation
MAP2	microtubule-associated protein 2
miRNPs	microRNA particles
mRNP	messenger ribonucleoprotein
PABP	poly (A)-binding protein
P bodies	processing bodies
PCR	polymerase chain reaction
RISC	RNA-induced silencing complex
RNG105	RNA-binding protein RNA granule 105
TLS	translocation liposarcoma protein
3'UTR	3'untranslated region
ZBP1	zipcode-binding protein 1
ZFR	zinc-finger protein

References

Aakalu, G., Smith, W.B., Nguyen, N., Jiang, C. and Schuman, E.M. (2001) Dynamic visualization of local protein synthesis in hippocampal neurons. Neuron, 30: 489–502.

Ainger, K., Avossa, D., Morgan, F., Hill, S.J., Barry, C., Barbarese, E. and Carson, J.H. (1993) Transport and localization of exogenous myelin basic protein mRNA microinjected into oligodendrocytes. J. Cell Biol., 123: 431–441.

Alarcon, J.M., Hodgman, R., Theis, M., Huang, Y.S., Kandel, E.R. and Richter, J.D. (2004) Selective modulation of some forms of schaffer collateral-CA1 synaptic plasticity in mice with a disruption of the CPEB-1 gene. Learn. Mem., 11: 318–327.

Anderson, P. and Kedersha, N. (2006) RNA granules. J. Cell Biol., 172: 803–808.

Angenstein, F., Evans, A.M., Ling, S.C., Settlage, R.E., Ficarro, S., Carrero-Martinez, F.A., Shabanowitz, J., Hunt, D.F. and Greenough, W.T. (2005) Proteomic characterization of messenger ribonucleoprotein complexes bound to nontranslated or translated poly(A) mRNAs in the rat cerebral cortex. J. Biol. Chem., 280: 6496–6503.

Ashraf, S.I., McLoon, A.L., Sclarsic, S.M. and Kunes, S. (2006) Synaptic protein synthesis associated with memory is regulated by the RISC pathway in Drosophila. Cell, 124: 191–205.

Bailey, C.H., Kandel, E.R. and Si, K. (2004) The persistence of long-term memory: a molecular approach to self-sustaining changes in learning-induced synaptic growth. Neuron, 44: 49–57.

Banko, J.L., Hou, L. and Klann, E. (2004) NMDA receptor activation results in PKA- and ERK-dependent Mnk1 activation and increased eIF4E phosphorylation in hippocampal area CA1. J. Neurochem., 91: 462–470.

Banko, J.L., Hou, L., Poulin, F., Sonenberg, N. and Klann, E. (2006) Regulation of eukaryotic initiation factor 4E by converging signaling pathways during metabotropic glutamate receptor-dependent long-term depression. J. Neurosci., 26: 2167–2173.

Banko, J.L., Poulin, F., Hou, L., DeMaria, C.T., Sonenberg, N. and Klann, E. (2005) The translation repressor 4E-BP2 is critical for eIF4F complex formation, synaptic plasticity, and memory in the hippocampus. J. Neurosci., 25: 9581–9590.

Bannai, H., Fukatsu, K., Mizutani, A., Natsume, T., Iemura, S., Ikegami, T., Inoue, T. and Mikoshiba, K. (2004) An RNA-interacting protein, SYNCRIP (heterogeneous nuclear ribonuclear protein Q1/NSAP1) is a component of mRNA granule transported with inositol 1,4,5-trisphosphate receptor type 1 mRNA in neuronal dendrites. J. Biol. Chem., 279: 53427–53434.

Barbee, S.A., Estes, P.S., Cziko, A.M., Hillebrand, J., Luedeman, R.A., Coller, J.M., Johnson, N., Howlett, I.C., Geng, C., Ueda, R., Brand, A.H., Newbury, S.F., Wilhelm, J.E., Levine, R.B., Nakamura, A., Parker, R. and Ramaswami, M. (2006) Staufen- and FMRP-containing neuronal RNPs are structurally and functionally related to somatic P bodies. Neuron, 52: 997–1009.

Bassell, G.J., Singer, R.H. and Kosik, K.S. (1994) Association of poly(A) mRNA with microtubules in cultured neurons. Neuron, 12: 571–582.

Blichenberg, A., Schwanke, B., Rehbein, M., Garner, C.C., Richter, D. and Kindler, S. (1999) Identification of a cis-acting dendritic targeting element in MAP2 mRNAs. J. Neurosci., 19: 8818–8829.

Breitwieser, W., Markussen, F.H., Horstmann, H. and Ephrussi, A. (1996) Oskar protein interaction with Vasa represents an essential step in polar granule assembly. Genes Dev., 10: 2179–2188.

Brendel, C., Rehbein, M., Kreienkamp, H.J., Buck, F., Richter, D. and Kindler, S. (2004) Characterization of Staufen 1 ribonucleoprotein complexes. Biochem. J., 384: 239–246.

Brosius, J. and Tiedge, H. (2001) Neuronal BC1 RNA: intracellular transport and activity-dependent modulation. Results Probl. Cell Differ., 34: 129–138.

Brown, V., Jin, P., Ceman, S., Darnell, J.C., O'Donnell, W.T., Tenenbaum, S.A., Jin, X., Feng, Y., Wilkinson, K.D., Keene, J.D., Darnell, R.B. and Warren, S.T. (2001) Microarray identification of FMRP-associated brain mRNAs and altered mRNA translational profiles in fragile X syndrome. Cell, 107: 477–487.

Bruckenstein, D.A., Lein, P.J., Higgins, D. and Fremeau, R.T., Jr. (1990) Distinct spatial localization of specific mRNAs in cultured sympathetic neurons. Neuron, 5: 809–819.

Bullock, S.L., Nicol, A., Gross, S.P. and Zicha, D. (2006) Guidance of bidirectional motor complexes by mRNA cargoes through control of dynein number and activity. Curr. Biol., 16: 1447–1452.

Burgin, K.E., Waxham, M.N., Rickling, S., Westgate, S.A., Mobley, W.C. and Kelly, P.T. (1990) In situ hybridization histochemistry of Ca^{2+}/calmodulin-dependent protein kinase in developing rat brain. J. Neurosci., 10: 1788–1798.

Bushati, N. and Cohen, S.M (2007) microRNA functions. Annu. Rev. Cell. Dev. Biol., 23: 175–205.

Cao, X., Yeo, G., Muotri, A.R., Kuwabara, T. and Gage, F.H. (2006) Noncoding RNAs in the mammalian central nervous system. Annu. Rev. Neurosci., 29: 77–103.

Carson, J.H., Kwon, S. and Barbarese, E. (1998) RNA trafficking in myelinating cells. Curr. Opin. Neurobiol., 8: 607–612.

Caudy, A.A., Myers, M., Hannon, G.J. and Hammond, S.M. (2002) Fragile X-related protein and VIG associate with the RNA interference machinery. Genes Dev., 16: 2491–2496.

Ceci, M., Gaviraghi, C., Gorrini, C., Sala, L.A., Offenhauser, N., Marchisio, P.C. and Biffo, S. (2003) Release of eIF6 (p27BBP) from the 60S subunit allows 80S ribosome assembly. Nature, 426: 579–584.

Ceman, S., O'Donnell, W.T., Reed, M., Patton, S., Pohl, J. and Warren, S.T. (2003) Phosphorylation influences the translation state of FMRP-associated polyribosomes. Hum. Mol. Genet., 12: 3295–3305.

Chen, L., Yun, S.W., Seto, J., Liu, W. and Toth, M. (2003) The fragile X mental retardation protein binds and regulates a novel class of mRNAs containing U rich target sequences. Neuroscience, 120: 1005–1017.

Chicurel, M.E., Terrian, D.M. and Potter, H. (1993) mRNA at the synapse: analysis of a synaptosomal preparation enriched in hippocampal dendritic spines. J. Neurosci., 13: 4054–4063.

Chowdhury, S., Shepherd, J.D., Okuno, H., Lyford, G., Petralia, R.S., Plath, N., Kuhl, D., Huganir, R.L. and Worley, P.F. (2006) Arc/Arg3.1 interacts with the endocytic machinery to regulate AMPA receptor trafficking. Neuron, 52: 445–459.

Coller, J. and Parker, R. (2005) General translational repression by activators of mRNA decapping. Cell, 122: 875–886.

Costa-Mattioli, M., Gobert, D., Harding, H., Herdy, B., Azzi, M., Bruno, M., Bidinosti, M., Ben Mamou, C., Marcinkiewicz, E., Yoshida, M., Imataka, H., Cuello, A.C., Seidah, N., Sossin, W., Lacaille, J.C., Ron, D., Nader, K. and Sonenberg, N. (2005) Translational control of hippocampal synaptic plasticity and memory by the eIF2alpha kinase GCN2. Nature, 436: 1166–1173.

Cracco, J.B., Serrano, P., Moskowitz, S.I., Bergold, P.J. and Sacktor, T.C. (2005) Protein synthesis-dependent LTP in isolated dendrites of CA1 pyramidal cells. Hippocampus, 15: 551–556.

Cristofanilli, M., Iacoangeli, A., Muslimov, I.A. and Tiedge, H. (2006) Neuronal BC1 RNA: microtubule-dependent dendritic delivery. J. Mol. Biol., 356: 1118–1123.

Dahm, R., Kiebler, M. and Macchi, P. (2007) RNA localisation in the nervous system. Semin. Cell Dev. Biol., 18: 216–223.

Darnell, J.C., Mostovetsky, O. and Darnell, R.B. (2005) FMRP RNA targets: identification and validation. Genes Brain Behav., 4: 341–349.

Davis, L., Banker, G.A. and Steward, O. (1987) Selective dendritic transport of RNA in hippocampal neurons in culture. Nature, 330: 477–479.

Donai, H., Sugiura, H., Ara, D., Yoshimura, Y., Yamagata, K. and Yamauchi, T. (2003) Interaction of Arc with CaM kinase II and stimulation of neurite extension by Arc in neuroblastoma cells expressing CaM kinase II. Neurosci. Res., 47: 399–408.

Dostie, J., Mourelatos, Z., Yang, M., Sharma, A. and Dreyfuss, G. (2003) Numerous microRNPs in neuronal cells containing novel microRNAs. RNA, 9: 180–186.

Dynes, J.L. and Steward, O. (2007) Dynamics of bidirectional transport of Arc mRNA in neuronal dendrites. J. Comp. Neurol., 500: 433–447.

Eberwine, J., Belt, B., Kacharmina, J.E. and Miyashiro, K. (2002) Analysis of subcellularly localized mRNAs using in situ hybridization, mRNA amplification, and expression profiling. Neurochem. Res., 27: 1065–1077.

Eberwine, J., Miyashiro, K., Kacharmina, J.E. and Job, C. (2001) Local translation of classes of mRNAs that are targeted to neuronal dendrites. Proc. Natl. Acad. Sci. U.S.A., 98: 7080–7085.

Elvira, G., Massie, B. and DesGroseillers, L. (2006a) The zinc-finger protein ZFR is critical for Staufen 2 isoform specific nucleocytoplasmic shuttling in neurons. J. Neurochem., 96: 105–117.

Elvira, G., Wasiak, S., Blandford, V., Tong, X.K., Serrano, A., Fan, X., del Rayo Sanchez-Carbente, M., Servant, F., Bell, A.W., Boismenu, D., Lacaille, J.C., McPherson, P.S., DesGroseillers, L. and Sossin, W.S. (2006b) Characterization of an RNA granule from developing brain. Mol. Cell. Proteomics, 5: 635–651.

Ethell, I.M. and Pasquale, E.B. (2005) Molecular mechanisms of dendritic spine development and remodeling. Prog. Neurobiol., 75: 161–205.

Fujii, R., Okabe, S., Urushido, T., Inoue, K., Yoshimura, A., Tachibana, T., Nishikawa, T., Hicks, G.G. and Takumi, T. (2005) The RNA binding protein TLS is translocated to dendritic spines by mGluR5 activation and regulates spine morphology. Curr. Biol., 15: 587–593.

Fujii, R. and Takumi, T. (2005) TLS facilitates transport of mRNA encoding an actin-stabilizing protein to dendritic spines. J. Cell Sci., 118: 5755–5765.

Garner, C.C., Tucker, R.P. and Matus, A. (1988) Selective localization of messenger RNA for cytoskeletal protein MAP2 in dendrites. Nature, 336: 674–677.

Gu, W., Pan, F., Zhang, H., Bassell, G.J. and Singer, R.H. (2002) A predominantly nuclear protein affecting cytoplasmic localization of beta-actin mRNA in fibroblasts and neurons. J. Cell Biol., 156: 41–51.

Hirokawa, N. and Takemura, R. (2005) Molecular motors and mechanisms of directional transport in neurons. Nat. Rev. Neurosci., 6: 201–214.

Hoek, K.S., Kidd, G.J., Carson, J.H. and Smith, R. (1998) hnRNP A2 selectively binds the cytoplasmic transport sequence of myelin basic protein mRNA. Biochemistry, 37: 7021–7029.

Hu, J.Y., Meng, X. and Schacher, S. (2003) Redistribution of syntaxin mRNA in neuronal cell bodies regulates protein expression and transport during synapse formation and long-term synaptic plasticity. J. Neurosci., 23: 1804–1815.

Huang, F., Chotiner, J.K. and Steward, O. (2005) The mRNA for elongation factor 1alpha is localized in dendrites and translated in response to treatments that induce long-term depression. J. Neurosci., 25: 7199–7209.

Huang, Y.S., Carson, J.H., Barbarese, E. and Richter, J.D. (2003) Facilitation of dendritic mRNA transport by CPEB. Genes Dev., 17: 638–653.

Huttelmaier, S., Zenklusen, D., Lederer, M., Dictenberg, J., Lorenz, M., Meng, X., Bassell, G.J., Condeelis, J. and Singer, R.H. (2005) Spatial regulation of beta-actin translation by Src-dependent phosphorylation of ZBP1. Nature, 438: 512–515.

Jin, P., Zarnescu, D.C., Ceman, S., Nakamoto, M., Mowrey, J., Jongens, T.A., Nelson, D.L., Moses, K. and Warren, S.T. (2004) Biochemical and genetic interaction between the fragile X mental retardation protein and the microRNA pathway. Nat. Neurosci., 7: 113–117.

Jonson, L., Vikesaa, J., Krogh, A., Nielsen, L.K., Hansen, T.V., Borup, R., Johnsen, A.H., Christiansen, J. and Nielsen, F.C. (2007) Molecular composition of IMP1 ribonucleoprotein granules. Mol. Cell. Proteomics, 6: 798–811.

Kanai, Y., Dohmae, N. and Hirokawa, N. (2004) Kinesin transports RNA: isolation and characterization of an RNA-transporting granule. Neuron, 43: 513–525.

Kang, H. and Schuman, E.M. (1996) A requirement for local protein synthesis in neurotrophin-induced hippocampal synaptic plasticity. Science, 273: 1402–1406.

Keene, J.D. and Tenenbaum, S.A. (2002) Eukaryotic mRNPs may represent posttranscriptional operons. Mol. Cell, 9: 1161–1167.

Kelleher, R.J., III, Govindarajan, A., Jung, H.Y., Kang, H. and Tonegawa, S. (2004a) Translational control by MAPK signaling in long-term synaptic plasticity and memory. Cell, 116: 467–479.

Kelleher, R.J., III, Govindarajan, A. and Tonegawa, S. (2004b) Translational regulatory mechanisms in persistent forms of synaptic plasticity. Neuron, 44: 59–73.

Kiebler, M.A. and Bassell, G.J. (2006) Neuronal RNA granules: movers and makers. Neuron, 51: 685–690.

Kiebler, M.A. and DesGroseillers, L. (2000) Molecular insights into mRNA transport and local translation in the mammalian nervous system. Neuron, 25: 19–28.

Kim, K.C. and Kim, H.K. (2006) Role of Staufen in dendritic mRNA transport and its modulation. Neurosci. Lett., 397: 48–52.

Kindler, S., Wang, H., Richter, D. and Tiedge, H. (2005) RNA transport and local control of translation. Annu. Rev. Cell Dev. Biol., 21: 223–245.

Kislauskis, E.H., Zhu, X. and Singer, R.H. (1994) Sequences responsible for intracellular localization of beta-actin messenger RNA also affect cell phenotype. J. Cell Biol., 127: 441–451.

Kleiman, R., Banker, G. and Steward, O. (1990) Differential subcellular localization of particular mRNAs in hippocampal neurons in culture. Neuron, 5: 821–830.

Kloc, M., Zearfoss, N.R. and Etkin, L.D. (2002) Mechanisms of subcellular mRNA localization. Cell, 108: 533–544.

Knowles, R.B., Sabry, J.H., Martone, M.E., Deerinck, T.J., Ellisman, M.H., Bassell, G.J. and Kosik, K.S. (1996) Translocation of RNA granules in living neurons. J. Neurosci., 16: 7812–7820.

Kohrmann, M., Luo, M., Kaether, C., DesGroseillers, L., Dotti, C.G. and Kiebler, M.A. (1999) Microtubule-dependent recruitment of Staufen-green fluorescent protein into large RNA-containing granules and subsequent dendritic transport in living hippocampal neurons. Mol. Biol. Cell, 10: 2945–2953.

Kosik, K.S. (2006) The neuronal microRNA system. Nat. Rev. Neurosci., 7: 911–920.

Kremerskothen, J., Kindler, S., Finger, I., Veltel, S. and Barnekow, A. (2006) Postsynaptic recruitment of dendrin depends on both dendritic mRNA transport and synaptic anchoring. J. Neurochem., 96: 1659–1666.

Krichevsky, A.M., King, K.S., Donahue, C.P., Khrapko, K. and Kosik, K.S. (2003) A microRNA array reveals extensive regulation of microRNAs during brain development. RNA, 9: 1274–1281.

Krichevsky, A.M. and Kosik, K.S. (2001) Neuronal RNA granules: a link between RNA localization and stimulation-dependent translation. Neuron, 32: 683–696.

Lagos-Quintana, M., Rauhut, R., Yalcin, A., Meyer, J., Lendeckel, W. and Tuschl, T. (2002) Identification of tissue-specific microRNAs from mouse. Curr. Biol., 12: 735–739.

Langford, G.M. (2002) Myosin-V, a versatile motor for short-range vesicle transport. Traffic, 3: 859–865.

Larocque, D., Pilotte, J., Chen, T., Cloutier, F., Massie, B., Pedraza, L., Couture, R., Lasko, P., Almazan, G. and Richard, S. (2002) Nuclear retention of MBP mRNAs in the quaking viable mice. Neuron, 36: 815–829.

Le Hir, H., Gatfield, D., Braun, I.C., Forler, D. and Izaurralde, E. (2001) The protein Mago provides a link between splicing and mRNA localization. EMBO Rep., 2: 1119–1124.

Leung, K.M., van Horck, F.P., Lin, A.C., Allison, R., Standart, N. and Holt, C.E. (2006) Asymmetrical beta-actin mRNA translation in growth cones mediates attractive turning to netrin-1. Nat. Neurosci., 9: 1247–1256.

Lewejohann, L., Skryabin, B.V., Sachser, N., Prehn, C., Heiduschka, P., Thanos, S., Jordan, U., Dell'Omo, G., Vyssotski, A.L., Pleskacheva, M.G., Lipp, H.P., Tiedge, H., Brosius, J. and Prior, H. (2004) Role of a neuronal small non-messenger RNA: behavioural alterations in BC1 RNA-deleted mice. Behav. Brain Res., 154: 273–289.

Liu, J., Hu, J.Y., Wu, F., Schwartz, J.H. and Schacher, S. (2006) Two mRNA-binding proteins regulate the distribution of syntaxin mRNA in *Aplysia* sensory neurons. J. Neurosci., 26: 5204–5214.

Lugli, G., Larson, J., Martone, M.E., Jones, Y. and Smalheiser, N.R. (2005) Dicer and eIF2c are enriched at postsynaptic densities in adult mouse brain and are modified by neuronal activity in a calpain-dependent manner. J. Neurochem., 94: 896–905.

Lynch, M.A. (2004) Long-term potentiation and memory. Physiol. Rev., 84: 87–136.

Macchi, P., Brownawell, A.M., Grunewald, B., DesGroseillers, L., Macara, I.G. and Kiebler, M.A. (2004) The brain-specific double-stranded RNA-binding protein Staufen2: nucleolar accumulation and isoform-specific exportin-5-dependent export. J. Biol. Chem., 279: 31440–31444.

Mallardo, M., Deitinghoff, A., Muller, J., Goetze, B., Macchi, P., Peters, C. and Kiebler, M.A. (2003) Isolation and characterization of Staufen-containing ribonucleoprotein particles from rat brain. Proc. Natl. Acad. Sci. U.S.A., 100: 2100–2105.

Martel, C., Macchi, P., Furic, L., Kiebler, M.A. and Desgroseillers, L. (2006) Staufen1 is imported into the nucleolus via a bipartite nuclear localization signal and several modulatory determinants. Biochem. J., 393: 245–254.

Martin, K.C. and Zukin, R.S. (2006) RNA trafficking and local protein synthesis in dendrites: an overview. J. Neurosci., 26: 7131–7134.

Mazroui, R., Huot, M.E., Tremblay, S., Filion, C., Labelle, Y. and Khandjian, E.W. (2002) Trapping of messenger RNA by fragile X mental retardation protein into cytoplasmic granules induces translation repression. Hum. Mol. Genet., 11: 3007–3017.

Michael, W.M., Eder, P.S. and Dreyfuss, G. (1997) The K nuclear shuttling domain: a novel signal for nuclear import and nuclear export in the hnRNP K protein. EMBO J., 16: 3587–3598.

Miki, T., Takano, K. and Yoneda, Y. (2005) The role of mammalian Staufen on mRNA traffic: a view from its nucleocytoplasmic shuttling function. Cell Struct. Funct., 30: 51–56.

Miller, S., Yasuda, M., Coats, J.K., Jones, Y., Martone, M.E. and Mayford, M. (2002) Disruption of dendritic translation of CaMKIIalpha impairs stabilization of synaptic plasticity and memory consolidation. Neuron, 36: 507–519.

Miyashiro, K., Dichter, M. and Eberwine, J. (1994) On the nature and differential distribution of mRNAs in hippocampal neurites: implications for neuronal functioning. Proc. Natl. Acad. Sci. U.S.A., 91: 10800–10804.

Miyashiro, K.Y., Beckel-Mitchener, A., Purk, T.P., Becker, K.G., Barret, T., Liu, L., Carbonetto, S., Weiler, I.J., Greenough, W.T. and Eberwine, J. (2003) RNA cargoes associating with FMRP reveal deficits in cellular functioning in Fmr1 null mice. Neuron, 37: 417–431.

Moccia, F., Frost, C., Berra-Romani, R., Tanzi, F. and Adams, D.J. (2004) Expression and function of neuronal nicotinic ACh receptors in rat microvascular endothelial cells. Am. J. Physiol. Heart Circ. Physiol., 286: H486–H491.

Mohr, E. (1999) Subcellular RNA compartmentalization. Prog. Neurobiol., 57: 507–525.

Mokin, M., Lindahl, J.S. and Keifer, J. (2006) Immediate-early gene-encoded protein Arc is associated with synaptic delivery of GluR4-containing AMPA receptors during in vitro classical conditioning. J. Neurophysiol., 95: 215–224.

Monné, L. (1948) Functioning of the cytoplasm. Adv. Enzymol., 8: 1–69.

Monshausen, M., Gehring, N.H. and Kosik, K.S. (2004) The mammalian RNA-binding protein Staufen2 links nuclear and cytoplasmic RNA processing pathways in neurons. Neuromolecular Med., 6: 127–144.

Mori, Y., Imaizumi, K., Katayama, T., Yoneda, T. and Tohyama, M. (2000) Two *cis*-acting elements in the 3′ untranslated region of alpha-CaMKII regulate its dendritic targeting. Nat. Neurosci., 3: 1079–1084.

Muslimov, I.A., Banker, G., Brosius, J. and Tiedge, H. (1998) Activity-dependent regulation of dendritic BC1 RNA in hippocampal neurons in culture. J. Cell Biol., 141: 1601–1611.

Muslimov, I.A., Nimmrich, V., Hernandez, A.I., Tcherepanov, A., Sacktor, T.C. and Tiedge, H. (2004) Dendritic transport and localization of protein kinase Mzeta mRNA: implications for molecular memory consolidation. J. Biol. Chem., 279: 52613–52622.

Muslimov, I.A., Santi, E., Homel, P., Perini, S., Higgins, D. and Tiedge, H. (1997) RNA transport in dendrites: a cis-acting targeting element is contained within neuronal BC1 RNA. J. Neurosci., 17: 4722–4733.

Muslimov, I.A., Titmus, M., Koenig, E. and Tiedge, H. (2002) Transport of neuronal BC1 RNA in Mauthner axons. J. Neurosci., 22: 4293–4301.

Ohashi, S., Kobayashi, S., Omori, A., Ohara, S., Omae, A., Muramatsu, T., Li, Y. and Anzai, K. (2000) The single-stranded DNA- and RNA-binding proteins pur alpha and pur beta link BC1 RNA to microtubules through binding to the dendrite-targeting RNA motifs. J. Neurochem., 75: 1781–1790.

Oleynikov, Y. and Singer, R.H. (2003) Real-time visualization of ZBP1 association with beta-actin mRNA during transcription and localization. Curr. Biol., 13: 199–207.

Ouyang, Y., Rosenstein, A., Kreiman, G., Schuman, E.M. and Kennedy, M.B. (1999) Tetanic stimulation leads to increased accumulation of Ca^{2+}/calmodulin-dependent protein kinase II via dendritic protein synthesis in hippocampal neurons. J. Neurosci., 19: 7823–7833.

Palacios, I.M. (2002) RNA processing: splicing and the cytoplasmic localisation of mRNA. Curr. Biol., 12: R50–R52.

Palacios, I.M. and St Johnston, D. (2001) Getting the message across: the intracellular localization of mRNAs in higher eukaryotes. Annu. Rev. Cell Dev. Biol., 17: 569–614.

Paquin, N., Menade, M., Poirier, G., Donato, D., Drouet, E. and Chartrand, P. (2007) Local activation of yeast ASH1 mRNA translation through phosphorylation of Kdh1p by the casein kinase Yck1p. Mol. Cell, 26: 795–809.

Pfeiffer, B.E. and Huber, K.M. (2006) Current advances in local protein synthesis and synaptic plasticity. J. Neurosci., 26: 7147–7150.

Poon, M.M., Choi, S.H., Jamieson, C.A., Geschwind, D.H. and Martin, K.C. (2006) Identification of process-localized mRNAs from cultured rodent hippocampal neurons. J. Neurosci., 26: 13390–13399.

Rial Verde, E.M., Lee-Osbourne, J., Worley, P.F., Malinow, R. and Cline, H.T. (2006) Increased expression of the immediate-early gene arc/arg3.1 reduces AMPA receptor-mediated synaptic transmission. Neuron, 52: 461–474.

Rocak, S. and Linder, P. (2004) DEAD-box proteins: the driving forces behind RNA metabolism. Nat. Rev. Mol. Cell Biol., 5: 232–241.

Rook, M.S., Lu, M. and Kosik, K.S. (2000) CaMKIIalpha 3′ untranslated region-directed mRNA translocation in living neurons: visualization by GFP linkage. J. Neurosci., 20: 6385–6393.

Ross, A.F., Oleynikov, Y., Kislauskis, E.H., Taneja, K.L. and Singer, R.H. (1997) Characterization of a beta-actin mRNA zipcode-binding protein. Mol. Cell. Biol., 17: 2158–2165.

Schratt, G.M., Nigh, E.A., Chen, W.G., Hu, L. and Greenberg, M.E. (2004) BDNF regulates the translation of a select group of mRNAs by a mammalian target of rapamycin-phosphatidylinositol 3-kinase-dependent pathway during neuronal development. J. Neurosci., 24: 7366–7377.

Schratt, G.M., Tuebing, F., Nigh, E.A., Kane, C.G., Sabatini, M.E., Kiebler, M. and Greenberg, M.E. (2006) A brain-specific microRNA regulates dendritic spine development. Nature, 439: 283–289.

Schuman, E.M., Dynes, J.L. and Steward, O. (2006) Synaptic regulation of translation of dendritic mRNAs. J. Neurosci., 26: 7143–7146.

Shepherd, J.D., Rumbaugh, G., Wu, J., Chowdhury, S., Plath, N., Kuhl, D., Huganir, R.L. and Worley, P.F. (2006) Arc/Arg3.1 mediates homeostatic synaptic scaling of AMPA receptors. Neuron, 52: 475–484.

Shiina, N., Shinkura, K. and Tokunaga, M. (2005) A novel RNA-binding protein in neuronal RNA granules: regulatory machinery for local translation. J. Neurosci., 25: 4420–4434.

Siomi, H. and Dreyfuss, G. (1995) A nuclear localization domain in the hnRNP A1 protein. J. Cell Biol., 129: 551–560.

Snay-Hodge, C.A., Colot, H.V., Goldstein, A.L. and Cole, C.N. (1998) Dbp5p/Rat8p is a yeast nuclear pore-associated DEAD-box protein essential for RNA export. EMBO J., 17: 2663–2676.

Sossin, W.S. and DesGroseillers, L. (2006) Intracellular trafficking of RNA in neurons. Traffic, 7: 1581–1589.

St Johnston, D. (2005) Moving messages: the intracellular localization of mRNAs. Nat. Rev. Mol. Cell Biol., 6: 363–375.

Steward, O. (1983) Polyribosomes at the base of dendritic spines of central nervous system neurons — their possible role in synapse construction and modification. Cold Spring Harb. Symp. Quant. Biol., 48(Pt. 2): 745–759.

Steward, O. and Falk, P.M. (1986) Protein-synthetic machinery at postsynaptic sites during synaptogenesis: a quantitative study of the association between polyribosomes and developing synapses. J. Neurosci., 6: 412–423.

Steward, O. and Halpain, S. (1999) Lamina-specific synaptic activation causes domain-specific alterations in dendritic immunostaining for MAP2 and CAM kinase II. J. Neurosci., 19: 7834–7845.

Steward, O. and Levy, W.B. (1982) Preferential localization of polyribosomes under the base of dendritic spines in granule cells of the dentate gyrus. J. Neurosci., 2: 284–291.

Steward, O. and Schuman, E.M. (2003) Compartmentalized synthesis and degradation of proteins in neurons. Neuron, 40: 347–359.

Steward, O., Wallace, C.S., Lyford, G.L. and Worley, P.F. (1998) Synaptic activation causes the mRNA for the IEG Arc to localize selectively near activated postsynaptic sites on dendrites. Neuron, 21: 741–751.

Steward, O. and Worley, P. (2002) Local synthesis of proteins at synaptic sites on dendrites: role in synaptic plasticity and memory consolidation? Neurobiol. Learn. Mem., 78: 508–527.

Sutton, M.A. and Schuman, E.M. (2006) Dendritic protein synthesis, synaptic plasticity, and memory. Cell, 127: 49–58.

Sutton, M.A., Wall, N.R., Aakalu, G.N. and Schuman, E.M. (2004) Regulation of dendritic protein synthesis by miniature synaptic events. Science, 304: 1979–1983.

Takano, K., Miki, T., Katahira, J. and Yoneda, Y. (2007) NXF2 is involved in cytoplasmic mRNA dynamics through interactions with motor proteins. Nucleic Acids Res., 35: 2513–2521.

Tang, S.J., Reis, G., Kang, H., Gingras, A.C., Sonenberg, N. and Schuman, E.M. (2002) A rapamycin-sensitive signaling pathway contributes to long-term synaptic plasticity in the hippocampus. Proc. Natl. Acad. Sci. U.S.A., 99: 467–472.

Thomas, K.L., Laroche, S., Errington, M.L., Bliss, T.V. and Hunt, S.P. (1994) Spatial and temporal changes in signal transduction pathways during LTP. Neuron, 13: 737–745.

Thomas, M.G., Martinez Tosar, L.J., Loschi, M., Pasquini, J.M., Correale, J., Kindler, S. and Boccaccio, G.L. (2005) Staufen recruitment into stress granules does not affect early mRNA transport in oligodendrocytes. Mol. Biol. Cell, 16: 405–420.

Tiruchinapalli, D.M., Oleynikov, Y., Kelic, S., Shenoy, S.M., Hartley, A., Stanton, P.K., Singer, R.H. and Bassell, G.J. (2003) Activity-dependent trafficking and dynamic localization of zipcode binding protein 1 and beta-actin mRNA in dendrites and spines of hippocampal neurons. J. Neurosci., 23: 3251–3261.

Tongiorgi, E., Righi, M. and Cattaneo, A. (1997) Activity-dependent dendritic targeting of BDNF and TrkB mRNAs in hippocampal neurons. J. Neurosci., 17: 9492–9505.

Tseng, S.S., Weaver, P.L., Liu, Y., Hitomi, M., Tartakoff, A.M. and Chang, T.H. (1998) Dbp5p, a cytosolic RNA helicase, is required for poly(A)$^+$ RNA export. EMBO J., 17: 2651–2662.

Tsokas, P., Grace, E.A., Chan, P., Ma, T., Sealfon, S.C., Iyengar, R., Landau, E.M. and Blitzer, R.D. (2005) Local protein synthesis mediates a rapid increase in dendritic elongation factor 1A after induction of late long-term potentiation. J. Neurosci., 25: 5833–5843.

Tsokas, P., Ma, T., Iyengar, R., Landau, E.M. and Blitzer, R.D. (2007) Mitogen-activated protein kinase upregulates the dendritic translation machinery in long-term potentiation by controlling the mammalian target of rapamycin pathway. J. Neurosci., 27: 5885–5894.

Tzingounis, A.V. and Nicoll, R.A. (2006) Arc/Arg3.1: linking gene expression to synaptic plasticity and memory. Neuron, 52: 403–407.

Villace, P., Marion, R.M. and Ortin, J. (2004) The composition of Staufen-containing RNA granules from human cells indicates their role in the regulated transport and translation of messenger RNAs. Nucleic Acids Res., 32: 2411–2420.

Wang, H., Iacoangeli, A., Lin, D., Williams, K., Denman, R.B., Hellen, C.U. and Tiedge, H. (2005) Dendritic BC1 RNA in translational control mechanisms. J. Cell Biol., 171: 811–821.

Weighardt, F., Biamonti, G. and Riva, S. (1995) Nucleo-cytoplasmic distribution of human hnRNP proteins: a search for the targeting domains in hnRNP A1. J. Cell Sci., 108(Pt. 2): 545–555.

Wells, D.G. (2006) RNA-binding proteins: a lesson in repression. J. Neurosci., 26: 7135–7138.

Wells, D.G., Dong, X., Quinlan, E.M., Huang, Y.S., Bear, M.F., Richter, J.D. and Fallon, J.R. (2001) A role for the cytoplasmic polyadenylation element in NMDA receptor-regulated mRNA translation in neurons. J. Neurosci., 21: 9541–9548.

Weston, A. and Sommerville, J. (2006) Xp54 and related (DDX6-like) RNA helicases: roles in messenger RNP assembly, translation regulation and RNA degradation. Nucleic Acids Res., 34: 3082–3094.

Wheeler, G., Ntounia-Fousara, S., Granda, B., Rathjen, T. and Dalmay, T. (2006) Identification of new central nervous system specific mouse microRNAs. FEBS Lett., 580: 2195–2200.

Willis, D., Li, K.W., Zheng, J.Q., Chang, J.H., Smit, A., Kelly, T., Merianda, T.T., Sylvester, J., van Minnen, J. and Twiss, J.L. (2005) Differential transport and local translation of cytoskeletal, injury-response, and neurodegeneration protein mRNAs in axons. J. Neurosci., 25: 778–791.

Wu, L., Wells, D., Tay, J., Mendis, D., Abbott, M.A., Barnitt, A., Quinlan, E., Heynen, A., Fallon, J.R. and Richter, J.D. (1998) CPEB-mediated cytoplasmic polyadenylation and the regulation of experience-dependent translation of alpha-CaMKII mRNA at synapses. Neuron, 21: 1129–1139.

Yao, J., Sasaki, Y., Wen, Z., Bassell, G.J. and Zheng, J.Q. (2006) An essential role for beta-actin mRNA localization and translation in Ca^{2+}-dependent growth cone guidance. Nat. Neurosci., 9: 1265–1273.

Yedavalli, V.S., Neuveut, C., Chi, Y.H., Kleiman, L. and Jeang, K.T. (2004) Requirement of DDX3 DEAD box RNA helicase for HIV-1 Rev-RRE export function. Cell, 119: 381–392.

Yoshimura, A., Fujii, R., Watanabe, Y., Okabe, S., Fukui, K. and Takumi, T. (2006) Myosin-Va facilitates the accumulation of mRNA/protein complex in dendritic spines. Curr. Biol., 16: 2345–2351.

Yuste, R. and Bonhoeffer, T. (2001) Morphological changes in dendritic spines associated with long-term synaptic plasticity. Annu. Rev. Neurosci., 24: 1071–1089.

Zalfa, F., Achsel, T. and Bagni, C. (2006) mRNPs, polysomes or granules: FMRP in neuronal protein synthesis. Curr. Opin. Neurobiol., 16: 265–269.

Zhang, H.L., Eom, T., Oleynikov, Y., Shenoy, S.M., Liebelt, D.A., Dictenberg, J.B., Singer, R.H. and Bassell, G.J. (2001) Neurotrophin-induced transport of a beta-actin mRNP complex increases beta-actin levels and stimulates growth cone motility. Neuron, 31: 261–275.

Zhong, J., Zhang, T. and Bloch, L.M. (2006) Dendritic mRNAs encode diversified functionalities in hippocampal pyramidal neurons. BMC Neurosci., 7: p. 17.

CHAPTER 4

Cap-dependent translation initiation and memory

Jessica L. Banko[1] and Eric Klann[2],*

[1]Department of Molecular Medicine, University of South Florida — Health, 12901 Bruce B. Downs Boulevard, MDC 61, Tampa, FL, USA
[2]Center for Neural Science, New York University, 4 Washington Place, Room 809, New York, NY, USA

Abstract: It is widely accepted that changes in gene expression contribute to enduring modifications of synaptic strength and are required for long-term memory. This is an exciting, wide-open area of research at this moment, one of those areas where it is clear that important work is underway but where the surface has just been scratched in terms of our understanding. Much attention has been given to the mechanisms of gene transcription; however, the regulation of transcription is only one route of manipulating gene expression. Regulation of mRNA translation is another route, and is the ultimate step in the control of gene expression, enabling cells to regulate protein production without altering mRNA synthesis or transport. One of the main advantages of this mechanism over transcriptional control in the nucleus lies in the fact that it endows local sites with independent decision-making authority, a consideration that is of particular relevance in neurons with complex synapto-dendritic architecture. There are a growing number of groups that are taking on the challenge of identifying the mechanisms responsible for regulating the process of mRNA translation during synaptic plasticity and memory formation. In this chapter we will discuss what has been discovered with regard to the localization and regulation of mRNA translation during specific types of neuronal activity in the mammalian central nervous system. The data are most complete for cap-dependent translation; therefore, particular attention will be paid to the machinery that initiates cap-dependent translation and its regulation during synaptic plasticity as well as the behavioral phenotypes consequent to its dysregulation.

Keywords: protein synthesis; translation initiation; long-term potentiation; long-term depression; learning and memory; eIF4E

Regulation of translation

Traditionally, the translation of an mRNA into a cognate protein proceeds in the three sequential steps of initiation, elongation, and termination. Regulation can occur at any of these steps, but initiation is typically rate limiting and thus often a target for regulation. Translation initiation itself also is divided into three steps. First, the initiator Met – $tRNA_i^{Met}$ must bind to the small ribosomal subunit, forming the 43S preinitiation complex. Then, the 43S complex must bind to an mRNA in the 5′ untranslated region and locate the initiation codon to form the 48S preinitiation complex. Finally, the large ribosomal subunit is added to generate a translation-competent ribosome that is now able to proceed with translation elongation. All three steps in the initiation of translation involve soluble proteins, most of which are referred to as initiation factors. The reader is

*Corresponding author. Tel.: +1 (212) 992-9769;
Fax: +1 (212) 995-4011; E-mail: eklann@cns.nyu.edu

referred to Mathews et al. (2007): "Translational Control in Biology and Medicine" for a complete discussion regarding the specifics of the biochemical regulation of mRNA translation.

Two principal pathways are available for attachment of the small ribosome subunit to the 5′ untranslated region in neurons. The first, termed cap-dependent translation initiation, relies on the fact that eukaryotic mRNAs are co-transcriptionally modified by attachment of an inverted, methylated guanine moiety to produce the 5′-terminal cap structure m^7GpppN (where N is the first transcribed nucleotide). The cap serves as an anchoring point for a cap-binding protein complex that can mediate recruitment of the small subunit of the ribosome to the extreme 5′ end of the mRNA. A second pathway uses complex secondary structure elements in the RNA called internal ribosomal entry segments (IRES) to recruit small ribosomal subunits either via direct RNA–ribosome contacts or indirectly via initiation factors that can bind both the IRES and the ribosome (Stoneley and Willis, 2004). Initiation via this pathway does not rely on the existence of a cap structure on the mRNA, and is therefore termed cap-independent. Evidence for the simultaneous existence of both cap-dependent and cap-independent initiation pathways within neurons has been reported; however, the overwhelming majority of eukaryotic mRNA transcripts are translated in a cap-dependent manner; therefore, this chapter will focus only on this mode of translation.

Cap-dependent translation initiation factors

The group of initiation factors responsible for recruiting small ribosomal subunits to the mRNA terminus belong to the eukaryotic initiation factor 4 group (eIF4). eIF4E binds to the cap structure of the mRNA, thus locating the 5′ end. eIF4A is a DEAD-box RNA-dependent ATPase that can unwind RNA duplexes in vitro and therefore has been proposed to be an RNA helicase that functions to unwind regions of secondary and tertiary structure in the 5′ end of the mRNA to facilitate ribosome scanning, a process that is stimulated by eIF4B. Both eIF4E and eIF4A bind to eIF4G, which is thought to act as a multifaceted adapter protein that coordinates interactions among factors in the initiation apparatus. Together, eIFs 4E, 4A, and 4G are called the eIF4F initiation complex (Fig. 1). The proposed role of the eIF4F complex is to bind to the 5′ end of the mRNA, unwind any structures found there, and then facilitate the loading of the 43S complex onto the now unstructured 5′ untranslated region to enable scanning for the initiation codon. Because any break in the chain cap–eIF4E–eIF4G–eIF4A impairs cap-dependent translation, the assembly of this complex has been offered as a potential target for translational control during activity-dependent neuronal processes.

eIF4E: the rate limiting factor

eIF4E is frequently described as the least abundant translation factor although the evidence for this is not particularly strong, being limited to examination of a very limited range of cell types at a time when tools were only available to study a small number of initiation factors. However, rather than abundance per se, it is likely that the availability of eIF4E is critically important for the activity of the initiation process. Indeed, there are several translation factors that inhibit translation via the sequestration of eIF4E from eIF4F complex association. This family of proteins has been named the 4E-binding proteins (4E-BPs). Binding of the 4E-BPs to eIF4E occurs via the region centering around Trp73 of eIF4E. Importantly, this residue does not interfere with the ability of eIF4E to bind mRNA, but is also required for eIF4E to bind to eIF4G. Whether the complete eIF4F initiation complex can assemble on a given mRNA depends upon the partner of the eIF4E molecule that binds to its 5′ end. mRNAs bound to 4E-BP–eIF4E are rendered translationally silent due to the occlusion of the eIF4G binding domain on eIF4E by 4E-BP. Consistent with this model, overexpression of 4E-BPs in mammalian cells, or addition of 4E-BPs to translation extracts, results in the inhibition of cap-dependent, but not cap-independent, translation (Pause et al., 1994). In addition to the 4E-BPs, various organisms have several other 4E-binding partners that compete

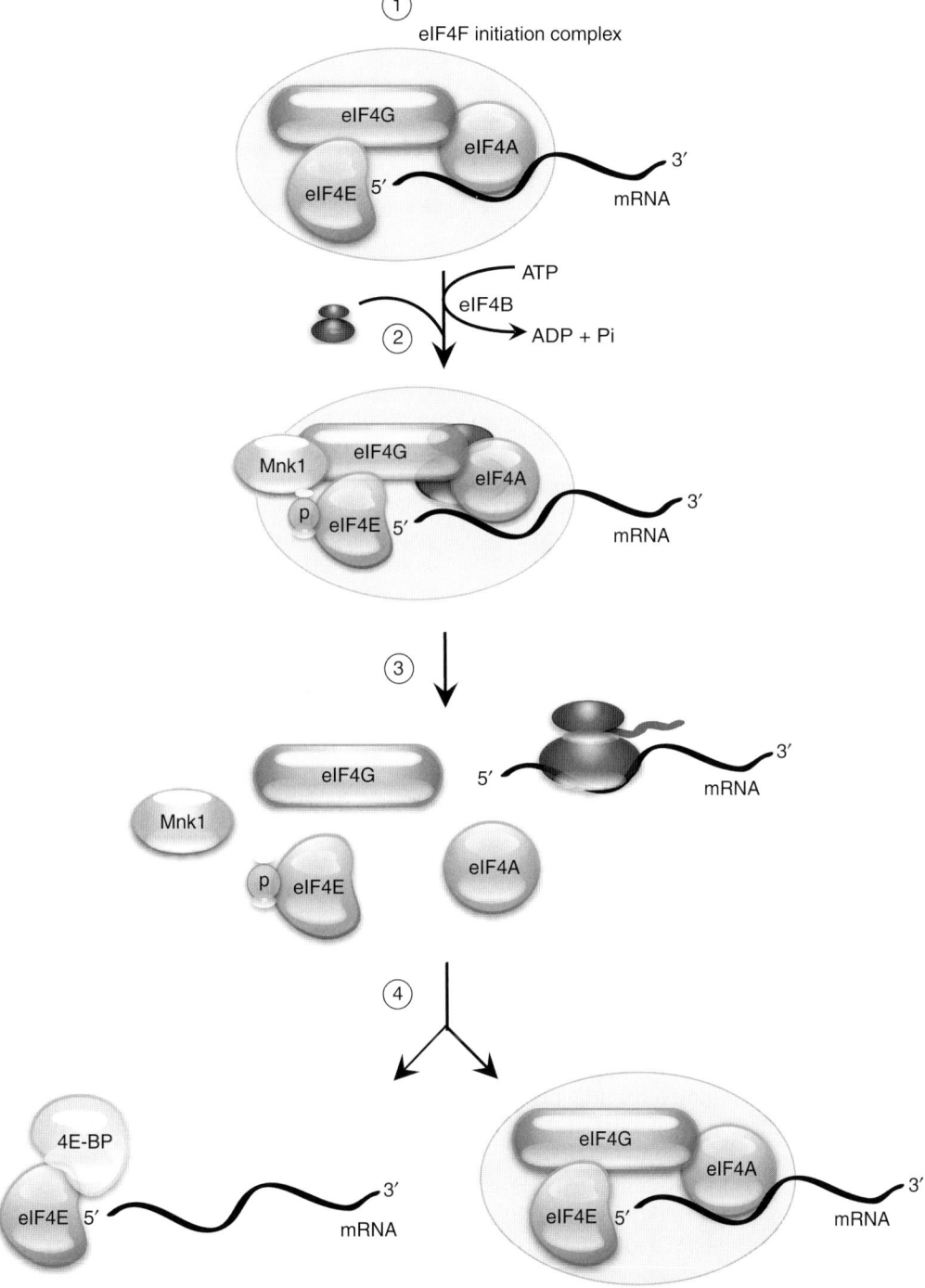

Fig. 1. Scheme for eIF4F recycling. (1) The eIF4F initiation complex is comprised of eIFs 4G, 4E, and 4A. Assembly of the eIF4F complex at the 5′ end of the mRNA targets a given mRNA for translation and recruits the ribosome. (2) Peptide elongation begins and Mnk1 phosphorylates eIF4E. (3) The eIF4F complex is released from the mRNA. (4) The released eIF4E is sequestered by 4E-BP or reassembles the eIF4F initiation complex for another round of translation.

with the formation of translationally active eIF4E–eIF4G protein complexes. These are generally larger proteins than the 4E-BPs, and can assume highly specialized roles during particular developmental stages. Examples of such proteins are Maskin in *Xenopus* (Cao and Richter, 2002) and Cup (Wilhelm et al., 2003) and Bicoid (Niessing et al., 2002) in *Drosophila*.

In higher animals, *Drosophila* and mammals, eIF4E is a phospho-protein. It is phosphorylated by the MAP kinase signal-integrating kinases, Mnk1 and Mnk2 at a single site in vivo, which is Ser209 in mammals (Flynn and Proud, 1995; Joshi et al., 1995). Studies with cell lines overexpressing Mnk cDNAs have shown that Mnk1 is strongly activated by growth factors, cellular stresses, and inflammatory cytokines via ERK and/or p38 MAPK, depending on the signaling context, whereas Mnk2 has a relatively high basal activity that is largely insensitive to changes in MAP kinase activity (Morley and McKendrick, 1997; Waskiewicz et al., 1997; Wang et al., 1998). The level of eIF4E phosphorylation in response to extracellular signals is determined by two factors. The first is the state of activation of the ERK or p38 MAP kinase pathways, which regulate the activity of Mnk1. The second is the quantity of eIF4F complexes, which bring together eIF4E and Mnk1 through their common binding partner, eIF4G.

The functional significance of eIF4E phosphorylation has been the subject of considerable interest within the field of translation research. Because stimuli that increase the rate of protein synthesis typically increase the state of phosphorylation of eIF4E, it was generally thought likely that phosphorylation would somehow activate eIF4E and perhaps increase its affinity for the mRNA cap. However, when the binding of eIF4E to cap analog (m^7GTP) was examined by fluorescence quenching, it was clear that phosphorylated eIF4E bound with lower affinity (2.5-fold difference) than the unphosphorylated protein (Scheper and Proud, 2002). Scheper et al. (2002) also examined binding of eIF4E to a capped oligonucleotide using a surface plasmon resonance approach (i.e. one carrying m^7GTP at its 5'-end). Arguably, this capped oligonucleotide more accurately resembled the physiological ligand of eIF4E, capped mRNA. In this case, phosphorylation of eIF4E again reduced its affinity for the ligand by approximately five-fold.

The binding constant for cap-binding by the eIF4E–eIF4G complex is approximately three orders of magnitude higher than for cap-binding by free eIF4E (Ptushkina et al., 1998). Importantly, the phosphorylation state of eIF4E does not alter the eIF4E–eIF4G interaction (Herbert et al., 2000; Knauf et al., 2001; Saghir et al., 2001); however, phosphorylation of eIF4E weakens its affinity for capped mRNA (Scheper and Proud, 2002). Based on these observations, a model has been proposed for how phosphorylation of eIF4E facilitates translational competence (Fig. 1). Phosphorylation of eIF4E by Mnk1 results in the release of the eIF4F complex from the mRNA and exposes the cap, allowing for the recruitment of a second eIF4E molecule and associated proteins. This facilitates the rapid loading of the next eIF4F initiation complex and ultimately the next ribosome onto the mRNA. Meanwhile, the released eIF4F proteins could be utilized by an additional mRNA molecule. This mechanism would serve to expedite ribosome loading and thus contribute to activation of translation initiation (Scheper and Proud, 2002). Operation of such a model would be consistent with the observations that pharmacological inhibition of eIF4E phosphorylation by the Mnk inhibitor CGP 57380 results in impaired polysome assembly, indicative of decreased recruitment of ribosomes onto the mRNA (Morley and Naegele, 2002). It should be noted that although polysome loading was compromised by CGP 57380, the overall level of protein synthesis was not reduced. Additional research is required to determine whether the transcripts undergoing translation before exposure to CGP 57380 were the same species as those undergoing translation during CGP 57380 exposure.

eIF4E phosphorylation during memory formation

From mollusks to mammals, memory can be divided into at least two phases: a transcription- and translation-independent phase that lasts minutes to hours termed short-term memory and a transcription- and translation-dependent phase that lasts days, weeks, or even longer termed long-term memory. For example, injection of the peptide elongation inhibitor anisomycin specifically impairs long-term memory in a well-characterized associative learning task, contextual fear conditioning. This task involves training rodents to learn to fear a new environment because of its temporal association with an aversive footshock. When exposed to the same environment (training context) at a later time, conditioned animals demonstrate the stereo-typical fear response, freezing. When mice received anisomycin injections immediately following training, freezing to the training context was normal 1 h later and impaired during memory tests the next day and 21 days later (Abel et al., 1997). This finding suggests that long-term memory, but not short-term memory, for fear conditioning requires the synthesis of new proteins.

Long-term memory formation for fear conditioning specifically elicits eIF4E phosphorylation. Kelleher et al. (2004) demonstrated that 30 min after training mice in a contextual fear conditioning paradigm, eIF4E phosphorylation at Ser209 was elevated in the hippocampus. There is a growing body of evidence suggesting that eIF4E is a critical molecule involved in the mRNA translation that supports the synaptic modifications that accompany memory formation. The specific consequences that interference with the regulation of eIF4E has on memory formation are discussed in detail later in this chapter.

Localization of translation in the neuron

Common cellular metabolism requires constitutive mRNA synthesis and translation; therefore, every neuron in the CNS by default contains the translation factors and the regulatory enzymes for cap-dependent translation initiation. Has there been a selective pressure to discriminate between the constitutive translation that occurs for normal cell maintenance and that which occurs during activity-induced synaptic modification and memory formation? On what basis could this be achieved? One means to achieve this discrimination is to localize the two translation processes in space and time.

Spatial localization of translation

The classical view of how new protein synthesis impacts activity-dependent neuronal function is that proteins are synthesized in the cell body and somehow are delivered to the activated synapses. In the late 1980s and early 1990s, however, it became increasingly clear that protein synthetic machinery and individual RNAs can be targeted to distant sites, sometimes at considerable distances from perikaryal somatic regions. Because evidence supporting this notion was emerging from diverse cell types including *Xenopus* oocytes, mammalian neurons, and glial cells, as well as from *Drosophila* embryos, researchers began to realize that a common principle might be at work. Today we understand that mRNAs are transported, localized, and locally translated in many eukaryotic cell types. These mechanisms are of fundamental importance in the regulation of gene expression as they allow cells to delegate control to autonomously acting local sites.

Using electron microscopy, Steward and Levy (1982) showed that the dendrites of neurons in the hippocampus contain polyribosomes that are preferentially associated with post-synaptic structures. Later studies showed hippocampal dendrites also possess the other protein and RNA components necessary for translation. Indeed, all of the cap-dependent translation initiation factors described above: eIF4E, 4E-BP, eIF4G, eIF4A, and Mnk1 have been observed in the dendritic layers and synaptic fractions of the hippocampus (Tang et al., 2002; Wang et al., 2002; Asaki et al., 2003; Banko et al., 2004). Numerous mRNAs have by now been shown to be localized in dendrites. Proteins encoded by these mRNAs belong to a variety of functional groups, including receptors such as the inositol 1,4,5-trisphosphate (InsP3) receptor type 1, several glutamate receptor subtypes, GABAA and glycine receptors, brain derived neurotrophic factor (BDNF) and its receptor TrkB, kinases such as the α subunit of Ca^{2+}/calmodulin-dependent protein kinase II (αCaMKII), cytoskeleton-associated proteins such as MAP2, and activity-regulated cytoskeleton-associated protein (Arc) and even translation factors themselves. It is estimated today that several hundred mRNAs are localized to dendrites.

Given that mRNA and translational machinery exists within dendrites, a logical hypothesis is that neurons transport specific mRNAs to dendrites to participate in the local production of new proteins that somehow modify the synapse. Compared to the alternative of making proteins in the soma and transporting them into the dendrites, local translation of dendritic mRNAs offers the potential of being faster and allowing more efficient spatial control: initiating translation on-site obviates the need to target proteins to specific synapses. Another possible reason for performing translation locally is that particular protein products may be much more difficult to transport than their mRNAs.

Temporal localization of translation

It is generally assumed that mRNAs are translationally repressed while en route to the dendrite. According to the above hypothesis, activity at a particular synapse could trigger translation of nearby, postsynaptically localized mRNAs, generating proteins that somehow modify the synapse. Evidence in support of the idea that certain forms of synaptic activity actually engage local translation is growing in abundance. For example, Kang and Schuman (1996) demonstrated that the neurotrophins, BDNF and neurotrophin 3, could elicit a long-lasting potentiation at Scahffer collateral synapses in hippocampal slices that required mRNA translation. Somatic mRNA translation could not account for what they observed because the potentiation and its dependence upon de novo translation persisted even when the cell bodies of both the presynaptic and postsynaptic neurons were cut off from their dendrites. Aakalu et al. (2001) then produced evidence that BDNF-induced synaptic potentiation could elicit local translation activation by demonstrating that translation of a green fluorescent protein (GFP)-based protein synthesis reporter was specifically and rapidly increased following the BDNF exposure.

Soon to follow were demonstrations that additional forms of synaptic plasticity within the CNS could also elicit local de novo translation. The late phase of the long-term potentiation (LTP) induced in the Schaffer collateral pathway by multiple tetani is perhaps the most well-characterized of them. Examination of this form of plasticity revealed that the spatio-temporal regulation of translation is even more intricate than activity-induced dendritic versus synaptic translation. It appears that regulation can take place at micro-domains within the dendrites themselves, perhaps even at specific synapses. For example, Bradshaw et al. (2003) demonstrated that when the translation inhibitor emetine was applied locally to the apical dendritic field of CA1 pyramidal cells in the hippocampus, tetanically induced late phase LTP (L-LTP) was impaired at apical but not at basal dendrites, and conversely when emetine was applied locally to the basal dendrites, L-LTP was impaired only at basal dendrites. Another interesting study that supports this hypothesis was reported by Steward et al. (1998) who examined the distribution of *Arc* mRNA and protein after activation of synapses in hippocampal slices. These researchers repetitively stimulated different subsets of synapses in hippocampal slices then looked at the distribution of newly synthesized *Arc* transcripts in the slices. They found that newly transcribed *Arc* mRNA concentrated in activated dendrites, bypassing non-activated dendrites of the same cells. At the same time, Arc protein also appeared in the activated dendrites, suggesting that local machinery translated the transported mRNAs.

An altogether distinct form of synaptic plasticity that elicits de novo translation is the LTP counterpart, long-term depression (LTD; Huber et al., 2000, 2001). Similar to the studies described above, LTD induced by pharmacological activation of the group I metabotropic glutamate receptors (mGluR) with DHPG in the Schaffer collateral pathway can be blocked by translation inhibitors and can be induced even when the cell bodies of CA1 are excised from their dendrites. Prior to these initial reports characterizing mGluR-LTD, it had been demonstrated that mRNA complementary to human *FMR1* and murine *Fmr1* transcripts moves into close association with polyribosomes isolated from rat cortex and undergoes rapid translation during the first 5 min after stimulation of mGluR1 with DHPG (Weiler et al., 1997). The protein product of these transcripts is the fragile X mental retardation protein (FMRP). FMRP functions as a translation repressor for a specialized population of mRNAs and has subsequently been shown to play an important role in mGluR-LTD.

Taken together, the characterization of these various forms of synaptic plasticity not only support the notion that extra-somatic mRNA translation is elicited by synaptic activity, but have spurred a new field of investigation into the specific mechanisms linking synaptic activity and local translation regulation mechanisms. In the following section we will explore what has recently been discovered regarding the mechanisms of translation initiation regulation during synaptic plasticity and the behavioral phenotypes consequent to its dysregulation.

Activity-induced regulation of translation initiation machinery

Disruptions in the regulatory mechanisms of mRNA translation are the basis for human cognitive disorders such as Fragile X syndrome (FXS) and vanishing white matter disease. The etiology of both diseases can be traced to mutations in individual genes that encode for translation regulatory proteins (Leegwater et al., 2001; Antar and Bassell, 2003). Furthermore, alteration in the regulation of translation initiation has been observed in bipolar disorder (Carter, 2007a) and schizophrenia (Carter, 2007b; Ogawa et al., 2005). Neuroscientists can glean a general knowledge of regulatory pathways governing translation initiation from the fields of cellular metabolism and cancer biology, but in truth, only recently have we begun to shed light on how synaptic activation specifically couples to local translation during specific patterns of neuronal activity. Interestingly, several labs have demonstrated that targeted regulation of eIF4E and its associated proteins accompanies several forms of synaptic plasticity (Fig. 2).

BDNF induces local activation of cap-dependent translation initiation machinery

It took several years after the initial report that BDNF-induced synaptic potentiation required local mRNA translation for investigators to elucidate a mechanism that coupled BDNF receptor activation with translation. Investigators drew upon what had been formerly demonstrated by the field of cellular metabolism as a signaling pathway that leads to the initiation of translation and looked to see whether BDNF could elicit activation of the same signaling cascade in neurons.

It had been well-established in vitro that binding of the 4E-BPs to eIF4E is reversible and dependent on the phosphorylation status of 4E-BP. Under conditions in which cells are not actively growing and the requirement for translation is reduced, the 4E-BPs are dephosphorylated and bind to eIF4E. However, under active growth conditions the 4E-BPs exist in a hyperphosphorylated state that prevents efficient binding to eIF4E and promotes formation of the eIF4F initiation complex (Pause et al., 1994). The pathway leading to 4E-BP phosphorylation involves the phosphoinositide-3 (PI3) kinase and its downstream effectors, the serine-threonine kinase Akt and the mammalian target of rapamycin (mTOR). Phosphorylation of 4E-BP occurs in a hierarchical manner on several sites; however, Thr37/46 appears critical for determining 4E-BP's affinity for eIF4E (for reviews, see Sonenberg and Gingras, 1998; Raught and Gingras, 1999).

Phosphorylation of Ser209 of eIF4E is also increased in response to a variety of growth-promoting conditions that include serum treatment of cells, growth factors, phorbol esters, and in some cell types, insulin. Where tested, these effects appear to be mediated via the ERK2–Mnk1 pathway. Certain cytokines and stressful conditions also increase the phosphorylation of eIF4E via the p38 MAP kinase pathway (for review, see Sonenberg and Gingras, 1998).

Takei et al. (2001, 2004) found that BDNF stimulated local translation initiation cascades resulting in 4E-BP Thr37/46 and eIF4E Ser209 phosphorylation. They demonstrated that BDNF

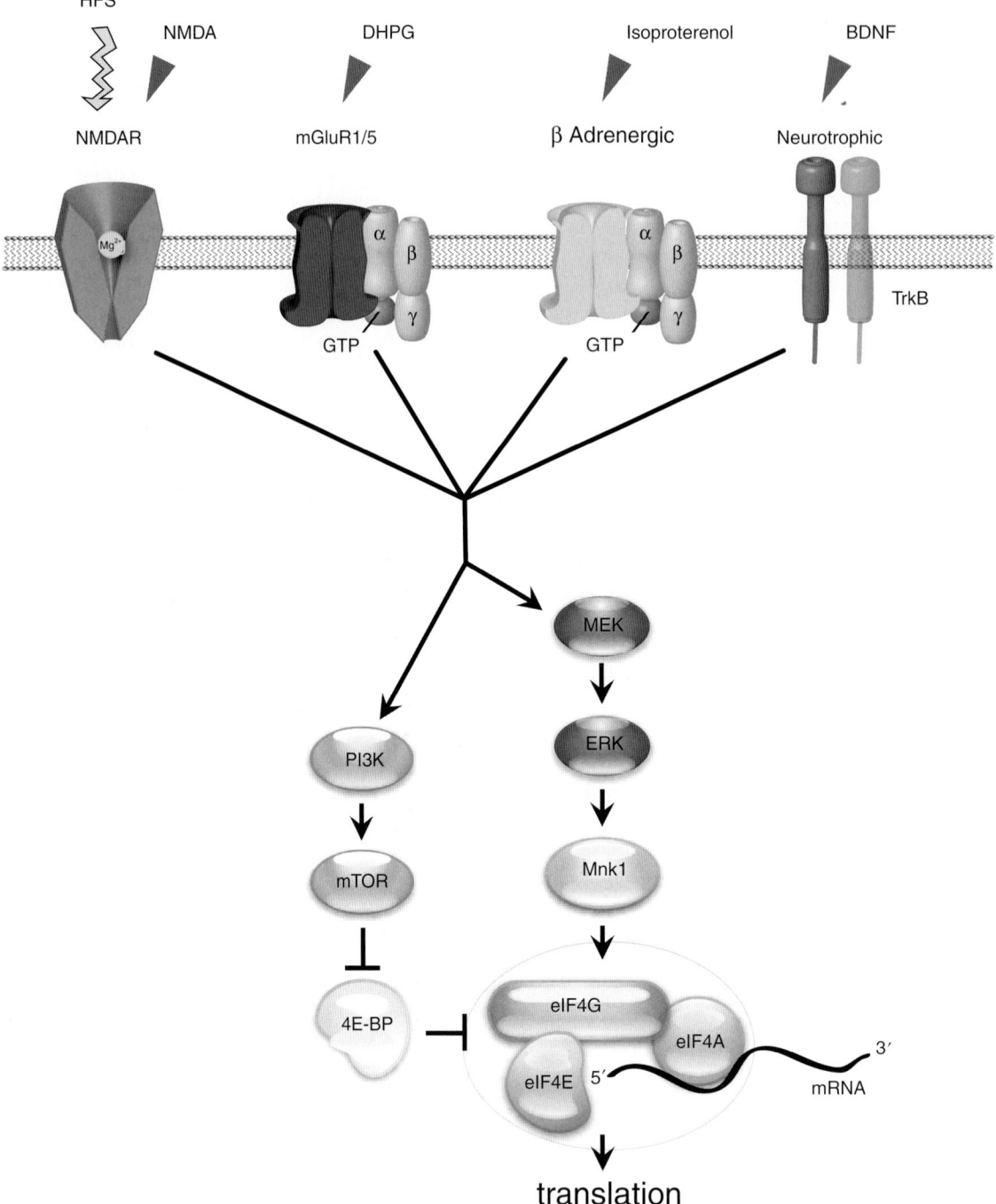

Fig. 2. Signal transduction pathways that regulate cap-dependent translation initiation machinery. At least three different classes of receptors in the hippocampus are coupled to signal transduction cascades that modulate local cap-dependent translation initiation machinery. Each of these receptor types can induce long-lasting synaptic plasticity at the Schaffer collateral pathway in the hippocampus that is translation-dependent. eIF4E serves as a point of signal integration between the PI3K–mTOR–4E-BP and MEK–ERK–Mnk1 pathways.

elicited mTOR activation and 4E-BP phosphorylation in isolated dendrites that could be inhibited by either inhibitors of PI3 kinase or mTOR. They also demonstrated that eIF4E phosphorylation was dependent upon ERK. Expanding on this initial observation, Smart et al. (2003) went on to demonstrate that BDNF resulted in the increased expression and local redistribution of eIF4E protein. They observed eIF4E translocation to synaptically localized mRNA granules following BDNF exposure. These observations strongly suggest that eIF4E participates in local translation initiation during BDNF-mediated synaptic potentiation.

Tetanic stimulation induces local activation of cap-dependent translation initiation machinery

The story is even more compelling for the more classically studied tetanically induced LTP. As detailed in earlier chapters, LTP can be described as having distinct phases based on the requirement for de novo gene expression. Early phase LTP (E-LTP) can be induced with a single tetanus of high frequency stimulation (HFS, 100 Hz for 1 s). E-LTP lasts 1–3 h and is characterized by post-translational modifications to pre-existing proteins. Multiple tetani of HFS induces L-LTP, which can endure several hours to days and is dependent upon transcription and translation for expression. Both forms of tetanically induced LTP are NMDA receptor-dependent in the Schaffer collateral pathway of the hippocampus. Activation of NMDA receptors engages multiple signaling cascades that could theoretically lie upstream of local translation initiation. For example, NMDA receptor activation leads to the activation of PKC, PKA, and ERK, which have all been shown to result in enhanced translation in multiple tissue systems.

Our group initially looked at the upstream signaling components of translation initiation pharmacologically in the hippocampal slice preparation. We determined that NMDA receptor activation from acute NMDA exposure resulted in local ERK-dependent activation of Mnk1 and phosphorylation of eIF4E (Banko et al., 2004). Although both PKC and PKA were involved in the coupling of NMDA receptors to ERK, we found that this was not the case for the coupling of NMDA receptors to Mnk1 and eIF4E. Pharmacological inhibition of PKA blocked not only NMDA-induced activation of ERK, but also the activation of Mnk1 and the increase in eIF4E phosphorylation. In contrast, we found that pharmacological inhibition of PKC blocked neither the NMDA-induced activation of Mnk1 nor the NMDA-induced increase in eIF4E phosphorylation. These data indicate that PKA, but not PKC, is required for the NMDAR-dependent activation of the ERK–Mnk1–eIF4E signaling cascade. It is possible that the PKC-dependent pool of activated ERK is involved in the coupling of NMDA to other translational machinery or the coupling of other neurotransmitter receptors to Mnk1 and eIF4E.

Consistent with our findings following pharmacological activation of NMDA receptors, we also observed that a single tetanus of HFS resulted in ERK-dependent increased activation of Mnk1 and eIF4E phosphorylation minutes after delivery. Additionally, we also found that phosphorylation of 4E-BP was increased in response to a single tetanus of HFS, which was accompanied by an increase in eIF4F complex formation (Banko et al., 2005; Gelinas et al., 2007). It was somewhat surprising that the signaling mechanisms that activate cap-dependent translation initiation machinery were engaged by a single tetanus of HFS given that the E-LTP induced by this protocol is not dependent upon protein synthesis for expression.

Not surprisingly, multiple tetani of HFS that induce L-LTP resulted in even further increases in Mnk1 activation, eIF4E phosphorylation, and eIF4F complex formation when compared to the single tetanus-induced E-LTP (Banko et al., 2005). Combined with the knowledge that multiple tetani of HFS produces L-LTP, these findings suggest the presence of a preset activation threshold for the initiation machinery that must be met before translation-dependent LTP is induced. In support of this notion, we found that when pairing the E-LTP-inducing stimuli, which on its own results in activation of the cap-dependent initiation machinery but does not engage sufficient translation activity to enable L-LTP, with either: (1) genetic manipulations that result in increased basal association of the eIF4F complex, or (2) additional synaptic stimulation that also results in a modest activation of the cap-dependent initiation machinery, E-LTP can be converted to L-LTP. In both of the above conditions the level of activation of the cap-dependent initiation machinery that is achieved when the conditions are paired is significantly greater than what is achieved through a single tetani of HFS in a wild type (WT) mouse. Although, we have found that more is not always better.

eIF4E as a signal integrator for the mTOR and ERK pathways: a common theme in multiple forms of synaptic plasticity in the hippocampus

In addition to receptor tyrosine kinases and voltage-gated ion channels, G-protein coupled receptors also mediate synaptic plasticity in the hippocampus and can likewise engage the activation of the cap-dependent initiation machinery. We have examined two different forms of translation-dependent synaptic plasticity mediated by G-protein coupled receptors, mGluR-LTD and β-adrenergic-facilitated LTP, to elucidate the signaling pathways that couple them to the cap-dependent initiation machinery. Interestingly, we found that the same signaling cascades are employed to subserve both translation-dependent forms of plasticity despite their opposing output in terms of synaptic strength.

Huber et al. (2000) showed that cap-dependent translation plays a critical role during mGluR-LTD. They demonstrated that addition of exogenous cap analog, which competes with endogenous transcripts for the eIF4 group of machinery, was sufficient to block the expression of mGluR-LTD. We subsequently probed the signaling cascades that engaged the cap-dependent initiation machinery after induction of mGluR-LTD to determine whether mechanisms similar to those exhibited following BDNF receptor and NMDA receptor activation were at play. Indeed, we demonstrated that chemical induction of mGluR-LTD with the group I agonist DHPG could trigger the local activation of an ERK–Mnk1–eIF4E signaling pathway, as well as a PI3K–mTOR–4E-BP–eIF4E signaling pathway, thereby increasing the quantity of the eIF4F complex (Banko et al., 2006).

Similar results were observed when we probed the signaling cascades that engaged the cap-dependent initiation machinery after induction of β-adrenergic receptor-induced L-LTP. This form of plasticity is achieved when an E-LTP inducing stimulation of a single tetanus of HFS for 1s is paired with exposure to β-adrenergic receptor agonist isoproterenol. Like L-LTP induced by multiple tetani of HFS, this LTP requires local protein synthesis, and the kinase activity of ERK and mTOR (Gelinas and Nguyen, 2005; Gelinas et al., 2007). We found that β-adrenergic receptor-induced L-LTP triggered the activation of the ERK–Mnk1–eIF4E signaling pathway, activation of the mTOR–4E-BP–eIF4E signaling pathway, and an increase in the quantity of eIF4F complex (Gelinas et al., 2007).

It is apparent that both the ERK and mTOR pathways are activated concomitantly during mGluR-LTD and β-adrenergic receptor-induced L-LTP and that they each signal to cap-dependent translation initiation factors. Recent evidence suggests that during HFS-induced LTP there is a cooperative crosstalk between these two signaling cascades (Tsokas et al., 2007), but we have not indisputably demonstrated a cooperative cross talk between the two during mGluR-LTD. We propose that eIF4E serves as the point of signal integration for the convergence of the two pathways based on the following observations (Banko et al., 2006; Gelinas et al., 2007; and unpublished results). First, phosphorylation of the critical Thr37/46 residues of 4E-BP is sensitive to inhibition of mTOR, but not ERK. Second, eIF4E–eIF4G complex association is sensitive to inhibition of mTOR, but not ERK. Third, Mnk1 activation is sensitive to inhibition of ERK, but not mTOR. Fourth, and finally, eIF4E phosphorylation is reduced by either inhibition of mTOR or ERK.

Our working model for how eIF4E serves as a point of signal integration for the convergence of the two pathways proposes that the activity-induced activation of the PI3K–Akt–mTOR pathway results in phosphorylation of 4E-BP2 Thr37/46, thereby liberating eIF4E. eIF4E is then free to bind eIF4G and form the eIF4F initiation complex, which initiates translation. Meanwhile, activity-induced activation of ERK elicits activation of Mnk1. The assembled eIF4F complex provides the necessary scaffold to facilitate Mnk1-dependent phosphorylation of eIF4E, a process that strongly correlates with increased mRNA translation (Fig. 2).

Lessons learned from the 4E-BP2 knockout mouse

Curiously, the regulation of 4E-BP phosphorylation following multiple tetani of HFS to induce L-LTP did not follow the same trend as observed for the other cap-dependent initiation machinery. We did not observe a further increase in 4E-BP2 phosphorylation when compared to the single tetanus that induces E-LTP (Fig. 4A). This observation lead us to examine more closely the role that 4E-BP plays in regulating the translational events that support synaptic plasticity. The results discussed in this section have been published in the following four reports: Banko et al. (2005, 2006, 2007) and Gelinas et al. (2007).

We began to investigate the role of 4E-BP by examining a mutant mouse generated to lack the 4E-BP isoform preferentially expressed in the mouse brain, 4E-BP2. The 4E-BP2 knockout (4E-BP2 KO) mice are identical to their WT littermates with regard to total protein levels of eIF4E and its kinase Mnk1, neuroanatomy and dendritic ultrastructure, basal synaptic transmission, and short-term synaptic plasticity. The primary biochemical difference between 4E-BP2 KO mice and their WT littermates is in basal eIF4F complex association. The genetic elimination of 4E-BP2 as a binding partner for eIF4E enables more eIF4E to bind to eIF4G under basal conditions. Although this appears to not have an effect on global translation, it may contribute to the plasticity phenotypes we observe in these animals (Fig. 3).

Our initial hypothesis was that increased levels of basal eIF4F complex association would facilitate translation-dependent forms of synaptic plasticity, and to a certain extent this was what we observed (Fig. 3). mGluR-LTD was enhanced in the 4E-BP2 KO mice. The enhanced LTD remained sensitive to inhibition of ERK activity; however, was immune to inhibition of mTOR, lending support to our model that offers eIF4E as the point of signal integration for the convergence of the two pathways (Fig. 2).

We also found that the absence of 4E-BP2 allowed a stimulus that normally elicits only E-LTP in WT mice, to elicit a facilitated LTP that resembles WT L-LTP (Fig. 3) in its duration, kinetic pattern of sensitivity to inhibitors of transcription and translation, and dependence on ERK activation. Interestingly, when we examined the levels of eIF4F complex formation in the 4E-BP2 KO slices during the facilitated LTP we found that they were comparable to what had been observed during WT L-LTP (Fig. 4D).

Levels of eIF4F complex formation following L-LTP in WT mice can be mimicked by pairing an E-LTP-inducing, single tetanus of HFS with pharmacological stimulation that co-recruits the cap-dependent initiation machinery. β-adrenergic receptor activation in WT mice results in similar eIF4F complex association as WT E-LTP does, but elicits only nominal and transient facilitation of the postsynaptic response (Figs. 3 and 4). Recall that earlier we described how pairing the single tetanus of HFS with the β-adrenergic receptor agonist isoproterenol converts WT E-LTP to L-LTP. This paired paradigm also elicits a combinatorial increase in eIF4F complex association that exceeds the levels following either condition independently and resembles the increases observed during tetanically induced L-LTP (Fig. 4F).

Conversely, isoproterenol exposure paired with test pulses delivered at 0.05 Hz to slices obtained from 4E-BP2 KO mice elicits enduring LTP that rivals the facilitated LTP achieved with a single tetanus of HFS in magnitude and kinetics (Fig. 3). These findings bring us back to the above hypothesis that a preset activation threshold exists for the cap-dependent translation initiation machinery that once reached, gates induction of translation-dependent LTP.

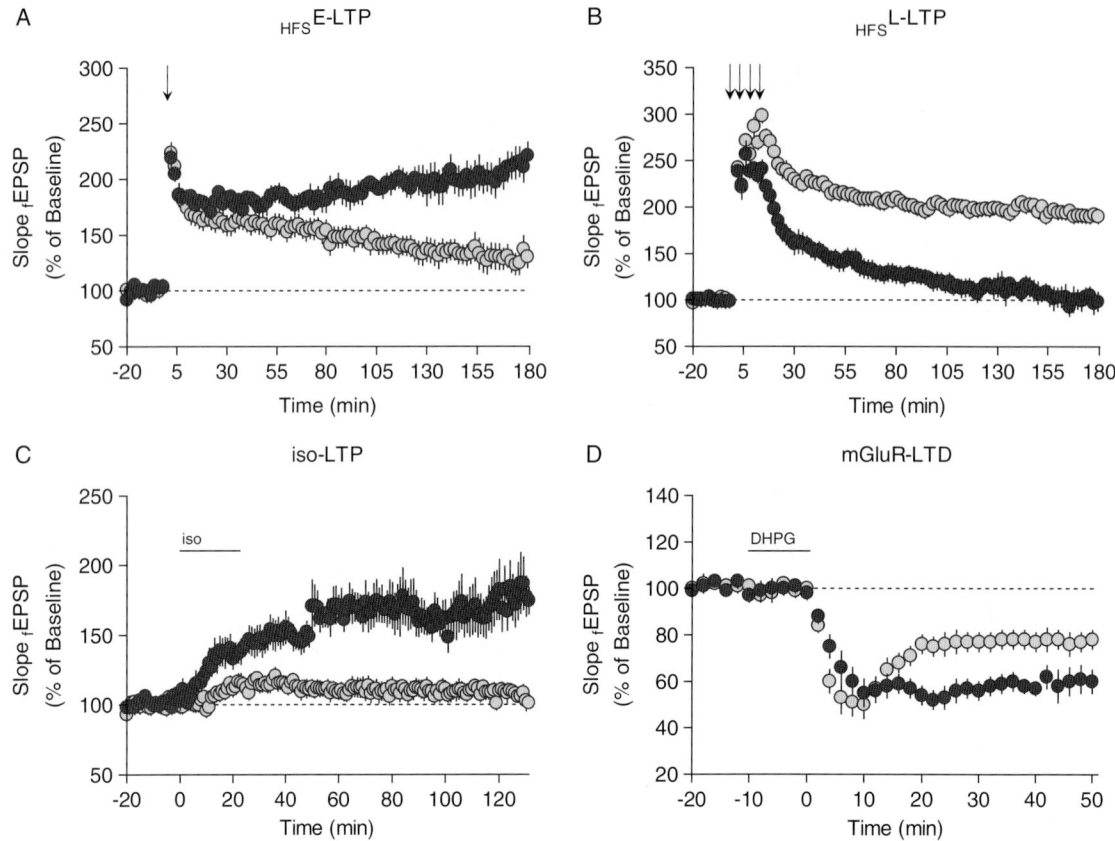

Fig. 3. 4E-BP2 knockout electrophysiology phenotypes. Gray circles represent WT measurements and black circles represent 4E-BP2 KO measurements of field extracellular post synaptic potentials at the Schaffer collateral synapse. (A) One train of HFS (100 Hz, 1 s) produces robust and long-lasting LTP in 4E-BP2 KO hippocampal slices. (B) Four trains of HFS (100 Hz, 1 s, 5 min ITI) fails to produce L-LTP in 4E-BP2 KO hippocampal slices. (C) Pharmacological stimulation of β-adrenergic receptors with isoproterenol (1 μM, 25 min) produces long lasting LTP in 4E-BP2 KO hippocampal slices. Adapted from Banko et al. (2005, 2006) and Gelinas et al. (2007). (D) Pharmacological stimulation of the group I mGluRs with DHPG (50 μM, 10 min) produces enhanced LTD in 4E-BP2 KO hippocampal slices.

Consequences of too much eIF4F complex association

We observed a pattern with regard to eIF4F complex association between the 4E-BP2 KO and WT mice (Fig. 4D). The quantity of basal eIF4F complex in the 4E-BP2 KO mice was elevated to a degree that resembled eIF4F levels during WT E-LTP. E-LTP-inducing stimulation resulted in eIF4F levels in the 4E-BP2 KO mice that resembled those during WT L-LTP, and induced synaptic potentiation that resembled WT L-LTP (Fig. 3). Finally, L-LTP-inducing stimulation further increased eIF4F levels in the 4E-BP2 KO mice such that they were approximately 1.5-fold greater than those observed during WT L-LTP, yet failed to elicit L-LTP (Fig. 3).

Recall the earlier model of how the eIF4F complex facilitates the initiation of translation by promoting small ribosome association with the 5' end of the mRNA and the role that phosphorylation of eIF4E plays in that process (Fig. 1). Interestingly, when we compared the ratio of phosphorylated eIF4E to total eIF4E in the eIF4F complex following the different patterns of stimulation, we found that the ratio remained

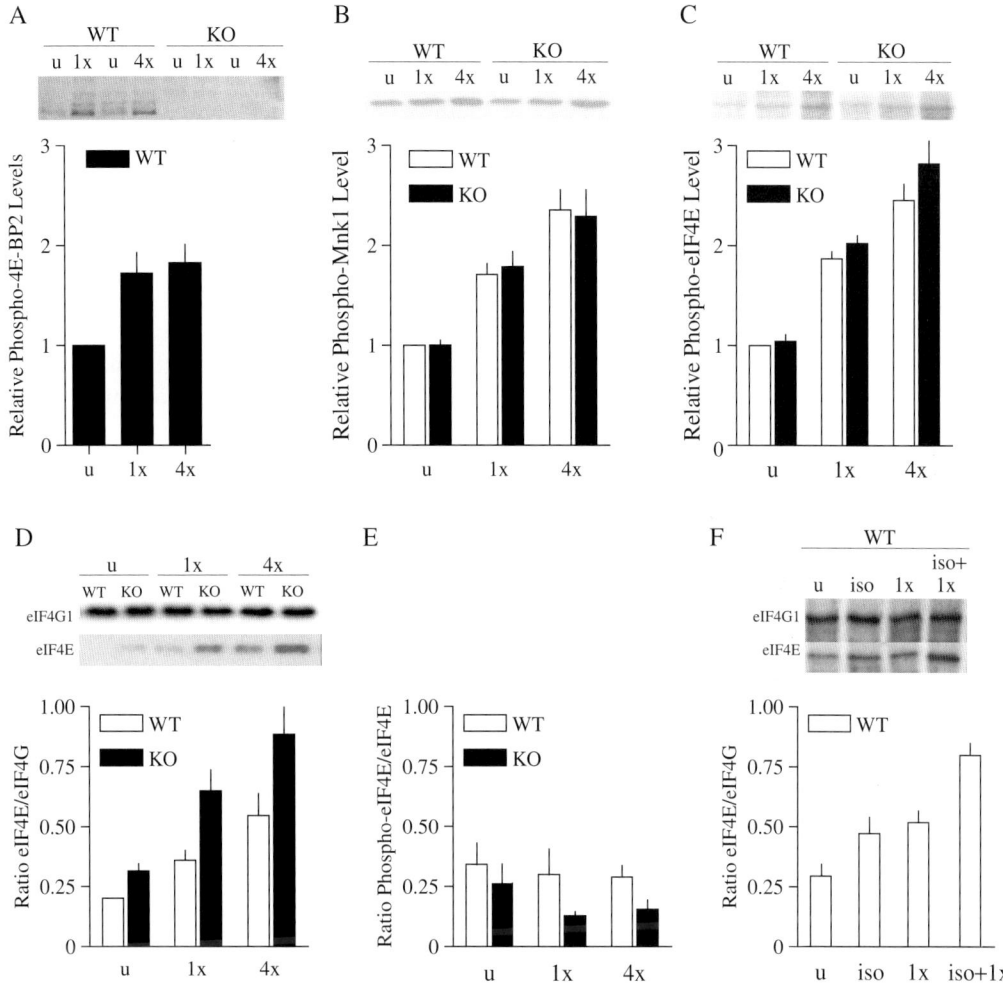

Fig. 4. Biochemical regulation of the cap-dependent machinery. (A–E) Quantitative Western blot analysis of hippocampal area CA1 from wild type (WT and 4E-BP2 knockout (KO) slices before (unstimulated: u), and 5 min following induction of either HFS-induced E-LTP (one train HFS 100 Hz, 1 s: 1×) or HFS-induced L-LTP (four trains HFS 100 Hz, 1 s, inter-train interval 5 min: 4×). (A) Phospho-4E-BP (Thr37/46) immunoreactivity, (B) Phospho-Mnk1 (Thr) immunoactivity, (C) Phospho-eIF4E (Ser209) immunoreactivity, (D) eIF4E/eIF4G1 coimmunoprecipitation, and (E) Ratio of Phospho-eIF4E to total eIF4E in the eIF4G1 immunoprecipitates. (F) eIF4E/eIF4G1 coimmunoprecipitation analysis of hippocampal area CA1 from wild type (WT) before (unstimulated: u), and 5 min following isoproterenol exposure alone (1 μM, 25 min: iso) or during isoproterenol-induced LTP [isoproterenol (1 μM, 25 min) followed by one train HFS: iso + HFS]. Adapted from Banko et al. (2005) and Gelinas et al. (2007).

constant in the WT mice, but decreased following LTP-inducing stimulation in the 4E-BP2 KO mice (Fig. 4E). Overall, WT mice experience coordinated increases in Mnk1 activation, eIF4E phosphorylation, eIF4F complex formation, and demonstrate homeostasis with regard to the ratio of phosphorylated eIF4E in the eIF4F complex despite dose-dependent elevations. On the other hand, 4E-BP2 KO mice do not experience coordinated increases in Mnk1 activation, eIF4E phosphorylation, or eIF4F complex formation. The levels of Mnk1 activation and eIF4E phosphorylation in response to LTP-inducing stimulation are the same as those found in WT mice, yet the eIF4F complex association is two-fold greater and the ratio of phosphorylated eIF4E in the eIF4F complex is reduced by 1/2 (Fig. 4). Several important conclusions can be drawn from these observations. First, eIF4F complex association is regulated during WT LTP in a manner consistent with increasing local translation activity. Second, excessive eIF4F complex quantities can be correlated with destabilization of LTP. Third, the phosphorylation state of the eIF4F complex is subject to activity-dependent homeostatic control. Fourth, apart from 4E-BP2, there are additional proteins that we have yet to identify that regulate the ability of eIF4E to join the eIF4F complex in response to synaptic activity.

It is clear that there is a tight homeostatic control that governs activity-dependent eIF4F complex regulation. The absence of additional 4E-BP2 phosphorylation during L-LTP may represent a mechanism that could have evolved to prevent excessive activity-induced eIF4E liberation. The reserve of unphosphorylated 4E-BP2 could recover the additional eIF4E liberated by these unidentified eIF4E regulatory proteins, thereby ensuring that the amount of eIF4F complex association achieved does not exceed the levels which could be appropriately regulated by Mnk1. The consequences of excessive unphosphorylated eIF4F complex could include compromised mRNA selectivity. It is not too difficult to imagine a case where deregulating the initiation of translation could be detrimental to synaptic plasticity because synthesis of a protein that destabilizes synaptic function might ensue. Our findings suggest that L-LTP induction in WT mice may be reliant on a mechanism of homeostatic control over the quantity of eIF4F complex that is governed by 4E-BP2 and the phosphorylation state of the complex that is governed by Mnk1.

Elimination of 4E-BP2 disrupts learning and memory in multiple brain regions

The aggregation of our results reveals explicit biochemical and genetic evidence delineating a role for the regulation of eIF4F complex formation in the stabilization of multiple forms of synaptic plasticity. The sheer complexity of the signaling cascades that govern the control of translation initiation suggest that there could be multiple pathways, converging and diverging, responsible for the regulation of translation initiation factors during each type of learning and memory. Examination of the relative contribution of each individual translation factor involved in learning and memory, as well as memory-independent behaviors, is a necessary step toward elucidating these translation regulatory mechanisms. It is likely that similar mechanisms stabilize the synaptic modifications that subserve learning and memory formation. Indeed, it has been shown that eIF4E phosphorylation accompanies memory formation of contextual fear conditioning as mentioned earlier (Kelleher et al., 2004). We performed a battery of tests on the 4E-BP2 knockout mice aimed toward probing a broad range of behavioral phenotypes that employ multiple brain regions (Table 1).

The disruption in the ability of the 4E-BP2 KO animals to express L-LTP in response to multiple tetani of HFS in the Schaffer collateral pathway compelled us to examine hippocampus-dependent learning and memory formation in the 4E-BP2 KO animals. We trained mice on the hidden platform version of the Morris water maze, a hippocampus-dependent spatial learning task that measures the

Table 1. 4E-BP2 knockout behavior phenotypes

Behavior	Phenotype
Open field	↔
Hot plate	↔
Passive avoidance	↔
Light–dark	↓
Light–light	↓
Elevated plus maze	↔
Tube test	↔
Rotarod	↓
Contextual fear conditioning (2 h)	↔
Contextual fear conditioning (24 h)	↓
Cued fear conditioning (2 h)	↔
Cued fear conditioning (24 h)	↔
Water maze (visual platform)	↔
Water maze (hidden platform)	↓
Spontaneous alternation T-maze	↓
Conditioned taste aversion	↑

Source: Table compiled from data reported in Banko et al. (2005, 2007).

ability of an animal to learn and remember the relationship between multiple distal cues and the location of the platform. During acquisition, WT and 4E-BP2 KO mice showed a decrease in escape latency and path length across days, indicating learning of the platform position. However, despite the day-to-day improvement, the 4E-BP2 KO mice displayed significantly higher escape latencies and longer path lengths than control mice overall. These results indicate that the 4E-BP2 KO mice exhibit a spatial learning impairment. We subsequently confirmed that the spatial learning deficit was not attributable to poor vision, motivation, or swimming ability. Moreover, during the probe test, the 4E-BP2 KO mice did not display a preference for the pool quadrant in which the platform was located during training and crossed the exact site where the platform was located during training fewer times than their WT littermates. These findings demonstrate compromised spatial learning and memory in the 4E-BP2 KO mice.

We utilized a conditioned fear paradigm to determine how the lack of 4E-BP2 affected associative learning and memory formation. Animals learn to fear a new environment or an emotionally neutral CS (tone) because of its temporal association with an aversive US (footshock). When exposed to the same training context (contextual conditioning) or tone (cued conditioning) at a later time, conditioned animals demonstrate the stereo-typical fear response, freezing. The 4E-BP2 KO mice and WT mice responded comparably to the training paradigm and showed similar acquisition of short-term memory for the context and the CS, indicating that mice lacking 4E-BP2 have normal associative learning and unaltered short-term fear-associative memory. There also was no difference between genotypes tested with the CS the next day. In contrast, 4E-BP2 knockout mice showed a significant reduction in long-term memory for the context. Contextual conditioning requires coordinated function of the amygdala and the hippocampus, whereas cued conditioning relies upon only the amygdala. Given that only long-term memory for the context was disrupted in the 4E-BP2 KO mice, we conclude that elimination of 4E-BP2 in the amygdala is not insurmountable for amygdala-dependent memory formation, but disrupts hippocampus-dependent long-term memory formation. We then extended these findings of compromised hippocampus-dependent long-term memory by demonstrating compromised hippocampus-dependent working memory with the spontaneous alternation T-maze.

Interestingly, we did find one hippocampus-dependent learning and memory task that was spared in the 4E-BP2 KO mice: passive avoidance learning and memory. One possible explanation for why this particular hippocampus-dependent process is not disrupted while spatial learning and associative contextual learning are impaired is that the passive avoidance paradigm is a less challenging test. Because of the constellation of cues that the mice must take in to navigate the water maze or learn a specific context, these learning tasks historically have been considered fairly challenging. In contrast, passive avoidance requires discrimination only on the basis of light versus dark. Overall, the hippocampus-dependent memory phenotypes, taken together with our observations of altered synaptic plasticity in the Schaffer collateral pathway of the 4E-BP2 KO mice, strongly suggest that loss of 4E-BP2 results in reduced functional capacity of the hippocampus for learning and memory.

The amygdala, septohippocampal system, medial hypothalamus, central periaqueductal gray, and frontal and cingulate cortices are important brain structures involved in the regulation of anxiety and fear. Concordant with our findings that the 4E-BP2 knockout mice were normal with respect to the cued component of the conditioned fear test, additional anxiety-related analyses, including light/dark exploration and elevated plus maze, did not reveal consistent anti-anxiety phenotypes in the 4E-BP2 KO mice. Interestingly, the chamber-based exploratory tasks did suggest that the 4E-BP2 KO mice favored exploration of a novel context based on first experience rather than light versus dark.

Motor skill learning circuits include the sensorimotor cortex, cerebellum, and the basal ganglia. Consolidation of rotating rod motor skill memory is sensitive to the disruption of de novo translation; therefore, we examined the ability of the 4E-BP2 KO mice to acquire and perform the motor skills associated with the rotating rod task. Although we observed no evidence for compromised locomotor activity with the open field test or water maze, we found that not only did naïve 4E-BP2 KO mice perform poorly on their initial rotating rod training day; they failed to demonstrate significant improvement over three consecutive days of training. These results are consistent with the notion that 4E-BP2 is important for complex motor skill performance.

Conditioned taste aversion: the Holy Grail of the 4E-BP2 knockout mouse

Amidst the myriad of tests in which the 4E-BP2 KO mice performed poorly, there was one that proved our initial naïve hypothesis that 4E-BP2 elimination would enhance cognition: conditioned taste aversion (CTA). CTA represents a robust form of associative learning in which a normally appetizing taste becomes aversive following its association with gastric distress. Learning and memory of CTA requires a neural circuit that includes the brain stem, pons, insular cortex, and amygdala; however, the role of the hippocampus in this task is not well defined. Interestingly, translation has been shown to be regulated during both the initiation and elongation phases following novel taste learning in the insular cortex (Belelovsky et al., 2005). The 4E-BP2 KO mice demonstrated a natural aversion to quinine and a natural preference for saccharin. Remarkably, 4E-BP2 KO mice demonstrated a significantly enhanced memory for CTA by avoiding the saccharin and NaCl solutions to a higher degree than WT mice following a one trial pairing of saccharin or NaCl with LiCl. Because amygdala-dependent learning was unaffected in the 4E-BP2 KO mice, these results suggest that in contrast to the hippocampus, shifting the scale towards enhanced translation initiation in the insular cortex improves learning and memory. This observation represents the first example of cognitive enhancement due to the genetic elimination of 4E-BP2.

Translation regulation and disease

Mutations in genes that result in altered translation control have been shown to result in several neurological diseases and mental retardation syndromes. FXS is the most common inherited disease causing mental retardation, affecting approximately 1:4000 males and 1:8000 females (Turner et al., 1996). The syndrome is characterized by moderate to severe mental retardation, mild facial dysmorphism, and macro-orchidism. FXS typically results from the expansion of a CGG repeat sequence in the 5′ untranslated region of fragile X mental retardation gene *FMR1*. The trinucleotide repeat expansion and subsequent hypermethylation cause transcriptional silencing of the *FMR1* gene, resulting in the loss of the FMRP (Darnell et al., 2001; O'Donnell and Warren, 2002; Jin and Warren, 2003). FMRP is an mRNA binding protein thought to regulate translation of specific mRNAs, including its own mRNA (Feng et al., 1997; Brown et al., 2001; Darnell et al., 2001; Zhang et al., 2001; Miyashiro et al., 2003). Previous studies have shown that FMRP is colocalized with polyribosomes in the neuronal soma as well as in dendritic spines, which suggests the possibility that FMRP is involved in local dendritic protein synthesis (Stefani et al., 2004). As mentioned earlier, it has been demonstrated that mRNA complementary to human *FMR1* and mouse *Fmr1* transcripts moves into close association with polyribosomes isolated from rat cortex and undergoes rapid translation during the first 5 of DHPG-induced mGluR-LTD (Weiler et al., 1997).

Another translation factor gaining interest is the initiation factor eIF2. The guanine nucleotide exchange factor eIF2B catalyzes the exchange of GDP for GTP on eIF2. GTP-bound eIF2 is then able to associate with the specific initiator methionyl-tRNA and form an eIF2–GTP–tRNA complex that is necessary for the ribosome to initiate translation. After the ribosome scans the mRNA and recognizes the AUG codon, the GTP is hydrolyzed and the stable eIF2–GDP complex is released. A guanine nucleotide exchange reaction is required to form eIF2–GTP to initiate the translation of another mRNA transcript. This process is regulated by phosphorylation of eIF2, which is mediated by the four kinases HRI, PKR, GCN2, and PERK. Phosphorylated eIF2 competes with non-phosphorylated eIF2 for a binding site on eIF2B. However, phosphorylated eIF2 inhibits eIF2B activity and subsequently decreases the rate of eIF2–GTP formation and the availability of the complex for translation initiation (for review, see Dever, 1999, 2002). Many genes reportedly associated with both schizophrenia and bipolar disorder code for proteins related to the control of the eIF2 kinases (IL1B, IL1RN, MTHFR, TNF, ND4, NDUFV2, XBP1; for review, see Carter, 2007a). Moreover, mutations in any of five eIF2B subunit genes are responsible for vanishing white matter disease, which provokes severe loss of oligodendrocytes and astrocytes in early life (Leegwater et al., 2001).

There are several pieces of evidence that suggest that altered translational control, and more specifically, upregulated mTOR signaling, might underlie abnormal long-term memory and other behaviors associated with autism spectrum disorder (ASD). The first piece of evidence that we will discuss was an exciting series of findings made by Kwon et al. (2006). Mutations in phosphatase and tensin homolog on chromosome ten (PTEN) have been reported in autistic individuals with macrocephaly (Butler et al., 2005). Kwon et al. (2006) generated a mouse that had a neuron-specific deletion of PTEN, which normally dephosphorylates PIP3, thereby negatively modulating Akt signaling. The investigators found that the PTEN-deficient mice had excessive Akt/mTOR signaling. More importantly, these authors also observed that the PTEN-deficient mice exhibited impaired long-term memory, abnormal social interactions, and exaggerated responses to sensory stimuli, autistic-like behaviors that also have been observed in Fmr1 knockout mice. Although these investigators did not determine whether blocking the excessive Akt/mTOR signaling could reverse the long-term memory and behavior phenotypes, their findings are consistent with the idea that excessive translation might result in autistic-like behaviors.

The second piece of evidence that is consistent with the idea that excessive translation via upregulated mTOR signaling might play a role in ASD deals with the tuberous sclerosis complex (TSC) gene products *TSC1* and *TSC2*. TSCs are heterodimers that normally function to repress mTOR activity. Individuals with TSC mutations have a number of nervous system disorders, including autism (Wiznitzer, 2004). Mice and rats with TSC mutations have a variety of phenotypes, including upregulated mTOR signaling, as well as altered synaptic plasticity and behavior (Govindarajan et al., 2005; von der Brelie et al., 2006; Waltereit et al., 2006). In addition, ablation of either TSC1 or TSC2 results in altered neuronal morphology (Tavazoie et al., 2005). Although there are no reports as of yet for autistic-like behavior in TSC mutant rodents, it is reasonable to propose that TSC mutations contribute to mental retardation and autism observed in patients with tuberous sclerosis.

Future directions

Large gaps exist upstream and downstream of the local initiation process. For example, little is known about the players involved in nucleus-to-dendrite mRNA translocation and ensuring mRNA stability once it arrives. Activity-induced targets for translation control downstream of initiation remain largely unexplored. Perhaps most important, the identities of the newly synthesized proteins that mediate the enduring synaptic modifications which subserve learning and memory remain largely unknown. These areas will certainly attract more attention, in part because it has been established that multiple neurological diseases are causally related to RNA localization and/or translation at synaptic sites. To fully understand these diseases, it will be necessary that we first arrive at a clear picture of these underlying molecular mechanisms.

Navigating the road to the synapse

How does the mRNA arrive at the synapse to begin with? Is there a dendritic localization signal that destines a given mRNA for transport to the dendritic arbor? Sequences within the mRNA account for mRNA localization. These sequences are referred to as localization elements (LEs), zipcodes and targeting elements (TEs) and are recognized by RNA binding proteins. Dendritic targeting elements (DTEs) have been identified in the untranslated region of a number of neuronal mRNAs like MAP2 (Mayford et al., 1996), αCaMKII (Mori et al., 2000), and β-actin (Zhang et al., 2001). However, such DTEs appear to be quite diverse in length, sequence, relative position, and they may also encode different destination sites within dendritic arborizations. We are therefore not yet able to predict whether a given mRNA is destined for synapto-dendritic localization based on sequence. What is urgently needed is a physical description, in conjunction with a functional dissection, of code-carrying RNA sequences and/or motifs that specify targeting.

Several potential mRNA-binding proteins have been identified that recognize these TEs. These proteins include the MARTA (−1 and −2) proteins and zipcode-binding protein (ZBP1; Rehbein et al., 2000; Tiruchinapalli et al., 2003). The interaction results in the formation of ribonucleotide protein (RNP) complexes, which travel along cytoskeletal filaments with the help of motor proteins. A physical description of RNA target-encoding motifs will also be essential for us to understand how transport RNPs are assembled and delivered to their final destination sites. The dendritic delivery of some of these mRNAs is dependent on neuronal activity. Synaptic activation in vivo strongly upregulates *Arc/arg3.1* gene expression (Lyford et al., 1995) and results in the selective, NMDA receptor–dependent recruitment of corresponding transcripts to dendritic segments in which synapses had been stimulated (Steward and Worley, 2001). Similarly, epileptogenic stimuli in vivo produce increased brain-derived BDNF mRNA levels, coupled with an NMDA receptor-dependent transcript accumulation in proximal dendritic segments in hippocampus (Tongiorgi et al., 2004). A common requirement in these pathways would be retrograde synapse-to-nucleus signaling. The nature of such a signal, and the mechanism by which it is transduced to the nuclear machinery, will be a major challenge for future work.

Are there targets for translational control downstream of initiation?

This chapter has mainly focused on trans-acting proteins that regulate local translation initiation. Elongation may also be a target for local translational control, if not as frequently as initiation. One such target is eukaryotic elongation factor 2 (eEF2). eEF2 is a GTP-binding protein that mediates the translocation of peptidyl-tRNA from the A site to the P site on the ribosome. Phosphorylation of eEF2 by eEF2 kinase, a Ca^{2+}/calmodulin-dependent enzyme, results in an inhibition of eEF2 activity and a general reduction of peptide elongation. Several studies have been reported that implicate eEF2 phosphorylation in synaptic plasticity. In the amphibian tectum, phosphorylation of eEF2 is dependent on the activation of NMDARs and can be induced by visual stimulation (Scheetz et al., 1997). The same group found that stimulation of synaptoneurosomes from rat superior colliculus also results in increased eEF2 phosphorylation, which was correlated with increased synthesis of αCaMKII (Scheetz et al., 2000). It was reported that glutamate enhances the phosphorylation of eEF2 via NMDA receptors in cultured cortical neurons (Marin et al., 1997). Phosphorylation of eEF2 results in reduced overall protein synthesis in eukaryotes (Nairn et al., 2001), but may enhance translation of some mRNAs in developing neurons (Scheetz et al., 2000). Indeed, Chotiner et al. (2003) demonstrated that chemically induced LTP elicits an inhibition of de novo total protein synthesis, coincident with increased phosphorylation of eEF2 but increased protein levels of Arc and Fos. It has been suggested that eEF2 may contribute to activity-induced, compartment-specific translation regulation by silencing translation at non-synaptic sites (Kanhema et al., 2006). Investigation into this hypothesis and others centered upon the lifecycle of local translation control is necessary to complete our picture of how synapses coordinate local translation.

What is often overlooked is the mRNA lifecycle itself and whether it is a target for regulation. The transition of an mRNA from active translation to being committed for degradation is a potential site for activity-dependent regulation. Interestingly, the 5′ cap structure not only serves as an anchoring site for the translation initiation complex, but also serves as an anchoring site for Argonaute (Ago) proteins. MicroRNAs (miRNAs) bind to Ago proteins and inhibit translation or promote degradation of mRNA targets (Kiriakidou et al., 2007). eIF4E itself may also play a role in this process. It has been demonstrated that eIF4E and one of its interaction partners, eIF4E-transporter (eIF4E-T), are components of mammalian P bodies, cytoplasmic foci enriched in 5′→3′ mRNA degrading enzymes (Andrei et al., 2005). Future research in the field of mRNA stability and degradation may address how protein synthesis is differentially regulated during different forms of synaptic plasticity.

What is the identity of the synaptic plasticity proteins?

An important question that has not been investigated to date on a broad scale is the identity of the newly synthesized proteins. This is a tougher problem because identification of proteins in dendrites does not indicate their site of synthesis. Furthermore, it is still unclear how many different mRNAs are present in dendrites. Current lists suggest hundreds of distinct mRNAs comprise the local mRNA population. Determination of the local mRNA population with conventional in situ hybridization techniques is critical, but falls short of getting at which proteins are synthesized in response to which patterns of synaptic stimulation. Given the preponderance of evidence that eIF4F is regulated during multiple forms of synaptic plasticity in a manner consistent with activating translation, it may present a possible solution for identification of the plasticity proteins. eIF4F–mRNA complexes could be isolated from synaptic fractions following specific patterns of stimulation and the mRNAs identified using microarrays or PCR-based strategies. Similar strategies using FMRP have been successful (Brown et al., 2001).

Inarguably, determining the identity of the proteins synthesized in response to synaptic activation will shed light on the mechanisms responsible for the enduring synaptic modifications the subserve learning and memory.

Acknowledgments

We thank Edwin Weeber, PhD for critically reading this chapter and providing assistance with graphic design.

References

Aakalu, G., Smith, W.B., Nguyen, N., Jiang, C. and Schuman, E.M. (2001) Dynamic visualization of local protein synthesis in hippocampal neurons. Neuron, 30: 489–502.

Abel, T., Nguyen, P.V., Barad, M., Deuel, T.A., Kandel, E.R. and Bourtchouladze, R. (1997) Genetic demonstration of a role for PKA in the late phase of LTP and in hippocampus-based long-term memory. Cell, 88: 615–626.

Andrei, M.A., Ingelfinger, D., Heintzmann, R., Achsel, T., Rivera-Pomar, R. and Luhrmann, R. (2005) A role for eIF4E and eIF4E-transporter in targeting mRNPs to mammalian processing bodies. RNA, 11: 717–727.

Antar, L.N. and Bassell, G.J. (2003) Sunrise at the synapse: the FMRP mRNP shaping the synaptic interface. Neuron, 37: 555–558.

Asaki, C., Usuda, N., Nakazawa, A., Kametani, K. and Suzuki, T. (2003) Localization of translational components at the ultramicroscopic level at postsynaptic sites of the rat brain. Brain Res., 972: 168–176.

Banko, J.L., Hou, L. and Klann, E. (2004) NMDA receptor activation results in PKA- and ERK-dependent Mnk1 activation and increased eIF4E phosphorylation in hippocampal area CA1. J. Neurochem., 91: 462–470.

Banko, J.L., Hou, L., Poulin, F., Sonenberg, N. and Klann, E. (2006) Regulation of eukaryotic initiation factor 4E by converging signaling pathways during metabotropic glutamate receptor-dependent long-term depression. J. Neurosci., 26: 2167–2173.

Banko, J.L., Poulin, F., Hou, L., DeMaria, C.T., Sonenberg, N. and Klann, E. (2005) The translation repressor 4E-BP2 is critical for eIF4F complex formation, synaptic plasticity, and memory in the hippocampus. J. Neurosci., 25: 9581–9590.

Banko, J.L., Merhav, M., Stern, E., Sonenberg, N., Rosenblum, K. and Klann, E. (2007) Behavioral alterations in mice lacking the translation repressor 4E-BP2. Neurobiol. Learn. Mem., 87: 248–256.

Belelovsky, K., Elkobi, A., Kaphzan, H., Nairn, A.C. and Rosenblum, K. (2005) A molecular switch for translational control in taste memory consolidation. Eur. J. Neurosci., 22: 2560–2568.

Bradshaw, K.D., Emptage, N.J. and Bliss, T.V. (2003) A role for dendritic protein synthesis in hippocampal late LTP. Eur. J. Neurosci., 18: 3150–3152.

Brown, V., Jin, P., Ceman, S., Darnell, J.C., O'Donnell, W.T., Tenenbaum, S.A., Jin, X., Feng, Y., Wilkinson, K.D., Keene, J.D., Darnell, R.B. and Warren, S.T. (2001) Microarray identification of FMRP-associated brain mRNAs and altered mRNA translational profiles in fragile X syndrome. Cell, 107: 477–487.

Butler, M.G., Dasouki, M.J., Zhou, X.P., Talebizadeh, Z., Brown, M., Takahashi, T.N., Miles, J.H., Wang, C.H., Stratton, R., Pilarski, R. and Eng, C. (2005) Subset of individuals with autism spectrum disorders and extreme macrocephaly associated with germline PTEN tumour suppressor gene mutations. J. Med. Genet., 42: 318–321.

Cao, Q. and Richter, J.D. (2002) Dissolution of the maskin–eIF4E complex by cytoplasmic polyadenylation and poly(A)-binding protein controls cyclin B1 mRNA translation and oocyte maturation. EMBO J., 21: 3852–3862.

Carter, C.J. (2007a) EIF2B and oligodendrocyte survival: where nature and nurture meet in bipolar disorder and schizophrenia? Schizophr. Bull., 33: 1343–1353.

Carter, C.J. (2007b) Multiple genes and factors associated with bipolar disorder converge on growth factor and stress activated kinase pathways controlling translation initiation: implications for oligodendrocyte viability. Neurochem. Int., 50: 461–490.

Chotiner, J.K., Khorasani, H., Nairn, A.C., O'Dell, T.J. and Watson, J.B. (2003) Adenylyl cyclase-dependent form of chemical long-term potentiation triggers translational regulation at the elongation step. Neuroscience, 116: 743–752.

Darnell, J.C., Jensen, K.B., Jin, P., Brown, V., Warren, S.T. and Darnell, R.B. (2001) Fragile X mental retardation protein targets G quartet mRNAs important for neuronal function. Cell, 107: 489–499.

Dever, T.E. (1999) Translation initiation: adept at adapting. Trends Biochem. Sci., 24: 398–403.

Dever, T.E. (2002) Gene-specific regulation by general translation factors. Cell, 108: 545–556.

Feng, Y., Absher, D., Eberhart, D.E., Brown, V., Malter, H.E. and Warren, S.T. (1997) FMRP associates with polyribosomes as an mRNP, and the I304N mutation of severe fragile X syndrome abolishes this association. Mol. Cell., 1: 109–118.

Flynn, A. and Proud, C.G. (1995) Serine 209, not serine 53, is the major site of phosphorylation in initiation factor eIF-4E in serum-treated Chinese hamster ovary cells. J. Biol. Chem., 270: 21684–21688.

Gelinas, J.N., Banko, J.L., Hou, L., Sonenberg, N., Weeber, E.J., Klann, E. and Nguyen, P.V. (2007) ERK and mTOR signaling couple beta-adrenergic receptors to translation initiation machinery to gate induction of protein synthesis-dependent LTP. J. Biol. Chem., 282: 27527–27535.

Gelinas, J.N. and Nguyen, P.V. (2005) Beta-adrenergic receptor activation facilitates induction of a protein synthesis-dependent late phase of long-term potentiation. J. Neurosci., 25: 3294–3303.

Govindarajan, B., Brat, D.J., Csete, M., Martin, W.D., Murad, E., Litani, K., Cohen, C., Cerimele, F., Nunnelley, M., Lefkove, B., Yamamoto, T., Lee, C. and Arbiser, J.L. (2005) Transgenic expression of dominant negative tuberin through a strong constitutive promoter results in a tissue-specific tuberous sclerosis phenotype in the skin and brain. J. Biol. Chem., 280: 5870–5874.

Herbert, T.P., Kilhams, G.R., Batty, I.H. and Proud, C.G. (2000) Distinct signalling pathways mediate insulin and phorbol ester-stimulated eukaryotic initiation factor 4F assembly and protein synthesis in HEK 293 cells. J. Biol. Chem., 275: 11249–11256.

Huber, K.M., Kayser, M.S. and Bear, M.F. (2000) Role for rapid dendritic protein synthesis in hippocampal mGluR-dependent long-term depression. Science, 288: 1254–1257.

Huber, K.M., Roder, J.C. and Bear, M.F. (2001) Chemical induction of mGluR5- and protein synthesis-dependent long-term depression in hippocampal area CA1. J. Neurophysiol., 86: 321–325.

Jin, P. and Warren, S.T. (2003) New insights into fragile X syndrome: from molecules to neurobehaviors. Trends Biochem. Sci., 28: 152–158.

Joshi, B., Cai, A.L., Keiper, B.D., Minich, W.B., Mendez, R., Beach, C.M., Stepinski, J., Stolarski, R., Darzynkiewicz, E. and Rhoads, R.E. (1995) Phosphorylation of eukaryotic protein synthesis initiation factor 4E at Ser-209. J. Biol. Chem., 270: 14597–14603.

Kang, H. and Schuman, E.M. (1996) A requirement for local protein synthesis in neurotrophin-induced hippocampal synaptic plasticity. Science, 273: 1402–1406.

Kanhema, T., Dagestad, G., Panja, D., Tiron, A., Messaoudi, E., Havik, B., Ying, S.W., Nairn, A.C., Sonenberg, N. and Bramham, C.R. (2006) Dual regulation of translation initiation and peptide chain elongation during BDNF-induced LTP in vivo: evidence for compartment-specific translation control. J. Neurochem., 99: 1328–1337.

Kelleher, III, R.J., Govindarajan, A., Jung, H.Y., Kang, H. and Tonegawa, S. (2004) Translational control by MAPK signaling in long-term synaptic plasticity and memory. Cell, 116: 467–479.

Kiriakidou, M., Tan, G.S., Lamprinaki, S., De Planell-Saguer, M., Nelson, P.T. and Mourelatos, Z. (2007) An mRNA m7G cap binding-like motif within human Ago2 represses translation. Cell, 129: 1141–1151.

Knauf, U., Tschopp, C. and Gram, H. (2001) Negative regulation of protein translation by mitogen-activated protein kinase-interacting kinases 1 and 2. Mol. Cell. Biol., 21: 5500–5511.

Kwon, C.H., Luikart, B.W., Powell, C.M., Zhou, J., Matheny, S.A., Zhang, W., Li, Y., Baker, S.J. and Parada, L.F. (2006) Pten regulates neuronal arborization and social interaction in mice. Neuron, 50: 377–388.

Leegwater, P.A., Vermeulen, G., Konst, A.A., Naidu, S., Mulders, J., Visser, A., Kersbergen, P., Mobach, D., Fonds, D., van Berkel, C.G., Lemmers, R.J., Frants, R.R., Oudejans, C.B., Schutgens, R.B., Pronk, J.C. and van der Knaap, M.S. (2001) Subunits of the translation initiation factor eIF2B are mutant in leukoencephalopathy with vanishing white matter. Nat. Genet., 29: 383–388.

Lyford, G.L., Yamagata, K., Kaufmann, W.E., Barnes, C.A., Sanders, L.K., Copeland, N.G., Gilbert, D.J., Jenkins, N.A., Lanahan, A.A. and Worley, P.F. (1995) Arc, a growth factor and activity-regulated gene, encodes a novel cytoskeleton-associated protein that is enriched in neuronal dendrites. Neuron, 14: 433–445.

Marin, P., Nastiuk, K.L., Daniel, N., Girault, J.A., Czernik, A.J., Glowinski, J., Nairn, A.C. and Premont, J. (1997) Glutamate-dependent phosphorylation of elongation factor-2 and inhibition of protein synthesis in neurons. J. Neurosci., 17: 3445–3454.

Mathews, M.B., Sonenberg, N., and Hershey, J.W.B. (Eds.) (2007) Translational control in biology and medicine. Cold Spring Harbor Laboratory Press, New York.

Mayford, M., Baranes, D., Podsypanina, K. and Kandel, E.R. (1996) The 3′-untranslated region of CaMKII alpha is a cis-acting signal for the localization and translation of mRNA in dendrites. Proc. Natl. Acad. Sci. U.S.A., 93: 13250–13255.

Miyashiro, K.Y., Beckel-Mitchener, A., Purk, T.P., Becker, K.G., Barret, T., Liu, L., Carbonetto, S., Weiler, I.J., Greenough, W.T. and Eberwine, J. (2003) RNA cargoes associating with FMRP reveal deficits in cellular functioning in Fmr1 null mice. Neuron, 37: 417–431.

Mori, Y., Imaizumi, K., Katayama, T., Yoneda, T. and Tohyama, M. (2000) Two cis-acting elements in the 3′ untranslated region of alpha-CaMKII regulate its dendritic targeting. Nat. Neurosci., 3: 1079–1084.

Morley, S.J. and McKendrick, L. (1997) Involvement of stress-activated protein kinase and p38/RK mitogen-activated protein kinase signaling pathways in the enhanced phosphorylation of initiation factor 4E in NIH 3T3 cells. J. Biol. Chem., 272: 17887–17893.

Morley, S.J. and Naegele, S. (2002) Phosphorylation of eukaryotic initiation factor (eIF) 4E is not required for de novo protein synthesis following recovery from hypertonic stress in human kidney cells. J. Biol. Chem., 277: 32855–32859.

Nairn, A.C., Matsushita, M., Nastiuk, K., Horiuchi, A., Mitsui, K., Shimizu, Y. and Palfrey, H.C. (2001) Elongation factor-2 phosphorylation and the regulation of protein synthesis by calcium. Prog. Mol. Subcell. Biol., 27: 91–129.

Niessing, D., Blanke, S. and Jackle, H. (2002) Bicoid associates with the 5′-cap-bound complex of caudal mRNA and represses translation. Genes Dev., 16: 2576–2582.

O'Donnell, W.T. and Warren, S.T. (2002) A decade of molecular studies of fragile X syndrome. Annu. Rev. Neurosci., 25: 315–338.

Ogawa, F., Kasai, M. and Akiyama, T. (2005) A functional link between Disrupted-In-Schizophrenia 1 and the eukaryotic translation initiation factor 3. Biochem. Biohys. Res. Commun., 338: 771–776.

Pause, A., Belsham, G.J., Gingras, A.C., Donze, O., Lin, T.A., Lawrence, Jr., J.C. and Sonenberg, N. (1994) Insulin-dependent stimulation of protein synthesis by phosphorylation of a regulator of 5′-cap function. Nature, 371: 762–767.

Ptushkina, M., von der Haar, T., Vasilescu, S., Frank, R., Birkenhager, R. and McCarthy, J.E. (1998) Cooperative modulation by eIF4G of eIF4E-binding to the mRNA 5′ cap in yeast involves a site partially shared by p. 20. EMBO J., 17: 4798–4808.

Raught, B. and Gingras, A.C. (1999) eIF4E activity is regulated at multiple levels. Int. J. Biochem. Cell. Biol., 31: 43–57.

Rehbein, M., Kindler, S., Horke, S. and Richter, D. (2000) Two trans-acting rat-brain proteins, MARTA1 and MARTA2, interact specifically with the dendritic targeting element in MAP2 mRNAs. Brain Res. Mol. Brain Res., 79: 192–201.

Saghir, A.N., Tuxworth, Jr., W.J., Hagedorn, C.H. and McDermott, P.J. (2001) Modifications of eukaryotic initiation factor 4F (eIF4F) in adult cardiocytes by adenoviral gene transfer: differential effects on eIF4F activity and total protein synthesis rates. Biochem. J., 356: 557–566.

Scheetz, A.J., Nairn, A.C. and Constantine-Paton, M. (1997) N-methyl-d-aspartate receptor activation and visual activity induce elongation factor-2 phosphorylation in amphibian tecta: a role for N-methyl-d-aspartate receptors in controlling protein synthesis. Proc. Natl. Acad. Sci. U.S.A., 94: 14770–14775.

Scheetz, A.J., Nairn, A.C. and Constantine-Paton, M. (2000) NMDA receptor-mediated control of protein synthesis at developing synapses. Nat. Neurosci., 3: 211–216.

Scheper, G.C. and Proud, C.G. (2002) Does phosphorylation of the cap-binding protein eIF4E play a role in translation initiation? Eur. J. Biochem., 269: 5350–5359.

Scheper, G.C., van Kollenburg, B., Hu, J., Luo, Y., Goss, D.J. and Proud, C.G. (2002) Phosphorylation of eukaryotic initiation factor 4E markedly reduces its affinity for capped mRNA. J. Biol. Chem., 277: 3303–3309.

Smart, F.M., Edelman, G.M. and Vanderklish, P.W. (2003) BDNF induces translocation of initiation factor 4E to mRNA granules: evidence for a role of synaptic microfilaments and integrins. Proc. Natl. Acad. Sci. U.S.A., 100: 14403–14408.

Sonenberg, N. and Gingras, A.C. (1998) The mRNA 5′ cap-binding protein eIF4E and control of cell growth. Curr. Opin. Cell. Biol., 10: 268–275.

Stefani, G., Fraser, C.E., Darnell, J.C. and Darnell, R.B. (2004) Fragile X mental retardation protein is associated with translating polyribosomes in neuronal cells. J. Neurosci., 24: 7272–7276.

Steward, O. and Levy, W.B. (1982) Preferential localization of polyribosomes under the base of dendritic spines in granule cells of the dentate gyrus. J. Neurosci., 2: 284–291.

Steward, O., Wallace, C.S., Lyford, G.L. and Worley, P.F. (1998) Synaptic activation causes the mRNA for the IEG Arc to localize selectively near activated postsynaptic sites on dendrites. Neuron, 21: 741–751.

Steward, O. and Worley, P.F. (2001) Selective targeting of newly synthesized Arc mRNA to active synapses requires NMDA receptor activation. Neuron, 30: 227–240.

Stoneley, M. and Willis, A.E. (2004) Cellular internal ribosome entry segments: structures, *trans*-acting factors and regulation of gene expression. Oncogene, 23: 3200–3207.

Takei, N., Kawamura, M., Hara, K., Yonezawa, K. and Nawa, H. (2001) Brain-derived neurotrophic factor enhances neuronal translation by activating multiple initiation processes: comparison with the effects of insulin. J. Biol. Chem., 276: 42818–42825.

Takei, N., Inamura, N., Kawamura, M., Namba, H., Hara, K., Yonezawa, K. and Nawa, H. (2004) Brain-derived neurotrophic factor induces mammalian target of rapamycin-dependent local activation of translation machinery and protein synthesis in neuronal dendrites. J. Neurosci., 24: 9760–9769.

Tang, S.J., Reis, G., Kang, H., Gingras, A.C., Sonenberg, N. and Schuman, E.M. (2002) A rapamycin-sensitive signaling pathway contributes to long-term synaptic plasticity in the hippocampus. Proc. Natl. Acad. Sci. U.S.A., 99: 467–472.

Tavazoie, S.F., Alvarez, V.A., Ridenour, D.A., Kwiatkowski, D.J. and Sabatini, B.L. (2005) Regulation of neuronal morphology and function by the tumor suppressors Tsc1 and Tsc2. Nat. Neurosci., 8: 1727–1734.

Tiruchinapalli, D.M., Oleynikov, Y., Kelic, S., Shenoy, S.M., Hartley, A., Stanton, P.K., Singer, R.H. and Bassell, G.J. (2003) Activity-dependent trafficking and dynamic localization of zipcode binding protein 1 and beta-actin mRNA in dendrites and spines of hippocampal neurons. J. Neurosci., 23: 3251–3261.

Tongiorgi, E., Armellin, M., Giulianini, P.G., Bregola, G., Zucchini, S., Paradiso, B., Steward, O., Cattaneo, A. and Simonato, M. (2004) Brain-derived neurotrophic factor mRNA and protein are targeted to discrete dendritic laminas by events that trigger epileptogenesis. J. Neurosci., 24: 6842–6852.

Tsokas, P., Ma, T., Iyengar, R., Landau, E.M. and Blitzer, R.D. (2007) Mitogen-activated protein kinase upregulates the dendritic translation machinery in long-term potentiation by controlling the mammalian target of rapamycin pathway. J. Neurosci., 27: 5885–5894.

Turner, G., Webb, T., Wake, S. and Robinson, H. (1996) Prevalence of fragile X syndrome. Am. J. Med. Genet., 64: 196–197.

von der Brelie, C., Waltereit, R., Zhang, L., Beck, H. and Kirschstein, T. (2006) Impaired synaptic plasticity in a rat model of tuberous sclerosis. Eur. J. Neurosci., 23: 686–692.

Waltereit, R., Welzl, H., Dichgans, J., Lipp, H.P., Schmidt, W.J. and Weller, M. (2006) Enhanced episodic-like memory and kindling epilepsy in a rat model of tuberous sclerosis. J. Neurochem., 96: 407–413.

Wang, H., Iacoangeli, A., Popp, S., Muslimov, I.A., Imataka, H., Sonenberg, N., Lomakin, I.B. and Tiedge, H. (2002) Dendritic BC1 RNA: functional role in regulation of translation initiation. J. Neurosci., 22: 10232–10241.

Wang, X., Flynn, A., Waskiewicz, A.J., Webb, B.L., Vries, R.G., Baines, I.A., Cooper, J.A. and Proud, C.G. (1998) The phosphorylation of eukaryotic initiation factor eIF4E in response to phorbol esters, cell stresses, and cytokines is mediated by distinct MAP kinase pathways. J. Biol. Chem., 273: 9373–9377.

Waskiewicz, A.J., Flynn, A., Proud, C.G. and Cooper, J.A. (1997) Mitogen-activated protein kinases activate the serine/threonine kinases Mnk1 and Mnk2. EMBO J., 16: 1909–1920.

Weiler, I.J., Irwin, S.A., Klintsova, A.Y., Spencer, C.M., Brazelton, A.D., Miyashiro, K., Comery, T.A., Patel, B., Eberwine, J. and Greenough, W.T. (1997) Fragile X mental retardation protein is translated near synapses in response to neurotransmitter activation. Proc. Natl. Acad. Sci. U.S.A., 94: 5395–5400.

Wilhelm, J.E., Hilton, M., Amos, Q. and Henzel, W.J. (2003) Cup is an eIF4E binding protein required for both the translational repression of Oskar and the recruitment of Barentsz. J. Cell. Biol., 163: 1197–1204.

Wiznitzer, M. (2004) Autism and tuberous sclerosis. J. Child. Neurol., 19: 675–679.

Zhang, H.L., Eom, T., Oleynikov, Y., Shenoy, S.M., Liebelt, D.A., Dictenberg, J.B., Singer, R.H. and Bassell, G.J. (2001) Neurotrophin-induced transport of a beta-actin mRNP complex increases beta-actin levels and stimulates growth cone motility. Neuron, 31: 261–275.

CHAPTER 5

Translational control of gene expression: a molecular switch for memory storage

Mauro Costa-Mattioli and Nahum Sonenberg*

Department of Biochemistry and McGill Cancer Center, 3655 Promendde Sir William Osler, McGill University, Montréal, QC H3G 1Y6, Canada

Abstract: A critical requirement for the conversion of the labile short-term memory (STM) into the consolidated long-term memory (LTM) is new gene expression (new mRNAs and protein synthesis). The first clues to the molecular mechanisms of the switch from short-term to LTM emerged from studies on protein synthesis in different species. Initially, it was shown that LTM can be distinguished from STM by its susceptibility to protein synthesis inhibitors. Later, it was found that long-lasting synaptic changes, which are believed to be a key cellular mechanism by which information is stored, are also dependent on new protein synthesis. Although the role of protein synthesis in memory was reported more than 40 years ago, recent molecular, genetic, and biochemical studies have provided fresh insights into the molecular mechanisms underlying these processes. In this chapter, we provide an overview of the role of translational control by the eIF2α signaling pathway in long-term synaptic plasticity and memory consolidation.

Keywords: translational control; synaptic plasticity; long-term memory; GCN2; eIF2α

Overview of translation initiation in eukaryotes

Translational control is an important mechanism by which cells govern gene expression, providing a rapid response by the cell without invoking nuclear pathways for mRNA synthesis and transport. In systems with little or no transcriptional control (e.g., reticulocytes, sea urchin eggs, *Drosophila* early embryogenesis, and oocytes), translation is the major mode of regulation of gene expression (Mathews et al., 2007a). Initiation is the rate-limiting step of translation and the main target of control. This regulation primarily involves the reversible phosphorylation of key eukaryotic initiation factors (eIFs). Phosphorylation of several initiation factors (eIF2B, eIF3, eIF4B, eIF4E, eIF4G) positively correlates with increased translation rates, whereas phosphorylation of other eIFs (e.g., eIF2α) results in inhibition of translation and suppression of cell growth (Raught and Gingras, 2007). The fundamentals and regulatory mechanisms of eukaryotic protein synthesis have been reviewed in a recent book by Mathews et al. (2007b). The key events in initiation are: (i) formation of the 43S ribosomal preinitiation complex, (ii) binding of the mRNA to the 43S complex, (iii) a start codon selection, (iv) 80S complex formation, and (v) recycling of eIF2 to generate a new ternary complex [eIF2 · Met − tRNA$_i^{Met}$ · GTP] (Fig. 1).

Ribosome recruitment to the mRNA is mediated by the eIF3 and eIF4 group of eIFs and occurs by one of two mechanisms: a cap-dependent

*Corresponding author. Tel.: +1 (514) 398-7274/5;
Fax: +1 (514) 398-1287; E-mail: nahum.sonenberg@mcgill.ca

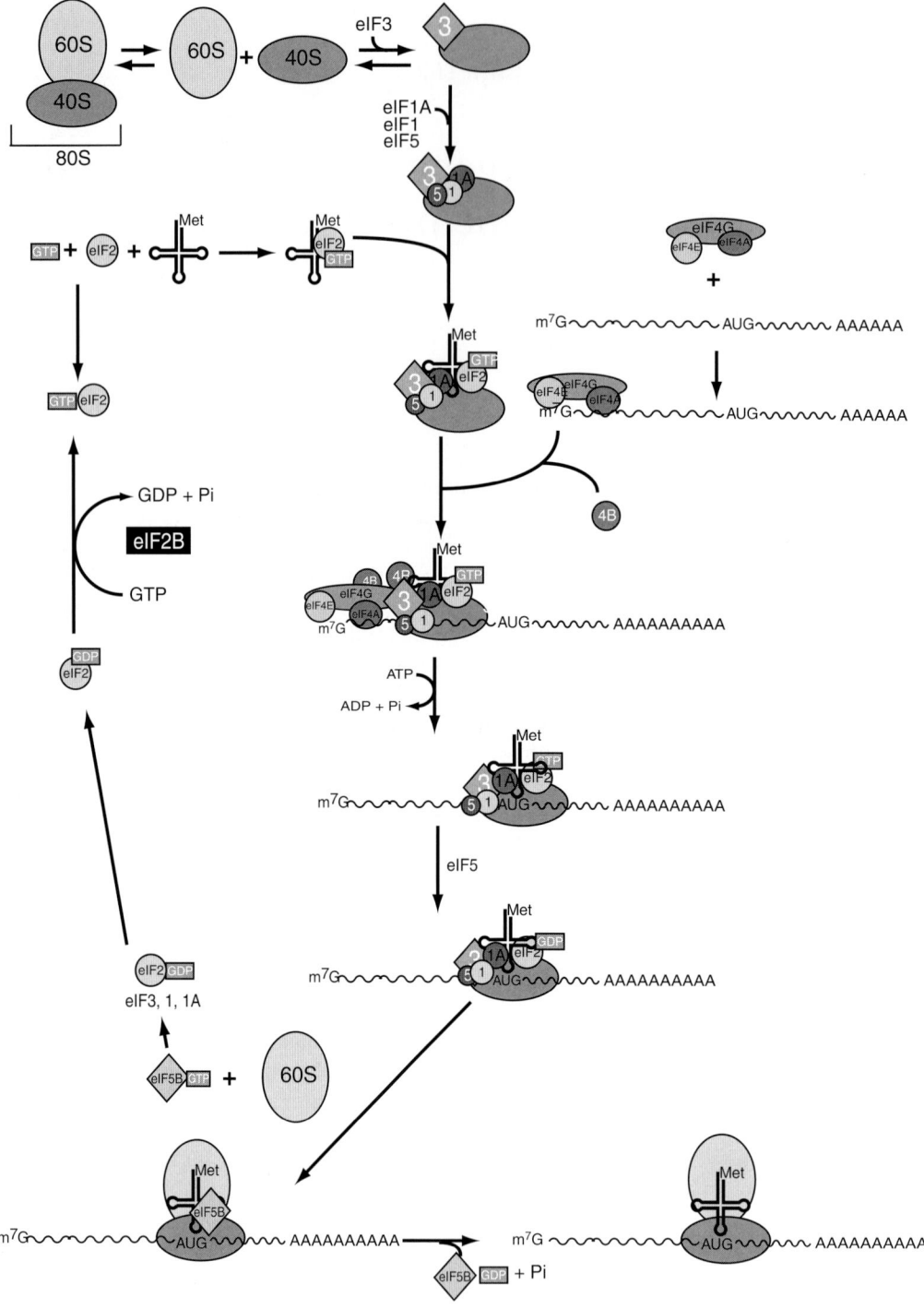

Fig. 1. Schematic representation of the translation initiation pathway in eukaryotes. Eukaryotic translation initiation factors are indicated as color-coded circles. (See Color Plate 5.1 in color plate section.)

(Pestova et al., 2007, p. 887) or a cap-independent mechanism, the latter involving internal recruitment of the ribosome to the mRNA 5′ UTR (Doudna and Sarnow, 2007, p. 864; Elroy-Stein and Merrick, 2007, p. 863). In eukaryotes, initiation on most mRNAs is thought to occur via a cap-dependent process. Indeed, the 5′ cap structure (m7GpppN, where m is a methyl group and N is any nucleotide), which is present at the 5′ end of all nuclear-transcribed eukaryotic mRNAs, is the first mRNA structure that is recognized by eIFs (Fig. 1). The protein complex involved in this step is eIF4F, which consists of: (i) eIF4E, the cap-binding protein responsible for directing eIF4F to the mRNA cap structure; (ii) eIF4A, an RNA helicase required to unwind local secondary structure to facilitate access of the 43S ribosomal complex to the mRNA template; and (iii) eIF4GI or eIF4GII (encoded by two different genes), modular scaffolding proteins that bridge the mRNA to the ribosome through interactions with eIF3, the largest and most complex eIF. The mammalian eIF3 contains 13 subunits that are designated eIF3a to eIF3m (Hinnebusch, 2006). In contrast, the yeast eIF3 version contains orthologs of only five mammalian eIF3 subunits eIF3a, eIF3b, eIF3c, eIF3g, and eIF3j. These five subunits are considered the core eIF3. eIF3 interacts with the 40S ribosomal subunit, thus serving as a link between the mRNA–eIF4F complex and the ribosome (Pestova et al., 2007).

Once bound to the mRNA, the 43S complex is thought to scan the 5′ UTR, supported by ATP hydrolysis, until the appropriate AUG start codon is encountered (Fig. 1). Because eIF4E is the least abundant of all the initiation factors, the mRNA recruitment step is rate-limiting (Duncan et al., 1987).

Regulation of translation initiation

Translational control of protein synthesis is generally achieved by changes in the phosphorylation state of eIFs or their regulators (see below). Two main targets for regulation are (i) the phosphorylation of eIF2α that regulates the exchange of GDP for GTP on eIF2 and (ii) the formation of eIF4F that controls the recruitment of the mRNA to the ribosome. Our review will focus on the translational control by the eIF2α signaling pathway. For a more detailed review on the role of the eIF4F complex and its regulators, eIF4E-binding proteins (4E-BPs) in translational control of long-lasting plasticity and learning and memory, see Chapter 10.

Translation regulation by phosphorylation of eIF2α

Translational control by phosphorylation of eIF2α is one of the best characterized translational mechanisms in eukaryotic cells. eIF2 consists of three subunits: a large γ subunit and two smaller α and β subunits. eIF2 binds both GTP and the Met – $tRNA_i^{Met}$ to form a ternary complex. eIF2 associates with the small ribosomal subunit in its GTP-bound form. Like other GTP-binding proteins, eIF2 alternates between its GTP-bound state and its GDP-bound state. GTP is hydrolyzed when the initiator AUG is engaged by the ribosome to produce eIF2 in the GDP-bound state. Exchange of GDP for GTP on eIF2 is catalyzed by the pentameric guanine nucleotide exchange factor eIF2B, and is required to reconstitute a functional ternary complex for a new round of translation initiation (Hinnebusch et al., 2007; Pestova et al., 2007). eIF2α is critical for the modulation of eIF2's activity. Phosphorylation of the α subunit on Ser51 decreases general translation initiation (Dever, 2002) by blocking the GDP–GTP exchange reaction and reducing the dissociation rate of eIF2 from eIF2B (Fig. 2). Since the cellular levels of eIF2B are much lower than the levels of eIF2, even when only a fraction (about 20–30%) of eIF2α is phosphorylated and complexed with eIF2B, the GTP–GDP exchange process is inhibited. Paradoxically this also results in stimulation of translation of a subset of mRNAs that contain upstream open reading frames (uORFs) (Dever et al., 2007; Hinnebusch et al., 2007; Jackson et al., 2007) (Fig. 2).

The molecular mechanism underlying this selective translation was extensively studied in the general amino acid control response in the yeast *S. cerevisiae* (Hinnebusch et al., 2007). Amino acid starvation of yeast cells increases translation of GCN4 mRNA, a process strongly dependent on the activation of the eIF2α kinase Gcn2p and

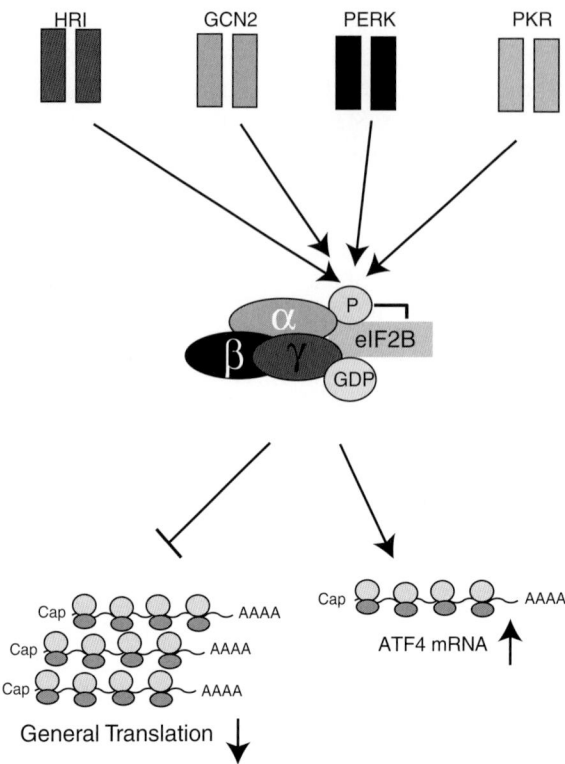

Fig. 2. Schematic representation of the eIF2α signaling pathway. The four eIF2α kinases (GCN2, PERK, PKR, and HRI), whose activity is regulated by different stress signals, phosphorylate Ser51 on the α subunit of eIF2. Phosphoryation of eIF2α leads to inhibition of general translation but it stimulates translation of ATF4 mRNA. (See Color Plate 5.2 in color plate section.)

phosphorylation of eIF2α. The specific induction of GCN4 translation upon eIF2α phosphorylation is mediated by four short open reading frames in the leader of GCN4 mRNA. When amino acids are available, scanning ribosomes translate these short ORFs but dissociate from the mRNA before reaching the authentic GCN4 start codon. In contrast, under starvation conditions, eIF2α phosphorylation by GCN2 inhibits eIF2B, thus causing a fraction of the scanning 40S subunits to form active translational complexes only after they bypassed the upstream ORFs, and allowing initiation at the proper GCN4 start codon (Hinnebusch et al., 2004, 2007).

In mammals, the translation of the Gcn4's metazoan counterpart, ATF4 (CREB2), is enhanced by eIF2α phosphorylation (Harding et al., 2000; Lu et al., 2004; Vattem and Wek, 2004). ATF4 contains two uORFs which are highly conserved across species. Both uORFs contribute in a different manner to ATF4 mRNA translation. The 5′-proximal uORF1, which is shorter than uORF2, is a positive element required to enhance translation of ATF4 mRNA in response to high eIF2α phosphorylation levels. In contrast, uORF2 overlaps the ATF4 ORF, rendering ribosomes that translate it unable to access the authentic ATF4 initiation site (Fig. 3). Similar to GCN4 mRNA translation, lowering the concentration of ternary complex (by phosphorylation of eIF2α) increases the probability that any rescanning 40S subunit will acquire the ternary complex at the ATF4 initiation site (Lu et al., 2004; Vattem and Wek, 2004).

eIF2α kinases

In mammalian cells, there are four known Ser/Thr protein kinases for which eIF2α is the major

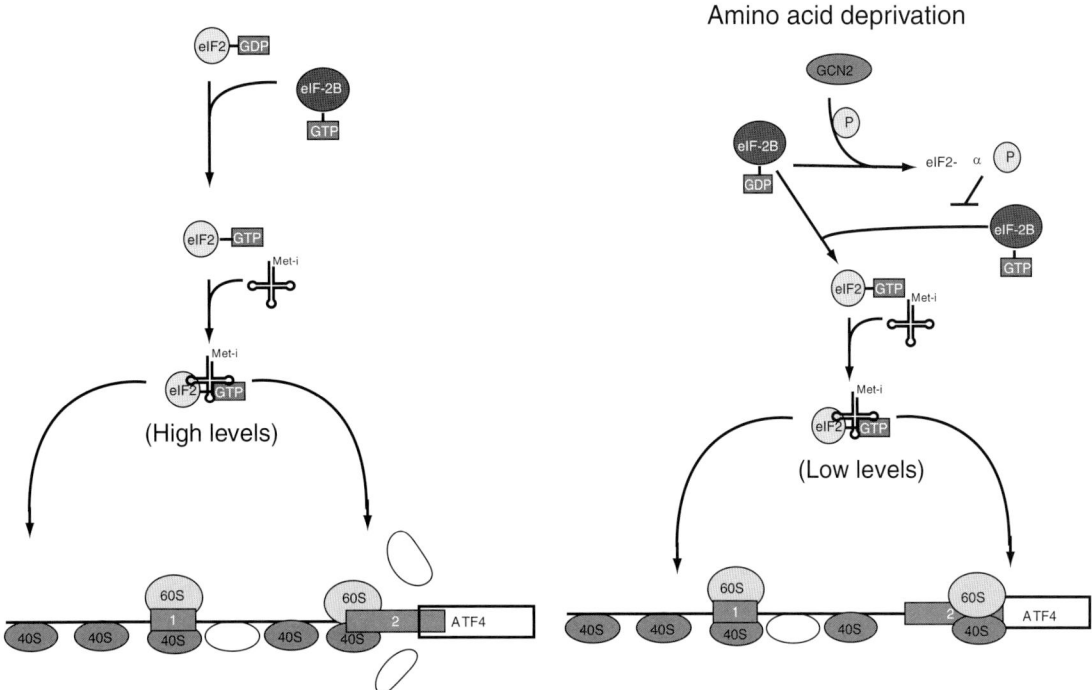

Fig. 3. A model for ATF4 mRNA translation. Schematic diagram of the 5' untranslated region of ATF4 mRNA. The open reading frames (ORFs) are shown as green boxes, and the ATF4 mRNA authentic ORF as an open rectangule. Under normal growing conditions (left panel) the 40S ribosome initiates at ORF1 and reinitiates at ORF2. Under amino acid deprivation conditions (right panel), due to a low concentration of ternary complex, 40S ribosomes conditions failed to reinitiate at ORF2 but reinitiated instead at the authentic ORF. (See Color Plate 5.3 in color plate section.)

substrate. They include the double-stranded (ds) RNA-activated protein kinase (PKR), the hemin-regulated inhibitor kinase (HRI), the pancreatic eIF2α or the PKR-endoplasmic reticulum (ER)-related kinase (PEK/PERK), and the general control non-derepressible kinase (GCN2). The eIF2α kinases share a conserved kinase domain (Dever et al., 2007) and their ability to respond to different stimuli is due to the presence of regulatory domains. For instance, double-stranded RNA (dsRNA), heme deficiency, misfolded proteins in the ER, and amino acid deprivation activate PKR, HRI, PERK, and GCN2, respectively.

With the exception of HRI whose level of expression is very low, all the eIF2α kinases are significantly expressed in the brain (Meurs et al., 1990; Chen et al., 1991; Chong et al., 1992; Crosby et al., 1994; Mellor et al., 1994; Shi et al., 1998; Berlanga et al., 1999; Harding et al., 1999; Sood et al., 2000). We will describe the most salient aspects of GCN2 since it plays a critical role in synaptic transmission, learning, and memory. GCN2 is the ancestral eIF2α kinase and is present in all eukaryotes from yeast to mammals (Hinnebusch et al., 2004; Dever et al., 2007).

GCN2 is activated under conditions of amino acid deprivation via the accumulation of uncharged tRNA. GCN2 has five domains. At the N-terminal, it contains a charged region which binds GCN1 and is required in vivo for activation of the kinase domain. This domain is followed by a pseudokinase domain, the eIF2α kinase domain, a domain related to histydyl-tRNA synthetase (HisRS), which includes a sequence (motif 2) interacting with all the deacylated tRNAs with similar affinity, and a carboxy-terminal domain which enhances tRNA binding, dimerizes, and mediates binding to ribosomes. In contrast to

PKR and PERK, which are monomers and their kinase activation requires dimerization, GCN2 is thought to be a constitutive dimer. In addition, the HisRS interacts with both the kinase domain and the carboxy-terminal domain. These inter-domain interactions are believed to keep the kinase in an inactive state. It is believed that the uncharged tRNAs, which increase in response to amino acid deprivation, bind to HisRS domain and cause the release of these inhibitory inter-domains interacting, thus activating the kinase.

GCN2 is also activated by other stresses such as UV irradiation, high salinity, glucose limitation, and rapamycin (Deng et al., 2002; Narasimhan et al., 2004). Interestingly, these stresses cannot activate a mutant GCN2 which lacks the m2 motifs, indicating that uncharged tRNA must be the main activator of GCN2.

Pharmacologic evidence that translation regulates long-term synaptic plasticity and memory

Short-term and long-term memory

The idea about two memory systems (STM and LTM) has emerged from the study of patients with memory impairments. A classic in the medical literature is the case of a patient, H.M., who suffered from seizures due to a head injury in a bicycle accident when he was 9 years-old (Scoville and Milner, 1957). To relieve his intractable seizures, neurosurgeons performed a bilateral surgical excision of the medial temporal region. As a result of the surgery H.M. exhibited a severe impairment in LTM but STM was intact. Further studies on other species have supported the same distinction between STM and LTM (Scoville and Milner, 1957).

Protein synthesis and behavioral learning

A similar idea about STM and LTM has emerged from the study of protein synthesis inhibitors (puromycin, anysomicin, emetine, acetoxycycloheximide, cycloheximide, and rapamycin) in memory formation. These studies provided the first molecular clues about the distinction between these two processes. The initial studies by Flexner et al. (1963), Agranoff et al. (Agranoff and Klinger, 1964), and Squire and Barondes (Davis and Squire, 1984), which showed that protein synthesis inhibitors block declarative memory, revolutionized the memory and cognition field. Further behavioral studies in different species strengthened the notion that protein synthesis inhibitors block specifically long-term memory (LTM) formation whereas short-term memory (STM) is spared. Interestingly, protein synthesis inhibitors are effective when given immediately before or after training. Indeed, when they are applied 1 h or later after training LTM is not affected (Davis and Squire, 1984).

Studies on the gill-withdrawal reflex of the marine mollusk *Aplysia californica* revealed similar results. A single stimulus to the tail gives rise to a protein synthesis independent, short-lasting sensitization (minutes to hours). In contrast, repetition of such a stimulus elicits LTM sensitization that can last days to week and requires the synthesis of new proteins (Pinsker et al., 1973; Carew et al., 1983; Frost et al., 1985; Castellucci et al., 1989). Taken together these data indicate that protein synthesis during or shortly after training is required for long-term but not STM.

Protein synthesis and long-lasting plasticity

Repeated activity strengthens synaptic connections between brain cells. This process, known as long-term potentiation (LTP), is believed to be a key cellular mechanism by which information is stored (Bliss and Collingridge, 1993; Malenka and Nicoll, 1999; Kandel, 2001). The first evidence that new protein synthesis is required for long-lasting LTP came from in vivo LTP experiments (Krug et al., 1984). Further studies on hippocampal slices in vitro have shown that two distinguished temporal phases of LTP were based on the sensitivity to protein synthesis inhibitors (Kelleher et al., 2004; Klann and Dever, 2004). Like memory, LTP exhibits two temporally distinct phases: early LTP (E-LTP), which depends on modification of pre-existing proteins; and late LTP (L-LTP), which requires transcription and synthesis of new proteins (Silva et al., 1998; Kelleher et al., 2004; Klann and Dever, 2004; Sutton and Schuman, 2006; Costa-Mattioli et al., 2007b). E-LTP is typically induced by a single train of high-frequency (tetanic) stimulation and lasts only 1–2 h. In contrast, L-LTP is generally induced by several repetitions of such stimulations (typically four tetanic trains separated by 5–10 min) and lasts many hours.

Consistent with these data, Montarolo and colleagues (Montarolo et al., 1986) showed that in *Aplysia* sensory neurons protein synthesis inhibitors blocked serotonin-mediated long-term facilitation but not short-term facilitation. In parallel to the behavior and plasticity experiment in vertebrates, the protein synthesis inhibitors were only effective if given around the time of the serotonin application.

At the molecular level, co-activation of pre- and post-synaptic neurons leads to the transcription of plasticity-related genes whose mRNAs are either subsequently translated in the cell body or transported to synapses on the dendrites where they are locally translated. The newly synthesized proteins are somehow captured by specifically "tagged" synapses, those most recently active (Frey and Morris, 1997; Martin et al., 1997). This is inferred from the finding that strongly stimulated synapses can enable weakly stimulated ones to generate L-LTP (reviewed by Morris and Frey, 1999; Martin and Kosik, 2002). Translation is also regulated locally (independently of changes in gene transcription) at stimulated synapses. The local synthesis model is supported by the presence of ribosomes and mRNAs in, or close to, dendritic spines. This process has been extensively reviewed (Steward and Schuman, 2001, 2003; Sutton and Schuman, 2006).

Genetic evidence that translation regulates long-term synaptic plasticity and memory

Identification of GCN2 as regulator of learning and memory and characterization of GCN2 knockout mice

Though we knew that memory consolidation requires new protein synthesis, the molecular mechanisms by which translation controls these processes remained obscure. GCN2 has several interesting features: GCN2-mediated phosphorylation of eIF2α suppresses general translation and selectively stimulates the translation of ATF4 (Dever et al., 2007; Pestova et al., 2007; Ron and Harding, 2007). Interestingly, ATF4 and its homologs are repressors of long-lasting synaptic plasticity and memory formation in diverse phyla (Bartsch et al., 1995; Abel et al., 1998; Kandel, 2001; Chen et al., 2003). More importantly, GCN2 is the major eIF2 kinase in the brain. Its mRNA is enriched in the brain of flies (Santoyo et al., 1997) and mammals (Berlanga et al., 1999; Sood et al., 2000), especially in the hippocampus (Costa-Mattioli et al., 2005). Thus, the well-documented requirement for translation in modulating synaptic activity and memory, together with strong evidence linking eIF2α phosphorylation and translational control to ATF4 activity (Lu et al., 2004; Vattem and Wek, 2004), raised the intriguing possibility of a role for GCN2 in regulating synaptic plasticity, learning, and memory. To explore this possibility, a GCN2 knockout mouse was generated. The GCN2 gene was mutated by deleting the essential exon 12 (Costa-Mattioli et al., 2005). In addition, splicing of exon 11 to exon 13 was predicted to disrupt the reading frame of the mRNA and introduce multiple stop codons which destabilize the mRNA. GCN2 was absent in the hippocampus of GCN2 knockouts as determined by two different antibodies which recognize the amino and carboxy-terminal of the protein. The GCN2 knockout mice are viable, fertile, and develop normally. In the hippocampus of GCN2 knockouts, both eIF2α phosphorylation and the memory repressor ATF4 are reduced (Costa-Mattioli et al., 2005).

We first examined synaptic plasticity at synapses on CA1 pyramidal cells by recording in stratum radiatum extracellular excitatory postsynaptic potentials evoked by electrical stimulation of the Schaffer collateral pathway. Basal synaptic transmission (i.e., input–output curves, fiber volley amplitude, and pair pulse facilitation) was not altered in GCN2 knockout mice. Surprisingly, a protocol that usually elicits an E-LTP, which is independent of gene expression (translation and transcription), elicited a typical gene expression-dependent L-LTP in hippocampal slices from GCN2 knockout mice (Costa-Mattioli et al., 2005). By three independent criteria, the LTP generated in GCN2 knockout mice is indistinguishable from the normal L-LTP induced by four trains of stimulation: it is (i) dependent on gene expression, (ii) dependent on PKA, and (iii) immune to depotentiation. The ability to convert a stimulus that normally leads to a short-term change to a long-term change in plasticity has been seen in numerous genetic manipulations in several species (Bartsch et al., 1995; Yin et al., 1995; Malleret et al., 2001; Barco et al., 2002; Genoux et al., 2002; Chen et al., 2003), and is invariably due to a decrease in the threshold for activating gene expression. For instance, mice with enhanced CREB activity (Barco et al., 2002) or those expressing a dominant negative for ATF4 (Chen et al., 2003) are remarkable examples of this principle: as in the GCN2 knockout mice, a single train of high-frequency stimulation was sufficient to elicit a sustained L-LTP in these mice. In this regard, it is noteworthy that in the basal state, GCN2 knockout mice exhibit decreased ATF4 levels. Thus, a plausible interpretation is that the facilitated LTP elicited in GCN2 knockout mice is associated with a lower threshold for activation of gene expression. Therefore, the effect of GCN2 on long-lasting changes in plasticity could be mediated through modulating ATF4/CREB activity. However, to our surprise, a L-LTP inducing protocol, such as four trains at 100 Hz or forskokin, elicited an impaired L-LTP in the GCN2 knockout hippocampal slices.

Deleting GCN2 affects LTP but not long-term depression (LTD), another well-characterized form of synaptic plasticity, induced by either low-frequency stimulation (LFS) or incubation with DHPG, an agonist of group I mGluRs (Costa-Mattioli et al., 2005). Therefore, it is possible that GCN2 does not play any role in protein synthesis-dependent LTD or the signaling pathway which regulates GCN2 activity is not altered by LTD-inducing protocols.

Is the altered synaptic efficacy in the hippocampal neurons of GCN2 knockout mice manifested at the behavioral level? To answer this question, mice were first subjected to Pavlovian fear conditioning. Pairing tone presentations with a foot shock in a particular environment leads to both auditory and contextual fear conditioning. Contextual fear conditioning is a hippocampus-dependent process in which LTM for the context is established following a single training session. Auditory fear conditioning on the other hand, which associates a tone (CS) with the foot shock (US), is dependent on the amygdala but not the hippocampus (Fanselow and LeDoux, 1999; LeDoux, 2000). Both contextual and cue fear conditioning are dependent on new protein synthesis (Bourtchouladze et al., 1998; Schafe et al., 1999).

GCN2 knockout mice exhibited reduced "freezing" when tested after a retention delay of 24 h and 10 days after training, indicating that hippocampus-dependent fear memories are impaired in these mice. By contrast, when GCN2 knockout mice were evaluated in auditory fear conditioning, they exhibited normal associative memory for the tone. These data indicate that the lack of GCN2 selectively affects hippocampal-dependent memories but not amygdala-dependent auditory fear conditioning.

To further assess the role of GCN2 in hippocampal-dependent memories, wild-type (WT) and GCN2 knockout mice were subjected to the hidden-platform version of the Morris water maze, a hippocampus-dependent reference memory task, in which a mouse must find a platform (using visual spatial clues) hidden under opaque water in a pool (Morris et al., 1982). Using a conventional training protocol (three trials per day), spatial learning of GCN2 knockout mice was impaired, as determined by escape latencies and quadrant occupancy (Costa-Mattioli et al., 2005). Since a short-lasting LTP protocol induced a robust gene-expression-dependent LTP in GCN2 knockout, mice were trained in a weak training protocol (once per day). Remarkably, when the GCN2 knockout mice were given a weak training, they exhibited enhanced spatial learning and memory (Costa-Mattioli et al., 2005). In parallel to the electrophysiological findings in hippocampal slices of GCN2 knockout mice, where weak stimulation elicited L-LTP instead of E-LTP, but strong stimulation failed to evoke the expected L-LTP, a weak training protocol induced an enhanced memory but a stronger protocol impaired memory consolidation in these mice. These data provide the first genetic evidence that translational control is critical for long-lasting synaptic plasticity and memory consolidation.

GCN2 in the brain regulates selection of balanced diet

Omnivorous animals such as rats reject diets lacking essential amino acids. Selection of such a balanced diet plays an important role in human evolution. It has been reported that neurons of the apical periform cortex, which project to appropriate feeding neuronal circuits, are activated by intracellular indispensable amino acids (Haberly and Price, 1978). The apical periform cortex appears to be critical for such an adversive response, because bilateral lesion of this region abolishes the bias against diets lacking essential amino acids (Gietzen, 1993). Interestingly, consumption of an imbalanced diet increases the phosphorylation of eIF2α in neurons of the apical periform cortex (Gietzen et al., 2004). Recently, two independent groups showed that GCN2 is responsible for this basic mechanism of functional stress (Hao et al., 2005; Maurin et al., 2005). Both groups showed that WT mice reject diet lacking threonine and leucine whereas GCN2 knockout mice consumed equal amount of balanced and imbalanced diet.

Consistent with these data, injection of amino alcohol threoninol, an inhibitor of a single tRNA synthetase into the apical periform cortex caused a decrease in food intake of a balanced diet (Hao et al., 2005). These data indicated that uncharged tRNA, which as amino acid deprivation activates GCN2, is the signal that triggers the feeding response. Thus, GCN2 appears to be responsible for the increased phosphorylation of eIF2α in neurons of the apical periform cortex. An increase in eIF2α phosphorylation was observed in brain section from WT mice feed with an imbalanced diet whereas the same diet did not alter eIF2α phosphorylation levels in GCN2 knockout brain slices (Hao et al., 2005; Maurin et al., 2005). It will be interesting to learn whether in WT mice either fed with a diet lacking amino acids or injected with amino alcohol threoninol in WT mice, the activity of GCN2 (phosphorylation) also is up-regulated in the apical periform cortex. Indeed, consumption of a diet lacking indispensable amino acids led to an eIF2α phosphorylation in the liver (Anthony et al., 2004; Maurin et al., 2005).

Taken together, these data indicate that GCN2 senses the imbalance in amino acids and activates down-stream signaling pathways which lead to the behavioral rejection of a diet lacking amino acids.

Important question remains with respect to the role of GCN2-mediated eIF2α in sensing amino acid deficiency in mammalian periform cortex. Is ATF4 the downstream target responsible for this or the decrease in translation associated with an increased eIF2α phosphorylation? Does local activation of the other eIF2α kinases, such as PERK or PKR, in the apical periform cortex, recapitulates the phenotype observed in WT fed with a diet lacking amino acids? Finally, it will be interesting to investigate whether other protein kinases also contribute to this feeding behavior.

A master switch for the conversion from short-term to long-term synaptic plasticity and memory formation

Consolidation of long-term memories requires the expression of new genes (Squire, 1987). Thus, if new gene expression is the rate-limiting step necessary to strengthen existing synaptic connections between neurons, how is this process turned on? If one could find such a mechanism and switch it on, then stimulation that normally elicits only E-LTP and STM should lead to L-LTP and LTM. This was the goal of our research. In diverse phyla, a key component in memory formation is the transcription factor CREB (cAMP responsive element-binding protein). CREB is regulated by phosphorylation of its serine 133, and is also under the control of the repressor protein ATF4 (CREB-2) (Bartsch et al., 1995; Silva et al., 1998; Kandel, 2001; Chen et al., 2003), which in turn is regulated at the level of mRNA translation. As described above, phosphorylation of eIF2α

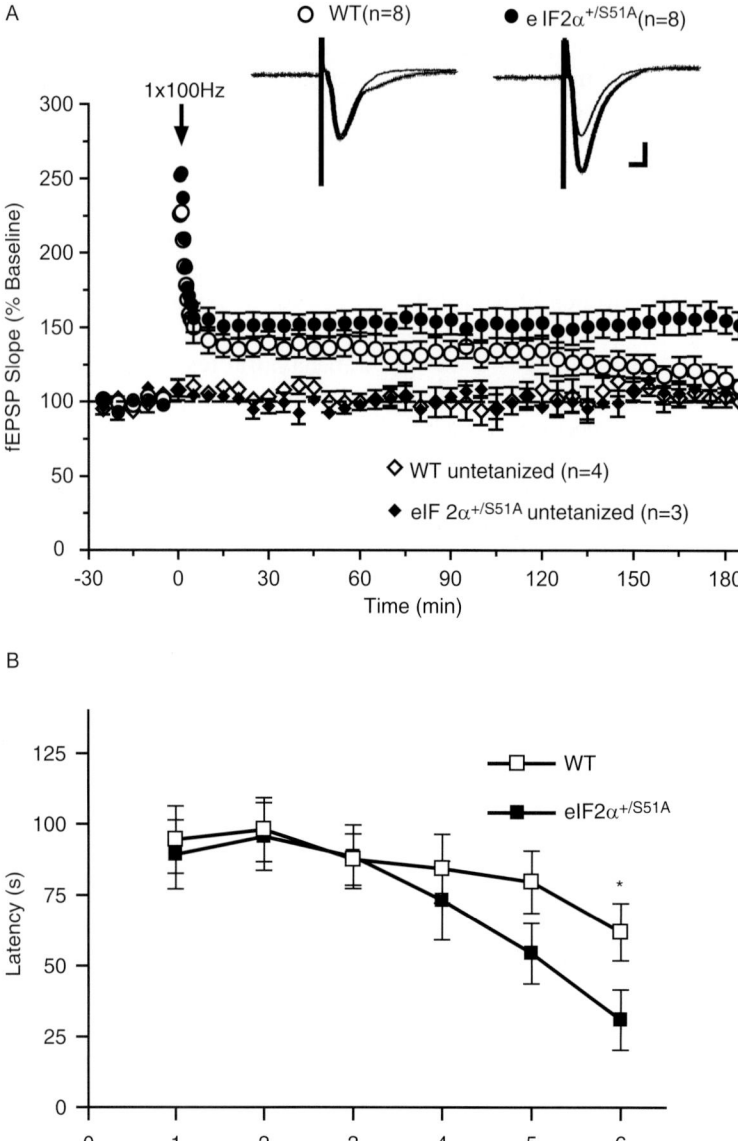

Fig. 4. Decreased eIF2α phosphorylation facilitates L-LTP and long-term spatial memory consolidation. (A) A single train of high-frequency stimulation elicits an enhanced LTP in hippocampal slices from eIF2α$^{+/S51A}$ mice. (B) eIF2α$^{+/S51A}$ mice exhibited and enhanced long-term spatial memory with a weak training protocol in the Morris Water Maze.

suppresses general translation, and selectively stimulates the translation of ATF4 (Dever et al., 2007; Ron and Harding, 2007). Thus, eIF2α phosphorylation regulates two fundamental processes essential for the conversion from short-term to long-term synaptic plasticity and memory: (a) de novo protein synthesis and (b) CREB-mediated gene expression via ATF4. According to recent evidence, eIF2α phosphorylation is tightly correlated with neuronal activity. We and others have shown that L-LTP-inducing protocols decrease eIF2α phosphorylation (Takei et al., 2001; Costa-Mattioli et al., 2005). In keeping with a role in memory formation, eIF2α phosphorylation is reduced when rats are trained in

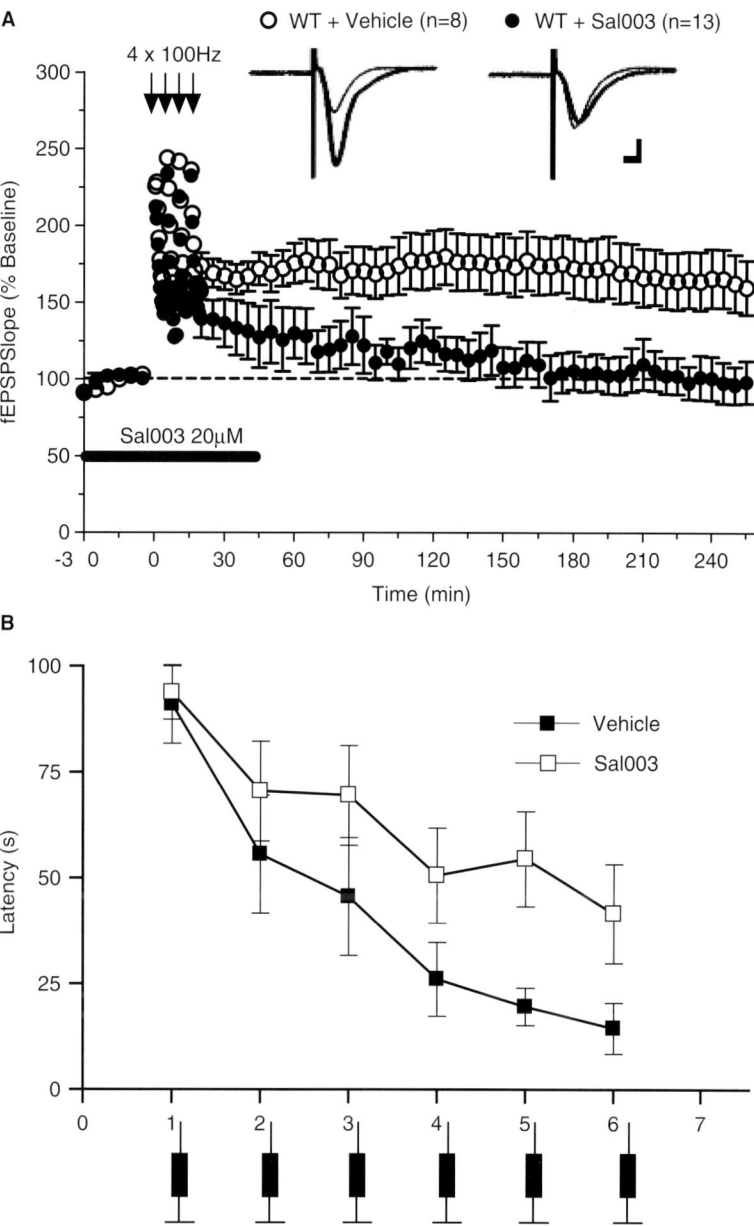

Fig. 5. Increased eIF2α phosphorylation impairs L-LTP and spatial memory consolidation. (A) The induction of L-LTP is blocked by Sal003, an inhibitor of the eIF2α phosphatases. (B) Infusion of Sal003 into the hippocampus immediately after training blocks memory consolidation. Dark syringes refer to either vehicle or Sal003 infusions across groups.

a Pavlovian-fear conditioning task (Costa-Mattioli et al., 2007b). In addition, in mice lacking GCN2 (the main eIF2α kinase in the mammalian brain) eIF2α levels are reduced and synaptic plasticity and memory are altered (Costa-Mattioli et al., 2005). We therefore predicted that reduced eIF2α phosphorylation would facilitate gene expression, L-LTP induction, and LTM storage.

We investigated long-lasting synaptic plasticity and memory in eIF2α heterozygous mutant mice (eIF2α$^{+/S51A}$) in which eIF2α phosphorylation and ATF4 levels are decreased. As expected, LTP was elicited more readily in hippocampal slices from these mice: stimulation that normally induces a short-lasting E-LTP in WT mice elicited a sustained, gene expression-dependent L-LTP in hippocampal slices from eIF2α$^{+/S51A}$ mice (Fig. 4A) (Costa-Mattioli et al., 2007a). Furthermore, mice lacking GCN2, in which eIF2α phosphorylation is reduced (Costa-Mattioli et al., 2005), and those expressing an inducible inhibitor of ATF4 (Chen et al., 2003) had similar phenotypes. In agreement with the enhanced LTP in hippocampal slices, eIF2α$^{+/S51A}$ mice have an enhanced memory and lower threshold for learning in several behavioral tasks, such as the Morris water maze (Fig. 4B), associative fear conditioning, and conditioned taste aversion. These data strongly support the notion that reduced phosphorylation of eIF2α facilitates the expression of genes required for long-lasting synaptic plasticity and memory consolidation.

We reasoned that, if dephosphorylation of eIF2α is essential for memory consolidation, preventing the decrease in eIF2α phosphorylation that normally occurs during memory formation should inhibit not only gene expression but also L-LTP and LTM. To test this prediction, we applied Sal003, a compound which suppresses eIF2α dephosphorylation (Boyce et al., 2005). As expected, the Sal003-mediated increase in eIF2α phosphorylation resulted in inhibition of general translation and selective increase in translation of ATF4 mRNA. Furthermore repeated tetanic stimulation induced only short-lasting LTP in Sal003-treated hippocampal slices from WT mice (Fig. 5A) (Costa-Mattioli et al., 2007a). In further electrophysiological tests, L-LTP elicited in slices from ATF4 knockouts was resistant to Sal003, confirming that the increase in ATF4 levels mediates the inhibitory action of Sal003. Moreover, locally injected Sal003 increased eIF2α phosphorylation in the hippocampus of WT mice and impaired their learning and memory in the water maze (Fig. 5B) and during contextual fear conditioning (Costa-Mattioli et al., 2007a).

The ability to enhance memory formation by decreasing the levels of repressors of gene expression appears to be a widely conserved mechanism, from sea slugs to mammals. After injecting antibodies against the *Aplysia* homolog of ATF4, ApCREB2, into sensory neurons, a single pulse of 5-HT, which normally induces only short-term facilitation (lasting minutes), is sufficient to evoke a gene expression-dependent long-term facilitation that lasts beyond one day (Bartsch et al., 1995).

Our findings thus reveal a crucial step in mnemonic processing: the phosphorylation of a single site on eIF2α determines whether a STM process is transformed into a long-lasting one, through modulation of gene expression. A better understanding of the molecular basis of memory should lead to improved therapy of memory loss, whether associated with aging or, more devastatingly, with Alzheimer-type dementia.

Summary

Significant advances in studies of translational control of synaptic plasticity and memory formation have emerged in the last few years. GCN2-mediated phosphorylation of eIF2α and signaling downstream is an ancient signaling pathway, which is critical for the regulation of various biological processes. Recent well-integrated multidisciplinary approaches (molecular biology, genetics, electrophysiology, and behavior) have revealed the crucial role of eIF2α phosphorylation in synaptic plasticity and memory, thus providing new insights into the molecular mechanisms underlying synaptic plasticity and memory formation.

Acknowledgments

We thank Kresimir Krnjevic for comments on the manuscript. This work was supported by a Team Grant from the Canadian Institute of Health Research (CIHR) to M.C.-M and N.S. and a Howard Hughes Medical Institute (HHMI) grant to N.S.

References

Abel, T., Martin, K.C., Bartsch, D. and Kandel, E.R. (1998) Memory suppressor genes: inhibitory constraints on the storage of long-term memory. Science, 279: 338–341.

Agranoff, B.W. and Klinger, P.D. (1964) Puromycin effect on memory fixation in the goldfish. Science, 146: 952–953.

Anthony, T.G., McDaniel, B.J., Byerley, R.L., McGrath, B.C., Cavener, D.R., McNurlan, M.A. and Wek, R.C. (2004) Preservation of liver protein synthesis during dietary leucine deprivation occurs at the expense of skeletal muscle mass in mice deleted for eIF2 kinase GCN2. J. Biol. Chem., 279: 36553–36561.

Barco, A., Alarcon, J.M. and Kandel, E.R. (2002) Expression of constitutively active CREB protein facilitates the late phase of long-term potentiation by enhancing synaptic capture. Cell, 108: 689–703.

Bartsch, D., Ghirardi, M., Skehel, P.A., Karl, K.A., Herder, S.P., Chen, M., Bailey, C.H. and Kandel, E.R. (1995) *Aplysia* CREB2 represses long-term facilitation: relief of repression converts transient facilitation into long-term functional and structural change. Cell, 83: 979–992.

Berlanga, J.J., Santoyo, J. and De Haro, C. (1999) Characterization of a mammalian homolog of the GCN2 eukaryotic initiation factor 2alpha kinase. Eur. J. Biochem., 265: 754–762.

Bliss, T.V. and Collingridge, G.L. (1993) A synaptic model of memory: long-term potentiation in the hippocampus. Nature, 361: 31–39.

Bourtchouladze, R., Abel, T., Berman, N., Gordon, R., Lapidus, K. and Kandel, E.R. (1998) Different training procedures recruit either one or two critical periods for contextual memory consolidation, each of which requires protein synthesis and PKA. Learn. Mem., 5: 365–374.

Boyce, M., Bryant, K.F., Jousse, C., Long, K., Harding, H.P., Scheuner, D., Kaufman, R.J., Ma, D., Coen, D.M., Ron, D. and Yuan, J. (2005) A selective inhibitor of eIF2alpha dephosphorylation protects cells from ER stress. Science, 307: 935–939.

Carew, T.J., Hawkins, R.D. and Kandel, E.R. (1983) Differential classical conditioning of a defensive withdrawal reflex in *Aplysia californica*. Science, 219: 397–400.

Castellucci, V.F., Blumenfeld, H., Goelet, P. and Kandel, E.R. (1989) Inhibitor of protein synthesis blocks long-term behavioral sensitization in the isolated gill-withdrawal reflex of *Aplysia*. J. Neurobiol., 20: 1–9.

Chen, A., Muzzio, I.A., Malleret, G., Bartsch, D., Verbitsky, M., Pavlidis, P., Yonan, A.L., Vronskaya, S., Grody, M.B. Cepeda, I., et al. (2003) Inducible enhancement of memory storage and synaptic plasticity in transgenic mice expressing an inhibitor of ATF4 (CREB-2) and C/EBP proteins. Neuron, 39: 655–669.

Chen, J.J., Throop, M.S., Gehrke, L., Kuo, I., Pal, J.K., Brodsky, M. and London, I.M. (1991) Cloning of the cDNA of the heme-regulated eukaryotic initiation factor 2 alpha (eIF-2 alpha) kinase of rabbit reticulocytes: homology to yeast GCN2 protein kinase and human double-stranded-RNA-dependent eIF-2 alpha kinase. Proc. Natl. Acad. Sci. U.S.A., 88: 7729–7733.

Chong, K.L., Feng, L., Schappert, K., Meurs, E., Donahue, T.F., Friesen, J.D., Hovanessian, A.G. and Williams, B.R. (1992) Human p68 kinase exhibits growth suppression in yeast and homology to the translational regulator GCN2. EMBO J., 11: 1553–1562.

Costa-Mattioli, M., Gobert, D., Harding, H., Herdy, B., Azzi, M., Bruno, M., Bidinosti, M., Ben Mamou, C., Marcinkiewicz, E., Yoshida, M., et al. (2005) Translational control of hippocampal synaptic plasticity and memory by the eIF2alpha kinase GCN2. Nature, 436: 1166–1173.

Costa-Mattioli, M., Gobert, D., Stern, E., Gamache, K., Colina, R., Cuello, C., Sossin, W., Kaufman, R., Pelletier, J., Rosenblum, K., et al. (2007a) eIF2alpha phosphorylation bidirectionally regulates the switch from short- to long-term synaptic plasticity and memory. Cell, 129: 195–206.

Costa-Mattioli, M., Sonenberg, N. and Klann, E. (2007b) Translational control mechanism of synaptic plasticity and memory. In: Sweatt J.D. (Ed.), Learning and Memory: A Comprehensive Reference. Elsevier. (in press).

Crosby, J.S., Lee, K., London, I.M. and Chen, J.J. (1994) Erythroid expression of the heme-regulated eIF-2 alpha kinase. Mol. Cell. Biol., 14: 3906–3914.

Davis, H.P. and Squire, L.R. (1984) Protein synthesis and memory: a review. Psychol. Bull., 96: 518–559.

Deng, J., Harding, H.P., Raught, B., Gingras, A.C., Berlanga, J.J., Scheuner, D., Kaufman, R.J., Ron, D. and Sonenberg, N. (2002) Activation of GCN2 in UV-irradiated cells inhibits translation. Curr. Biol., 12: 1279–1286.

Dever, T.E. (2002) Gene-specific regulation by general translation factors. Cell, 108: 545–556.

Dever, T.E., Dar, A.C. and Sicheri, F. (2007) The eIF2alpha kinases. In: Mathews M.B., Sonenberg N. and Hershey J.W.B. (Eds.), Translational Control in Biology and Medicine. Cold Spring Harbor Laboratory Press, Cold Spring Harbor, NY, pp. 319–345.

Doudna, J.A. and Sarnow, P. (2007) Translation initiation by viral internal ribosome entry sites. In: Mathews M.B., Sonenberg N. and Hershey J.W.B. (Eds.), Translational Control in Biology and Medicine. Cold Spring Harbor Laboratory Press, Cold Spring Harbor, NY, pp. 129–154.

Duncan, R., Milburn, S.C. and Hershey, J.W. (1987) Regulated phosphorylation and low abundance of HeLa cell initiation factor eIF-4F suggest a role in translational control. Heat shock effects on eIF-4F. J. Biol. Chem., 262: 380–388.

Elroy-Stein, O. and Merrick, W.C. (2007) Translation initiation via cellular internal ribosome entry sites. In: Mathews M.B., Sonenberg N. and Hershey J.W.B. (Eds.), Translational Control in Biology and Medicine. Cold Spring Harbor Laboratory Press, Cold Spring Harbor, NY, pp. 155–172.

Fanselow, M.S. and LeDoux, J.E. (1999) Why we think plasticity underlying Pavlovian fear conditioning occurs in the basolateral amygdala. Neuron, 23: 229–232.

Flexner, J.B., Flexner, L.B. and Stellar, E. (1963) Memory in mice as affected by intracerebral puromycin. Science, 141: 57–59.

CHAPTER 6

Regulation of hippocampus-dependent memory by cyclic AMP-dependent protein kinase

Ted Abel[1] and Peter V. Nguyen[2],*

[1]University of Pennsylvania, Department of Biology, Biological Basis of Behavior Program, Philadelphia, PA 19104, USA
[2]University of Alberta School of Medicine, Departments of Physiology and Psychiatry, Centre for Neuroscience, Edmonton, AB T6G 2H7, Canada

Abstract: The hippocampus is crucial for the consolidation of new declarative long-term memories. Genetic and behavioral experimentation have revealed that several protein kinases are critical for the formation of hippocampus-dependent long-term memories. Cyclic-AMP dependent protein kinase (PKA) is a serine–threonine kinase that has been strongly implicated in the expression of specific forms of hippocampus-dependent memory. We review evidence that PKA is required for hippocampus-dependent memory in mammals, and we highlight some of the proteins that have been implicated as targets of PKA. Future directions and open questions regarding the role of PKA in memory storage are also described.

Keywords: memory; learning; cyclic AMP-dependent protein kinase; hippocampus; synaptic plasticity; long-term potentiation

Introduction

Protein kinases modulate a plethora of important processes, including synaptic plasticity, learning, and memory. Multiple chemical neurotransmitters, hormones, and other signaling substances use cyclic adenosine 3′,5′ monophosphate (cAMP) as an intracellular second messenger. The principal target for cAMP in mammalian cells is cAMP-dependent protein kinase (PKA), which is ubiquitously expressed and mediates intracellular signal transduction. Pioneering work by Earl Sutherland identified cAMP as the first intracellular second messenger (Sutherland and Rall, 1957; Sutherland et al., 1965). Subsequently, Edwin Krebs, Paul Greengard, and their colleagues purified PKA from rabbit skeletal muscle (Walsh et al., 1968; Reimann et al., 1971), and they showed that PKA activity was stimulated by cAMP (Miyamoto et al., 1968, 1969; Walsh et al., 1968; Beavo et al., 1974). Other advances, made possible by genetic, molecular, and cell biological techniques, have shed light on the molecular characteristics, dynamics, and functional plurality of the PKA holoenzyme (reviewed by McKnight et al., 1988; Beebe, 1994; Skalhegg and Tasken, 2000). It is now well established that PKA regulates many biological processes through its phosphorylation of proteins. Also, phosphatases such as protein phosphatase 1 and the calcium-regulated phosphatase, calcineurin, can dephosphorylate proteins that had been phosphorylated by PKA, thus allowing PKA signaling events to be reversible (Fig. 1).

*Corresponding author. Tel.: +1 780 492 8163;
Fax: +1 780 492 8915; E-mail: Peter.Nguyen@ualberta.ca

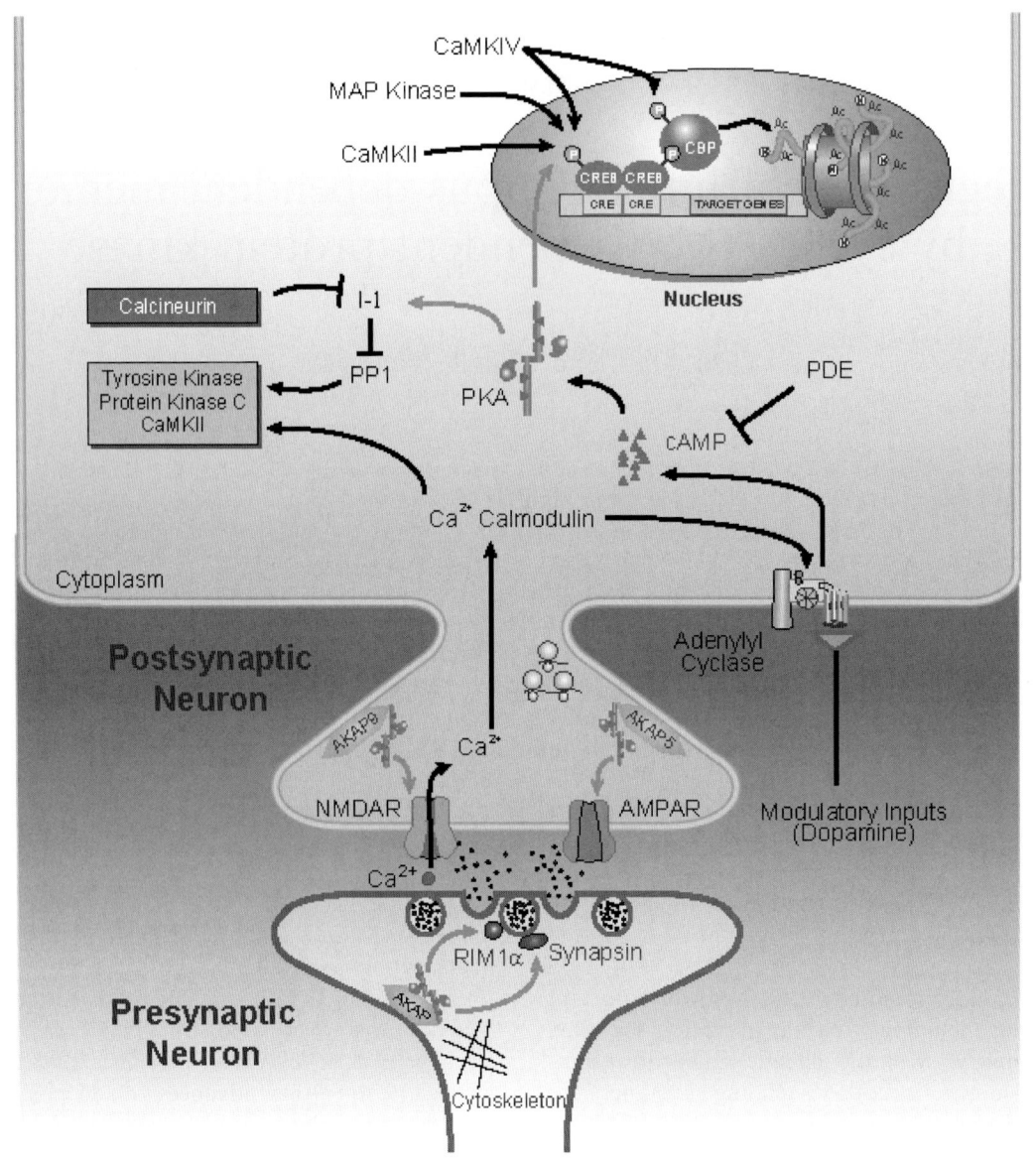

Fig. 1. The cAMP/PKA signaling pathway critically regulates molecular components underlying long-term potentiation and long-term memory. In the postsynaptic neuron, PKA targets include NMDA and AMPA receptors, inhibitor-1 (I-1) and CREB. In the presynaptic terminal, PKA targets include RIM1α and synapsin. (See Color Plate 6.1 in color plate section.)

A fundamental process that is modulated by PKA is synaptic transmission, which can be modified by the electrical activity of a neuron — a process termed "activity-dependent synaptic plasticity" (reviewed by Nguyen and Woo, 2003). Because synaptic plasticity involves long-lasting modifications of intercellular signaling in the nervous system, it plays significant roles in regulating learning and memory (Castellucci et al., 1970; Kupfermann et al., 1970; McKernan and Shinnick-Gallagher, 1997; Abraham et al., 2002; Whitlock et al., 2006). Not surprisingly, certain types of long-lasting synaptic plasticity, such as long-term potentiation (LTP) and long-term depression (LTD), are believed to underlie some forms of learning and memory in the mammalian brain (Bliss and Lomo, 1973; Dudek and Bear, 1992; Martin et al., 2000; Abraham et al., 2002; Whitlock et al., 2006).

Cyclic AMP/PKA signaling has been shown to be pivotal for specific types of long-term synaptic plasticity and for long-term memory (Fig. 1), as demonstrated in the landmark studies of learning and memory in *Aplysia* and *Drosophila* (reviewed by Burrell and Sahley, 2001). In particular, electrophysiological and behavioral studies of the marine snail, *Aplysia californica*, have revealed a requirement for cAMP/PKA signaling in the establishment of short- and long-lasting forms of synaptic plasticity, learning, and memory (Kandel, 2001). These studies demonstrated an important role for PKA in mediating short-term synaptic facilitation (by reversible phosphorylation of ion channels), and long-term facilitation and long-term memory for behavioral sensitization (by modulation of gene expression and protein synthesis) (Kandel, 2001). Studies in *Drosophila* have also been instrumental in defining the role of PKA in learning and memory (reviewed by McGuire et al., 2005; Davis, 1996). The learning mutant, *dunce*, was the first learning mutant isolated in *Drosophila*, and *dunce* encodes a cAMP phosphodiesterase. Similarly, mutations in adenylyl cyclase (*rutabaga*) and mutations in PKA itself impair learning and memory in *Drosophila*. Overall, these landmark experiments have laid much of the conceptual foundations for subsequent research on the roles of PKA in learning and memory in the mammalian brain. In this article, the roles of PKA in hippocampus-dependent memory are reviewed. We focus only on the mammalian hippocampus, because of its pivotal roles in the consolidation of spatial and non-spatial long-term memories. We discuss studies that have shown a requirement for PKA signaling in the hippocampus. We then consider some of the downstream targets of PKA, emphasizing studies that have identified requirements for PKA-mediated phosphorylation and signaling in hippocampal LTP, a form of synaptic plasticity that has been linked to memory formation (Abel et al., 1997; Gruart et al., 2006; Whitlock et al., 2006).

cAMP-dependent protein kinase (PKA)

The mammalian PKA family consists of four regulatory (R) subunits (RIα, RIβ, RIIα, RIIβ) and three catalytic (C) subunits (Cα, Cβ, Cγ). Each subunit is encoded by a unique gene (McKnight et al., 1988; Doskeland et al., 1993) and they are all expressed in the mammalian brain (Cadd and McKnight, 1989). Two major isozymes of PKA, termed type I (with RIα and RIβ dimers) and type II (with RIIα and RIIβ dimers), have been characterized and were initially identified based on their patterns of chromatographic elution (Tasken et al., 1993; Francis and Corbin, 1999; Skalhegg and Tasken, 2000). In the absence of cAMP, PKA is an inactive tetrameric holoenzyme composed of two R subunits bound to two C subunits. Each regulatory subunit contains two tandem cAMP binding sites, a high-affinity site and a low-affinity site (Taylor et al., 1990). Sequential and cooperative binding of cAMP to these two sites releases monomeric C subunits (Taylor et al., 1990; Gibbs et al., 1992). These dissociated C subunits can then phosphorylate serine and threonine residues on numerous proteins. In essence, the regulatory subunits inhibit the phosphotransferase activities of the catalytic subunits.

Properties of memory

Learning, the change in behavior as a result of experience, and memory, the retention of changes in behavior that result from learning, have been of great interest to neurobiologists. The psychological study of memory began over 100 years ago with the experiments of Herman Ebbinghaus; since then, researchers have sought to define the principles underlying memory storage. This work has defined three major properties of memory storage (Milner et al., 1998). First, the study of patients like H. M. revealed that there are multiple memory systems that function in distinct but overlapping ways. Thus, declarative or explicit memory, the conscious memory of facts and events, is mediated by the hippocampal memory system, whereas non-declarative or implicit memory, involving unconscious learning of emotional tasks or skills and habits, is mediated by other brain systems such as the amygdala and striatum . Secondly, memory consists of specific temporal phases, including short-term and long-term memory. As we will see in our discussion here of the cAMP/PKA signaling pathway, these phases of memory can be mediated by distinct molecular pathways. Thirdly, memory consists of specific stages. Memory is acquired during training, consolidated in the period following training into a more stable form, and then retrieved when needed. Following retrieval, memory can be modified by processes of reconsolidation and extinction. In this review, we will focus on the role of the cAMP/PKA signaling pathway in hippocampus-dependent forms of memory in rodents.

The role of the cAMP/PKA system in spatial memory

In rodents, the quintessential hippocampus-dependent behavioral task is spatial learning and memory in the hidden platform version of the Morris water maze. In this task, animals learn to use the distal cues located in the room to find the location of a submerged platform during repeated trials. Performance is measured by an increase in latency to find the platform during training trials and memory is tested in a probe (or transfer) test in which the platform is removed from the pool and animals swim freely. During this probe test, the spatial preference of the animals is measured by a tracking device and spatial learning is reflected in a spatial bias to this search pattern: well-trained animals spend most of their time swimming in the vicinity of the platform location. This form of spatial learning depends on the dorsal hippocampus and it is NMDA receptor-dependent (Morris et al., 1982, 1986; Tsien et al., 1996). Importantly, the water maze can be configured in many different ways: with a visible platform that is shifted from trial to trial to test for procedural memory, or with a hidden platform that is shifted from day to day to test for working memory. As typically configured with a positionally fixed hidden platform, the water maze tests spatial reference memory. In addition to the water maze, spatial memory can also be tested in the radial arm maze and in the Y maze.

What is the role of the cAMP/PKA signaling pathway in spatial memory? This issue has been difficult to address genetically because of redundancy in the genes encoding various components of the cAMP/PKA signaling pathway, particularly PKA itself. In addition, the large number of training trials in the water maze makes region-specific pharmacological manipulation difficult. Initial work in the mammalian hippocampus defined a role for cAMP and PKA in synaptic plasticity (reviewed in Nguyen and Woo, 2003). The first demonstration of a role for cAMP in spatial memory came in 1995 when Daniel Storm and colleagues examined the effects of mutations in the gene encoding type I adenylyl cyclase on behavior (Wu et al., 1995). They found that mutant mice lacking type I adenylyl cyclase had reductions in the levels of calcium-stimulated cAMP synthesis and exhibited spatial memory deficits in the hidden platform version of the water maze (Wu et al., 1995). Studies of similar conventional null mutations in the genes encoding specific regulatory and catalytic subunits of PKA were difficult to interpret because of compensatory changes in the expression of genes encoding other PKA subunits in these mutant mice (Brandon et al., 1997). Also, the behavioral phenotypes observed in mutant mice is sensitive to genetic background (Howe et al., 2002). To define the role of PKA in learning and memory in rodents, Abel et al. (1997) created R(AB) transgenic mice expressing a dominant negative form of the regulatory subunit of PKA. By using a transgenic approach, PKA can be inhibited in specific cell types at particular times, thereby reducing potential compensatory effects. Furthermore, the R(AB) transgene will interfere broadly with type I PKA within the cell types in which it is expressed. In the initial R(AB) transgenics, the CaMKIIα promoter restricted expression to postnatal excitatory neurons within the forebrain. These transgenic mice have provided the clearest demonstration that PKA activity is critical for spatial memory, as these mice exhibited normal initial learning in the hidden platform version of the water maze, but showed deficits in memory in retrieval tests. Biochemical studies of PKA activation following training have revealed that spatial learning activates PKA in the hippocampus. Interestingly, such activation is observed early in learning at a time when PKA inhibition does not impair performance, suggesting that PKA activation during training sets into motion a biochemical cascade that leads to memory storage (Vazquez et al., 2000; Mizuno et al., 2002). Such a cascade may likely involve the activation of ERK (extracellular signal-regulated protein kinases) and the transcription factor, CREB (cAMP response element binding protein).

Few studies have probed the effects of pharmacological inhibitors of PKA on spatial learning, perhaps because of the assumption that PKA would need to be inhibited during each training trial. Surprisingly, a single infusion of H-89, an inhibitor of PKA, into area CA1 of the dorsal hippocampus after the last trial on the third day of training significantly impaired spatial memory (Sharifzadeh et al., 2005). Clearly, additional pharmacological studies are needed to define the role that PKA might play earlier in acquisition or during retrieval, reconsolidation, and extinction of spatial memory. As always with pharmacological studies, it is important to note that the effects of inhibitors are rarely specific (Davies et al., 2000).

The role of the cAMP/PKA system in contextual conditioning

Experiments with spatial memory clearly establish the role of the cAMP/PKA pathway in hippocampus-dependent memory. However, because of the repeated training trials necessary for spatial learning, these tasks do not allow researchers to precisely define the role of this signaling pathway in short- and long-term memory. Since 1890, when William James made the distinction between short-term and long-term memory (James, 1890), researchers have sought to determine the relationship between these processes. Using single-trial learning tasks such as contextual fear conditioning and step-down inhibitory avoidance, researchers can define the mechanisms underlying short-term and long-term memory in the mammalian brain. In contextual fear conditioning, animals are exposed to a novel context and receive a mild footshock. A single 2- to 3-min training trial is sufficient for rodents to exhibit fear when re-exposed to the conditioning context. Such a fear response is measured most frequently by immobility ("freezing"). Contextual fear conditioning is sensitive to lesions of the amygdala and hippocampus (reviewed by Maren, 2001). Importantly, cued fear conditioning, in which animals are exposed to a tone and a footshock, is sensitive to lesions of the amygdala but not to hippocampal lesions, providing a good behavioral control for experiments aimed at addressing the impact of manipulations on hippocampal function.

The first evidence that PKA played a role in contextual fear conditioning came from studies of R(AB) transgenic mice (Abel et al., 1997; Bourtchouladze et al., 1998). These mutant mice exhibited selective deficits in long-term memory for contextual fear conditioning. Short-term memory for contextual fear as well as cued fear conditioning were intact. More recent R(AB) transgenic lines have used the tetracycline system for conditional regulation in adulthood, and these studies have confirmed that genetic inhibition of neuronal PKA in adulthood selectively impairs contextual long-term memory (Isiegas et al., 2006). Mice lacking the calcium-responsive forms of adenylyl cyclase also exhibit selective long-term memory deficits in contextual conditioning and passive avoidance, but the time course of these deficits appears to be distinct from that observed in the PKA mutant mice, perhaps because of differences in experimental procedures or genetic backgrounds of the different mutant mice (Wong et al., 1999). Although most research has focused on aversive conditioning, recent studies using novel object recognition have also revealed an important role for cAMP signaling in long-term memory in this non-aversive task (Pineda et al., 2004; Wang et al., 2004). Interestingly, studies in invertebrates (*Drosophila*, *Aplysia*) have found that the cAMP/PKA pathway is involved in short-term and long-term memory. In contrast, the majority of the studies in vertebrate systems have found that the cAMP/PKA signaling pathway plays a selective role in long-term memory. The reasons for this distinction are unclear.

Intra-ventricular or intra-hippocampal injection of PKA inhibitors has been shown to impair long-term contextual memory (Bourtchouladze et al., 1998; Wallenstein et al., 2002; Ahi et al., 2004). In experiments using step-down inhibitory avoidance, Izquierdo and colleagues have also found that inhibitors of PKA administered into hippocampal area CA1 selectively impair long-term memory (Bernabeu et al., 1997; Vianna et al., 2000; Quevedo et al., 2004). In rodents, inhibitors of protein synthesis impair long-term memory, similar to inhibitors of PKA, suggesting that PKA plays a role in activating the biochemical mechanisms that engage protein synthesis and transcription.

Much research has focused on the idea that PKA targets transcriptional machinery (Fig. 1). One target of PKA of great interest has been CREB, and CREB phosphorylation is increased after fear conditioning, as is expression of a CRE-driven reporter gene (Impey et al., 1998; Stanciu et al., 2001; Athos et al., 2002; Mizuno et al., 2002). It is important to note in this context that PKA is one of several kinases that can phosphorylate CREB on serine 133, and identifying which kinase is most critical under specific behavioral circumstances remains a challenge (Athos et al., 2002; Sindreu et al., 2007). The analysis of CREB mutant mice, however, has not been as clearly in support of a selective role for CREB in long-term memory as one might have expected, perhaps because of compensatory changes in other CRE-binding proteins in CREB mutant mice (Hummler et al., 1994) or because of effects of genetic background (Bourtchouladze et al., 1994; Gass et al., 1998; Bucan and Abel, 2002; Graves et al., 2002; Balschun et al., 2003). Indeed, the data appear to be more consistent with the idea that a CRE-binding protein related to CREB is involved in long-term memory storage, because the clearest phenotypes have been observed in genetically modified mice expressing dominant-negative forms of CREB (Kida et al., 2002; Pittenger et al., 2002), which effectively interfere with many proteins binding to CRE sites, and in CBP mutant mice, which also interfere with the action of a number of transcription factors (Oike et al., 1999; Bourtchouladze et al., 2003; Alarcon et al., 2004; Korzus et al., 2004; Wood et al., 2005, 2006).

The study of the molecular mechanisms underlying memory storage has highlighted important questions about the nature of memory. First, what is the relationship between long-term and short-term memory? Is short-term memory a process on the way towards long-term memory or are these two forms of memory mediated by independent, parallel processes? Most treatments that impair short-term memory also impair long-term memory, suggesting that in many cases they are serial processes. However, under certain circumstances, treatments can lead to impairments in short-term memory without impairing long-term memory, implying that these forms of memory can be mediated by distinct, parallel mechanisms (Emptage and Carew, 1993; Izquierdo et al., 1998; Sherff and Carew, 2004). Secondly, is memory consolidation a single process set into motion by training or does it have distinct phases? Here the data from studies of the role of PKA in long-term memory storage suggest that there are at least two time periods during which PKA and protein synthesis are required during memory consolidation (Bernabeu et al., 1997; Bourtchouladze et al., 1998). Finally, what is the behavioral role of cAMP/PKA signaling in the hippocampus and how does this compare to the role of this signaling pathway in other brain regions?

It is interesting to note that genetically modified mice in which the PKA pathway has been manipulated exhibited selective deficits in contextual fear conditioning, with intact cued fear conditioning (Abel et al., 1997; Isiegas et al., 2006). This suggests that the cAMP/PKA signaling pathway has distinct roles in the hippocampus and the amygdala, because direct pharmacological manipulation of PKA in the amygdala impairs cued fear conditioning (Schafe and LeDoux, 2000). In addition, it may be that the hippocampus forms a representation of the context during training whereas the amygdala may be the site of the association of the conditioned and unconditioned stimuli (CS and US) (Maren, 2001; Keeley et al., 2006). If this is the case, then there may be fundamental biochemical differences in the ways in which circuits in the hippocampus and amygdala are activated by training. Future experiments will be required to investigate these issues.

The cAMP/PKA pathway as a target of cognition-enhancing drugs

Our discussion has focused on the impact of impairments in cAMP/PKA signaling, but researchers have also investigated the behavioral effects of increased activity in the cAMP/PKA pathway. Pharmacological experiments have revealed a time window of 3–6 h after training when treatment with drugs that directly or indirectly activate PKA can enhance memory storage (Bernabeu et al., 1997). Blockade of the degradation of cAMP by treatment with inhibitors of phosphodiesterases such as the PDE4 inhibitor, rolipram, enhances memory (Barad et al., 1998). Interestingly, these PDE inhibitors can reverse the memory impairments found in aged mice (Bach et al., 1999) and in mouse models of mental retardation (Bourtchouladze et al., 2003) and Alzheimer's Disease (Vitolo et al., 2002; Gong et al., 2004). Although the benefits of pharmacological blockade of PDE for memory storage have been known for some time, the use of this approach in human patients is still in its nascent stages (Rose et al., 2005).

Studies of genetic manipulations designed to increase levels of cAMP have revealed that it is important to distinguish between manipulations that increase cAMP *levels* and those that increase cAMP *signals*. Genetic manipulations that increase cAMP *signals* (such as overexpression of calcium-responsive adenylyl cyclase) lead to memory enhancements like those observed with PDE inhibitors, whereas genetic manipulations that increase basal cAMP *levels* lead to memory impairments (Pineda et al., 2004; Wang et al., 2004; Bourtchouladze et al., 2006).

Synaptic tagging

Neurons typically receive inputs from thousands of synaptic contacts. However, L-LTP is input-specific (Andersen et al., 1977; Nguyen et al., 1994). To preserve the input specificity of L-LTP, a mechanism to mark, or "tag", active synapses has been proposed to allow newly synthesized gene products to be captured and utilized at appropriately activated synapses (Sossin, 1996; Frey and Morris, 1997; Schuman, 1997). Frey and Morris (1997) first provided evidence for the synaptic tag theory in the rat hippocampus. They proposed that proteins synthesized in response to long-term synaptic changes at one set of synapses could be captured and utilized by other synapses to express L-LTP if a synaptic tag is generated by appropriate synaptic activity. In accordance with this idea, they found that transient potentiation resulting from weaker synaptic activation could be prolonged to resemble L-LTP if paired with established protein synthesis-dependent L-LTP at separate synaptic inputs (Frey and Morris, 1997).

Interestingly, low-frequency stimulation (LFS) can impair L-LTP subsequently induced at the previously activated synapses (homosynaptic inhibition), as well as at other synapses converging on the same postsynaptic cells (heterosynaptic inhibition) (Young and Nguyen, 2005). Both homosynaptic and heterosynaptic inhibition of L-LTP by prior LFS required protein phosphatase activity (Young et al., 2006), and LFS impaired signaling through the cAMP/PKA pathway (Young et al., 2006). More importantly, pharmacological or genetic inhibition of PKA impaired synaptic capture, and pharmacological activation of cAMP/PKA signaling was sufficient to generate a synaptic tag that enabled persistent synaptic facilitation (Young et al., 2006). Thus, PKA has a key role in synaptic tagging that may be a novel control point in the consolidation of L-LTP, and perhaps, of long-term memory. PKA-mediated signaling can be constrained by prior episodes of synaptic activity to regulate future L-LTP expression and the spatiotemporal integration of multiple synaptic events. Because there is evidence that a behavioral analog of synaptic tagging and capture may occur during spatial exploration of a novel environment (Moncada and Viola, 2007), it will be of interest to determine, by in vivo inhibition of PKA, whether PKA also has a role in producing this modification of behavioral learning and memory.

Downstream substrates for PKA in synaptic plasticity and memory storage

The cAMP/PKA signaling cascade has garnered much attention because of its critical requirement for both long-term memory and for specific forms of long-term synaptic plasticity that are believed to underlie memory storage. The precise molecular mechanisms underlying these processes have not been fully elucidated, but many proteins that are important mediators of synaptic plasticity are regulated by cAMP/PKA signaling (Fig. 1; reviewed by Nguyen and Woo, 2003). Here, we discuss several key proteins that are targets of PKA and that have been implicated in mediating the induction and expression of hippocampal LTP, an activity-dependent enhancement of synaptic strength that is believed to be a cellular mechanism for information storage in the mammalian brain (Bliss and Collingridge, 1993; Martin et al., 2000; Gruart et al., 2006; Whitlock et al., 2006). These proteins include NMDA (N-methyl-D-aspartate) and AMPA (alpha-amino-3-hydroxy-5-methyl-4-isoxazole propionate) glutamatergic receptors, RIM1α, inhibitor-1 (I-1, an inhibitor of protein phosphatase-1), and the transcription factor, CREB (Fig. 1). We highlight these substrates to underscore the notion that PKA signaling is important for the expression of hippocampus-dependent forms of memory because it can modify specific signal transduction molecules that are critical for hippocampal LTP, a type of synaptic plasticity that has been strongly correlated with memory storage.

NMDA receptors (NMDARs)

In rat (Collingridge et al., 1983) and mouse (Tsien et al., 1996; Woo et al., 2003) hippocampal area CA1, LTP induction requires activation of NMDARs. Pharmacological block of brain NMDARs impairs spatial learning and attenuates hippocampal LTP in rats (Morris et al., 1986). Genetic knockout of NMDARs in area CA1 per se impairs spatial learning in the Morris water maze and blocks induction of LTP (Tsien et al., 1996). At the molecular level, PKA is coupled directly to NMDARs through an A-kinase anchoring protein (AKAP), yotiao, which permits modulation of NMDAR channel activity by PKA (Westphal et al., 1999). Recent studies by Suzanne Zukin and her colleagues have shown that pharmacological blockers of PKA reduce both the calcium permeability of NMDARs and NMDAR-mediated calcium signals in active dendritic spines of rodent hippocampal neurons (Skeberdis et al., 2006). The same blockers of PKA attenuated LTP at Schaeffer collateral-CA1 synapses (Skeberdis et al., 2006), thus linking PKA-dependent synaptic plasticity to calcium signaling in spines and to calcium permeability of NMDARs. Interestingly, this regulation of NMDAR currents by PKA was more prominent in neurons cultured from immature animals, suggesting developmental regulation of PKA modulation of these currents (Skeberdis et al., 2006). Thus, PKA phosphorylates the NMDAR or another associated protein, thereby altering the size of the NMDA channel pore to modify calcium permeability. This mechanism would regulate the amount of calcium influx into hippocampal neurons, which is a critical determinant of the type of plasticity (LTP or LTD) that is expressed. PKA regulation of calcium permeability can thus serve as a critical "front line" gate for controlling the strength of downstream intracellular signaling leading to induction and long-term expression of LTP. It is unclear whether this modulation of NMDAR calcium influx by PKA per se is critical for hippocampus-dependent memory, but given the established links between NMDARs, PKA, and hippocampal memory (Tsien et al., 1996; Abel et al., 1997), this idea is a testable one that will require more exact mutations of the NMDAR channel complex to selectively modify NMDAR calcium permeability in the intact animal.

AMPA receptors (AMPARs)

AMPA receptors (AMPARs) are targets of PKA (and other kinases), and PKA-mediated phosphorylation of these receptors has been implicated in the expression of hippocampal LTP. These receptors are composed of four subunits (GluR1–GluR4), which combine in different ratios to form functional receptors (Hollmann and Heinemann, 1994;

Dingledine et al., 1999). Alteration of subunit compositions of AMPARs is a key mechanism for regulating synaptic strength (Liu and Cull-Candy, 2000) through control of trafficking of these receptors (Shi et al., 2001). Furthermore, reversible phosphorylation of serine residues in specific subunits is critically linked to the expression of distinct forms of synaptic plasticity. Serine residues 831 and 845, located in the intracellular C-terminal domain of GluR1, are phosphorylated by CaMKII (Barria et al., 1997; Mammen et al., 1997) and PKA (Roche et al., 1996), respectively.

The first demonstrations that PKA can regulate AMPAR currents in hippocampal neurons were accomplished by intracellular or extracellular application of modifiers of PKA activity (Greengard et al., 1991; Wang et al., 1991). AMPA/kainate receptor whole-cell currents in dissociated hippocampal neurons were potentiated by intracellular application of catalytic subunits of PKA (Wang et al., 1991), and this same treatment increased the opening frequency and open times of these channels (Greengard et al., 1991), thereby providing a molecular basis for the enhancement of whole-cell EPSC amplitudes seen following bath applications of PKA activators (Wang et al., 1991; Duffy and Nguyen, 2003).

It is noteworthy that direct infusion of membrane-impermeant PKI (a specific peptide inhibitor of PKA) into CA1 pyramidal neurons of mouse hippocampal slices blocks maintenance of LTP, and that infusion of catalytic subunits of PKA into the same CA1 neurons is sufficient to elicit long-lasting synaptic potentiation of AMPAR-mediated excitatory postsynaptic currents (Duffy and Nguyen, 2003). Also, spatial "enrichment" of mice facilitates both PKA-dependent potentiation of AMPAR-mediated synaptic responses in area CA1 and hippocampus-dependent contextual fear memory (Duffy et al., 2001), suggesting that PKA-dependent synaptic plasticity is linked to improved performance on a hippocampal memory task (cf. Abel et al., 1997).

Lee et al. (2000) have shown that induction of homosynaptic LTD in area CA1 of naïve hippocampal slices elicited dephosphorylation of serine-845 (the PKA site) of GluR1, whereas induction of LTD in previously potentiated slices dephosphorylated serine-831 (the CaMKII site). Conversely, if LTP was induced at previously depressed synapses, serine-845 was phosphorylated, whereas LTP induction at naïve synapses elicited serine-831 phosphorylation (Lee et al., 2000). Interestingly, hippocampal learning (inhibitory avoidance performance) is associated with LTP-like increases in synaptic strength in the rat hippocampus and with GluR1 phosphorylation patterns similar to those seen after tetanus-induced LTP (Whitlock et al., 2006). These findings are consistent with the general concept that protein phosphorylation favors LTP, whereas dephosphorylation favors LTD. The results also demonstrate that different phosphorylation sites on a single receptor subunit are responsible for opposite polarities of synaptic plasticity.

AMPA receptor trafficking also plays an important role in hippocampal synaptic plasticity (Song and Huganir, 2002; Derkach et al., 2007), experience-dependent synaptic strengthening in cortical circuits (Clem and Barth, 2006), and associative learning (Rumpel et al., 2005). Recycling and sorting of AMPARs are activity-dependent, and they can be regulated by the phosphorylation state of the receptor (Carroll et al., 1999; Shi et al., 1999; Hayashi et al., 2000; Lu et al., 2001). Indeed, phosphorylation and dephosphorylation of AMPARs have been implicated in hippocampal LTD (Kameyama et al., 1998; Lee et al., 1998) and in LTP (reviewed by Soderling and Derkach, 2000), which in turn, involve the removal (Carroll et al., 1999; Man et al., 2000) or addition (Shi et al., 1999; Hayashi et al., 2000) of synaptic AMPARs, respectively. By expressing a recombinant construct of a specific subunit tagged with green fluorescent protein, Malinow and colleagues have defined some of the characteristics of AMPAR insertion (Shi et al., 1999, 2001; Hayashi et al., 2000), including a requirement for PKA-dependent phosphorylation of AMPAR subunits during synaptic plasticity (Esteban et al., 2003). The latter study showed that PKA controls the synaptic delivery of GluR4 subunits in rat hippocampal neurons, by relieving a retentive process that inhibits subunit incorporation at synapses. Interestingly, the effects of PKA are not restricted to GluR4: incorporation of GluR1, which requires CaMKII, was also blocked by inhibition of PKA (Esteban et al., 2003). Phosphorylation of serine-845 by PKA promotes AMPAR insertion and reduces AMPAR endocytosis (Man et al., 2007). More importantly, GluR1 is incorporated into synapses in a PKA-dependent manner during LTP, and stable LTP required PKA-mediated phosphorylation of serine-845 in the GluR1 subunit (Esteban et al., 2003). It should be noted, however, that ser-845 is phosphorylated under basal conditions (Lee et al., 2000); this is consistent with the observation that acute PKA activity is not necessary for induction of early stages of LTP (Huang and Kandel, 1994; but see Otmakhova et al., 2000; Yasuda et al., 2003).

In addition to AMPAR phosphorylation, AMPAR synthesis can increase several hours after L-LTP induction, and this synthesis, along with late LTP (L-LTP), were both attenuated by inhibitors of either PKA or transcription (Nayak et al., 1998). Thus, increased AMPA receptor synthesis is another mechanism by which PKA may cause L-LTP, which is known to be PKA-dependent. Overall, AMPAR incorporation at hippocampal synapses, resulting from increased synthesis and insertion, requires PKA, and it may explain the PKA-dependence of some types of LTP.

RIM1α and presynaptic LTP

The mossy fiber (MF) pathway, which connects granule cells of the dentate gyrus to pyramidal neurons of area CA3, displays a form of LTP (MF-LTP) that is expressed presynaptically and is independent of NMDA receptor activation (for a review, see Nguyen and Woo, 2003). Activation of PKA is required for MF-LTP (Huang et al., 1994; Weisskopf et al., 1994). One presynaptic substrate for PKA during MF-LTP is RIM1α, a presynaptic active zone protein that binds to Rab3A, a synaptic vesicle protein also implicated in MF-LTP (Castillo et al., 2002). RIM1α knockout mice display poor MF-LTP (Castillo et al., 2002), and impaired spatial learning on the Morris water maze (Powell et al., 2004). The presynaptic localization of RIM1α positions it for a role in vesicular docking and fusion, processes that are logically vital for presynaptic expression of MF-LTP. Thus, MF-LTP is one form of synaptic plasticity that requires a functional interface between synaptic vesicles and active zones, mediated through binding between Rab3A and RIM1α.

Interestingly, LTP in other brain regions also display a requirement for RIM1α and, by inference, involvement of presynaptic PKA activation. In the cerebellum, proof that PKA phosphorylation of RIM1α can elicit LTP was obtained by Lonart et al. (2003). LTP in RIM1α mutant mice was rescued by presynaptic expression of RIM1α, whereas introduction of mutant RIM1α lacking the PKA phosphorylation site impaired LTP in wildtype neurons (Lonart et al., 2003). Also, in the Schaffer collateral pathway of CA1, genetic knockout of RIM1α impaired L-LTP but not E-LTP, and this impairment of L-LTP may result from defective synaptic capture of gene products (Huang et al., 2005). Thus, PKA-mediated phosphorylation of RIM1α may impact numerous processes during synaptic plasticity and possibly at multiple sites. In a broader framework, a challenge for future research is to elucidate how the pre- and postsynaptic contributions of PKA are regulated to orchestrate the multiple molecular events that underlie PKA-dependent synaptic plasticity in distinct brain structures.

CREB and CRE-mediated gene expression

Hippocampal L-LTP, but not E-LTP (early LTP), requires transcription (Nguyen et al., 1994). Activation of PKA may cause L-LTP by regulating gene expression. PKA can modify transcription by phosphorylating several different transcription factors, one of which is CREB. CREB modulates transcription of genes containing cAMP response elements (CRE) in their promoters (Montminy, 1997; Lonze and Ginty, 2002). Following stimulation that leads to an increase in cytosolic cAMP, C subunits of PKA translocate into the nucleus to phosphorylate serine-133 on CREB (Bacskai et al., 1993). This phosphorylation can initiate transcription of CRE-associated genes (Yamamoto et al., 1988; Gonzalez and Montminy, 1989).

Studies in *Aplysia* (Dash et al., 1990; Bartsch et al., 1995; Martin et al., 1997) and *Drosophila* (Tully, 1991; Yin et al., 1994) demonstrated that members of the CREB family of transcription factors are necessary for long-lasting forms of synaptic plasticity and for long-term memory. In mice, it is unclear whether CREB activation is necessary for L-LTP. Bourtchouladze et al. (1994) found that knockout of the alpha and delta isoforms of CREB impaired LTP induced by just one 100-Hz train. Balschun et al. (2003) found normal LTP induced by three trains of 100-Hz stimulation in slices from CREB mutants, and Pittenger et al. (2002) similarly found intact multi-train LTP in mice expressing a dominant negative inhibitor of CREB. Thus, the requirement for CREB in L-LTP in the mammalian hippocampus may involve regulatory mechanisms that are more complex than previously hypothesized from studies on invertebrates. Genetic backgrounds of CREB mutant mice may also play key roles in determining the synaptic phenotypes of these animals.

An important question is whether PKA-mediated activation of CREB is critical for L-LTP. Direct evidence that PKA is critical for CREB-mediated changes in gene expression that may be linked to L-LTP comes from a study by Matsushita et al. (2001), in which specific blockade of *nuclear* PKA activity, by targeted delivery of PKI (a specific PKA inhibitor) to the nucleus, resulted in deficient CA1 L-LTP with a corresponding reduction in CREB phosphorylation (Matsushita et al., 2001). Thus, L-LTP in area CA1 of mouse hippocampal slices involves, but may not *require*, CREB-mediated transcription that requires nuclear PKA activation. Furthermore, whether CREB itself, or a related CRE-binding protein, is the effector for PKA-dependent L-LTP remains to be determined as the antibodies that bind to phosphorylated CREB also recognize phosphorylated forms of closely related proteins.

Inhibitor-1 and regulation of protein phosphatase-1 (PP-1)

Protein phosphatases can interact with the cAMP signaling pathway. Activation of dopaminergic D1 receptors, which are positively coupled to cAMP production, leads to phophorylation of DARPP-32 (dopamine and cAMP-regulated phosphoprotein, 32 kDa), an isoform of I-1 that is an endogenous inhibitor of PP-1 (Huang and Glinsmann, 1976; Hemmings et al., 1984; Greengard et al., 1999). I-1 and PP-1 have been implicated in E-LTP by studies suggesting that activation of the cAMP/PKA pathway may permit expression of E-LTP by suppressing PP-1's inhibition of E-LTP (Blitzer et al., 1995, 1998). PKA is thought to phosphorylate I-1, thereby enabling I-1 to inactivate PP-1; this would then allow E-LTP to be expressed (Blitzer et al., 1995, 1998). Indeed, stimulation that induced LTP triggered cAMP-dependent phosphorylation of I-1 and decreased PP-1 activity (Blitzer et al., 1998). Conversely, dephosphorylation of I-1 by calcineurin is thought to enable expression of LTD under conditions when calcium binding to calmodulin can activate calcineurin, which is known to be required for LTD (Mulkey et al., 1994). Calcineurin thus appears to act as an inhibitory constraint on LTP, because genetic inhibition of calcineurin enhances expression of E-LTP (Malleret et al., 2001).

The gating model of PKA/PP-1 regulation of E-LTP was also extended to L-LTP. R(AB) transgenic mice display reduced hippocampal PKA activity and deficient L-LTP in area CA1 (Abel et al., 1997; Woo et al., 2000). L-LTP in hippocampal slices from these mice was rescued by acute application of PP-1 inhibitors (Woo et al., 2002), indicating that PP-1 acts downstream of PKA to impair L-LTP. Similarly, pharmacological inhibition of PP-1 in R(AB) mutant slices also rescued forskolin-induced synaptic facilitation, which was defective in mutants (Woo et al., 2002). Interestingly, pharmacological inhibition of PP-1 had no effect on E-LTP induced by a single 100-Hz train in mutant slices (Woo et al., 2002). Thus, suppression of PP-1 by PKA selectively regulates L-LTP in mice. These collective data show that PP-1 acts as an inhibitory constraint on L-LTP, and that genetic reduction of PKA activity impairs L-LTP by enhancing this inhibitory constraint.

Some evidence indicates that I-1 is not required for hippocampal spatial learning and memory (Allen et al., 2000). Genetic knockout of I-1 did not impair performance on the Morris water maze, but it did attenuate LTP at lateral perforant path synapses (but not at Schaeffer collateral-CA1 synapses) (Allen et al., 2000). Thus, some caution is required in attributing a critical role for I-1 in specific types of hippocampal learning and memory, and not all hippocampal synapses may rely on I-1-mediated regulation of LTP.

The role of PKA in shaping the essence of memory

Our review has highlighted multiple roles for PKA in hippocampal memory. However, much remains unexplored. For example, what are the specific roles of PKA in the distinct phases of memory processing? These phases include acquisition, consolidation, reconsolidation, and retrieval of memories, and their mechanistic definitions are essential goals that must be achieved before researchers can fully grasp the essence of memory in the mammalian brain. Reversible pharmacological inactivation of hippocampal synaptic activity, by blockade of AMPAR-mediated transmission, has revealed that synaptic activity is required for encoding, consolidation, and retrieval of spatial memories in rats (Riedel et al., 1999). Reversible, genetic reduction of hippocampal PKA activity might definitively shed light on the roles of PKA in these important temporal phases of memory processing (see Isiegas et al., 2006, for evidence of PKA's inhibitory role in extinction), but more refinement is needed to achieve increasingly rapid temporal and spatial control of transgene expression to definitively resolve the contributions of PKA to the phases of memory processing.

We also need to explore the functional role of specific forms of PKA. Broadly speaking, PKA is present as RI or RII containing tetramers, referred to as type I or type II PKA respectively. These forms of PKA have different affinities for cAMP and they are localized to specific subcellular locations by AKAPs (reviewed in Bauman et al., 2004). Identifying isoform specific functions of PKA and the function of specific PKA-containing AKAP complexes is a challenge, but selective pharmacological tools are becoming available to examine these questions (e.g., Gold et al., 2006). Appropriate genetic approaches, such as the expression of specific inhibitors or the conditional mutation of appropriate genes, may also provide effective resolutions to these questions. Most of the current genetic experiments use broad-based inhibitors of either PKA activity or PKA anchoring. Identifying the specific PKA complex involved in memory storage will be critical for identifying the relevant PKA targets among the many that have been found to date.

PKA may selectively control specific aspects of memory processing. If so, this might lead to novel strategies, involving pharmacological, environmental, or genetic manipulations of PKA signaling, for alleviating memory disorders that arise from deficits of one or more of these temporal stages of memory processing. However, such strategies must be tempered by the fact that PKA activation in several brain structures, such as the prefrontal cortex and nucleus accumbens, can impair, rather than improve, cognition (Arnsten et al., 2005).

Central for further advancing our understanding of the roles of PKA in memory is the need for more research on PKA's roles in regulating synaptic plasticity in multiple brain structures. The history of this field is rich with examples of how the elucidation of mechanisms of synaptic plasticity can enhance our understanding of the cellular and molecular bases of learning and memory. Understanding the essence of memory requires a clear grasp of the integrative nature of memory processing, which involves coordinated communication between many distinct brain structures and subregions. Thus, the role of PKA in extrahippocampal synaptic physiology and cognition is ripe for future investigation. There is evidence that PKA is important for synaptic plasticity in brain regions outside of the hippocampus, such as prefrontal cortex (Jay et al., 1998), amygdala (Huang and Kandel, 1998), neostriatum (Colwell and Levine, 1995), thalamus (Castro-Alamancos and Calcagnotto, 1999), visual cortex (Hensch et al., 1998), and cerebellum (Salin et al., 1996; Chen and Regehr, 1997; Chavis et al., 1998; Linden and Ahn, 1999). Another important goal will be to distinguish between PKA's role as a *mediator*, *versus* its role as a *modulator*, of memory function. PKA may act as a mediator by directly controlling memory functions, whereas a modulatory role for PKA will be evident if PKA modifies the efficacies of other signaling pathways and neurotransmitter systems to alter memory dynamics.

Most, if not all, studies on hippocampal memory and synaptic plasticity are correlative because it is very difficult to establish causality between a molecular event, such as PKA activation, and a more complex process, such as synaptic potentiation or memory. At the molecular level, a critical requirement that needs to be satisfied before a definitive causal relationship between PKA activation and memory is established is direct visualization of PKA activation during memory processing. Apart from in vitro cellular imaging of PKA activity (Zhang et al., 2001; Gervasi et al., 2007), it will also be critical to refine techniques for in vivo imaging of kinase expression or activities. Such methods may be used to assess the causal relationships between PKA activity and specific behaviors, such as spatial and contextual learning. Some progress has been achieved in this domain, particularly with the emergence of two-photon imaging of neural activity in freely moving animals (Helmchen et al., 2001), but further fine-tuning is necessary to accommodate the stringent requirements for stable and specific monitoring of signal transduction in neural circuits in vivo.

Another technical approach that may yield important insights on the roles of PKA in behavior is in vivo recording of neural network activity patterns. Simultaneous recording of electrical activity in two or more interconnected brain regions in awake, freely moving animals is an important step towards elucidating the roles of identified signaling molecules in complex brain functions such as perception, learning, and memory. The relative paucity of data on multi-regional network activity patterns in genetically modified mice may now be remedied by the application of multi-electrode methods that can record oscillatory firing patterns in awake rodents, including mice (Buzsáki et al., 2003). It will be vital to apply these approaches to genetically modified PKA mutant mice (Abel et al., 1997) to directly assess the roles of PKA in regulating multi-region brain activity. Overall, the integrated application of methods from molecular and systems neuroscience will reveal novel roles for intracellular signaling in general, and for PKA signaling in particular, in learning and memory.

Acknowledgments

Our research was supported by the Canadian Institutes of Health Research (PVN), the Alberta Heritage Foundation for Medical Research (PVN), the National Institutes of Health (TA), and the Human Frontiers Science Program (TA). PVN thanks the University of Pennsylvania Provost's Distinguished International Scholars Program for supporting a sabbatical visit during which this article was written. We thank Ted Huang (University of Pennsylvania) for his help with constructing the figure. We thank S. Belleville, V. Castellucci, M. Dumas, J. C. Lacaille, and W. Sossin for making the "Essence of Memory" meeting in Montreal both memorable and fruitful. We apologize to colleagues whose research we could not discuss because of space limitations.

References

Abel, T., Nguyen, P.V., Barad, M., Deuel, T.A.S., Kandel, E.R. and Bourtchuladze, R. (1997) Genetic demonstration of a role for PKA in the late phase of LTP and in hippocampus-based long-term memory. Cell, 88: 615–626.

Abraham, W.C., Logan, B., Greenwood, J.M. and Dragunow, M. (2002) Induction and experience-dependent consolidation of stable long-term potentiation lasting months in the hippocampus. J. Neurosci., 22: 9626–9634.

Ahi, J., Radulovic, J. and Spiess, J. (2004) The role of hippocampal signaling cascades in consolidation of fear memory. Behav. Brain Res., 149: 17–31.

Alarcon, J.M., Malleret, G., Touzani, K., Vronskaya, S., Ishii, S., Kandel, E.R. and Barco, A. (2004) Chromatin acetylation, memory, and LTP are impaired in CBP(+/−) Mice: a model for the cognitive deficit in Rubinstein–Taybi syndrome and its amelioration. Neuron, 42: 947–959.

Allen, P.B., Hvalby, O., Jensen, V., Errington, M.L., Ramsay, M., Chaudry, F.A., Bliss, T.V.P., Storm-Mathisen, J., Morris, R.G.M., Andersen, P. and Greengard, P. (2000) Protein phosphatase-1 regulation in the induction of LTP: heterogeneous molecular mechanisms. J. Neurosci., 20: 3537–3543.

Andersen, P., Sundberg, S.H., Sveen, O. and Wigstrom, H. (1977) Specific long-lasting potentiation of synaptic transmission in hippocampal slices. Nature, 266: 736–737.

Arnsten, A.F.T., Ramos, B.P., Birnbaum, S.G. and Taylor, J.R. (2005) Protein kinase A as a therapeutic target for memory disorders: rationale and challenges. Trends Neurosci., 11: 121–128.

Athos, J., Impey, S., Pineda, V.V., Chen, X. and Storm, D.R. (2002) Hippocampal CRE-mediated gene expression is required for contextual memory formation. Nat. Neurosci., 5: 1119–1120.

Bach, M.E., Barad, M., Son, H., Zhuo, M., Lu, Y.F., Shih, R., Mansuy, I., Hawkins, R.D. and Kandel, E.R. (1999) Age-related defects in spatial memory are correlated with defects in the late phase of hippocampal long-term potentiation in vitro and are attenuated by drugs that enhance the cAMP signaling pathway. Proc. Natl. Acad. Sci. U.S.A., 96: 5280–5285.

Bacskai, B.J., Hochner, B., Mahaut-Smith, M., Adams, S.R., Kaang, B.K., Kandel, E.R. and Tsien, R.Y. (1993) Spatially resolved dynamics of cAMP and protein kinase A subunits in *Aplysia* sensory neurons. Science, 260: 222–226.

Balschun, D., Wolfer, D.P., Gass, P., Mantamadiotis, T., Welzl, H., Schutz, G., Frey, J.U. and Lipp, H.P. (2003) Does cAMP response element-binding protein have a pivotal role in hippocampal synaptic plasticity and hippocampus-dependent memory? J. Neurosci., 23: 6304–6314.

Barad, M., Bourtchouladze, R., Winder, D.G., Golan, H. and Kandel, E. (1998) Rolipram, a type IV-specific phosphodiesterase inhibitor, facilitates the establishment of long-lasting long-term potentiation and improves memory. Proc. Natl. Acad. Sci. U.S.A., 95: 15020–15025.

Barria, A., Derkach, V. and Soderling, T. (1997) Identification of the Ca^{2+}/calmodulin-dependent protein kinase II regulatory phosphorylation site in the alpha-amino-3-hydroxyl-5-methyl-4-isoxazole-propionate-type glutamate receptor. J. Biol. Chem., 272: 32727–32730.

Bartsch, D., Ghirardi, M., Skehel, P.A., Karl, K.A., Herder, S.P., Chen, M., Bailey, C.H. and Kandel, E.R. (1995) *Aplysia* CREB2 represses long-term facilitation: relief of repression converts transient facilitation into long-term functional and structural change. Cell, 83: 979–992.

Bauman, A.L., Goehring, A.S. and Scott, J.D. (2004) Orchestration of synaptic plasticity through AKAP signaling complexes. Neuropharmacology, 46: 299–310.

Beavo, J.A., Bechtel, P.J. and Krebs, E.G. (1974) Activation of protein kinase by physiological concentrations of cyclic AMP. Proc. Natl. Acad. Sci. U.S.A., 71: 3580–3583.

Beebe, S.J. (1994) The cAMP-dependent protein kinases and cAMP signal transduction. Semin. Cancer Biol., 5: 285–294.

Bernabeu, R., Bevilaqua, L., Ardenghi, P., Bromberg, E., Schmitz, P., Bianchin, M., Izquierdo, I. and Medina, J.H. (1997) Involvement of hippocampal cAMP/cAMP-dependent protein kinase signaling pathways in a late memory consolidation phase of aversively motivated learning in rats. Proc. Natl. Acad. Sci. U.S.A., 94: 7041–7046.

Bliss, T.V. and Collingridge, G.L. (1993) A synaptic model of memory: long-term potentiation in the hippocampus. Nature, 361: 31–39.

Bliss, T.V. and Lomo, T. (1973) Long-lasting potentiation of synaptic transmission in the dentate area of the anaesthetized rabbit following stimulation of the perforant path. J. Physiol., 232: 331–356.

Blitzer, R., Wong, T., Nouranifar, R., Iyengar, R. and Landau, E. (1995) Postsynaptic cAMP pathway gates early LTP in hippocampal CA1 region. Neuron, 15: 1403–1414.

Blitzer, R.D., Connor, J.H., Brown, G.P., Wong, T., Shenolikar, S., Iyengar, R. and Landau, E.M. (1998) Gating of CaMKII by cAMP-regulated protein phosphatase activity during LTP. Science, 280: 1940–1942.

Bourtchouladze, R., Abel, T., Berman, N., Gordon, R., Lapidus, K. and Kandel, E.R. (1998) Different training procedures for contextual memory in mice can recruit either one or two critical periods for memory consolidation that require protein synthesis and PKA. Learn. Mem., 5: 365–374.

Bourtchouladze, R., Frenguelli, B., Blendy, J., Cioffi, D., Schütz, G. and Silva, A.J. (1994) Deficient long-term memory in mice with a targeted mutation of the cAMP-responsive element-binding protein. Cell, 79: 59–68.

Bourtchouladze, R., Lidge, R., Catapano, R., Stanley, J., Gossweiler, S., Romashko, D., Scott, R. and Tully, T. (2003) A mouse model of Rubinstein–Taybi syndrome: defective long-term memory is ameliorated by inhibitors of phosphodiesterase 4. Proc. Natl. Acad. Sci. U.S.A., 100: 10518–10522.

Bourtchouladze, R., Patterson, S.L., Kelly, M.P., Kreibich, A., Kandel, E.R. and Abel, T. (2006) Chronically increased Gs alpha signaling disrupts associative and spatial learning. Learn. Mem., 13: 745–752.

Brandon, E.P., Idzerda, R.L. and McKnight, G.S. (1997) PKA isoforms, neural pathways, and behavior: making the connection. Curr. Opin. Neurobiol., 7: 397–403.

Bucan, M. and Abel, T. (2002) The mouse: genetics meets behaviour. Nat. Rev. Genet., 3: 114–123.

Burrell, B.D. and Sahley, C.L. (2001) Learning in simple systems. Curr. Opin. Neurobiol., 11: 757–764.

Buzsáki, G., Buhl, D.L., Harris, K.D., Csicsvari, J., Czeh, B. and Morozov, A. (2003) Hippocampal network patterns of activity in the mouse. Neuroscience, 116: 201–211.

Cadd, G. and McKnight, G.S. (1989) Distinct patterns of cAMP-dependent protein kinase gene expression in mouse brain. Neuron, 3: 71–79.

Carroll, R.C., Lissin, D.V., von Zastrow, M., Nicoll, R.A. and Malenka, R.C. (1999) Rapid redistribution of glutamate receptors contributes to long-term depression in hippocampal cultures. Nat. Neurosci., 2: 454–460.

Castellucci, V., Pinsker, H., Kupfermann, I. and Kandel, E.R. (1970) Neuronal mechanisms of habituation and dishabituation of the gill-withdrawal reflex in *Aplysia*. Science, 167: 1745–1748.

Castillo, P.E., Schoch, S., Schmitz, F., Sudhof, T.C. and Malenka, R.C. (2002) RIM1alpha is required for presynaptic long-term potentiation. Nature, 415: 327–330.

Castro-Alamancos, M.A. and Calcagnotto, M.E. (1999) Presynaptic long-term potentiation in corticothalamic synapses. J. Neurosci., 19: 9090–9097.

Chavis, P., Mollard, P., Bockaert, J. and Manzoni, O. (1998) Visualization of cyclic AMP-regulated presynaptic activity at cerebellar granule cells. Neuron, 20: 773–781.

Chen, C. and Regehr, W.G. (1997) Mechanism of cAMP-mediated enhancement at a cerebellar synapse. J. Neurosci., 17: 8687–8694.

Clem, R.L. and Barth, A. (2006) Pathway-specific trafficking of native AMPARs by in vivo experience. Neuron, 49: 663–670.

Collingridge, G.L., Kehl, S.J. and McLennan, H. (1983) Excitatory amino acids in synaptic transmission in the Schaffer collateral-commissural pathway of the rat hippocampus. J. Physiol., 334: 33–46.

Colwell, C.S. and Levine, M.S. (1995) Excitatory synaptic transmission in neostriatal neurons: regulation by cyclic AMP-dependent mechanisms. J. Neurosci., 15: 1704–1713.

Dash, P.K., Hochner, B. and Kandel, E.R. (1990) Injection of the cAMP-responsive element into the nucleus of *Aplysia* sensory neurons blocks long-term facilitation. Nature, 345: 718–721.

Davies, S.P., Reddy, H., Caivano, M. and Cohen, P. (2000) Specificity and mechanism of action of some commonly used protein kinase inhibitors. Biochem. J., 351: 95–105.

Davis, R.L. (1996) Physiology and biochemistry of *Drosophila* learning mutants. Physiol. Rev., 76: 299–317.

Derkach, V.A., Oh, M.C., Guire, E.S. and Soderling, T.R. (2007) Regulatory mechanisms of AMPA receptors in synaptic plasticity. Nat. Rev. Neurosci., 8: 101–113.

Dingledine, R., Borges, K., Bowie, D. and Traynelis, S.F. (1999) The glutamate receptor ion channels. Pharmacol. Rev., 51: 7–61.

Doskeland, S.O., Maronde, E. and Gjertsen, B.T. (1993) The genetic subtypes of cAMP-dependent protein kinase — functionally different or redundant? Biochim. Biophys. Acta, 1178: 249–258.

Dudek, S.M. and Bear, M.F. (1992) Homosynaptic long-term depression in area CA1 of hippocampus and effects of N-methyl-D-aspartate receptor blockade. Proc. Natl. Acad. Sci. U.S.A., 89: 4363–4367.

Duffy, S.N., Craddock, K.J., Abel, T. and Nguyen, P.V. (2001) Environmental enrichment modifies the PKA-dependence of hippocampal LTP and improves hippocampus-dependent memory. Learn. Mem., 8: 26–34.

Duffy, S.N. and Nguyen, P.V. (2003) Postsynaptic application of a peptide inhibitor of cAMP-dependent protein kinase blocks expression of long-lasting synaptic potentiation in hippocampal neurons. J. Neurosci., 23: 1142–1150.

Emptage, N.J. and Carew, T.J. (1993) Long-term synaptic facilitation in the absence of short-term facilitation in *Aplysia* neurons. Science, 262: 253–256.

Esteban, J.A., Shi, S.H., Wilson, C., Nuriya, M., Huganir, R.L. and Malinow, R. (2003) PKA phosphorylation of AMPA receptor subunits controls synaptic trafficking underlying plasticity. Nat. Neurosci., 6: 136–143.

Francis, S.H. and Corbin, J.D. (1999) Cyclic nucleotide-dependent protein kinases: intracellular receptors for cAMP and cGMP action. Crit. Rev. Clin. Lab. Sci., 36: 275–328.

Frey, U. and Morris, R.G. (1997) Synaptic tagging and long-term potentiation. Nature, 385: 533–536.

Gass, P., Wolfer, D.P., Balschun, D., Rudolph, D., Frey, U., Lipp, H. and Schutz, G. (1998) Deficits in memory tasks of mice with CREB mutations depend on gene dosage. Learn. Mem., 5: 274–288.

Gervasi, N., Hepp, R., Tricoire, L., Zhang, J., Lambolez, B., Paupardin-Tritsch, D. and Vincent, P. (2007) Dynamics of protein kinase A signaling at the membrane, in the cytosol, and in the nucleus of neurons in mouse brain slices. J. Neurosci., 27: 2744–2750.

Gibbs, C.S., Knighton, D.R., Sowadski, J.M., Taylor, S.S. and Zoller, M.J. (1992) Systematic mutational analysis of cAMP-dependent protein kinase identifies unregulated catalytic subunits and defines regions important for the recognition of the regulatory subunit. J. Biol. Chem., 267: 4806–4814.

Gold, M.G., Lygren, B., Dokurno, P., Hoshi, N., McConnachie, G., Tasken, K., Carlson, C.R., Scott, J.D. and Barford, D. (2006) Molecular basis of AKAP specificity for PKA regulatory subunits. Mol. Cell., 24: 383–395.

Gong, B., Vitolo, O.V., Trinchese, F., Liu, S., Shelanski, M. and Arancio, O. (2004) Persistent improvement in synaptic and cognitive functions in an Alzheimer mouse model after rolipram treatment. J. Clin. Invest., 114: 1624–1634.

Gonzalez, G.A. and Montminy, M.R. (1989) Cyclic AMP stimulates somatostatin gene transcription by phosphorylation of CREB at serine 133. Cell, 59: 675–680.

Graves, L., Dalvi, A., Lucki, I., Blendy, J.A. and Abel, T. (2002) Behavioral analysis of CREB alphadelta mutation on a B6/129 F1 hybrid background. Hippocampus, 12: 18–26.

Greengard, P., Allen, P.B. and Nairn, A.C. (1999) Beyond the dopamine receptor: the DARPP-32/protein phosphatase-1 cascade. Neuron, 23: 435–447.

Greengard, P., Jen, J., Nairn, A. and Stevens, C.F. (1991) Enhancement of the glutamate response by cAMP-dependent protein kinase in hippocampal neurons. Science, 253: 1135–1138.

Gruart, A., Munoz, M.D. and Delgado-Garcia, J.M. (2006) Involvement of the CA3–CA1 synapse in the acquisition of associative learning in behaving mice. J. Neurosci., 26: 1077–1087.

Hayashi, Y., Shi, S.H., Esteban, J.A., Piccini, A., Poncer, J.C. and Malinow, R. (2000) Driving AMPA receptors into synapses by LTP and CaMKII: requirement for GluR1 and PDZ domain interaction. Science, 287: 2262–2267.

Helmchen, F., Fee, M.S., Tank, D.W. and Denk, W. (2001) A miniature head-mounted two-photon microscope. high-resolution brain imaging in freely moving animals. Neuron, 31: 903–912.

Hemmings, H.C., Jr., Nairn, A.C. and Greengard, P. (1984) DARPP-32, a dopamine- and adenosine 3′:5′-monophosphate-regulated neuronal phosphoprotein. II. Comparison of the kinetics of phosphorylation of DARPP-32 and phosphatase inhibitor 1. J. Biol. Chem., 259: 14491–14497.

Hensch, T.K., Gordon, J.A., Brandon, E.P., McKnight, G.S., Idzerda, R.L. and Stryker, M.P. (1998) Comparison of plasticity in vivo and in vitro in the developing visual cortex of normal and protein kinase A RIbeta-deficient mice. J. Neurosci., 18: 2108–2117.

Hollmann, M. and Heinemann, S. (1994) Cloned glutamate receptors. Annu. Rev. Neurosci., 17: 31–108.

Howe, D.G., Wiley, J.C. and McKnight, G.S. (2002) Molecular and behavioral effects of a null mutation in all PKA C beta isoforms. Mol. Cell Neurosci., 20: 515–524.

Huang, F.L. and Glinsmann, W.H. (1976) Separation and characterization of two phosphorylase phosphatase inhibitors from rabbit skeletal muscle. Eur. J. Biochem., 70: 419–426.

Huang, Y.Y. and Kandel, E.R. (1994) Recruitment of long-lasting and protein kinase A-dependent long-term potentiation in the CA1 region of hippocampus requires repeated tetanization. Learn. Mem., 1: 74–82.

Huang, Y.Y. and Kandel, E.R. (1998) Postsynaptic induction and PKA-dependent expression of LTP in the lateral amygdala. Neuron, 21: 169–178.

Huang, Y.Y., Li, X.C. and Kandel, E.R. (1994) cAMP contributes to mossy fiber LTP by initiating both a covalently mediated early phase and macromolecular synthesis-dependent late phase. Cell, 79: 69–79.

Huang, Y.Y., Zakharenko, S.S., Schoch, S., Kaeser, P.S., Janz, R., Sudhof, T.C., Siegelbaum, S.A. and Kandel, E.R. (2005) Genetic evidence for a protein-kinase-A-mediated presynaptic component in NMDA-receptor-dependent forms of long-term synaptic potentiation. Proc. Natl. Acad. Sci. U.S.A., 102: 9365–9370.

Hummler, E., Cole, T., Blendy, J., Ganss, R., Aguzzi, A., Schmid, W., Beermann, F. and Schulz, G. (1994) Targeted mutation of the CREB gene: compensation within the CREB/ATF family of transcription factors. Proc. Natl. Acad. Sci. U.S.A., 91: 5647–5651.

Impey, S., Smith, D.M., Obrietan, K., Donahue, R., Wade, C. and Storm, D.R. (1998) Stimulation of cAMP response element (CRE)-mediated transcription during contextual learning. Nat. Neurosci., 1: 595–601.

Isiegas, C., Park, A., Kandel, E.R., Abel, T. and Lattal, K.M. (2006) Transgenic inhibition of neuronal protein kinase A activity facilitates fear extinction. J. Neurosci., 26: 12700–12707.

Izquierdo, I., Barros, D.M., Mello e Souza, T., de Souza, M.M., Izquierdo, L.A. and Medina, J.H. (1998) Mechanisms for memory types differ. Nature, 393: 635–636.

James, W. (1890) The Principles of Psychology. Holt, New York.

Jay, T.M., Gurden, H. and Yamaguchi, T. (1998) Rapid increase in PKA activity during long-term potentiation in the hippocampal afferent fibre system to the prefrontal cortex in vivo. Eur. J. Neurosci., 10: 3302–3306.

Kameyama, K., Lee, H.K., Bear, M.F. and Huganir, R.L. (1998) Involvement of a postsynaptic PKA substrate in expression of homosynaptic LTD. Neuron, 21: 1163–1175.

Kandel, E.R. (2001) The molecular biology of memory storage: A dialogue between genes and synapses [Nobel Lecture]. Science, 294: 1030–1038.

Keeley, M.B., Wood, M.A., Isiegas, C., Stein, J., Hellman, K., Hannenhalli, S. and Abel, T. (2006) Differential transcriptional response to nonassociative and associative components of classical fear conditioning in the amygdala and hippocampus. Learn. Mem., 13: 135–142.

Kida, S., Josselyn, S.A., de Ortiz, S.P., Kogan, J.H., Chevere, I., Masushige, S. and Silva, A.J. (2002) CREB required for the stability of new and reactivated fear memories. Nat. Neurosci., 5: 348–355.

Korzus, E., Rosenfeld, M.G. and Mayford, M. (2004) CBP histone acetyltransferase activity is a critical component of memory consolidation. Neuron, 42: 961–972.

Kupfermann, I., Castellucci, V., Pinsker, H. and Kandel, E. (1970) Neuronal correlates of habituation and dishabituation of the gill-withdrawal reflex in *Aplysia*. Science, 167: 1743–1745.

Lee, H., Barbarosie, M., Kameyama, K., Huganir, R.L. and Bear, M.F. (2000) Regulation of distinct AMPA receptor phosphorylation sites during bidirectional synaptic plasticity. Nature, 405: 955–959.

Lee, H., Kameyama, K., Huganir, R.L. and Bear, M.F. (1998) NMDA induces long-term synaptic depression and dephosphorylation of the GluR1 subunit of AMPA receptors in hippocampus. Neuron, 21: 1151–1162.

Linden, D.J. and Ahn, S. (1999) Activation of presynaptic cAMP-dependent protein kinase is required for induction of cerebellar long-term potentiation. J. Neurosci., 19: 10221–10227.

Liu, S.G. and Cull-Candy, G.G. (2000) Synaptic activity at calcium-permeable AMPA receptors induces a switch in receptor subtype. Nature, 405: 454–458.

Lonart, G., Schoch, S., Kaeser, P.S., Larkin, C.J., Sudhof, T.C. and Linden, D.J. (2003) Phosphorylation of RIM1alpha by PKA triggers presynaptic long-term potentiation at cerebellar parallel fiber synapses. Cell, 115: 49–60.

Lonze, B.E. and Ginty, D.D. (2002) Function and regulation of CREB family transcription factors in the nervous system. Neuron, 35: 605–623.

Lu, W., Man, H., Ju, W., Trimble, W.S., MacDonald, J.F. and Wang, Y.T. (2001) Activation of synaptic NMDA receptors induces membrane insertion of new AMPA receptors and LTP in cultured hippocampal neurons. Neuron, 29: 243–254.

Malleret, G., Haditsch, U., Genoux, D., Jones, M.W., Bliss, T.V., Vanhoose, A.M., Weitlauf, C., Kandel, E.R., Winder, D.G. and Mansuy, I.M. (2001) Inducible and reversible enhancement of learning, memory, and long-term potentiation by genetic inhibition of calcineurin. Cell, 104: 675–686.

Mammen, A.L., Kameyama, K., Roche, K.W. and Huganir, R.L. (1997) Phosphorylation of the alpha-amino-3-hydroxy-5-methylisoxazole4-propionic acid receptor GluR1 subunit by calcium/calmodulin-dependent kinase II. J. Biol. Chem., 272: 32528–32533.

Man, H.Y., Lin, J.W., Ju, W.H., Ahmadian, G., Liu, L., Becker, L.E., Sheng, M. and Wang, Y.T. (2000) Regulation of AMPA receptor-mediated synaptic transmission by clathrin-dependent receptor internalization. Neuron, 25: 649–662.

Man, H.Y., Sekine-Aizawa, Y. and Huganir, R.L. (2007) Regulation of AMPA receptor trafficking through PKA phosphorylation of the GluR1 subunit. Proc. Natl. Acad. Sci. U.S.A., 104: 3579–3584.

Maren, S. (2001) Neurobiology of Pavlovian fear conditioning. Annu. Rev. Neurosci., 24: 897–931.

Martin, K.C., Casadio, A., Zhu, H., E, Y., Rose, J.C., Chen, M., Bailey, C.H. and Kandel, E.R. (1997) Synapse-specific long-term facilitation of *Aplysia* sensory to motor synapses: a function for local protein synthesis in memory storage. Cell, 91: 927–938.

Martin, S., Grimwood, P. and Morris, R.G.M. (2000) Synaptic plasticity and memory: an evaluation of the hypothesis. Annu. Rev. Neurosci., 23: 649–711.

Matsushita, M., Tomizawa, K., Moriwaki, A., Li, ST., Terada, H. and Matsui, H. (2001) A high-efficiency protein transduction system demonstrating the role of PKA in long-lasting LTP. J. Neurosci., 21: 6000–6007.

McGuire, S.E., Deshazer, M. and Davis, R.L. (2005) Thirty years of olfactory learning and memory research in *Drosophila melanogaster*. Prog. Neurobiol., 76: 328–347.

McKernan, M.G. and Shinnick-Gallagher, P. (1997) Fear conditioning induces a lasting potentiation of synaptic currents in vitro. Nature, 390: 607–611.

McKnight, G.S., Clegg, C.H., Uhler, M.D., Chrivia, J.C., Cadd, G.G., Correll, L.A. and Otten, A.D. (1988) Analysis of the cAMP-dependent protein kinase system using molecular genetic approaches. Rec. Prog. Horm. Res., 44: 307–335.

Milner, B., Squire, L.R. and Kandel, E.R. (1998) Cognitive neuroscience and the study of memory. Neuron, 20: 445–468.

Miyamoto, E., Kuo, J.F. and Greengard, P. (1968) Adenosine 3′,5′-monophosphate-dependent protein kinase from brain. Science, 165: 63–65.

Miyamoto, E., Kuo, J.F. and Greengard, P. (1969) Cyclic nucleotide-dependent protein kinases. 3. Purification and properties of adenosine 3′,5′-monophosphate-dependent protein kinase from bovine brain. J. Biol. Chem., 244: 6395–6402.

Mizuno, M., Yamada, K., Maekawa, N., Saito, K., Seishima, M. and Nabeshima, T. (2002) CREB phosphorylation as a molecular marker of memory processing in the hippocampus for spatial learning. Behav. Brain Res., 133: 135–141.

Moncada, D. and Viola, H. (2007) Induction of long-term memory by exposure to novelty requires protein synthesis: evidence for a behavioral tagging. J. Neurosci., 27: 7476–7481.

Montminy, M. (1997) Transcriptional regulation by cyclic AMP. Annu. Rev. Biochem., 66: 807–822.

Morris, R.G.M., Anderson, E., Lynch, G.S. and Baudry, M. (1986) Selective impairment of learning and blockade of long-term potentiation by an *N*-methyl-D-aspartate receptor antagonist, AP5. Nature, 319: 774–776.

Morris, R.G.M., Garrud, P., Rawlins, J.N.P. and O'Keefe, J. (1982) Place navigation impaired in rats with hippocampal lesions. Nature, 297: 681–683.

Mulkey, R.M., Endo, S., Shenolikar, S. and Malenka, R.C. (1994) Involvement of a calcineurin/inhibitor-1 phosphatase cascade in hippocampal long-term depression. Nature, 369: 486–488.

Nayak, A., Zastrow, D.J., Lickteig, R., Zahniser, N.R. and Browning, M.D. (1998) Maintenance of late-phase LTP is accompanied by PKA-dependent increase in AMPA receptor synthesis. Nature, 394: 680–683.

Nguyen, P.V., Abel, T. and Kandel, E.R. (1994) Requirement of a critical period of transcription for induction of a late phase of LTP. Science, 265: 1104–1107.

Nguyen, P.V. and Woo, N.H. (2003) Regulation of hippocampal synaptic plasticity by cyclic AMP-dependent protein kinases. Prog. Neurobiol., 71: 401–437.

Oike, Y., Hata, A., Mamiya, T., Kaname, T., Noda, Y., Suzuki, M., Yasue, H., Nabeshima, T., Araki, K. and Yamamura, K. (1999) Truncated CBP protein leads to classical Rubinstein–Taybi syndrome phenotypes in mice: implications for a dominant-negative mechanism. Hum. Mol. Genet., 8: 387–396.

Otmakhova, N.A., Otmakhov, N., Mortenson, L. and Lisman, J.E. (2000) Inhibition of the cAMP pathway decreases early LTP at CA1 hippocampal synapses. J. Neurosci., 20: 4446–4451.

Pineda, V.V., Athos, J.I., Wang, H., Celver, J., Ippolito, D., Boulay, G., Birnbaumer, L. and Storm, D.R. (2004) Removal of G(ialpha1) constraints on adenylyl cyclase in the hippocampus enhances LTP and impairs memory formation. Neuron, 41: 153–163.

Pittenger, C., Huang, Y.Y., Paletzki, R.F., Bourtchouladze, R., Scanlin, H., Vronskaya, S. and Kandel, E.R. (2002) Reversible inhibition of CREB/ATF transcription factors in region CA1 of the dorsal hippocampus disrupts hippocampus-dependent spatial memory. Neuron, 34: 447–462.

Powell, C.M., Schoch, S., Monteggia, L., Barrot, M., Matos, M.F., Feldmann, N., Sudhof, T.C. and Nestler, E.J. (2004) The presynaptic active zone protein RIM1alpha is critical for normal learning and memory. Neuron, 42: 143–153.

Quevedo, J., Vianna, M.R., Martins, M.R., Barichello, T., Medina, J.H., Roesler, R. and Izquierdo, I. (2004) Protein synthesis, PKA, and MAP kinase are differentially involved in short- and long-term memory in rats. Behav. Brain Res., 154: 339–343.

Reimann, E.M., Walsh, D.A. and Krebs, E.G. (1971) Purification and properties of rabbit skeletal muscle adenosine 3′,5′-monophosphate-dependent protein kinases. J. Biol. Chem., 246: 1986–1995.

Riedel, G., Micheau, J., Lam, A.G., Roloff, E., Martin, S.J., Bridge, H., Hoz, L., Poeschel, B., McCulloch, J. and Morris, R.G. (1999) Reversible neural inactivation reveals hippocampal participation in several memory processes. Nat. Neurosci., 2: 898–905.

Roche, K.W., O'Brien, R.J., Mammen, A.L., Bernhardt, J. and Huganir, R.L. (1996) Characterization of multiple phosphorylation sites on the AMPA receptor GluR1 subunit. Neuron, 16: 1179–1188.

Rose, G.M., Hopper, A., De Vivo, M. and Tehim, A. (2005) Phosphodiesterase inhibitors for cognitive enhancement. Curr. Pharm. Des., 11: 3329–3334.

Rumpel, S., LeDoux, J., Zador, A. and Malinow, R. (2005) Postsynaptic receptor trafficking underlying a form of associative learning. Science, 308: 83–88.

Salin, P.A., Malenka, R.C. and Nicoll, R.A. (1996) Cyclic AMP mediates a presynaptic form of LTP at cerebellar parallel fiber synapses. Neuron, 16: 797–803.

Schafe, G.E. and LeDoux, J.E. (2000) Memory consolidation of auditory Pavlovian fear conditioning requires protein synthesis and protein kinase A in the amygdala. J. Neurosci., 20: p. RC96.

Schuman, E.M. (1997) Synapse specificity and long-term information storage. Neuron, 18: 339–342.

Sharifzadeh, M., Sharifzadeh, K., Naghdi, N., Ghahremani, M.H. and Roghani, A. (2005) Posttraining intrahippocampal infusion of a protein kinase AII inhibitor impairs spatial memory retention in rats. J. Neurosci. Res., 79: 392–400.

Sherff, C.M. and Carew, T.J. (2004) Parallel somatic and synaptic processing in the induction of intermediate-term and long-term synaptic facilitation in *Aplysia*. Proc. Natl. Acad. Sci. U.S.A., 101: 7463–7468.

Shi, S., Hayashi, Y., Esteban, J.A. and Malinow, R. (2001) Subunit-specific rules governing AMPA receptor trafficking to synapses in hippocampal pyramidal neurons. Cell, 105: 331–343.

Shi, S.-H., Hayashi, Y., Petralia, R.S., Zaman, S.H., Wenthold, R.J., Svoboda, K. and Malinow, R. (1999) Rapid spine delivery and redistribution of AMPA receptors after synaptic NMDA receptor activation. Science, 284: 1811–1816.

Sindreu, C.B., Scheiner, Z.S. and Storm, D.R. (2007) Ca^{2+}-stimulated adenylyl cyclases regulate ERK-dependent activation of MSK1 during fear conditioning. Neuron, 53: 79–89.

Skalhegg, B.S. and Tasken, K. (2000) Specificity in the cAMP/PKA signaling pathway. Differential expression, regulation, and subcellular localization of subunits of PKA. Front. Biosci., 5: D678–D693.

Skeberdis, V.A., Chevaleyre, V., Lau, C.G., Goldberg, J.H., Pettit, D.L., Suadicani, S.O., Lin, Y., Bennett, M.V., Yuste, R., Castillo, P.E. and Zukin, R.S. (2006) Protein kinase A regulates calcium permeability of NMDA receptors. Nat. Neurosci., 9: 501–510.

Soderling, T.R. and Derkach, V.A. (2000) Postsynaptic protein phosphorylation and LTP. Trends Neurosci., 23: 75–80.

Song, I. and Huganir, R.L. (2002) Regulation of AMPA receptors during synaptic plasticity. Trends Neurosci., 25: 578–588.

Sossin, W.S. (1996) Mechanisms for the generation of synapse specificity in long-term memory: the implications of a requirement for transcription. Trends Neurosci., 19: 215–218.

Stanciu, M., Radulovic, J. and Spiess, J. (2001) Phosphorylated cAMP response element binding protein in the mouse brain after fear conditioning: relationship to Fos production. Brain Res. Mol. Brain Res., 94: 15–24.

Sutherland, E.W., Oye, I. and Butcher, R.W. (1965) The action of epinephrine and the role of the adenyl cyclase system in hormone action. Rec. Prog. Horm. Res., 21: 623–646.

Sutherland, E.W. and Rall, T.W. (1957) The properties of an adenine ribonucleotide produced with cellular particles, ATP, Mg, and epinephrine or glucagons. J. Am. Chem. Soc., 79: p. 3608.

Tasken, K., Skalhegg, B.S., Solberg, R., Andersson, K.B., Taylor, S.S., Lea, T., Blomhoff, H.K., Jahnsen, T. and Hansson, V. (1993) Novel isozymes of cAMP-dependent protein kinase exist in human cells due to formation of RI alpha–RI beta heterodimeric complexes. J. Biol. Chem., 268: 21276–21283.

Taylor, S.S., Buechler, J.A. and Yonemoto, W. (1990) cAMP-dependent protein kinase: framework for a diverse family of regulatory enzymes. Annu. Rev. Biochem., 59: 971–1005.

Tsien, J.Z., Huerta, P.T. and Tonegawa, S. (1996) The essential role of hippocampal CA1 NMDA receptor-dependent synaptic plasticity in spatial memory. Cell, 87: 1327–1338.

Tully, T. (1991) Physiology of mutations affecting learning and memory in *Drosophila* — the missing link between gene product and behavior. Trends Neurosci., 14: 163–164.

Vazquez, S.I., Vazquez, A. and Pena de Ortiz, S. (2000) Different hippocampal activity profiles for PKA and PKC in spatial discrimination learning. Behav. Neurosci., 114: 1109–1118.

Vianna, M.R., Izquierdo, L.A., Barros, D.M., Ardenghi, P., Pereira, P., Rodrigues, C., Moletta, B., Medina, J.H. and Izquierdo, I. (2000) Differential role of hippocampal cAMP-dependent protein kinase in short- and long-term memory. Neurochem. Res., 25: 621–626.

Vitolo, O.V., Sant'Angelo, A., Costanzo, V., Battaglia, F., Arancio, O. and Shelanski, M. (2002) Amyloid beta-peptide inhibition of the PKA/CREB pathway and long-term potentiation: reversibility by drugs that enhance cAMP signaling. Proc. Natl. Acad. Sci. U.S.A., 99: 13217–13221.

Wallenstein, G.V., Vago, D.R. and Walberer, A.M. (2002) Time-dependent involvement of PKA/PKC in contextual memory consolidation. Behav. Brain Res., 133: 159–164.

Walsh, D.A., Perkins, J.P. and Krebs, E.G. (1968) An adenosine 3′,5′-monophosphate-dependant protein kinase from rabbit skeletal muscle. J. Biol. Chem., 243: 3763–3765.

Wang, H., Ferguson, G.D., Pineda, V.V., Cundiff, P.E. and Storm, D.R. (2004) Overexpression of type-1 adenylyl cyclase in mouse forebrain enhances recognition memory and LTP. Nat. Neurosci., 7: 635–642.

Wang, L.Y., Salter, M.W. and MacDonald, J.F. (1991) Regulation of kainate receptors by cAMP-dependent protein kinase and phosphatases. Science, 253: 1132–1135.

Weisskopf, M.G., Castillo, P.E., Zalutsky, R.A. and Nicoll, R.A. (1994) Mediation of hippocampal mossy fiber long-term potentiation by cyclic AMP. Science, 265: 1878–1882.

Westphal, R.S., Tavalin, S.J., Lin, J.W., Alto, N.M., Fraser, I.D., Langeberg, L.K., Sheng, M. and Scott, J.D. (1999) Regulation of NMDA receptors by an associated phosphatase-kinase signaling complex. Science, 285: 93–96.

Whitlock, J.R., Heynen, A.J., Shuler, M.G. and Bear, M.F. (2006) Learning induces long-term potentiation in the hippocampus. Science, 313: 1093–1097.

Wong, S.T., Athos, J., Figueroa, X.A., Pineda, V.V., Schaefer, M.L., Chavkin, C.C., Muglia, L.J. and Storm, D.R. (1999) Calcium-stimulated adenylyl cyclase activity is critical for hippocampus-dependent long-term memory and late phase LTP. Neuron, 23: 787–798.

Woo, N.H., Abel, T. and Nguyen, P.V. (2002) Genetic and pharmacological demonstration of a role for cyclic AMP-dependent protein kinase-mediated suppression of protein phosphatases in gating the expression of late LTP. Eur. J. Neurosci., 16: 1871–1876.

Woo, N.H., Duffy, S.N., Abel, T. and Nguyen, P.V. (2000) Genetic and pharmacological demonstration of differential recruitment of cAMP-dependent protein kinases by synaptic activity. J. Neurophysiol., 84: 2739–2745.

Woo, N.H., Duffy, S.N., Abel, T. and Nguyen, P.V. (2003) Temporal spacing of synaptic stimulation critically modulates the dependence of LTP on cyclic AMP-dependent protein kinase. Hippocampus, 13: 293–300.

Wood, M.A., Attner, M.A., Oliveira, A.M., Brindle, P.K. and Abel, T. (2006) A transcription factor-binding domain of the coactivator CBP is essential for long-term memory and the expression of specific target genes. Learn. Mem., 13: 609–617.

Wood, M.A., Kaplan, M.P., Park, A., Blanchard, E.J., Oliveira, A.M., Lombardi, T.L. and Abel, T. (2005) Transgenic mice expressing a truncated form of CREB-binding protein (CBP) exhibit deficits in hippocampal synaptic plasticity and memory storage. Learn. Mem., 12: 111–119.

Wu, Z.L., Thomas, S.A., Villacres, E.C., Xia, Z., Simmons, M.L., Chavkin, C., Palmiter, R.D. and Storm, D.R. (1995) Altered behavior and long-term potentiation in type 1 adenylyl cyclase mutant mice. Proc. Natl. Acad. Sci. U.S.A., 92: 220–224.

Yamamoto, K.K., Gonzalez, G.A., Biggs, W.H., III and Montminy, M.R. (1988) Phosphorylation-induced binding and transcriptional efficacy of nuclear factor CREB. Nature, 334: 494–498.

Yasuda, H., Barth, A.L., Stellwagen, D. and Malenka, R.C. (2003) A developmental switch in the signaling cascades for LTP induction. Nat. Neurosci., 6: 15–16.

Yin, J.C.P., Wallach, J.S., Del Vecchio, M., Wilder, E.L., Zhou, H., Quinn, W.G. and Tully, T. (1994) Induction of a dominant negative CREB transgene specifically blocks long-term memory in *Drosophila*. Cell, 79: 49–58.

Young, J.Z., Isiegas, C., Abel, T. and Nguyen, P.V. (2006) Metaplasticity of late-LTP: a critical role for PKA in synaptic tagging. Eur. J. Neurosci., 23: 1784–1794.

Young, J.Z. and Nguyen, P.V. (2005) Homosynaptic and heterosynaptic inhibition of synaptic tagging and capture of LTP by prior synaptic activity. J. Neurosci., 25: 7221–7231.

Zhang, J., Ma, Y., Taylor, S. and Tsien, R.Y. (2001) Genetically encoded reporters of PKA activity reveal impact of substrate tethering. Proc. Natl. Acad. Sci. U.S.A., 98: 14997–15002.

CHAPTER 7

'Synaptic tagging' and 'cross-tagging' and related associative reinforcement processes of functional plasticity as the cellular basis for memory formation

Sabine Frey and Julietta U. Frey[*]

Leibniz Institute for Neurobiology, Brenneckestrasse 6, D-39118 Magdeburg, Germany

Abstract: We focus on new properties of cellular and network processes of memory formation involving 'synaptic tagging' and 'cross-tagging' during long-term potentiation (LTP) and long-term depression (LTD) as well as associative heterosynaptic interactions, the latter of which are characterized by a time-window of about 1 h. About 20 years ago we showed for the first time that the maintenance of LTP, like memory storage, depends on intact protein synthesis and thus consists of at least two temporal phases. Later, similar properties for LTD were shown by our own and other laboratories. Here we describe the requirements for the induction of the transient early-LTP/LTD and of the protein synthesis-dependent late-LTP/LTD. Late-LTP/LTD depend on the associative activation of heterosynaptic inputs, i.e. the synergistic activation of glutamatergic and modulatory reinforcing inputs within specific, effective time-windows during their induction. The induction of late-LTP/LTD is characterized by novel, late-associative properties such as 'synaptic tagging', 'cross-tagging' and 'late-associative reinforcement'. All of these phenomena require the associative setting of synaptic tags as well as the availability of plasticity-related proteins (PRPs) and they are restricted to functional dendritic compartments, in general. 'Synaptic tagging' guarantees input specificity, 'cross-tagging' determines the interaction between LTP and LTD in a neuron, and thus both are required for the specific processing of afferent signals for the establishment of late-LTP/LTD. 'Late-associative reinforcement' describes a process where early-LTP/LTD by the co-activation of modulatory inputs can be transformed into late-LTP/LTD in activated synapses where a tag is set. Recent experiments in the freely moving rat revealed a number of modulatory brain structures involved in the transformation of early-plasticity events into long-lasting ones. Further to this, we have characterized time-windows and activation patterns to be effective in the reinforcement process. Studies using a combined electrophysiological and behavioural approach revealed the physiological relevance of these reinforcement processes, which is also supported by fMRI studies in humans, which led to the hypothesis outlined here on cellular and system memory-formation by late-associative heterosynaptic interactions at the cellular level during functional plasticity events.

Keywords: protein synthesis-dependent late-LTP; protein synthesis-dependent late-LTD; memory formation; synaptic tagging; reinforcement; late associativity; cross-tagging

[*]Corresponding author. Tel.: +49-391-6263422;
Fax: +49-391-6263421; E-mail: frey@ifn-magdeburg.de

There is growing interest in the phenomena of functional plasticity such as long-term potentiation (LTP) and long-term depression (LTD), the cellular models of activity-dependent, long-lasting changes of synaptic transmission in the vertebrate brain. Bliss and co-workers (for review, see Bliss and Gardner-Medwin, 1973; Bliss and Lomo, 1973) first described LTP in the dentate region of the hippocampus. The best-studied form of LTP as well as the one which can be induced in the intact adult animal has been referred to as an 'associative, NMDA-receptor-dependent LTP'. The induction of LTP by brief high-frequency trains activates the glutamatergic and modulatory, heterosynaptic inputs required for its prolonged maintenance. This particular form of LTP requires the activation of various kinases as well as the synthesis of plasticity-related proteins (PRPs) and thus consists of phases (Krug et al., 1984; Frey et al., 1988; Reymann et al., 1988a, b; Otani et al., 1989; Matthies et al., 1990b; Bliss and Collingridge, 1993). Similar properties were also shown for LTD in the healthy adult vertebrate hippocampus (Bear and Malenka, 1994; Huber et al., 2000; Manahan-Vaughan et al., 2000; Sajikumar and Frey, 2004a). While other forms of synaptic plasticity are important, we shall hereafter discuss only the associative, NMDA receptor- and protein synthesis-dependent forms and refer to them, for simplicity, as 'LTP' or 'LTD'. Because of its long duration, input specificity and associative properties, LTP as well as LTD is widely used as a model for the investigation of basic, cellular memory mechanisms.

Matthies (1974) developed his hypothesis of neuronal mechanisms of memory formation in the mammalian brain: the assumed phases of short-term, intermediate and long-term memory — their different time-courses and decay-times, different biological correlates and different sensitivities to interventions — reflect corresponding cellular properties at the synaptic, synaptosomal and nuclear level required for the formation of a memory trace (Matthies, 1974, 1989). Learning experiments revealed a biphasic occurrence of protein synthesis. It was suggested that the early phase of protein synthesis represents the synthesis of regulatory proteins. The regulatory proteins then control the formation of target proteins which, finally, remodel neuronal connectivity/efficiency during late stages of memory formation.

LTP and LTD as elementary cellular memory models

Currently synaptic LTP and LTD are the only cellular identified correlates which could underlie memory. Learning and memory, however, are complex processes involving more than just local synaptic processes. Conventionally, LTP and LTD are properties of a single synapse or synapse population in a restricted, artificial circuit activated by tetanization, which serves to investigate mechanisms of synaptic efficacy. Thus, studying LTP/LTD at the synaptic or small network level can only contribute to the better basic understanding of the formation of elementary memory traces but not of memory taking part in an organism's behaviour. Memory formation in the organism represents the property of several circuits including a multi-level, distributed storage system, which enables the adaptation of an individual organism to changes in its environment. The transfer and test of results obtained in cellular studies into behaviour should therefore be a must. The study of distinct associative properties of LTP and LTD and their combination with behaviour could lead to a better understanding of basic learning mechanisms as well as serve to design special learning experiments to verify the cellular properties in a complex system. Here we review evidence which indicates that LTP and LTD share a number of similar properties with memory consolidation such as requiring the synthesis of PRPs induced in an associative and heterosynaptic way. Furthermore, we have shown first that both processes can associatively interact within effective time-windows thus sharing information within a single cell — a phenomenon which we have named 'cross-tagging'. Finally, we have recently shown that heterosynaptic pathways which are required for the reinforcement or consolidation of memory can be assigned to specific behavioural content. These associative reinforcement processes must take place within specific time-windows ('late-associativity') around LTP-induction and can represent emotional and/or cognitive aspects.

Consolidation of LTP and LTD: requirement for protein synthesis

In 1984 Manfred Krug and colleagues (Krug et al., 1984) demonstrated for the first time that the maintenance of hippocampal LTP beyond 3–6 h in the dentate gyrus (DG) in vivo can be blocked by anisomycin, a reversible translation inhibitor. The role of protein synthesis for the maintenance of LTP has since then been replicated by various other laboratories (e.g., Frey et al., 1988; Otani et al., 1989; Fazeli et al., 1993; Osten et al., 1996; Nayak et al., 1998). Taken together, these first results led us to the hypothesis that LTP has stages or phases like memory consolidation and that the synthesis of proteins is required for the long-lasting maintenance of synaptic changes in efficacy (Krug et al., 1984; Frey et al., 1988; Reymann et al., 1988a, b; Matthies et al., 1990b).

Although the multiple phase model of LTP is now widely accepted, questions were raised whether the different time-course and duration of LTP reflects only quantitative differences depending on tetanization strength. Indeed, experimental protocols used to elicit consolidation in LTP are usually based on increasing the intensity of inputs. For example, one 100 Hz tetanization is used for the induction of early-LTP and three repeated tetani are required to elicit a late-LTP (Reymann et al., 1985; Huang and Kandel, 1994). Later studies revealed, however, that a single tetanus can also lead to late-LTP if its stimulation intensity and the number of stimuli per tetanus are sufficiently high (Bortolotto and Collingridge, 2000; Sajikumar et al., 2005a). Obviously, the stimulation strength determines whether heterosynaptic inputs become activated by field stimulation which in turn has been shown to be required to induce late-LTP. The involvement of additional transmitters (see below) and protein synthesis shows that a new quality of cellular interaction for consolidation is reached when late-LTP is induced. Similar properties were also found for LTD. We prefer to distinguish between different phases of LTP/LTD instead of distinguishing just the induction and maintenance process. The mechanisms which enable the late stages also have to be induced, probably as early as those underlying early-LTP/LTD. Interestingly, late-LTP/LTD can, in certain circumstances, be induced very shortly after the expression of the enhanced synaptic change that constitutes early-LTP/LTD or even in the event of cross-tagging, the late plasticity event can be induced by the opposite early plasticity event. In the latter case the "cross-tagged/captured" input would result in late-LTP/LTD without necessarily expression of early-LTP/LTD, if a tag was set in this input (see below). This tag captures the PRPs from the other strong input which results in an immediate late-LTP/LTD without early-LTP/LTD being necessarily expressed. Thus, early-LTP/LTD might be a staging-post on the way to a longer-lasting change; alternatively, it could be a separate or parallel phase of LTP/LTD expression that is not required for late-LTP/LTD. Further evidence for a mechanistically separate early phase comes from findings demonstrating that after saturated LTP has been induced, an additional potentiation can be evoked by turning down the test intensity. Normally, it was thought that the induction of LTP precludes the further induction of LTP in the same input. We had shown in 1995 (Frey et al., 1995), however, that in a time-window of approximately 3 h, i.e. during a period where early-LTP is expressed in these inputs, only a short-term potentiation (STP) can be induced, whereas later, beyond 4 h, a full-length LTP can be generated by further tetani. Thus, only during the maintenance of early-LTP is the further induction of longer-lasting LTP precluded. If, however, early-LTP was transformed into late-LTP, the same synapses further allowed LTP induction again, so the induction of LTP was not precluded any more. It remains to be investigated whether LTD follows similar rules. Even if the same principal expression mechanism as, e.g. an increase in AMPA receptor function were used for different LTP phases, different mechanisms seem to underlie the different phases of LTP and LTD. In general, recent findings suggest that mechanistic differentiation in parallel as well as sequential processes is somewhat misleading. For late phases to occur a distinct interplay between most of these processes seems to be required (see also later: 'synaptic tagging').

Local protein synthesis inhibition in the entorhinal cortex demonstrated that the presynaptic somata whose axons form the perforant path stimulating the granular cells in the DG are not the critical locus of protein synthesis for late-LTP. Only if anisomycin was applied into the DG was late-LTP blocked (Otani and Abraham, 1989). A still-open question is whether the molecular signals for initiating translation and the protein synthesis itself are localized at the nearby dendrites or distributed between dendrites and remote cell bodies. Strong stimulation that is used to induce long-lasting LTP certainly engages somatic signalling cascades. Evidence comes from our study where LTP was induced in dendritic stumps of CA1 pyramidal cells of hippocampal slices in vitro (Frey et al., 1989b). In these experiments the cell-body layer, the major site of transcription and translation, was surgically removed from the apical dendrites. The isolated dendrites revealed a pronounced early-LTP in the field-EPSP, similar to that in intact slices. The potentiation decreased, however, after about 3–4 h, thus showing the same lack of late-LTP as observed in complete CA1 neurons after inhibition of protein synthesis with anisomycin. These results indicate that the mechanisms responsible for late-LTP are located in postsynaptic compartments and somatic factors are needed for NMDA-receptor-dependent late-LTP in the CA1 region. If somatic signalling or even transcription is necessary for the synthesis, the question arises of how the input-specificity of late-LTP is achieved without elaborate protein-trafficking. This important question will be discussed later in the chapter on 'synaptic tagging'. It becomes evident, however, that the picture with respect to the locus of protein synthesis is more complicated. Local application of the protein synthesis inhibitor emetine to either basal or apical dendrites revealed a specific contribution of protein synthesis in dendritic compartments irrespective of the contribution of somatic protein synthesis (Bradshaw et al., 2003). Anisomycin and cycloheximide block LTP already after 1 h, even in isolated dendrites (Cracco et al., 2005; Vickers et al., 2005). A theta-frequency stimulation applied to the Schaffer collateral pathway can also induce a form of late-LTP (Huang and Kandel, 2005) that is maintained even in isolated CA1 dendrites and most likely involves GABA-ergic processes at a higher degree than that observed after conventional high-frequency stimulation and that observed in the intact animal (Buzsáki, 1997; Chapman et al., 1998; Perez et al., 1999). Translation of pre-existing mRNAs present in isolated dendrites (Steward and Schuman, 2001) might be sufficient to mediate potentiation or is necessary in addition to proteins or mRNA transported from the soma (for RNA see below), at least to maintain partially early-LTP in distinct preparations. Besides postsynaptic protein synthesis, a late NMDA-receptor-sensitive increase of presynaptic vesicle proteins was also described (Lynch et al., 1994). It should be mentioned that besides local protein function in dendrites and presynaptic terminals the release of newly synthesized proteins into the extracellular space might also be important for late-LTP (Duffy et al., 1981; Fazeli et al., 1988; Pang and Lu, 2004). Similar processes can be expected for late-LTD although more thorough analyses are required. In general, all the above findings might find their explanations in the need for local dendritic protein synthesis, by drug site effects (see below), or paradigm/preparation/incubation-dependence (Sajikumar et al., 2005a).

The identification of the classes and functions of plasticity proteins, which might themselves be receptors destined for insertion into the cell membrane or factors like kinases that enhance or inhibit the activity of existing molecular entities, is ongoing. If macromolecules are necessary for consolidation, the question arises whether similar if not identical molecules (e.g., PRP) are required in all synapses, a topic discussed later in more detail. The foregoing results are consistent with the finding that the incorporation of radioactive-labelled amino acids into cytosomal proteins of hippocampal neurons is elevated for 1 h immediately after tetanization (e.g., Duffy et al., 1981; Fazeli et al., 1988; Bullock et al., 1990; Frey et al., 1991a). This transient enhancement of protein synthesis coincides with the time-window after tetanization during which the inhibition of protein synthesis prevents the initiation of late-LTP/LTD (Frey et al., 1988; Manahan-Vaughan et al., 2000; Sajikumar and Frey, 2003, 2004a). Learning experiments have shown a second peak in the elevation of leucine incorporation 8 h after training (see Fig. 1a) (Lössner et al., 1982). Later experiments supported these results for hippocampal LTP, revealing an increase in radioactive-labelled proteins shortly after tetanization and a subsequent processing and incorporation into synaptosomal membranes and postsynaptic densities 8 h after tetanization (Bullock et al., 1990; Frey et al., 1991a). Although the first stage of protein synthesis takes place immediately after tetanization, further post-translational processing of proteins during early-LTP is required for the induction of late-LTP. Recent confocal imaging analysis revealed that the dendritic synthesis of a probe protein and its fast degradation are a prerequisite of late-LTP (Karpova et al., 2006). Furthermore, in addition to processing of functional proteins it can be assumed that late-LTP/LTD induction also leads to the synthesis of regulatory proteins involved in the long-lasting regulation of PRPs via gene expression.

Consolidation of LTP/LTD: role for transcription?

Mammalian long-term memory involves the regulation of mRNA-synthesis (for review, see Matthies, 1974; Matthies et al., 1990a; Alkon et al., 1991). In contrast to memory early studies of LTP in intact animals in vivo have shown no influence of the mRNA-synthesis inhibitor actinomycin D (Otani et al., 1989). Later work from the laboratory of Kandel (Nguyen et al., 1994) suggested that a distinct LTP in hippocampal slices in vitro was prevented after 1–3 h when the inhibitors were applied during tetanization. A side-effect of the drugs on mechanisms involved in LTP-generation could not be excluded. Since the question of whether mRNA synthesis takes place in LTP/LTD is crucial for understanding the mechanisms underlying late-LTP/LTD, we performed experiments not subject to these shortcomings (Frey et al., 1993b, 1996; Manahan-Vaughan et al., 2000). The effect of the mRNA-synthesis inhibitor actinomycin D as well as 5,6-dichloro-1-β-D-ribofuranosyl benzimidazole (DRB) was studied in two different systems, the CA1 in vitro and the DG in vivo, and for LTP as well as for LTD in the CA1 region in vivo. To avoid unspecific effects of the drugs on LTP/LTD-induction, the irreversible drugs were applied sufficiently before tetanization/low-frequency stimulation. Although we found that the population spike was affected by the drugs in the LTP studies (where both the population spike and the field-EPSP had been analysed), the more relevant factor for synaptic efficacy — the EPSP — showed, if any, only a delayed weak effect after about 6–8 h during LTP (Fig. 1c) or had no effect in LTD (Manahan-Vaughan et al., 2000).

Unfortunately, animals treated with irreversible mRNA-synthesis inhibitors can only be investigated for about 12 h before baseline potentials decline by unspecific drug effects, and the animals often die after 24 h. Thus, in the intact animal and in hippocampal slices, LTP as well as LTD can be maintained at least for the first 8 h, i.e. with inhibited mRNA — but an intact protein synthesis machinery (Sajikumar et al., 2005a). In other words, the protein synthesis-dependent first 8 h of LTP/LTD are maintained by pre-existing mRNA whereas later stages may require additional regulation of PRPs by gene expression. In line with these observations, a recent study by Vickers et al. (2005) showed that late-LTP in slices was prevented by translational inhibitors after about 1–2 h whereas mRNA-synthesis inhibitors were only effective in blocking LTP starting after about 4–6 h. This suggests the requirement of gene expression to maintain the plasticity event. Whether, it is only required to refill stores of "housekeeping" proteins including the PRPs remains an open question. The above results let us suggest, however, that the regulation of such "housekeeping" proteins instead of transcription-dependent functionally new proteins seems to be more comprehensible in maintaining long-lasting plasticity events. If the latter is true, the search for new "genes" and proteins after LTP/LTD-induction, and maybe also learning, is useless, and the complex regulation of existing ones should be investigated instead in future studies.

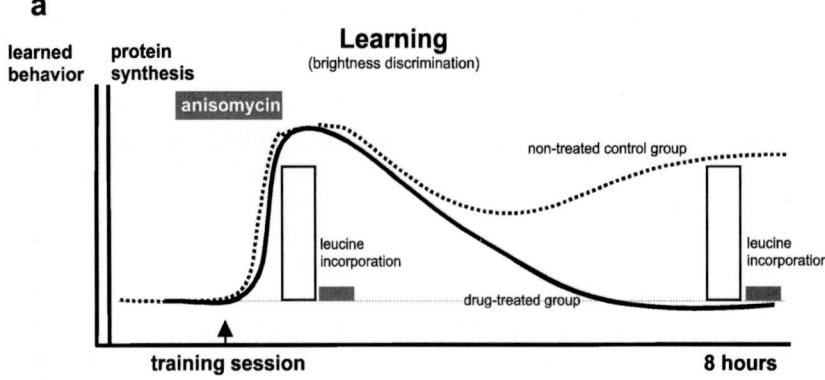
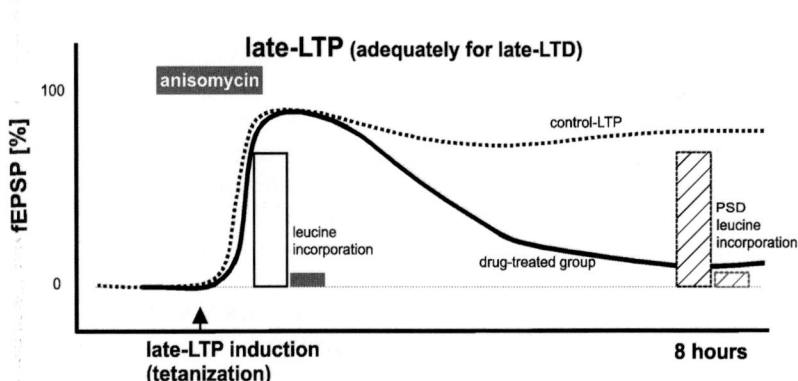
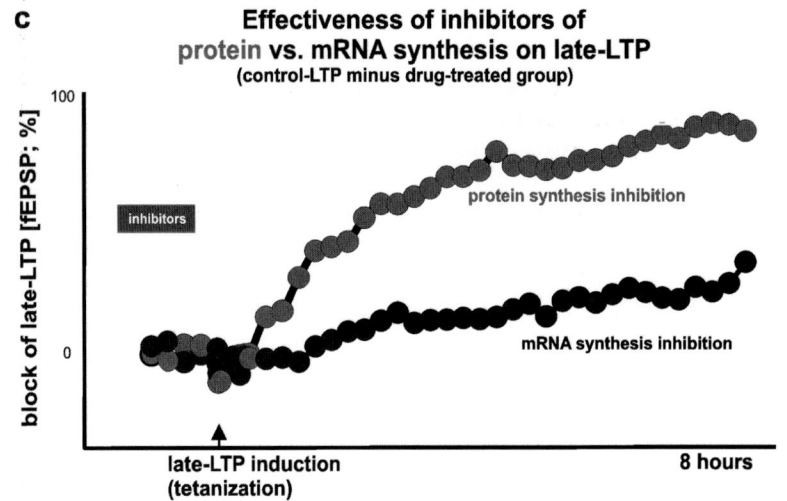

LTP/LTD-induction by glutamate receptors

After the distinguishing of early- and late-LTP/LTD on the basis of their different sensitivity to protein synthesis inhibition the question arises which transmitter receptors and second messengers are involved in the induction of early- and late-LTP as well as -LTD. Since we focus in this review on processes of late-LTP/LTD we want to make brief mention that co-activation of NMDA- and AMPA receptors during tetanic stimulation or low-frequency-stimulation and a concurrent influx of Ca^{2+} into the postsynaptic cell via NMDA-receptor operated channels is necessary for LTP as well as LTD induction in the intact animal (Herron et al., 1986; Davis et al., 1992; Manahan-Vaughan et al., 2000). These processes guarantee the induction and maintenance of short-term potentiation/depression(?) (STP/STD) and early-LTP/LTD (for review see Reymann and Frey, 2007). For late-LTP/LTD to occur, however, further non-glutamatergic associative interactions must take place during their induction processes.

Non-glutamatergic, neuromodulatory and "reinforcing" transmitters required for late plasticity events

Late-LTP in CA1 and DG, as well as late-LTD, can be modulated by inhibitors and activators of aminergic, or metabotropic glutamatergic receptors as well as the action of opioids coupled to the cAMP cascade (for review, see Frey, 1997, 2001; Sajikumar and Frey, 2004a; Navakkode et al., 2005, 2007). We therefore suggest that late-LTP/LTD requires concomitant activation of different transmitter systems (Frey et al., 1989a, 1990, 1991b; Matthies et al., 1990b; Frey, 1997; Frey and Morris, 1998a; Sajikumar and Frey, 2004a). The typical LTP/LTD experiment, involving a brief period of high frequency stimulation (or, with intracellular recording, pairing of pre- and postsynaptic activation) for LTP or a short period of low-frequency stimulation in the case of LTD, overlooks the more likely situation in the behaving organism, where the activation of a population of glutamatergic synapses of an individual neuronal cell is likely to be accompanied by a dynamic activation of non-glutamatergic heterosynaptic inputs. The tetanization/low-frequency stimulation used in LTP/LTD studies, involving simultaneous field activation of hundreds of fibres, activates more than one kind of neurotransmitter input, and it is that cooperative action of inputs that induces late-LTP/LTD. We therefore favour the notion that a time-dependent convergence of two or more events is required for late-LTP/LTD.

Given the effects of dopamine on learning (Flood et al., 1980), we investigated first whether dopamine is simultaneously released during LTP induction in the apical dendrites of CA1-neurons in the hippocampus, and whether it is required for late-LTP and late-LTD there (Frey et al., 1989a, 1990, 1991b; Sajikumar and Frey, 2004a). The apical dendrites of the CA1 in the hippocampus are innervated by dopaminergic fibres that course through the mesolimbic pathway (Baulac et al.,

Fig. 1. The consolidation of a memory trace: the requirement for protein synthesis (adapted from Frey (1997)). (a) During a brightness discrimination task a distinct memory is formed (dotted line) lasting for at least 8 h. If the reversible protein synthesis inhibitor anisomycin was applied during training a memory phase beyond 4 h was blocked (full line). Measurement of incorporation of radioactive-labelled amino acids (boxes) revealed a biphasic increase immediately after training and 8 h later. These enhancements of protein synthesis were blocked if anisomycin was given during training (grey boxes). (b) Similar properties as in learning can be described during the consolidation of hippocampal CA1-LTP as well as LTD. Strong tetanization can induce late-LTP or distinct low-frequency stimulation late-LTD, respectively, with a duration of 8 h (dotted line) which can be blocked by protein synthesis inhibitors if applied during or shortly after the induction (filled line). Similarly to learning during LTP an increase of incorporation of amino acids into soluble proteins could be detected shortly after tetanization (open black box; Frey et al., 1991a) whereas an increase of amino acid incorporation into the structure of the postsynaptic density was detected 8 h after LTP induction (scattered black box; Bullock et al., 1990). Both effects were abolished if anisomycin was applied during tetanization (grey boxes). (c) Represents the effectiveness of protein- (grey symbols) vs. mRNA- (black symbols) synthesis inhibitors on LTP measured as fEPSP in hippocampal CA1. Protein synthesis inhibitors prevented LTP after about 4 h, i.e. prevented late-LTP whereas mRNA-inhibitors exerted their action only at later time-points of late late-LTP. The graphs schematically represent the change of the fEPSP related to pretetanic values.

1986), and there is evidence of the expression of the D5 receptor in CA1 pyramidal cells. The D5 receptor is related to the D1 dopamine receptor that is coupled to adenylyl cyclase. In addition to glutamate, and possibly other neurotransmitters, dopamine levels increase during conventional LTP induction in the above region (Frey et al., 1990). To determine whether dopamine might be the additional activator necessary for late-LTP and/or late-LTD, we showed that it plays a crucial role in the initiation of the mechanisms responsible for late-LTP as well as late-LTD in the apical dendrites of hippocampal CA1 neurons (Sajikumar and Frey, 2004a; Navakkode et al., 2007). When specific inhibitors of the dopaminergic D1 and D2 receptors were administered during tetanization or LTD-induction, late-LTP/LTD was prevented (Frey et al., 1990, 1991b; Frey, 1997; Sajikumar and Frey, 2004a; Navakkode et al., 2007). Our results suggest that the influences of aminergic transmitters are not just modulatory. We have demonstrated that activation of dopaminergic inputs to hippocampal CA1 neurons plays a critical role, perhaps acting as a cellular switch to establish the late phase of LTP or LTD, since the time-course of LTP/LTD decay after dopamine antagonists is similar to that seen after anisomycin. It was shown that repeated application or transient application (Huang and Kandel, 1995; Frey and Morris, 1998a; Sajikumar and Frey, 2004a) of a D1/D5 receptor agonist in CA1 synapses in adult animals induces a delayed potentiation or depression whose time-courses are similar to those found after administration of dopamine. Interestingly, whether LTP or LTD is induced depends only on relatively small differences in the concentration of applied dopamine. Thus, repeated administration of 10 µM dopamine revealed late-LTD whereas 50 µM resulted in late-LTP if measured in apical CA1 dendrites (Sajikumar and Frey, 2004a). For the late event to occur we have recently shown that a synergistic glutamatergic input during dopamine application (Navakkode et al., 2007) is required which points to analogous properties observed during electrical inductions of the late event. There it was shown that a threefold tetanization resulted in transient increased cAMP-levels which were also dependent on NMDA-receptor function (Frey et al., 1993a). The fact that small differences in the concentration of the modulatory transmitter can result in opposite plasticity events suggests that one and the same synapse (or synapse population representing a functional input) can express both forms of the late form of either LTP or LTD, whereas PRPs can represent a pool of process-specific (such as PKMζ for LTP) and process-unspecific molecules, unable to result in LTP or LTD, however, without the adequate process-specific tag.

Our findings on the requirement of dopamine for late-LTP in apical dendrites of hippocampal CA1 neurons have recently been supported by studies from the Morris laboratory in which some of the above experiments were reinvestigated (O'Carroll and Morris, 2004). Furthermore, similar results to those with dopamine have been obtained by the application of norepinephrine in the DG (Stanton and Sarvey, 1985).

As mentioned above late-LTP induction causes the transient formation of cAMP. Cyclic-AMP activates cAMP-dependent protein kinase A (PKA) or other cAMP-dependent processes. The short-term elevation of cAMP was blocked by a D1/D5 antagonist, SCH 23390, and the NMDA-receptor blocker AP-5. Our observations reveal that application of PKA inhibitors is ineffective in blocking STP and early-LTP but they do block late-LTP (Frey et al., 1993a; Matthies and Reymann, 1993). Transient application of a membrane-permeable cAMP analog or of other PKA activators initiated a delayed potentiation that seemed to simulate late-LTP (Frey et al., 1993a; Pockett et al., 1993; Selbach et al., 1997). These results suggest that the cAMP-dependent activation of PKA is required for late-LTP and thus PKA-dependent processes are specifically involved in the synthesis of PRPs, which in turn are required for the expression of late-LTP.

Phases of LTP and LTD

All of the above experiments point to the multi-stage model of LTP which we proposed first in the 1980s for LTP and then later for LTD (for review, see Matthies et al., 1990b; Frey, 2001; Reymann and Frey, 2007). The late stages require an associative and heterosynaptic induction involving different transmitter systems. The main message is that LTP/LTD consolidation takes place not only as merely as a function of more-of-the-same (i.e., transmitter), but also of coincidence of several transmitter inputs with several second messengers coupled to them and the coincidence of the specific plasticity state at other synapses of the same neuron. This includes the time-dependent plasticity state of the neuron as a whole as we will see later on. Only in dependence on these functional states does the consolidation of LTP or LTD take place. In general, this corresponds with the natural situation, in which much of the information that we consolidate as long-term memory is distinguished from the non-consolidated one by virtue of context and association involving many brain structures rather than just by the intensity at a single synapse. In sum, we had first proposed that LTP is not an unitary phenomenon, but consists of at least three phases: (i) STP, a decremental phase dependent on NMDA receptor activation and Ca^{2+}/calmodulin, (ii) early-LTP, an intermediate phase which depends on phosphorylation of PKC- and CaMK-substrates, and (iii) late-LTP, a long-lasting component which requires the synthesis of proteins (Reymann et al., 1988b; Reymann and Frey, 2007). Similar events — involving the same and different molecular key-players — can be suggested for LTD (Manahan-Vaughan et al., 2000; Sajikumar and Frey, 2004a; Navakkode et al., 2005). As discussed before, there is some indication of an additional late phase (at least for LTP), which depends on gene transcription regulating mRNA synthesis for PRPs (Fig. 1; see also Krug et al., 1984; Frey et al., 1988; Reymann et al., 1988a, b; Otani et al., 1989; Matthies et al., 1990b; Bliss and Collingridge, 1993).

Input-specificity of late-LTP/LTD via 'synaptic tagging' and by associative properties

How pre-existing or newly synthesized PRPs interact with specific, activated synapses expressing LTP (or, also LTD, e.g. see Sajikumar and Frey, 2004a), but not with non-activated synapses, is fundamental to the synapse-specificity thought to be critical for information-processing and memory formation. We explain synaptic input-specificity by the concept of 'synaptic tagging', in which newly synthesized PRPs activated by heterosynaptic interactions bind to recently potentiated, glutamatergic 'tagged' synapses, thus maintaining LTP and input-specificity (Frey and Morris, 1997, 1998a, b). We had also shown that stimulation that normally leads to early-LTP could also induce late-LTP if a separate pathway had been strongly tetanized within a specific time-window. Thus tetanization of a pathway can induce a LTP with variable persistence as a function of the prior history of activation of the neuron (Frey and Morris, 1998b). The tag is transiently active with an expected half-life of about 30 min in the intact animal. It has also been suggested that the PRPs are characterized by a specific, relatively short half-life of about 1–2 h (for review, see Korz and Frey, 2004; Sajikumar et al., 2005a). Only in the event when both processes, the synapse-specific tag as well as PRPs, are available, can the two interact and transform early- into late-LTP at the stimulated synapses (see Fig. 2; Frey and Morris, 1998b). The existence of tag- and PRP-dynamics therefore determines an effective, functionally important time-window during which a normally transient form of functional plasticity can be transformed into a long-lasting one. 'Synaptic tagging' can therefore also explain how a strong tetanus, producing protein synthesis-dependent LTP (late-LTP) in one pathway, can prolong the potentiation in an independent, weakly tetanized pathway that would normally have produced only early-LTP with a tag set (Frey and Morris, 1998b).

The late-associative time-window is not only determined by the mechanistic half-life of tags and PRPs. Synaptic tags or tag-complexes can not only be passively deactivated by cellular degradational processes but also actively through distinct electrical stimulation, such as by depotentiating stimuli within 5–10 min after the setting of the tag by specific stimulation (Stäubli and Scafidi, 1999; Sajikumar and Frey, 2004b). We are currently studying the functional relevance of these processes in the intact animal. With respect to

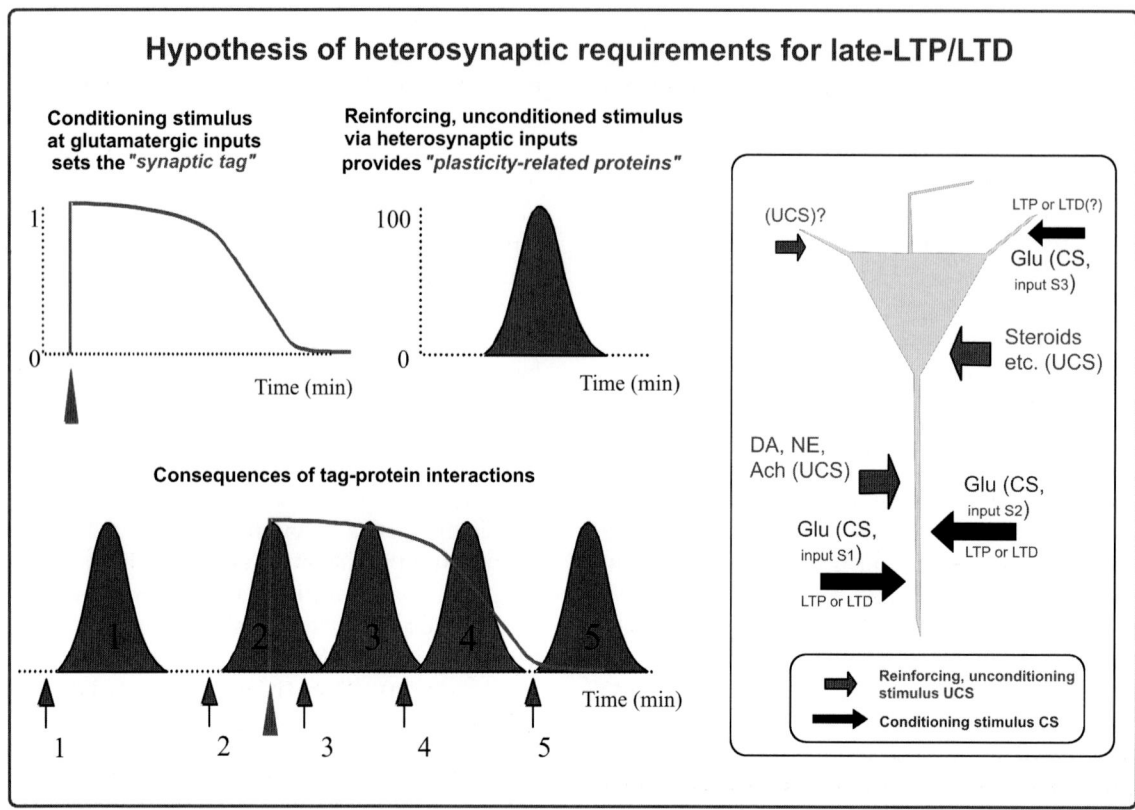

Fig. 2. Hypothesis of a heterosynaptic induction of long-lasting plastic changes. The left part shows specific properties of the interaction of setting a synaptic tag and the availability of plasticity-related proteins (PRPs). Our studies revealed that most likely glutamatergic inputs are required to set the synaptic tag in addition to maintaining early-LTP/LTD (conditioning stimulus, CS) and modulatory inputs regulate the synthesis of PRPs (although with synergistic interactions with NMDA-receptor function). We have identified dopaminergic, noradrenergic and cholinergic inputs as reinforcing, unconditioning stimuli (UCS) which seems to be substructure-dependent. We hypothesize, however, that other molecules can also act as UCS (e.g., steroids during stress, etc.). Our current hypothesis of synaptic tagging required for late-LTP/LTD is as follows: The induction of early-LTP/LTD on a glutamatergic CS pathway (blue arrow symbol) sets a tag with a probability of 1 and this happens immediately. The tag lasts a period of time and then decays in a probabilistic manner. We could identify that the activation of the tag is transient and decays within 30–60 min. Stimulation of a non-glutamatergic reinforcing, unconditioning input (UCS) either via a strong tetanus/strong low-frequency stimulation or via direct stimulation of extrahippocampal, modulatory systems, induces the synthesis and distribution to appropriate postsynaptic sites of putative PRPs with which the tag interacts to stabilize (or permissively enable) LTP/LTD. The lower left part shows the possible dynamics of heterosynaptic tag–protein interactions. Induction of early-LTP/LTD on a CS pathway is shown at the centre of the time diagram (blue arrow). Five separate cases are then considered. Case 1 involves the distribution and decay of macromolecule availability a long time before the tag is set. There can then be no interaction. Case 2 results in the peak availability of plasticity-proteins coinciding with the setting of a tag on the other pathway. Case 3 involves the setting of a tag and the subsequent synthesis and distribution of plasticity-proteins. In case 4, the interval is longer and only a limited interaction can occur, resulting in a smaller proportion of synapses becoming stabilized. Finally, in case 5, the interval is too long for any effective interaction to occur. These dynamics may represent the more native possibilities of interstructural interactions important for the induction of long-lasting plastic neuronal changes, schematically shown on the right part of the figure (DA: dopamine, NE: norepinephrine, Glu: glutamate, ACh: acetylcholine). Modified from Frey and Morris (1998b). Our later work suggests that glutamatergic inputs can transiently store sensory information under distinct circumstances. If the system then decides that this information is system-relevant, it can influence the duration of the normally transient storage of that sensory information by reinforcing heterosynaptic associative interactions within a time-window of about 30 min, i.e. the transient event can be "reinforced" into a long-term memory trace. The neuromodulatory input thereby regulates specifically the synthesis of PRPs. The right schema represents a CA1 pyramidal neuron with inputs representing the sensory information on glutamatergic synapses (conditioning stimulus, CS) and adequate neuromodulatory inputs (reinforcing, unconditioning stimulus, UCS) regulating and providing the availability of PRPs required for the transformation of the originally transient trace stored at CS-inputs into long-lasting memory traces. (See Color Plate 7.2 in color plate section.)

heterosynaptic, neuromodulatory requirements for inducing late-LTP/LTD, there is now growing evidence that in adult brains non-glutamatergic inputs, such as dopaminergic ones in the apical dendritic layer of CA1 pyramidal neurons, modulate the availability of PRPs in synergism with NMDA-receptor function, whereas the glutamatergic input alone mainly mediates the maintenance processes of LTP during early-LTP as well as the setting of the tag (for review, see Frey, 2001). Synergistic interactions with neuromodulatory inputs, however, during establishment of early-LTP as well as setting the tag can also not be ruled out.

During the last few years our results on tagging have been verified in various laboratories (e.g., Kauderer and Kandel, 2000; Barco et al., 2002, 2005; Dudek and Fields, 2002; Adams and Dudek, 2005; Young and Nguyen, 2005), and it was also shown that analogous tagging processes can occur in invertebrates' neurons (e.g., Martin et al., 1997). Furthermore, tagging was not only described for LTP but also LTD in mammals (Kauderer and Kandel, 2000; Sajikumar and Frey, 2004a; Sajikumar et al., 2005b) characterized by similar cellular and functional properties (Sajikumar and Frey, 2004a). Recently, the synaptic tagging model has been expanded to include functional interactions between LTP and LTD, referred to as 'cross-tagging' (Sajikumar and Frey, 2004a; Sajikumar et al., 2005b). 'Cross-tagging' describes the capability of late-LTP/late-LTD in one synaptic input (S1) to transform the opposite, protein synthesis-independent early-LTD/early-LTP in an independent synaptic input (S2) into its long-lasting form (Sajikumar and Frey, 2004a) (see also Fig. 3). Cross-tagging not only expands the repertoire of functional interactions between different synapses and afferent pathways, but also raises the following fundamental questions

(1) Cross-tagging requires that the transient synaptic tag must mark an activated synapse specifically for either LTP or LTD. Therefore the question arises, which molecules make tags LTP- or LTD-specific?
(2) Cross-tagging studies should allow searches for the specificity of the tag-complex within determined brain structures. We have recently shown that PRPs can be process-specific for LTP in apical dendrites of CA1 pyramidal neurons (i.e., protein kinase Mζ (PKMζ, a persistently active PKC isoform that is synthesized during LTP and required for its maintenance (Ling et al., 2002)), as well as process-unspecific or regulatory (such as the PDE4B3; see Sajikumar et al., 2005b). Can further cross-tagging studies also identify LTD-specific molecules and can differences between different brain (sub)structures be determined?
(3) If PRPs are not restricted to a local synaptic site one can ask whether their protein synthesis is restricted to local dendritic compartments (Martin et al., 2000; Steward and Schuman, 2001; Kelleher et al., 2004) or can the synthesis of PRPs affect distant synapses neuron-wide?
(4) Do neuromodulatory brain structures determine local, functional compartments including the specificity of tags and PRPs?

The latter questions address the fundamental issue as to whether distinct functional neuronal compartments exist, or if induction of a late plastic event affects the entire pyramidal neuron.

Identifying molecules mediating synaptic tags and the identity of PRPs within functional compartments

As mentioned above synaptic tags must mark an activated synapse specifically for either LTP or LTD. Which molecules then make tags LTP- or LTD-specific? Further, PRPs can be process-specific, -unspecific and regulatory ("process-specificity" means here its relation to the plasticity-type induced) (Navakkode et al., 2005; Sajikumar et al., 2005b). Can we therefore search for the specificity of the tag-complex by using cross-tagging paradigms? If PRPs are not restricted to a synapse, is their synthesis restricted to local dendritic compartments (Martin et al., 2000; Steward, 2002; Kelleher et al., 2004; Govindarajan et al., 2006) or can PRP-synthesis affect synapses neuron-wide? CA1 neurons receive their inputs in apical and basal dendrites from other hippocampal neurons and extrahippocampal structures such as the amygdala, ventral tegmental area and others (Amaral and Kurz, 1985; Alkon et al., 1991; Amaral and Witter, 1998; Pikkarainen et al., 1999; Lisman and Grace, 2005). The question arises how mechanisms of input-specificity, local dendritic protein synthesis and the interaction of different synaptic and modulatory inputs could interact in CA1 neurons to assure integrated information

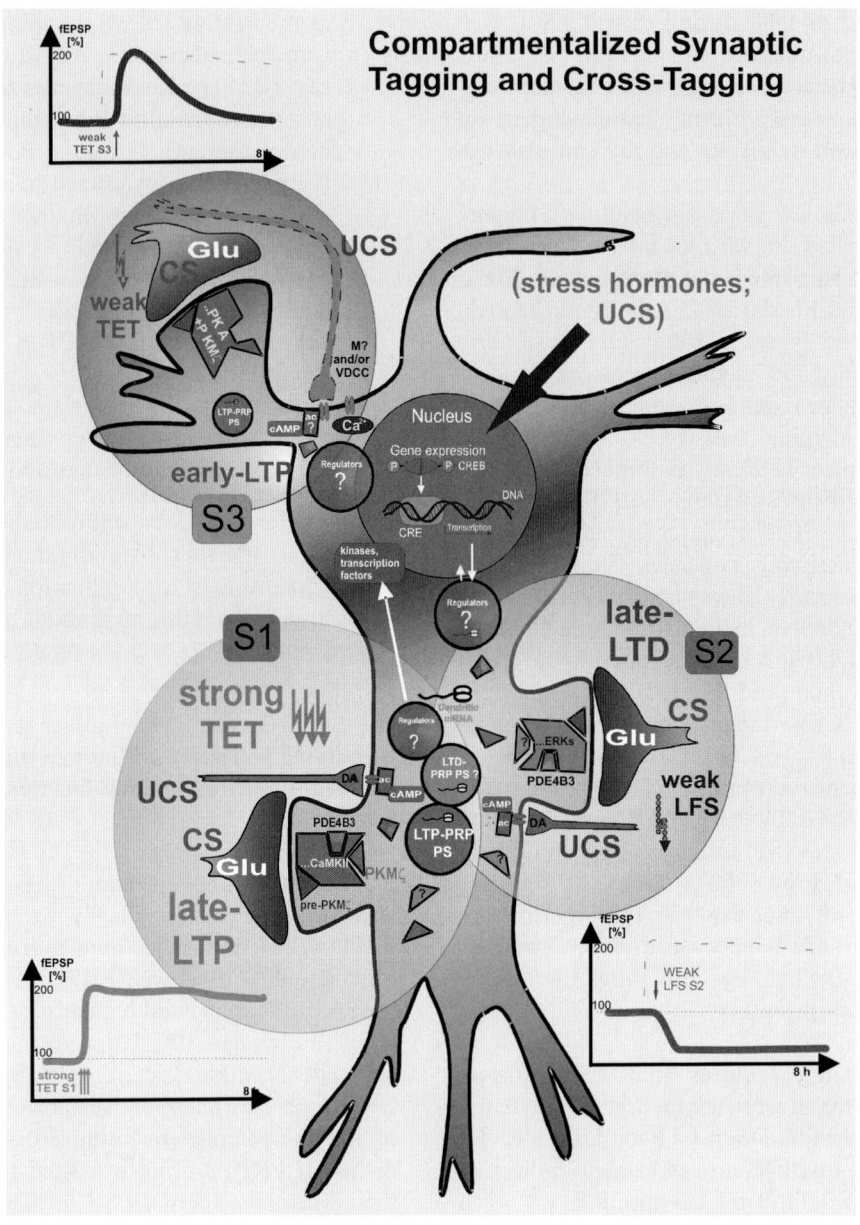

processing. The information from inputs containing spatial, contextual, relational information or information with system-relevant "decision" — content can be integratively processed at the target neuron, for instance, by processes of synaptic tagging or cross-tagging. We were now able to show using the 'cross-tagging'-paradigm (Sajikumar et al., 2007a) that the induction of late-LTP sets a process-specific synaptic tag that is mediated by calcium-calmodulin-dependent-kinase II (CaMKII), whereas induction of late-LTD sets its LTD-tags mediated by mitogen-activated protein kinases (MAPKs) in apical dendrites of hippocampal CA1. Interestingly, very similar results for MAPK-requirements were described for different forms of memory in Aplysia (Sharma et al., 2003). In basal dendrites (stratum oriens) we studied processes of synaptic LTP-tagging and found that MAPKs and CaMKII are unimportant for setting LTP-tags but either PKA or PKMζ mediate the setting of tags. Furthermore, we found that LTP or LTD induction is restricted to functional compartments in which PRP-synthesis occurs. Under the investigated conditions the PRPs appeared compartment-restricted. It still remains to be investigated, if LTD also depends on process-specific PRPs and if different inputs into pyramidal neurons as well as other brain structures are characterized by the expression of structure-specific key-proteins mediating the tag and/or PRP-interactions. One hint comes from recent work where it was shown that CaMKII-activity is not required for late-LTP in the DG (Cooke et al., 2006).

Fig. 3. Compartmentalized functional plasticity versus uncompartmentalization in a pyramidal CA1-neuron. The schematical pyramidal CA1-neuron represents our cellular hypothesis of compartmentalized synaptic tagging in the hippocampal CA1 region during cognitive information-processing (in contrast to uncompartmentalization by stress — see text). The induction of late-LTP (but also similarly late-LTD) in the apical synaptic input S1 activates postsynaptically, via a heterosynaptic synergistic interaction of glutamatergic (conditioning stimulus, CS) as well as dopaminergic receptors (modulatory input; unconditioning stimulus, UCS), a transient, protein synthesis-independent synaptic tag, which is made specific for LTP by CaMKII-dependent processes (dark red symbol at the postsynaptic site in S1) and the synthesis of a pool of PRPs (triangles, trapeziums). PRPs consist of LTP-specific proteins such as PKMζ (red triangles) and regulatory proteins such as PDE4B4 (red-blue symbols). Both the tag and the PRPs have a distinct half-life during that time-window an interaction of the two can occur, thus maintaining LTP in S1 for at least 8 h. If within this time-window of available PRPs within this dendritic compartment a second, independent synaptic input S2 in the apical dendritic compartment is stimulated to set a synaptic tag (e.g., by induction of early-LTD via a weak low-frequency stimulation (weak LFS)), the latter can benefit from the compartment-restricted availability of PRPs, thus transforming the normally transient plastic form into an enduring one, in the example shown, from early-LTD into late-LTD in S2 by processes of synaptic tagging, cross-tagging and capturing of the PRPs provided through input S1. LTD-tags in the apical CA1-dendrites however, are not dependent on CaMKII but rather on MAPK-activity. If within the same time-window at a basal dendritic compartment S3 an early-LTP is induced even with a synaptic tag set (here early-LTP by a weak tetanization in S3 (weak TET)), these tags cannot benefit from the compartmentalized availability of PRPs in the apical dendrites, thus resulting in a transient expression of LTP, in early-LTP. Synaptic tagging or better cross-compartmental tagging/capturing does not occur. Interestingly, the basal compartment is also characterized by different key-players responsible for setting the local tag-here, either PKA and/or PKMζ is required to mediate the tag. It remains unclear, whether basal plastic events require an activation of modulatory inputs and which molecules are the PRPs in basal dendritic compartments. For a prolonged maintenance of a plastic event beyond 8 h the expression of genes has been shown to be an obvious requirement, which is also schematically represented in the figure. Furthermore, under high stress or life-threatening situations (grey arrow on gene expression) an immediate gene expression can also occur, which results in an uncompartmentalization of the plasticity events, i.e. the transformation of early- into late-LTP in the transcompartments, as between S1 and S3 (see text and Sajikumar et al., 2007b). The graphs at the synaptic inputs S1, S2 and S3 schematically illustrate compartmentalized "synaptic tagging" in CA1 (with the exception of uncompartmentalization by stress hormones (see text). Induction of late-LTP by a strong tetanus (strong TET) in a synaptic input S1 in the apical dendrites leads to late-LTP (red curve). If a weak low-frequency stimulation (weak LFS) was applied within a distinct time-window to a separate synaptic input S2 in the apical dendrite the normally early-LTD was transformed into late-LTD by processes of synaptic cross-tagging (red curve in right lower graph). Weak TET to a synaptic input S3 in the basal dendrites — even if applied within the normally effective time-window of tagging within a dendritic compartment — revealed, however, that strong TET in apical S1 was ineffective in influencing early-LTP in S3 at basal dendrites (blue curve in the upper graph) during cognitive information-processing. Thus tagging is restricted to distinct dendritic compartments (stress however, is able to break the compartmentalization — see text). Modified from Frey and Morris (1998a); Reymann and Frey (2007); Sajikumar et al. (2007a). (See Color Plate 7.3 in color plate section.)

Late-associative properties of LTP and LTD: restriction of associative properties to neuronal compartments

In the light of the intact organism, an associative interaction between information from system-wide inputs to the hippocampus must be expected. Inputs containing spatial, contextual, relational information or information containing "system-relevant decisions" must therefore be associatively processed during the cellular formation of a memory trace within distinct neuronal nets. Such inputs can arrive as modulatory information integratively processed at the given neuronal compartment, e.g. by mechanisms of synaptic tagging and cross-tagging. Hippocampal CA1 neurons receive their inputs in apical and basal dendrites from hippocampal neurons and from various extrahippocampal structures, including modulatory inputs from the amygdala, ventral tegmental area and many other brain regions (Amaral and Witter, 1998). Apical and basal dendrites do not, however, just differ with respect to their innervation but also in their morphology and physiological properties (Leung and Shen, 1995; Pike et al., 2000). Thus many questions arise from these differences, such as whether system-relevant information processing and integration are restricted to single synapses, or to dendritic compartments, or to the whole neuron.

Recent results suggest that instead of the formation of a memory trace within a single synapse, functional synaptic populations within a dendritic compartment exist. These synaptic populations with their related intracellular messenger machinery could act as integrative units within neural networks (Engert and Bonhoeffer, 1997; Kelleher et al., 2004; Tonegawa, 2005; Govindarajan et al., 2006). Our hypothesis is that processes of synaptic tagging and cross-tagging should be locally restricted to control more efficiently and reliably systemic and organism-relevant information processing and computation (Frey, 2001; Polsky et al., 2004; Sajikumar et al., 2005b; Reymann and Frey, 2007). Until now these processes were mainly investigated in different synaptic inputs in the apical CA1-dendrites. Recently, Young and Nguyen (2005) have expanded this local restriction and described the first results of a negative interaction between stratum oriens stimulation and the availability to induce late-LTP in stratum radiatum of hippocampal pyramidal CA1 neurons. Dudek and Fields (2002) have shown that somatic action potentials can "prime" synaptic tagging at apical dendrites. In contrast to our hypothesis, the latter results suggest neuron-wide processes of synaptic tagging and cross-tagging. Recently, the Kandel-laboratory has shown that orthodromic induction of CA1-LTP is restricted to local, functional compartments in general. If, however, stimulation is strong enough, a distinct spread of plastic events between compartments is described (Alarcon et al., 2006). In parallel we replicated these results and have expanded these studies by showing that the induction of LTP as well as LTD and their interactions, such as cross-tagging, is also restricted to functional compartments (Sajikumar et al., 2007a), i.e. to neuronal parts receiving a distinct neuromodulatory input with some exceptions. With respect to the uncompartmentalization by very strong stimulation recently seen by Alarcon et al. (2006), we presented (Sajikumar et al., 2007b) one cellular pathway responsible for the uncompartmentalization of the normally localized plasticity processes by the action of rolipram, an inhibitor of type 4 phosphodiesterases. In contrast with compartment-restricted information-processing, uncompartmentalization requires immediate transcription. In the search for system relevance of compartmentalization versus uncompartmentalization we described firstly data which show that more cognitive information-processing in rats' behaviour may follow rules of compartmentalization, whereas stressful, more life-threatening, inputs abolish compartment-restricted information-processing involving transcription. Our findings allowed us to suggest that consolidation of processes which take place during the cognitive event most probably depend on local protein synthesis, whereas stress immediately induces gene expression in addition, resulting in a compartment-unspecific up-regulation of PRPs, providing the entire neuron with a higher level of "reactiveness". These data provide a specific functional cellular mechanism to respond differentially and effectively to behaviourally weighted inputs. Figure 3 represents compartmentalized/uncompartmentalized interactions of LTP and LTD in a hippocampal CA1 neuron.

Physiological, structural and behavioural reinforcement of early-plasticity events by activation of neuromodulatory inputs

We have shown that the induction of hippocampal late-LTP/LTD requires the associative activation of heterosynaptic inputs, such as the activation of glutamatergic and dopaminergic receptors in area CA1 or glutamatergic and noradrenergic or muscarinergic receptors in the DG (Frey, 2001). The structure-specific neuromodulatory inputs for LTD remain to be investigated. In the next part we refer therefore to LTP studies rather than LTD experiments, although first hints come from Manahan-Vaughan laboratory suggesting again very similar properties for LTD in the behaving animal (Kemp and Manahan-Vaughan, 2004, 2007; Lemon and Manahan-Vaughan, 2006). In the search for the physiological significance of the associative activation of two different inputs, however, it should be noted that dopaminergic neurons, for instance, are assumed to mediate, in part, the effect of biologically significant reinforcing stimuli during learning (Bischoff, 1986; Kemp and Manahan-Vaughan, 2004; Lemon and Manahan-Vaughan, 2006). Normal dopamine functioning also appears necessary for the establishment and maintenance of conditioned reinforcement, or incentive learning (Beninger, 1983). Thus, prolonged enhancement of synaptic efficacy during the processing of specific sensory information becomes understandable. Dopaminergic facilitation of learning is consistent with observations from Packard and White (1989), who demonstrated a role of dopaminergic receptors in memory facilitation on learning tasks sensitive to both hippocampal and caudate lesions. The question arises whether this requirement of the joint activation of heterosynaptic inputs, i.e. of extrahippocampal structures in addition to intrahippocampal processes, can be detected during an animal's behaviour and acquisition of information in combination with the induction of plasticity processes. So far, there are only a few experiments linking LTP as a physiological phenomenon involved in normal animal behaviour (see above). In these studies it was expected that specific behaviour might interfere with evoked population potential recordings, i.e. it was shown that novelty, for instance, can facilitate evoked potentials for a distinct time (Li et al., 2003; Straube et al., 2003b; Davis et al., 2004; Lemon and Manahan-Vaughan, 2006). Similar findings have been shown for LTD with respect to learning of novel objects in new locations (Kemp and Manahan-Vaughan, 2004). This approach requires that behaviour may drive a whole population of neurons within one direction. The question remains whether such a dramatic shift in the excitability of a well-known structure corresponds to a physiological processing of relevant information. Since 1995 we have tried different approaches taking into account the newly identified properties of late-LTP, i.e. its late associative properties. In a first approach, our laboratory evaluated the effect of behavioural and motivational states, which are known to affect learning, on LTP induced by strong or weak tetanic stimulation in the DG of freely moving rats (Seidenbecher et al., 1995, 1997). The strong tetanization produced a "saturated" LTP, which lasted more than 24 h and showed no significant difference between the behavioural states. LTP induced by weak tetanization, however, did differ across behavioural states. LTP which normally lasted only 5–7 h was clearly extended to more than 24 h if the water-deprived rats were allowed to drink during or up to 30 min after the tetanus. Similar results were obtained with application of foot shock instead of water. We concluded that both appetitive and aversive stimuli act within a certain time-window as reinforcers to consolidate LTP in a similar manner to that of memory formation. Under low stress conditions, such as during novelty detection or water delivery in water-deprived rats, the transformation of early-LTP into late-LTP in the DG depends on the activation of β-adrenergic receptors (Seidenbecher et al., 1997) and on protein synthesis (Straube et al., 2003a). It seems plausible that during our strong field stimulation, the transmitter the necessity to investigate processes in the CA1 region in the intact animal cannot be replaced by DG-recordings. Thus, we were also interested in CA1 experiments in vivo which would allow us to work in a similarly flexible manner as in vitro, i.e. to investigate associative interactions of modulatory inputs on different forms of LTP/LTD. During the last few years we have optimized stimulation and recording parameters for the CA1 which now have led to very reliable simultaneous recording of the field-EPSP and the population spike in freely moving animals for several weeks. To induce distinct forms of LTP we normally stimulate contralateral hippocampal substructures to avoid electrical-unspecific interactions with the ipsilateral site and to guarantee the stimulation of separate synaptic inputs (Hassan et al., 2006) and combine that with the activation of modulatory brain structures. To study whether CA1-early-LTP in the intact animal can also be transformed into late-LTP we have first studied the effect of the electrical activation of the ventral tegmental area (VTA, a structure which innervates the CA1 region with dopamine) at different time-points after tetanization of the CA1. Early-LTP was transformed into late-LTP in a protein synthesis and D1-receptor-dependent manner when the VTA was stimulated within 30 min after CA1-tetanization. These results in vivo resemble our in vitro results.

Behavioural reinforcement: identification of content and interaction of modulatory brain structures

Swim stress in the water-maze can result in the reinforcement of early- into late-LTP in the DG (Korz and Frey, 2003). This particular swim stress resulted in an enhanced phosphorylation of MAPK and CaMKII and the reinforcing effect on LTP by swim stress was mediated by mineralocorticoid receptors (MR) but not corticosterone or norepinephrine, muscarinergic or dopaminergic receptors. Our current working hypothesis is that strong emotional content such as high stress up-regulates the synthesis of PRPs in a more general way most likely through gene expression (for review, see Korz and Frey, 2004; see also Reymann and Frey, 2007).

Reinforcement, for instance, by novelty detection typically is a rapid process that occurs after a single exposure to the arousing situation and can be blocked by a brief pre-exposure to the stimuli making them familiar. The term "emotional tagging" has been introduced to characterize this form of LTP-reinforcement related to the rapid consolidation of relevant new information (Richter-Levin and Akirav, 2003). Recently, we have shown that early-LTP can be transformed into late-LTP when early-LTP is associated with holeboard training in a specific manner, i.e. critical time-windows within the learning curve determine the ability for reinforcement (Korz and Frey, 2004; Uzakov et al., 2005). During holeboard training, a significant improvement of learning and the consolidation of the reference memory after seven training trials can be detected. If the induction of early-LTP was related to that 7th trial (i.e., tetanization of the DG 15 min before the 7th trial) reference memory consolidation as well as LTP-reinforcement by holeboard-learning was detected, which was protein synthesis-dependent. These results suggest that the same newly synthesized proteins involved in the formation of reference memory are also required for LTP maintenance. Memory consolidation in this model, as well as LTP-reinforcement was dependent on the heterosynaptic activity of noradrenergic as well as dopaminergic inputs. Recently we were also able to describe a similar cognitive reinforcement of LTP by water-maze training after five training trials. After 10 trials, i.e. at a time-point where the reference memory is already consolidated, no reinforcement was observed. If, however, we removed the platform and the animals searched for it, a reinforcement of LTP appeared again (Korz and Frey, 2004). Support for the above results with respect to the modulation of LTP and LTD by dopamine-dependent behavioural aspects during various information processes comes from recent work by the Manahan-Vaughan laboratory (Lemon and Manahan-Vaughan, 2006; Kemp and Manahan-Vaughan, 2007).

Figure 5 schematically represents our current findings with respect to behavioural LTP-reinforcement.

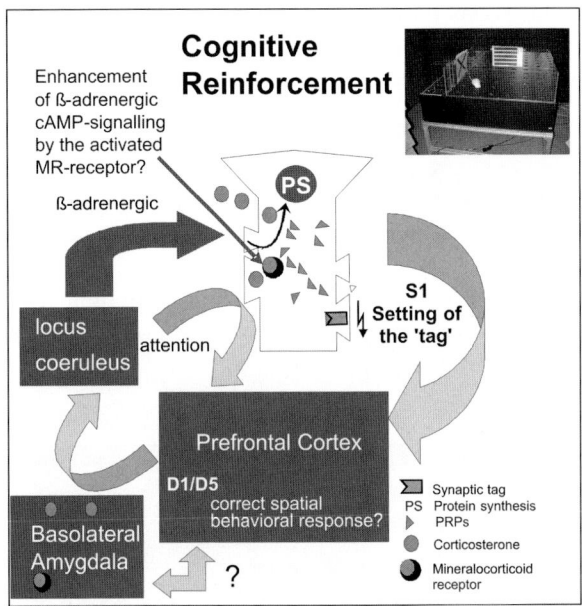
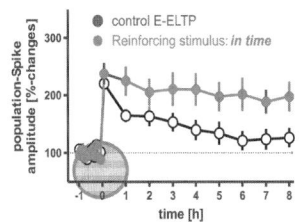
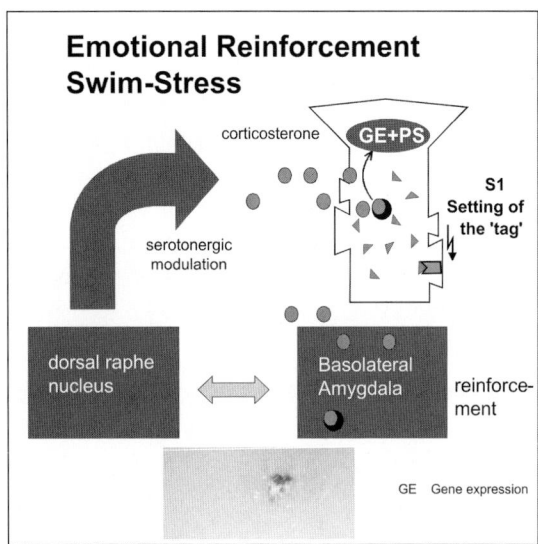

Fig. 5. Behavioural LTP-reinforcement. Hippocampal early-LTP in the DG (blue curve in the upper right graph, which sets a synaptic tag) can be reinforced into late-LTP (red curve) by the associative heterosynaptic stimulation of structure-specific, modulatory brain structures (provided in the lower schemes) during behavioural tasks. Like structural reinforcement distinct behavioural paradigms with different content can activate these modulatory structures to reinforce early-LTP into late-LTP. The modulatory brain structures regulate the synthesis of PRPs which then interact with the synaptic tags, thus expressing late-LTP. Emotional, high stress conditions (lower right scheme) revealed that a 2-min-swim in a water tank 15 min after a weak tetanus also prolongs early-LTP up to 24 h in naïve animals. The prolongation is dependent on gene expression, and protein synthesis but not on β-adrenergic activation. Through such a high stress paradigm, LTP is maintained by corticosterone-induced activation of mineralocorticoid receptors (MR). Because the activated MR-complex acts as a transcription factor (circles) and modulates the activity of other factors, resulting translational products may include PRPs (triangles) being involved in the maintenance of LTP. Indeed, this form of LTP-reinforcement requires immediate gene expression in addition to protein synthesis. LTP can also be reinforced by the mastering of a spatial task in a holeboard (lower left panel). In contrast to emotional reinforcement this cognitive reinforcement appears only after repetitive training in the same environment and is related to the formation of a long-lasting spatial reference memory, suggesting that proteins involved in memory consolidation also contribute to the maintenance of hippocampal LTP. This type of LTP-reinforcement does not require gene expression but involves protein synthesis. Different cellular signalling is involved in emotional and cognitive reinforcement of LTP as indicated by the boxes in each scheme. Adapted with permission from Korz and Frey (2004). (See Color Plate 7.5 in color plate section.)

Concept of 'cellular memory formation'

Finally, current results allow us to outline our hypothesis of compartmentalized memory consolidation at the cellular level (Frey, 2001; Korz and Frey, 2004; Sajikumar and Frey, 2004a; Sajikumar et al., 2005b; Reymann and Frey, 2007). We suggest that processing of encoding as well as consolidation of a memory trace takes place in functional compartments determined by late-associative interactions, i.e. synaptic tagging and synaptic cross-tagging including heterosynaptic requirements during its induction (Frey and Morris, 1998a; Sajikumar and Frey, 2004a). The functional relevance of compartmentalized cellular memory formation is thereby determined by the physiological relevance of the afferent information. As described previously (Frey, 2001), in the intact animal afferent information might be automatically and transiently stored at glutamatergic synapses (e.g., by processes of early-LTP or early-LTD), irrespective of their late, final relevance for the organism in this context. If, however, the system decides the information is relevant to be stored more permanently — within the time-window of about 30 min for late-associative interactions — the system can regulate in an associative manner the synthesis of PRPs within distinct functional neuronal compartments by modulatory inputs such as dopamine or other neuromodulators, or neuron-wide by stress hormones to transform the normally transient events into long-lasting ones and by transmitting the information from one to another set of neurons (by interactions between synapses during tagging). The latter would result in the fine-tuning or construction of existing or new neuronal networks containing and further processing the encoded and consolidated information. The uncompartmentalized regulation of PRPs by stress could also fulfil a complementary function and could affect the simultaneous consolidation of different compartment-specific memories by an arousal-dependent process.

Figure 6 summarizes our current concept of cellular memory formation in a pyramidal CA1-neuron including the consolidation of the memory trace in neuronal circuits.

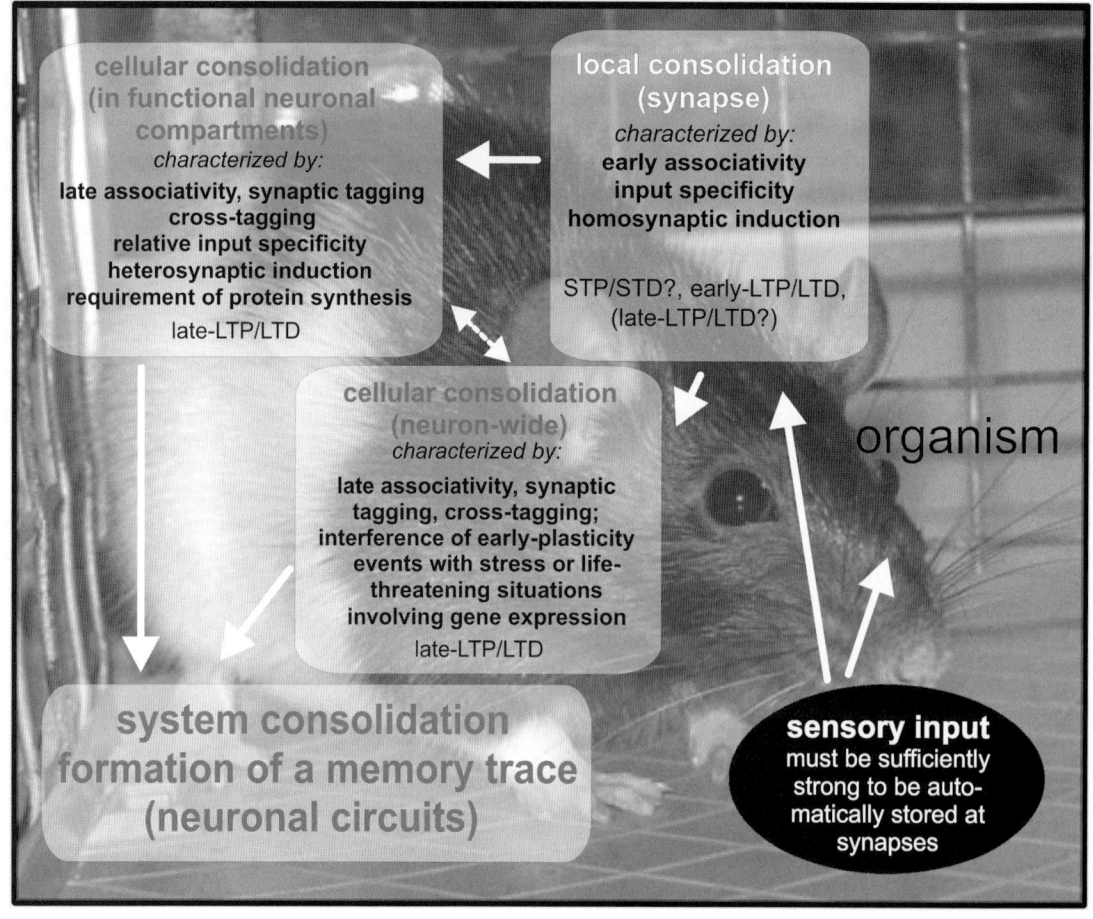

In general, we suggest that (see also Frey, 2001; Reymann and Frey, 2007) the induction of a late, protein synthesis-dependent form of functional plasticity can exert dramatic long-lasting effects on a population of synapses within specific functional dendritic compartments, if long-lasting, and with dramatic effects too on neuronal circuits. Hereby, the following specific rules can be described

(1) Specific afferent stimulation can induce either LTP or LTD at a specific set of synapses. The physiological meaning of the opposite way in storing information of LTP or LTD is still to be investigated. Local processes, however, guarantee the expression of only one form at a particular synapse over time by the possibility of setting a process-specific tag, i.e. CaMKII-dependent mechanisms for LTP, and ERK-dependent processes for LTD (Fig. 6 — local consolidation).

(2) If the afferent stimulation is strong enough to induce a more enduring form of LTP/LTD, in adult animals the heterosynaptic activation/regulation of the synthesis of PRPs is required, consisting of process-specific PRPs (e.g., PKMζ for LTP, and perhaps other molecules for LTD) and process-unspecific regulatory proteins (e.g., PDE4B3). Process-specific PRPs with a specific half-life can be captured by synaptic

Fig. 6. Consolidation of a memory trace. Schematic illustration of different localization levels where elements of memory consolidation by processes of functional plasticity may take place. Forms of LTP or LTD could be involved in the local and cellular processes required for system-wide consolidation. A sensory stimulation can result in the automatic and transient recording of the information at glutamatergic synapses (white). At the same time the information will be evaluated by the organism for its relevance (right arrow from the sensory information box into the organism). If the system decides the information is important and should be stored for longer periods of time, under non-life-threatening situations, a restricted, compartmentalized regulation of PRPs is achieved by the associative activation of specific neuromodulatory inputs within an effective time-window (e.g., by β-adrenergic inputs to DG-neurons or by dopaminergic inputs at apical CA1 dendrites). Depending on the task, more than one modulatory brain structure might be involved here, characterized by its own dynamics and associative cross-interactions with hippocampal neurons. Furthermore, depending on the state of the system prior to the sensory information it is also possible that PRPs are already available, i.e. the system is already primed by the heterosynaptic inputs resulting in the interaction with subsequent sensory information still transferring it into long-lasting memory (cellular consolidation in compartments — upper left box, grey). In contrast, the processing of very important information for the organism, like a life-threatening event with a high emotional content, requires storing of information in a large number of neuronal inputs reflected within several neuronal networks by loss of compartment-specificity within the given neuron. Basal dendrites in CA1 pyramidal cells with their close relation to the soma, and thus voltage-dependent-calcium-channel (VDCC)-dependent mechanisms or the activation of mineralocorticoid receptors which could act as transcription factors, may regulate a more general, soma-wide synthesis of PRPs, possibly via a direct activation of gene expression. These PRPs can then also be captured by synapses far away from the trigger of PRP-synthesis (Dudek and Fields, 2002; Adams and Dudek, 2005), and provide a cellular mechanism of a general cellular increase in the capacity to store relevant information at the cellular level (neuron-wide cellular consolidation, box in the middle). This is consistent with a partial spreading of tagging-processes from soma-near events to the apical dendrites (Alarcon et al., 2006; Sajikumar et al., 2007b). Furthermore, this could also explain the action of non-specifically acting stress hormones that up-regulate the synthesis of corticoid-receptor-dependent PRPs through a direct gene expression mechanism, which could provide the entire neuron with PRPs for a distinct period of time, which could then be captured at all activated synapses with a tag set. Such a mechanism was mentioned above in the description of the induction of PRP synthesis via swim-stress activation of corticoid-receptors which resulted in the reinforcement of early- into late-LTP (Korz and Frey, 2003; Sajikumar et al., 2007b). Compartmentalization and uncompartmentalization are most likely specific for distinct brain structures depending on their morphological organization/cyto-architecture and function. The more general regulation of neuron-wide changes of plasticity-capacity in CA1 pyramidal neurons by basal dendritic synaptic events points to such a mechanism. Thus, in that case, it could also be that in addition to the above-discussed modulatory mechanisms GABA-ergic neurons are specifically involved in the processing of information arriving at basal CA1-dendrites. For example, a fine-tuning of inhibition could lead to an enhanced depolarization of pyramidal cells through the activation of VDCCs located near the soma. Influx of calcium at this locus by GABA-ergic mechanisms could therefore also directly activate somatic protein synthesis and/or gene expression with a soma-wide impact on long-lasting plasticity events. See also the text and Reymann and Frey (2007). Long-lasting changes of synaptic efficacy at the cellular level affect also the neuronal networks in which the given neuron takes part. By mechanisms as yet unknown the cellular memory trace must then be stored and most likely also transferred to other neurons and brain areas resulting in the system's consolidation of the memory trace (lower left box).

tags within a time-window, and may further regulate the processing of proteins as well as the induction of gene expression. The latter could require, in addition, the action of process-unspecific regulatory proteins which would lead to the synthesis of mRNAs most likely specific for the compartment from which the induction of gene expression was generated (Steward et al., 1998) (Fig. 6 — cellular consolidation (in functional neuronal compartments)). Thus this mRNA would be transported to the local, restricted dendritic compartment under normal conditions, or — in conditions with a high emotional impact — this mRNA could provide the neuron with PRPs used neuron-wide (Fig. 6 — cellular consolidation (neuron-wide)) for the transformation of early- into late-plasticity. Under distinct circumstances, the compartmentalized restriction of information-processing can be abolished by the impact of life-threatening or stressful inputs (see above; Sajikumar et al., 2007b).

(3) Changes in neuronal plasticity as described in (2) can cause the formation of an enduring memory trace most likely involving many neurons and other brain structures. Future studies are required for a better understanding of the impact of functional plasticity on these processes (Fig. 6 — system consolidation).

(4) A time-window of about 30 min in intact animals or of about 1–2 h in slices maintained at a temperature of 32 °C (see Sajikumar et al., 2005a) was detected for cross-synaptic late associative interactions to occur. Within this time interval — which is longer than that conventionally considered for heterosynaptic associative processes — heterosynaptic processes can induce/regulate the availability of a pool of the above-mentioned local PRPs, which contains process-specific proteins (e.g., PKMζ) and process-non-specific proteins. The latter would be required for local processing at the synapses or the induction of a prolonged regulation of the availability of local sets of mRNAs by activation of gene expression. The latter two properties may require a system-wide evaluation of the information to be consolidated, including extrahippocampal brain structures to determine the information as sufficiently relevant to store it more permanently (Fig. 6 — system consolidation). In that case, the activation of heterosynaptic inputs within the above time interval is sufficient effectively to transform the transient into a long-lasting event by regulating the availability of PRPs. We suggest that there are different kinds of neuromodulators fulfilling different but specific functions by regulating the synthesis of PRPs. Thus, as mentioned above, general modulators like stress hormones may interfere with neuron-wide, compartment-unspecific PRP-synthesis whereas dopaminergic inputs in the hippocampal CA1 carrying relational or contextual information may regulate local protein synthesis at apical dendrites. Other, as yet unknown modulators could be required for the synthesis of PRPs in the basal dendritic compartments.

In sum, the described new properties of late-plastic events in hippocampal neurons support a new principle of cellular information storage, i.e. the idea of a "compartmentalized processing and storage of information" (Frey, 2001; Reymann and Frey, 2007). We provided the first experimental evidence for the existence of functional clusters during cellular information storage (Govindarajan et al., 2006) by studying processes of hippocampal long-lasting functional plasticity. Recently, we were also able to describe similar dopaminergic associative interactions with hippocampal function during novelty detection as well as reward-related hippocampus-dependent long-term memory formation in humans using functional magnetic resonance imaging (fMRI) (Schott et al., 2004; Wittmann et al., 2005). First, as we have discussed before, LTP in the hippocampus can be enhanced and prolonged by dopaminergic inputs from mesolimbic mid-brain structures such as the medial substantia nigra (MSN). This improved synaptic plasticity was now hypothesized to be associated with better memory consolidation in the hippocampus. We used a condition that reliably elicited a dopaminergic response, reward anticipation, to study the relationship between activity of dopaminergic mesolimbic mid-brain areas and hippocampal long-term memory consolidation formation in healthy human adults. Pictures of object drawings that predicted monetary reward were associated with stronger fMRI activity in reward-related brain areas including the MSN compared with non-reward-predicting pictures. Three weeks later, recollection and source memory were better for reward-predicting than for non-reward-predicting pictures. Functional MRI activity in the hippocampus and the MSN mid-brain was higher for reward-predicting pictures that were later recognized compared with later forgotten pictures. These data are consistent with the hypothesis that activation of dopaminergic mid-brain regions enhances hippocampus-dependent memory formation, possibly by enhancing consolidation. In the second case, as we have also shown earlier exploring a novel environment can facilitate subsequent hippocampal LTP in animals. We now reported a related behavioural enhancement in humans. In two separate experiments recollection and free recall, both measures of hippocampus-dependent memory formation, were enhanced for words studied after a 5-min exposure to unrelated novel, as opposed to familiar, images depicting indoor and outdoor scenes. With fMRI the enhancement was predicted by specific activity patterns observed during novelty-exposure, in brain regions known to be linked to attentional orienting to novel stimuli and perceptual processing of scenes (data submitted for publication).

These findings provide specific behavioural clues to enhancing hippocampus-dependent memory in humans in direct correlation with our findings at the cellular level during hippocampal functional plasticity.

Abbreviations

CA1	cornu ammonis 1
cAMP	cyclic amino-mono-phosphate
CaMK	calcium-calmodulin activated kinase
CS	conditioning stimulus
DA	dopamine
DG	dentate gyrus
fMRI	functional magnetic resonance imaging
EPSP	excitatory postsynaptic potential
GABA	gamma-amino-butyric acid
Glu	glutamate
LTD	long-term depression
LTP	long-term potentiation
MAPK	mitogen-activated kinase
mGluR	metabotropic glutamate receptor
mRNA	messenger ribonucleic acid
MSN	medial substantia nigra
NMDA	N-methyl-D-aspartate
PDE	phosphodiesterase
PKA	protein kinase A
PKM	protein kinase M
PRP	plasticity-related proteins
STD	short-term depression
STP	short-term potentiation
UCS	unconditioning stimulus

Acknowledgements

We are very grateful to our first teachers in the neuroscience of learning and memory, and to our friends and colleagues Hansjuergen Matthies and Manfred Krug.

References

Adams, J.P. and Dudek, S.M. (2005) On the late-phase of LTP: getting to the nucleus. Nat. Rev. Neurosci., 6: 737–743.

Alarcon, J.M., Barco, A. and Kandel, E.R. (2006) Capture of the late phase of long-term potentiation within and across the apical and basilar dendritic compartments of CA1 pyramidal neurons: synaptic tagging is compartment restricted. J. Neurosci., 26: 256–264.

Alkon, D.L., Amaral, D.G., Bear, M.F., Black, J., Carew, T.J., Cohen, N.J., Disterhoft, J.F., Eichenbaum, H., Golski, S., Gorman, L.K., Lynch, G., McNaughton, B.L., Mishkin, M., Moyer, J.R.J., Olds, J.L., Olton, D.S., Otto, T., Squire, L.R., Staubli, U., Thompson, L.T. and Wible, C. (1991) Learning and memory. Brain Res. Rev., 16: 193–220.

Amaral, D.G. and Kurz, J. (1985) An analysis of the origins of the cholinergic and noncholinergic septal projections to the hippocampal formation of the rat. J. Comp. Neurol., 240: 37–59.

Amaral, D.G. and Witter, M.P. (1998) Hippocampal formation. In: Paxinos G. (Ed.), The Rat Nervous System. Academic Press, San Diego, CA, pp. 443–493.

Barco, A., Alarcon, J.M. and Kandel, E.R. (2002) Expression of constitutively active CREB protein facilitates the late phase of long-term potentiation by enhancing synaptic capture. Cell, 108: 689–703.

Barco, A., Patterson, S., Alarcon, J.M., Gromova, P., Mata-Roig, M., Morozov, A. and Kandel, E.R. (2005) Gene expression profiling of facilitated L-LTP in VP16-CREB mice reveals that BDNF is critical for the maintenance of LTP and its synaptic capture. Neuron, 48: 123–137.

Baulac, M., Verney, C. and Berger, B. (1986) Dopaminergic innervation of parahippocampal and hippocampal regions in the rat. Rev. Neurol., 142: 895–905.

Bear, M.F. and Malenka, R.C. (1994) Synaptic plasticity: LTP and LTD. Curr. Opin. Neurobiol., 4: 389–399.

Beninger, R.J. (1983) The role of dopamine in locomotor activity and learning. Brain Res. Rev., 6: 173–196.

Bergado, J., Frey, S., Lopez, J., Almaguer-Melian, W. and Frey, J.U. (2007) Cholinergic afferents to the locus coeruleus and noradrenergic afferents to the medial septum mediate LTP-reinforcement in the dentate gyrus by stimulation of the amygdala. Neurobiol. Learn. Mem., 88: 331–341.

Bischoff, S. (1986) Mesohippocampal dopamine system: characterization, functional and clinical implications. In: Isaacson R.L. and Pribram K.H. (Eds.), The Hippocampus. Plenum Press, New York, pp. 1–32.

Bliss, T.V. and Gardner-Medwin, A.R. (1973) Long-lasting potentiation of synaptic transmission in the dentate area of the unanaesthetized rabbit following stimulation of the perforant path. J. Physiol. (Lond.), 232: 357–374.

Bliss, T.V. and Lomo, T. (1973) Long-lasting potentiation of synaptic transmission in the dentate area of the anaesthetized rabbit following stimulation of the perforant path. J. Physiol. (Lond.), 232: 331–356.

Bliss, T.V.P. and Collingridge, G.L. (1993) A synaptic model of memory: long-term potentiation in the hippocampus. Nature, 361: 31–39.

Bortolotto, Z.A. and Collingridge, G.L. (2000) A role for protein kinase C in a form of metaplasticity that regulates the induction of long-term potentiation at CA1 synapses of the adult rat hippocampus. Eur. J. Neurosci., 12: 4055–4062.

Bradshaw, K.D., Emptage, N.J. and Bliss, T.V. (2003) A role for dendritic protein synthesis in hippocampal late LTP. Eur. J. Neurosci., 18: 3150–3152.

Bullock, S., Lössner, B., Krug, M., Frey, S., Rose, S.P. and Matthies, H. (1990) Posttetanic long-term potentiation in rat dentate area increases postsynaptic 411B immunoreactivity. J. Neurochem., 55: 708–713.

Buzsáki, G. (1997) Functions for interneuronal nets in the hippocampus. Can. J. Physiol. Pharmacol., 75: 508–515.

Chapman, C.A., Perez, Y. and Lacaille, J.C. (1998) Effects of $GABA_A$ inhibition on the expression of long-term potentiation in CA1 pyramidal cells are dependent on tetanization parameters. Hippocampus, 8: 289–298.

Cooke, S.F., Wu, J., Plattner, F., Errington, M., Rowan, M., Peters, M., Hirano, A., Bradshaw, K.D., Anwyl, R., Bliss, T. and Giese, K.P. (2006) Autophosphorylation of {alpha}-CaMKII is not a general requirement for NMDA receptor-dependent LTP in the adult mouse. J. Physiol., 574: 805–818.

Cracco, J.B., Serrano, P., Moskowitz, S.I., Bergold, P.J. and Sacktor, T.C. (2005) Protein synthesis-dependent LTP in isolated dendrites of CA1 pyramidal cells. Hippocampus, 15: 551–556.

Davis, C.D., Jones, F.L. and Derrick, B.E. (2004) Novel environments enhance the induction and maintenance of long-term potentiation in the dentate gyrus. J. Neurosci., 24: 6497–6506.

Davis, S., Butcher, S.P. and Morris, R.G. (1992) The NMDA receptor antagonist D-2-amino-5-phosphonopentanoate (D-AP5) impairs spatial learning and LTP in vivo at intracerebral concentrations comparable to those that block LTP in vitro. J. Neurosci., 12: 21–34.

Dudek, S.M. and Fields, R.D. (2002) Somatic action potentials are sufficient for late-phase LTP-related cell signaling. Proc. Natl. Acad. Sci. U.S.A., 99: 3962–3967.

Duffy, C., Teyler, T.J. and Shashoua, V.E. (1981) Long-term potentiation in the hippocampal slice: evidence for stimulated secretion of newly synthesized proteins. Science, 212: 1148–1151.

Engert, F. and Bonhoeffer, T. (1997) Synapse specificity of long-term potentiation breaks down at short distances. Nature, 388: 279–284.

Fazeli, M.S., Corbet, J., Dunn, M.J., Dolphin, A.C. and Bliss, T.V.P. (1993) Changes in protein synthesis accompanying long-term potentiation in the dentate gyrus in vivo. J. Neurosci., 13: 1346–1353.

Fazeli, M.S., Errington, M.L., Dolphin, A.C. and Bliss, T.V. (1988) Long-term potentiation in the dentate gyrus of the anaesthetized rat is accompanied by an increase in protein efflux into push–pull cannula perfusates. Brain Res., 473: 51–59.

Flood, J.F., Smith, G.E. and Jarvik, M.E. (1980) A comparison of the effects of localized brain administration of catecholamine and protein synthesis inhibitors on memory processing. Brain Res., 197: 153–165.

Frey, J.U. (2001) Long-lasting hippocampal plasticity: cellular model for memory consolidation? In: Richter D. (Ed.), Cell Polarity and Subcellular RNA Localization. Springer-Verlag, Berlin-Heidelberg, pp. 27–40.

Frey, S., Bergado, J.A. and Frey, J.U. (2003) Modulation of late phases of long-term potentiation in rat dentate gyrus by stimulation of the medial septum. Neuroscience, 118: 1055–1062.

Frey, S., Bergado-Rosado, J., Seidenbecher, T., Pape, H.C. and Frey, J.U. (2001) Reinforcement of early long-term potentiation (early-LTP) in dentate gyrus by stimulation of the basolateral amygdala: heterosynaptic induction mechanisms of late-LTP. J. Neurosci., 21: 3697–3703.

Frey, S., Schweigert, C., Krug, M. and Lossner, B. (1991a) Long-term potentiation induced changes in protein synthesis of hippocampal subfields of freely moving rats: time-course. Biomed. Biochim. Acta, 50: 1231–1240.

Frey, U. (1997) Cellular mechanisms of long-term potentiation: late maintenance. In: Donahoe J.W. and Dorsel V.P. (Eds.), Neural Network Models of Cognition: Biobehavioral Foundations. Elsevier Science Press, Amsterdam, pp. 105–128.

Frey, U., Frey, S., Schollmeier, F. and Krug, M. (1996) Influence of actinomycin D, a RNA synthesis inhibitor, on long-term potentiation in rat hippocampal neurons in vivo and in vitro. J. Physiol., 490: 703–711.

Frey, U., Hartmann, S. and Matthies, H. (1989a) Domperidone, an inhibitor of the D2-receptor, blocks a late phase of an electrically induced long-term potentiation in the CA1-region in rats. Biomed. Biochim. Acta, 48: 473–476.

Frey, U., Huang, Y.-Y. and Kandel, E.R. (1993a) Effects of cAMP simulate a late stage of LTP in hippocampal CA1 neurons. Science, 260: 1661–1664.

Frey, U., Krug, M., Brödemann, R., Reymann, K. and Matthies, H. (1989b) Long-term potentiation induced in dendrites separated from rat's CA1 pyramidal somata does not establish a late phase. Neurosci. Lett., 97: 135–139.

Frey, U., Krug, M., Reymann, K.G. and Matthies, H. (1988) Anisomycin, an inhibitor of protein synthesis, blocks late phases of LTP phenomena in the hippocampal CA1 region in vitro. Brain Res., 452: 57–65.

Frey, U., Matthies, H. and Reymann, K.G. (1991b) The effect of dopaminergic D1 receptor blockade during tetanization on the expression of long-term potentiation in the rat CA1 region in vitro. Neurosci. Lett., 129: 111–114.

Frey, U. and Morris, R.G.M. (1997) Synaptic tagging and long-term potentiation. Nature, 385: 533–536.

Frey, U. and Morris, R.G.M. (1998a) Synaptic tagging: implications for late maintenance of hippocampal long-term potentiation. Trends Neurosci., 21: 181–188.

Frey, U. and Morris, R.G.M. (1998b) Weak before strong: dissociating synaptic tagging and plasticity-factor accounts of late-LTP. Neuropharmacology, 37: 545–552.

Frey, U., Schollmeier, K., Reymann, K.G. and Seidenbecher, T. (1995) Asymptotic hippocampal long-term potentiation in rats does not preclude additional potentiation at later phases. Neuroscience, 67: 799–807.

Frey, U., Schroeder, H. and Matthies, H. (1990) Dopaminergic antagonists prevent long-term maintenance of posttetanic LTP in the CA1 region of rat hippocampal slices. Brain Res., 522: 69–75.

Frey, U., Seidenbecher, T. and Krug, M. (1993b) The action of the RNA-synthesis inhibitor actinomycin D on a late stage of LTP in hippocampal neurons in vivo and in vitro. Soc. Neurosci. Abstr., 19: p. 912.

Govindarajan, A., Kelleher, R.J. and Tonegawa, S. (2006) A clustered plasticity model of long-term memory engrams. Nat. Rev. Neurosci., 7: 575–583.

Hassan, H., Frey, S. and Frey, J.U. (2006) Search for a two-input model for future investigations of 'synaptic tagging' in freely moving animals in vivo. J. Neurosci. Methods, 152: 220–228.

Herron, C.E., Lester, R.A., Coan, E.J. and Collingridge, G.L. (1986) Frequency-dependent involvement of NMDA receptors in the hippocampus: a novel synaptic mechanism. Nature, 322: 265–268.

Huang, Y.Y. and Kandel, E.R. (1994) Recruitment of long-lasting and protein kinase A-dependent long-term potentiation in the CA1 region of hippocampus requires repeated tetanization. Learn. Mem., 1: 74–82.

Huang, Y.-Y. and Kandel, E.R. (1995) D1/D5 receptor agonists induce a protein synthesis-dependent late potentiation in the CA1 region of the hippocampus. Proc. Natl. Acad. Sci. U.S.A., 92: 2446–2450.

Huang, Y.Y. and Kandel, E.R. (2005) Theta frequency stimulation induces a local form of late phase LTP in the CA1 region of the hippocampus. Learn. Mem., 12: 587–593.

Huber, K.M., Kayser, M.S. and Bear, M.F. (2000) Role for rapid dendritic protein synthesis in hippocampal mGluR-dependent long-term depression. Science, 288: 1254–1256.

Karpova, A., Sanna, P.P. and Behnisch, T. (2006) Involvement of multiple phosphatidylinositol 3-kinase-dependent pathways in the persistence of late-phase long-term potentiation expression. Neuroscience, 137: 833–841.

Kauderer, B.S. and Kandel, E.R. (2000) Capture of a protein synthesis-dependent component of long-term depression. Proc. Natl. Acad. Sci. U.S.A., 97: 13342–13347.

Kelleher, III, R.J., Govindarajan, A. and Tonegawa, S. (2004) Translational regulatory mechanisms in persistent forms of synaptic plasticity. Neuron, 44: 59–73.

Kemp, A. and Manahan-Vaughan, D. (2004) Hippocampal long-term depression and long-term potentiation encode different aspects of novelty acquisition. Proc. Natl. Acad. Sci. U.S.A., 101: 8192–8197.

Kemp, A. and Manahan-Vaughan, D. (2007) Hippocampal long-term depression: master or minion in declarative memory processes? Trends Neurosci., 30: 111–118.

Korz, V. and Frey, J.U. (2003) Stress-related modulation of hippocampal long-term potentiation in rats: involvement of adrenal steroid receptors. J. Neurosci., 23: 7281–7287.

Korz, V. and Frey, J.U. (2004) Emotional and cognitive reinforcement of rat hippocampal long-term potentiation by different learning paradigms. Neuron Glia Biol., 1: 253–261.

Krug, M., Lössner, B. and Ott, T. (1984) Anisomycin blocks the late phase of long-term potentiation in the dentate gyrus of freely moving rats. Brain Res. Bull., 13: 39–42.

Lemon, N. and Manahan-Vaughan, D. (2006) Dopamine D_1/D_5 receptors gate the acquisition of novel information through hippocampal long-term potentiation and long-term depression. J. Neurosci., 26: 7723–7729.

Leung, L.S. and Shen, B. (1995) Long-term potentiation at the apical and basal dendritic synapses of CA1 after local stimulation in behaving rats. J. Neurophysiol., 73: 1938–1946.

Li, S., Cullen, W.K., Anwyl, R. and Rowan, M.J. (2003) Dopamine-dependent facilitation of LTP induction in hippocampal CA1 by exposure to spatial novelty. Nat. Neurosci., 6: 526–531.

Ling, D.S., Benardo, L.S., Serrano, P.A., Blace, N., Kelly, M.T., Crary, J.F. and Sacktor, T.C. (2002) Protein kinase Mzeta is necessary and sufficient for LTP maintenance. Nat. Neurosci., 5: 295–296.

Lisman, J.E. and Grace, A.A. (2005) The hippocampal-VTA loop: controlling the entry of information into long-term memory. Neuron, 46: 703–713.

Lössner, B., Jork, R., Krug, M. and Matthies, H. (1982) Protein synthesis in rat hippocampus during training and stimulation experiments. In: Marsan C.A. and Matthies H. (Eds.), Neural Plasticity and Memory Formation. Ravens Press, New York, pp. 183–191.

Lynch, M.A., Voss, K.L., Rodriguez, J. and Bliss, T.V.P. (1994) Increase in synaptic vesicle proteins accompanies long-term potentiation in the dentate gyrus. Neuroscience, 60: 1–5.

Manahan-Vaughan, D., Kulla, A. and Frey, J.U. (2000) Requirement of translation but not transcription for the maintenance of long-term depression in the CA1 region of freely moving rats. J. Neurosci., 20: 8572–8576.

Martin, K.C., Barad, M. and Kandel, E.R. (2000) Local protein synthesis and its role in synapse-specific plasticity. Curr. Opin. Neurobiol., 10: 587–592.

Martin, K.C., Casadio, A., Zhu, H.X., Rose, J.C., Chen, M., Bailey, C.H. and Kandel, E.R. (1997) Synapse-specific, long-form facilitation of *Aplysia* sensory to motor synapses: a function for local protein synthesis in memory storage. Cell, 91: 927–938.

Matthies, H. (1974) The biochemical basis of learning and memory. Life Sci., 15: 2017–2031.

Matthies, H. (1989) Neurobiological aspects of learning and memory. Annu. Rev. Psychol., 40: 381–404.

Matthies, H., Frey, U., Krug, M., Pohle, W., Reymann, K.G. and Rüthrich, H. (1990a) Long-term synaptic potentiation and macromolecular changes in memory formation. Ergebn. Exp. Med., 51: 88–94.

Matthies, H., Frey, U., Reymann, K., Krug, M., Jork, R. and Schroeder, H. (1990b) Different mechanisms and multiple stages of LTP. Adv. Exp. Med. Biol., 268: 359–368.

Matthies, H. and Reymann, K.G. (1993) Protein kinase A inhibitors prevent the maintenance of hippocampal long-term potentiation. Neuroreport, 4: 712–714.

Navakkode, S., Sajikumar, S. and Frey, J.U. (2005) Mitogen-activated protein kinase-mediated reinforcement of hippocampal early long-term depression by the type IV-specific phosphodiesterase inhibitor rolipram and its effect on synaptic tagging. J. Neurosci., 25: 10664–10670.

Navakkode, S., Sajikumar, S. and Frey, J.U. (2007) Synergistic requirements for the induction of dopaminergic D1/D5-receptor-mediated LTP in hippocampal slices of rat CA1 in vitro. Neuropharmacology, 52: 1547–1554.

Nayak, A., Zastrow, D.J., Lickteig, R., Zahniser, N.R. and Browning, M.D. (1998) Maintenance of late-phase LTP is accompanied by PKA-dependent increase in AMPA receptor synthesis. Nature, 394: 680–683.

Nguyen, P.V., Abel, T. and Kandel, E.R. (1994) Requirement of a critical period of transcription for induction of a late phase of LTP. Science, 265: 1104–1107.

O'Carroll, C.M. and Morris, R.G.M. (2004) Heterosynaptic co-activation of glutamatergic and dopaminergic afferents is required to induce persistent long-term potentiation. Neuropharmacology, 47: 324–332.

Osten, P., Valsamis, L., Harris, A. and Sacktor, T.C. (1996) Protein synthesis-dependent formation of protein kinase Mzeta in long-term potentiation. J. Neurosci., 16: 2444–2451.

Otani, S. and Abraham, W.C. (1989) Inhibition of protein synthesis in the dentate gyrus, but not the entorhinal cortex, blocks maintenance of long-term potentiation in rats. Neurosci. Lett., 106: 175–180.

Otani, S., Marshall, C.J., Tate, W.P., Goddard, G.V. and Abraham, W.C. (1989) Maintenance of long-term potentiation in rat dentate gyrus requires protein synthesis but not messenger RNA synthesis immediately post-tetanization. Neuroscience, 28: 519–526.

Packard, M.G. and White, N.M. (1989) Memory facilitation produced by dopamine agonists: role of receptor subtype and mnemonic requirements. Pharmacol. Biochem. Behav., 33: 511–518.

Pang, P.T. and Lu, B. (2004) Regulation of late-phase LTP and long-term memory in normal and aging hippocampus: role of secreted proteins tPA and BDNF. Ageing Res. Rev., 3: 407–430.

Perez, Y., Chapman, C.A., Woodhall, G., Robitaille, R. and Lacaille, J.C. (1999) Differential induction of long-lasting potentiation of inhibitory postsynaptic potentials by theta patterned stimulation versus 100-Hz tetanization in hippocampal pyramidal cells in vitro. Neuroscience, 90: 747–757.

Pike, F.G., Goddard, R.S., Suckling, J.M., Ganter, P., Kasthuri, N. and Paulsen, O. (2000) Distinct frequency preferences of different types of rat hippocampal neurones in response to oscillatory input currents. J. Physiol., 529(Pt. 1): 205–213.

Pikkarainen, M., Rönkkö, S., Savander, V., Insausti, R. and Pitkänen, A. (1999) Projections from the lateral, basal, and accessory basal nuclei of the amygdala to the hippocampal formation in rat. J. Comp. Neurol., 403: 229–260.

Pockett, S., Slack, J.R. and Peacock, S. (1993) Cyclic AMP and long-term potentiation in the CA1 region of rat hippocampus. Neuroscience, 52: 229–236.

Polsky, A., Mel, B.W. and Schiller, J. (2004) Computational subunits in thin dendrites of pyramidal cells. Nat. Neurosci., 7: 621–627.

Reymann, K., Frey, U. and Matthies, H. (1988a) A multi-phase model of synaptic long-term potentiation in hippocampal CA1 neurones: protein kinase C activation and protein synthesis are required for the maintenance of the trace. In: Haas H.L. and Buzsaki G. (Eds.), Synaptic Plasticity in the Hippocampus. Springer, Berlin, Heidelberg, pp. 126–129.

Reymann, K.G. and Frey, J.U. (2007) The late maintenance of hippocampal LTP: requirements, phases, 'synaptic tagging', 'late-associativity' and implications. Neuropharmacology, 52: 24–40.

Reymann, K.G., Frey, U., Jork, R. and Matthies, H. (1988b) Polymyxin B, an inhibitor of protein kinase C, prevents the maintenance of synaptic long-term potentiation in hippocampal CA1 neurons. Brain Res., 440: 305–314.

Reymann, K.G., Malisch, R., Schulzeck, K., Brödemann, R., Ott, T. and Matthies, H. (1985) The duration of long-term potentiation in the CA1 region of the hippocampal slice preparation. Brain Res. Bull., 15: 249–255.

Richter-Levin, G. and Akirav, I. (2003) Emotional tagging of memory formation: in the search for neural mechanisms. Brain Res. Rev., 43: 247–256.

Sajikumar, S. and Frey, J.U. (2003) Anisomycin inhibits the late maintenance of long-term depression in rat hippocampal slices in vitro. Neurosci. Lett., 338: 147–150.

Sajikumar, S. and Frey, J.U. (2004a) Late-associativity, synaptic tagging, and the role of dopamine during LTP and LTD. Neurobiol. Learn. Mem., 82: 12–25.

Sajikumar, S. and Frey, J.U. (2004b) Resetting of 'synaptic tags' is time- and activity-dependent in rat hippocampal Ca1 in vitro. Neuroscience, 129: 503–507.

Sajikumar, S., Navakkode, S. and Frey, J.U. (2005a) Protein synthesis-dependent long-term functional plasticity: methods and techniques. Curr. Opin. Neurobiol., 15: 607–613.

Sajikumar, S., Navakkode, S. and Frey, J.U. (2007a) Identification of compartment- and process-specific molecules required for "synaptic tagging" during long-term potentiation and long-term depression in hippocampal CA1. J. Neurosci., 27: 5068–5080.

Sajikumar, S., Navakkode, S., Korz, V. and Frey, J.U. (2007b) Cognitive and emotional information processing: protein synthesis versus gene expression. J. Physiol. Aug. 16 [Epub ahead of print].

Sajikumar, S., Navakkode, S., Sacktor, T.C. and Frey, J.U. (2005b) Synaptic tagging and cross-tagging: the role of protein kinase Mzeta in maintaining long-term potentiation but not long-term depression. J. Neurosci., 25: 5750–5756.

Schott, B.H., Sellner, D.B., Lauer, C.J., Habib, R., Frey, J.U., Guderian, S., Heinze, H.J. and Duzel, E. (2004) Activation of midbrain structures by associative novelty and the formation of explicit memory in humans. Learn. Mem., 11: 383–387.

Seidenbecher, T., Balschun, D. and Reymann, K.G. (1995) Drinking after water deprivation prolongs "unsaturated" LTP in the dentate gyrus of rats. Physiol. Behav., 57: 1001–1004.

Seidenbecher, T., Reymann, K.G. and Balschun, D. (1997) A post-tetanic time window for the reinforcement of long-term potentiation by appetitive and aversive stimuli. Proc. Natl. Acad. Sci. U.S.A., 94: 1494–1499.

Selbach, O., Brown, R.E. and Haas, H.L. (1997) Long-term increase of hippocampal excitability by histamine and cyclic AMP. Neuropharmacology, 36: 1539–1548.

Sharma, S.K., Sherff, C.M., Shobe, J., Bagnall, M.W., Sutton, M.A. and Carew, T.J. (2003) Differential role of mitogen-activated protein kinase in three distinct phases of memory for sensitization in *Aplysia*. J. Neurosci., 23: 3899–3907.

Stanton, P.K. and Sarvey, J.M. (1985) Blockade of norepinephrine-induced long-lasting potentiation in the hippocampal dentate gyrus by an inhibitor of protein synthesis. Brain Res., 361: 276–283.

Stäubli, U. and Scafidi, J. (1999) Time-dependent reversal of long-term potentiation in area CA1 of the freely moving rat induced by theta pulse stimulation. J. Neurosci., 19: 8712–8719.

Steward, O. (2002) Local synthesis of proteins at synaptic sites on dendrites: role in synaptic plasticity and memory consolidation? Neurobiol. Learn. Mem., 78: 508–527.

Steward, O. and Schuman, E.M. (2001) Protein synthesis at synaptic sites on dendrites. Annu. Rev. Neurosci., 24: 299–325.

Steward, O., Wallace, C.S., Lyford, G.L. and Worley, P.F. (1998) Synaptic activation causes the mRNA for the IEG Arc to localize selectively near activated postsynaptic sites on dendrites. Neuron, 21: 741–751.

Straube, T. and Frey, J.U. (2003) Involvement of β-adrenergic receptors in protein synthesis-dependent late long-term potentiation (LTP) in the dentate gyrus of freely moving rats: the critical role of the LTP induction strength. Neuroscience, 119: 473–479.

Straube, T., Korz, V., Balschun, D. and Frey, J.U. (2003a) Requirement of β-adrenergic receptor activation and protein synthesis for LTP-reinforcement by novelty in rat dentate gyrus. J. Physiol., 552: 953–960.

Straube, T., Korz, V. and Frey, J.U. (2003b) Bidirectional modulation of long-term potentiation by novelty-exploration in rat dentate gyrus. Neurosci. Lett., 344: 5–8.

Tonegawa, S. (2005) The clustered plasticity hypothesis of memory engrams.

Uzakov, S., Frey, J.U. and Korz, V. (2005) Reinforcement of rat hippocampal LTP by holeboard training. Learn. Mem., 12: 165–171.

Vickers, C.A., Dickson, K.S. and Wyllie, D.J.A. (2005) Induction and maintenance of late-phase long-term potentiation in isolated dendrites of rat hippocampal CA1 pyramidal neurones. J. Physiol., 568: 803–813.

Wittmann, B.C., Schott, B.H., Guderian, S., Frey, J.U., Heinze, H.J. and Düzel, E. (2005) Reward-related fMRI activation of dopaminergic midbrain is associated with enhanced hippocampus-dependent long-term memory formation. Neuron, 45: 459–467.

Young, J.Z. and Nguyen, P.V. (2005) Homosynaptic and heterosynaptic inhibition of synaptic tagging and capture of long-term potentiation by previous synaptic activity. J. Neurosci., 25: 7221–7231.

CHAPTER 8

Synaptic plasticity in learning and memory: stress effects in the hippocampus

John G. Howland[1] and Yu Tian Wang[2,*]

[1]Neural Systems and Plasticity Research Group, Department of Psychology, University of Saskatchewan, Saskatoon, SK, S7N 5A5, Canada
[2]Department of Medicine and the Brain Research Centre, University of British Columbia, Vancouver, BC V6T 2B5, Canada

Abstract: Synaptic plasticity has often been argued to play an important role in learning and memory. The discovery of long-term potentiation (LTP) and long-term depression (LTD), the two most widely cited cellular models of synaptic plasticity, significantly spurred research in this field. Although correlative evidence suggesting a role for synaptic changes such as those seen in LTP and LTD in learning and memory has been gained in a number of studies, definitive demonstrations of a specific role for either LTP or LTD in learning and memory are lacking. In this review, we discuss a number of recent advancements in the understanding of the mechanisms that mediate LTP and LTD in the rodent hippocampus and focus on the use of subunit-specific N-methyl-D-aspartate receptor antagonists and interference peptides as potential tools to study the role of synaptic plasticity in learning and memory. By using the modulation of synaptic plasticity and hippocampal-dependent learning and memory by acute stress as an example, we review a large body of convincing evidence indicating that alterations in synaptic plasticity underlie the changes in learning and memory produced by acute stress.

Keywords: glutamate; NR2B; interference peptide; water maze; LTP; LTD; endocytosis; AMPA; NMDA; GluR2

Introduction: synaptic plasticity, learning, and memory

The theory postulating that changes at synapses within the brain underlie learning and memory (along with many other behavioral phenomena) was formalized by Donald Hebb in his seminal work *The Organization of Behavior: A Neuropsychological Theory* (Hebb, 1949). Since that time, strong support for his theory has been gained through a number of lines of research. Most important in this regard is likely the discovery of synaptic long-term potentiation (LTP) and long-term depression (LTD) within the mammalian hippocampus (Bliss and Lomo, 1973; Lynch et al., 1977; Dudek and Bear, 1992; Bliss and Collingridge, 1993; Bear and Abraham, 1996; Malenka and Nicoll, 1999). Intense interest has focused on these forms of synaptic plasticity as they have a number of properties that make them suitable as models for the synaptic changes that likely occur during learning and memory. The classic properties of LTP including input specificity, associative induction, and persistence are key examples of such properties

*Corresponding author. Tel.: +1 604 822 0398;
Fax: +1 604 822 7170; E-mail: ytwang@interchange.ubc.ca

DOI: 10.1016/S0079-6123(07)00008-8

and have been discussed extensively elsewhere (Bliss and Collingridge, 1993; Martin et al., 2000).

Since the discovery of LTP and LTD, rigorous experiments have been conducted that support the role of synaptic plasticity in learning and memory. Changes in synaptic plasticity have been observed for different types of memory which depend on discrete neural circuits including the hippocampus and amygdala (Martin et al., 2000; Sigurdsson et al., 2007). Most of these experiments have focused on the potential role of LTP-like plasticity in learning and memory using normal (Rogan et al., 1997; Pastalkova et al., 2006; Whitlock et al., 2006) or genetically modified rodents (Mayford et al., 1996; Tang et al., 1999), however, a lack of specific inhibitors for either LTP or LTD has hindered progress in determining the specific types of synaptic plasticity involved in various forms of learning and memory.

Importantly, numerous experiential factors, such as acute stress, have profound effects on learning and memory which are correlated with altered synaptic plasticity in relevant brain areas (Kim et al., 2006). A number of recent advancements in understanding the mechanisms through which stress affects synaptic plasticity have shed new light on how altered synaptic plasticity may affect behavior. The present review focuses on these advancements as an example of the critical role of synaptic plasticity in the biological basis of learning and memory. In this review, we begin by summarizing recent advancements in understanding the mechanisms of N-methyl-D-aspartate glutamate receptor (NMDAR)-mediated LTP and LTD in the CA1 area of the hippocampus and continue with a detailed discussion of number of recent studies that have examined the contribution of these mechanisms to the effects of acute stress on learning and memory. The implications of these findings for current theories of synaptic plasticity and memory are also discussed.

Mechanisms underlying synaptic plasticity in the hippocampus

Research into the mechanisms underlying LTP and LTD in the CA1 region of the hippocampus is especially vigorous (Malenka and Nicoll, 1999; Malenka and Bear, 2004). However, determining the specific alterations in synaptic plasticity that are critical for learning and memory has been hindered by a lack of specific inhibitors of LTP and LTD. The recent discovery of a number of compounds that may be suitable for specifically targeting LTP or LTD in behaving animals has provided potential new avenues for research in this area. The following section will review evidence regarding the usefulness of these compounds for understanding the potential link between synaptic plasticity and learning and memory.

Induction of LTP/LTD

It is generally well accepted that the induction of hippocampal CA1 homosynaptic LTP and LTD depends on the activation of NMDARs (Bliss and Collingridge, 1993; Malenka and Bear, 2004). NMDARs are heteromeric complexes of NR1 subunits, at least one type of NR2 subunits (NR2A-D), and NR3 (A or B) subunits in some areas (Cull-Candy et al., 2001; Paoletti and Neyton, 2007). NMDARs have a number of unique characteristics which make them particularly attractive as a molecular substrate mediating the induction of synaptic plasticity. For example, under conditions of low post-synaptic activity, NMDARs are blocked in a voltage-dependent manner by magnesium ions. When post-synaptic activity is high, such as under conditions suitable for producing plasticity, the post-synaptic membrane depolarizes enough to remove the magnesium block. Once activated, NMDARs are also highly permeable to calcium ions. Importantly, the post-synaptic influx of calcium is a critical step underlying both LTP and LTD, although the detailed mechanisms surrounding calcium influx that give rise to either LTP or LTD is still the subject of significant debate (Malenka and Bear, 2004).

Converging evidence supports the hypothesis that the subunit composition of NMDARs confers distinct roles of the receptors in normal and pathological brain function (Cull-Candy et al., 2001; Paoletti and Neyton, 2007). The development of NMDAR subunit-selective pharmacological agents such as NVP-AAM077 (Auberson et al.,

2002) for NR2A-containing receptors and Ro25-6981 (Mutel et al., 1998) for NR2B-containing receptors has made it possible to test this hypothesis. Several studies using these compounds in in vitro brain slices prepared from both young and adult rodents provide evidence for a critical role of NR2A-containing NMDAR activation in hippocampal CA1 LTP (Liu et al., 2004) and NR2B-containing NMDARs activation in the induction of hippocampal CA1 LTD (Liu et al., 2004; Woo et al., 2005; Yang et al., 2005; Izumi et al., 2006). Similar results have also been obtained in slices from the perirhinal cortex of young adult rats (Massey et al., 2004).

However, contradictory results have been reported by others (Hendricson et al., 2002; Berberich et al., 2005; Morishita et al., 2007). Since results both for and against a critical involvement of NR2B-containing receptors in LTD were independently obtained from more than one laboratory, it is possible that the subunit requirements for LTD are state-dependent phenomena and these contradictory results may be due in part to different conditions used in the in vitro studies. Additionally, the subunit specificity of NVP-AAM077 for NR2A- versus NR2B-containing receptors may be less than originally reported, especially when rat recombinant NMDA receptors are used for the binding assays (Frizelle et al., 2006). Therefore, further validation of the role of NR2A-containing receptors in synaptic plasticity awaits the development of a new generation of NR2A antagonists with better pharmacological subunit specificity. Determining the exact roles of NMDAR subunits in synaptic plasticity is also complicated by the potential existence of native NMDAR complexes of NR1/NR2A/NR2B subunits (Sheng et al., 1994). Such trimeric receptor complexes would likely have pharmacological properties that differ from either NR1/NR2A or NR1/NR2B dimeric complex. However, using quantitative immunoprecipitation techniques, Al-Hallaq et al. (2007) have recently reported that the majority of native NMDARs in the rat hippocampus are di-heteromeric receptors containing either NR1/NR2A or NR1/NR2B subunits.

Importantly, the subunit specificity of NMDAR-dependent LTP and LTD originally reported in in vitro studies has recently been corroborated by in vivo studies. Thus, within the narrow range of concentrations used, NVP-AAM077 and Ro25-6981 were shown to preferentially inhibit hippocampal CA1 LTP and LTD, respectively, in anesthetized rats (Fox et al., 2006). Although these findings remain to be confirmed by other laboratories and extended to other brain areas, such preferential inhibition of either LTP or LTD with NMDAR subunit-specific antagonists in vivo offers a valuable opportunity to use these drugs to begin to probe the specific roles of LTP and/or LTD in mediating some aspects of learning and memory (Duffy et al., 2007; Wong et al., 2007). A number of experiments designed to address this question will be further described in the section discussing the effects of stress on synaptic plasticity and learning and memory.

Expression of LTP/LTD

Although the activation of NMDARs is required for the induction of LTP and LTD (Malenka and Bear, 2004), the expression of these forms of synaptic plasticity is likely dependent on both presynaptic changes in transmitter release and post-synaptic changes in the α-amino-3-hydroxy-5-methylisoxazole-4-propionic acid subtype of glutamate receptors (AMPARs; Malenka and Nicoll, 1999; Collingridge et al., 2004; Malenka and Bear, 2004). AMPARs mediate the majority of the fast synaptic transmission in the mammalian brain and are generally expressed as tetramers composed of various patterns of four subunits (GluR1-4; Derkach et al., 2007). Although modifications to pre-existing AMPARs likely contribute to various forms of synaptic plasticity, considerable research also suggests receptor trafficking (rapid changes in the number of AMPARs) plays an important role in LTP and LTD (Collingridge et al., 2004; Derkach et al., 2007). Strong support for the insertion and endocytosis of AMPARs during LTP and LTD, respectively, has been gained using a number of methods in vitro (Ahmadian et al., 2004; Derkach et al., 2007). Clearly, a more thorough understanding of the molecular mechanisms underlying receptor trafficking in synaptic plasticity may have important relevance for determining the

the effects of stress depend on elevated levels of adrenal hormones; however, the data do not entirely support this hypothesis. Although blocking the activation of glucocorticoid receptors before or immediately following stress blocks the effects of stress on synaptic plasticity (Xu et al., 1998; Yang et al., 2004, 2005), adrenalectomizing rats prior to stress fails to block the disruptive effects of stress on LTP (Shors et al., 1990). Additionally, animals that can terminate administered shocks show elevated levels of corticosterone similar to those animals not able to terminate the shock, but do not show alterations in synaptic plasticity following stress (Shors et al., 1989). Thus, although elevated levels of corticosterone are a critical determinant of the effects of stress on synaptic plasticity, in some cases, the increase in glucocorticoid hormones must interact with other factors to enable stress to alter synaptic plasticity.

Substantial evidence also suggests that the effects of acute stress on synaptic plasticity are mediated by glutamatergic neurotransmission. For example, pretreatment with NMDA antagonists blocks the effects of stress on the induction of both LTP and LTD in the hippocampus (Kim et al., 1996). Interestingly, Yang and colleagues recently showed that exploration of a novel environment following acute stress reverses the expected effects of acute stress on both LTP and LTD (Yang et al., 2006). The novelty effects depend on activation of the cholinergic system and NMDARs, which in turn activate the protein phosphatase 2B and striatal-enriched tyrosine phosphatase. Thus, exposure to certain environmental stimuli which activate NMDARs also reverse the stress-induced changes in hippocampal synaptic plasticity.

Moreover, recent evidence suggests that stress may enable LTD in the CA1 region of the hippocampus by either enhancing the release of glutamate or blocking glutamate reuptake (Lowy et al., 1993, 1995; Yang et al., 2005; Wong et al., 2007). Either of these mechanisms allow for the activation of extra-synaptic NMDARs, which are thought to be comprised mostly of NR2B-containing NMDARs in the adult rodent CA1 region (Hardingham et al., 2002; Yang et al., 2005; Duffy et al., 2007; Wong et al., 2007). Antagonists for NR2B-containing NMDARs are effective at blocking the induction of LTD following stress (Yang et al., 2005; Fox et al., 2006; Wang et al., 2006; Wong et al., 2007), and may also reverse the stress-induced disruption of LTP (Wang et al., 2006). Thus, it appears that the LTD enabled by stress may share similar mechanisms to LTD induced in the hippocampus without stress in vitro (Liu et al., 2004; Woo et al., 2005; Yang et al., 2005; Izumi et al., 2006). Further support that stress-enabled LTD shares similar mechanisms to LTD induced without stress comes from a recent study reporting that administration of the Tat-GluR2$_{3Y}$, but not the Tat-GluR2$_{3A}$, peptide blocks stress-enabled CA1 LTD in young adult rats in vivo (Fox et al., 2007). Thus, it appears that the expression of stress-enabled LTD is also dependent on the clathrin-dependent endocytosis of GluR2-containing AMPARs.

What is the mechanism through which stress alters synaptic plasticity?

Significant progress has been made regarding the role of specific alterations in synaptic plasticity caused by acute stress in learning and memory. However, a number of important challenges remain. Most importantly, considerable debate remains regarding the mechanism by which acute stress alters synaptic plasticity in the hippocampus by promoting the induction of LTD and inhibiting the induction of LTP (Kim and Yoon, 1998; Abraham, 2004; Diamond et al., 2004, 2005; Huang et al., 2005). A number of lines of evidence suggest that stress and LTP share common molecular mechanisms and that stress may saturate LTP, thereby inhibiting its induction (Diamond et al., 2004, 2005; Huang et al., 2005). Correlational support for this hypothesis comes from studies showing a number of common effects between stress and the induction of LTP on the activation of immediate early genes, ionotropic glutamate receptors, and learning and memory (see Diamond et al., 2004 for specific references). Direct support for this hypothesis would be gained with a clear demonstration that acute stress increases synaptic potentials in a manner similar to LTP. One study (Sacchetti et al., 2001) provides such support by showing that contextual fear conditioning (which is inherently stressful) produces a long-lasting increase in the evoked CA1 response in hippocampal slices of fear conditioned rats. However, this finding has not been replicated and no significant changes in evoked responses from the hippocampus have been observed during or immediately following acute stress per se (Xu et al., 1997; Huang et al., 2005).

A second possibility is that stress exerts modulatory effects on synaptic plasticity in the hippocampus by altering the induction threshold for LTP and LTD (Kim and Yoon, 1998; Abraham, 2004; Huang et al., 2005). Such 'metaplastic' changes or alterations in the ability to induce different forms of synaptic plasticity by prior activity are an important characteristic of a number of neural circuits, including the CA1 region of the hippocampus (Abraham and Bear, 1996; Kirkwood et al., 1996; Mockett et al., 2002). The Bienenstock, Cooper, and Munro (BCM) model is a frequently cited computational model designed to explain the modifications that may occur in synaptic plasticity as a result of experience (Bienenstock et al., 1982). One of the central components of the model is that the threshold for plasticity (referred to as θ_m) in a given circuit is not fixed, but rather changes as a result of experience. Thus, as stress favors the induction of LTD over LTP, it could be hypothesized that this results from a rightward shift of θ_m. Direct support for the a shift in θ_m following stress has been difficult to obtain (Huang et al., 2005), although promising results have been reported following the pharmacological activation of glucocorticoid receptors (Coussens et al., 1997).

Effects of acute stress on hippocampal-based learning and memory

Acute stress has differential effects on learning and memory that depend on a number of factors including the type of task, timing of the stress, and sex of the subject (Joels et al., 2006; Kim et al., 2006; Shors, 2006). In the present review, we restrict our discussion to spatial memory retrieval deficits following acute stress because recent advancements in this area provide an excellent example of the consequences of specific changes in synaptic plasticity on learning and memory.

When rodents are trained in spatial memory tasks, such as the water maze or radial arm maze, acute stress does not significantly affect the ability of the animals to learn the task (Diamond et al., 1999; Kim et al., 2005) or the retrieval of hippocampal-independent reference memory (Diamond et al., 1996; Woodson et al., 2003). In contrast, acute stress disrupts the retrieval of hippocampal-dependent spatial memory whether the stress occurs before the learning (Diamond et al., 2006) or retrieval phase of the test (Diamond et al., 1996, 1999; de Quervain et al., 1998). The disruptive effects of acute stress on hippocampal-dependent memory retrieval are particularly robust and have also been demonstrated using a number of paradigms in human (Het et al., 2005; Kuhlmann et al., 2005) and rodent studies (Baker and Kim, 2002). Importantly, these effects can be mimicked by cortisol or corticosterone treatment (de Quervain et al., 1998, 2000, 2003; Het et al., 2005) and are modulated by a number of factors including levels of arousal and emotional valence of the stimuli (Kuhlmann et al., 2005; Kuhlmann and Wolf, 2006a, b). Interestingly, acute stress also enhances memory in aversive hippocampal-dependent tasks such as contextual fear conditioning and trace eye-blink conditioning (Beylin and Shors, 1998; Nijholt et al., 2004), thereby demonstrating that the effects of stress may differ in aversively motivated contexts.

The mechanisms that underlie the effects of stress on spatial memory retrieval have received considerable attention. In an important paper from McGaugh's group (de Quervain et al., 1998), spatial memory retrieval deficits were observed 30 min, but not 2 min or 4 h, after footshock stress. Thirty minutes following stress coincided with the peak level of circulating corticosterone, thereby supporting the notion that increases in corticosterone may cause the memory retrieval deficits. This hypothesis is further supported by the fact that pharmacologically inhibiting corticosterone synthesis blocks the memory retrieval deficits, while administration of corticosterone in the absence of stress induced retrieval deficits (de Quervain et al., 1998).

However, as with the effects of stress on hippocampal synaptic plasticity, increased release of corticosterone cannot fully account for the effects of acute stress on memory retrieval. In a well-designed experiment, Diamond and his colleagues exposed rats to either a cat or a sexually receptive female rat before spatial memory testing (Woodson et al., 2003). Although both stimuli aroused the rats and resulted in similar increases in corticosterone, only the group exposed to the cat displayed disrupted hippocampal-dependent memory. Therefore, similar to the effects of stress on synaptic plasticity, increased release of corticosterone is not sufficient to cause stress-induced memory retrieval deficits.

Roles of LTP and LTD in stress-induced spatial memory impairment

Other studies have examined the role altered hippocampal glutamatergic synaptic plasticity may play in acute stress-induced spatial memory retrieval impairments. As was reviewed above, acute stress induces a profound shift in the pattern of hippocampal synaptic plasticity by enabling the induction of LTD and blocking the induction of LTP. If the memory retrieval impairments are due to alterations in synaptic plasticity caused by stress, treatments which reverse the effects of stress on synaptic plasticity would be expected to also reverse the memory retrieval impairments. Strong support for this hypothesis has been gained from a number of studies. For example, NMDA antagonists block the enabling and inhibiting effects of acute stress on LTD and LTP, respectively (Kim et al., 1996). Blocking NMDARs with a broad spectrum antagonist (CPP) also reverses the disruptive effects of stress on spatial memory (Park et al., 2004), thereby supporting the conjecture that the disruptive effects of stress on memory retrieval may be a result of NMDAR-dependent alterations in hippocampal synaptic plasticity. It is worth noting that under normal conditions, administration of NMDAR antagonists to unstressed animals disrupts spatial memory retrieval but in this case, NMDA antagonism prevents the effects of stress on spatial memory retrieval by preserving the normal function of the hippocampus (i.e. blocking LTD and keeping it in an 'LTP' prone state; Diamond et al., 2005).

Although these studies implicate altered synaptic plasticity in the stress-induced impairment of memory, it is unclear whether specific alterations in either LTP or LTD caused by stress are responsible for the stress-induced impairment of spatial memory retrieval. In a recent study, we provide strong evidence for an essential and sufficient role of hippocampal LTD in mediating acute stress-induced impairment of spatial memory retrieval by specifically inhibiting LTD with the structurally and mechanistically distinct inhibitors Ro25-6981 and the Tat-GluR2$_{3Y}$ peptide and facilitating the induction of LTD by inhibiting glutamate uptake (Fig. 1; Wong et al., 2007). More specifically, administration of the specific NR2B subunit-containing NMDAR antagonist (Ro25-6981) reversed the disruptive effect of stress on spatial memory retrieval in a water maze task (Wong et al., 2007). Given that Ro25-6981 has been shown convincingly to block stress-enabled LTD (Yang et al., 2005; Fox et al., 2006; Wang et al., 2006; Wong et al., 2007) and reverse the disruption of LTP by stress (Wang et al., 2006), these data suggest that the activation of NR2B-containing receptors are critical for the effects of stress on synaptic plasticity and memory retrieval. Interestingly, the increases in hippocampal glutamate efflux observed after stress can be mimicked pharmacologically by local injections of the glutamate transporter inhibitor DL-TBOA (Wong et al., 2007). Under these conditions, LTD is readily induced in the hippocampus in vivo and spatial memory retrieval is disrupted. Similar to the effects of stress, both the alterations in synaptic plasticity and memory retrieval in TBOA-treated animals can be reversed with Ro25-6981 (Wong et al., 2007). Further support for the essential role of LTD in the spatial memory retrieval disruptions following acute stress is gained from experiments showing that administration of the Tat-GluR2$_{3Y}$ peptide, which specifically blocks AMPAR endocytosis and stress-enabled LTD in vivo (Fox et al., 2007), also blocks the disruptive effects of acute stress on spatial memory retrieval (Wong et al., 2007). Taken together, the results of these studies provide convincing support for the hypothesis that

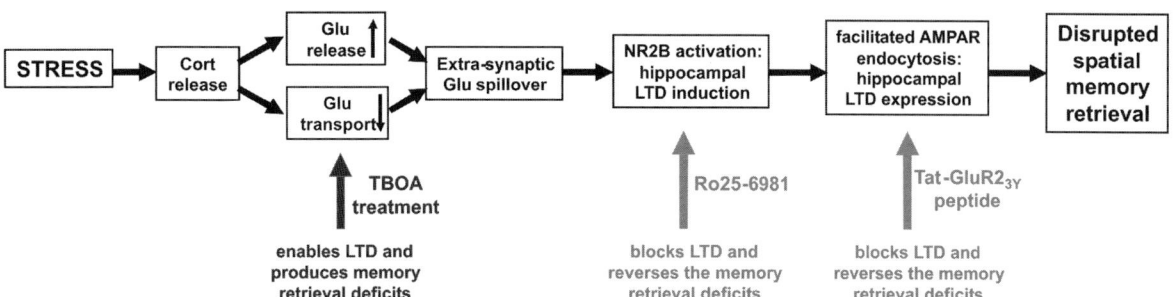

Fig. 1. Schematic describing the hypothetical mechanisms by which acute stress enables induction of long-term depression (LTD) in the hippocampus, thereby causing impaired spatial memory retrieval and the steps at which experimental treatments performed in Wong et al. (2007) interfere with this process. Acute stress causes the release of corticosterone (Cort) which then increases glutamate concentration in the synaptic cleft either through increased glutamate (Glu) release and/or decreased glutamate transport in the hippocampus. The increased glutamate concentration enables the induction of LTD via a spill-over activation of extra-synaptically localized NR2B-containing NMDARs, and hence the expression of LTD via facilitating the endocytosis of post-synaptic AMPARs, thereby leading to the disrupted spatial memory retrieval. The treatments used to experimentally induce and inhibit LTD are indicated in green and red, respectively. Figure used with permission, copyright (2007), National Academy of Sciences, U.S.A. (Wong et al., 2007). (See Color Plate 8.1 in color plate section.)

LTD-like changes in synaptic plasticity, involving NR2B-subunit containing NMDARs and AMPAR endocytosis, underlie the effects of acute stress on spatial memory retrieval (Fig. 1; Yang et al., 2005; Wang et al., 2006; Fox et al., 2007; Wong et al., 2007).

Understanding the detailed mechanisms of how corticosterone release following acute stress alters glutamate transmission and subsequent synaptic plasticity and behavior requires further investigation. While it is possible that these effects are mediated by the classic actions of corticosterone on gene expression, the time course of the effects make this unlikely. For example, in the experiments of de Quervain et al. (1998) and Wong et al. (2007), the disruptive effects of acute stress on memory retrieval were observed 30 min after the initiation of the stressor. A number of recent studies suggest that corticosteriods have rapid, likely non-genomic actions on glutamate transmission in the central nervous system that may be mediated by unidentified membrane-associated receptors (Di et al., 2003; Karst et al., 2005; Tasker et al., 2006). Thus, non-genomic effects of stress hormones may at least partially explain the effects of acute stress on synaptic plasticity and spatial memory retrieval. However, as previously discussed, such an explanation is complicated by data showing that glucocorticoid receptor antagonists reverse the effects of acute stress on synaptic plasticity (Xu et al., 1998) and increased corticosterone release is insufficient to disrupt memory retrieval (Woodson et al., 2003).

The results of these studies have important implications for theories of hippocampal-dependent learning and memory. In the case of spatial learning and memory, it is tempting to speculate that during the learning phase of a task such as the water maze, a memory for the platform location is formed by the potentiation of a subset of synapses in the hippocampus (Diamond et al., 2004). The activation of these synapses during subsequent retrieval allows the successful retrieval of the memory. As previously discussed, if acute stress is experienced immediately before retrieval, the memory is impaired (de Quervain et al., 1998; Wong et al., 2007). Given the points discussed above, it is possible that exposure to stress disrupted retrieval by saturating LTP in the hippocampus (Diamond et al., 2004, 2005). However, the recent results showing that specifically blocking the expression of LTD with the Tat-GluR2$_{3Y}$ peptide is sufficient to block the disruptive effects of stress on spatial memory retrieval refutes this hypothesis (Wong et al., 2007). Given that hippocampal CA1 LTD can be produced in an input specific manner, it is plausible that stress could depress only those synapses that were potentiated in the original learning episode. Alternatively, stress may "reset" the entire hippocampal network by depressing all synapses, whether or not they were potentiated during learning (Diamond et al., 2005). Finally, as previously discussed, acute stress may affect hippocampal metaplasticity and thereby alter the optimal balance of LTP and LTD within the hippocampal circuit for memory retrieval (Kim and Yoon, 1998; Huang et al., 2005). Although a number of possibilities remain regarding the exact mechanism that allows stress to disrupt spatial memory retrieval, it is clear that altered patterns of synaptic plasticity which specifically favor the induction of LTD in the hippocampus ultimately underlies these behavioral effects.

Conclusion

This review focused on recent research aimed at understanding the role of hippocampal synaptic plasticity in learning and memory. In the context of acute stress, we provided strong support for the hypothesis that distinct forms of synaptic plasticity underlie the effects of experience on learning and memory. In particular, recent results from our laboratory and others suggest that stress-enabled hippocampal LTD underlies the deficits in spatial memory retrieval commonly observed after acute stress (Wang et al., 2006; Fox et al., 2007; Wong et al., 2007). The conclusions drawn from these experiments were made possible by using two recently developed specific inhibitors for LTD. However, specific inhibitors of LTP are still lacking. Thus, efforts aimed at the development of additional compounds suitable for selectively inhibiting various forms of synaptic plasticity will be highly profitable for understanding the specific roles of synaptic plasticity in learning and memory.

Abbreviations

AMPAR	α-amino-3-hydroxy-5-methylisoxazole-4-propionic acid receptor
CT	carboxyl tail
HIV	human immunodeficiency virus
HPA	hypothalamic–pituitary–adrenal
LTD	long-term depression
LTP	long-term potentiation
NMDAR	N-methyl-D-aspartate receptor

Acknowledgments

John G. Howland is supported by fellowships from the Canadian Institutes of Health Research (CIHR) and the Michael Smith Foundation for Health Research (MSFHR). Yu Tian Wang is a Howard Hughes Medical Institute International Scholar, CIHR Investigator, MSFHR Senior Scholar, and Heart and Stroke Foundation of British Columbia and the Yukon Chair in Stroke Research.

References

Abraham, W.C. (2004) Stress-related phenomena. Hippocampus, 14: 675–676.

Abraham, W.C. and Bear, M.F. (1996) Metaplasticity: the plasticity of synaptic plasticity. Trends Neurosci., 19: 126–130.

Ahmadian, G., Ju, W., Liu, L., Wyszynski, M., Lee, S.H., Dunah, A.W., Taghibiglou, C., Wang, Y., Lu, J., Wong, T.P., Sheng, M. and Wang, Y.T. (2004) Tyrosine phosphorylation of GluR2 is required for insulin-stimulated AMPA receptor endocytosis and LTD. EMBO J., 23: 1040–1050.

Al-Hallaq, R.A., Conrads, T.P., Veenstra, T.D. and Wenthold, R.J. (2007) NMDA di-heteromeric receptor populations and associated proteins in rat hippocampus. J. Neurosci., 27: 8334–8343.

Auberson, Y.P., Allgeier, H., Bischoff, S., Lingenhoehl, K., Moretti, R. and Schmutz, M. (2002) 5-Phosphonomethylquinoxalinediones as competitive NMDA receptor antagonists with a preference for the human 1A/2A, rather than 1A/2B receptor composition. Bioorg. Med. Chem. Lett., 12: 1099–1102.

Baker, K.B. and Kim, J.J. (2002) Effects of stress and hippocampal NMDA receptor antagonism on recognition memory in rats. Learn. Mem., 9: 58–65.

Bear, M.F. and Abraham, W.C. (1996) Long-term depression in hippocampus. Annu. Rev. Neurosci., 19: 437–462.

Berberich, S., Punnakkal, P., Jensen, V., Pawlak, V., Seeburg, P.H., Hvalby, O. and Kohr, G. (2005) Lack of NMDA receptor subtype selectivity for hippocampal long-term potentiation. J. Neurosci., 25: 6907–6910.

Beylin, A.V. and Shors, T.J. (1998) Stress enhances excitatory trace eyeblink conditioning and opposes acquisition of inhibitory conditioning. Behav. Neurosci., 112: 1327–1338.

Bienenstock, E.L., Cooper, L.N. and Munro, P.W. (1982) Theory for the development of neuron selectivity: orientation specificity and binocular interaction in visual cortex. J. Neurosci., 2: 32–48.

Bliss, T.V. and Collingridge, G.L. (1993) A synaptic model of memory: long-term potentiation in the hippocampus. Nature, 361: 31–39.

Bliss, T.V. and Lomo, T. (1973) Long-lasting potentiation of synaptic transmission in the dentate area of the anaesthetized rabbit following stimulation of the perforant path. J. Physiol., 232: 331–356.

Brebner, K., Wong, T.P., Liu, L., Liu, Y., Campsall, P., Gray, S., Phelps, L., Phillips, A.G. and Wang, Y.T. (2005) Nucleus accumbens long-term depression and the expression of behavioral sensitization. Science, 310: 1340–1343.

Carroll, R.C., Beattie, E.C., Xia, H., Luscher, C., Altschuler, Y., Nicoll, R.A., Malenka, R.C. and von, Z.M. (1999) Dynamin-dependent endocytosis of ionotropic glutamate receptors. Proc. Natl. Acad. Sci. U.S.A., 96: 14112–14117.

Chung, H.J., Steinberg, J.P., Huganir, R.L. and Linden, D.J. (2003) Requirement of AMPA receptor GluR2 phosphorylation for cerebellar long-term depression. Science, 300: 1751–1755.

Collingridge, G.L., Isaac, J.T. and Wang, Y.T. (2004) Receptor trafficking and synaptic plasticity. Nat. Rev. Neurosci., 5: 952–962.

Coussens, C.M., Kerr, D.S. and Abraham, W.C. (1997) Glucocorticoid receptor activation lowers the threshold for NMDA-receptor-dependent homosynaptic long-term depression in the hippocampus through activation of voltage-dependent calcium channels. J. Neurophysiol., 78: 1–9.

Cull-Candy, S., Brickley, S. and Farrant, M. (2001) NMDA receptor subunits: diversity, development and disease. Curr. Opin. Neurobiol., 11: 327–335.

de Quervain, D.J., Henke, K., Aerni, A., Treyer, V., McGaugh, J.L., Berthold, T., Nitsch, R.M., Buck, A., Roozendaal, B. and Hock, C. (2003) Glucocorticoid-induced impairment of declarative memory retrieval is associated with reduced blood flow in the medial temporal lobe. Eur. J. Neurosci., 17: 1296–1302.

de Quervain, D.J., Roozendaal, B. and McGaugh, J.L. (1998) Stress and glucocorticoids impair retrieval of long-term spatial memory. Nature, 394: 787–790.

de Quervain, D.J., Roozendaal, B., Nitsch, R.M., McGaugh, J.L. and Hock, C. (2000) Acute cortisone administration impairs retrieval of long-term declarative memory in humans. Nat. Neurosci., 3: 313–314.

Derkach, V.A., Oh, M.C., Guire, E.S. and Soderling, T.R. (2007) Regulatory mechanisms of AMPA receptors in synaptic plasticity. Nat. Rev. Neurosci., 8: 101–113.

Di, S., Malcher-Lopes, R., Halmos, K.C. and Tasker, J.G. (2003) Nongenomic glucocorticoid inhibition via endocannabinoid release in the hypothalamus: a fast feedback mechanism. J. Neurosci., 23: 4850–4857.

Diamond, D.M., Campbell, A.M., Park, C.R., Woodson, J.C., Conrad, C.D., Bachstetter, A.D. and Mervis, R.F. (2006) Influence of predator stress on the consolidation versus retrieval of long-term spatial memory and hippocampal spinogenesis. Hippocampus, 16: 571–576.

Diamond, D.M., Fleshner, M., Ingersoll, N. and Rose, G.M. (1996) Psychological stress impairs spatial working memory: relevance to electrophysiological studies of hippocampal function. Behav. Neurosci., 110: 661–672.

Diamond, D.M., Fleshner, M. and Rose, G.M. (1994) Psychological stress repeatedly blocks hippocampal primed burst potentiation in behaving rats. Behav. Brain Res., 62: 1–9.

Diamond, D.M., Park, C.R., Campbell, A.M. and Woodson, J.C. (2005) Competitive interactions between endogenous LTD and LTP in the hippocampus underlie the storage of emotional memories and stress-induced amnesia. Hippocampus, 15: 1006–1025.

Diamond, D.M., Park, C.R., Heman, K.L. and Rose, G.M. (1999) Exposing rats to a predator impairs spatial working memory in the radial arm water maze. Hippocampus, 9: 542–552.

Diamond, D.M., Park, C.R. and Woodson, J.C. (2004) Stress generates emotional memories and retrograde amnesia by inducing an endogenous form of hippocampal LTP. Hippocampus, 14: 281–291.

Dudek, S.M. and Bear, M.F. (1992) Homosynaptic long-term depression in area CA1 of hippocampus and effects of N-methyl-D-aspartate receptor blockade. Proc. Natl. Acad. Sci. U.S.A., 89: 4363–4367.

Duffy, S., Labrie, V. and Roder, J.C. (2007) D-Serine augments NMDA-NR2B receptor-dependent hippocampal long-term depression and spatial reversal learning. Neuropsychopharmacology advance online publication 11 July 2007; doi:10.1038/sj.npp.1301486.

Fox, C.J., Russell, K., Titterness, A.K., Wang, Y.T. and Christie, B.R. (2007) Tyrosine phosphorylation of the GluR2 subunit is required for long-term depression of synaptic efficacy in young animals in vivo. Hippocampus, 17: 600–605.

Fox, C.J., Russell, K.I., Wang, Y.T. and Christie, B.R. (2006) Contribution of NR2A and NR2B NMDA subunits to bidirectional synaptic plasticity in the hippocampus in vivo. Hippocampus, 16: 907–915.

Foy, M.R., Stanton, M.E., Levine, S. and Thompson, R.F. (1987) Behavioral stress impairs long-term potentiation in rodent hippocampus. Behav. Neural Biol., 48: 138–149.

Frizelle, P.A., Chen, P.E. and Wyllie, D.J. (2006) Equilibrium constants for (R)-[(S)-1-(4-bromo-phenyl)-ethylamino]-(2,3-dioxo-1,2,3,4-tetrahydroquinoxalin-5-yl)-methyl]-phosphonic acid (NVP-AAM077) acting at recombinant NR1/NR2A and NR1/NR2B N-methyl-D-aspartate receptors: implications for studies of synaptic transmission. Mol. Pharmacol., 70: 1022–1032.

Hardingham, G.E., Fukunaga, Y. and Bading, H. (2002) Extrasynaptic NMDARs oppose synaptic NMDARs by triggering CREB shut-off and cell death pathways. Nat. Neurosci., 5: 405–414.

Hayashi, T. and Huganir, R.L. (2004) Tyrosine phosphorylation and regulation of the AMPA receptor by SRC family tyrosine kinases. J. Neurosci., 24: 6152–6160.

Hebb, D.O. (1949) The Organization of Behavior: A Neuropsychological Theory. Wiley, New York.

Hendricson, A.W., Miao, C.L., Lippmann, M.J. and Morrisett, R.A. (2002) Ifenprodil and ethanol enhance NMDA receptor-dependent long-term depression. J. Pharmacol. Exp. Ther., 301: 938–944.

Het, S., Ramlow, G. and Wolf, O.T. (2005) A meta-analytic review of the effects of acute cortisol administration on human memory. Psychoneuroendocrinology, 30: 771–784.

Huang, C.C., Yang, C.H. and Hsu, K.S. (2005) Do stress and long-term potentiation share the same molecular mechanisms? Mol. Neurobiol., 32: 223–235.

Izumi, Y., Auberson, Y.P. and Zorumski, C.F. (2006) Zinc modulates bidirectional hippocampal plasticity by effects on NMDA receptors. J. Neurosci., 26: 7181–7188.

Joels, M., Pu, Z., Wiegert, O., Oitzl, M.S. and Krugers, H.J. (2006) Learning under stress: how does it work? Trends Cogn. Sci., 10: 152–158.

Karst, H., Berger, S., Turiault, M., Tronche, F., Schutz, G. and Joels, M. (2005) Mineralocorticoid receptors are indispensable for nongenomic modulation of hippocampal glutamate transmission by corticosterone. Proc. Natl. Acad. Sci. U.S.A., 102: 19204–19207.

Kim, C.H., Chung, H.J., Lee, H.K. and Huganir, R.L. (2001) Interaction of the AMPA receptor subunit GluR2/3 with PDZ domains regulates hippocampal long-term depression. Proc. Natl. Acad. Sci. U.S.A., 98: 11725–11730.

Kim, J.J. and Diamond, D.M. (2002) The stressed hippocampus, synaptic plasticity and lost memories. Nat. Rev. Neurosci., 3: 453–462.

Kim, J.J., Foy, M.R. and Thompson, R.F. (1996) Behavioral stress modifies hippocampal plasticity through N-methyl-D-aspartate receptor activation. Proc. Natl. Acad. Sci. U.S.A., 93: 4750–4753.

Kim, J.J., Koo, J.W., Lee, H.J. and Han, J.S. (2005) Amygdalar inactivation blocks stress-induced impairments in hippocampal long-term potentiation and spatial memory. J. Neurosci., 25: 1532–1539.

Kim, J.J., Song, E.Y. and Kosten, T.A. (2006) Stress effects in the hippocampus: synaptic plasticity and memory. Stress, 9: 1–11.

Kim, J.J. and Yoon, K.S. (1998) Stress: metaplastic effects in the hippocampus. Trends Neurosci., 21: 505–509.

Kirkwood, A., Rioult, M.C. and Bear, M.F. (1996) Experience-dependent modification of synaptic plasticity in visual cortex. Nature, 381: 526–528.

Kuhlmann, S., Piel, M. and Wolf, O.T. (2005) Impaired memory retrieval after psychosocial stress in healthy young men. J. Neurosci., 25: 2977–2982.

Kuhlmann, S. and Wolf, O.T. (2006a) Arousal and cortisol interact in modulating memory consolidation in healthy young men. Behav. Neurosci., 120: 217–223.

Kuhlmann, S. and Wolf, O.T. (2006b) A non-arousing test situation abolishes the impairing effects of cortisol on delayed memory retrieval in healthy women. Neurosci. Lett., 399: 268–272.

Lee, H.K., Barbarosie, M., Kameyama, K., Bear, M.F. and Huganir, R.L. (2000) Regulation of distinct AMPA receptor phosphorylation sites during bidirectional synaptic plasticity. Nature, 405: 955–959.

Lee, H.K., Takamiya, K., Han, J.S., Man, H., Kim, C.H., Rumbaugh, G., Yu, S., Ding, L., He, C., Petralia, R.S., Wenthold, R.J., Gallagher, M. and Huganir, R.L. (2003) Phosphorylation of the AMPA receptor GluR1 subunit is required for synaptic plasticity and retention of spatial memory. Cell, 112: 631–643.

Lee, S.H., Liu, L., Wang, Y.T. and Sheng, M. (2002) Clathrin adaptor AP2 and NSF interact with overlapping sites of GluR2 and play distinct roles in AMPA receptor trafficking and hippocampal LTD. Neuron, 36: 661–674.

Liu, L., Wong, T.P., Pozza, M.F., Lingenhoehl, K., Wang, Y., Sheng, M., Auberson, Y.P. and Wang, Y.T. (2004) Role of NMDA receptor subtypes in governing the direction of hippocampal synaptic plasticity. Science, 304: 1021–1024.

Lowy, M.T., Gault, L. and Yamamoto, B.K. (1993) Adrenalectomy attenuates stress-induced elevations in extracellular glutamate concentrations in the hippocampus. J. Neurochem., 61: 1957–1960.

Lowy, M.T., Wittenberg, L. and Yamamoto, B.K. (1995) Effect of acute stress on hippocampal glutamate levels and spectrin proteolysis in young and aged rats. J. Neurochem., 65: 268–274.

Luscher, C., Xia, H., Beattie, E.C., Carroll, R.C., von, Z.M., Malenka, R.C. and Nicoll, R.A. (1999) Role of AMPA receptor cycling in synaptic transmission and plasticity. Neuron, 24: 649–658.

Lynch, G.S., Dunwiddie, T. and Gribkoff, V. (1977) Heterosynaptic depression: a postsynaptic correlate of long-term potentiation. Nature, 266: 737–739.

Malenka, R.C. and Bear, M.F. (2004) LTP and LTD: an embarrassment of riches. Neuron, 44: 5–21.

Malenka, R.C. and Nicoll, R.A. (1999) Long-term potentiation — a decade of progress? Science, 285: 1870–1874.

Man, H.Y., Lin, J.W., Ju, W.H., Ahmadian, G., Liu, L., Becker, L.E., Sheng, M. and Wang, Y.T. (2000) Regulation of AMPA receptor-mediated synaptic transmission by clathrin-dependent receptor internalization. Neuron, 25: 649–662.

Martin, S.J., Grimwood, P.D. and Morris, R.G. (2000) Synaptic plasticity and memory: an evaluation of the hypothesis. Annu. Rev. Neurosci., 23: 649–711.

Massey, P.V., Johnson, B.E., Moult, P.R., Auberson, Y.P., Brown, M.W., Molnar, E., Collingridge, G.L. and Bashir, Z.I. (2004) Differential roles of NR2A and NR2B-containing NMDA receptors in cortical long-term potentiation and long-term depression. J. Neurosci., 24: 7821–7828.

Mayford, M., Bach, M.E., Huang, Y.Y., Wang, L., Hawkins, R.D. and Kandel, E.R. (1996) Control of memory formation through regulated expression of a CaMKII transgene. Science, 274: 1678–1683.

McEwen, B.S. (1999) Stress and hippocampal plasticity. Annu. Rev. Neurosci., 22: 105–122.

McEwen, B.S. (2004) Protection and damage from acute and chronic stress: allostasis and allostatic overload and relevance to the pathophysiology of psychiatric disorders. Ann. N.Y. Acad. Sci., 1032: 1–7.

McEwen, B.S. (2005) Stressed or stressed out: what is the difference? J. Psychiatry Neurosci., 30: 315–318.

McEwen, B.S. and Sapolsky, R.M. (1995) Stress and cognitive function. Curr. Opin. Neurobiol., 5: 205–216.

Mesches, M.H., Fleshner, M., Heman, K.L., Rose, G.M. and Diamond, D.M. (1999) Exposing rats to a predator blocks primed burst potentiation in the hippocampus in vitro. J. Neurosci., 19: p. RC18.

Mockett, B., Coussens, C. and Abraham, W.C. (2002) NMDA receptor-mediated metaplasticity during the induction of long-term depression by low-frequency stimulation. Eur. J. Neurosci., 15: 1819–1826.

Morishita, W., Lu, W., Smith, G.B., Nicoll, R.A., Bear, M.F. and Malenka, R.C. (2007) Activation of NR2B-containing NMDA receptors is not required for NMDA receptor-dependent long-term depression. Neuropharmacology, 52: 71–76.

Mutel, V., Buchy, D., Klingelschmidt, A., Messer, J., Bleuel, Z., Kemp, J.A. and Richards, J.G. (1998) In vitro binding properties in rat brain of [3H]Ro 25-6981, a potent and selective antagonist of NMDA receptors containing NR2B subunits. J. Neurochem., 70: 2147–2155.

Nijholt, I., Farchi, N., Kye, M., Sklan, E.H., Shoham, S., Verbeure, B., Owen, D., Hochner, B., Spiess, J., Soreq, H. and Blank, T. (2004) Stress-induced alternative splicing of acetylcholinesterase results in enhanced fear memory and long-term potentiation. Mol. Psychiatry, 9: 174–183.

Paoletti, P. and Neyton, J. (2007) NMDA receptor subunits: function and pharmacology. Curr. Opin. Pharmacol., 7: 39–47.

Park, C.R., Fleshner, M. and Diamond, D.M. (2004) An NMDA antagonist can impair, protect or have no effect on memory depending on training parameters and stress at the time of retrieval. Soc. Neurosci. Abstr., 34: p. 776.22.

Pastalkova, E., Serrano, P., Pinkhasova, D., Wallace, E., Fenton, A.A. and Sacktor, T.C. (2006) Storage of spatial information by the maintenance mechanism of LTP. Science, 313: 1141–1144.

Reul, J.M. and de Kloet, E.R. (1985) Two receptor systems for corticosterone in rat brain: microdistribution and differential occupation. Endocrinology, 117: 2505–2511.

Rogan, M.T., Staubli, U.V. and LeDoux, J.E. (1997) Fear conditioning induces associative long-term potentiation in the amygdala. Nature, 390: 604–607.

Roozendaal, B. (2002) Stress and memory: opposing effects of glucocorticoids on memory consolidation and memory retrieval. Neurobiol. Learn. Mem., 78: 578–595.

Sacchetti, B., Lorenzini, C.A., Baldi, E., Bucherelli, C., Roberto, M., Tassoni, G. and Brunelli, M. (2001) Long-lasting hippocampal potentiation and contextual memory consolidation. Eur. J. Neurosci., 13: 2291–2298.

Sapolsky, R.M. (1992) Stress, the Aging Brain, and the Mechanisms of Neuron Death. MIT Press, Cambridge, MA.

Schwarze, S.R., Ho, A., Vocero-Akbani, A. and Dowdy, S.F. (1999) In vivo protein transduction: delivery of a biologically active protein into the mouse. Science, 285: 1569–1572.

Sheng, M., Cummings, J., Roldan, L.A., Jan, Y.N. and Jan, L.Y. (1994) Changing subunit composition of heteromeric NMDA receptors during development of rat cortex. Nature, 368: 144–147.

Shors, T.J. (2004) Learning during stressful times. Learn. Mem., 11: 137–144.

Shors, T.J. (2006) Stressful experience and learning across the lifespan. Annu. Rev. Psychol., 57: 55–85.

Shors, T.J., Levine, S. and Thompson, R.F. (1990) Effect of adrenalectomy and demedullation on the stress-induced impairment of long-term potentiation. Neuroendocrinology, 51: 70–75.

Shors, T.J., Seib, T.B., Levine, S. and Thompson, R.F. (1989) Inescapable versus escapable shock modulates long-term potentiation in the rat hippocampus. Science, 244: 224–226.

Sigurdsson, T., Doyere, V., Cain, C.K. and LeDoux, J.E. (2007) Long-term potentiation in the amygdala: a cellular mechanism of fear learning and memory. Neuropharmacology, 52: 215–227.

Tang, Y.P., Shimizu, E., Dube, G.R., Rampon, C., Kerchner, G.A., Zhuo, M., Liu, G. and Tsien, J.Z. (1999) Genetic enhancement of learning and memory in mice. Nature, 401: 63–69.

Tasker, J.G., Di, S. and Malcher-Lopes, R. (2006) Minireview: rapid glucocorticoid signaling via membrane-associated receptors. Endocrinology, 147: 5549–5556.

Wang, M., Yang, Y., Dong, Z., Cao, J. and Xu, L. (2006) NR2B-containing N-methyl-D-aspartate subtype glutamate receptors regulate the acute stress effect on hippocampal long-term potentiation/long-term depression in vivo. Neuroreport, 17: 1343–1346.

Wang, Y.T. and Linden, D.J. (2000) Expression of cerebellar long-term depression requires postsynaptic clathrin-mediated endocytosis. Neuron, 25: 635–647.

Whitlock, J.R., Heynen, A.J., Shuler, M.G. and Bear, M.F. (2006) Learning induces long-term potentiation in the hippocampus. Science, 313: 1093–1097.

Wong, T.P., Howland, J.G., Robillard, J.M., Ge, Y., Yu, W., Titterness, A.K., Brebner, K., Liu, L., Weinberg, J., Christie, B.R., Phillips, A.G. and Wang, Y.T. (2007) Hippocampal long-term depression mediates acute stress-induced spatial memory retrieval impairment. Proc. Natl. Acad. Sci. U.S.A., 104: 11471–11476.

Woo, N.H., Teng, H.K., Siao, C.J., Chiaruttini, C., Pang, P.T., Milner, T.A., Hempstead, B.L. and Lu, B. (2005) Activation of p75(NTR) by proBDNF facilitates hippocampal long-term depression. Nat. Neurosci., 8: 1069–1077.

Woodson, J.C., Macintosh, D., Fleshner, M. and Diamond, D.M. (2003) Emotion-induced amnesia in rats: working memory-specific impairment, corticosterone–memory correlation, and fear versus arousal effects on memory. Learn. Mem., 10: 326–336.

Xu, L., Anwyl, R. and Rowan, M.J. (1997) Behavioural stress facilitates the induction of long-term depression in the hippocampus. Nature, 387: 497–500.

Xu, L., Holscher, C., Anwyl, R. and Rowan, M.J. (1998) Glucocorticoid receptor and protein/RNA synthesis-dependent mechanisms underlie the control of synaptic plasticity by stress. Proc. Natl. Acad. Sci. U.S.A., 95: 3204–3208.

Yang, C.H., Huang, C.C. and Hsu, K.S. (2004) Behavioral stress modifies hippocampal synaptic plasticity through corticosterone-induced sustained extracellular signal-regulated kinase/mitogen-activated protein kinase activation. J. Neurosci., 24: 11029–11034.

Yang, C.H., Huang, C.C. and Hsu, K.S. (2005) Behavioral stress enhances hippocampal CA1 long-term depression through the blockade of the glutamate uptake. J. Neurosci., 25: 4288–4293.

Yang, C.H., Huang, C.C. and Hsu, K.S. (2006) Novelty exploration elicits a reversal of acute stress-induced modulation of hippocampal synaptic plasticity in the rat. J. Physiol., 577: 601–615.

CHAPTER 9

The role of the GluR-A (GluR1) AMPA receptor subunit in learning and memory

D.J. Sanderson[1,*], M.A. Good[2], P.H. Seeburg[3], R. Sprengel[3], J.N.P. Rawlins[1] and D.M. Bannerman[1,*]

[1]*Department of Experimental Psychology, University of Oxford, South Parks Road, Oxford, OX1 3UD, UK*
[2]*School of Psychology, Cardiff University, Tower Building, Park Place, PO Box 901, Cardiff, CF10 3YG, UK*
[3]*Max-Planck Institute of Medical Research, Department of Molecular Neurobiology, Jahnstrasse 29, D-69120 Heidelberg, Germany*

Abstract: It is widely believed that synaptic plasticity may provide the neural mechanism that underlies certain kinds of learning and memory in the mammalian brain. The expression of long-term potentiation (LTP) in the hippocampus, an experimental model of synaptic plasticity, requires the GluR-A subunit of the AMPA subtype of glutamate receptor. Genetically modified mice lacking the GluR-A subunit show normal acquisition of the standard, fixed-location, hidden-platform watermaze task, a spatial reference memory task that requires the hippocampus. In contrast, these mice are dramatically impaired on hippocampus-dependent, spatial working memory tasks, in which the spatial response of the animal is dependent on information in short-term memory. Taken together, these results argue for two distinct and independent spatial information processing mechanisms: (i) a GluR-A-independent associative learning mechanism through which a particular spatial response is gradually or incrementally strengthened, and which presumably underlies the acquisition of the classic watermaze paradigm and (ii) a GluR-A-dependent, non-associative, short-term memory trace which determines performance on spatial working memory tasks. These results are discussed in terms of Wagner's SOP model (1981).

Keywords: glutamate receptors; long-term potentiation; hippocampus; spatial memory; associative learning; working memory

Introduction

Over the years, studies of learning and memory processes in animals have employed a great many experimental approaches and techniques, and have used a range of different species. In the past 15 years the rapid advances in genetic engineering and molecular biological techniques have provided a unique and extremely powerful tool for analyzing these processes in rodents and, in particular, in mice (Chen and Tonegawa, 1997). The study of genetically modified mice has revealed dissociations in aspects of information processing that were hitherto unapparent on the basis of more traditional experimental approaches such as lesion studies.

The use of these transgenic approaches has been particularly successful for studying the contribution

*Corresponding authors. Tel.: +441865271377 or +441865271426; Fax: +441865310447; E-mail: david.sanderson@psy.ox.ac.uk or david.bannerman@psy.ox.ac.uk

DOI: 10.1016/S0079-6123(07)00009-X

that different glutamate receptors and their specific subunits make to distinct aspects of learning and memory. Glutamatergic neurotransmission underlies the majority of fast, synaptic neurotransmission in the central nervous system (CNS) and also plays a crucial role in plasticity processes by which the efficacy of synaptic connections between neurons can be strengthened or weakened. These plasticity mechanisms are widely considered to play an essential role in learning and memory (Hebb, 1949; Bliss and Lomo, 1973; Morris et al., 1986a; Martin et al., 2000).

Glutamatergic neurotransmission in the CNS is mediated through both metabotropic and ionotropic receptors. There are three main forms of ionotropic glutamate receptor — L-α-amino-3-hydroxy-5-methyl-4-isoxazelopropionate (AMPA), N-methyl-D-aspartate (NMDA) and kainate — characterised by their distinct anatomical localizations within the brain, differential sensitivity to a variety of pharmacological antagonists and their selective activation by various glutamate analogues, which provide the basis for their nomenclature (Wisden and Seeburg, 1993; Sprengel, 2006). Although the role of the NMDA receptor in learning and memory has been exhaustively studied and extensively described (Morris et al., 1990a; Martin et al., 2000; Martin and Morris, 2002), the contribution that AMPA receptors, and their individual subunits, make to these aspects of information processing has been less well studied, in part as a consequence of the lack of the appropriate pharmacological tools. Recent advances in genetic technologies now allow the importance of the individual receptor subunits to be assessed.

The functional significance of a particular receptor subunit or subtype will obviously depend upon the brain region and extended neural circuitry within which it is embedded. This chapter aims to describe and discuss the role of AMPA receptors, and in particular GluR-A (GluR1)-containing AMPA receptors, in hippocampus-dependent forms of learning and memory. GluR-A-containing AMPA receptors in other regions of the brain are likely to play an equally important role in other aspects of behaviour. For example, the role of the GluR-A subunit in encoding and/or retrieving the sensory-specific aspects of unconditioned stimuli has now been established (Mead and Stephens, 2003; Johnson et al., 2005; Rumpel et al., 2005), a phenotype that bears strong resemblance to the effects of cytotoxic, basolateral amygdala lesions (Blundell et al., 2001; Dwyer and Killcross, 2006). This chapter will, however, focus on the hippocampal memory system.

The hippocampus and spatial memory

Numerous lines of evidence have implicated the rodent hippocampus in spatial forms of learning and memory (O'Keefe and Nadel, 1978). For example, hippocampal lesions produce robust and reliable deficits on spatial memory tasks such as the Morris watermaze in both rats and mice (Morris et al., 1982, 1990b; Deacon et al., 2002). On the standard version of the task the animals are trained to find a hidden escape platform that remains in the same, fixed location on every trial. The rat or mouse starts from various starting points around the perimeter of the maze and has to use the extramaze cues located around the room to navigate to the platform. This version of the task is often described as a spatial reference memory task (Olton et al., 1979): the relationship between the spatial cues and the goal location, in this case the escape platform, is the same on every trial. If, for example, the escape platform is located in the north-east quadrant of the pool for a given rat or mouse, then for that animal the correct allocentric spatial response is always to go to the north-east quadrant.

Rats with hippocampal lesions are severely impaired on this task (Morris et al., 1982, 1990b) (see also Figs. 1 and 2). In particular, the dorsal subregion of the hippocampus appears to be crucial for normal spatial memory performance (Moser et al., 1995; Bannerman et al., 1999). Mice with hippocampal lesions are equally devastated (Deacon et al., 2002; Reisel et al., 2002). Hippocampal-lesioned animals take significantly longer (in terms of both escape latencies and pathlengths) to find the platform during training (Fig. 1). In addition, a probe trial conducted at the end of training, during which the platform is removed from the pool and the rat or mouse allowed to swim freely for 60 s, shows that whereas control

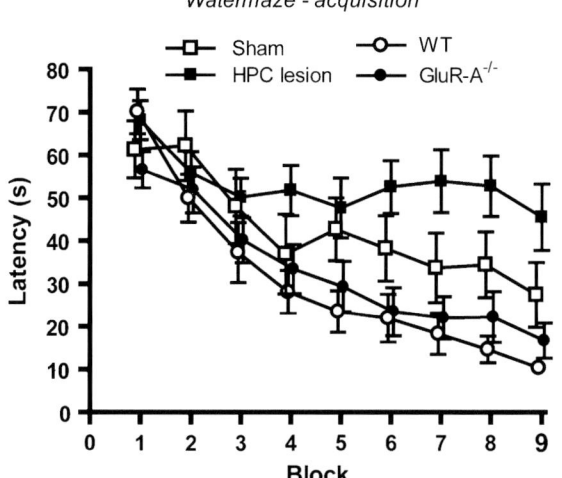

Fig. 1. Hippocampal lesions but not GluR-A deletion impair acquisition of the standard, fixed-location, hidden escape platform (spatial reference memory) version of the Morris watermaze task. Acquisition: Mean escape latency (±SEM) to find the platform across 9 days of training for Sham (white squares) and complete hippocampal-lesioned (black squares); both data taken from Deacon et al. (2002), and for wild-type (white circles) and GluR-A-knockout mice (black circles); both data taken from Reisel et al. (2002).

animals will spend most of their time searching in the area of the pool where the platform was previously located, rats or mice with hippocampal lesions show no such search preference for the training quadrant (Fig. 2). If the amount of time the animals spend in the four quadrants of the pool is plotted, the control animals show a clear preference for the training quadrant in which the platform had been located, whereas animals with complete or dorsal hippocampal lesions show no such preference (chance performance is 25%).

AMPA receptors and hippocampus-dependent spatial memory

A generic role for AMPARs in hippocampus-dependent spatial learning is well established (Riedel et al., 1999). Direct intra-hippocampal infusion of the AMPAR antagonist, LY326325, given during the training phase, disrupted acquisition of the standard version of the fixed-location, hidden-platform, spatial reference memory version

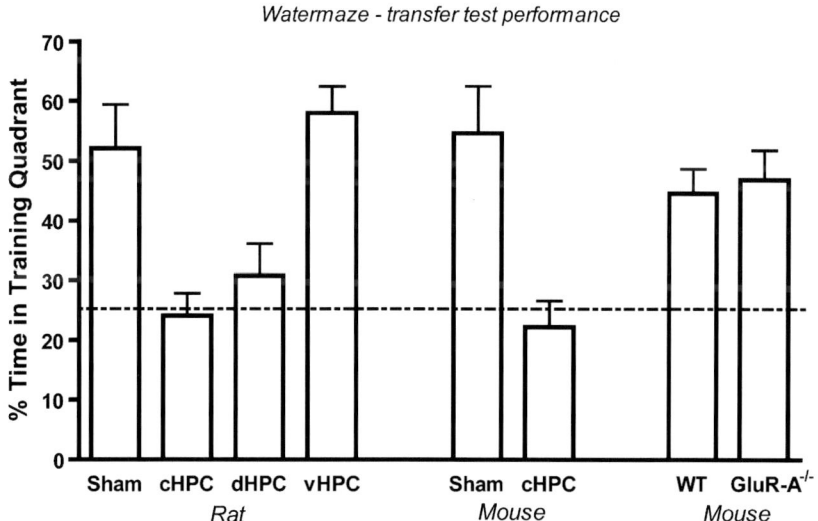

Fig. 2. Complete and dorsal hippocampal lesions but not GluR-A deletion impair transfer test performance following training on the standard, fixed location, hidden escape platform (spatial reference memory) version of the Morris watermaze task. Transfer test performance: Mean percent time in the training quadrant (±SEM) during a probe trial in which the platform is removed from the pool and the rats allowed to swim freely. Left: Performance of sham, complete (cHPC), dorsal (dHPC) and ventral (vHPC) lesioned rats (data taken from Bannerman et al., 1999); centre: performance of sham and complete (cHPC) hippocampal-lesioned mice (data taken from Deacon et al., 2002); right: performance of wild-type and GluR-A-knockout mice (data taken from Reisel et al., 2002). Broken line equates to chance performance (25%).

of the watermaze task. Similarly, the expression of previously acquired spatial information was also disrupted by AMPAR blockade. Animals that were trained on this version of the watermaze task under vehicle infusion conditions, but then subsequently received the AMPAR antagonist during the probe test to assess expression of this previously acquired spatial memory, were unable to successfully retrieve the correct platform location. This study, therefore, clearly demonstrated a role for hippocampal AMPARs in spatial memory.

The contribution that different AMPAR subunits make to hippocampal function was initially limited by the lack of subunit-selective ligands. The development of genetically modified mice in which individual AMPAR subunits can be selectively deleted or modified has now opened the door, allowing the contribution that these individual proteins make to various forms of learning and memory to be studied. The AMPAR subtype of excitatory amino acid glutamate receptor is a hetero-oligomeric protein complex consisting of combinations of four different kinds of subunits (GluR-A–GluR-D; GluR1–GluR4), each encoded by a separate gene: gria 1–4 (Wisden and Seeburg, 1993). In recent years, genetically modified mice lacking each of the individual GluR subunits have been generated and their behaviour has been studied across a range of learning and memory paradigms: GluR-B (Shimshek et al., 2006); GluR-C (Sanchis-Segura et al., 2006); GluR-D (Fuchs et al., 2007). Mice lacking the GluR-A subunit of the AMPA receptor have been of particular interest (Zamanillo et al., 1999). Studies of these mice have revealed important dissociations between distinct aspects of hippocampal information processing.

GluR-A knockout mice

Genetically modified, adult mice lacking the GluR-A AMPA receptor subunit exhibit fast synaptic transmission in the hippocampus, mediated by the receptors formed from the remaining AMPA receptor subunits GluR-B and GluR-C (Sommer et al., 1991), but are deficient in a rapidly expressed component of synaptic plasticity at CA3-to-CA1 synapses in the hippocampus (Zamanillo et al., 1999; Hoffman et al., 2002; Malinow and Malenka, 2002; Jensen et al., 2003). Behavioural studies of GluR-A knockout mice have revealed striking dissociations between different aspects of hippocampus-dependent information processing. GluR-A$^{-/-}$ mice, like animals with hippocampal lesions, display a robust and enduring spatial working memory deficit (Reisel et al., 2002; Schmitt et al., 2003, 2004b). Nevertheless, GluR-A$^{-/-}$ mice show no impairment on tests of spatial reference memory (Zamanillo et al., 1999; Reisel et al., 2002; Schmitt et al., 2003). This is in direct contrast to hippocampal-lesioned rodents (Morris et al., 1982; Deacon et al., 2002; Reisel et al., 2002) and rats with a intra-hippocampal AMPAR blockade following infusion of LY326325 (Riedel et al., 1999). Taken together these results suggest that there are separate and distinct molecular mechanisms within the hippocampus supporting spatial working and reference memory.

Spatial reference memory is GluR-A-independent

In contrast to the severe impairments seen with hippocampal lesions, or complete intra-hippocampal AMPAR blockade, GluR-A knockout mice were able to acquire the spatial reference memory watermaze task as well as wild-type controls (Zamanillo et al., 1999; Reisel et al., 2002). The knockout mice had similar escape latencies and pathlengths during acquisition (Fig. 1), and the performance of the wild-type and GluR-A$^{-/-}$ groups on the probe trial was indistinguishable (Fig. 2). Both groups had learned to an equal extent about the location of the platform, with wild-type and knockout mice showing a strong preference for the quadrant of the pool in which the platform had been located. Similar findings have been reported for genetically modified mice in which phosphorylation sites on the GluR-A subunit have been altered. These GluR-A point mutants also display intact spatial reference memory acquisition in the watermaze (Lee et al., 2003).

The observation that GluR-A knockout mice acquire spatial reference memory tasks as well as control mice was subsequently confirmed using an appetitively motivated radial maze task (Olton et al., 1979). The radial maze that we used consisted of six arms radiating out from a central platform like spokes on a wheel (Fig. 3). To assess spatial reference memory, three of the arms were baited with milk rewards and the other three were never baited. The identity of the baited and non-baited arms remained constant across all trials. Consequently the animal had to learn which arms are rewarded and which arms are never rewarded. The mice were placed on the central platform at the start of a trial and allowed to explore and enter arms until all three milk rewards had been collected. During this phase of the study animals were only allowed to enter each arm once. Once an arm had been visited, the door to that arm was closed and remained shut thereafter. Crucially, the maze was periodically rotated to prevent the mice from using intramaze cues. For this reference memory task, the spatial responses that the animals are required to make to collect the food rewards remain the same across all of the training trials. If an animal entered an arm that was never baited then that was scored as a reference memory error.

Mice with hippocampal lesions were unable to acquire the spatial reference memory radial maze task (Schmitt et al., 2003) (Fig. 3). They were unable to learn which arms were baited and which arms were never baited, and continued to enter the un-rewarded arms. Indeed, even after extensive training the hippocampal-lesioned mice showed no improvement. In contrast to hippocampal-lesioned mice, but not surprisingly in view of the previous watermaze result, the GluR-A knockout mice acquired the spatial reference memory component of the three from six radial maze task just as well as the control animals (Fig. 3). The mutant mice were perfectly capable of learning to avoid the never-baited arms, and both groups of mice made equivalent numbers of reference memory errors during acquisition. This result confirmed that hippocampus-dependent spatial reference memory is GluR-A-independent. In view of the widely held belief that hippocampal long-term potentization (LTP) underlies performance on hippocampus-dependent memory tasks, this result came as something of a surprise. It also suggested that AMPARs composed of other subunits were likely to be of crucial importance for spatial reference memory performance (Shimshek et al., 2006).

Spatial working memory is GluR-A-dependent

The radial maze task also allowed us to study hippocampus-dependent spatial working memory. On spatial working memory tasks, the correct spatial response that the animal must make varies

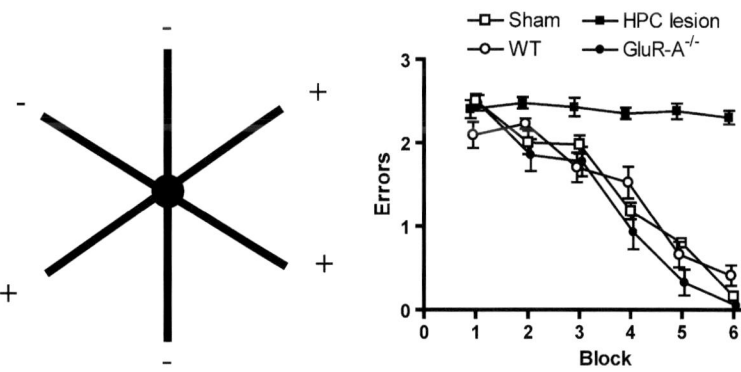

Fig. 3. Hippocampal lesions but not GluR-A deletion impair acquisition of the spatial reference memory component of the radial maze task. Left: Mice were trained to discriminate between three baited arms (+) and three non-baited arms (−) of a six-arm radial maze. Doors prevented mice re-entering arms they had already visited on that particular visit to the maze (i.e. of making working memory errors) during this stage of the experiment. Right: Mean number of reference memory errors per trial ($\pm$SEM) during six blocks of training (four trials per block). Sham-lesioned (white squares), wild-type (WT: white circles) and GluR-A-knockout mice (GluR-A$^{-/-}$: black circles) all acquired the task at an equivalent rate. Hippocampal-lesioned mice (HPC: black squares) were completely unable to learn which arms were baited and which arms were not baited (all data taken from Schmitt et al., 2003).

daily, pseudo-trial-unique arm presentations on the radial maze, and then also when conducting each trial on a different three-arm maze, each in a novel testing room. There was, however, no evidence that the magnitude of the spatial working memory impairment was alleviated in any way when potential proactive interference was reduced.

In further experiments, performance of experimentally naïve wild-type and GluR-A$^{-/-}$ mice was assessed during just a single trial of spontaneous alternation, thus completely eliminating any source of proactive interference. Under these conditions, the spatial non-matching to place, alternation task becomes a spatial novelty preference test. Performance on this test is dependent on the hippocampus (Fig. 8). During a single sample phase, the mice are allowed to explore the start arm and only one of the two goal arms. On the subsequent choice test, in which access to all three arms is now permitted, wild-type controls showed a preference for the novel arm over the familiar arms. Experimentally naïve GluR-A$^{-/-}$ mice, which had no prior experience of the maze or of that spatial environment, showed no preference for the novel or the familiar arms (Fig. 8). This result is important because it shows that the deficit in the knockout animals is not simply the result of an inability to overcome the potentially deleterious effects of proactive interference, arising from previous trials.

In summary, we have argued previously that the pattern of impaired spatial working memory combined with intact spatial reference memory performance might be explained by an increased sensitivity to proactive interference in the knockout animals. However, the results of the spatial novelty preference test suggest that the spatial working memory impairment in GluR-A$^{-/-}$ animals is not due to increased proactive interference, and that in these knockout mice there is a fundamental deficit in making the appropriate spatial response during these working memory tasks.

Two separable forms of hippocampal-dependent, spatial information processing

The spatial novelty preference test in experimentally naïve animals demonstrates that GluR-A$^{-/-}$ mice exhibit a one-trial spatial memory deficit that is independent of proactive interference. This result, of course, suggests the existence of two distinct, and totally independent, spatial information processing mechanisms that are differentially sensitive to GluR-A deletion: the first, a *GluR-A-dependent* mechanism that rapidly encodes or expresses information relating to a single spatial episode or experience; the other, a *GluR-A-independent* mechanism that encodes information about spatial locations that is relevant across many trials as typified by spatial reference memory acquisition.

This GluR-A-independent mechanism, by which the associative strength of spatial locations can be increased (and/or decreased) gradually over many trials, and which presumably underlies spatial reference memory acquisition during tasks such as the Morris watermaze or the radial maze (Zamanillo et al., 1999; Reisel et al., 2002; Schmitt et al., 2003), may, in certain respects, resemble the incremental process by which non-spatial stimuli acquire associative strength during simple conditioning procedures using discrete objects, odours or auditory cues, although importantly these two forms of associative learning do differ in that the former, but not the latter, is dependent on the hippocampus for its expression (e.g. Morris et al., 1982, 1986b). Nevertheless, although the expression of long-term spatial memories during performance on spatial reference memory tasks may depend upon an intact hippocampus, it has been suggested that these memories are in fact stored elsewhere in the cortex, and that the role of the hippocampus is not in memory formation per se, but with response selection and the appropriate expression of behaviour, on the basis of both contextual cues and information retrieved from memory (e.g. Virley et al., 1999; Gray and McNaughton, 2000; Honey and Good, 2000; Bannerman et al., 2004). So what then might be the specific role of GluR-A-dependent synaptic plasticity within the framework of a hippocampal system that mediates aspects of response selection and the expression of behaviour based on contextual information and information retrieved from memory?

et al., 1979). The radial maze that we used consisted of six arms radiating out from a central platform like spokes on a wheel (Fig. 3). To assess spatial reference memory, three of the arms were baited with milk rewards and the other three were never baited. The identity of the baited and non-baited arms remained constant across all trials. Consequently the animal had to learn which arms are rewarded and which arms are never rewarded. The mice were placed on the central platform at the start of a trial and allowed to explore and enter arms until all three milk rewards had been collected. During this phase of the study animals were only allowed to enter each arm once. Once an arm had been visited, the door to that arm was closed and remained shut thereafter. Crucially, the maze was periodically rotated to prevent the mice from using intramaze cues. For this reference memory task, the spatial responses that the animals are required to make to collect the food rewards remain the same across all of the training trials. If an animal entered an arm that was never baited then that was scored as a reference memory error.

Mice with hippocampal lesions were unable to acquire the spatial reference memory radial maze task (Schmitt et al., 2003) (Fig. 3). They were unable to learn which arms were baited and which arms were never baited, and continued to enter the un-rewarded arms. Indeed, even after extensive training the hippocampal-lesioned mice showed no improvement. In contrast to hippocampal-lesioned mice, but not surprisingly in view of the previous watermaze result, the GluR-A knockout mice acquired the spatial reference memory component of the three from six radial maze task just as well as the control animals (Fig. 3). The mutant mice were perfectly capable of learning to avoid the never-baited arms, and both groups of mice made equivalent numbers of reference memory errors during acquisition. This result confirmed that hippocampus-dependent spatial reference memory is GluR-A-independent. In view of the widely held belief that hippocampal long-term potentiation (LTP) underlies performance on hippocampus-dependent memory tasks, this result came as something of a surprise. It also suggested that AMPARs composed of other subunits were likely to be of crucial importance for spatial reference memory performance (Shimshek et al., 2006).

Spatial working memory is GluR-A-dependent

The radial maze task also allowed us to study hippocampus-dependent spatial working memory. On spatial working memory tasks, the correct spatial response that the animal must make varies

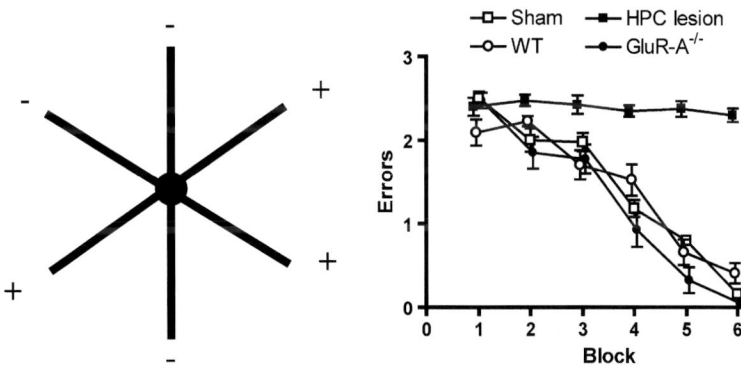

Fig. 3. Hippocampal lesions but not GluR-A deletion impair acquisition of the spatial reference memory component of the radial maze task. Left: Mice were trained to discriminate between three baited arms (+) and three non-baited arms (−) of a six-arm radial maze. Doors prevented mice re-entering arms they had already visited on that particular visit to the maze (i.e. of making working memory errors) during this stage of the experiment. Right: Mean number of reference memory errors per trial (±SEM) during six blocks of training (four trials per block). Sham-lesioned (white squares), wild-type (WT: white circles) and GluR-A-knockout mice (GluR-A$^{-/-}$: black circles) all acquired the task at an equivalent rate. Hippocampal-lesioned mice (HPC: black squares) were completely unable to learn which arms were baited and which arms were not baited (all data taken from Schmitt et al., 2003).

from trial to trial and is dependent on some conditional, trial-specific information that is relevant to that particular trial. The idea of distinct working memory and reference memory systems was first proposed by Honig (1978), and later elaborated upon by Olton et al. (1979). It is also worth pointing out, however, that the term 'working memory' has been used to reflect different processes by different researchers. For example, to many, and in particular researchers working with primates, working memory is a term often used to refer to a short-term, on-line memory supported by frontal lobe structures (Goldman-Rakic, 1987). In the present context, however, we use the term to refer to a flexible memory system in which conditional, trial-specific information is used to select between response options that are variably correct or incorrect. The performance on spatial working memory tasks is extremely sensitive to hippocampal lesions (Olton et al., 1979).

Once the animals had learned which radial maze arms were baited and which never baited, then the working memory component of the task was introduced (Schmitt et al., 2003). Food rewards were not replaced within a trial and the mice were now allowed to re-enter arms as often as they wished. So now the animals needed to keep track of which arms they had already visited on that particular trial for efficient performance. If an animal entered an arm that was normally baited, but which it had already entered on that trial (and had already received the milk reward), then that was scored as a working memory error. In addition, reference memory errors (entries into never-baited arms) were scored as before.

GluR-A$^{-/-}$ mice were less able to keep track of which arms they have already visited on a particular trial, and they made numerous working memory errors in which they re-visited arms that were no longer baited, whereas the wild-type mice made virtually none of these kinds of errors (Fig. 4). Importantly, despite the working memory impairment, the GluR-A$^{-/-}$ mice were, however, still able to remember which arms were never baited. They showed intact reference memory performance on the very same trials during which they were making numerous working memory-type errors. This, therefore, provides a within-subject and within-task demonstration of impaired

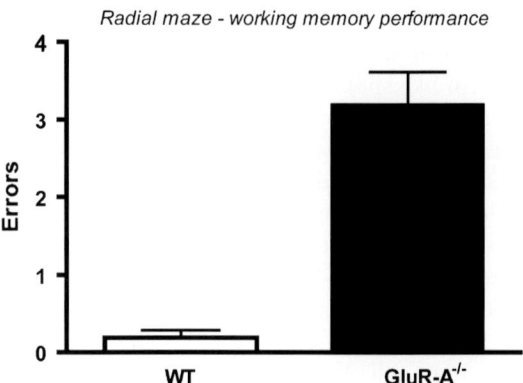

Fig. 4. GluR-A deletion impairs spatial working memory performance on the six-arm radial maze task. Mice were still rewarded in the same three arms of the maze and not rewarded in the three non-baited arms, but now they were allowed to re-enter arms as often as they liked, and rewards were not replaced within a trial. Mean number of working memory errors per trial ($\pm$SEM) for wild-type (white bar) and GluR-A-knockout mice (GluR-A$^{-/-}$: black bar; all data taken from Schmitt et al., 2003).

hippocampus-dependent spatial working memory but spared hippocampus-dependent spatial reference memory performance. As both reference and working memory components of the radial maze task involve using the same set of spatial cues, and presumably have the same sensorimotor and motivational demands, this dissociation argues very strongly for a specific role for GluR-A in hippocampus-dependent spatial working memory processes, and against an account based on a non-specific disruption of maze performance per se.

Spatial working memory on the T-maze

Spatial working memory performance can also be studied using a spatial non-matching to place or rewarded alternation task on a simple three-arm maze such as an elevated T-maze or Y-maze. As with the spatial working memory component of the radial maze task, the relationship between the spatial cues and the goal location (i.e. the milk reward) changes from trial to trial. The animal receives some information on a sample trial which acts as conditional information that the animal must then use to select the appropriate response on the subsequent choice trial. The conditional information is only relevant on that trial. On the T-maze-rewarded alternation task the rat or mouse is first forced into one arm of the maze during a sample run where it then receives a reward (Fig. 5). The mouse is then returned to the beginning of the start arm and is given a free choice of either arm. The mouse is rewarded for choosing the previously unvisited arm (i.e. for alternating). Many trials are run over a period of time with the goal arm visited on the sample run designated by a pseudorandom sequence. Successful performance on this task also depends on an intact hippocampus, and, in particular, on the dorsal hippocampus (Rawlins and Olton, 1982; Hock and Bunsey, 1998; Bannerman et al., 1999). Both mice and rats with complete and dorsal hippocampal lesions are dramatically impaired on this task and display chance levels of performance (50%) (Bannerman et al., 2002; Deacon et al., 2002) (Fig. 5). Similarly, the GluR-A knockout mice exhibit a devastating and lasting impairment. The performance of the knockout mice, like rats or mice with hippocampal lesions, remains steadfast at chance even after many trials (Reisel et al., 2002). The GluR-A$^{-/-}$ mice are unable to discriminate between the unvisited and visited arm from the previous sample trial (Fig. 5).

Genetic rescue of GluR-A-dependent spatial working memory

Spatial working memory performance in GluR-A$^{-/-}$ mice can be restored by the forebrain-specific expression of GluR-A subunits labeled with green-fluorescent protein (GFP) on an otherwise

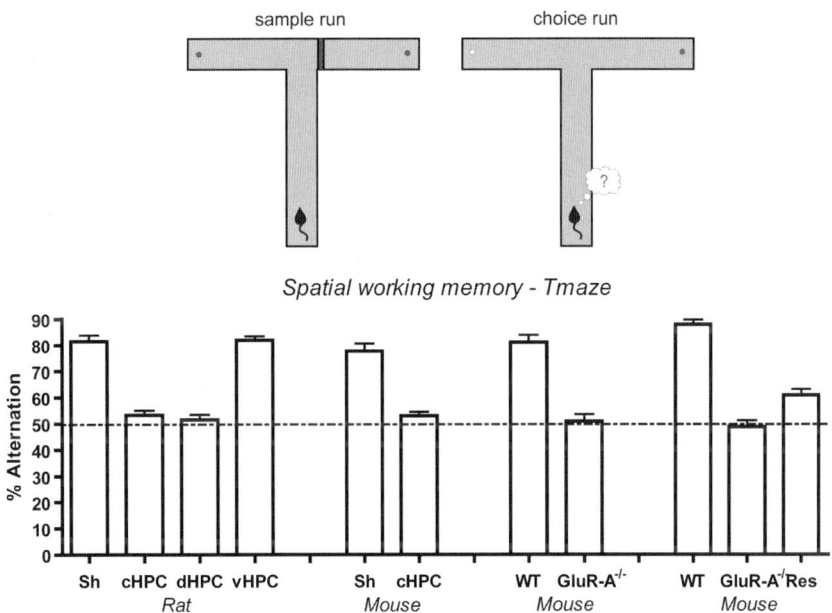

Fig. 5. Hippocampal lesions and GluR-A deletion both impair spatial working memory performance during non-matching to place testing (rewarded alternation) on the elevated T-maze. Top: During a sample run (left) mice are forced into either the left or right goal arm, according to a pseudorandom sequence, and receive a milk reward. During the choice run (right) the animals are required to go to the opposite (previously unvisited) goal arm for a second milk reward. Many trials are run over a period of time with the identity of the arm visited on the sample run varying pseudorandomly. Bottom: Mean percentage of trials on which the mouse alternated successfully ($\pm$SEM). Left: Performance of sham, complete (cHPC), dorsal (dHPC) and ventral (vHPC) lesioned rats (data taken from Bannerman et al., 2002); centre left: performance of sham and complete (cHPC) hippocampal-lesioned mice (unpublished); centre right: performance of wild-type and GluR-A-knockout mice (data taken from Reisel et al., 2002). Right: Performance of wild-type (WT), GluR-A-knockout mice (GluR-A$^{-/-}$) and GFP-labelled GluR-A rescued GluR-A-knockout mice (Res: data taken from Schmitt et al., 2005). Broken line equates to chance performance (50%).

GluR-A knockout background (Schmitt et al., 2005). This genetic rescue was under the control of the αCaMKII promoter and was especially pronounced in the CA1 subfield of the dorsal subregion of the hippocampus. The level of performance of the rescued mice on the spatial working memory component of the radial maze and during non-matching to place on the T-maze (Fig. 5) was intermediate between the performance of the wild-type controls and the full GluR-A knockout mice. Importantly, GFP-GluR-A expression also resulted in a significant rescue of hippocampal field LTP (Mack et al., 2001). Again the level of LTP in the rescued animals was intermediate between the wild-type and GluR-A$^{-/-}$ mice, thus further strengthening the correlation between this form of hippocampal synaptic plasticity and a rapid, flexible form of information processing that underlies spatial working memory performance.

GluR-A and performance on conditional learning tasks

On spatial working memory tests such as the rewarded alternation T-maze task, or the win-shift component of the radial maze task, trial-specific information is provided by the spatial location of the goal arm just visited on the sample run (T-maze), or the arms of the radial maze already entered on that particular visit to the maze. This trial-specific, response-generated spatial information acts as a conditional cue which the animal can then use to select the appropriate spatial response on that particular trial.

In a separate series of experiments, the trial-specific conditional information was provided by floor inserts placed in the start arm of the maze (Schmitt et al., 2004a). For example, for a given mouse, when there was a white perspex floor insert in the start arm of the maze then the animal had to go to the left goal arm for a food reward (Fig. 6, left-hand side). When there was a black, wire mesh floor insert in the start arm of the maze then the mouse had to go to the right goal arm to get its food reward (Murray and Ridley, 1999). Mice with hippocampal lesions found this task insoluble (Fig. 6). They remained at chance levels of performance throughout testing. When wild-type and GluR-A$^{-/-}$ mice were compared on this conditional task, there was also a complete impairment in the knockout animals, resembling that seen in the lesioned mice (Fig. 6). Thus, GluR-A$^{-/-}$ mice also found the task insoluble.

Importantly, in a second experiment we found that if the floor inserts extended throughout the entire maze, including not only the start arm but also both of the goal arms (see Fig. 6, right-hand side), then both mice with hippocampal lesions and GluR-A$^{-/-}$ mice were perfectly able to acquire the task (Fig. 6). The same learning rules applied, but now the relevant floor insert was also present at the location where the animal received the reward. Separate groups of mice were trained on this version of the task. For these animals there was no spatio-temporal discontiguity between the floor insert and the place where the animal was rewarded. Under these conditions there was no impairment in either lesioned or knockout mice. Both hippocampal-lesioned and GluR-A$^{-/-}$ mice acquired this task as well as their respective control groups.

Taken together, these results show that hippocampal GluR-A is essential when the animal is required to bridge a spatio-temporal discontiguity between the conditional cue and the place where the reward is experienced (Rawlins, 1985). In other words, hippocampal GluR-A is required when the appropriate spatial response is selected rapidly and flexibly on the basis of a cue representation that is retrieved from recent, short-term memory.

Non-spatial hippocampus-dependent tasks

The deficit in GluR-A$^{-/-}$ mice is not solely restricted to tasks in which the animal is required to make a spatial response. There is increasing evidence that the hippocampus may be important for the temporal encoding of non-spatial, as well as spatial, information (Meck et al., 1984; Young and McNaughton, 2000; Fortin et al., 2002; Kesner et al., 2002). For example, mice or rats with hippocampal lesions are impaired on a non-spatial, operant differential reinforcement of low rates of

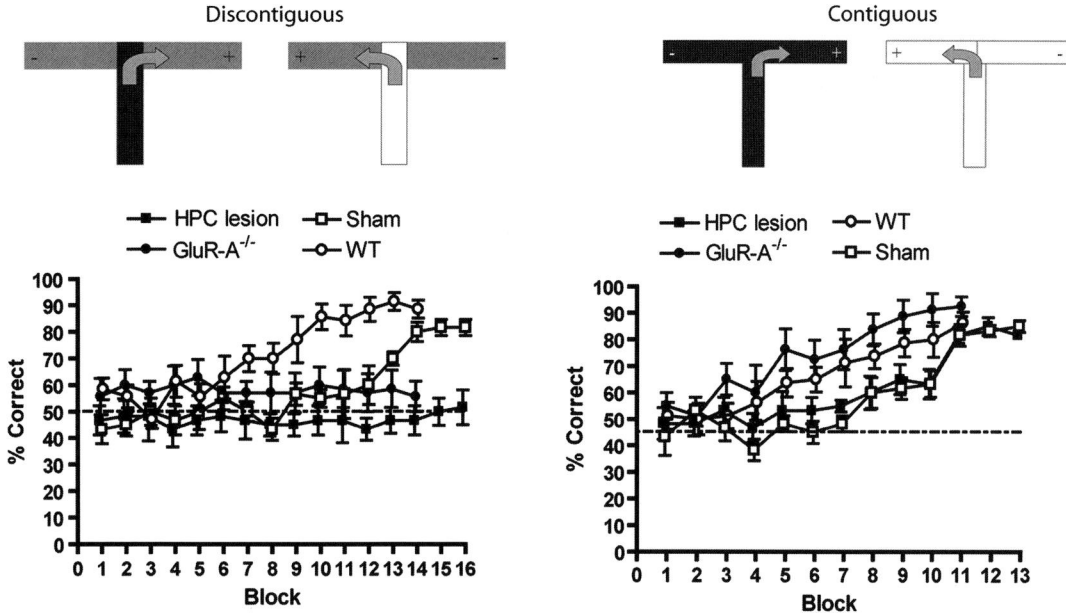

Fig. 6. Hippocampal lesions and GluR-A deletion both impair a conditional discrimination T-maze task in which inserts (black wire mesh vs. white Perspex) present on the floor of the maze indicate which goal arm contains reward, but only when there is a spatiotemporal discontiguity between the floor inserts and the place where the reward is delivered. Left: Discontiguous task. The floor inserts are only present in the start arm of the maze. The presence of the black/wire mesh insert indicates that the food reward is in the right goal arm. The presence of the white Perspex insert indicates that the food reward is in the left goal arm. Hippocampal lesions and GluR-A deletion both impair acquisition of the Discontiguous task. Mean percent correct choices ($\pm$SEM) during blocks of 10 trials for Sham-lesioned (white squares), hippocampal-lesioned mice (HPC: black squares), wild-type (WT: white circles) and GluR-A-knockout mice (GluR-A$^{-/-}$: black circles). Right: Contiguous task. Now the same conditional rules apply but the floor inserts extend throughout the entire maze. All mice acquired the task. There were no differences between Sham- and hippocampal-lesioned mice, or between wild-type and GluR-A-knockout mice (all data taken from Schmitt et al., 2004a).

responding (DRL) task (Clark and Isaacson, 1965; Jarrard and Becker, 1977; Johnson et al., 1977; Boitano et al., 1980; Sinden et al., 1986; Reisel et al., 2005). In this task, animals are trained in an operant chamber to press a lever to obtain a food reward. They are then required to wait a prescribed period of time before pressing the lever again to obtain subsequent food rewards. If the animal responds prematurely during this time period, then the animal is not rewarded and the timer resets to zero. This task therefore requires the animal to switch between withholding and making a response contingent upon the elapse of some minimum period of time (the DRL requirement) since the last response.

Hippocampal-lesioned mice were less able to withhold responding during a DRL-10 requirement (i.e. when the animals had to wait 10 s between lever presses), and were thus less efficient than sham operated controls. In contrast, GluR-A$^{-/-}$ mice did not differ from wild-type mice when the DRL requirement was 10 s. However, they were significantly impaired when the DRL requirement was increased to 15 s. This result was similar to that previously observed with D-AP5-treated rats (Tonkiss et al., 1988). Animals receiving the competitive NMDAR antagonist displayed a robust impairment, although the deficit was also less pronounced than that seen in rats with hippocampal lesions (Sinden et al., 1986; Tonkiss et al., 1988). GluR-A$^{-/-}$ mice therefore display impairments on both spatial and non-spatial tasks which require a rapid, flexible memory system that enables the animal to select between potentially equally valid behavioural options (or response alternatives) that have previously each been associated with reward and non-reward.

NMDA receptors, LTP and spatial memory

Our work with GluR-A$^{-/-}$ mice has therefore established a specific role for a GluR-A-dependent form of synaptic plasticity in a particular aspect of hippocampal-dependent information processing. The demonstration that it was still possible to induce some LTP in these GluR-A$^{-/-}$ mice, using a theta-burst stimulation paradigm, leaves open the possibility that GluR-A-independent forms of synaptic plasticity may be sufficient to support other aspects of hippocampal information processing, such as acquisition of spatial reference memory tasks (Hoffman et al., 2002). In this regard, it is worth pointing out that the residual, GluR-A-independent LTP observed in GluR-A$^{-/-}$ mice, was blocked by the NMDAR antagonist, D-AP5 (Hoffman et al., 2002). This of course might suggest that if the residual synaptic plasticity does underlie performance on spatial reference memory tasks, then that performance should be sensitive to manipulations of NMDAR function.

The precise contribution that NMDARs make to spatial memory is still a matter of debate (see Bannerman et al., 2006, for discussion). Traditionally, this issue has been explored using pharmacological approaches. In a seminal study, Morris et al. (1986a) showed that experimentally naïve rats receiving chronic intra-cerebroventricular infusion of the NMDAR antagonist, D-AP5, were impaired in acquiring the standard spatial reference memory watermaze task. However, subsequent studies have shown that if the rats are given watermaze pre-training as normal animals, prior to being tested with NMDA antagonists in a novel spatial environment, then the drug-treated animals are in fact perfectly capable of acquiring the spatial reference memory task (Bannerman et al., 1995; Saucier and Cain, 1995). This suggests that NMDARs are not required to form a spatial representation of the environment to incrementally form an association between a particular spatial location and a goal (e.g. obtain a food reward or escape from water), or to use the spatial representation to navigate from a variable start location to a fixed location, hidden escape platform.

In contrast, a role for NMDARs in hippocampus-dependent, spatial working memory tasks has been well supported by pharmacological studies. There are reports of spatial working memory deficits with AP5 on a number of tasks (Caramanos and Shapiro, 1994; Lee and Kesner, 2002; Day et al., 2003; Bast et al., 2005). For example, D-AP5-treated rats are impaired on the spatial, delayed non-matching to position (rewarded alternation) task on the elevated T-maze, the same task on which we see such a striking deficit in the GluR-A$^{-/-}$ mice (Tonkiss and Rawlins, 1991). D-AP5 also impairs performance on a spatial working memory, matching to place version of the watermaze task in which the platform moves to a novel position on each day of testing (Steele and Morris, 1999).

NMDAR mutants and spatial memory

More recently, the development of various genetically modified mouse lines has furthered the study of NMDARs and their role in spatial memory. The NMDAR is a heteromeric complex comprising an obligatory NR1 receptor subunit and variable combinations of NR2 subunits, of which NR2A and NR2B predominate in hippocampal pyramidal cells in the adult mouse brain.

In agreement with the results obtained from GluR-A$^{-/-}$ mice, mice in which the NR1 subunit of the NMDA receptor has been selectively deleted from the CA3 subfield of the hippocampus displayed normal acquisition of the standard reference memory version of the watermaze task (Nakazawa et al., 2002), but were impaired on a spatial working memory, matching to position version (Nakazawa et al., 2003). In contrast, CA1-specific NR1 knockout mice have been reported to display poorer spatial reference memory acquisition in the watermaze (Tsien et al., 1996). It should also be noted, however, that these CA1 NR1 knockout mice were also impaired during performance of a visible platform control task (a task that does not require an intact hippocampus (Morris et al., 1986b), although by the end of training they were performing as well as the controls. This might suggest a more general disruption of watermaze performance rather than a spatial learning impairment per se, and may also suggest that the NR1 deletion may be less specific to the hippocampus than was hoped for.

NR2 subunit selective knockout mice have also been produced. Mice with unconditional deletion of the NR2B subunits die perinatally (Kutsuwada et al., 1996), but NR2A knockout mice are viable. These NR2A$^{-/-}$ mice have been reported to show a mild impairment in the Morris watermaze, although they clearly learned a substantial amount about the spatial location of the platform. Therefore, there are questions as to how robust and reliable this deficit actually is, and whether it generalises to other tests of spatial reference memory (Bannerman et al., 2006).

Dentate-gyrus-specific NR1 knockout selectively impairs spatial working memory

We have also recently generated genetically modified mice that are deficient in the NR1 subunit of the NMDAR, specifically and exclusively within the dentate gyrus of the hippocampal formation (Niewoehner et al., 2007) (Fig. 7). Mice carrying floxed NR1 alleles were generated and tTA-induced Cre expression was used to destroy the NR1 gene, specifically in dentate granule cells (Krestel et al., 2004) (Fig. 7A). In situ hybridization and immunohistochemical analyses revealed a selective deletion of the NR1 subunit in the dentate gyrus, whereas there was no evidence of any reduction in NR1 subunit levels in any of the other hippocampal subfields, or anywhere else in the brain (Fig. 7B, C). These $NR1^{\Delta DG}$ mice displayed severely impaired LTP in both medial and lateral perforant path inputs to the DG, whereas LTP was unchanged in CA3-to-CA1 cell synapses in hippocampal slices.

We assessed the behaviour of the $NR1^{\Delta DG}$ mice using the same three from six radial maze task that we had previously used to characterise the GluR-A$^{-/-}$ mice, allowing us to investigate both spatial reference memory and spatial working memory (Schmitt et al., 2003). $NR1^{\Delta DG}$ and wild-type mice acquired the spatial reference memory component of the radial maze task at the same rate, and both groups were able to learn to discriminate between the always-baited and never-baited arms (Fig. 7D). However, when the working memory component was introduced, the $NR1^{\Delta DG}$ mice were less able to keep track of which arms they had already visited on a particular trial, and made significantly more spatial working memory errors than their wild-type counterparts (Fig. 7E). At the same time, however, they were still able to avoid the non-baited arms. This pattern of sparings and impairments is the same as that seen with the GluR-A$^{-/-}$ mice, although the magnitude of the spatial working memory impairment in the dentate-specific knockout is much smaller than that observed in the constitutive, wholebrain GluR-A knockout.

Is the spatial working memory impairment in GluR-A$^{-/-}$ mice related to frontal or hippocampal processing deficits?

So far we have worked under the assumption that the spatial working memory phenotype in mice with a constitutive, wholebrain GluR-A deletion is hippocampal in origin. Although we have clearly demonstrated that an exclusively hippocampus-specific manipulation of LTP (the dentate gyrus-specific deletion of the NR1 subunit of the NMDAR) can give rise to a behavioural phenotype in which impaired spatial working memory co-exists with normal spatial reference memory (Nakazawa et al., 2002, 2003; Niewoehner et al., 2007), it is of course the case that we cannot rule out the possibility that deletion of GluR-A in a brain area, or brain areas, other than the hippocampus could also contribute to the observed dissociation between impaired spatial working memory and spared spatial reference memory.

Historically, a role for the prefrontal cortex in spatial working memory has been repeatedly claimed (for review, see Kolb, 1984). Older studies using traditional, non-fibre-sparing lesions (e.g. aspiration lesions) often resulted in spatial alternation deficits. In contrast, more recent studies employing cytotoxic, fibre-sparing techniques have found that lesions of medial prefrontal cortex (mPFC) do not always reliably affect non-matching to place performance in rodents, and where deficits do exist they are generally mild and transient in nature (Shaw and Aggleton, 1993; Aggleton et al., 1995; Sanchez-Santed et al., 1997;

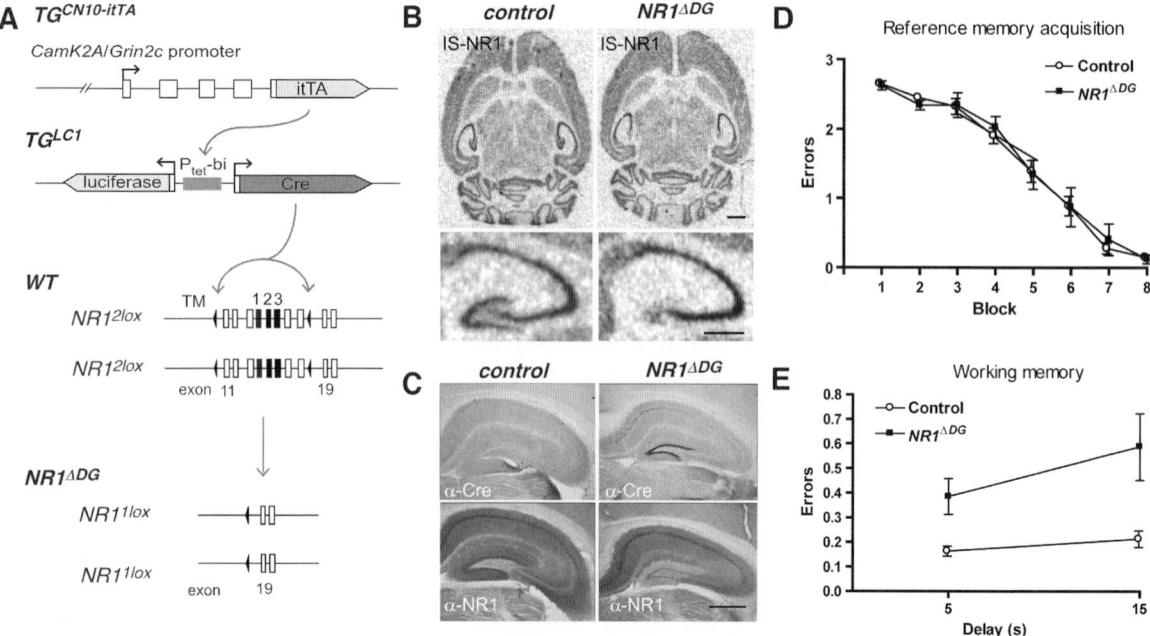

Fig. 7. Dentate gyrus-specific deletion of the NR1 subunit of the NMDA receptor results in a selective impairment of spatial working memory but has no effect on spatial reference memory. (A) Depletion of functional NMDAR in $NR1^{\Delta DG}$ mice. Genetic elements used for cell-specific NR1 gene deletion. Expression of itTA from the $CamK2A/Grin2c$ hybrid promoter (transgene from line $TG^{CN10-itTA}$) drives Cre-recombinase expression by activation of the bidirectional Ptet$_{bi}$ promoter of the luciferase/Cre tet-responder transgene from line (TG^{LC1}). Cre-recombinase excises the loxP (black triangle) flanked introns 11–18 (white boxes) of the gene targeted modified $NR1^{2lox}$ alleles. Exons encoding membrane regions 1–3 (TM1-3) are given in black. In cells with active Cre the $NR1^{2lox}$ alleles are converted to $NR1^{1lox}$ alleles. (B) In situ hybridisation of littermate control animals (left) and DG-specific NR1 deletion mice ($NR1^{\Delta DG}$) (right) with NR1-specific probe. The lower panels give a zoom of the hippocampus from the respective upper panels (scale bars: 1 mm). (C) Cre immunohistochemistry (top panels) of littermate $NR1^{2lox}$ (control, left) and $NR1^{\Delta DG}$ mice (right) and with anti-NMDAR1 antibody (bottom panels) (scale bar: 1 mm). (D) $NR1^{\Delta DG}$ mice display normal spatial reference memory acquisition on a 3/6 radial arm maze task. Mean ($\pm$SEM) number of reference memory errors per trial (maximum of 3) for control (white circles; n = 16) and $NR1^{\Delta DG}$ mice (black squares; n = 8; doors prevented working memory errors in this phase of the experiment). Each block consisted of four trials. (E) $NR1^{\Delta DG}$ mice display impaired spatial working memory performance. Mean ($\pm$SEM) number of working memory errors per trial (averaged over 24 trials) during assessment of working memory performance on the task (doors no longer prevented working memory errors). The inter-choice interval was either 5 or 15 s (all data reconstructed from Niewoehner et al., 2007).

Delatour and Gisquet-Verrier, 2000; Dias and Aggleton, 2000; Deacon et al., 2003; Walton et al., 2003; Kellendonk et al., 2006).

More recent studies have clearly shown that rodents with mPFC lesions are perfectly capable of learning the spatial non-match to place rule (e.g. Touzani et al., 2007), and can perform perfectly well even at longer delays (Gisquet-Verrier and Delatour, 2006). Indeed, Gisquet-Verrier and Delatour (2006, p. 585) have suggested that the mPFC "is not directly involved in the short-term maintenance of specific information but is implicated when changes, such as the sudden introduction of a delay or exposure to unexpected interfering events, alter the initial situation" (see also Touzani et al., 2007). This is in obvious contrast to the robust and lasting spatial working memory impairments that have been observed following lesions of the septo-hippocampal formation, or following GluR-A deletion (e.g. Deacon et al., 2002; Reisel et al., 2002).

Understanding the psychological basis of the impairment in GluR-A$^{-/-}$ mice

To summarise, GluR-A$^{-/-}$ mice exhibit a profound and lasting impairment of hippocampus-dependent spatial working memory but, at the same time, display unimpaired hippocampus-dependent spatial reference memory. The GluR-A$^{-/-}$ mice are *as* impaired as hippocampal-lesioned mice on spatial working memory tasks, but at the same time, they are *as* proficient as control animals on tests of spatial reference memory. These findings suggest that distinct molecular pathways support different aspects of hippocampal information processing. They also pose a conundrum: Why does an impaired spatial working memory system not impinge upon spatial reference memory performance? For example, why does a memory of what happened on the previous trial not aid the mice on the present trial during spatial reference memory acquisition on the radial maze or Y-maze? Similarly, why does a memory of where the animal has just searched not help it be more efficient in finding the platform during spatial reference memory performance in the fixed-location, hidden platform, watermaze task? Any psychological account of the role of GluR-A in hippocampal function not only needs to explain the spatial working memory impairment in the GluR-A$^{-/-}$ mice but, equally importantly, must also explain why this impairment has no consequences for spatial reference memory acquisition or performance.

Are GluR-A$^{-/-}$ mice more susceptible to proactive interference?

We have previously suggested that the pattern of impaired spatial working memory but spared spatial reference memory observed with GluR-A$^{-/-}$ mice could be accounted for by an increased susceptibility to within-session proactive interference (Schmitt et al., 2004a; Bannerman et al., 2006). Proactive interference arises when previous reinforcement or response histories conflict with the response requirements of the current trial. An increased susceptibility to proactive interference would be expected to result in impairments on spatial working memory tasks where responding is governed by information relevant to only one trial. For example, during spatial non-matching to place (rewarded alternation) testing on the T-maze, a memory of a previous trial's different sample or choice response might interfere with memory for

the present sample, or with the generation of the appropriate choice of response to it. However, in contrast, no impairment would be expected on spatial reference memory tasks where previous and current schedules of reinforcement or responding are congruent. Any tendency to recall correct responses on previous trials would simply increase the likelihood of making another, identical, correct response. We have previously suggested that GluR-A-dependent processing may contribute to reducing interference from trial to trial on working memory tasks (Schmitt et al., 2004a).

Furthermore, we argued that an inability to encode or monitor the temporal sequence of events or episodes in order to discriminate between other similar or related events (encoding what-happened-when) could result in an increased susceptibility to proactive interference and, thus, in principle explain the pattern of impaired spatial working memory and spared spatial reference memory in GluR-A$^{-/-}$ mice. As previously mentioned, there is increasing evidence that the hippocampus may play a role in the processing of temporal information, with respect to non-spatial, as well as spatial, stimuli. For example, previous studies have shown that rats with hippocampal lesions are unable to encode the temporal sequence in which they are exposed to a series of odours (Fortin et al., 2002; see also Kesner et al., 2002; Charles et al., 2004). Our recent demonstration that GluR-A$^{-/-}$ mice exhibit a DRL impairment (Reisel et al., 2005), suggested that GluR-A-dependent synaptic plasticity may contribute to a temporal aspect of hippocampal information processing.

If the role of GluR-A-dependent synaptic plasticity is concerned with reducing the influence of proactive interference, perhaps by underlying the encoding of the temporal sequence of events, then spatial working memory performance in GluR-A$^{-/-}$ mice should improve if proactive interference from prior trials is radically reduced or eliminated. In a recent study we therefore assessed spatial working memory performance in GluR-A$^{-/-}$ mice under a range of interference conditions (Sanderson et al., 2007). In a first experiment, spatial working memory was assessed during non-matching to place testing in wild-type and GluR-A$^{-/-}$ mice, both during testing with daily, pseudo-trial-unique arm presentations on the radial maze, and then also when conducting each trial on a different three-arm maze, each in a novel testing room. There was, however, no evidence that the magnitude of the spatial working memory impairment was alleviated in any way when potential proactive interference was reduced.

In further experiments, performance of experimentally naïve wild-type and GluR-A$^{-/-}$ mice was assessed during just a single trial of spontaneous alternation, thus completely eliminating any source of proactive interference. Under these conditions, the spatial non-matching to place, alternation task becomes a spatial novelty preference test. Performance on this test is dependent on the hippocampus (Fig. 8). During a single sample phase, the mice are allowed to explore the start arm and only one of the two goal arms. On the subsequent choice test, in which access to all three arms is now permitted, wild-type controls showed a preference for the novel arm over the familiar arms. Experimentally naïve GluR-A$^{-/-}$ mice, which had no prior experience of the maze or of that spatial environment, showed no preference for the novel or the familiar arms (Fig. 8). This result is important because it shows that the deficit in the knockout animals is not simply the result of an inability to overcome the potentially deleterious effects of proactive interference, arising from previous trials.

In summary, we have argued previously that the pattern of impaired spatial working memory combined with intact spatial reference memory performance might be explained by an increased sensitivity to proactive interference in the knockout animals. However, the results of the spatial novelty preference test suggest that the spatial working memory impairment in GluR-A$^{-/-}$ animals is not due to increased proactive interference, and that in these knockout mice there is a fundamental deficit in making the appropriate spatial response during these working memory tasks.

Two separable forms of hippocampal-dependent, spatial information processing

The spatial novelty preference test in experimentally naïve animals demonstrates that GluR-A$^{-/-}$ mice exhibit a one-trial spatial memory deficit that is independent of proactive interference. This result, of course, suggests the existence of two distinct, and totally independent, spatial information processing mechanisms that are differentially sensitive to GluR-A deletion: the first, a *GluR-A-dependent* mechanism that rapidly encodes or expresses information relating to a single spatial episode or experience; the other, a *GluR-A-independent* mechanism that encodes information about spatial locations that is relevant across many trials as typified by spatial reference memory acquisition.

This GluR-A-independent mechanism, by which the associative strength of spatial locations can be increased (and/or decreased) gradually over many trials, and which presumably underlies spatial reference memory acquisition during tasks such as the Morris watermaze or the radial maze (Zamanillo et al., 1999; Reisel et al., 2002; Schmitt et al., 2003), may, in certain respects, resemble the incremental process by which non-spatial stimuli acquire associative strength during simple conditioning procedures using discrete objects, odours or auditory cues, although importantly these two forms of associative learning do differ in that the former, but not the latter, is dependent on the hippocampus for its expression (e.g. Morris et al., 1982, 1986b). Nevertheless, although the expression of long-term spatial memories during performance on spatial reference memory tasks may depend upon an intact hippocampus, it has been suggested that these memories are in fact stored elsewhere in the cortex, and that the role of the hippocampus is not in memory formation per se, but with response selection and the appropriate expression of behaviour, on the basis of both contextual cues and information retrieved from memory (e.g. Virley et al., 1999; Gray and McNaughton, 2000; Honey and Good, 2000; Bannerman et al., 2004). So what then might be the specific role of GluR-A-dependent synaptic plasticity within the framework of a hippocampal system that mediates aspects of response selection and the expression of behaviour based on contextual information and information retrieved from memory?

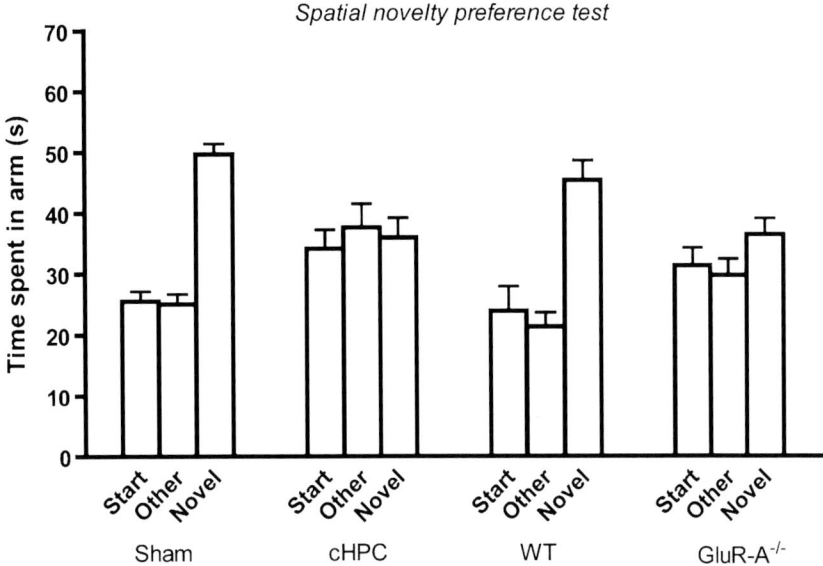

Fig. 8. Hippocampal lesions and GluR-A deletion both impair performance on a spatial novelty preference test. Sham and wild-type (WT) mice exhibit a preference for a previously unexposed (Novel) arm of a Y-maze over two familiar arms to which they have previously been exposed (Start and Other). GluR-A knockout mice (GluR-A$^{-/-}$) and hippocampal-lesioned mice (cHpc) did not show a significant preference for the novel arm (data taken from Sanderson et al., 2007).

Hippocampal GluR-A and priming in short-term memory

We have suggested elsewhere that alternation behaviour during rewarded, non-matching to place, or during spontaneous alternation on the T-maze, or win-shift behaviour in the radial maze task, might be best considered in terms of a short-term habituation process (Sanderson et al., 2007). One account of short-term habituation is that a recently experienced stimulus is rendered less surprising because a representation of that stimulus remains active for a period of time (a refractory period) before decaying to an inactive state. Thus, if this representation is still active in short-term memory when the stimulus is again presented, it will receive a reduced level of processing. One consequence of this process is that the unconditioned response elicited by the stimulus will be reduced. Thus, if an animal is exposed to a particular spatial environment (e.g. an arm of the maze), and then after a short interval is allowed to choose between this familiar location and a new spatial location (a different arm of the maze), it follows from the above analysis that the new location will elicit a stronger exploratory response.

This idea was propounded in a formal model of associative learning by Wagner (1981). In this model, a stimulus is assumed to excite a mnemonic representation (a node) that consists of a set of elements. These elements normally reside in an inactive state (I, or long-term memory) but may be activated into one of two states, A1 and A2 (Fig. 9). The A1 state is considered a primary state, where stimuli are the focus of attention or in a state of rehearsal. The A2 state is likened to a stimulus at the margin of attention. It is assumed that (i) a given element of a stimulus cannot be in two states at the same time, (ii) that elements in the A2 state cannot move directly into the A1 state and (iii) that an element in the A2 state is less capable of generating responding than one that is in the A1 state. Wagner proposes that the only route by which representational elements may enter the A1 state is by presenting the stimulus itself. In contrast there are two routes into the A2 state, which may result from either self-generated or associative priming.

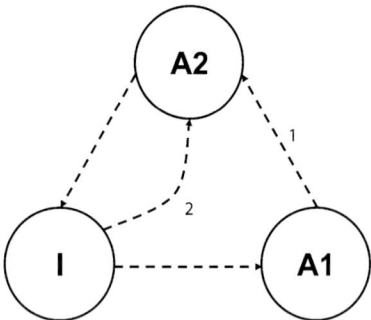

Fig. 9. The states of activation, which elements of a memory can reside, and the permissible transitional routes between states, according to Wagner (1981). Presentation of a stimulus leads to a proportion of its elements transferring from the inactive state (I) to the A1 state. Elements then decay to a secondary activation state, A2, before returning to an inactive state (I). Elements that are active in the A2 state cannot return to the A1 state on subsequent presentation of the stimulus, thus leading to reduced A1 activity. A2 state activity can occur due to the recent presentation of a stimulus (self-generated priming) (see arrow 1). Also, presentation of a stimulus leads to A2 state activation of elements of other stimuli with which it is associated (retrieval-generated priming) (see arrow 2).

It is assumed that on the first occasion in which a novel stimulus is presented, a significant proportion of its representational elements are activated into the A1 state. These elements then decay into the A2 state (a process referred to as self-generated priming), and then eventually into the inactive state. If the same stimulus is applied after a sufficiently short period of time, then the elements currently in the A2 state will be unavailable for activation into the A1 state, and therefore there will be fewer elements in the A1 state. As a consequence, there follows a corresponding decline in the amplitude of the response elicited by A1 processing with each subsequent presentation of the stimulus. In other words, with repeated presentations the stimulus becomes less surprising and is processed (rehearsed) less effectively in short-term memory.

The alternative (retrieval-generated priming) route depends upon previously formed associations. Positive associations are formed when elements representing cues are co-activated in the A1 state. Thus, if a stimulus is repeatedly presented in one context, it will form an association with the context. Under these conditions, the contextual cue is able to prime activation of the associated elements into the A2 state. Importantly, it is assumed that associative priming can only elicit an A2, and not A1, state of activation. Accordingly, therefore, habituation may occur at least in part by the context priming the elements of the target stimulus into the A2 state and thereby limiting the number of elements available for activation into the A1 state when the target stimulus itself is then presented. This feature of the model is able to explain the characteristics of long-term habituation that occur over several training sessions.

Recent evidence suggests that the hippocampus may play an important role in not only retrieval-generated priming (Honey et al., 1998; Honey and Good, 2000), but also self-generated priming. For example, Marshall et al. (2004) presented normal rats and rats with lesions of the hippocampus with a visual stimulus, V1, followed a by a second visual stimulus, V2. Shortly after, test trials were scheduled in which V1 and V2 were simultaneously presented. Normal rats oriented towards (approached and made contact with) V1 rather than V2. According to the priming account, this pattern of responding emerges because the more recent presentation of V2 resulted in a proportion of its elements being active in the A2 state on its subsequent presentation. In contrast, elements representing V1 had time to decay from A2 into the inactive state, and thus, on subsequent presentations, V1 was able to activate more of its elements to the A1 state. Rats with hippocampal lesions failed to show this effect and, if anything, showed orienting towards the more recently presented of the two visual cues. One implication of these results is that the hippocampus contributes to short-term memory processes that limit the extent to which a stimulus can undergo processing in memory. And if one accepts that the hippocampus contributes to short-term habituation then it seems reasonable to suggest that hippocampal synaptic plasticity mechanisms may also contribute to these processes.

Accordingly, it is tempting to speculate that the maintenance of a stimulus representation in the A2 state for a short period of time after the presentation of a specific stimulus (i.e. self-generated priming) may rely upon GluR-A-dependent synaptic plasticity. If this were the case, GluR-A deletion would increase the rate at which elements of a cue presented recently decay into an inactive state (I) from the A2 state. One consequence of this processing deficit is that performance elicited by A2 processing will be impaired in GluR-A$^{-/-}$ mice (e.g. as in spontaneous and rewarded alternation). Whether GluR-A receptors also contribute to the maintenance of associatively primed elements in the A2 state remains an issue for further investigation. However, it is worth noting that GluR-A$^{-/-}$ mice show normal acquisition of conditioned responding in several standard associative learning tasks (Mead and Stephens, 2003; Johnson et al., 2005), and in spatial reference memory tasks (Reisel et al., 2002; Schmitt et al., 2003). This could be taken as evidence that associative learning and retrieval-generated priming are independent of GluR-A receptor function, at least under some conditions. Although this account is in its infancy it offers several predictions. For example, it makes the novel prediction that short-term habituation will be impaired in GluR-A$^{-/-}$ mice but that these mice will nevertheless show evidence of long-term habituation. This prediction is currently being investigated.

Conclusions

In summary, we have found that GluR-A receptors play a distinctive role in a specific subset of hippocampal-dependent information processing mechanisms. The finding that GluR-A deletion impairs spatial working memory whilst sparing spatial reference memory demonstrates that whilst these two types of learning are both dependent on the hippocampus, they are independent of each other at the molecular level. One possible account of this dissociation is that deficits in spatial working memory reflect impairments in specific short-term memory processes, and that associative learning (including spatial reference memory acquisition) is supported by a parallel, GluR-A-independent, mechanism. The role of GluR-A-dependent synaptic plasticity may therefore be to maintain the representation of a stimulus in a memory state that renders it less surprising if it is subsequently presented a second time (self-generated priming). The role that GluR-A plays in other forms of priming, such as retrieval-generated priming and hippocampus-dependent occasion-setting, requires further investigation (Schmitt et al., 2004a).

Finally, it has been shown that increases in synaptic efficacy following LTP induction involve the rapid insertion of GluR-A-containing AMPARs into the post-synaptic membrane, and that these receptors are then gradually and continually replaced by GluR-B/GluR-C-containing AMPARs (see Malinow and Malenka, 2002, for review). This could very reasonably be taken to imply that hippocampal memory depends on a strictly serial storage process, in which GluR-B/GluR-C containing AMPARs could only support lasting memory formation following the prior establishment of changes in synaptic weights involving GluR-A-containing AMPARs. The observation that a profound spatial working memory deficit does not induce any difficulty in spatial reference memory acquisition in GluR-A$^{-/-}$ mice clearly contradicts such a hypothesis. It thus seems that these different memory systems must be capable, at least in principle, of functioning in parallel rather than having to function in series.

References

Aggleton, J.P., Neave, N., Nagle, S. and Sahgal, A. (1995) A comparison of the effects of medial prefrontal, cingulate cortex, and cingulum bundle lesions on tests of spatial memory: evidence of a double dissociation between frontal and cingulum bundle contributions. J. Neurosci., 15: 7270–7281.

Bannerman, D.M., Deacon, R.M., Offen, S., Friswell, J., Grubb, M. and Rawlins, J.N. (2002) Double dissociation of function within the hippocampus: spatial memory and hyponeophagia. Behav. Neurosci., 116: 884–901.

Bannerman, D.M., Good, M.A., Butcher, S.P., Ramsay, M. and Morris, R.G. (1995) Distinct components of spatial learning revealed by prior training and NMDA receptor blockade. Nature, 378: 182–186.

Bannerman, D.M., Rawlins, J.N. and Good, M.A. (2006) The drugs don't work-or do they? Pharmacological and transgenic studies of the contribution of NMDA and GluR-A-containing AMPA receptors to hippocampal-dependent memory. Psychopharmacology, 188: 552–566.

Bannerman, D.M., Rawlins, J.N., McHugh, S.B., Deacon, R.M., Yee, B.K., Bast, T., Zhang, W.N., Pothuizen, H.H. and Feldon, J. (2004) Regional dissociations within the hippocampus — memory and anxiety. Neurosci. Biobehav. Rev., 28: 273–283.

Bannerman, D.M., Yee, B.K., Good, M.A., Heupel, M.J., Iversen, S.D. and Rawlins, J.N. (1999) Double dissociation of function within the hippocampus: a comparison of dorsal, ventral, and complete hippocampal cytotoxic lesions. Behav. Neurosci., 113: 1170–1188.

Bast, T., da Silva, B.M. and Morris, R.G. (2005) Distinct contributions of hippocampal NMDA and AMPA receptors to encoding and retrieval of one-trial place memory. J. Neurosci., 25: 5845–5856.

Bliss, T.V. and Lomo, T. (1973) Long-lasting potentiation of synaptic transmission in the dentate area of the anaesthetized rabbit following stimulation of the perforant path. J. Physiol., 232: 331–356.

Blundell, P., Hall, G. and Killcross, S. (2001) Lesions of the basolateral amygdala disrupt selective aspects of reinforcer representation in rats. J. Neurosci., 21: 9018–9026.

Boitano, J.J., Dokla, C.P., Mulinski, P., Misikonis, S. and Kaluzynski, T. (1980) Effects of hippocampectomy in an incremental-step DRL paradigm. Physiol. Behav., 25: 273–278.

Caramanos, Z. and Shapiro, M.L. (1994) Spatial memory and N-methyl-D-aspartate receptor antagonists APV and MK-801: memory impairments depend on familiarity with the environment, drug dose, and training duration. Behav. Neurosci., 108: 30–43.

Charles, D.P., Gaffan, D. and Buckley, M.J. (2004) Impaired recency judgments and intact novelty judgments after fornix transection in monkeys. J. Neurosci., 24: 2037–2044.

Chen, C. and Tonegawa, S. (1997) Molecular genetic analysis of synaptic plasticity, activity-dependent neural development, learning, and memory in the mammalian brain. Annu. Rev. Neurosci., 20: 157–184.

Clark, C.V. and Isaacson, R.L. (1965) Effect of bilateral hippocampal ablation on Drl performance. J. Comp. Physiol. Psychol., 59: 137–140.

Day, M., Langston, R. and Morris, R.G. (2003) Glutamate-receptor-mediated encoding and retrieval of paired-associate learning. Nature, 424: 205–209.

Deacon, R.M., Bannerman, D.M., Kirby, B.P., Croucher, A. and Rawlins, J.N. (2002) Effects of cytotoxic hippocampal lesions in mice on a cognitive test battery. Behav. Brain Res., 133: 57–68.

Deacon, R.M., Penny, C. and Rawlins, J.N. (2003) Effects of medial prefrontal cortex cytotoxic lesions in mice. Behav. Brain Res., 139: 139–155.

Delatour, B. and Gisquet-Verrier, P. (2000) Functional role of rat prelimbic-infralimbic cortices in spatial memory: evidence for their involvement in attention and behavioural flexibility. Behav. Brain Res., 109: 113–128.

Dias, R. and Aggleton, J.P. (2000) Effects of selective excitotoxic prefrontal lesions on acquisition of nonmatching- and matching-to-place in the T-maze in the rat: differential involvement of the prelimbic-infralimbic and anterior cingulate cortices in providing behavioural flexibility. Eur. J. Neurosci., 12: 4457–4466.

Dwyer, D.M. and Killcross, S. (2006) Lesions of the basolateral amygdala disrupt conditioning based on the retrieved representations of motivationally significant events. J. Neurosci., 26: 8305–8309.

Fortin, N.J., Agster, K.L. and Eichenbaum, H.B. (2002) Critical role of the hippocampus in memory for sequences of events. Nat. Neurosci., 5: 458–462.

Fuchs, E.C., Zivkovic, A.R., Cunningham, M.O., Middleton, S., Lebeau, F.E., Bannerman, D.M., Rozov, A., Whittington, M.A., Traub, R.D., Rawlins, J.N. and Monyer, H. (2007) Recruitment of parvalbumin-positive interneurons determines hippocampal function and associated behavior. Neuron, 53: 591–604.

Gisquet-Verrier, P. and Delatour, B. (2006) The role of the rat prelimbic/infralimbic cortex in working memory: not involved in the short-term maintenance but in monitoring and processing functions. Neuroscience, 141: 585–596.

Goldman-Rakic, P.S. (1987) Circuitry of primate prefrontal cortex and regulation of behaviour by representational memory. In: Plum F. (Ed.), Handbook of Physiology: The Nervous System. American Physiology Society, Bethesda, MD, pp. 373–417.

Gray, J.A. and McNaughton, N. (2000) The Neuropsychology of Anxiety. Oxford University Press, Oxford.

Hebb, D.O. (1949) The Organisation of Behaviour. Wiley, New York.

Hock, B.J. and Bunsey, M.D. (1998) Differential effects of dorsal and ventral hippocampal lesions. J. Neurosci., 18: 7027–7032.

Hoffman, D.A., Sprengel, R. and Sakmann, B. (2002) Molecular dissection of hippocampal theta-burst pairing potentiation. Proc. Natl. Acad. Sci. U.S.A., 99: 7740–7745.

Honey, R.C. and Good, M. (2000) Associative components of recognition memory. Curr. Opin. Neurobiol., 10: 200–204.

Honey, R.C., Watt, A. and Good, M. (1998) Hippocampal lesions disrupt an associative mismatch process. J. Neurosci., 18: 2226–2230.

Honig, W.K. (1978) Studies of working memory in the pigeon. In: Hulse S.H., Fowler H. and Honig W.K. (Eds.), Cognitive Processes in Animal Behavior. Lawrence Erlbaum, New Jersey, pp. 211–248.

Jarrard, L.E. and Becker, J.T. (1977) The effects of selective hippocampal lesions on DRL behavior in rats. Behav. Biol., 21: 393–404.

Jensen, V., Kaiser, K.M., Borchardt, T., Adelmann, G., Rozov, A., Burnashev, N., Brix, C., Frotscher, M., Andersen, P., Hvalby, O., Sakmann, B., Seeburg, P.H. and Sprengel, R. (2003) A juvenile form of postsynaptic hippocampal long-term potentiation in mice deficient for the AMPA receptor subunit GluR-A. J. Physiol., 553: 843–856.

Johnson, A.W., Bannerman, D.M., Rawlins, N.P., Sprengel, R. and Good, M.A. (2005) Impaired outcome-specific devaluation of instrumental responding in mice with a targeted deletion of the AMPA receptor glutamate receptor 1 subunit. J. Neurosci., 25: 2359–2365.

Johnson, C.T., Olton, D.S., Gage, F.H., III and Jenko, P.G. (1977) Damage to hippocampus and hippocampal connections: effects on DRL and spontaneous alternation. J. Comp. Physiol. Psychol., 91: 508–522.

Kellendonk, C., Simpson, E.H., Polan, H.J., Malleret, G., Vronskaya, S., Winiger, V., Moore, H. and Kandel, E.R. (2006) Transient and selective overexpression of dopamine D2 receptors in the striatum causes persistent abnormalities in prefrontal cortex functioning. Neuron, 49: 603–615.

Kesner, R.P., Gilbert, P.E. and Barua, L.A. (2002) The role of the hippocampus in memory for the temporal order of a sequence of odors. Behav. Neurosci., 116: 286–290.

Kolb, B. (1984) Functions of the frontal cortex of the rat: a comparative review. Brain Res., 320: 65–98.

Krestel, H.E., Shimshek, D.R., Jensen, V., Nevian, T., Kim, J., Geng, Y., Bast, T., Depaulis, A., Schonig, K., Schwenk, F., Bujard, H., Hvalby, O., Sprengel, R. and Seeburg, P.H. (2004) A genetic switch for epilepsy in adult mice. J. Neurosci., 24: 10568–10578.

Kutsuwada, T., Sakimura, K., Manabe, T., Takayama, C., Katakura, N., Kushiya, E., Natsume, R., Watanabe, M., Inoue, Y., Yagi, T., Aizawa, S., Arakawa, M., Takahashi, T., Nakamura, Y., Mori, H. and Mishina, M. (1996) Impairment of suckling response, trigeminal neuronal pattern formation, and hippocampal LTD in NMDA receptor epsilon 2 subunit mutant mice. Neuron, 16: 333–344.

Lee, H.K., Takamiya, K., Han, J.S., Man, H., Kim, C.H., Rumbaugh, G., Yu, S., Ding, L., He, C., Petralia, R.S., Wenthold, R.J., Gallagher, M. and Huganir, R.L. (2003) Phosphorylation of the AMPA receptor GluR1 subunit is required for synaptic plasticity and retention of spatial memory. Cell, 112: 631–643.

Lee, I. and Kesner, R.P. (2002) Differential contribution of NMDA receptors in hippocampal subregions to spatial working memory. Nat. Neurosci., 5: 162–168.

Mack, V., Burnashev, N., Kaiser, K.M., Rozov, A., Jensen, V., Hvalby, O., Seeburg, P.H., Sakmann, B. and Sprengel, R. (2001) Conditional restoration of hippocampal synaptic potentiation in Glur-A-deficient mice. Science (New York, N.Y.), 292: 2501–2504.

Malinow, R. and Malenka, R.C. (2002) AMPA receptor trafficking and synaptic plasticity. Ann. Rev. Neurosci., 25: 103–126.

Marshall, V.J., McGregor, A., Good, M. and Honey, R.C. (2004) Hippocampal lesions modulate both associative and nonassociative priming. Behav. Neurosci., 118: 377–382.

Martin, S.J., Grimwood, P.D. and Morris, R.G. (2000) Synaptic plasticity and memory: an evaluation of the hypothesis. Ann. Rev. Neurosci., 23: 649–711.

Martin, S.J. and Morris, R.G. (2002) New life in an old idea: the synaptic plasticity and memory hypothesis revisited. Hippocampus, 12: 609–636.

Mead, A.N. and Stephens, D.N. (2003) Selective disruption of stimulus-reward learning in glutamate receptor gria1 knockout mice. J. Neurosci., 23: 1041–1048.

Meck, W.H., Church, R.M. and Olton, D.S. (1984) Hippocampus, time, and memory. Behav. Neurosci., 98: 3–22.

Morris, R.G., Anderson, E., Lynch, G.S. and Baudry, M. (1986a) Selective impairment of learning and blockade of long-term potentiation by an N-methyl-D-aspartate receptor antagonist, AP5. Nature, 319: 774–776.

Morris, R.G., Davis, S. and Butcher, S.P. (1990a) Hippocampal synaptic plasticity and NMDA receptors: a role in information storage? Philos. Trans. R. Soc. Lond., 329: 187–204.

Morris, R.G., Garrud, P., Rawlins, J.N. and O'Keefe, J. (1982) Place navigation impaired in rats with hippocampal lesions. Nature, 297: 681–683.

Morris, R.G., Hagan, J.J. and Rawlins, J.N. (1986b) Allocentric spatial learning by hippocampectomised rats: a further test of the "spatial mapping" and "working memory" theories of hippocampal function. Q. J. Exp. Psychol., 38: 365–395.

Morris, R.G., Schenk, F., Tweedie, F. and Jarrard, L.E. (1990b) Ibotenate lesions of hippocampus and/or subiculum: dissociating components of allocentric spatial learning. Eur J. Neurosci., 2: 1016–1028.

Moser, M.B., Moser, E.I., Forrest, E., Andersen, P. and Morris, R.G. (1995) Spatial learning with a minislab in the dorsal hippocampus. Proc. Natl. Acad. Sci. U.S.A., 92: 9697–9701.

Murray, T.K. and Ridley, R.M. (1999) The effect of excitotoxic hippocampal lesions on simple and conditional discrimination learning in the rat. Behav. Brain Res., 99: 103–113.

Nakazawa, K., Quirk, M.C., Chitwood, R.A., Watanabe, M., Yeckel, M.F., Sun, L.D., Kato, A., Carr, C.A., Johnston, D., Wilson, M.A. and Tonegawa, S. (2002) Requirement for hippocampal CA3 NMDA receptors in associative memory recall. Science (New York, N.Y.), 297: 211–218.

Nakazawa, K., Sun, L.D., Quirk, M.C., Rondi-Reig, L., Wilson, M.A. and Tonegawa, S. (2003) Hippocampal CA3 NMDA receptors are crucial for memory acquisition of one-time experience. Neuron, 38: 305–315.

Niewoehner, B., Single, F.N., Hvalby, O., Jensen, V., Borgloh, S.M., Seeburg, P.H., Rawlins, J.N., Sprengel, R. and Bannerman, D.M. (2007) Impaired spatial working memory but spared spatial reference memory following functional loss of NMDA receptors in the dentate gyrus. Eur. J. Neurosci., 25: 837–846.

O'Keefe, J. and Nadel, L. (1978) The Hippocampus as a Cognitive Map. Clarendon Press, Oxford.

Olton, D.S., Becker, J.T. and Handelman, G.H. (1979) Hippocampus, space, and memory. Behav. Brain Sci., 2: 313–365.

Rawlins, J.N. and Olton, D.S. (1982) The septo-hippocampal system and cognitive mapping. Behav. Brain Res., 5: 331–358.

Rawlins, J.N.P. (1985) Associations across time: the hippocampus as a temporary memory store. Behav. Brain Sci., 8: 479–528.

Reisel, D., Bannerman, D.M., Deacon, R.M., Sprengel, R., Seeburg, P.H. and Rawlins, J.N. (2005) GluR-A-dependent synaptic plasticity is required for the temporal encoding of nonspatial information. Behav. Neurosci., 119: 1298–1306.

Reisel, D., Bannerman, D.M., Schmitt, W.B., Deacon, R.M., Flint, J., Borchardt, T., Seeburg, P.H. and Rawlins, J.N. (2002) Spatial memory dissociations in mice lacking GluR1. Nat. Neurosci., 5: 868–873.

Riedel, G., Micheau, J., Lam, A.G., Roloff, E.L., Martin, S.J., Bridge, H., de Hoz, L., Poeschel, B., McCulloch, J. and Morris, R.G. (1999) Reversible neural inactivation reveals hippocampal participation in several memory processes. Nat. Neurosci., 2: 898–905.

Rumpel, S., LeDoux, J., Zador, A. and Malinow, R. (2005) Postsynaptic receptor trafficking underlying a form of associative learning. Science (New York, N.Y.), 308: 83–88.

Sanchez-Santed, F., de Bruin, J.P., Heinsbroek, R.P. and Verwer, R.W. (1997) Spatial delayed alternation of rats in a T-maze: effects of neurotoxic lesions of the medial prefrontal cortex and of T-maze rotations. Behav. Brain Res., 84: 73–79.

Sanchis-Segura, C., Borchardt, T., Vengeliene, V., Zghoul, T., Bachteler, D., Gass, P., Sprengel, R. and Spanagel, R. (2006) Involvement of the AMPA receptor GluR-C subunit in alcohol-seeking behavior and relapse. J. Neurosci., 26: 1231–1238.

Sanderson, D.J., Gray, A., Simon, A., Taylor, A.M., Deacon, R.M., Seeburg, P.H., Sprengel, R., Good, M.A., Rawlins, J.N. and Bannerman, D.M. (2007) Deletion of glutamate receptor-A (GluR-A) AMPA receptor subunits impairs one-trial spatial memory. Behav. Neurosci., 121: 559–569.

Saucier, D. and Cain, D.P. (1995) Spatial learning without NMDA receptor-dependent long-term potentiation. Nature, 378: 186–189.

Schmitt, W.B., Arianpour, R., Deacon, R.M., Seeburg, P.H., Sprengel, R., Rawlins, J.N. and Bannerman, D.M. (2004a) The role of hippocampal glutamate receptor-A-dependent synaptic plasticity in conditional learning: the importance of spatiotemporal discontiguity. J. Neurosci., 24: 7277–7282.

Schmitt, W.B., Deacon, R.M., Reisel, D., Sprengel, R., Seeburg, P.H., Rawlins, J.N. and Bannerman, D.M. (2004b) Spatial reference memory in GluR-A-deficient mice using a novel hippocampal-dependent paddling pool escape task. Hippocampus, 14: 216–223.

Schmitt, W.B., Deacon, R.M., Seeburg, P.H., Rawlins, J.N. and Bannerman, D.M. (2003) A within-subjects, within-task demonstration of intact spatial reference memory and impaired spatial working memory in glutamate receptor-A-deficient mice. J. Neurosci., 23: 3953–3959.

Schmitt, W.B., Sprengel, R., Mack, V., Draft, R.W., Seeburg, P.H., Deacon, R.M., Rawlins, J.N. and Bannerman, D.M. (2005) Restoration of spatial working memory by genetic rescue of GluR-A-deficient mice. Nat. Neurosci., 8: 270–272.

Shaw, C. and Aggleton, J.P. (1993) The effects of fornix and medial prefrontal lesions on delayed non-matching-to-sample by rats. Behav. Brain Res., 54: 91–102.

Shimshek, D.R., Jensen, V., Celikel, T., Geng, Y., Schupp, B., Bus, T., Mack, V., Marx, V., Hvalby, O., Seeburg, P.H. and Sprengel, R. (2006) Forebrain-specific glutamate receptor B deletion impairs spatial memory but not hippocampal field long-term potentiation. J. Neurosci., 26: 8428–8440.

Sinden, J.D., Rawlins, J.N., Gray, J.A. and Jarrard, L.E. (1986) Selective cytotoxic lesions of the hippocampal formation and DRL performance in rats. Behav. Neurosci., 100: 320–329.

Sommer, B., Kohler, M., Sprengel, R. and Seeburg, P.H. (1991) RNA editing in brain controls a determinant of ion flow in glutamate-gated channels. Cell, 67: 11–19.

Sprengel, R. (2006) Role of AMPA receptors in synaptic plasticity. Cell Tissue Res., 326: 447–455.

Steele, R.J. and Morris, R.G. (1999) Delay-dependent impairment of a matching-to-place task with chronic and intrahippocampal infusion of the NMDA-antagonist D-AP5. Hippocampus, 9: 118–136.

Tonkiss, J., Morris, R.G. and Rawlins, J.N. (1988) Intraventricular infusion of the NMDA antagonist AP5 impairs performance on a non-spatial operant DRL task in the rat. Exp. Brain Res. (Experimentelle Hirnforschung), 73: 181–188.

Tonkiss, J. and Rawlins, J.N. (1991) The competitive NMDA antagonist AP5, but not the non-competitive antagonist MK801, induces a delay-related impairment in spatial working memory in rats. Experimental Brain Res. (Experimentelle Hirnforschung), 85: 349–358.

Touzani, K., Puthanveettil, S.V. and Kandel, E.R. (2007) Consolidation of learning strategies during spatial working memory task requires protein synthesis in the prefrontal cortex. Proc. Natl. Acad. Sci. U.S.A., 104: 5632–5637.

Tsien, J.Z., Huerta, P.T. and Tonegawa, S. (1996) The essential role of hippocampal CA1 NMDA receptor-dependent synaptic plasticity in spatial memory. Cell, 87: 1327–1338.

Virley, D., Ridley, R.M., Sinden, J.D., Kershaw, T.R., Harland, S., Rashid, T., French, S., Sowinski, P., Gray, J.A., Lantos, P.L. and Hodges, H. (1999) Primary CA1 and conditionally immortal MHP36 cell grafts restore conditional discrimination learning and recall in marmosets after excitotoxic lesions of the hippocampal CA1 field. Brain, 122(Pt. 12): 2321–2335.

Wagner, A.R. (1981) SOP: a model of automatic memory processing in animal behavior. In: Spear N.E. and Miller R.R. (Eds.), Information Processing in Animals: Memory Mechanisms. Erlbaum, Hillsdale, NJ, pp. 5–47.

Walton, M.E., Bannerman, D.M., Alterescu, K. and Rushworth, M.F. (2003) Functional specialization within medial frontal cortex of the anterior cingulate for evaluating effort-related decisions. J. Neurosci., 23: 6475–6479.

Wisden, W. and Seeburg, P.H. (1993) Mammalian ionotropic glutamate receptors. Curr. Opin. Neurobiol., 3: 291–298.

Young, B. and McNaughton, N. (2000) Common firing patterns of hippocampal cells in a differential reinforcement of low rates of response schedule. J. Neurosci., 20: 7043 7051.

Zamanillo, D., Sprengel, R., Hvalby, O., Jensen, V., Burnashev, N., Rozov, A., Kaiser, K.M., Koster, H.J., Borchardt, T., Worley, P., Lubke, J., Frotscher, M., Kelly, P.H., Sommer, B., Andersen, P., Seeburg, P.H. and Sakmann, B. (1999) Importance of AMPA receptors for hippocampal synaptic plasticity but not for spatial learning. Science (New York, N.Y.), 284: 1805–1811.

CHAPTER 10

Synaptic remodeling, synaptic growth and the storage of long-term memory in *Aplysia*

Craig H. Bailey[2,3,*] and Eric R. Kandel[1,2,3]

[1]*Howard Hughes Medical Institute, New York, NY 10032, USA*
[2]*Department of Neuroscience, College of Physicians and Surgeons of Columbia University, New York State Psychiatric Institute, New York, NY 10032, USA*
[3]*Kavli Institute for Brain Sciences, 1051 Riverside Drive, New York, NY 10032, USA*

Abstract: Synaptic remodeling and synaptic growth accompany various forms of long-term memory. Storage of the long-term memory for sensitization of the gill-withdrawal reflex in *Aplysia* has been extensively studied in this respect and is associated with the growth of new synapses by the sensory neurons onto their postsynaptic target neurons. Recent time-lapse imaging studies of living sensory-to-motor neuron synapses in culture have monitored both functional and structural changes simultaneously so as to follow remodeling and growth at the same specific synaptic connections continuously over time and to examine the functional contribution of these learning-related structural changes to the different time-dependent phases of memory storage. Insights provided by these studies suggest the synaptic differentiation and growth induced by learning in the mature nervous system are highly dynamic and often rapid processes that can recruit both molecules and mechanisms used for de novo synapse formation during development.

Keywords: active zone; apCAM; *Aplysia*; cell adhesion molecules; learning and memory; long-term memory; long-term sensitization; nascent synapse; presynaptic facilitation; presynaptic terminal; silent synapse; structural changes; synapse formation; synaptic growth; synaptic plasticity; synaptic remodeling

Introduction

Studies of simple forms of implicit memory in higher invertebrates and more complex forms of explicit memory in mammals have found that the storage of long-term memory is represented at the cellular level by activity-dependent modulation of both the function and the structure of specific synaptic connections (Kandel, 2001). Although a number of molecular components that underlie the functional changes associated with memory storage have been characterized, little is known about how these are regulated by and coupled to the signaling pathways that give rise to the synaptic structural changes. This in turn raises two questions central to an understanding of the mechanisms that underlie long-term memory: (1) does the alteration in synaptic strength that accompanies memory storage result from a structural change in

*Corresponding author. Tel.: +1 (212) 543-5404; Fax: +1 (212) 543-5797; E-mail: chb1@columbia.edu

pre-existing connections — for example, from the conversion of nonfunctional (silent) synapses to functional synapses — from the addition of newly formed functional synapses, or from perhaps both? (2) how closely do the mechanisms and signaling interactions that regulate alterations in the structure of the synapse that are induced by learning resemble those that govern de novo synaptogenesis and the fine-tuning of synaptic connections during development?

In this chapter, we address these questions by focusing on molecular and structural studies of long-term memory storage in *Aplysia*. We begin by examining the remodeling and growth of identified sensory neuron synapses that accompany long-term sensitization — an elementary form of implicit memory. We then turn to recent in vitro studies of the sensory-to-motor neuron synapse reconstituted in dissociated cell culture and consider the cellular and molecular events that give rise to these learning-related structural changes and their functional contribution to the different temporal phases of long-term memory storage. Finally, we outline some of the insights that have been provided by these studies of synaptic remodeling and synaptic growth in *Aplysia* and discuss how molecules and mechanisms important for synapse formation during the development of the nervous system may be reutilized in the adult for the purposes of synaptic plasticity and memory storage.

Functional architecture of the synapse

The interaction between neurons is largely restricted to specialized cellular sites where one nerve cell comes into close apposition with its target cell. This junction is called the synapse, a term introduced by Charles Sherrington in 1897. Although the concept of the synapse was originally framed in physiological terms, it was also realized that there had to be a stable physical structure mediating the function of each synapse. Morphological support for this idea was first provided by Sherrington's contemporary, Santiago Ramon y Cajal, who demonstrated that all synapses had two conserved elements: a presynaptic terminal and a postsynaptic target site. Ramon y Cajal also inferred the existence of a third element, the synaptic cleft, the space between the presynaptic and postsynaptic elements. Collectively, these three components of the synapse underlie the ability of neurons to communicate with one another, the process of synaptic transmission.

Modern studies of the central nervous system (CNS), beginning with those of Palay and Palade, have revealed that chemical synapses are asymmetric sites of cell–cell contact designed for rapid and repetitive signaling between neurons. The presynaptic compartment contains a highly specialized and restricted region, known as the active zone, where synaptic vesicles preferentially dock and fuse with the presynaptic membrane. The fine structure of the active zone consists of an electron-dense meshwork of cytoskeletal filaments and associated proteins at the plasma membrane, embedded with clusters of synaptic vesicles, which is contiguous with the electron-dense postsynaptic density (PSD) of the target neuron. The cytoskeletal matrix associated with the active zone (CAZ) contains a large family of scaffolding and signaling molecules. These are thought to play a fundamental role in the formation and organization of sites along the presynaptic membrane where neurotransmitter is released, maintaining the presynaptic active zone in full alignment with the PSD, and regulating the mobilization of synaptic vesicles and the refilling of transmitter release sites in the presynaptic compartment.

Directly apposed to the active zone, and separated from it by a distance of typically 10–50 nm, is a postsynaptic membrane specialization — the PSD — that serves as a molecular apparatus for both the reception and transduction of the chemical information released by the presynaptic neuron. The PSD consists of a dense network of cytoskeletal filaments that extends across the synaptic cleft to the presynaptic active zone as well as into the cytoplasm of the postsynaptic compartment where it is also associated with a family of scaffolding and regulatory proteins. This postsynaptic matrix serves to anchor and cluster neurotransmitter receptors, ion channels regulated by the receptors and cell adhesion molecules (CAMs) in the postsynaptic membrane and can recruit a variety of signaling cascades that link these structures with the cytoskeleton to coordinate electrical and more enduring cellular responses. The close apposition of the pre- and postsynaptic compartments as well as the precise structural alignment of the molecular components for transmitter release and transmitter reception across the synaptic cleft facilitates the efficacy of synaptic transmission.

Until approximately three decades ago chemical synapses were thought to convey information in only one direction — from the presynaptic to the postsynaptic neuron. It now is clear that synaptic transmission is a bidirectional and self-modifiable form of cell–cell communication (Jessell and Kandel, 1993). The bidirectional nature of signaling across synapses has been demonstrated in biophysical studies of synaptic transmission, and it also is evident in the assembly of synapses during development and during activity-dependent plasticity of synapses in the mature brain. The relative contributions of both the pre- and postsynaptic neuron and their reciprocal signaling interactions is consistent with the current view of intercellular communication that incorporates the biology of nerve cells and, specifically, signaling in the nervous system, into the broader field of cell biology. This emerging appreciation provides a conceptual framework for the interpretation of learning-related changes in the structure of the synapse.

Synaptic plasticity and memory storage

Modern behavioral and biological studies have revealed that memory is not a unitary faculty of the mind but consists of distinct families of mental processes that can be grouped into at least two general categories, each with its own rules (Polster et al., 1991; Squire and Zola-Morgan, 1991). Explicit or declarative memory is the conscious recall of knowledge about people, places and things, and is particularly well developed in the vertebrate brain. Implicit or nondeclarative memory is memory for motor and perceptual skills as well as other tasks and is expressed through performance, without conscious recall of past experience. Implicit memory includes simple associative forms of memory such as classical and operant conditioning, and nonassociative forms such as sensitization and habituation. Explicit and implicit memory have been localized to different neural systems within the brain (Milner, 1985; Polster et al., 1991). Explicit memory is critically dependent on structures in the medial temporal lobe of the cerebral cortex, including the hippocampal formation. Implicit memory is a family of different processes that are represented in a number of brain systems including the cerebellum, the striatum, the amygdala and in the simplest cases, the sensory and motor pathways recruited during the learning process for particular perceptual or motor skills. As a result, implicit memory can be studied in a variety of simple reflex systems, including those of higher invertebrates, whereas explicit memory can best (and perhaps only) be studied in mammals.

Two experimental model systems have been extensively studied as representative examples of these two forms of memory storage: sensitization in the marine snail *Aplysia californica* as an example of implicit memory, and spatial memory formation in the rodent hippocampus as an example of explicit memory. In both model systems, the elementary events that underlie synaptic plasticity, the ability of neurons to modulate the strength of their synapses in response to extra- or intracellular cues, are thought to be fundamental for both the fine-tuning of synaptic connections during development, as well as for behavioral learning and memory storage in the adult organism. Cell biological and molecular studies in both *Aplysia* and the hippocampus suggest that activity-dependent modulation of synaptic function and synaptic structure is an important mechanism by which information is encoded, processed and stored within the brain (Kandel, 2001; Bliss et al., 2003).

For both implicit and explicit memory, two general types of storage mechanisms have been described: short-term memory lasting minutes and long-term memory lasting days, weeks or longer. This temporal distinction in behavior is reflected in specific forms of synaptic plasticity that underlie each form of behavioral memory as well as specific molecular requirements for each of these two forms of synaptic plasticity. The short-term forms involve the covalent modifications of pre-existing proteins by a variety of kinases (e.g., PKA and MAPK) and are expressed as alterations in the effectiveness of pre-existing connections. By contrast, in addition to the kinases recruited during the short-term forms, the long-term forms also require CREB-mediated gene expression, new mRNA and protein synthesis and are often associated with the growth of new synaptic connections (Bailey et al., 1996). For both implicit and explicit memory storage, the synaptic growth is thought to represent the final and self-sustaining cellular change that stabilizes the long-term process (Bailey and Kandel, 1993). Moreover, recent studies in *Aplysia* and mammals have provided evidence that activity-dependent remodeling of pre-existing synapses and the growth of new synapses might play an important role in the storage of information at both the level of individual synaptic connections as well as in more complex neuronal networks by modulating and perhaps reconfiguring the activity of the neural network in which this occurs (Bailey et al., 2004; Lamprecht and LeDoux, 2004).

Although each chemical synapse consists of two precisely aligned, tightly adherent, highly specialized and functionally coupled anatomical components, most studies of the structural changes that accompany long-lasting forms of synaptic plasticity have focused on either the presynaptic or postsynaptic compartment. For example, learning-related remodeling and growth of the presynaptic compartment have been most extensively studied in the sensory neurons of the gill-withdrawal reflex of *Aplysia* following long-term sensitization — a simple form of implicit memory storage. Conversely, activity-dependent structural changes in the postsynaptic compartment have been the major focus of studies on dendritic spines in the hippocampus during long-term potentiation (LTP) — a more complex form of explicit memory storage in mammals. Historically, this dichotomy reflects, at least in part, the unique experimental advantages of each model system as well as the underlying hypotheses regarding the mechanisms that give rise to these different forms of plasticity.

The significance of learning-related changes in either the structure of pre-existing synapses or changes in the number of synapses must ultimately be considered in a functional context, that is, the contribution of a specific structural modification to the change in effectiveness of that synaptic connection. When viewed in this light, it is readily apparent that studies of the structural changes associated with synaptic plasticity should consider each synaptic contact in its entirety and recognize that functionally relevant changes in the morphology of the presynaptic compartment are likely to be accompanied by reciprocal changes in the morphology of the postsynaptic compartment. Of the changes in pre-existing synapses, the most reliable and potentially best suited for correlation with changes in synaptic effectiveness are those that involve reorganization of the active zone and associated cytoskeletal matrix (CAZ) in the presynaptic compartment and the PSD in the postsynaptic compartment. Any change in the number, size, continuity or shape of one of these focal and highly specialized regions of the synapse should be reflected by a comparable remodeling in its cognate partner. Similarly, in order to modulate synaptic strength, an increase or decrease in the number of synapses must consist of parallel changes in both the presynaptic transmitter release machinery and the postsynaptic receptive apparatus. Indeed, alterations in the number and structure of both the presynaptic and postsynaptic compartments have been found to accompany long-term sensitization in *Aplysia* and LTP in mammals, and an increasing body of evidence now suggests that coordinated pre- and postsynaptic mechanisms may underlie each form of memory storage (Hawkins et al., 2006).

Growth of new sensory neuron synapses and the persistence of long-term sensitization

The CNS of *Aplysia* contains only approximately 20,000 large and frequently identifiable nerve cells, clustered into 9 major ganglia. The ability to identify many of the individual neurons of this nervous system and record their activity has made it possible to define the major components of the neuronal circuits of specific behaviors, and to delineate the critical sites and underlying mechanisms used to store memory-related representations.

The mechanisms contributing to implicit memory storage have been most extensively studied for sensitization of the gill-withdrawal reflex in *Aplysia* (Kandel, 2001). Sensitization is an elementary type of nonassociative learning, a form of learned fear, by which an animal learns about the properties of a single noxious stimulus. When a light touch is applied to the siphon of the snail, the snail responds by withdrawing its gill and siphon. This response is enhanced when the animal is given a noxious, sensitizing stimulus. As with other forms of defensive behaviors, the memory for sensitization of the withdrawal reflex is graded, and repeated tail shocks lead to a longer-lasting memory: a single tail shock produces short-term sensitization that lasts for minutes, whereas repeated tail shocks given at spaced intervals produce long-term sensitization that lasts for up to several weeks (Frost et al., 1985).

The simplicity of the neuronal circuit underlying this behavioral modification — including direct monosynaptic connections between identified mechanoreceptor sensory neurons and their follower cells (Castellucci et al., 1970) — has allowed reduction of the analysis of the short- and long-term memory for sensitization to the cellular and molecular level. This monosynaptic sensory to motor neuron connection, which is thought to be glutamatergic, can be reconstituted in dissociated cell culture. Despite its simplification, this in vitro model system reproduces what is observed in the whole animal during behavioral training. In this simplified culture preparation tail shocks are replaced with brief applications of serotonin (5-HT), a modulatory transmitter normally released by sensitizing stimuli in the intact animal (Montarolo et al., 1986). A single brief application of 5-HT produces a short-term change in synaptic effectiveness (short-term facilitation) much as does a single tail shock, whereas repeated and spaced applications produce changes in synaptic strength that can last for more than a week (long-term facilitation or LTF). A component of the increase in synaptic strength observed during both the short- and long-term changes is due, in each case, to enhanced release of transmitter by the sensory neuron onto its follower cells, and is accompanied by an increase in the excitability of the sensory neuron attributable to the depression of specific sets of potassium channels (Klein and Kandel, 1980; Frost et al., 1985; Hochner et al., 1986; Montarolo et al., 1986; Dale et al., 1987; Scholz and Byrne, 1987).

Despite this superficial similarity, the short-term cellular changes differ fundamentally from the long-term changes in at least two important ways: (1) the long-term change, but not the short-term change, requires new protein synthesis (Schwartz et al., 1971; Montarolo et al., 1986; Castellucci et al., 1989) and (2) the long-term but not the short-term process involves a structural change (Bailey and Chen, 1983, 1988a, b, 1989). Long-term sensitization is associated with the growth of new synaptic connections by the sensory neurons onto their follower cells.

In the early 1980s, studies in *Aplysia* first began to explore the morphological basis of the synaptic plasticity that might underlie the transition from short-term to long-term memory. By combining selective intracellular labeling techniques with the analysis of serial thin sections and transmission electron microscopy, complete reconstructions of unequivocally identified sensory neuron synapses were quantitatively analyzed from both control and behaviorally modified animals. The storage of long-term memory for sensitization (lasting several weeks) was accompanied by a family of distinct structural changes at identified sensory neuron synapses. These changes reflected a learning-induced remodeling of the functional architecture of presynaptic sensory neuron varicosities at two different levels of synaptic organization: (1) alterations in focal regions of membrane specialization of the synapse that mediate transmitter release — the number, size and vesicle complement of sensory neuron active zones were larger in sensitized animals than in controls (Bailey and Chen, 1983, 1988b) and (2) a growth process that appeared similar to synaptogenesis during development and led to a pronounced increase in the total number of presynaptic varicosities per sensory neuron (Bailey and Chen, 1988a). Thus, sensory neurons from long-term sensitized animals exhibited a twofold increase in the number of synaptic varicosities, as well as an enlargement in the linear extent of each neuron's axonal arbor when compared to sensory neurons from untrained animals (Fig. 1).

To determine which class of structural changes at sensory neuron synapses might contribute to the retention of long-term sensitization, Bailey and Chen (1989) compared the time course for each morphological change with the behavioral duration of the memory. They found that not all of the structural changes persisted as long as the memory. The increase in the size and synaptic vesicle complement of sensory neuron active zones present 24 h following the completion of behavioral training returned to control levels when tested 1 week later. These data indicated that, insofar as this relatively transient modulation of active zone size and associated synaptic vesicles is one of the structural mechanisms underlying long-term sensitization, it is associated with the initiation and early expression of the long-term process and not with its persistence. By contrast, the duration of changes in varicosity and active zone number, which persisted unchanged for at least 1 week and were partially reversed at the end of 3 weeks, paralleled the behavioral time course of memory storage indicating that only the increase in the number of sensory neuron synapses contributes to the stable maintenance of long-term sensitization. These results directly linked a change in synaptic structure to a long-lasting behavioral memory and suggested that the morphological alterations could represent an anatomical substrate for memory consolidation. In addition, the finding that some components of the learning-induced changes in synaptic architecture were transient whereas others endured suggested that not all of these modifications were regulated synchronously. At the structural level, the sensory neuron appears to have multiple mechanisms and parameters of plasticity available to it. Thus, during the later phases of long-term memory storage for sensitization, although there are more synapses, each individual synapse may recruit all of the mechanisms of plasticity that were present before training.

Unlike the extensive anatomical changes observed at sensory neuron synapses following long-term training, the structural correlates of short-term memory in *Aplysia* (lasting minutes to hours rather than days to weeks) are far less pronounced (Bailey and Chen, 1988c). For example, the decrease in the strength of the sensory to motor neuron connection that accompanies short-term habituation is not associated with any detectable alterations in either the number of sensory neuron presynaptic varicosities or the number of active zones within the presynaptic varicosities. Nor does it alter the size of active zones or the total number of synaptic vesicles within the presynaptic varicosity. Rather, there is a reduction in the number of vesicles that are docked at the active zones and thus there are fewer packets of transmitter ready to be released.

Taken together, these initial morphological studies of short- and long-term memory in *Aplysia* began to suggest a clear difference in the nature, extent and time course of changes in the functional architecture of the synapse that may underlie memories of differing durations. The transient durations of short-term memories involving covalent modifications of pre-existing proteins (proteins that turn over slowly) are accompanied only by modest structural rearrangements that appear to be restricted to shifts in the proximity of synaptic vesicle populations contiguous to the release site. By contrast, the prolonged durations of long-term memories depend on altered gene expression and the synthesis of new proteins and are associated with more substantial and potentially more enduring structural alterations that are reflected by frank changes in both the number of synaptic contacts and their active zone morphology.

These studies also demonstrated, for the first time, that learning-induced structural changes could be detected at the level of specific, identified synaptic connections known to be critically involved in the behavioral modification and

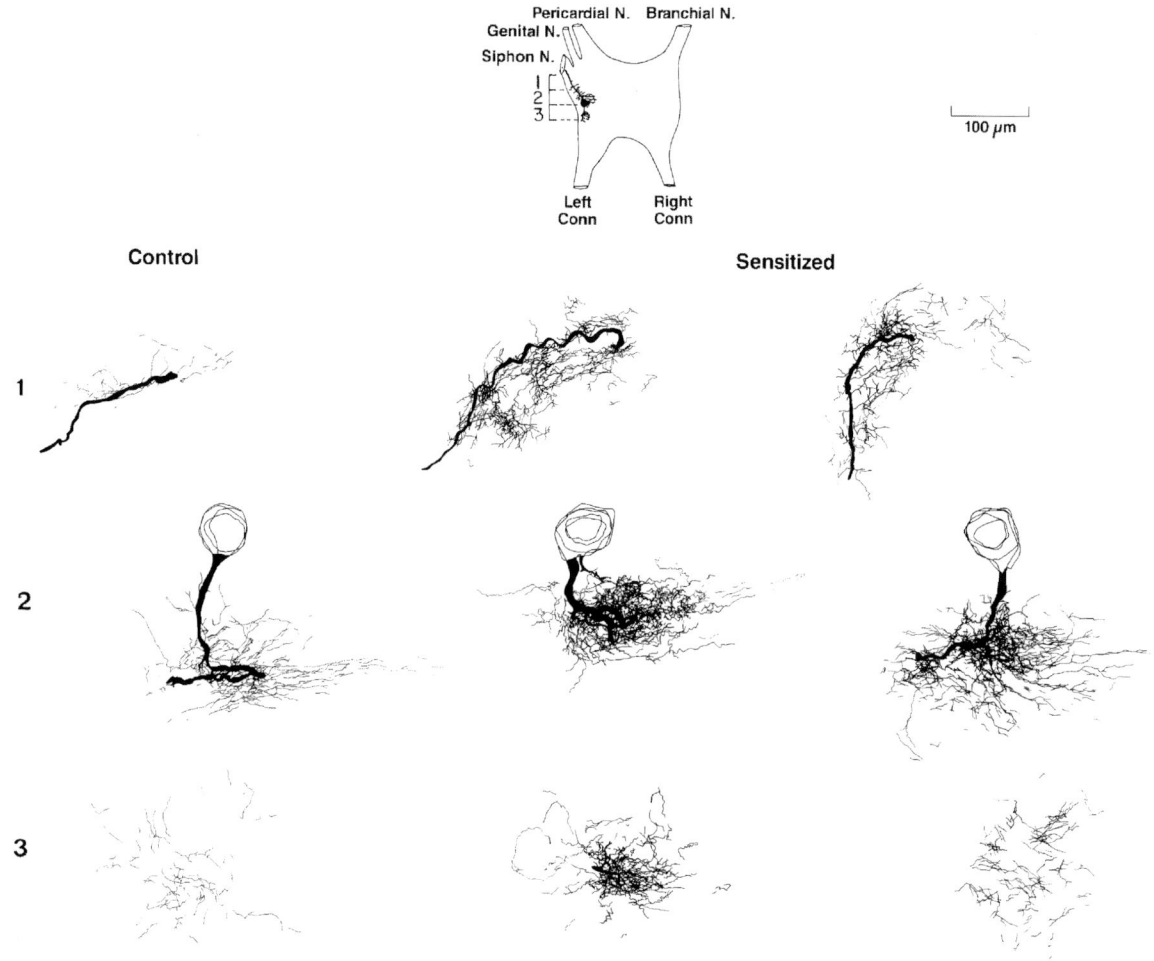

Fig. 1. Growth of sensory neurons induced by long-term sensitization in *Aplysia*. Serial reconstruction of identified sensory neurons labeled with horseradish peroxidase (HRP, Bailey et al., 1979) from long-term sensitized and control animals. Total extent of the neuropil arbors of sensory neurons from one control (untrained) and two long-term-sensitized animals are shown. In each case, the rostral (row 3) to caudal (row 1) extent of the arbor is divided roughly into thirds. Each panel was produced by the superimposition of camera lucida tracings of all HRP-labeled processes present in 17 consecutive slab-thick Epon sections and represents a linear segment through the ganglion of roughly 340 μm. For each composite, ventral is up, dorsal is down, lateral is to the left, and medial is to the right. By examining images across each row (rows 1, 2, and 3), the viewer is comparing similar regions of each sensory neuron. In all cases, the axonal arbor of long-term-sensitized cells is markedly increased compared to cells from control (untrained) animals and parallels the concomitant increase in the number of sensory neuron presynaptic varicosities. (From Bailey and Chen, 1988a.)

provided evidence for an intriguing notion — that active zones are plastic rather than immutable components of the synapse. Even elementary forms of learning can remodel the basic anatomical scaffolding of the neuron, in this case by altering the number and organization of transmitter release sites in the presynaptic compartment, to modulate the functional expression of synaptic connections. In addition, complete serial reconstructions of identified sensory neuron varicosities in untrained (naïve) animals revealed that approximately 60% of these presynaptic terminals lacked a structurally detectable active zone suggesting the possibility of nascent synapses in the adult brain. The extent to which learning and memory can convert these immature, and presynaptically silent synapses into mature and functionally competent synaptic connections is discussed below. Finally, these initial studies in *Aplysia* suggested that the growth of new sensory neuron synapses may represent the final and perhaps most stable phase of long-term memory storage, and raised the possibility that the stability of the long-term process might be achieved, at least in part, because of the relative stability of synaptic structure.

Subsequent studies by Wainwright et al. (2002) have examined the effects of different sensitization training protocols on the structure of sensory neurons in the pleural ganglion mediating the tail-siphon withdrawal reflex in *Aplysia*. A 4-day training period produced a robust and localized outgrowth of both neuritic processes and presynaptic varicosities in these sensory neurons observed 24 h after the end of training. These structural changes were consistent with the previous results of Bailey and Chen (1988a) in siphon sensory neurons. By contrast, 1 day of sensitization training, which can induce both long-term behavioral sensitization and synaptic facilitation (Frost et al., 1985; Cleary et al., 1998), was not associated with a comparable outgrowth of tail sensory neurons indicating that storage of the memory for sensitization induced by the 1 day of training and the spaced 4 day protocols are likely to recruit different mechanisms. These investigators have also reported a dissociation of morphological and physiological changes associated with long-term sensitization of the tail-siphon withdrawal reflex in *Aplysia* (Wainwright et al., 2004). The behavioral effects of long-term sensitization training were restricted to the trained side of the animal as were changes in the strength of sensory to motor connections. By contrast, long-term training produced varicosity formation on both sides of the animal. Interestingly, on the trained side, this outgrowth was reflected by an increase in the number of putative contacts with follower neurons as well as with an increase in synaptic strength and behavioral enhancement suggesting a causal relationship between these changes.

The long-lasting growth of new synaptic connections between sensory neurons and their follower cells during long-term sensitization can be reconstituted in sensory–motor neuron co-cultures by five repeated applications of 5-HT (Glanzman et al., 1990; Bailey et al., 1992b) as well as induced in the intact ganglion by the intracellular injection of cAMP, a second messenger activated by 5-HT (Nazif et al., 1991). In culture, the synaptic growth can be correlated with the long-term (24–72 h) enhancement in synaptic effectiveness and depends on the presence of an appropriate target cell similar to the synapse formation that occurs during development.

Time-lapse imaging reveals LTF is associated with presynaptic activation of silent synapses and growth of new functional synapses

In most model learning systems, the functional contribution of the structural changes that accompany long-lasting forms of synaptic plasticity remains largely unknown. In particular one would like to know if changes in the number or structure of synaptic connections induced by learning are functionally effective and capable of contributing to the storage of long-term memory. Both technical and experimental limitations prevented the earlier studies in *Aplysia* discussed above from examining whether the increase in synaptic strength during long-term sensitization resulted from the conversion of pre-existing but nonfunctional (silent) synapses to active synapses, or from the addition of newly formed functional synapses, or perhaps both. To address these issues directly, more recent in vitro studies of the sensory-to-motor neuron synapse in *Aplysia* culture have monitored both functional and structural changes simultaneously so as to follow remodeling and growth at the same specific synaptic varicosities continuously over time and to examine the functional contribution of these presynaptic structural changes to the different time-dependent phases of LTF.

Toward that end, Kim et al. (2003) combined time-lapse confocal imaging of individual presynaptic varicosities of sensory neurons labeled with three different fluorescent markers: the whole cell marker *Alexa-594*, and two presynaptic marker proteins: *synaptophysin-eGFP* which monitors changes in the distribution of synaptic vesicles within individual varicosities and *synapto-PHluorin* (synPH), a monitor of active transmitter release sites (Miesenbock et al., 1998). They found that repeated pulses of 5-HT induce two temporally, morphologically and molecularly distinct classes of presynaptic changes: (1) the rapid activation of silent presynaptic terminals through the filling of pre-existing empty varicosities with synaptic vesicles, which requires translation but not transcription and (2) the generation of new synaptic varicosities which occurs more slowly and requires both transcription and translation. The enrichment of pre-existing but empty varicosities with synaptophysin is completed within 3–6 h, parallels intermediate-term facilitation and accounts for approximately 32% of the newly activated synapses evident at 24 h. By contrast, the new sensory neuron varicosities, which account for 68% of the newly activated synapses at 24 h, do not form until 12–18 h after exposure to 5 pulses of 5-HT. The rapid activation of silent presynaptic terminals suggests that in addition to its role in LTF, this modification of pre-existing synapses may also contribute to the intermediate phase of synaptic plasticity and memory storage (Ghirardi et al., 1995; Mauelshagen et al., 1996; Sutton et al., 2001) (Fig. 2).

This temporal analysis of the synaptic remodeling and growth that underlie the development of LTF served to bring the structural changes into register with the physiological correlates of the different phases of long-term synaptic plasticity. For example, the different time windows for the two classes of presynaptic structural changes — a rapid enrichment of pre-existing empty varicosities with synaptic vesicle proteins and a slower generation of new varicosities — appear to be consistent with the onset and duration of intermediate-term and LTF, respectively. Since intermediate-term facilitation lasts for about 3–6 h whereas LTF first appears at 10–15 h and lasts for days (Ghirardi et al., 1995; Mauelshagen et al., 1996), these two classes of presynaptic changes may represent distinct structural correlates of the two phases of memory storage (Sutton et al., 2001). This idea is further supported by the finding that the 5-HT-induced clustering of synaptic vesicles into pre-existing empty varicosities is blocked by the inhibition of de novo protein synthesis, which has previously been shown to attenuate both intermediate-term facilitation and the corresponding intermediate phase of memory (Ghirardi et al., 1995; Sutton et al., 2001).

To test this hypothesis, Kim et al. (2003) developed a reduced 5-HT protocol to induce selectively facilitation in the intermediate-term time domain without inducing LTF (see also Ghirardi et al., 1995). They found that the isolated intermediate-term facilitation is also accompanied by the redistribution and clustering of synaptophysin-eGFP into empty varicosities at 0.5 and 3 h similar to what occurs when intermediate- and LTF are recruited together. However, the presynaptic structural changes induced by the reduced 5-HT protocol differed from those induced by the long-term protocol in at least two ways. First, there was no growth of new sensory neuron varicosities in the isolated intermediate phase. Second, unlike the filling of pre-existing empty varicosities during the intermediate-term phase induced by the long-term protocol, the newly filled varicosities did not persist for 24 h and were unaffected by inhibitors of protein synthesis suggesting that the structural remodeling activated by the reduced 5-HT protocol involved only a simple rearrangement of pre-existing synaptic components. This may reflect a fundamental difference in the molecular mechanisms that underlie the two 5-HT protocols. Although both protocols induce intermediate-term facilitation, the long-term protocol may recruit additional molecular events (including the machinery for translational activation) required to set up the long-term phase, perhaps by stabilizing the intermediate phase.

To quantitatively assess the contribution of these two distinct classes of presynaptic structural changes to LTF, Kim et al. (2003) monitored the functional state of individual sensory neuron varicosities in living cells before and 24 h after 5 pulses of 5-HT using the activity-sensitive fusion protein synPH. With synPH, release of labeled synaptic vesicles yields an increase in fluorescence

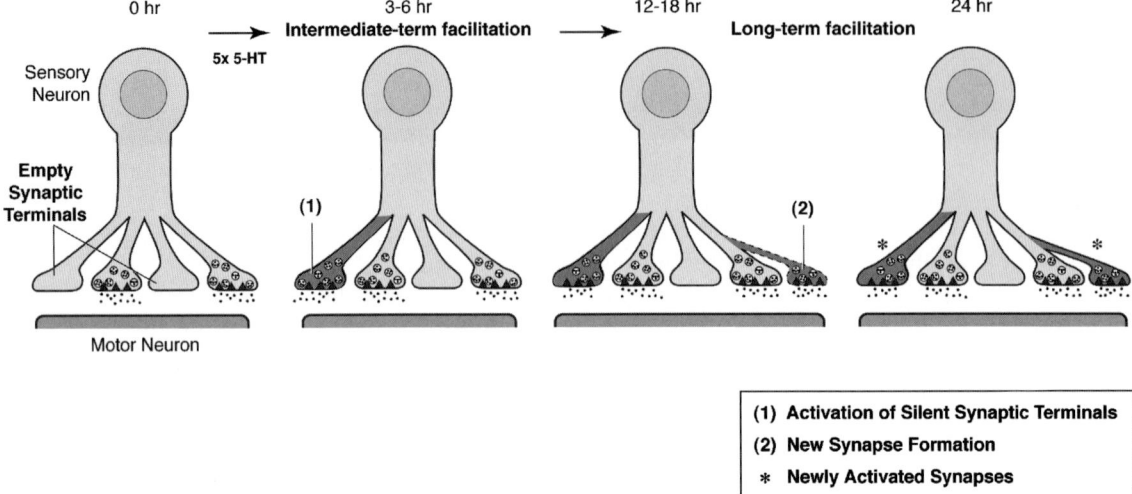

Fig. 2. Time course and functional contribution of two distinct presynaptic structural changes associated with intermediate- and long-term facilitation in *Aplysia*. Repeated pulses of 5-HT in sensory to motor neuron co-cultures trigger two distinct classes of presynaptic structural changes: (1) the rapid clustering of synaptic vesicles to pre-existing silent sensory neuron varicosities (3–6 h) and (2) the slower generation of new sensory neuron synaptic varicosities (12–18 h). The resultant newly filled and newly formed varicosities are functionally competent (capable of evoked transmitter release) and contribute to the synaptic enhancement that underlies LTF. The rapid filling and activation of silent presynaptic terminals at 3 h suggests that, in addition to its role in LTF, this modification of pre-existing varicosities may also contribute to the intermediate phase of synaptic plasticity. Red triangles represent transmitter release sites (active zones). (Modified from Kim et al., 2003.) (See Color Plate 10.2 in color plate section.)

due to the externalization of pHluorin to a more basic exterior medium, which returns to basal levels by the re-acidification of synaptic vesicles following endocytosis in a Ca^{2+}-dependent fashion (Sankaranarayanan and Ryan, 2001). When expressed in *Aplysia* sensory neurons, depolarization by bath application of KCl leads to the evoked exocytotic release of synaptic vesicles within individual varicosities as indicated by an increase in the fluorescence signal of synPH as has been previously reported in cultured hippocampal neurons (Miesenbock et al., 1998).

In this fashion, Kim et al. (2003) were able to monitor continuously over time individual 5-HT-induced structural changes at the same sensory neuron varicosities and to examine directly the functional contribution of two distinct classes of presynaptic structural changes to the different temporal phases of LTF. They found that 24 h after repeated pulses of 5-HT there was a 59% increase in the total number of synPH-active sensory neuron varicosities. These real time experiments on living synapses in culture were remarkably consistent with the earlier electron microscopic studies of Bailey and Chen (1983) which had demonstrated that long-term sensitization in the intact animal is accompanied by an increase of 65% in the number of sensory neuron varicosities that contained fully differentiated presynaptic active zones (transmitter release sites) when compared to varicosities from untrained (naïve) animals. The enrichment of pre-existing but empty varicosities accounted for approximately 32% of these newly activated synapses at 24 h, whereas newly formed varicosities accounted for approximately 68%. These results suggested that both classes of structural changes — the presynaptic activation of pre-existing silent synapses and the growth of new functional synapses — appear to contribute to the synaptic enhancement that characterizes LTF at 24 h.

Previous studies have shown that specific 5-HT protocols (Sun and Schacher, 1998; Casadio et al., 1999) or experimental manipulation in *Aplysia*

(Hatada et al., 2000; Udo et al., 2005) can induce LTF at 24 h without the formation of new varicosities. How might such an increase in synaptic strength persist for 24 h in the absence of synaptic growth? The results of Kim and associates suggest that additional modifications of pre-existing connections including the activation of previously silent synapses, may play a role in the initial phases of synaptic maintenance and highlight the fact that there are likely to be multiple types of structural mechanisms that can contribute to LTF at 24 h (Schacher et al., 1997; Bailey et al., 2000; Sutton and Carew, 2000).

Of the two classes of presynaptic structural plasticity induced by 5-HT in culture, synaptic growth appears to contribute more to the synaptic enhancement present at 24 h than does the activation of pre-existing silent synapses. It will be of interest to see if the functional contribution by newly formed synapses increases with time when the growth process is more fully developed and memory storage is likely to be more stable. This would be consistent with the earlier studies in the intact animal outlined above, which have shown that only the increases in the number of sensory neuron varicosities and active zones persist for several weeks in parallel with the behavioral duration of the memory, as well as more recent work in culture which has demonstrated that synaptic growth plays a more prominent role in the expression of the later phases of LTF (Martin et al., 1997b; Casadio et al., 1999).

The activation of silent synapses also seems to play a major role in LTP in the hippocampus. Although in mammals the term refers to a very specific molecular configuration found in synapses in different regions of the CNS of vertebrates (Malinow et al. 2000; Malinow and Malenka, 2002). In this case, the term *silent synapse* refers to excitatory glutamatergic synapses whose postsynaptic membrane contains NMDARs but no AMPARs. The lack of AMPAR-mediated signaling renders these synapses inactive, or "silent", under normal conditions. Synaptic stimulation activates these silent synapses through the insertion of AMPARs into the postsynaptic membrane, a phenomenon some times referred to as *AMPAfication*. Calcium/calmodulin-dependent protein kinase II (CaMKII) plays a critical role in this process. Once this kinase is activated by high frequency stimulation, it phosphorylates AMPARs or associated proteins, triggering their insertion into the postsynaptic membrane. The synapse is then no longer silent and postsynaptic responses are, by consequence, enhanced.

Remodeling of the presynaptic actin network for learning-related synaptic growth requires activation of Cdc42-mediated signaling pathways

Actin is enriched in both the pre- and postsynaptic compartment of neurons (Matus, 2000). Although the activity-dependent modulation of actin dynamics at postsynaptic spines has been well documented, the extent and role of actin regulation in presynaptic terminals is less clear (Colicos et al., 2001; Antonova et al., 2001). During development, reorganization of actin in growth cones has been shown to play an important role in axonal path finding (Yuan et al., 2003). However, in mature neurons, it has been suggested that the presynaptic actin network may function more as an intracellular scaffold rather than providing a propulsive force that could contribute directly to the type of rapid structural remodeling reported for postsynaptic dendritic spines (Sankaranarayanan et al., 2003).

In *Aplysia*, repeated applications of 5-HT that lead to LTF induce a slower and more persistent alteration in the dynamics of the presynaptic actin network leading to the remodeling and growth of sensory neuron synapses. Both the 5-HT-induced enrichment of synaptic proteins in pre-existing varicosities and the formation of new and functionally competent sensory neuron varicosities during LTF involve an activity-dependent rearrangement of the presynaptic actin cytoskeleton (Udo et al., 2005). These findings in *Aplysia* are consistent with previous reports that structural remodeling of synapses in response to physiological activity requires the reorganization of actin (Colicos et al., 2001; Huntley et al., 2002) and that the inhibition of actin function blocks synapse formation and interferes with long-term synaptic plasticity (Hatada et al., 2000; Krucker et al., 2000; Zhang and Benson, 2001). Furthermore, several synaptic proteins such as synapsin can bind to the presynaptic actin cytoskeleton and participate in synaptic vesicle trafficking (Humeau et al., 2001) that may contribute to the 5-HT-induced enrichment of pre-existing varicosities observed during LTF.

One attractive molecular candidate for the 5-HT-induced reorganization of the cytoskeleton at sensory neuron presynaptic varicosities is the Rho-family of small GTPases, which can modulate actin polymerization in response to extracellular signals and can be regulated by neuronal activity in vivo (Luo et al., 1996; Hall, 1998). In *Aplysia*, Udo et al. (2005) found that the application of toxin B, a general inhibitor of the Rho family, blocks 5-HT-induced LTF, as well as growth of new synapses in sensory–motor neuron co-cultures. Moreover, repeated pulses of 5-HT selectively induce the spatial and temporal regulation of the activity of only one of the small GTPases-Cdc42-at a subset of sensory neuron presynaptic varicosities. The activation of ApCdc42 induced by 5-HT is dependent on both the P13K and PLC pathways and, in turn, recruits the downstream effectors PAK (p21-Cdc42/Rac-activated kinase) and N-WASP (neuronal Wiskott–Aldrich syndrome protein) to regulate the presynaptic actin network. This initial molecular cascade leads to the outgrowth of filopodia, some of which represent morphological precursors for the growth of new sensory neuron varicosities associated with the storage of LTF.

5-HT-induced internalization of apCAM: a preliminary and permissive step for initiation of learning-related synaptic growth

Cell adhesion molecules (CAMs) are cell surface glycoproteins that mediate cell-to-cell and cell-to-extracellular matrix adhesions. In the CNS, CAMs are involved in cell migration, neurite outgrowth and more recently have been shown to participate in synapse formation during development (Scheiffele, 2003; Washbourne et al., 2004) as well as various forms of learning-related synaptic plasticity in the adult brain (Martin and Kandel, 1996; Fields and Itoh, 1996; Benson et al., 2000). Some of the first evidence for a role of CAMs during learning and memory came from studies of an immunoglobulin-related CAM in *Aplysia*, designated apCAM, which is homologous to NCAM in vertebrates and Fasciclin II in *Drosophila* (Mayford et al., 1992). Following the application of 5-HT or cAMP, there is a decrease in the expression of apCAM and this occurs in a transcriptionally dependent manner. Furthermore, imaging of fluorescently labeled MAbs to apCAM indicates that not only is there a decrease in the level of expression but that even pre-existing protein is lost from the surface membrane of the sensory neurons within 1 h after the addition of 5-HT. This transient modulation by 5-HT of CAMs, therefore, may represent one of the early molecular steps required for initiating learning-related growth of synaptic connections. Indeed, blocking the expression of the antigen by MAb causes defasciculation, a step that appears to precede synapse formation during development in *Aplysia* (Keller and Schacher, 1990).

To examine the mechanisms that underlie the 5-HT-induced down-regulation of apCAM and, in particular, how these relate to the initiation of synaptic growth, Bailey et al. (1992a) combined thin-section electron microscopy with immuno-labeling using a gold-conjugated MAb specific to apCAM. They found that a 1 h application of 5-HT led to a 50% decrease in the density of gold-labeled apCAM complexes at the surface membrane of the sensory neuron. This down-regulation was particularly prominent at adherent processes of the sensory neurons and was achieved by a heterologous, protein synthesis-dependent activation of the endosomal pathway, leading to internalization and apparent degradation of apCAM. As is the case for the down-regulation at the level of expression, the 5-HT-induced internalization of apCAM can be simulated by cAMP. Concomitant with the down-regulation of apCAM, Hu et al. (1993) further demonstrated that, as part of this coordinated program for endocytosis, 5-HT and cAMP also induce an increase in the number of coated pits and coated vesicles in the sensory neurons and an increase in the expression of the light chain of clathrin (apClathrin). Since the apClathrin light chain contains the important functional domains of both LCa and LCb of mammalian clathrin thought to be essential for the coated pit assembly and disassembly, the increase in clathrin may be an important component in the activation of the endocytic cycle required for the internalization of apCAM.

The learning-induced internalization of apCAM is thought to have at least two major structural consequences: (1) disassembly of homophilically associated fascicles of the sensory neurons (defasciculation), a process that may destabilize adhesive contacts normally inhibiting growth, and (2) endocytic activation that may lead to a redistribution of membrane components to sites where new synapses form. Thus, aspects of the initial steps in the learning-related growth of synaptic connections that is a hallmark of the long-term process may eventually be understood in the context of a novel and targeted form of receptor-mediated endocytosis.

These initial effects of 5-HT on the remodeling of the surface and internal membrane systems of sensory neurons in *Aplysia* bear a striking similarity to the morphological changes induced in non-neuronal systems by growth factors, which suggests that modulatory transmitters important for learning, such as 5-HT, may serve a double function. In addition to producing a transient regulation of the excitability of neurons, with repeated or prolonged exposure they may also produce an action comparable to that of a growth factor, which results in a more persistent regulation of the architecture of the neuron.

To further define the mechanisms whereby 5-HT leads to apCAM down-regulation, Bailey et al. (1997) used epitope tags to examine the fate of the two apCAM isoforms (transmembrane and GPI-linked) and found that only the transmembrane form (TM-apCAM) is internalized (Fig. 3). This internalization was blocked by overexpression of TM-apCAM with a point mutation in the two MAPK phosphorylation consensus sites, as well as by injection of a specific MAPK antagonist into sensory neurons. These data suggest that activation of the MAPK pathway is important for the internalization of TM-apCAM and may represent one of the initial and perhaps permissive stages of learning-related synaptic growth in *Aplysia*. In addition, the differential down-regulation of the GPI-linked and transmembrane forms of apCAM raised the interesting possibility that learning-related synaptic growth in the adult is initiated by an activity-dependent recruitment of specific isoforms of CAMs, similar to the modulation of cell-surface receptors during the fine-tuning of synaptic connections in the developing nervous system. One consequence of isoform recruitment is that it would allow neuronal activity to regulate the surface expression of each isoform, a process that might take on additional functional significance if these surface molecules were distributed differentially along the three-dimensional extent of the neuron and, thus may provide a regulatory unit capable of acting sequentially at multiple cytoplasmic and plasma membrane sites during the early inductive phases of the long-term process.

Han et al. (2004) have examined more closely the relationship between the 5-HT-induced down-regulation of TM-apCAM and synaptic growth by overexpressing various HA-epitope tagged recombinant apCAMs in *Aplysia* sensory neurons. They found that overexpression of TM-apCAM, but not the GPI-linked isoform of apCAM, blocked both LTF as well as the associated increase in the number of sensory neuron varicosities. By interrupting the adhesive function of apCAM with an anti-HA antibody, this inhibition of LTF induced by the overexpression of TM-apCAM was restored. Moreover, LTF could be completely blocked by overexpression of the cytoplasmic tail portion of apCAM alone. These studies indicated that the extracellular domain of TM-apCAM has an inhibitory function that is neutralized by internalization to induce LTF and suggested that the cytoplasmic domain provides an interactive platform for both signal transduction and the internalization machinery.

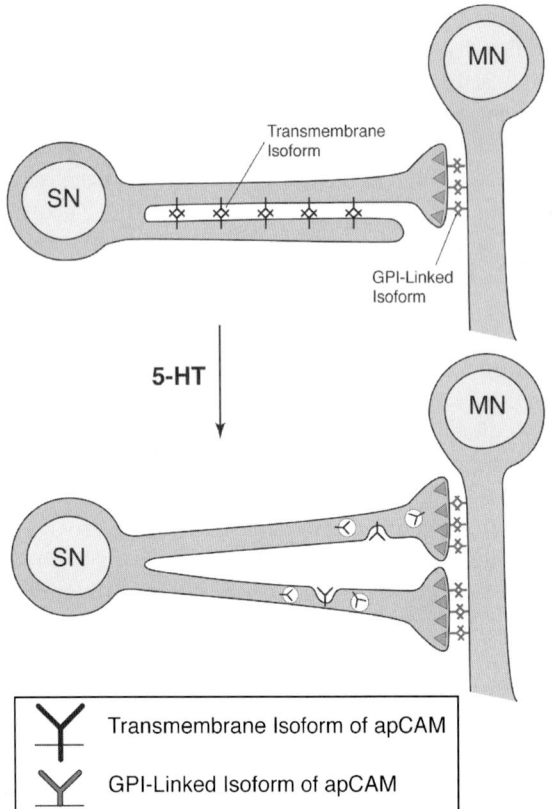

Fig. 3. Regional specific down-regulation of the transmembrane isoform of apCAM. This model is based on the assumption that the relative concentration of the GPI-linked versus transmembrane isoforms of apCAM is highest at points of synaptic contact between the sensory neuron and motor neuron and reflects the results of studies done in dissociated cell culture. Thus, previously established connections might remain intact following exposure to 5-HT since they would be held in place by the adhesive, homophilic interactions of the GPI-linked isoforms and the process of outgrowth from sensory neuron axons would be initiated by down-regulation of the transmembrane form at extrasynaptic sites of membrane apposition. In the intact ganglion, the axons of sensory neurons are likely to fasciculate not only with other sensory neurons but also with the processes of other neurons and perhaps even glia. One of the attractive features of this model is that the mechanism for down-regulation is intrinsic to the sensory neurons. Thus, even if some of the sensory neuron axonal contacts in the intact ganglion were heterophilic in nature, i.e., with other neurons or glia, we would still expect the selective internalization of apCAM at the sensory neuron surface membrane at these sites of heterophilic apposition to destabilize adhesive contacts and to facilitate disassembly. (From Bailey et al., 1997.) (See Color Plate 10.3 in color plate section.)

Nuclear translocation of apCAM-associated protein (CAMAP) activates presynaptic gene transcription for induction of LTF

Lee et al. (2007) have examined the 5-HT-induced signaling interactions mediated by the cytoplasmic domain of TM-apCAM and found an additional, and novel role for this CAM in synapse-specific forms of long-lasting plasticity. LTF at the sensory to motor neuron synapse requires the activation of CREB1 in the nucleus of the sensory neuron (Bartsch et al., 1998). Activated CREB1 induces the transcription factor ApC/EBP that in turn acts on downstream genes encoding proteins important for synaptic growth and the stable maintenance of LTF (Alberini et al., 1994). As discussed above, an initial step, thought to be permissive, for the initiation of learning-related growth is the clathrin-mediated internalization and consequent down-regulation of TM-apCAM.

To examine directly how the internalization of TM-apCAM is related to the initiation of transcription, Lee et al. (2007) first looked for molecules that could bind to the cytoplasmic tail of TM-apCAM and cloned an CAMAP by yeast two-hybrid screening. They found that 5-HT signaling at the synapse activates PKA which in turn phosphorylates CAMAP to induce the dissociation of CAMAP from apCAM and that this dissociation is a prerequisite for the internalization of apCAM. The 5-HT-induced dissociated CAMAP is subsequently translocated to the nucleus of the sensory neurons. In the nucleus, CAMAP acts as a transcriptional co-activator for CREB1 that is essential for the activation of ApC/EBP required for the initiation of LTF. Combined, these data suggest that CAMAP is one of the retrograde signals from the synapse to the nucleus where it acts as a co-regulator of the presynaptic gene expression associated with the induction of LTF in *Aplysia*. In addition, these findings demonstrate the importance, for learning-related synaptic plasticity, of signal propagation into the nucleus from the surface membrane of activated synaptic sites mediated by a molecule directly interacting with a cell surface adhesion molecule and suggest a novel presynaptic molecular mechanism to turn on the gene transcription required for long-term memory.

An overall view

Perhaps the most striking finding in the biology of memory storage is that long-term memory involves transcription in the nucleus and structural changes at the synapse. The structural changes associated with the storage of long-term memory can be grouped into two general categories: remodeling of pre-existing synapses and growth of new synapses (Greenough and Bailey, 1988; Bailey and Kandel, 1993; Yuste and Bonhoeffer, 2001; Bailey et al., 2004; Lamprecht and LeDoux, 2004). Despite an increasing body of evidence for changes in the number or structure of synaptic connections and long-term memory, it has so far proven difficult to follow individual structural changes at the same synapse over time and to relate directly this remodeling to physiological function and memory storage.

Recent studies have found that activity-dependent remodeling of pre-existing synapses and the growth of new synaptic connections occurs in the mammalian CNS (Buchs and Muller, 1996; Engert and Bonhoeffer, 1999; Maletic-Savatic et al., 1999; Toni et al., 1999; Colicos et al., 2001; De Paola et al., 2003). However, in the mammalian brain these structural changes are difficult to study because the effects are often modest. Moreover, the specific role of this structural plasticity remains unclear because the functional contribution of individual synapses to memory processes in these more complex neuronal networks is not yet well defined (Lamprecht and LeDoux, 2004; Hayashi and Majewska, 2005; Segal, 2005). For example, although the generation and enlargement of dendritic spines has been associated with the production of LTP and synaptic activity in organotypic hippocampal slices (Matsuzaki et al., 2004; Nagerl et al., 2004) and acute slices of neonatal animals (Zhou et al., 2004), these structural changes are much more subtle in the adult brain (Lang et al., 2004). In adults there is only a modest production of new spines (Zuo et al., 2005), and learning-related plasticity seems to rely more on subcellular changes than on anatomical changes. Thus, neuronal activity regulates the transport of polysomes from dendritic shafts to active spines (Ostroff et al., 2002), as well as the trafficking of neurotransmitter receptors (Malinow and Malenka, 2002).

By contrast, in *Aplysia* the learning-induced structural changes that accompany long-term sensitization in vivo and LTF in vitro are robust, highly reproducible and easy to study and can be shown to be both functionally effective and capable of contributing to memory storage. Time-lapse imaging studies of the sensory to motor neuron synapse in culture have revealed that LTF is accompanied by two temporally and morphologically distinct classes of presynaptic structural change: the rapid activation of silent pre-existing varicosities by filling with synaptic vesicles and the slower growth of new functional varicosities. These findings, the first to be made on individually identified presynaptic varicosities, suggest that the duration of the changes in synaptic effectiveness that accompany memory storage may be reflected by the differential regulation of two fundamentally disparate forms of presynaptic compartment: (1) nascent (silent) varicosities that can be rapidly and reversibly remodeled into active transmitter release sites and (2) mature, more stable and functionally competent varicosities that following long-term training may undergo a process of fission to form new stable synaptic contacts.

Recent live imaging studies in the mammalian CNS have reported a comparable remodeling and differentiation of the presynaptic compartment in developing synapses (for review, see Ziv and Garner, 2004). For example, the establishment of functional transmitter release sites in cultured hippocampal neurons can occur in less than 1 h (Ahmari et al., 2000; Friedman et al., 2000). This short delay is similar to the 5-HT-induced enrichment and subsequent activation of pre-existing silent varicosities in *Aplysia*, which also occurs very rapidly — initial changes in the recruitment of synaptic vesicle proteins to empty presynaptic varicosities can be detected as early as 30 min after the completion of 5-HT training.

How is this rapid differentiation of nascent presynaptic compartments achieved? One model of active zone assembly suggests proteins that comprise the cytoskeletal matrix of the active zone (CAZ) are packaged into transport vesicles for delivery and fusion with the plasma membrane at nascent synaptic contacts (Zhai et al., 2001; Shapira et al., 2003). These transport vesicles contain multiple molecular components including Piccolo and Bassoon — important for assembly and structural maintenance of the active zone — as well as other CAZ scaffolding molecules implicated in synaptic vesicle exocytosis but typically not synaptic vesicle proteins. The rapid remodeling and presynaptic activation of nascent sensory neuron varicosities induced by 5-HT in *Aplysia* culture is consistent with this molecular model for active zone assembly in developing synapses of the mammalian CNS. Moreover, the apparent heterogeneity in the content of these mobile pre-assembled packets — some of which contain synaptic vesicle proteins and the others which contain components required for assembly of the active zone (Friedman et al., 2000; Zhai et al., 2001) could explain why more than half of the sensory neuron varicosities enriched in only synaptophysin following 5-HT-induced LTF were not functional.

The second general class of learning-related presynaptic structural change associated with LTF in *Aplysia* involves a slower generation of new and functionally effective sensory neuron varicosities. How are these new varicosities formed? Time-lapse imaging of individual sensory neuron varicosities indicates that the 5-HT-induced recruitment of synaptic proteins to a pre-existing varicosity leads directly to both an enrichment of these presynaptic constituents as well as to an overall increase in the size of the varicosity. The presynaptic remodeling and growth is followed by the apparent division or splitting of a subset of these pre-existing varicosities (Hatada et al., 2000; Kim et al., 2003; Udo et al., 2005). This physical process may lead to the budding off of components of the active zone and associated synaptic vesicle cluster from each pre-existing presynaptic compartment that could then serve as a nucleation site to seed the formation of a new varicosity. Aspects of these structural transformations of the presynaptic compartment that precede learning-related synaptic growth in *Aplysia* have also been reported at developing synapses in mammals. For example, imaging studies of the early stages of synapse formation have shown that presynaptic sites formed immediately after initial contact of axonal and dendritic processes are highly unstable. Moreover, even apparent mature presynaptic sites are relatively unstable as occasionally "orphan release sites" break off from fully formed boutons and either migrate to adjacent presynaptic sites or participate in the formation of completely new ones (Friedman et al.; 2000; Hopf et al., 2002; Krueger et al., 2003).

Taken together, these time-lapse observations of living synapses indicate that differentiation and growth of the presynaptic compartment, either induced by learning in the mature nervous system or as mechanistic steps during development, are highly dynamic and rapid processes that can recruit both pre-existing proteins as well as pre-assembled synaptic components. The instability of nascent presynaptic compartments during the early stages of development is characterized by the dispersion of mobile packets of synaptic components and a renewal of their migration once the transient pre- and postsynaptic contacts breakup. As the neurons mature, an increasing proportion of these initial contacts develop into more stable and fully functional presynaptic terminals (Ziv and Garner, 2004). Results in the *Aplysia* sensory to motor neuron culture preparation further indicate that even some of these mature presynaptic contacts can be selectively destabilized during learning and memory leading to the generation of new and functionally competent synaptic varicosities by processes that appear similar to those that govern developmental synaptogenesis.

Conclusions

Over the past two decades, it has become clear that synaptic growth and the formation of new synapses are hallmarks of long-term, learning-related synaptic plasticity. The morphological correspondence between the studies of long-term sensitization in *Aplysia* and LTP in the mammalian hippocampus indicates that learning resembles a process of growth and neuronal differentiation across a broad segment of the animal kingdom and suggests that learning-related synapse formation may be a highly conserved feature for the storage of both implicit and explicit forms of long-term memory. Such changes are likely to reflect the recruitment by environmental stimuli of developmental processes that are latent or inhibited in the fully differentiated neuron. Recent studies of the sensory to motor neuron connection in *Aplysia* have begun to characterize the cellular and molecular events that underlie the structural changes induced by learning at the level of individual identified synapses. This in turn has suggested the synaptic remodeling and synaptic growth that accompany learning and memory storage in the adult brain may reutilize mechanisms important for de novo synapse formation and the fine-tuning of synaptic connections during development of the nervous system.

Acknowledgments

Research in this review was supported in part by National Institutes of Health grant MH37134 (to C.H.B.), the Howard Hughes Medical Institute (to E.R.K.), and the Kavli Institute for Brain Sciences.

References

Ahmari, S.E., Buchanan, J. and Smith, S.J. (2000) Assembly of presynaptic active zones from cytoplasmic transport packets. Nat. Neurosci., 3: 445–451.

Alberini, C.M., Ghirardi, M., Metz, R. and Kandel, E.R. (1994) C/EBP is an immediate-early gene required for the consolidation of long-term facilitation in *Aplysia*. Cell, 76: 1099–1114.

Antonova, I., Arancio, O., Trillat, A.-C., Wang, H.G., Zablow, L., Udo, H., Kandel, E.R. and Hawkins, R.D. (2001) Rapid increase in clusters of presynaptic proteins at onset of long-lasting potentiation. Science, 294: 1547–1550.

Bailey, C.H., Bartsch, D. and Kandel, E.R. (1996) Toward a molecular definition of long-term memory storage. Proc. Natl. Acad. Sci. U.S.A., 93: 13445–13452.

Bailey, C.H. and Chen, M. (1983) Morphological basis of long-term habituation and sensitization in *Aplysia*. Science, 220: 91–93.

Bailey, C.H. and Chen, M. (1988a) Long-term memory in *Aplysia* modulates the total number of varicosities of single identified sensory neurons. Proc. Natl. Acad. Sci. U.S.A., 85: 2373–2377.

Bailey, C.H. and Chen, M. (1988b) Long-term sensitization in *Aplysia* increases the number of presynaptic contacts onto the identified gill motor neuron L7. Proc. Natl. Acad. Sci. U.S.A., 85: 9356–9359.

Bailey, C.H. and Chen, M. (1988c) Morphological basis of short-term habituation in *Aplysia*. J. Neurosci., 8: 2452–2459.

Bailey, C.H. and Chen, M. (1989) Time course of structural changes at identified sensory neuron synapses during long-term sensitization in *Aplysia*. J. Neurosci., 9: 1774–1780.

Bailey, C.H., Chen, M., Keller, F. and Kandel, E.R. (1992a) Serotonin-mediated endocytosis of apCAM: an early step of learning-related synaptic growth in *Aplysia*. Science, 25: 645–649.

Bailey, C.H., Giustetto, M., Zhu, H., Chen, M. and Kandel, E.R. (2000) A novel function for serotonin-mediated short-term facilitation in *Aplysia*: conversion of a transident cell-wide homosynaptic Hebbian plasticity into a persistent, protein synthesis-independent synapse-specific enhancement. Proc. Natl. Acad. Sci. U.S.A., 97: 11581–11586.

Bailey, C.H., Kaang, B.K., Chen, M., Marin, C., Lim, C.S., Casadio, A. and Kandel, E.R. (1997) Mutation in the phosphorylation sites of MAP kinase blocks learning-related internalization of apCAM in *Aplysia* sensory neurons. Neuron, 18: 913–924.

Bailey, C.H. and Kandel, E.R. (1993) Structural changes accompanying memory storage. Annu. Rev. Physiol., 55: 397–426.

Bailey, C.H., Kandel, E.R. and Si, K. (2004) The persistence of long-term memory: a molecular approach to self-sustaining changes in learning-induced synaptic growth. Neuron, 44: 49–57.

Bailey, C.H., Montarolo, P.G., Chen, M., Kandel, E.R. and Schacher, S. (1992b) Inhibitors of protein and RNA synthesis block the structural changes that accompany long-term heterosynaptic plasticity in the sensory neurons of *Aplysia*. Neuron, 9: 749–758.

Bailey, C.H., Thompson, E.B., Castellucci, V.F. and Kandel, E.R. (1979) Ultrastructure of the synapses of sensory neurons that mediate the gill-withdrawal reflex in *Aplysia*. J. Neurocytol., 8: 415–444.

Bartsch, D., Casadio, A., Karl, K.A., Serodio, P. and Kandel, E.R. (1998) CREB1 encodes a nuclear activator, a repressor, and a cytoplasmic modulator that form a regulatory unit critical for long-term facilitation. Cell, 95: 211–223.

Benson, D.L., Schnapp, L.M., Shapiro, L. and Huntley, G.W. (2000) Making memories stick: cell adhesive molecules in synaptic plasticity. Trends Cell Biol., 10: 473–480.

Bliss, T.V., Collingridge, G.L. and Morris, R.G. (2003) Introduction. Long-term potentiation and structure of the issue. Philos. Trans. R. Soc. Lond. B Biol. Sci., 358: 607–611.

Buchs, P.A. and Muller, D. (1996) Induction of long-term potentiation is associated with major ultrastructural changes of activated synapses. Proc. Natl. Acad. Sci. U.S.A., 93: 8040–8045.

Casadio, A., Martin, K.C., Giustetto, M., Zhu, H., Chen, M., Bartsch, D., Bailey, C.H. and Kandel, E.R. (1999) A transient, neuron-wide form of CREB-mediated long-term facilitation can be stabilized at specific synapses by local protein synthesis. Cell, 99: 221–237.

Castellucci, V., Pinsker, H., Kupfermann, I. and Kandel, E.R. (1970) Neuronal mechanisms of habituation and dishabituation of the gill-withdrawal reflex in Aplysia. Science, 167: 1745–1748.

Castellucci, V.F., Blumenfeld, H., Goelet, P. and Kandel, E.R. (1989) Inhibitor of protein synthesis blocks long-term behavioral sensitization in the isolated gill-withdrawal reflex of Aplysia. Science, 220: 91–93.

Cleary, L.J., Lee, W.L. and Byrne, J.H. (1998) Cellular correlates of long-term sensitization in Aplysia. J. Neurosci., 18: 5988–5998.

Colicos, M.A., Collins, B.E., Sailor, M.J. and Goda, Y. (2001) Remodeling of synaptic actin induced by photoconductive stimulation. Cell, 107: 605–616.

Dale, N., Kandel, E.R. and Schacher, S. (1987) Serotonin produces long-term changes in the excitability of Aplysia sensory neurons in culture that depend on new protein synthesis. J. Neurosci., 7: 2232–2238.

De Paola, V., Arber, S. and Caroni, P. (2003) AMPA receptors regulate dynamic equilibrium of presynaptic terminals in mature hippocampal networks. Nat. Neurosci., 6: 491–500.

Engert, F. and Bonhoeffer, T. (1999) Dendritic spine changes associated with hippocampal long-term synaptic plasticity. Nature, 399: 66–70.

Fields, R.D. and Itoh, K. (1996) Neural cell adhesion molecules in activity-dependent development and synaptic plasticity. Trends Neurosci., 19: 473–480.

Friedman, H.V., Bresler, T., Garner, C.C. and Ziv, N.E. (2000) Assembly of new individual excitatory synapses: time course and temporal order of synaptic molecule recruitment. Neuron, 27: 57–69.

Frost, W.N., Castellucci, V.F., Hawkins, R.D. and Kandel, E.R. (1985) Monosynaptic connections made by the sensory neurons of the gill- and siphon-withdrawal reflex in Aplysia participates in the storage of long-term memory for sensitization. Proc. Natl. Acad. Sci. U.S.A., 82: 8266–8269.

Ghirardi, M., Montarolo, P.G. and Kandel, E.R. (1995) A novel intermediate stage in the transition between short- and long-term facilitation in the sensory to motor neuron synapse of Aplysia. Neuron, 14: 413–420.

Glanzman, D.L., Kandel, E.R. and Schacher, S. (1990) Target-dependent structural changes accompanying long-term synaptic facilitation in Aplysia neurons. Science, 249: 779–802.

Greenough, W.T. and Bailey, C.H. (1988) The anatomy of a memory: convergence of results across a diversity of tests. Trends Neurosci., 11: 142–147.

Hall, A. (1998) Rho GTPases and the actin cytoskeleton. Science, 279: 509–514.

Han, J.H., Lim, Y.S., Kandel, E.R. and Kaang, B.K. (2004) Role of Aplysia cell adhesion molecules during 5-HT-induced long-term functional and structural changes. Learn. Mem., 11: 421–435.

Hatada, Y., Wu, F., Sun, Z.Y., Schacher, S. and Goldberg, D.J. (2000) Presynaptic morphological changes associated with long-term synaptic facilitation are triggered by actin polymerization at preexisting varicosities. J. Neurosci., 20: p. RC82.

Hawkins, R.D., Kandel, E.R. and Bailey, C.H. (2006) Molecular mechanisms of memory storage in Aplysia. Biol. Bull., 210: 174–191.

Hayashi, Y. and Majewska, A.K. (2005) Dendritic spine geometry: functional implication and regulation. Neuron, 46: 529–532.

Hochner, B., Klein, M., Schacher, S. and Kandel, E.R. (1986) Additional components in the cellular mechanisms of presynaptic facilitation contributes to behavioral dishabituation in Aplysia. Proc. Natl. Acad. Sci. U.S.A., 83: 8794–8798.

Hopf, F.W., Walters, J., Mehta, S. and Smith, S.J. (2002) Stability and plasticity of developing synapses in hippocampal neuronal cultures. J. Neurosci., 22: 775–781.

Hu, Y., Barzilai, A., Chen, M., Bailey, C.H. and Kandel, E.R. (1993) 5-HT and cAMP induce the formation of coated pits and vesicles and increase the expression of clathrin light chain in sensory neurons of Aplysia. Neuron, 10: 921–929.

Humeau, Y., Doussau, F., Vitiello, F., Greengard, P., Benfenati, F. and Poulain, B. (2001) Synapsin controls both reserve and releasable synaptic vesicle pools during neuronal activity and short-term plasticity in Aplysia. J. Neurosci., 21: 4195–4206.

Huntley, G.W., Benson, D.L. and Colman, D.R. (2002) Structural remodeling of the synapse in response to physiological activity. Cell, 108: 1–4.

Jessell, T.M. and Kandel, E.R. (1993) Synaptic transmission: a bidirectional and self-modifiable form of cell-cell communication. Cell 72/Neuron 10, 1–30.

Kandel, E.R. (2001) The molecular biology of memory storage: a dialogue between genes and synapses. Science, 294: 1030–1038.

Keller, Y. and Schacher, S. (1990) Neuron-specific membrane glycoproteins promoting neurite fasciculation in Aplysia californica. J. Cell Biol., 111: 2637–2650.

Klein, M. and Kandel, E.R. (1980) Mechanism of calcium current modulation underlying presynaptic facilitation and behavioral sensitization in Aplysia. Proc. Natl. Acad. Sci. U.S.A., 77: 6912–6916.

Kim, J.-H., Udo, H., Li, H.-L., Youn, T.Y., Chen, M., Kandel, E.R. and Bailey, C.H. (2003) Presynaptic activation of silent synapses and growth of new synapses contribute to intermediate and long-term facilitation in Aplysia. Neuron, 40: 151–165.

Krucker, T., Siggins, G.R. and Halpain, S. (2000) Dynamic actin filaments are required for stable long-term potentiation

(LTP) in area CA1 of the hippocampus. Proc. Natl. Acad. Sci. U.S.A., 97: 6856–6861.

Krueger, S.R., Kolar, A. and Fitzsimonds, R.M. (2003) The presynaptic release apparatus is functional in the absence of dendritic contact and highly mobile within isolated axons. Neuron, 40: 945–957.

Lamprecht, R. and LeDoux, J. (2004) Structural plasticity and memory. Nat. Rev. Neurosci., 5: 45–54.

Lang, C., Barco, A., Zablow, L., Kandel, E.R., Siegelbaum, S.A. and Zakharenko, S.S. (2004) Transient expansion of synaptically connected dendritic spines upon induction of hippocampal long-term potentiation. Proc. Natl. Acad. Sci. U.S.A., 101: 16665–16670.

Lee, S.-H., Lim, C.-S., Park, H., Lee, J.-A., Han, J.-H., Kim, H., Cheang, Y.-H., Lee, S.-H., Lee, Y.-S., Ko, H.-G., Jang, D.-H., Kim, H., Miniaci, M.C., Bartsch, D., Kim, E., Bailey, C.H., Kandel, E.R. and Kaang, B.-K. (2007) Nuclear translocation of CAM-associated protein activates transcription for long-term facilitation in *Aplysia*. Cell, 129: 801–812.

Luo, L., Hensch, T.K., Ackerman, L., Barbel, S., Jan, L.Y. and Jan, Y.N. (1996) Differential effects of the Rac GTPase on Purkinje cell axons and dendritic trunks and spines. Nature, 379: 837–840.

Maletic-Savatic, M., Maliriow, R. and Svoboda, K. (1999) Rapid dendritic morphogenesis in CA1 hippocampal dendrites induced by synaptic activity. Science, 283: 1923–1927.

Malinow, R. and Malenka, R.C. (2002) AMPA receptor trafficking and synaptic plasticity. Annu. Rev. Neurosci., 25: 103–126.

Malinow, R., Mainen, Z.F. and Hayashi, Y. (2000) LTP mechanisms: from silence to four-lane traffic. Curr. Opin. Neurobiol., 10: 352–357.

Martin, E.C., Casadio, A., Zhu, H., Yaping, E., Rose, J., Chen, M., Bailey, C.H. and Kandel, E.R. (1997) Synapse-specific long-term facilitation of *Aplysia* sensory somatic synapses: a function for local protein synthesis memory storage. Cell, 91: 927–938.

Martin, K.C. and Kandel, E.R. (1996) Cell adhesion molecules, CREB and the formation of new synaptic connections during development and learning. Neuron, 17: 567–570.

Matsuzaki, M., Honkura, N., Ellis-Davies, G.C.R. and Kasai, H. (2004) Structural basis of long-term potentiation in single dendritic spines. Nature, 429: 761–766.

Matus, A. (2000) Actin-based plasticity in dendritic spines. Science, 290: 754–758.

Mauelshagen, J., Parker, G.R. and Carew, T.J. (1996) Dynamics of induction and expression of long-term synaptic facilitation in *Aplysia*. J. Neurosci., 16: 7099–7108.

Mayford, M., Barzilai, A., Keller, F., Schacher, S. and Kandel, E.R. (1992) Modulation of an NCAM-related adhesion molecule with long-term synaptic plasticity in *Aplysia*. Science, 256: 638–644.

Miesenbock, G., De Angelis, D.A. and Rothman, J.E. (1998) Visualizing secretion and synaptic transmission with pH-sensitive green fluorescent proteins. Nature, 394: 192–195.

Milner, B. (1985) Memory and the human brain. In: Shafto M. (Ed.), How We Know. Harper and Rowe, San Francisco, CA; Acad. Sci. U.S.A., 95, 1864–1869.

Montarolo, P.G., Goelet, P., Castellucci, V.F., Morgan, J., Kandel, E.R. and Schacher, S. (1986) A critical period for macromolecular synthesis in long-term heterosynaptic facilitation in *Aplysia*. Science, 234: 1249–1254.

Nagerl, U.V., Eberhorn, N., Cambridge, S.B. and Bonhoeffer, T. (2004) Bidirectional activity-dependent morphological plasticity in hippocampal neurons. Neuron, 44: 759–767.

Nazif, F.A., Byrne, J.H. and Cleary, L.J. (1991) cAMP induces long-term morphological changes in sensory neurons of *Aplysia*. Brain Res., 539: 324–327.

Ostroff, L.E., Fiala, J.C., Allwardt, B. and Harris, K.M. (2002) Polyribosomes redistribute from dendritic shafts into spines with enlarged synapses during LTP in developing rat hippocampal slices. Neuron, 35: 535–545.

Polster, M.R., Nadel, L. and Schachter, D.L. (1991) Cognitive neuroscience: an analysis of memory: a historical perspective. J. Cogn. Neurosci., 3: 95–116.

Sankaranarayanan, S., Atluri, P.P. and Ryan, T.A. (2003) Actin has a molecular scaffolding, not propulsive, role in presynaptic function. Nat. Neurosci., 6: 127–135.

Sankaranarayanan, S. and Ryan, T.A. (2001) Calcium accelerates endocytosis of vSNAREs at hippocampal synapses. Nat. Neurosci., 4: 129–136.

Schacher, S., Wu., F. and Sun, Z.-Y. (1997) Pathway-specific synaptic plasticity: activity-dependent enhancement and suppression of long-term facilitation at converging inputs on a single target. J. Neurosci., 17: 597–606.

Scheiffele, P. (2003) Cell-cell signaling during synapse formation in the CNS. Annu. Rev. Neurosci., 26: 485–508.

Scholz, K.P. and Byrne, J.H. (1987) Long-term sensitization in *Aplysia*: biophysical correlates in tail sensory neurons. Science, 235: 685–687.

Schwartz, H., Castellucci, V.F. and Kandel, E.R. (1971) Functions of identified neurons and synapses in abdominal ganglion of *Aplysia* in absence of protein synthesis. J. Neurophysiol., 34: 9639–9653.

Segal, M. (2005) Dendritic spines and long-term plasticity. Nat. Rev. Neurosci., 6: 277–284.

Shapira, M., Zhai, R.G., Dresbach, T., Bresler, T., Torres, V.I., Gundelfinger, E.D., Ziv, N.E. and Garner, C.C. (2003) Unitary assembly of presynaptic active zones from piccolo-bassoon transport vesicles. Neuron, 38: 237–252.

Squire, L.R. and Zola-Morgan, S. (1991) The medial temporal lobe memory system. Science, 253: 1380–1386.

Sun, Z.Y. and Schacher, S. (1998) Binding of serotonin to receptors at multiple sites is required for structural plasticity accompanying long-term facilitation of *Aplysia* sensorimotor synapses. J. Neurosci., 18: 3991–4000.

Sutton, M.A. and Carew, T.J. (2000) Parallel molecular pathways mediated expression of distinct forms of intermediate-term facilitation at tail sensory-motor synapses in *Aplysia*. Neuron, 26: 219–231.

Sutton, M.A., Masters, S.E., Bagnall, M.W. and Carew, T.J. (2001) Molecular mechanisms underlying a unique intermediate phase of memory in *Aplysia*. Neuron, 31: 143–154.

Toni, N., Buchs, P.A., Nikonenko, I., Bron, C.R. and Muller, D. (1999) LTP promotes formation of multiple spine synapses between a single axon terminal and a dendrite. Nature, 402: 421–425.

Udo, H., Jin, I., Kim, J-H., Li, H-L., Youn, T., Hawkins, R.D., Kandel, E.R. and Bailey, C.H. (2005) Serotonin-induced regulation of the actin network for learning-related synaptic growth requires CdC42, N-WASP and PAK in *Aplysia* sensory neurons. Neuron, 45: 887–901.

Wainwright, M.L., Byrne, J.H. and Cleary, L.J. (2004) Dissociation of morphological and physiological changes associated with long-term memory in *Aplysia*. J. Neurophysiol., 92: 2628–2632.

Wainwright, M.L., Zhang, H., Byrne, J.H. and Cleary, L.J. (2002) Localized neuronal outgrowth induced by long-term sensitization training in *Aplysia*. J. Neurosci., 22: 4132–4141.

Washbourne, P., Dityatev, A., Scheiffele, P., Biederer, T., Weiner, J.A., Christopherson, K.S. and El-Husseini, A. (2004) Cell adhesion molecules in synapse formation. J. Neurosci., 24: 9244–9249.

Yuan, X.B., Jin, M., Xu, X., Song, Y.Q., Wu, C.P., Poo, M.M. and Duan, S. (2003) Signaling and crosstalk of GTPases in mediating axon guidance. Nat. Cell Biol., 5: 38–45.

Yuste, R. and Bonhoeffer, T. (2001) Morphological changes in dendritic spines associated with long-term synaptic plasticity. Annu. Rev. Neurosci., 24: 1071–1108.

Zhai, R.G., Vardinon-Friedman, H., Cases-Langhoff, C., Becker, B., Gundelfinger, E.D., Ziv, N.E. and Garner, C.C. (2001) Assembling the presynaptic active zone: a characterization of an active one precursor vesicle. Neuron, 29: 131–143.

Zhang, W. and Benson, D.L. (2001) Stages of synapse development defined by dependence on F-actin. J. Neurosci., 21: 5169–5181.

Zhou, Q., Homma, K.J. and Poo, M.M. (2004) Shrinkage of dendritic spines associated with long-term depression of hippocampal synapses. Neuron, 44: 749–757.

Ziv, N.E. and Garner, C.G. (2004) Cellular mechanisms of presynaptic assembly. Nat. Rev. Neurosci., 5: 385–399.

Zuo, Y., Lin, A., Chang, P. and Gan, W.B. (2005) Development of long-term dendritic spine stability in diverse regions of cerebral cortex. Neuron, 46: 181–189.

CHAPTER 11

Spine dynamics and synapse remodeling during LTP and memory processes

M. De Roo, P. Klauser, P. Mendez Garcia, L. Poglia and D. Muller*

Department of Neuroscience, Centre Médical Universitaire, 1211 Geneva 4, Switzerland

Abstract: While changes in the efficacy of synaptic transmission are believed to represent the physiological bases of learning mechanisms, other recent studies have started to highlight the possibility that a structural reorganization of synaptic networks could also be involved. Morphological changes of the shape or size of dendritic spines or of the organization of postsynaptic densities have been described in several studies, as well as the growth and formation following stimulation of new protrusions. Confocal in vivo imaging experiments have further revealed that dendritic spines undergo a continuous turnover and replacement process that may vary as a function of development, but can be markedly enhanced by sensory activation or following brain damage. The implications of these new aspects of plasticity for learning and memory mechanisms are discussed.

Keywords: dendritic spines; synaptic plasticity; structural remodeling; hippocampus; rat

Introduction

A main approach to understanding learning and memory mechanisms has been to examine how information storage occurs at the level of a synaptic network. The answer to this question is now widely accepted to involve the contribution of properties of synaptic plasticity and in particular, properties such as long-term potentiation (LTP) and long-term depression (LTD; Bliss and Collingridge, 1993; Cooke and Bliss, 2006). These properties, through a lasting modification of the efficacy of synaptic transmission, are believed to durably alter the integration of synaptic responses in target neurons and consequently their probability to fire in response to a specific input. This in turn will re-shape the population of cells activated by the input, providing therefore a mechanism which, at the level of a neuronal network, fulfills the criteria expected for a learning process. Indeed, much evidence indicates that these properties of plasticity are intimately associated to learning and memory mechanisms (Cooke and Bliss, 2006). LTP and LTD are induced by patterns of activity that are physiologically relevant and enhanced during learning in regions critical for memory processes such as the hippocampus (Whitlock et al., 2006). Also many examples of pharmacological or genetic studies in rodents have shown correlations between the capacity to express these properties and the performance in behavioral learning tasks (Tang et al., 1999; Cui et al., 2004).

These results have thus strongly stimulated research on the cellular and molecular mechanisms underlying these properties of plasticity. In

*Corresponding author. Tel.: +41 22 379 5434; Fax: +41 22 379 5452;
E-mail: Dominique.Muller@medecine.unige.ch

particular, the identification of the locus (pre- or postsynaptic) and nature of the events responsible for the lasting increase in synaptic efficacy has represented an important issue and a continuous subject of debate (Malinow and Malenka, 2002). While the answer is likely to be complex, much evidence indicates that modifications in the expression of postsynaptic glutamate receptors represent a major mechanism accounting for the changes in synaptic efficacy (Malinow and Malenka, 2002; Nicoll, 2003). Currently, the most accepted view proposes that NMDA receptor activation during high-frequency stimulation leads to a rise in calcium in dendritic spines which in turn activates signaling cascades, among which protein kinases, such as calcium/calmodulin dependent protein kinase II or protein kinase C, are likely to play important roles (Lisman et al., 2002). Through a sequence of molecular events that are not yet fully understood these signaling cascades probably affect the expression of specific subunits of AMPA receptors at the synapse and thus the sensitivity to glutamate of the postsynaptic structure (Malinow and Malenka, 2002). A main mechanism contributing to synaptic enhancement thus probably involves a rapid regulation of receptor expression at the synapse.

There are however also other mechanisms that have been reported in association with synaptic plasticity and in particular many studies have examined the possibility that the structural organization of the synapse or the number of synapses could be modified (Yuste and Bonhoeffer, 2001). Spines have indeed been shown to be dynamic structures (Matus, 2000) that can be replaced at a rather high rate in young animals and with some plasticity that remains even in the adult. The possibility that structural changes are associated with LTP or LTD remains therefore an interesting and pertinent issue. We review here some of the evidence supporting this hypothesis and discuss these results in the context of new data showing spine plasticity and turnover.

Morphological changes associated to synaptic plasticity

Ultrastructural analyses of spine synapses under conditions of plasticity have revealed many changes in the shape or size of spine synapses going from enlargements of the spine head, formation of protrusions, including spinules, or modifications in the convexity or concavity of spines (Lee et al., 1980; Fifkova and Anderson, 1981; Chang and Greenough, 1984; Desmond and Levy, 1986; Chang et al., 1991). However, a very frequent finding that has been associated with changes in synaptic activity or with plasticity is the occurrence of perforated synapses, which on single sections appear as synapses with a discontinuity in their PSD (Calverley and Jones, 1990). Three-dimensional reconstructions further show that the size of perforated synapses is usually considerably increased and their shape may actually vary from macular aspects to fully segmented PSDs with separate release zones (Harris and Stevens, 1989; Geinisman, 1993). Perforated synapses might thus represent examples of highly functional synapses. Changes in the proportion of perforated synapses have been reported under multiple conditions: these include (i) in vivo situations with animals stimulated with LTP trains or following kindlings, or rats raised in a complex environment, or submitted to learning or training procedures (Greenough et al., 1978), (ii) development or regeneration models in association with synaptogenesis (Nieto-Sampedro et al., 1982; Carlin and Siekevitz, 1983; Harris et al., 1992) and (iii) in vitro situations following LTP induction or epilepsy (Geinisman et al., 1993; Buchs and Muller, 1996; Stewart et al., 2005). Overall the formation of perforated synapses is likely to be related to changes in synaptic functions, although many aspects of their properties remain unclear. It is still unknown what is the dynamic of perforated synapses, how long they maintain their organization, are they only transient as suggested by some in vitro studies (Toni et al., 1999), what are the release properties and the level of synaptic efficacy expressed at perforated synapses? Answers to these questions will be required in order to better understand their role in plasticity.

In addition to perforated synapses, a few other morphological changes have also been described. These include spines with multiple synapses, multisynapse boutons in which the presynaptic terminals contacts several spines and bifurcating spines found in association with an increase in spine density (Trommald et al., 1996; Collin et al., 1997; Toni et al., 1999, 2001). These changes have been reported both under in vitro situations following LTP induction or short ischemic conditions, and also under in vivo situations following learning or training protocols. Overall these changes have generally been suggested to reflect synaptogenesis mechanisms.

Morphological remodeling of activated spines

Most of the morphological changes reported above under plasticity conditions were in fact observed following large-scale analyses of synapses using EM approaches. A major drawback of this type of approach is related to the impossibility to really appreciate how exactly they were related to activity, whether they concerned mainly or exclusively activated synapses and what was their evolution over time. Only a few studies developed approaches to try to address this issue. In previous work in hippocampal slice culture, we used a calcium precipitation methodology to try to identify activated synapses and then determine their morphology over the next 2 h using EM and 3D reconstruction. While this technique clearly allowed to detect a population of spines with calcium precipitate, it was not fully devoid of biases, since there was some level of background staining, and furthermore precipitate often tended to be localized in the spine apparatus and thus in larger spines. However, despite these caveats, the approach revealed several interesting findings. In particular, activated spines observed 30 min after stimulation were characterized by larger spine heads and larger PSDs and exhibited a greater proportion of perforated synapses (Buchs and Muller, 1996). Interestingly however these changes were not stable and found to reverse to control values after 2 h (Toni et al., 1999). They were thus interpreted as suggesting a fast remodeling of synaptic membranes with extension of the receptor zone, the whole process being detectable for about 30–60 min following LTP induction.

The recent development of confocal and 2-photon imaging applied to in vitro preparations has made it possible to further investigate the behavior of activated spine synapses following activity or LTP induction. Calcium transient in spines were reported to affect their morphology (Korkotian and Segal, 2001) and recent work by Matsuzaki et al. (2004) showed that induction of LTP through release of caged glutamate was associated with a lasting increase in the size of the spine head, an observation also made following chemical induction of LTP (Kopec et al., 2006). Other studies however failed to observe such changes, at least on a short time scale (Bagal et al., 2005), and Lang et al. (2004) only found a transient expansion of stimulated spines. In contrast, EM studies have regularly strengthened the good correlation existing between spine head size, PSD size and level of receptor expression (Harris et al., 1992; Nusser et al., 1998). Thus, although many data point to an increase in size of dendritic spines and of their PSD following LTP induction, further work will be required to determine whether these changes accompany or, on the other hand, are necessary for the expression of the potentiation. New approaches combining confocal and 3D EM studies might provide new means to address this question.

Synaptogenesis associated to plasticity

A second very interesting morphological correlate of LTP mechanisms is the possibility that plasticity also involves the growth of new protrusions and formation of new synapses. Work by several laboratories provided evidence for such a process in association with LTP stimulation (Engert and Bonhoeffer, 1999; Maletic-Savatic et al., 1999; Toni et al., 1999). Such changes were mainly reported under in vitro conditions, but also following sensory stimulation in the cortex of adult animals (Trachtenberg et al., 2002; Holtmaat et al., 2006). Under these conditions, new spines can be formed de novo within periods of 30–60 min following stimulation and 3D EM studies suggest that some of them at least may indeed become mature synaptic contacts (Toni et al., 1999). Quantitatively, the phenomenon is clearly significant since up to 1–2% of new protrusions have been reported within 2–3 h following LTP induction protocols. Interestingly also, while new protrusions can be formed following LTP induction, the reverse appears to occur following application of LTD protocols (Nagerl et al., 2004). Thus protrusion growth and spine elimination are two opposite facets accompanying bidirectional plasticity of excitatory transmission.

An important issue with regard to these activity-dependent mechanisms of spine formation is whether they ultimately result in the formation of stable, functional synapses and whether this process does contribute to the synaptic enhancement characterizing LTP. The first new spines formed upon LTP induction or other stimulation protocols have mainly been reported 20–30 min after stimulation and their number appeared to then progressively increase over the next few hours (Jourdain et al., 2003; Engert and Bonhoeffer, 1999). This result thus clearly indicates that new protrusions cannot directly account for the synaptic potentiation that is detectable as early as a few minutes after stimulation. This does not exclude however that new spines formed after LTP induction could contribute at later time points to the enhancement of synaptic responses. For this, newly formed spines should become functional synapses and involve similar presynaptic axons as the ones activated during high-frequency stimulation. Our previous 3D EM analysis of stimulated synapses suggested that this could indeed happen at least in some particular cases, since we found examples of duplicated spine synapses, 2 h after LTP induction, where two spines from the same dendrite contacted the same presynaptic partner (Toni et al., 1999). This situation is however not likely to be the rule for two main reasons. In a recent study of spine turnover, we found that a great proportion of newly formed spines are in fact likely to disappear rather quickly, and that their stabilization requires expression of a PSD which usually becomes detectable on new spines only between 5 and 19 h after formation of the protrusion (De Roo et al., 2008). This was also confirmed in a recent study (Nagerl et al., 2007) that reveals that the formation of morphologically functional synapses, as analyzed by 3D EM reconstruction of newly formed spines after LTP induction, is a process that takes time. New spines only a few hours old only rarely formed synapses, while they more consistently expressed features of functional synapses after 15–19 h. This is also in line with previous in vivo data that indicated that a few days were required for the formation of morphologically mature synapses (Knott et al., 2006). Thus all together, these studies strongly suggest that, while synaptogenesis is likely to take place after LTP induction and result in the formation of mature synapses, these synapses are unlikely to account for the synaptic enhancement associated with LTP or at least not within the first hours or days.

Mechanisms of plasticity-induced synaptogenesis

If synaptogenesis and synaptic enhancement are two separate mechanisms induced by the same patterns of activity, an important issue then is to understand how they are regulated. Experiments show that both of them require NMDA receptor activation and depend on calcium influx in the postsynaptic cell. A further link between protrusion growth and potentiation mechanisms concerns the molecular events contributing their regulation. In particular, protein kinases, such as PKC, and also CaMKII, which play a critical role for the functional changes in synaptic strength, appear to regulate protrusion formation (Jourdain et al., 2003; Pilpel and Segal, 2004). It is therefore likely that the same mechanisms that control expression of receptors might also sub-serve changes in protrusion numbers. The downstream effectors of these changes remain however yet unknown.

Another interesting question is whether these new protrusions grow in an undefined manner or whether their growth is more directed towards, for example, pre-existing and possibly activated boutons. An interesting morphological feature which might bring an answer to these questions is the frequent observation under conditions of synaptogenesis of multiple synapse boutons where several spines contact the same terminal. Special cases of multiple synapse boutons in which two spines arising from the same dendrite and contacting the same terminal were found following LTP induction in hippocampal slice cultures (Toni et al., 1999).

Another study has further shown an increase in multiple synapse boutons under in vivo conditions following learning protocols (Geinisman et al., 2001). Finally, studies of spine dynamics in the somatosensory cortex showed that spines initially grow without a synaptic contact and then get in touch with a presynaptic partner which, in many cases, already formed a synapse with another spine (Knott et al., 2006; Toni et al., 2007). This was also the case of the new spines reconstructed after LTP induction (Nagerl et al., 2007). This therefore clearly suggest a directed mechanism where new protrusions probably grow towards existing boutons and there get in competition with the synapses already established by other spines. Accordingly, by stimulating spine growth, LTP mechanisms might in fact promote competition between spines and a process of input selection which might be reflected by the observation of multiple synapse boutons.

Spine turnover and synapse formation mechanisms

An important new aspect of the mechanisms of spine dynamics was brought by experiments of repetitive confocal imaging in living mice. Studies by several groups provided evidence that dendritic spines do undergo some sort of turnover and that there exist a process of continuous growth and elimination of spines (Grutzendler et al., 2002; Trachtenberg et al., 2002). Although there has been some debate about the magnitude of this phenomenon, an issue that may be related to the approach used for imaging live neurons (Xu et al., 2007), current evidence indicates that spine turnover varies greatly during development, affecting as many as 10–15% of spines per 24 h in very young animals. Later on spines become progressively more and more stable with only less than a few percent of spines undergoing replacement in adult tissue (Zuo et al., 2005). Furthermore this process appears to vary in different cortical regions and even show some cell-type specificity (Holtmaat et al., 2005). Finally some data suggest that this spine turnover process could also be affected by sensory activity and thus promote remodeling of synaptic networks (Holtmaat et al., 2006).

This phenomenon of spine turnover is particularly important in young hippocampal slices cultures, a model that is very often used for studying plasticity and learning-related mechanisms. Through repetitive imaging of the same cells over several days in hippocampal slice cultures, we recently found that the rate of spine turnover affects about 20% of all spines in 15-days-old cultures, but only about 10% after 25 days in vitro (De Roo et al., 2008). These values, which are quite close to those obtained in very young animals, suggest that spine turnover retains similar properties in vitro than in vivo and particularly its developmental dependency. Another interesting aspect was that this rate of spine turnover is actually underestimated by the use of long observation intervals. It turns out that in 15 DIV slice cultures, the basal rate of protrusion formation reaches values in the order of 2% of all spines per hour, which represents hundreds of new protrusions per day and per neuron. The reason why spine density remains nevertheless stable is that most of these new protrusions do in fact disappear fairly quickly and only a small proportion of them become stable spine synapses (Fig. 1). Curiously, we found, in agreement with other in vivo studies, that filopodia, usually considered as precursors of spine synapses, only exceptionally lead to formation of stable contacts (Zuo et al., 2005). Protrusions are thus generated at a high rate, but only a fraction of them become finally stable spines. We also found that this required a process of maturation that lasted about 24 h. During this period, new spines usually grew in size and started to express a PSD after about 5 h (Fig. 1) (Knott et al., 2006; De Roo et al., 2008). Interestingly, expression of this PSD was activity-dependent, since blockade of AMPA and NMDA receptors prevented its expression and reduced the probability of the spine to be stabilized. All together these experiment suggested a model in which development of synaptic networks proceed through an extensive, non specific growth of dendritic protrusions, followed by the stabilization of a small number of spines. This appears to involve the expression of a PSD through mechanisms that are driven by synaptic activity.

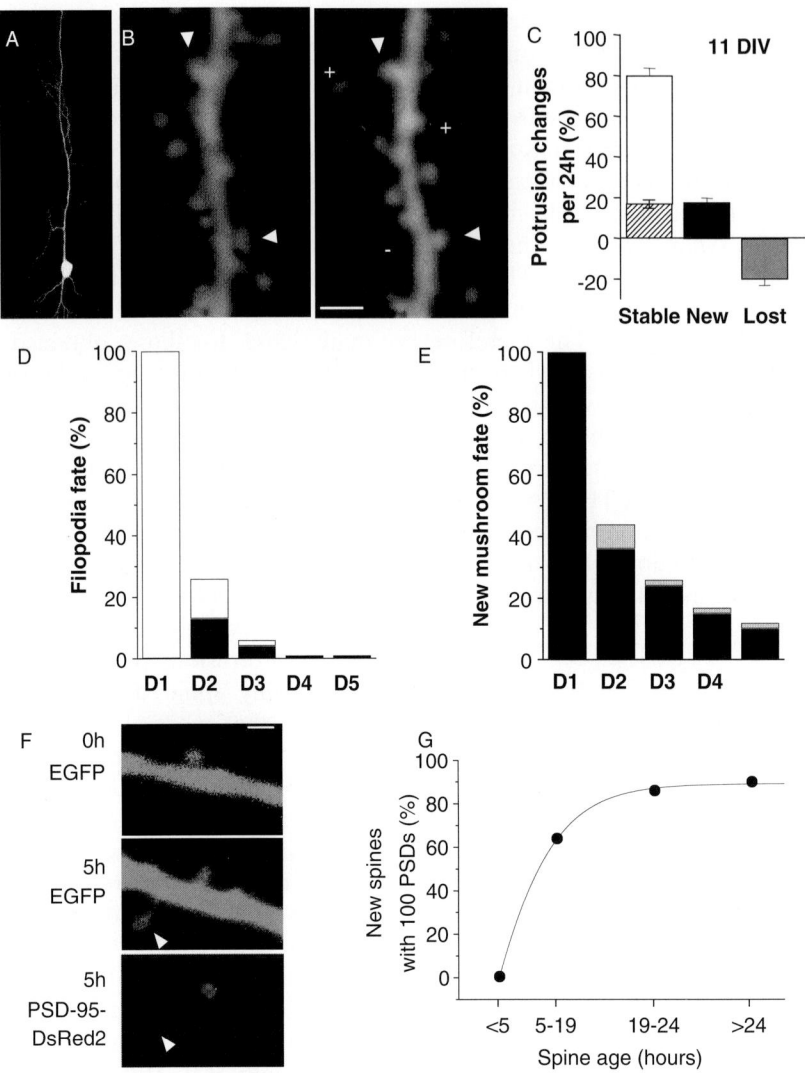

Fig. 1. Spine dynamics in hippocampal organotypic slice cultures. (A) EGFP transfected CA1 pyramidal neuron. (B) Repetitive imaging of a dendritic segment at 24 h interval reveals the occurrence of new and lost protrusions. (C) Summary of the proportion of stable spines (open column), which include spines exhibiting changes in morphology (dashed column), of newly formed (black column) and disappearing (gray column) spines. (D) Stability over 5 days of newly formed filopodia. Note that most of them disappear within 1–2 days and only exceptionally lead to the formation of a stable spine. (E) Stability of newly formed spines. (F) Illustration of a newly formed spine (age < 5 h; arrow head, middle panel) which do not express PSD-95-DsRed2 (arrow head, lower panel). (G) Time course of PSD-95-DsRed2 expression in newly formed spines. (See Color Plate 11.1 in color plate section.)

Memory and synapse formation mechanisms

The interesting implication of these new data is that in young cortical structures, development of synaptic contacts on a given neuron occurs through a mechanism of trial and error where most of generated protrusions are in fact rapidly eliminated. In this process a factor that seems to be important to stabilize new protrusions is the presence of an active terminal in the vicinity. The glutamate so released could drive the expression of a PSD on the new spine and thus promote its stabilization. Overall, these experiments also point to the rather low level of stability and high rate of replacement of dendritic spines in young tissue. This clearly raises issues with regard to mechanisms of learning and memory under such situations and emphasizes the likely importance of synapse remodeling for these processes. Indeed in preliminary work, we found that an important effect of LTP induction in slice cultures is to enhance the basal rate of protrusion formation over a period of several days, leading to an increased and lasting competition between spines. These observations thus clearly point to the role of spine pruning and replacement as a major mechanism susceptible to affect the organization of synaptic networks during development. LTP in this context might have an important function by promoting and enhancing the phenomenon. Theoretical work does indeed suggest that spine remodeling could represent an extremely powerful mechanism for processing and storing information (Mel, 2002; Chklovskii et al., 2004).

Thus, while Hebbian mechanisms affecting synaptic efficacy certainly represent a powerful tool for quickly modifying signal integration in a given network, it is not unlikely that the long-term maintenance of memory traces could also depend on network reorganization mechanisms associated to synaptic plasticity (Feldman and Brecht, 2005). In this sense, changes in synaptic function and remodeling of the network could represent two different, but inter-related aspects of the mechanisms underlying learning and memory formation.

Conclusion

Since the discovery that LTP induction in hippocampus not only affects synaptic efficacy but also promotes growth of new protrusions, evidence has continued to accumulate indicating that spines are indeed dynamic structures that may undergo continuous replacement throughout life. Although it appears that these dynamic properties are mainly expressed in young developing cortex, a capacity for structural plasticity is clearly retained in adult animals (Grutzendler et al., 2002; Trachtenberg et al., 2002; Holtmaat et al., 2005; Majewska et al., 2006). Although only a few percent of total spines seem to be regularly replaced in older tissue, this still represents a significant capacity for reorganization of synaptic networks. Furthermore these properties might be reactivated under specific circumstances where plasticity is required, for example, following ischemic damage or neurodegenerative diseases (Grutzendler and Gan, 2006). Accordingly the capacity of LTP to promote spine growth and synapse remodeling during several days following stimulation might in fact reveal a key mechanism for the acquisition and then long-term maintenance of new information, providing a new mechanistic link between forms of neuronal activity and the development and plasticity of synaptic circuits. Thus in addition to the Hebbian concept of changes in synaptic strength, it becomes more and more apparent that activity-dependent remodeling of synaptic networks represents another major aspect of the changes induced by specific patterns of neuronal activity.

Acknowledgment

This work was supported by the Swiss Science Foundation and the European project Promemoria.

References

Bagal, A.A., Kao, J.P., Tang, C.M. and Thompson, S.M. (2005) Long-term potentiation of exogenous glutamate responses at single dendritic spines. Proc. Natl. Acad. Sci. U.S.A., 40: 14434–14439.

Bliss, T.V. and Collingridge, G.L. (1993) A synaptic model of memory: long-term potentiation in the hippocampus. Nature, 6407: 31–39.

Buchs, P.A. and Muller, D. (1996) Induction of long-term potentiation is associated with major ultrastructural changes of activated synapses. Proc. Natl. Acad. Sci. U.S.A., 15: 8040–8045.

Calverley, R.K. and Jones, D.G. (1990) Contributions of dendritic spines and perforated synapses to synaptic plasticity. Brain Res. Brain Res. Rev., 3: 215–249.

Carlin, R.K. and Siekevitz, P. (1983) Plasticity in the central nervous system: do synapses divide? Proc. Natl. Acad. Sci. U.S.A., 11: 3517–3521.

Chang, F.L. and Greenough, W.T. (1984) Transient and enduring morphological correlates of synaptic activity and efficacy change in the rat hippocampal slice. Brain Res., 1: 35–46.

Chang, P.L., Isaacs, K.R. and Greenough, W.T. (1991) Synapse formation occurs in association with the induction of long-term potentiation in two-year-old rat hippocampus in vitro. Neurobiol. Aging, 5: 517–522.

Chklovskii, D.B., Mel, B.W. and Svoboda, K. (2004) Cortical rewiring and information storage. Nature, 7010: 782–788.

Collin, C., Miyaguchi, K. and Segal, M. (1997) Dendritic spine density and LTP induction in cultured hippocampal slices. J. Neurophysiol., 3: 1614–1623.

Cooke, S.F. and Bliss, T.V. (2006) Plasticity in the human central nervous system. Brain, Pt. 7: 1659–1673.

Cui, Z., Wang, H., Tan, Y., Zaia, K.A., Zhang, S. and Tsien, J.Z. (2004) Inducible and reversible NR1 knockout reveals crucial role of the NMDA receptor in preserving remote memories in the brain. Neuron, 5: 781–793.

De Roo, M., Klauser, P., Mendez, P., Poglia, L. and Muller, D. (2008) Activity-dependent PSD formation and stabilization of newly formed spines in hippocampal slice cultures. Cereb. Cortex, 18: 151–161.

Desmond, N.L. and Levy, W.B. (1986) Changes in the numerical density of synaptic contacts with long-term potentiation in the hippocampal dentate gyrus. J. Comp. Neurol., 4: 466–475.

Engert, F. and Bonhoeffer, T. (1999) Dendritic spine changes associated with hippocampal long-term synaptic plasticity. Nature, 6731: 66–70.

Feldman, D.E. and Brecht, M. (2005) Map plasticity in somatosensory cortex. Science, 5749: 810–815.

Fifkova, E. and Anderson, C.L. (1981) Stimulation-induced changes in dimensions of stalks of dendritic spines in the dentate molecular layer. Exp. Neurol., 2: 621–627.

Geinisman, Y. (1993) Perforated axospinous synapses with multiple, completely partitioned transmission zones: probable structural intermediates in synaptic plasticity. Hippocampus, 4: 417–433.

Geinisman, Y., Berry, R.W., Disterhoft, J.F., Power, J.M. and Van der Zee, E.A. (2001) Associative learning elicits the formation of multiple-synapse boutons. J. Neurosci., 15: 5568–5573.

Geinisman, Y., de Toledo-Morrell, L., Morrell, F., Heller, R.E., Rossi, M. and Parshall, R.F. (1993) Structural synaptic correlate of long-term potentiation: formation of axospinous synapses with multiple, completely partitioned transmission zones. Hippocampus, 4: 435–445.

Greenough, W.T., West, R.W. and DeVoogd, T.J. (1978) Subsynaptic plate perforations: changes with age and experience in the rat. Science, 4372: 1096–1098.

Grutzendler, J. and Gan, W.B. (2006) Two-photon imaging of synaptic plasticity and pathology in the living mouse brain. NeuroRx, 4: 489–496.

Grutzendler, J., Kasthuri, N. and Gan, W.B. (2002) Long-term dendritic spine stability in the adult cortex. Nature, 6917: 812–816.

Harris, K.M., Jensen, F.E. and Tsao, B. (1992) Three-dimensional structure of dendritic spines and synapses in rat hippocampus (CA1) at postnatal day 15 and adult ages: implications for the maturation of synaptic physiology and long-term potentiation. J. Neurosci., 7: 2685–2705.

Harris, K.M. and Stevens, J.K. (1989) Dendritic spines of CA 1 pyramidal cells in the rat hippocampus: serial electron microscopy with reference to their biophysical characteristics. J. Neurosci., 8: 2982–2997.

Holtmaat, A., Wilbrecht, L., Knott, G.W., Welker, E. and Svoboda, K. (2006) Experience-dependent and cell-type-specific spine growth in the neocortex. Nature, 7096: 979–983.

Holtmaat, A.J., Trachtenberg, J.T., Wilbrecht, L., Shepherd, G.M., Zhang, X., Knott, G.W. and Svoboda, K. (2005) Transient and persistent dendritic spines in the neocortex in vivo. Neuron, 2: 279–291.

Jourdain, P., Fukunaga, K. and Muller, D. (2003) Calcium/calmodulin-dependent protein kinase II contributes to activity-dependent filopodia growth and spine formation. J. Neurosci., 33: 10645–10649.

Knott, G.W., Holtmaat, A., Wilbrecht, L., Welker, E. and Svoboda, K. (2006) Spine growth precedes synapse formation in the adult neocortex in vivo. Nat. Neurosci., 9: 1117–1124.

Kopec, C.D., Li, B., Wei, W., Boehm, J. and Malinow, R. (2006) Glutamate receptor exocytosis and spine enlargement during chemically induced long-term potentiation. J. Neurosci., 7: 2000–2009.

Korkotian, E. and Segal, M. (2001) Spike-associated fast contraction of dendritic spines in cultured hippocampal neurons. Neuron, 3: 751–758.

Lang, C., Barco, A., Zablow, L., Kandel, E.R., Siegelbaum, S.A. and Zakharenko, S.S. (2004) Transient expansion of synaptically connected dendritic spines upon induction of hippocampal long-term potentiation. Proc. Natl. Acad. Sci. U.S.A., 47: 16665–16670.

Lee, K.S., Schottler, F., Oliver, M. and Lynch, G. (1980) Brief bursts of high-frequency stimulation produce two types of structural change in rat hippocampus. J. Neurophysiol., 2: 247–258.

Lisman, J., Schulman, H. and Cline, H. (2002) The molecular basis of CaMKII function in synaptic and behavioural memory. Nat. Rev. Neurosci., 3: 175–190.

Majewska, A.K., Newton, J.R. and Sur, M. (2006) Remodeling of synaptic structure in sensory cortical areas in vivo. J. Neurosci., 11: 3021–3029.

Maletic-Savatic, M., Malinow, R. and Svoboda, K. (1999) Rapid dendritic morphogenesis in CA1 hippocampal dendrites induced by synaptic activity. Science, 5409: 1923–1927.

Malinow, R. and Malenka, R.C. (2002) AMPA receptor trafficking and synaptic plasticity. Annu. Rev. Neurosci., 103–126.

Matsuzaki, M., Honkura, N., Ellis-Davies, G.C. and Kasai, H. (2004) Structural basis of long-term potentiation in single dendritic spines. Nature, 6993: 761–766.

Matus, A. (2000) Actin-based plasticity in dendritic spines. Science, 5492: 754–758.

Mel, B.W. (2002) Have we been hebbing down the wrong path? Neuron, 2: 175–177.

Nagerl, U.V., Eberhorn, N., Cambridge, S.B. and Bonhoeffer, T. (2004) Bidirectional activity-dependent morphological plasticity in hippocampal neurons. Neuron, 5: 759–767.

Nagerl, U.V., Kostinger, G., Anderson, J.C., Martin, K.A. and Bonhoeffer, T. (2007) Protracted synaptogenesis after activity-dependent spinogenesis in hippocampal neurons. J. Neurosci., 30: 8149–8156.

Nicoll, R.A. (2003) Expression mechanisms underlying long-term potentiation: a postsynaptic view. Philos. Trans. R. Soc. Lond. B Biol. Sci., 1432: 721–726.

Nieto-Sampedro, M., Hoff, S.F. and Cotman, C.W. (1982) Perforated postsynaptic densities: probable intermediates in synapse turnover. Proc. Natl. Acad. Sci. U.S.A., 18: 5718–5722.

Nusser, Z., Lujan, R., Laube, G., Roberts, J.D., Molnar, E. and Somogyi, P. (1998) Cell type and pathway dependence of synaptic AMPA receptor number and variability in the hippocampus. Neuron, 3: 545–559.

Pilpel, Y. and Segal, M. (2004) Activation of PKC induces rapid morphological plasticity in dendrites of hippocampal neurons via Rac and Rho-dependent mechanisms. Eur. J. Neurosci., 12: 3151–3164.

Stewart, M.G., Medvedev, N.I., Popov, V.I., Schoepfer, R., Davies, H.A., Murphy, K., Dallerac, G.M., Kraev, I.V. and Rodriguez, J.J. (2005) Chemically induced long-term potentiation increases the number of perforated and complex postsynaptic densities but does not alter dendritic spine volume in CA1 of adult mouse hippocampal slices. Eur. J. Neurosci., 12: 3368–3378.

Tang, Y.P., Shimizu, E., Dube, G.R., Rampon, C., Kerchner, G.A., Zhuo, M., Liu, G. and Tsien, J.Z. (1999) Genetic enhancement of learning and memory in mice. Nature, 6748: 63–69.

Toni, N., Buchs, P.A., Nikonenko, I., Bron, C.R. and Muller, D. (1999) LTP promotes formation of multiple spine synapses between a single axon terminal and a dendrite. Nature, 6760: 421–425.

Toni, N., Buchs, P.A., Nikonenko, I., Povilaitite, P., Parisi, L. and Muller, D. (2001) Remodeling of synaptic membranes after induction of long-term potentiation. J. Neurosci., 16: 6245–6251.

Toni, N., Teng, E.M., Bushong, E.A., Aimone, J.B., Zhao, C., Consiglio, A., van Praag, H., Martone, M.E., Ellisman, M.H. and Gage, F.H. (2007) Synapse formation on neurons born in the adult hippocampus. Nat. Neurosci., 6: 727–734.

Trachtenberg, J.T., Chen, B.E., Knott, G.W., Feng, G., Sanes, J.R., Welker, E. and Svoboda, K. (2002) Long-term in vivo imaging of experience-dependent synaptic plasticity in adult cortex. Nature, 6917: 788–794.

Trommald, M., Hulleberg, G. and Andersen, P. (1996) Long-term potentiation is associated with new excitatory spine synapses on rat dentate granule cells. Learn. Mem., 2–3: 218–228.

Whitlock, J.R., Heynen, A.J., Shuler, M.G. and Bear, M.F. (2006) Learning induces long-term potentiation in the hippocampus. Science, 5790: 1093–1097.

Xu, H.T., Pan, F., Yang, G. and Gan, W.B. (2007) Choice of cranial window type for in vivo imaging affects dendritic spine turnover in the cortex. Nat. Neurosci., 5: 549–551.

Yuste, R. and Bonhoeffer, T. (2001) Morphological changes in dendritic spines associated with long-term synaptic plasticity. Annu. Rev. Neurosci., 24: 1071–1089.

Zuo, Y., Lin, A., Chang, P. and Gan, W.B. (2005) Development of long-term dendritic spine stability in diverse regions of cerebral cortex. Neuron, 2: 181–189.

SECTION II

Systems Approaches to the Essence of Memory

… CHAPTER 12

The age of plasticity: developmental regulation of synaptic plasticity in neocortical microcircuits

Arianna Maffei and Gina Turrigiano*

Department of Biology and Center for Behavioral Genomics, Brandeis University, Waltham, MA 02454, USA

Abstract: Proper wiring of neural circuits during development depends on both molecular cues that guide connectivity and activity-dependent mechanisms that use patterned activation to adjust the strength and number of synaptic connections. In this chapter, we discuss some of the plasticity mechanisms underlying the experience-dependent rewiring of visual cortical microcircuits focusing on layer 4 of rat primary visual cortex. The microcircuit in layer 4 has the ability to regulate its excitability by shifting the balance between excitatory and inhibitory synaptic transmission in an experience-dependent manner. Early in postnatal development (shortly after eye opening), visual deprivation activates several forms of homeostatic plasticity that cooperate to adjust layer 4 excitability to compensate for reduced sensory drive. In contrast, during the classical sensitive period for rodent visual system plasticity, this homeostatic response is replaced by mechanisms that reduce the responsiveness of deprived cortex. We discuss this developmentally regulated switch in plasticity within layer 4 and how this might depend on the maturation of excitatory and inhibitory monosynaptic connections. Based on our published data, we propose inhibitory plasticity as an important player in circuit refinement that can contribute both to the compensatory forms of circuit plasticity in the early stages of development and to the pathological loss of function induced by visual deprivation during the critical period.

Keywords: critical period; synaptic plasticity; homeostatic; inhibitory interneurons; visual cortex

It is known that neocortical circuits in the rodent undergo a major period of maturation in the few weeks after birth. Circuit formation initially occurs under control of molecular cues (Tessier-Lavigne and Goodman, 1996; Charron and Tessier-Lavigne, 2005) which guide migration of the various neuronal types to the correct layers (Nadarajah et al., 2001, 2003), and also guide axons to form synaptic contacts with the correct targets (Goodman and Shatz, 1993; Bishop et al., 2000, 2003). During this phase a very large number of synapses are formed (Rakic et al., 1986), some of which are active very early in development (Rumpel et al., 1998). The initial patterns of connectivity that are set up through these molecular processes are not fully functional, but must be refined in a use-dependent manner. This process of refinement and maturation is activity-dependent and is thought to be shaped in large part by changes in synaptic strength and connectivity (Katz and Shatz, 1996; Catalano and Shatz, 1998; Inan and Crair, 2007). Many forms of

*Corresponding author. Tel.: +1-781-736-2684; Fax: +1-781-736-3107; E-mail: Turrigiano@brandeis.edu

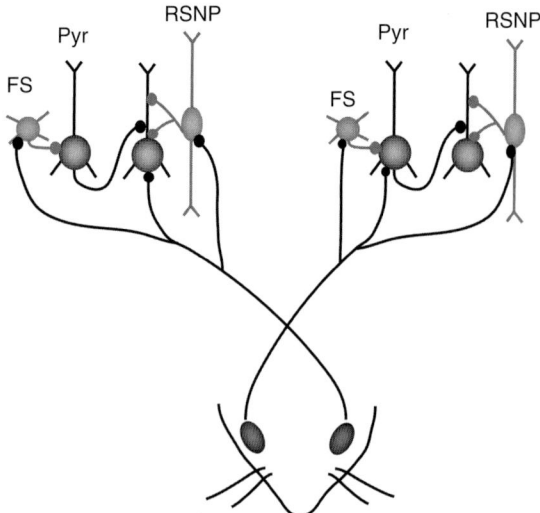

Fig. 1. Layer 4 in rat primary visual cortex. The sketch represents the neuron types and their synaptic connections in rat primary visual cortex. Pyr, star pyramidal neurons; FS, small GABAergic basket cells with multipolar morphology; RSNP, bipolar GABAergic neurons whose firing shows frequency adaptation in response to depolarizing current steps. The extent of the direct thalamic drive onto both types of inhibitory neurons is unknown; however, both are likely to receive feedforward input from thalamic afferents (Hajós et al., 1997; Sun et al., 2006).

plasticity are expressed at cortical synapses (Turrigiano et al., 1998; Maffei et al., 2004, 2006; Malenka and Bear, 2004), but we still know little about when and how these mechanisms are engaged during development or in response to particular forms of sensory experience. Here we will review the development and plasticity of synaptic transmission between excitatory and inhibitory cortical neurons, focusing on recent work from our lab on the microcircuitry in layer 4 of rat primary visual cortex (Fig. 1).

Plasticity and circuit excitability

One of the main principles of circuit wiring is captured by the phrase "cells that fire together wire together," meaning that synapses between excitatory neurons whose firing is correlated will become stronger and those between neurons whose firing is uncorrelated will become weaker (Brown et al., 1990). This principle leads to the maintenance of strong inputs and the loss of weak inputs. These correlation-based forms of synaptic plasticity are known as "Hebbian" after Donald Hebb, and include several forms of long-term potentiation (LTP) and depression (LTD). Besides being involved in circuit wiring and maturation, these long-lasting modifications of synaptic strength are thought to be a major substrate for learning and memory (Palm, 1982; Malenka and Bear, 2004).

It has been suggested that LTP and LTD might be destabilizing for the neural circuit because increased synaptic strength between two excitatory neurons will increase the correlation between them, which will further strengthen the connection in a positive feedback cycle. On the other hand, if LTD is induced, the correlation between two neurons will be reduced and drive a further reduction in synaptic strength, potentially leading to loss of synaptic contacts. In the absence of mechanisms designed to stabilize network excitability, LTP can thus initiate a positive feedback loop that will lead to excessive excitability (Abbott and Nelson, 2000; Turrigiano and Nelson, 2000). Under these conditions, neurons are bound to respond strongly even to spontaneous background activity and will become unable to discriminate meaningful differences in the structure of incoming signals. Despite the need for Hebbian synaptic modification for learning, memory, and circuit wiring, it is very important that the network "gain" (the relationship between input and output magnitude) is maintained. Homeostatic forms of plasticity have been proposed to contribute to neural circuit stability and function by endowing neurons with the ability to readjust their excitability globally in response to changes in network activity (Turrigiano, 1999). Homeostatic changes are thought to preserve the relative differences in strength between excitatory synaptic inputs previously induced by Hebbian plasticity, but allow neurons to bring their excitability back into an optimal working range, thus maintaining their ability to discriminate fine differences in the structure of incoming signals (Turrigiano and Nelson, 2004). Synaptic scaling is one of these forms of plasticity. It globally scales up or down the strength of excitatory synapses onto one neuron by the same multiplicative factor depending on the level of network activity (Turrigiano et al., 1998; Watt et al., 2000). It has been shown to act by regulating the number of glutamatergic receptors on the postsynaptic terminals while not affecting the probability of neurotransmitter release (Turrigiano et al., 1998; Wierenga et al., 2005). Neurons have also been shown to adjust their excitability by modulating their voltage-dependent conductance (Desai et al., 1999; Golowasch et al., 1999; Desai, 2003).

All of these forms of plasticity have been shown to occur at excitatory synapses; however, neuronal circuits also contain inhibitory interneurons with a variety of morphologies and firing properties (Kawaguchi and Kubota, 1997). Evidence is now accumulating for their role in circuit development (Hollrigel et al., 1998; Cohen et al., 2001) and for inhibitory synapses being plastic (Komatsu and Iwakiri, 1993; Holmgren and Zilberter, 2001; Maffei et al., 2006), properties that would make them important contributors to circuit refinement and function. The combination of changes in synaptic strength and excitability for both excitatory and inhibitory neurons would endow neural networks with a wide variety of strategies for information storage and experience-dependent rewiring.

Plasticity in the visual system

Visual cortex is one of the best-studied models for activity-dependent circuit wiring and experience-dependent plasticity. Most mammals are born with their eyes closed, and therefore the initial activity-dependent wiring events are driven by spontaneous retinal activity (Del Rio and Feller, 2006). After eye opening, however, patterned visual input is thought to be the main drive leading to maturation of the visual cortical circuit (Tagawa et al., 2005; Smith and Trachtenberg, 2007). Impairment of visual experience for long periods of time can retard the maturation of visual cortical circuitry and lead to irreversible loss of function (Blasdel and Pettigrew, 1978). However, the forms of plasticity that contribute to these changes are not fully understood. Both Hebbian and homeostatic forms of plasticity have been implicated in the changes induced by altered visual experience during development (Rittenhouse et al., 1999; Desai et al., 2002). Most of what is known about the effect of experience in visual cortex comes from experiments measuring neuronal firing in response to visual stimulation in anesthetized animals which were deprived of visual input by monocular deprivation (MD) during a developmental period known as the "critical period" for visual cortical plasticity. The critical period is a time when neurons in binocular cortex show a shift in ocular dominance toward the open eye in response to deprivation of the other eye (Hubel and Wiesel, 1970). The shift is thought to occur because the neurons activated by the deprived eye respond to visual flashes with lower firing rates than those driven by the open eye (Mioche and Singer, 1989; Frenkel and Bear, 2004). The interpretation proposed to explain this loss of visual responsiveness is that monosynaptic LTD had been induced at thalamocortical and intracortical excitatory synapses onto neurons activated by the deprived eye (Rittenhouse et al., 1999; Crozier et al., 2007). Such reduction of synaptic strength would result in lower firing rates and decreased visual responsiveness (Mioche and Singer 1989; Frenkel and Bear 2004).

The critical period has traditionally been considered to be the moment in cortical development when synapses are most plastic and are most strongly modified by experience (Hubel and Wiesel, 1970). Recently published data has provided new evidence for the ability of visual cortical synapses to be very plastic also during what is often referred to as the "pre-critical" period, which starts right at eye opening (in the rat at p14) (Maffei et al., 2004; Feller and Scanziani, 2005; Smith and Trachtenberg, 2007). The experience-dependent rewiring during these stages involve quite different sets of synaptic modifications and raise a new series of questions about developmental regulation of multiple forms of synaptic plasticity and their role in cortical function (Maffei et al., 2004, 2006).

Experience-dependent plasticity in the absence of competition

The majority of studies on visual cortical plasticity were designed to investigate the development of ocular dominance columns and their plasticity in binocular cortex, where inputs from the two eyes compete with each other during development (Hubel and Wiesel, 1970; Rittenhouse et al., 1999; Smith and Trachtenberg, 2007). Circuit refinement in binocular cortex likely involves both competitive and non-competitive synaptic plasticity mechanisms, but it is difficult to tease apart their relative contribution. The effects of experience on microcircuits in monocular cortex of rodents, where competitive mechanisms are less pronounced, have recently been undertaken by us (Maffei et al., 2004, 2006). Our analysis is focused on the effect of development and sensory deprivation at excitatory and inhibitory synapses in the monocular region of rat visual cortex. Rat primary visual cortex is composed of a small binocular visual area and a much larger monocular portion driven only by the contralateral eye. MD affects only the circuit contralateral to the closed eye, leaving the other hemisphere unaffected. This is the ideal preparation to directly compare deprived and control hemispheres within the same animal, and to probe the contribution of non-competitive plasticity mechanisms in circuit refinement.

One of our major findings was that sensory deprivation induces dramatic microcircuit rewiring in monocular cortex, leading to pronounced changes in network excitability. This is achieved by readjusting the balance of excitation and inhibition in the deprived microcircuit through both Hebbian and homeostatic forms of plasticity. The effects of visual experience on the cortical circuit are very complex and developmentally regulated: different forms of plasticity were induced depending on the age of the animal at the time of deprivation (Maffei et al., 2004, 2006). Below, we discuss in turn the changes induced by visual deprivation during the pre-critical period (between eye opening and postnatal day 18) and during the critical period (after about postnatal day 19).

Pre-critical period plasticity within layer 4

To study the effect of experience-dependent plasticity during the pre-critical period, we sutured one eye (MD) between p14 (just before eye opening) and p17 (before the opening of the classical critical period), and obtained patch clamp recordings from coronal slices containing monocular cortex (data are summarized in Figs. 2 and 3).

To determine the specific synapses involved in regulating the balance between excitation and inhibition, we performed paired recordings between visually identified neurons. Four neurons were recorded simultaneously and the presence of synaptic connections between them was assessed by firing putative presynaptic neurons with precisely timed depolarizing steps in current clamp and recording the putative postsynaptic neurons in voltage clamp. Quadruple patch clamp allows us to test 12 putative connections at a time, increasing the probability of finding connected pairs. Connection probability was calculated measuring the ratio of found versus tested pairs (Figs. 2 and 3; CP). In addition, we measured the probability that two neurons were reciprocally connected by measuring the ration of reciprocally connected pairs versus the total number of pairs we found for any specific connection tested (Figs. 2 and 3; RCP).

MD induced a doubling in amplitude and connection probability for monosynaptic connections between star pyramidal neurons (Fig. 2b; p17, EPSC; layer 4 pyramidal neurons in V1 are named star pyramids because of their lack of complex arborization of their apical dendrite). Detailed analysis of synaptic responses showed no significant changes in the coefficient of variation (Fig. 2c; p17, CV) and failure rates (Fig. 2c; p17, % failure), parameters that have been shown to correlate with changes in release probability at the presynaptic terminal, suggesting that MD induces increased EPSC amplitude mostly through changes at the postsynaptic terminals. A small, but significant change was also observed in the steady-state short-term depression of EPSC amplitude in response to trains of presynaptic spikes at 20 Hz, suggesting a small presynaptic change that would not be enough to explain the large change in EPSC amplitude (Maffei et al., 2004; Fig. 2c; p17, SS STD). The data are consistent with previous published results showing that MD induces synaptic scaling of miniature EPSC (mEPSC) in

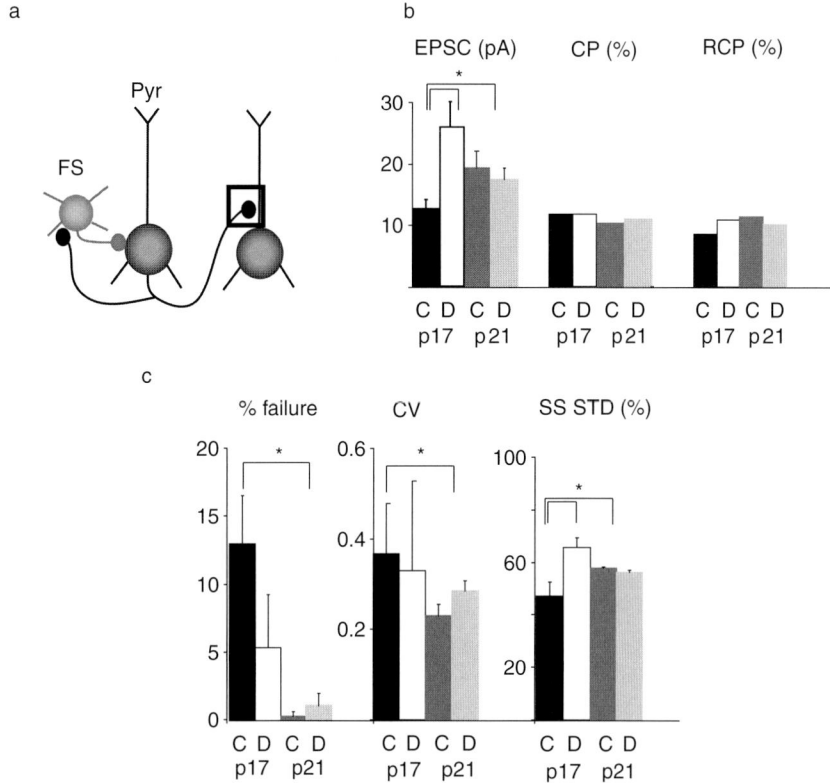

Fig. 2. Development and plasticity of star pyramidal connections (diagrammed in a). The box indicates the synapse examined in the rest of the plots. (b) Effect of development and experience-dependent plasticity on amplitude (EPSC), connection probability (CP), and recurrent connectivity (RCP) for monosynaptic connections between star pyramidal neurons in control (C) and deprived (D) conditions at p17 (black bar for control: C; white bar for deprived: D) and at p21 (dark gray bar for control: C; light gray bar for deprived: D). (c) Bar plot showing effects of development and monocular deprivation on failure rates (% failure), coefficient of variation (CV), and steady-state short-term depression (SS STD) of EPSC from connected pairs of star pyramids [p17 (black bars for control: C; white bars for deprived: D); p21 (dark gray bars for control: C; light gray bars for deprived: D)]. Data are presented as average and standard error. The asterisks mark significant changes.

star pyramids from layer 4 during the pre-critical period (Desai et al., 2002).

Changes in inhibitory synaptic transmission onto star pyramids also contributed to the shift in the excitation/inhibition balance (Fig. 3; p17). Two separate populations of inhibitory neurons provide most inhibitory drive in layer 4 and they both were affected by MD. One of them is composed of regular-spiking non-pyramidal neurons (RSNP; see Fig. 1) which show regular firing pattern in response to depolarizing current steps and have bipolar morphology. The global effect of MD on these inhibitory synapses was to maintain the overall level of inhibition by reducing their connection probability onto star pyramids to half of control and increasing the strength of inhibitory postsynaptic currents without changes in failure rates and coefficient of variation (CV) (Maffei et al., 2004). A very different set of changes was observed when we measured strength and synaptic properties for fast-spiking (FS) inhibitory interneurons (Fig. 3a–c; p17). These interneurons have multipolar basket-like morphology and show fast non-adapting firing in response to depolarizing current steps. The strength of their IPSC onto star pyramids was significantly reduced by MD with no changes in connection probability (Fig. 3b; p17, IPSC, CP). CV and failure rates were significantly

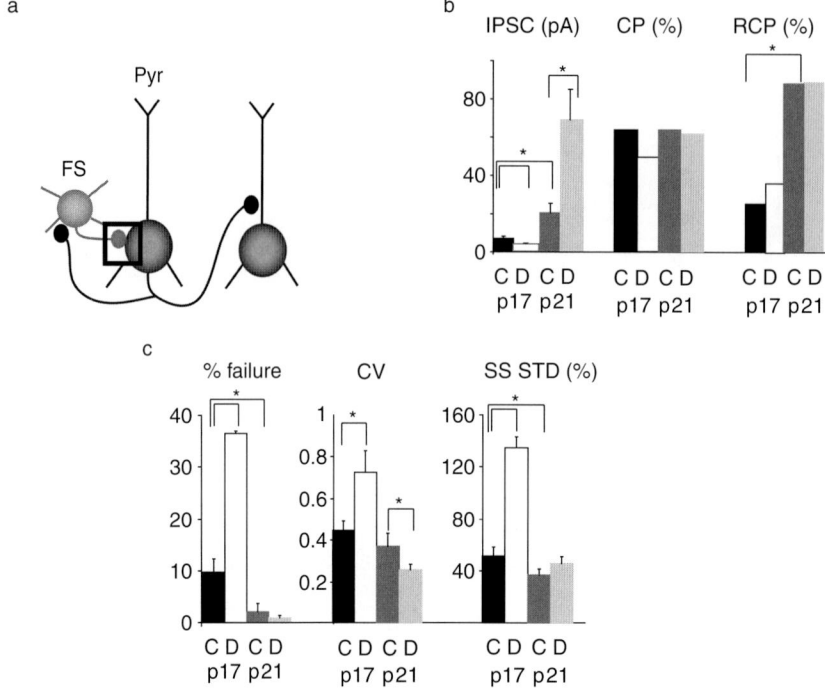

Fig. 3. Development and plasticity of inhibitory synapses between fast-spiking (FS) and star pyramidal neurons (diagrammed in a). (b) Data summarizing developmental and experience-dependent changes for amplitude (IPSC), connection probability (CP), and recurrent connectivity (RCP) for FS to star pyramidal neurons connections in control (C, black bars) and deprived (D, white bars) conditions at p17 and p21 (dark gray bar for control: C; light gray bar for deprived: D). (c) Bar plot showing changes in failure rates (% failure), coefficient of variation (CV), and steady-state short-term depression (SS STD) for IPSC measured from FS to star pyramids connected pairs [p17 (black bars for control: C; white bars for deprived: D); p21 (dark gray bars for control: C; light gray bars for deprived: D)]. Data are presented as average and standard error. The asterisks mark significant changes.

increased, suggesting a decrease in presynaptic release probability as the main mechanism for the reduction in IPSC amplitude (Fig. 3c; p17, CV, % failure). One striking effect of MD at these synapses was a switch in short-term plasticity: they showed depression in response to trains of presynaptic spikes at 20 Hz in the control hemisphere, but became strongly facilitating in response to the same trains of stimulation in the deprived hemisphere so that the steady-state IPSC amplitude in the train was not significantly different between control and deprived conditions (Maffei et al., 2004; Fig. 3c; p17, SS STD). One possible functional consequence of these changes is that, while at low levels of network excitability weak inhibition would boost the excitability of the layer 4 microcircuit, when network excitability becomes too high a facilitating FS to star pyramid IPSC will recruit additional inhibition and prevent runaway excitation from completely destabilizing it. The global effect of the deprivation on the layer 4 circuit at this developmental stage was to increase the excitability of the local microcircuit. This occurred through a shift in the balance between excitatory and inhibitory inputs onto star pyramids to favor excitation, therefore increasing spontaneous firing rates (Maffei et al., 2004; Fig. 4; left panel).

The complex set of changes we observed in layer 4 microcircuit appears to be a compensatory (or homeostatic) response. It contributes to maintaining network function in the face of reduced sensory drive and is mediated by the induction of specific forms of plasticity at different synapse depending on the role each of them plays within the microcircuit. In our experimental condition,

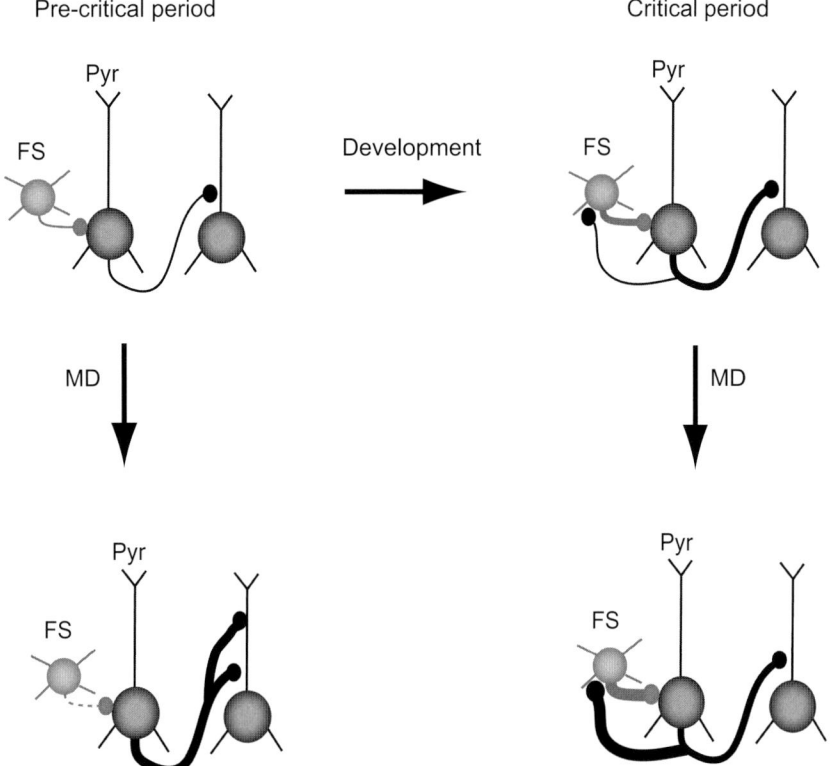

Fig. 4. Summary showing the significant changes induced by development and experience-dependent plasticity at excitatory and inhibitory synapses composing the cortical microcircuit in layer 4 of rat primary visual cortex.

neurons from the deprived hemisphere have never been primed by visual stimulation because MD began before eye opening. Homeostatic plasticity might allow the layer 4 microcircuit to remain in a functional, "receptive" state while waiting to be activated by sensory input.

An alternative interpretation proposed to explain the effects of deprivation, which begun before eye opening, is that visual deprivation prevents the maturation of both excitatory and inhibitory circuits in the deprived hemisphere, maintaining it in an underdeveloped state (Blasdel and Pettigrew, 1978; Bartoletti et al., 2004). We will discuss this hypothesis in the next paragraphs where new evidence for developmental changes in layer 4 microcircuit will be presented together with experience-dependent changes. What remains clear is that contrary to what has traditionally been thought, the pre-critical period is a very plastic phase of development when the basis for proper circuit wiring and plasticity are being set. Furthermore, experience dramatically affects circuit wiring and plasticity even in the absence of competition between inputs from the two eyes, suggesting that some of the early events of sensory-dependent rewiring might be needed to adjust the balance between excitation and inhibition of the deprived neurons so that the events triggered by competition may take place.

Developmental changes in synaptic properties within cortical layer 4

Most of the published data on experience-dependent visual system plasticity has found a loss of visual responsiveness in vivo in neurons activated by the deprived eye. This is in apparent contrast with our data on early deprivation in layer 4, where MD increases excitability and thus would be expected to increase visual responsiveness. This discrepancy is likely because most visual deprivation studies have been conducted later in life, during the classical critical period for ocular dominance plasticity, whereas our experiments were performed during the pre-critical period (Mioche and Singer, 1989; Frenkel and Bear, 2004; Maffei et al., 2004). If there are developmentally regulated changes in connectivity, synaptic strength, and synaptic plasticity within the microcircuit in layer 4, then visual deprivation may have very different effects depending on when it is performed during development. The classically defined critical period is the time during development in which a shift in ocular dominance is observed in binocular cortex in many mammals in response to monocular alteration of visual inputs. In the rat, it begins around p18 and ends around p45 (Fagiolini et al., 1994). There are just a few days between eye opening and the beginning of the critical period, but such short period of visual experience during a very plastic moment in visual cortical development might produce maturational changes in the cortical microcircuit, which might then induce a different pattern of rewiring in response to MD.

To examine developmental changes in layer 4 microcircuitry between the pre-critical and critical period, we compared strength and connectivity of excitatory and inhibitory synapses in layer 4 taken in control rats at p17 (end of pre-critical period) and at p21 (beginning of critical period). These few days were sufficient to dramatically alter synaptic strength and connectivity in this cortical layer. Monosynaptic connection between star pyramids more than doubled in amplitude (Fig. 2b; EPSC, p21: 251% of p17; $p<0.05$) and their CV (Fig. 2c; CV, p21: 64% of p17; $p<0.006$) and failure rates decreased significantly (Fig. 2c; % failures, p21: $2.00\pm1.73\%$, p17: $9.72\pm2.45\%$; $p<0.006$), while their connection probability and the probability of finding recurrent connections remained unchanged (Fig. 2b; CP and RCP; p21: 79 out of 739, 10.7%; p17: 25 out of 236, 10.6%). Excitatory synapses therefore became stronger and more reliable as presynaptic release probability increases (Maffei et al., 2004, 2006; Fig. 2; p17, p21).

What are the mechanisms underlying these developmental changes in synaptic strength? The increased strength of EPSCs seems to recapitulate the effect of MD before eye opening. However, the mechanisms through which the increase in EPSC amplitude is achieved during normal development and in response to visual deprivation are fundamentally different: the developmental change appears to be driven by presynaptic changes (such as increased release probability), while MD induces postsynaptic changes that closely resemble synaptic scaling (Turrigiano et al., 1998; Desai et al., 2002). LTP of some excitatory synapses in neocortex has been shown to depend on presynaptic mechanisms (Hardingham et al., 2007), suggesting that this developmental strengthening could be due to a presynaptic form of LTP. Furthermore, the increase in connection probability between pyramidal neurons after MD might be due to an activity-dependent increase in formation of new connections or the prevention of visually driven pruning that might normally occur between p14 and p17. Since connection probability does not get reduced developmentally between p17 and p21 (Fig. 2b; CP and RCP, p17, p21), the pruning idea seems less likely, although further experiments are needed to measure connection probability at these synapses before eye opening in order to confirm this hypothesis.

In addition to developmental changes at excitatory synapses, dramatic changes in connectivity and synaptic strength were observed also at inhibitory synapses between FS neurons and star pyramids between p17 and p21 (Fig. 3a–c; p17, p21). There was a threefold increase in IPSC amplitude (Fig. 3b; IPSC, p21: 282.1% of p17, $p<0.01$) without any significant changes in CV, failure rates, or steady-state depression (Fig. 3c; $p<0.32, p<0.76, p<0.15$; Fig. 2; p17, p21). One of the most striking changes in the inhibitory feedback loop between FS neurons and star pyramids during these few days of development is a fourfold higher chance of finding recurrent connections from star pyramids back onto FS neurons (Fig. 3b; RCP). At p17, these connections would occur for only 28.5% of the FS to star pyramid pairs, whereas at p21 they were found in 82.3% of the pairs found (Fig. 3b; RCP, p17, p21; $p<0.01\ \chi^2$), while their strength was not changed (p21: 19.9±8.3 pA; p17: 15.4±4.8 pA; $p=0.48$). FS neurons have been shown to receive direct thalamic drive. Our developmental data on connection probability also show that by the beginning of the critical period this population of inhibitory interneurons receives a strong recurrent excitatory drive within layer 4 (Fig. 4). The increase in amplitude and connectivity of the inhibitory feedback loop between FS neurons and star pyramids is consistent with recent data suggesting that visual stimulation triggers the maturation of GABAergic synapses (Fagiolini and Hensch, 2000). The authors in fact suggest that the maturation of the inhibitory circuit determines the closure of the critical period for visual cortical plasticity; moreover, they indicate the synapse between FS and star pyramidal neurons as the one involved in determining the duration of this plastic period (Katagiri et al., 2007).

As previously mentioned, one of the hypotheses proposed to explain the effects of MD initiated before eye opening is that visual cortical circuit is maintained in an immature state. In light of our data about cortical circuit rewiring during development and after MD, this hypothesis does not seem to be valid. In fact excitatory synapses increase in strength, recapitulating one of the changes driven by development at these synapses, but on top of this their recurrent connectivity is also increased. The increase in connectivity seems to be specifically dependent on decreased visual drive because connection probability does not change during development. Moreover, inhibitory synapses are modified quite differently by development and by visual deprivation, suggesting that the changes in the inhibitory circuit depend on the role that specific types of interneurons play in layer 4 microcircuit. The visual input sets in motion a very complex set of changes in synaptic strength and connectivity at excitatory and inhibitory synapses that are likely to contribute to the maturation of cortical microcircuits. Instead of freezing the cortical circuit in an immature state, the deprivation of sensory drive induces a unique set of experience-dependent changes that contribute to rewire the microcircuit in layer 4.

Critical period plasticity within cortical layer 4

Over the past 50 years, it was proposed that changes at the level of the cortical excitatory and inhibitory circuits might determine the complex set of events underlying experience-dependent plasticity (Hensch and Fagiolini, 2005). It was shown that visual deprivation during the critical period induces profound functional changes in the visual responsiveness of cortical neurons. The neurons activated by the deprived eye show reduced responsiveness to visually evoked potentials, loss of orientation selectivity, and loss of visual acuity (White et al., 2001; Prusky and Douglas, 2002; Frenkel and Bear, 2004). One of the models proposed to explain these dramatic effects is that visual deprivation induces homosynaptic LTD at excitatory cortical synapses (Rittenhouse et al., 1999), therefore depressing the ability of cortical neurons to respond to visual inputs. In addition, it was also shown that changes in cortical inhibition play an important role in regulating the beginning and duration of the window for plasticity in visual cortex (Fagiolini and Hensch, 2000), and that cortical infusion of GABA blockers restored visual responsiveness to the deprived neurons (Reiter and Stryker, 1988; Duffy et al., 1976; but see also Sillito et al., 1981). At the cellular level, it was shown that the beginning of the maturation of perisomatic inhibition was the permissive stage for the opening of the critical period for visual cortical plasticity and the complete maturation of these synapses is involved in determining the end of this plastic stage of plasticity (Huang et al., 1999; Chattopadhyaya et al., 2004). Our experimental paradigm allowed for a fine analysis of these hypotheses, thanks to the possibility of directly measuring strength, connectivity, and plasticity of both excitatory and inhibitory unitary synapses within layer 4.

If MD induced LTD at cortical excitatory synapses, we would expect that the amplitude of monosynaptic connections in connected pairs of star pyramids in the deprived hemisphere would be smaller than at pairs in the control hemisphere. We performed MD between p18 and p21 (the beginning of the critical period), and recorded monosynaptic currents in connected pairs of star pyramidal neurons in control and deprived hemispheres (Maffei et al., 2006). To our surprise, the amplitude of unitary EPSCs within layer 4 remained unchanged (Fig. 2b; EPSC, p21). This is in contrast with the proposed idea that MD induces homosynaptic depression at excitatory synapses and suggests that changes at the level of inhibitory transmission might be involved in determining the decreased responsiveness of excitatory neurons to visual stimulation. Their short-term dynamics and ability to undergo LTD were also unaffected. However, the overall excitability of the layer 4 circuit was indeed reduced by MD, requiring further circuit analysis in order to understand how this might be explained (Maffei et al., 2006). We turned our attention to the inhibitory circuit whose connectivity and synapses were profoundly affected by development (Fig. 3a–c).

First, we measured possible changes at excitatory synapses between star pyramids and FS neurons. Their connectivity dramatically increased after eye opening, and possible changes in the excitatory drive onto this type of interneurons might have important repercussions on the overall excitability of the microcircuit in layer 4. Indeed their strength increased threefold following deprivation. There was also a significant increase in steady-state depression in response to presynaptic spiking at 20 Hz and no significant change in connection probability (Maffei et al., 2006). These data show that the neuronal type targeted by star pyramid synapses has an important part in determining the change in synaptic strength. When the target of synaptic contacts from star pyramids was an FS cell, excitatory synapses were potentiated following MD, whereas when the target was another star pyramid, there was no change in excitatory synaptic strength. These results illustrate the important point that synaptic changes within the cortical microcircuit are highly cell-type specific.

In addition, when measuring the amplitude of IPSC from FS to star pyramids, we found that they were also increased in amplitude by about threefold (Fig. 3b; IPSC, p21). There was a significant reduction in CV, but no changes in short-term plasticity, connection probability, or IPSC reversal potential (Fig. 3b, c; CV, SS STD, CP; Maffei et al., 2006). Non-stationary fluctuation analysis showed that increased IPSC amplitude depends upon a significant increase in number of open channels at the peak of the IPSC, suggesting that visual deprivation induced either an increase in quantal content or an increase in the number of postsynaptic GABA receptors, or both.

This dramatic increase in the strength of inhibitory feedback within layer 4 should make it much harder to propagate sensory input through the layer, and may be a major contributor to the loss of visual responsiveness of neurons activated by the deprived eye. We found that a form of LTP of GABAergic transmission (LTPi; Maffei et al., 2006) can be induced at this synapse, and that the expression characteristics of this LTPi and of the MD-induced change in inhibition were the same.

We tested the hypothesis that the increase in inhibition induced by MD is due to an experience-dependent induction of LTPi at FS to star pyramid synapses. In support of this idea, our experiments showed that further LTPi induction was occluded after MD. Further investigation on the properties of LTPi suggested that it is successfully induced only when presynaptic and postsynaptic firing are uncorrelated (Maffei et al., 2006). LTPi induction, therefore, has unique induction rules, and would be predicted to be activated by changes in sensory input that de-correlated FS and star pyramid activity.

This is a different form of inhibitory plasticity than that described in hippocampal preparations (Woodin et al., 2003) in which it depends on a de-correlation of FS and star pyramidal neuron firing, and it does not determine changes in chloride reversal potential (measured in perforated patch; Maffei et al., 2006), excluding the involvement of modulation of chloride pumps.

Experience-dependent rewiring is regulated by cortical development

Taken together, the above experiments on pre-critical and critical period plasticity suggest that there is a developmentally regulated switch in experience-dependent plasticity mechanisms within layer 4, which leads to very different changes in the layer 4 microcircuit depending on when the activity deprivation is initiated. Early in development, the circuit in layer 4 is able to respond homeostatically to brief visual deprivation by increasing its excitability, suggesting that the gain of the circuit is increased in order to compensate for reduced visual drive. This ability to compensate is lost by the beginning of the critical period. Visual deprivation performed at this time in fact induces a strengthening of the inhibitory feedback loop between FS and star pyramids, thus reducing layer 4 excitability. The maturation of both excitatory and inhibitory transmission during the few days between eye opening and the beginning of the critical period may be instrumental in determining the forms of plasticity that are present within layer 4 and its ability to rewire itself in response to altered sensory experience.

It has been recently shown that between eye opening and the beginning of the critical period, $GABA_A$ receptors change their subunits composition and that around p21 the composition of the receptor reaches its mature configuration (Heinen et al., 2004). NMDA receptors are also undergoing a major subunit rearrangement between eye opening and the beginning of the critical period (Cao et al., 2000). Such shifts in NMDA and $GABA_A$ subunits composition might contribute to the developmental changes in inhibitory plasticity that we have observed following MD in layer 4. In addition, developmental changes in the wiring of the microcircuit in layer 4 likely contribute to differences in plasticity following visual deprivation, since they will affect how activity propagates through the network and thus may influence both correlation-based and homeostatic plasticity. For example, the increase in recurrent connectivity between FS and star pyramids affects dramatically the excitability of FS neurons which during the critical period receive two major sources of excitatory drive, from thalamic neurons as well as from star pyramids within layer 4.

This opens the possibility that the changes in the circuit triggered by eye opening will determine the switch between homeostatic and non-homeostatic rewiring of layer 4 circuit. In addition, the density of GABAergic synapses and their positioning onto star pyramidal neurons make LTPi a possible player in determining experience-dependent changes also in the presence of competition between the inputs from the two eyes. In fact there is evidence that GABAergic synapses are positioned near the site of thalamic input on star pyramids (Beaulieau et al., 1994). An interesting possibility is that such localized positioning could provide a precise increase in inhibition localized at synapses driven by the closed eye, therefore contributing to a depression of thalamocortical synaptic efficacy driven by the deprived eye.

Is homeostatic plasticity still induced during the critical period?

The experience-dependent changes we observed suggest that the layer 4 microcircuit loses the ability to respond homeostatically to a decrease in sensory drive. This is potentially a source of instability for a cortical circuit, which will not be able to achieve the optimal balance between excitation and inhibition required for sensory processing. Desai et al. (2002) showed that during the critical period, brief periods of visual deprivation by intraocular TTX injection could not induce synaptic scaling in layer 4, but could induce scaling of AMPA miniature EPSCs in layer 2/3 pyramidal neurons. These results suggest that the cortical microcircuit in layer 2/3 becomes the site of homeostatic plasticity during the critical period, endowing this microcircuit with the ability to produce compensatory responses when layer 4 is overwhelmed by inhibition. New experiments are needed to investigate whether visual deprivation during the critical period induces shifts in the balance between excitation and inhibition, thus affecting the excitability of layer 2/3 in a compensatory manner.

References

Abbott, L. and Nelson, S. (2000) Synaptic plasticity: taming the beast. Nat. Neurosci., 3(Suppl.): 1178–1183.

Bartoletti, A., Medini, P., Berardi, N. and Maffei, L. (2004) Environmental enrichment prevents effects of dark-rearing in the rat visual cortex. Nat. Neurosci., 7: 215–216.

Beaulieau, C., Campistron, G. and Crevier, C. (1994) Quantitative aspects of GABA circuitry in the primary visual cortex of the adult rat. J. Comp. Neurol., 339: 559–572.

Bishop, K.M., Garel, S., Nakagawa, Y., Rubenstein, J. and O'Leary, D. (2003) Emx1 and Emx2 cooperate to regulate cortical size, lamination, neuronal differentiation, development of cortical efferents, and thalamocortical pathfinding. J. Comp. Neurol., 457: 345–360.

Bishop, K., Goudreau, G. and O'Leary, D. (2000) Regulation of area identity in the mammalian neocortex by Emx2 and Pax6. Science, 288: 344–349.

Blasdel, G. and Pettigrew, J. (1978) Effect of prior visual experience on cortical recovery from the effects of unilateral eyelid suture in kittens. Brain Res., 274: p. 601.

Brown, T., Kairiss, E. and Keenan, C. (1990) Hebbian synapses: biophysical mechanisms and algorithms. Annu. Rev. Neurosci., 13: 475–511.

Cao, Z., Liu, L., Lickey, M. and Gordon, B. (2000) Development of NR1, NR2A and NR2B mRNA in NR1 immunoreactive cells of rat visual cortex. Brain Res., 868: 296–305.

Catalano, S. and Shatz, C. (1998) Activity-dependent cortical target selection by thalamic axons. Science, 281: 559–562.

Charron, F. and Tessier-Lavigne, M. (2005) Novel brain wiring functions for classical morphogens: a role as graded positional cues in axon guidance. Development, 132: 2251–2262.

Chattopadhyaya, B., Di Cristo, G., Higashiyama, H., Knott, G., Kuhlman, S., Welker, E. and Huang, Z. (2004) Experience and activity-dependent maturation of perisomatic GABAergic innervation in primary visual cortex during a postnatal critical period. J. Neurosci., 24: 9598–9611.

Cohen, A., Lin, D. and Coulter, D. (2001) Protracted postnatal development of inhibitory synaptic transmission in rat hippocampal area CA1 neurons. J. Neurophysiol., 84: 2465–2476.

Crozier, R., Wang, Y., Liu, C. and Bear, M. (2007) Deprivation-induced synaptic depression by distinct mechanisms in different layers of mouse visual cortex. Proc. Natl. Acad. Sci. U.S.A., 104: 1383–1388.

Del Rio, T. and Feller, M. (2006) Early retinal activity and visual circuit development. Neuron, 52: 221–222.

Desai, N. (2003) Homeostatic plasticity in the CNS: synaptic and intrinsic forms. J. Physiol. (Paris), 97: 391–402.

Desai, N., Cudmore, R., Nelson, S. and Turrigiano, G. (2002) Critical periods for experience-dependent synaptic scaling in visual cortex. Nat. Neurosci., 5: 783–789.

Desai, N., Rutherford, L. and Turrigiano, G. (1999) Plasticity in the intrinsic excitability of cortical pyramidal neurons. Nat. Neurosci., 2: 515–520.

Duffy, F.H., Burchfield, J.L. and Conway, J.L. (1976) Bicuculline reversal of deprivation amblyopia in the cat. Nature, 260: 256–257.

Fagiolini, M. and Hensch, T. (2000) Inhibitory threshold for critical-period activation in primary visual cortex. Nature, 404: 183–186.

Fagiolini, M., Pizzorusso, T., Berardi, N., Domenici, L. and Maffei, L. (1994) Functional postnatal development of the rat primary visual cortex and the role of visual experience: dark rearing and monocular deprivation. Vis. Res., 34: 709–720.

Feller, M.B. and Scanziani, M. (2005) A precritical period for plasticity in visual cortex. Curr. Opin. Neurobiol. 15: 94–100.

Frenkel, M. and Bear, M. (2004) How monocular deprivation shifts ocular dominance in visual cortex of young mice. Neuron, 44: 917–923.

Golowasch, J., Casey, M., Abbott, L. and Marder, E. (1999) Network stability from activity-dependent regulation of neuronal conductances. Neural. Comput., 11: 1079–1096.

Goodman, C. and Shatz, C.J. (1993) Developmental mechanisms that generate precise patterns of neuronal connectivity. Cell, 72: 77–98.

Hajós, F., Staiger, J.F., Halasy, K., Freund, T.F. and Zilles, Z. (1997) Geniculo-cortical afferents form synaptic contacts with vasoactive intestinal polypeptide (VIP) immunoreactive neurons of the rat visual cortex. Neurosci. Lett., 228: 179–182.

Hardingham, N., Hardingham, G., Fox, K. and Jack, J. (2007) Presynaptic efficacy directs normalization of synaptic strength in layer 2/3 rat neocortex after paired activity. J. Neurophysiol., 97: 2965–2975.

Heinen, K., Bosman, L., Spijker, S., van Pelt, J., Smit, A., Voorn, P., Baker, R. and Brussaard, A. (2004) GABAA receptor maturation in relation to eye opening in the rat visual cortex. Neuroscience, 124: 161–171.

Hensch, T. and Fagiolini, M. (2005) Excitatory-inhibitory balance and critical period plasticity in developing visual cortex. Prog. Brain. Res., 147: 115–124.

Hollrigel, G., Ross, S. and Soltesz, I. (1998) Temporal patterns and depolarizing actions of spontaneous $GABA_A$ receptor activation in granule cells of the early postnatal dentate gyrus. J. Neurophysiol., 80: 2340–2351.

Holmgren, C. and Zilberter, Y. (2001) Coincident spiking activity induces long-term changes in inhibition of neocortical pyramidal cells. J. Neurosci., 21: 8270–8277.

Huang, Z., Kirkwood, A., Pizzorusso, T., Porciatti, V., Morales, B., Bear, M., Maffei, L. and Tonegawa, S. (1999) BDNF regulates the maturation of inhibition and the critical period of plasticity in mouse visual cortex. Cell, 98: 739–755.

Hubel, D. and Wiesel, T. (1970) The period of susceptibility to the physiological effects of unilateral eye closure in kittens. J. Physiol., 206: 419–436.

Inan, M. and Crair, M. (2007) Development of cortical maps: perspectives from the barrel cortex. Neuroscientist, 13: 49–61.

Katagiri, H., Fagiolini, M. and Hensch, T. (2007) Optimization of somatic inhibition at critical period onset in mouse visual cortex. Neuron, 53: 805–812.

Katz, L.C. and Shatz, C.J. (1996) Synaptic activity and the construction of cortical circuits. Science, 274: 1133–1138.

Kawaguchi, Y. and Kubota, Y. (1997) GABAergic cell subtypes and their synaptic connections in rat frontal cortex. Cereb. Cortex, 7: 476–486.

Komatsu, Y. and Iwakiri, M. (1993) Long-term modification of inhibitory synaptic transmission in developing visual cortex. Neuroreport, 4: 907–910.

Maffei, A., Nataraj, K., Nelson, S. and Turrigiano, G. (2006) Potentiation of cortical inhibition by visual deprivation. Nature, 443: 81–84.

Maffei, A., Nelson, S. and Turrigiano, G. (2004) Selective reconfiguration of layer 4 visual cortical circuitry by visual deprivation. Nat. Neurosci., 7: 1353–1359.

Malenka, R. and Bear, M. (2004) LTP and LTD: an embarrassment of riches. Neuron, 44: 5–21.

Mioche, L. and Singer, W. (1989) Chronic recordings from single sites of kitten striate cortex during experience-dependent modifications of receptive-field properties. J. Neurophysiol., 62: 185–197.

Nadarajah, B., Alifragis, P., Wong, R. and Parnavelas, J. (2003) Neuronal migration in the developing cerebral cortex: observations based on real-time imaging. Cereb. Cortex, 13: 607–611.

Nadarajah, B., Brunstrom, J., Grutzendler, J., Wong, R. and Pearlman, A. (2001) Two modes of radial migration in early development of the cerebral cortex. Nat. Neurosci., 4: 143–150.

Palm, G. (1982) Neural Assemblies: An Alternative Approach. Springer-Verlag, Berlin, Heidelberg.

Prusky, G. and Douglas, R. (2002) Developmental plasticity of mouse visual acuity. Eur. J. Neurosci., 17: 167–173.

Rakic, P., Bourgeois, J., Eckenhoff, M., Zecevic, N. and Goldman-Rakic, P. (1986) Concurrent overproduction of synapses in diverse regions of the primate cerebral cortex. Science, 232: 232–235.

Reiter, H. and Stryker, M. (1988) Neural plasticity without postsynaptic action potentials: less-active inputs become dominant when kitten visual cortical cells are pharmacologically inhibited. Proc. Natl. Acad. Sci. U.S.A., 85: 3623–3627.

Rittenhouse, C., Shouval, H., Paradiso, M. and Bear, M. (1999) Monocular deprivation induces homosynaptic long-term depression in visual cortex. Nature, 397: 347–350.

Rumpel, S., Hatt, H. and Gottmann, K. (1998) Silent synapses in the developing rat visual cortex: evidence for postsynaptic expression of synaptic plasticity. J. Neurosci., 18: 8863–8874.

Sillito, A., Kemp, J. and Blakemore, C. (1981) The role of GABAergic inhibition in the cortical effects of monocular deprivation. Nature, 291: 318–320.

Smith, S. and Trachtenberg, J. (2007) Experience-dependent binocular competition in the visual cortex begins at eye opening. Nat. Neurosci., 10: 370–375.

Sun, Q., Huguenard, J. and Prince, D. (2006) Barrel cortex microcircuits: thalamocortical feedforward inhibition in spiny stellate cells is mediated by a small number of fast-spiking interneurons. J. Neurosci., 24: 1219–1230.

Tagawa, Y., Kanold, P., Majdan, M. and Shatz, C. (2005) Multiple periods of functional ocular dominance plasticity in mouse visual cortex. Nat. Neurosci., 8: 380–388.

Tessier-Lavigne, M. and Goodman, C.S. (1996) The molecular biology of axon guidance. Science, 274: 1123–1133.

Turrigiano, G. (1999) Homeostatic plasticity in neuronal networks: the more things change, the more they stay the same. Trends Neurosci., 22: 221–227.

Turrigiano, G., Leslie, K., Desai, N., Rutherford, L. and Nelson, S. (1998) Activity-dependent scaling of quantal amplitude in neocortical neurons. Nature, 391: 892–896.

Turrigiano, G. and Nelson, S. (2000) Hebb and homeostasis in neuronal plasticity. Curr. Opin. Neurobiol., 10: 358–364.

Turrigiano, G. and Nelson, S. (2004) Homeostatic plasticity in the developing nervous system. Nat. Rev. Neurosci., 5: 97–107.

Watt, A., van Rossum, M., MacLeod, K., Nelson, S. and Turrigiano, G. (2000) Activity coregulates quantal AMPA and NMDA currents at neocortical synapses. Neuron, 26: 659–670.

White, L., Coppola, D. and Fitzpatrick, D. (2001) The contribution of sensory experience to the maturation of orientation selectivity in ferret visual cortex. Nature, 411: 1049–1052.

Wierenga, C., Ibata, K. and Turrigiano, G. (2005) Postsynaptic expression of homeostatic plasticity at neocortical synapses. J. Neurosci., 25: 2895–2905.

Woodin, M., Ganguly, K. and Poo, M. (2003) Coincident pre- and postsynaptic activity modifies GABAergic synapses by postsynaptic changes in Cl⁻ transporter activity. Neuron, 39: 807–820.

CHAPTER 13

Differential mechanisms of transmission and plasticity at mossy fiber synapses

Chris J. McBain*

Laboratory of Cellular and Synaptic Neurophysiology, Program in Developmental Neurobiology, National Institute of Child Health and Human Development, National Institutes of Health, Building 35, Bethesda, MD 20892, USA

Abstract: The last few decades have seen the hippocampal formation at front and center in the field of synaptic transmission. However, much of what we know about hippocampal short- and long-term plasticity has been obtained from research at one particular synapse; the Schaffer collateral input onto principal cells of the CA1 subfield. A number of recent studies, however, have demonstrated that there is much to be learned about target-specific mechanisms of synaptic transmission by study of the lesser known synapse made between the granule cells of the dentate gyrus; the so-called mossy fiber synapse, and its targets both within the hilar region and the CA3 hippocampus proper. Indeed investigation of this synapse has provided an embarrassment of riches concerning mechanisms of transmission associated with feedforward excitatory and inhibitory control of the CA3 hippocampus. Importantly, work from a number of labs has revealed that mossy fiber synapses possess unique properties at both the level of their anatomy and physiology, and serve as an outstanding example of a synapse designed for target-specific compartmentalization of synaptic transmission. The purpose of the present review is to highlight several aspects of this synapse as they pertain to a novel mechanism of bidirectional control of synaptic plasticity at mossy fiber synapses made onto hippocampal stratum lucidum interneurons. It is not my intention to pour over all that is known regarding the mossy fiber synapse since many have explored this topic exhaustively in the past and interested readers are directed to other fine reviews (Henze et al., 2000; Urban et al., 2001; Lawrence and McBain, 2003; Bischofberger et al., 2006; Nicoll and Schmitz, 2005).

Keywords: hippocampus; local circuit inhibitory interneuron; long-term depression; glutamate receptors; plasticity; mossy fiber; mGluR7

Anatomy of the granule cell mossy fiber axon

Before discussing the physiological properties of the mossy fiber synapse, a short description of the granule cell anatomy and in particular the mossy fiber axon is warranted. The granule cell is the principal cell of the dentate gyrus and releases the neurotransmitter glutamate (Spruston and McBain, 2006). Granule cells typically possess small, ovoid cell bodies with a single apical, and conical dendritic tree, which extends into the molecular layer and terminates close to the hippocampal fissure. Compared to other principal cells of the hippocampal formation the granule cell is relatively small, possessing a somata approximately 10 μm in diameter and about 18 μm

---

*Corresponding author. Tel.: +1 301 402 4778;
Fax: +1 301 402 4777; E-mail: mcbainc@mail.nih.gov

along its long axis. The number of dendritic branches is highly variable but the total dendritic length is significantly shorter than their CA1 and CA3 pyramidal neuron counterparts. The axon of the granule cell emerges close to the basal pole of the somata where it passes unmyelinated into the hilus and the CA3 hippocampus.

Based solely on their anatomy these axons are unique among hippocampal and cortical neurons (and indeed throughout the mammalian central nervous system) in that they form anatomically specialized synapses depending on the nature of their postsynaptic targets. Indeed the anatomy of the granule cell axons was first noted as peculiar by the anatomist Sala and then subsequently described in greater detail by Ramon y Cajal. In his landmark original work *Textura del sistema del hombre y los vertebrados* (1899 and 1904) Cajal remarked: "In well impregnated material from the 8-day old rabbit or the newborn Guinea-pig, one can easily see that these varicosities are triangular or stellate masses of cytoplasm with angles that give rise either to short, thick, diverging processes or to thin, rather long filaments ending in a varicosity. Although perhaps less striking this appearance is definitely reminiscent of certain fibers in the cerebellum that we referred to as mossy fibers". Cajal's early appreciation of the anatomy of the granule cell axon is evident in his Figure 479 from his original text and reproduced here as Fig. 1A.

Typically, each granule cell gives rise to a single unmyelinated axonal fiber approximately 0.2 μm in diameter, with a total length (including collaterals) of more than 3 mm (Claiborne et al., 1986; Frotscher et al., 1991; Acsády et al., 1998). These axons form a number of local collaterals, which innervate the numerous cell types present within the hilar subfield. After leaving the hilar region, the mossy fiber axon contains few, if any further branch points and projects to the apical and in some cases basal dendrites of CA3 pyramidal neurons (Claiborne et al., 1986). The mossy-fiber projection to CA3 forms a tight bundle running parallel to the CA3 pyramidal cell body layer in a region corresponding to approximately the first 100 μm of the CA3 pyramidal neuron apical dendrite, the stratum lucidum, so-called because of its translucent appearance under the light microscope (Fig. 1B).

Three basic types of mossy fiber presynaptic terminal exist along the entire length of the main axon: large mossy terminals (4–10 μm in diameter), filopodial extensions (0.5–2.0 μm) that project from the large mossy boutons, and small en passant varicosities (0.5–2.0 μm) (Amaral, 1979; Claiborne et al., 1986; Frotscher et al., 1991; Acsády et al., 1998) (Fig. 1C, D). Both of these latter, small terminals are somewhat larger than the typical varicosities found in axons of CA3 pyramidal neurons (Acsády et al., 1998). The total number of large mossy terminals along a single axon is about 10 in the hilar region and about 12 (range 10–18) in the CA3 region, with the terminals being distributed somewhat evenly spaced (approximately every 150 μm) along the mossy fiber axon (Acsády et al., 1998). The large mossy fiber terminals, which contain up to 35 individual release sites envelop the postsynaptic thorny excrescences of CA3 pyramidal neuron dendrites. In contrast the mossy cells constitute the principal target of the large mossy terminals in the hilus. Although large mossy fiber boutons do not typically innervate GABAergic interneurons, they have on occasion been observed to directly contact the dendrites of stratum lucidum basket cell interneurons in Guinea pig (Frotscher, 1985). More typical however, the small filopodial extensions, and en passant terminals emerging from the parent axon, predominantly innervate only inhibitory interneuron targets in the stratum lucidum (Acsády et al., 1998). These terminals form single, often perforated, asymmetric synapses on the cell bodies, dendrites, and spines of GABAergic inhibitory interneurons.

In 1998, Lazlo Acsády, then in Gyorgyi Buzsaki's Lab, made the important observation that the total number of filopodial extensions and small, en passant synapses outnumbered the large mossy fiber terminals by approximately 10-fold (Acsády et al., 1998). Specifically, a single granule cell typically gives rise to 7–12 large mossy boutons in the hilus and 11–18 boutons in the CA3 region. In contrast, the number of small terminals (en passant and filopodial) was

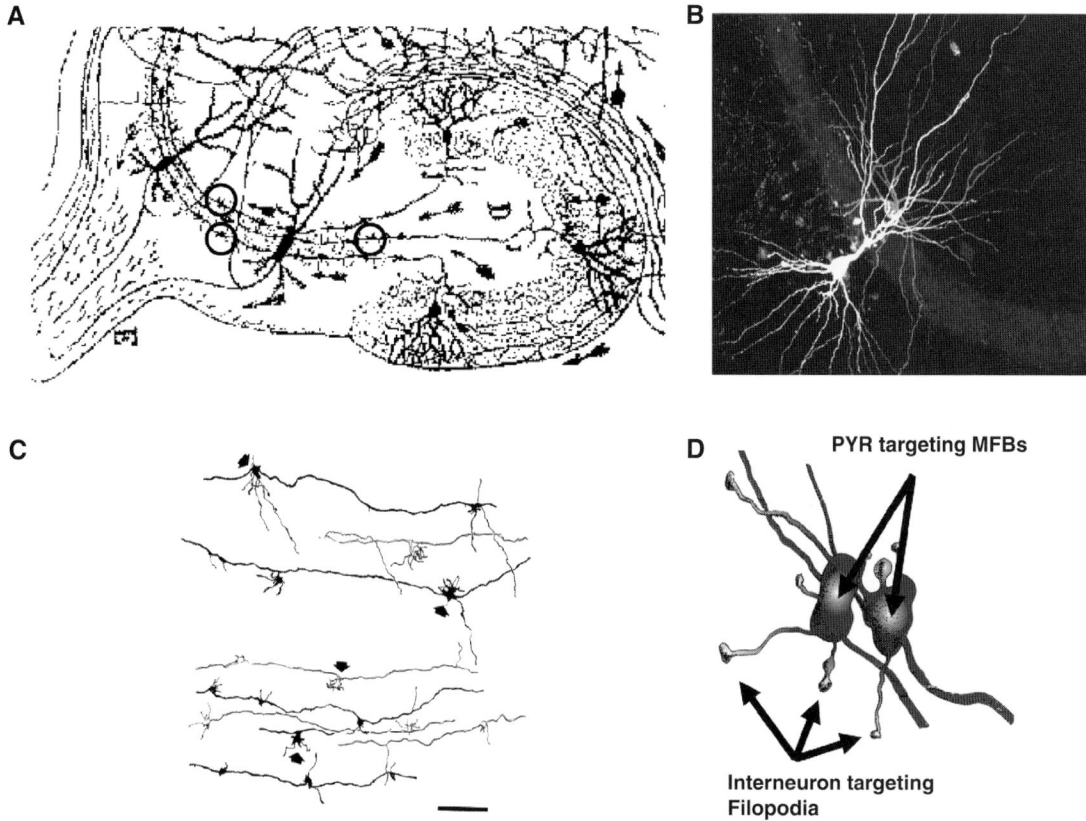

Fig. 1. Anatomy of the mossy fiber–CA3 synapse. (A) Enlargement of Ramon y Cajal's classic drawing of the hippocampal formation to illustrate the mossy fiber pathway as visualized by his Golgi impregnation technique. Three individual mossy fiber boutons are circled to illustrate that even at this level of resolution Cajal could appreciate the anatomical structure of the mossy fiber bouton as being something more than a typical swelling or en passant synapse. (B) Triple fluorescence image shows the juxtaposition of a CA3 pyramidal cell, a stratum lucidum interneuron, and the MF pathway (stained for calbindin immunoreactivity). (C) Camera lucida reconstruction of a Golgi-impregnated mossy fiber axon illustrating individual boutons and their associated filopodial extensions (indicated by arrows) C. A cartoon rendering of reconstructed mossy fiber boutons, which innervate pyramidal cells and their filopdial extensions which typically innervate the smooth dendrites of local circuit inhibitory interneurons. Figures have been modified and reproduced with permission from: Panel A, Ramon y Cajal (1904, 1995); Panel C, Amaral and Dent (1981); Panel D, Acsády et al. (1998); Panel B, from K Pelkey and C.J. McBain, unpublished observation.

considerably higher both in the hilus (120–150) and the CA3 region (40–50). This circuit arrangement is such that each large mossy fiber bouton in the CA3 subfield gives rise to approximately 2.5 small bulbous ending filopodial extensions, indicating that an average granule cell mossy fiber axon in the CA3 subfield has 25–35 filopodial extensions associated with ∼10 large boutons (Fig. 1D). Since the larger mossy fiber boutons primarily target CA3 pyramidal cells and filopodial extensions target the smooth dendrites of local circuit inhibitory interneurons this would suggest that, if based solely on the total number of actual synapses, the primary targets of dentate gyrus granule cells are inhibitory interneurons (Acsády et al., 1998). This anatomical segregation of terminal types is unprecedented throughout the mammalian central nervous systems and suggests that it may serve to provide a functional specialization of synaptic output.

Basic properties of mossy fiber-inhibitory interneuron transmission

Without prior electrophysiological knowledge, based on these anatomical properties alone, it is possible to speculate that the physiological properties of each synapse type within this synaptic arrangement would differ, since the large bouton alone has 20–35 release sites compared to the single release site of each filopodial extension. In fact, electrophysiological experiments have largely confirmed the hypothesis that the large and small synaptic specializations of the mossy fibers are functionally distinct (Henze et al., 2000; Urban et al., 2001). Individual mossy fiber release sites onto principal cells have a low initial release probability ($P_r = 0.01–0.05$) (Jonas et al., 1993). However, because of the large number of release sites per bouton this endows these presynaptic terminals with a high degree of short-term- and frequency-dependent facilitation of transmission (Salin et al., 1996; Toth et al., 2000; Lawrence et al., 2004). In contrast, mossy fiber synapses onto interneurons in stratum lucidum possess a greater than one order of magnitude higher initial release probability (0.1–0.5) (Lawrence et al., 2004). This relatively high release probability endows these filopodial synapses with either mild facilitation or depression in response to brief stimulus trains (Toth et al., 2000).

Consideration of these different initial release probabilities at the mossy fiber bouton and filopodia synapse highlights an important feature of this synapse, namely that dentate gyrus granule cells firing at different frequencies will differentially impact the CA3 network function. The higher probability of release and higher potency of each release site at the filopodial synapse onto inhibitory interneurons (Lawrence et al., 2004), coupled to their interconnectivity, will tend to provide a robust feedforward inhibitory drive to all downstream targets at low (<0.5 Hz) in vivo discharge rates of dentate granule cells (Jung and McNaughton, 1993). This would suggest that at low frequencies at least, the output of the mossy fiber pathway largely drives a net inhibition onto the CA3 network. This is a hypothesis that has been well appreciated from in vivo recordings where synchronous activation of dentate granule cells in vivo during dentate spikes results in a net inhibition of CA3 pyramidal cell firing (Bragin et al., 1995a, b; Penttonen et al., 1997). However, the relative release dynamics change at higher dentate granule-cell firing frequencies (Salin et al., 1996; Toth et al., 2000; Henze et al., 2002). Data from in vivo recordings demonstrate that when the animal moves into the dentate granule cell's place field, cells fire in short, high-frequency bursts during the in-field discharge (Henze et al., 2002). This high frequency granule cell discharge transiently alters the network dynamics of mossy fiber transmission such that pyramidal cells are now robustly excited due to the massive facilitation of mossy fiber transmission at events late in the epoch. In contrast, since the mossy fiber–interneuron synapse can only weakly facilitate or often depresses in response to such high-frequency afferent drive, this results in an erosion of mossy fiber drive of inhibitory interneuron targets. The net effect is to provide an expansion of the CA3 recurrent excitability and the triggering of action potential discharge within the CA3 pyramidal cell population (Henze et al., 2002; Lawrence and McBain, 2003; Mori et al., 2004). Thus, the dynamics of excitation and inhibition within the CA3 network will shift depending on the firing frequency of the dentate granule cells.

Two types of AMPA/NMDAR populate mossy fiber–interneuron synapses

One important aspect of mossy fiber synapses onto stratum lucidum interneurons is the nature of the AMPA and NMDA receptors that populated the postsynaptic sites. We have demonstrated that in rat hippocampus a continuum of AMPA receptor types exist at these synapses (Toth et al., 2000; Lei and McBain, 2002) (Fig. 2). Of particular interest, at one end of the continuum there are MF–interneuron synapses that comprise GluR2-lacking, Ca^{2+}-permeable (CP-) AMPA receptors. These CP AMPA receptors lack the GluR2 subunit, show inward rectifying current–voltage relationships and are blocked by exogenous polyamine containing toxins (Fig. 3). These

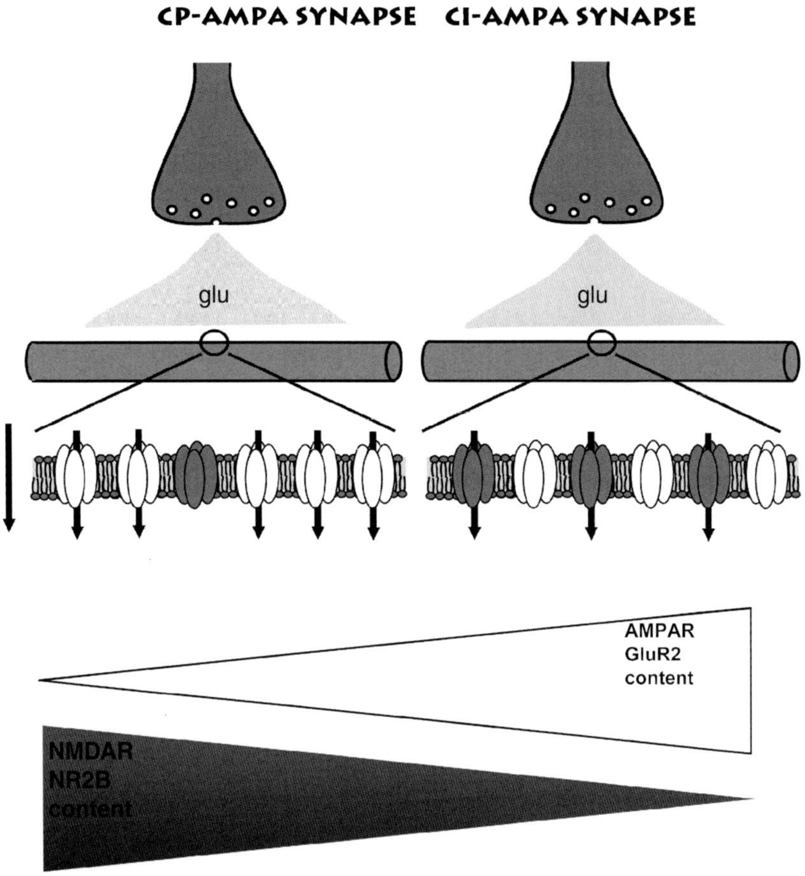

CP-AMPA SYNAPSE   CI-AMPA SYNAPSE

Fig. 2. A continuum of glutamate receptors exist at mossy fiber–interneuron synapses. A schematic illustrating the typical synapses that make up the population of mossy fiber–interneuron connections. Typically, GluR2-lacking Ca^{2+}-permeable (CP) AMPA receptors are found at synapses also populated with NR2B-containing NMDA receptors. At the other end of the continuum is a population of synapses that contain GluR2-containing, CP-AMPA receptors together with NR2B-lacking NMDA receptors. Figure modified and reproduced with permission from Bischofberger and Jonas (2002).

CP-AMPARs typically occur at synapses that also possess NMDA receptors containing the NR2B subunit (Fig. 2). At the other end of the continuum, GluR2-lacking, Ca^{2+}-impermeable (CI-) AMPARs are typically found associated with NR2B lacking NMDARs (Lei and McBain, 2002) (Fig. 2). These CI-AMPARs possess linear current voltage relationships and are largely resistant to exogenous polyamine toxins (Fig. 3). The ratios of these two synapse types show some developmental regulation, with CP-AMPA receptor containing synapses modestly predominating (62% of measured synapses) early in development. However, it is important to appreciate that there is not a complete conversion of one synapse type to another on developmental maturation, at least at time points up to postnatal day 40 (Lei and McBain, 2002). At postnatal day 40, approximately 45% of mossy fiber–interneuron synapses are comprised of CI-AMPARs, whereas only 37% are comprised CP-AMPARs. The remaining ~20% of synapses are a mixed population that possess intermediate rectification ratios, AMPARs with low sensitivity to polyamine block and NMDARs with intermediate sensitivity to the NR2B preferring antagonist ifenprodil (Figs. 2 and 3). At this time it is unclear what rules dictate the expression of each type of AMPA and NMDA receptor and how

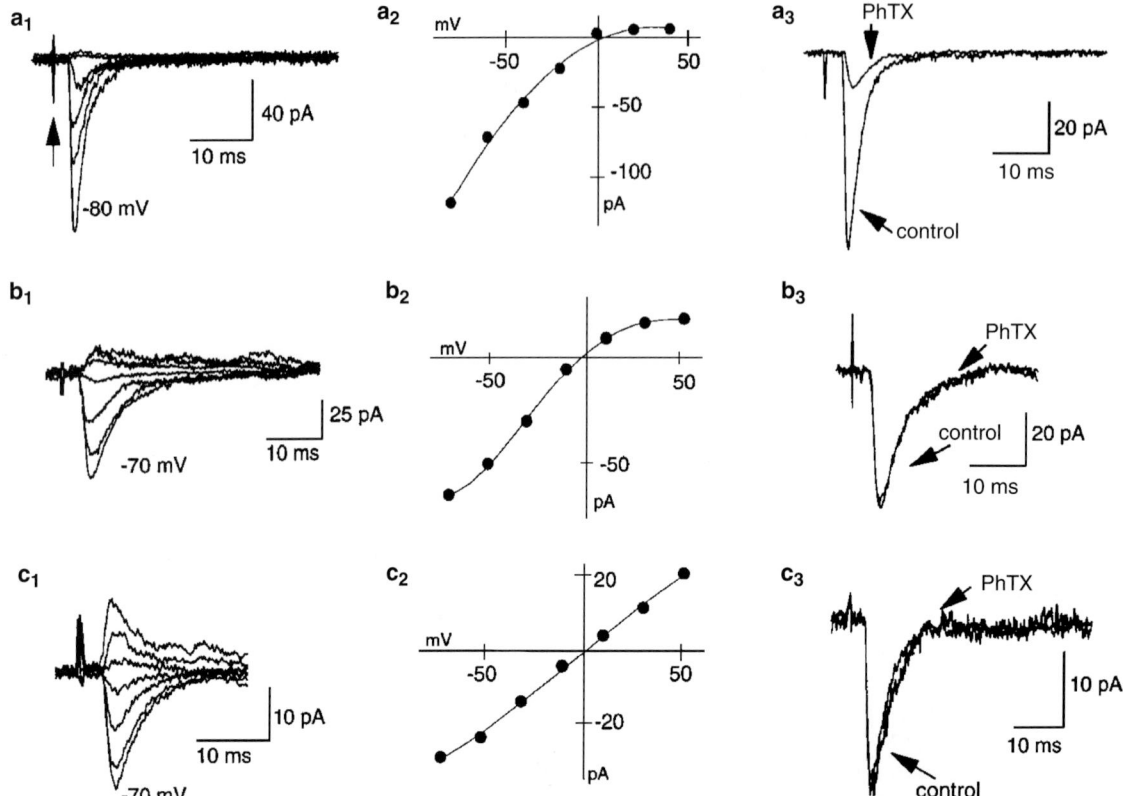

Fig. 3. Rectification properties and PhTx sensitivity of EPSCs on stratum lucidum interneurons. Examples of mossy-fiber-evoked EPSCs onto three stratum lucidum interneurons (left panels). Low-intensity stimulation of mossy fibers evokes EPSCs that show a continuum of I–V relationships (center panels) from either strongly rectifying to essentially linear. Five EPSCs were evoked at each potential (20-mV increments) and averaged. The stimulus artifact is indicated by the arrow. (a, b) Two cells where EPSCs demonstrated strong and moderate inward rectification (center), respectively. The rectification indices of these two cells were 0.01 (a) and 0.12 (b). Of the two sets of evoked EPSCs, only the one with strong inward rectification (a) was blocked by the polyamine toxin PhTx-433 (10 μM, right panels). (c) In contrast EPSCs with linear current–voltage relationships (rectification index of 0.88) are insensitive to PhTx (right). Figure reproduced from Toth and McBain (1998).

these synapse types map onto their presynaptic granule cell afferents. Although not as rigorously determined, the ratio of mossy fiber–CP-AMPAR: CI-AMPAR synapses in the mouse is significantly higher across all ages tested than that observed in the rat.

Some differences in the basic features of transmission at CP-AMPAR and CI-AMPAR synapses exist (Toth et al., 2000; Walker et al., 2002). One of two important features worthy of further consideration is the presence of an entirely postsynaptic mechanism of short-term plasticity at CP-AMPARs, which results from the use-dependent unblock of intracellular polyamines from the intracellular pore of the channel (Rozov et al., 1998; Rozov and Burnashev, 1999; Toth et al., 2000). Traditionally most mechanisms of short-term synaptic plasticity are considered presynaptic in origin and typically arise from changes in the release probability due to accumulation of residual calcium loads, or exhaustion of the readily releasable pool of neurotransmitter. However, CP-AMPA receptors lacking the GluR2 subunit posses an unexpected and entirely postsynaptic form of short-term plasticity due to their susceptibility to block by intracellular polymines (Rozov et al., 1998; Rozov and Burnashev, 1999).

Within the pore of AMPA receptors a single amino acid, the so-called Q/R site not only dictates their Ca permeability but also influences their sensitivity to block by both endogenous intracellular polyamines such as spermine and spermidine as well as externally applied polyamine toxins such as philanthotoxin (Bowie and Mayer, 1995; Kamboj et al., 1995; Koh et al., 1995; Isaac et al., 2007). The interaction with intracellular polyamines is in part voltage-dependent, with block being greatest at depolarized potentials. It is this polyamine block that is responsible for the highly rectifying current voltage relationship of GluR2-lacking AMPA receptors. The affinity of polyamines for their intracellular binding site within the AMPA receptor is such that it provides a tonic block of native Ca-permeable AMPA receptors. Relief of this block is use- and voltage-dependent (Bowie and Mayer, 1995) and typically requires multiple receptor activations (i.e. multiple synaptic events) to force the polyamine from the inner vestibule of the channel. This use-dependence results in an apparent increased current flow due to the progressive unblock of the pore and imparts a novel mechanism of short-term facilitation to CP-AMPA receptors. Because of the voltage-dependence of the polyamine block, facilitation is greatest when the cell is depolarized and is almost completely absent at resting or hyperpolarized membrane potentials. Such a mechanism is observed at mossy fiber-CP AMPARs interneuron synapses (Toth et al., 2000). Indeed short trains (five events) of synaptically evoked currents were observed to induced facilitation when measured at $-20\,mV$ but not at resting potential. This facilitation is absent when polyamines are omitted from the intracellular recording solution or when endogenous polyamines are chelated with ATP.

A second important feature that differentiates activity through CP- versus CI-AMPAR synapses is the magnitude of temporal summation observed at either synapse when measured under current clamp conditions (Lei and McBain, 2002). At CI-AMPAR containing synapses the presence of NR2B-lacking NMDARs endows synapses with a distinct temporal feature when compared with CP-AMPAR containing synapses (which contain NR2B-containing NMDARs). The presence of the larger amplitude NR2B-lacking NMDA receptor-mediated excitatory postsynaptic potentials (EPSPs) at CI-AMPAR synapses endows these synapses with a broad depolarizing envelope and significant temporal summation during repetitive stimulation when measured at resting potentials (Lei and McBain, 2002). This long-lasting depolarization results in repeated action potential firing and a reduced action potential precision. In contrast, temporal summation at CP-AMPAR/NR2B-containing NMDARs is modest. Synaptic events triggered by NR2B-containing NMDARs possess a slower time course, but their small amplitude renders them unable to significantly influence transmission at these synapses at resting membrane potentials (Bischofberger and Jonas, 2002; Lei and McBain, 2002). Consequently, repetitive activation of CP-AMPAR synapses results in rapid and brief synaptic potentials (Lei and McBain, 2002; Walker et al., 2002) that trigger single action potentials at their peak with little jitter. Consistent with a modest role for NMDA receptors at this synapse type, addition of an NMDA receptor antagonist has little impact on the temporal envelope underlying repetitive activation and the temporal precision of action potential initiation is largely unchanged (Lei and McBain, 2002). This scenario suggests that CP-AMPAR containing synapses are designed for precision timing and rapid synaptic transmission (Lawrence and McBain, 2003; Jonas et al., 2004) whereas robust depolarization and multiple action potentials with no requirement for precision timing is a feature of mossy fiber–CI-AMPAR interneuron synapses.

Mossy fiber-inhibitory interneuron plasticity

In the 1990s, Maccaferri and McBain published several papers that suggested that excitatory synapses onto inhibitory interneurons lacked the NMDAR-dependent forms of long-term potentiation (LTP) that were observed at excitatory synapses onto principal cells (Maccaferri and

McBain, 1995, 1996; for reviews, see McBain and Maccaferri, 1997; McBain and Maccaferri, 1997). Our failure to observe the most widely studied form of postsynaptic expressed LTP at synapses onto interneurons drove us to consider whether *presynaptically* expressed forms of long-lasting plasticity were similarly absent at excitatory synapses onto inhibitory interneurons. To this end, we turned our attention to the mossy fiber synapse where a well-established form of presynaptically expressed NMDAR-*independent* form of LTP had been described (Nicoll and Schmitz, 2005) (Figs. 4 and 5). Our hypothesis was that if this form of LTP was expressed as a change in presynaptic transmitter release probability, then these changes should distribute to all presynaptic terminals, regardless of the postsynaptic target type. It was also at this time that we became aware of the mossy fiber bouton versus filopodial arrangements described by Acsády et al. (1998) (Fig. 1); indeed this circuit arrangement fueled the experiments that were to follow. Given that the filopodial extensions originate from the large MF terminal, it might be assumed that any change in presynaptic release probability arising from LTP in the large MF bouton would distribute

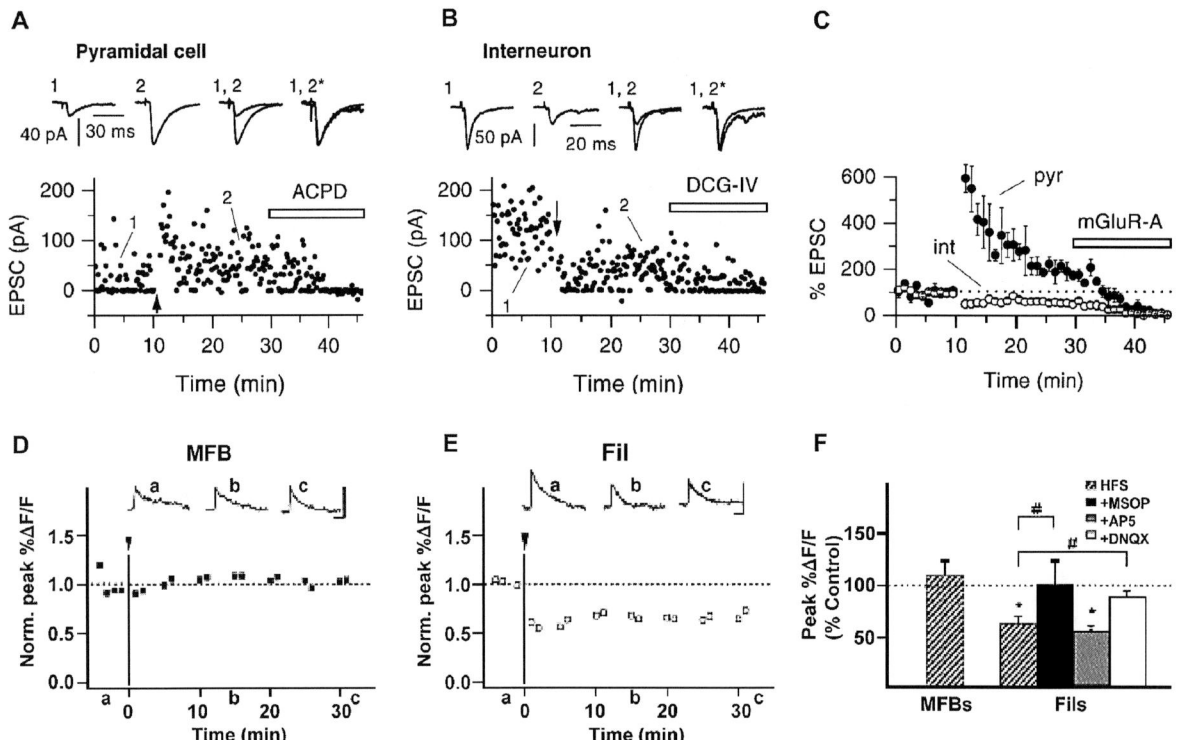

Fig. 4. Target-cell specific presynaptic LTD in the hippocampal MF pathway. A–C Representative single recording examples (A, B) and group data (C) illustrating divergent forms of NMDA receptor independent presynaptic plasticity at MF-PYR and MF-SLIN synapses following HFS (arrows). MF-PYR synapses undergo LTP while mossy fiber filopodial–interneuron synapses undergo LTD. Traces above are averaged EPSCs obtained at the times indicated, and are also shown scaled to each other (*). Application of an mGluR agonist (ACPD or DCG-IV) at the end of each recording confirms the MF origin of evoked synaptic events. Arrows in each panel reflect the time point for delivery of the induction protocol. D–F: Presynaptic two-photon Ca^{2+} imaging reveals compartmentalized VGCC regulation at MFB and Fil release sites. HFS (arrow and line) does not alter MFB CaTs that are undergoing LTP (D, F) but produces LTD of filopodial CaTs (E, F). The HFS-induced LTD of filopodial–interneuron synaptic CaTs is prevented by antagonizing presynaptic mGluR7 with MSOP and also by blocking postsynaptic CP-AMPARs with DNQX but is not affected by NMDAR inhibition with AP5 (F). Figures have been modified and reproduced Panels A–C from Maccaferri et al. (1998), Panels D–F from Pelkey et al. (2006).

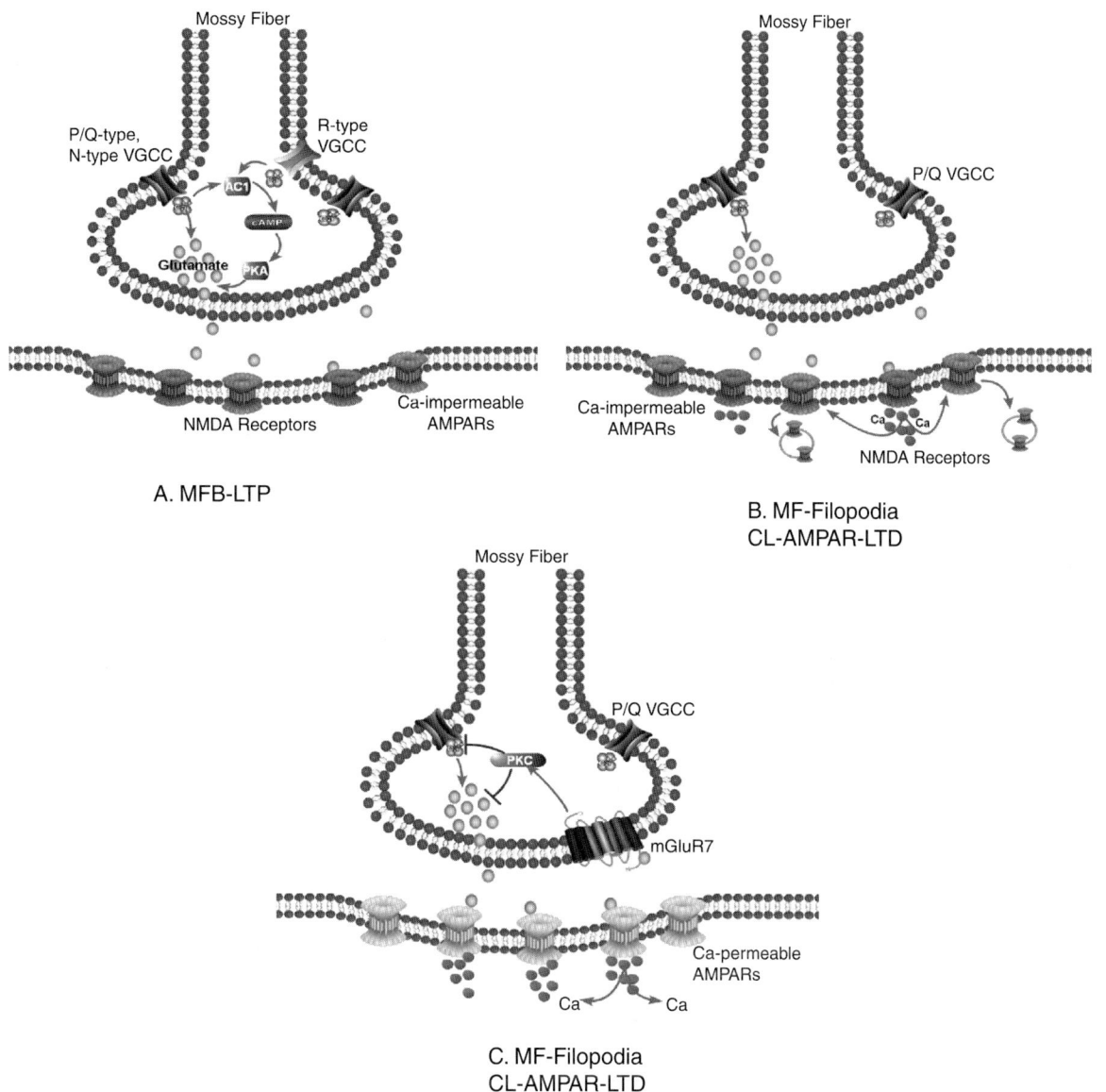

Fig. 5. Target-specific plasticity at mossy fiber synapses. Three schematics to illustrate the prevailing mechanisms underlying the three forms of plasticity at mossy fiber synapses onto pyramidal cells (Panel A) and inhibitory interneurons (Panels B and C). (A) A high-frequency non-associative induction paradigm produces LTP at mossy fiber bouton-pyramidal cell synapses that is NMDAR-independent and has a presynaptic locus of expression. Expression is thought to involve a Ca^{2+}-dependent increase in neurotransmitter release via a mechanism involving adenylyl cyclase 1 (AC1), cAMP, and protein kinase A (PKA) modulation of the release machinery. B and C: At filopodial synapses onto mossy fiber-CA3 interneuron a non-associative induction paradigm induces two forms of long-term depression. (B) Long-term depression at mossy fiber–interneuron synapses comprised of Ca^{2+}-impermeable (CI-) AMPA receptor synapses has a postsynaptic locus of induction and expression, is NMDA receptor dependent, and involves an endocytosis of surface AMPA receptors reminiscent of NMDAR-dependent LTD observed at principal cell synapses. (C) Long-term depression at mossy fiber–interneuron synapses comprised of CP-AMPA receptors has a requirement for a postsynaptic increase in intracellular Ca^{2+} but is expressed presynaptically. This form of LTD is NMDA receptor independent, requires activation of presynaptic mGluR7 and downstream PKC-dependent cascades to reduce transmitter release probability via a reduction of the voltage-gated Ca^{2+} transient generated by P/Q Ca^{2+} channels.

evenly to all synapses, resulting in a potentiation of transmission at both MF-pyramidal cell and interneuron synapses.

MF-principal cell LTP arises via an adenylyl cyclase-cAMP-dependent mechanism involving PKA phosphorylation of Rab3/Rim1a to ultimately strengthen synaptic transmitter release (Castillo et al., 1997, 2002; Nicoll and Schmitz, 2005) (Fig. 5). Although there are arguments for and against a requirement for a postsynaptic induction locus (Yeckel et al., 1999; Nicoll and Schmitz, 2005) almost all camps agree that expression of mossy fiber-pyramidal cell LTP proceeds as an increase in initial release probability of transmitter glutamate release. In marked contrast, the same high-frequency induction protocol that causes LTP at principal cell synapses induces two distinct forms of long-term depression (LTD) at mossy fiber–interneuron synapses (Figs. 4 and 5) (Maccaferri et al., 1998; Lei and McBain, 2004; Pelkey et al., 2005). One can immediately appreciate that this differential, strengthening of synapses onto CA3 pyramidal cells combined with a simultaneous weakening of transmission onto stratum lucidum interneurons will have important consequences for the dynamics of the CA3 circuit.

When these results were originally published (McBain and Maccaferri, 1997) we were unaware that two distinct mossy fiber–interneuron synapse types existed; i.e. synapses formed with either CI-AMPARs or CP-AMPARs (Figs. 2 and 3). However once we had characterized more fully the presence of CI- versus CP-AMPARs at mossy fiber–interneuron synapses, Toth et al. (2000) and Lei and McBain (2002) revisited these original experiments to establish whether common mechanisms underlay LTD at either synapse type. Indeed high-frequency stimulation of mossy fiber afferents induced LTD at both mossy fiber–CP-AMPAR and CI-AMPAR synapse types. Although the induction locus was postsynaptic in both cases, and blocked by inclusion of BAPTA in the recording pipette, the locus of LTD expression was presynaptic at CP-AMPAR but *postsynaptic* at CI-AMPARs (Figs. 4 and 5) (Lei and McBain, 2004). Importantly the mechanisms underlying either form of LTD were also distinct. LTD at CI-AMPARs is NMDAR dependent and blocked by the antagonist, AP5, and by inclusion of intracellular BAPTA indicating a role for Ca^{2+}-elevation via NMDA receptor activation. Moreover, parameters consistent with changes in presynaptic transmitter release, paired pulse ratio, coefficient of variance (CV), failure rate, and the magnitude of block by the low-affinity glutamate receptor antagonist γ-DGG were all unchanged after LTD of CI-AMPA receptor-mediated transmission. In addition, this form of LTD relies on an AP-2 dependent internalization of AMPAR by a mechanism similar to that seen at other synapses expressing NMDAR-dependent LTD (Lei and McBain, 2004).

At mossy fiber CP-AMPAR-interneuron synapses, LTD is NMDAR-*independent*. Moreover, this form of depression is accompanied by changes in the paired pulse ratio, failure rate, and the CV, all indices consistent with changes in presynaptic transmitter release probability. However, none of these parameters by themselves definitively prove a presynaptic locus. To explore the presynaptic locus of LTD at CP-AMPAR synapses in more detail, Lei and McBain (2004) took advantage of the low-affinity glutamate receptor antagonist γ-DGG. This compound has been used at a number of synapses to probe for changes in the transmitter release profile under a variety of conditions (Liu et al., 1999; Wadiche and Jahr, 2001). At CP-AMPAR synapses, the magnitude of γ-DGG block of the evoked EPSC was increased following LTD expression, indicative of a lower concentration of transmitter release following depression. These data strongly support the hypothesis that LTD at CP-AMPARs proceeds as a change in glutamate release probability either by alterations in multivesicular release or for example by converting full fusion vesicular release events into partial fusion, thus reducing the cleft glutamate concentration. Independent support of this presynaptic locus of expression came from experiments using postsynaptic delivery of peptides (pep2m) designed to interfere with AMPA receptor internalization. In contrast to CI-AMPARs, infusion of pep2m into the postsynaptic compartment had no effect of either basal synaptic transmission or the ability to induce LTD at CP-AMPAR synapses (Lei and McBain, 2004).

Taken together these data highlight two important aspects of mossy fiber transmission onto inhibitory interneurons. First, mossy fiber–interneuron synapses are constructed from two distinct types of postsynaptic glutamate receptor, GluR2-lacking CP-AMPA receptors and GluR2-containing, CI AMPA receptors. Importantly, distinct forms of long-lasting plasticity are generated at these synapses, that appear to be largely dictated by the nature of the AMPA receptor present on the postsynaptic side. Interestingly, in synapses that contain a mixture of either CP-AMPARs and CI-AMPARs, i.e. cells with only mild sensitivity to philanthotoxin or are weakly rectifying, it appears that LTD defaults to the presynaptic mechanism (Lei and McBain, 2004).

Before discussing this presynaptic form of LTD in greater detail, it is worthwhile pausing to emphasize that these data suggest that a common non-associative induction paradigm triggers a presynaptic form of strengthening at mossy fiber bouton synapses onto pyramidal cells yet simultaneously triggers a long-lasting depression at filopodial synapses onto inhibitory interneurons. This observation strengthens the hypothesis that the anatomical arrangement of filopodia emerging from the parent mossy fiber bouton may serve to function as a solution for a compartmentalization of a biochemical mechanism (for further discussion, see Pelkey and McBain, 2007). Importantly, these data suggest that each synapse type can independently regulate activity-induced changes in synaptic strength using distinct signaling cascades. Evidence in support of this hypothesis comes from our original observation that the adenylyl cyclase activator, forskolin, which has been used repeatedly as a exogenous means to trigger LTP at mossy fiber-CA3 synapses, has no impact on synaptic strength at the filopodial–interneuron synapse under naïve conditions (Maccaferri et al., 1998). This indicates that some aspect of the cAMP-dependent cascade present in the mossy fiber bouton may be absent from the filopodial extension or that the target(s) of the adenylyl cyclase-activated cascade may be inactive or sequestered.

mGluR7 functions as a trigger for bidirectional plasticity at the mossy fiber–interneuron synapses

The above discussion highlighted that the filopodial extensions and the parent mossy fiber bouton function independently with regard to their respective transmitter release probabilities, short-term mechanism of plasticity, and their differential sensitivity to high-frequency stimulation. This might suggest that the mechanism underlying transmitter release at each terminal is controlled by different synaptic machinery. At the large mossy fiber bouton, transmitter exocytosis is controlled by the high threshold P/Q (Cav2.1) and N-type (Cav2.2) voltage-gated calcium channels (VGCCs), with some evidence for a minor role for R-type VGCCs (Bischofberger et al., 2006; Nicoll and Schmitz, 2005). We considered it possible that the filopodial extensions may utilize a different compliment of VGCCs to trigger transmitter release. However, both evoked mossy fiber–interneuron synaptic events as well as presynaptic Ca^{2+} transients were similarly sensitive to the P/Q and N-type Ca^{2+} channel blockers ω-agatoxin IVa and ω-conotoxin GVIa. These data suggest that at the level of transmitter release, the filopodial extensions and the large parent mossy fiber bouton utilize the same VGCC types (Pelkey et al., 2006).

One important clue regarding the control of transmitter release/plasticity at the filopodial extensions versus the mossy fiber bouton came from the study of Shigemoto et al. (1997). In an elegant study of the distribution of metabotropic glutamate receptors throughout the hippocampal formation, they demonstrated that the Group III mGluR7b isoform was expressed exclusively within the mossy fiber pathway. Surprisingly, mGluR7b was largely restricted to the small filopodial and en passant terminals made onto smooth dendrites of inhibitory interneurons and was largely absent at the larger mossy fiber boutons. Given that many mGluRs function as auto- or heterosynaptic receptors to regulate presynaptic release probability we set out to determine whether mGluR7 performed a similar function at the filopodial synapse onto interneurons (Pelkey et al., 2005). Surprisingly, brief application of the Group III agonist, L-AP4, at a concentration capable of activating mGluR7, induced a long-lasting, chemical depression of excitatory transmission at mossy fiber–CP-AMPAR synapses. This chemical LTD is virtually identical to that induced by high-frequency stimulation at the same synapse; L-AP4 caused an increase in the paired pulse ratio, increased CV, and failure rate suggesting a presynaptic locus of expression for this chemical LTD. To our surprise, application of L-AP4 in the absence of synaptic transmission fails to trigger the chemical LTD and instead only a reversible form of short-term depression is observed. Like high-frequency induced LTD, inclusion of the calcium chelator BAPTA in the intracellular solution prevented chemical LTD induction. This suggests that similar to high-frequency stimulation-induced LTD, postsynaptic activity is also required for induction of the chemical-induced form of LTD. How postsynaptic activation/induction triggers presynaptic LTD expression either driven by L-AP4 or HFS is presently unclear.

LTD induced by either exogenous application of L-AP4, or high-frequency stimulation, results in a persistent depression of the presynaptic voltage-gated calcium channel signal monitored by two-photon microscopy (Fig. 4D–F). This persistent depression of the presynaptic Ca^{2+} transient arises through a preferential depression of P/Q voltage-gated Ca^{2+} channels and is blocked by an antagonist of mGluR7 (MSOP) (Pelkey et al., 2006). Importantly, blockade of P/Q channels prevents both the depression of synaptic events and calcium transients by either L-AP4 or high-frequency stimulation indicating that the remaining functional N-type channels play no (or little) role in this form of LTD. A similar application of L-AP4 at the mossy fiber bouton synapses made onto CA3 pyramidal cells failed to result in a persistent depression of the presynaptic Ca^{2+} transient, consistent with an absence of mGluR7 at these terminals. Moreover, although there likely exist other Group III mGluRs at mossy fiber-pyramidal cell synapses (Shigemoto et al., 1997), their activation triggers only a short-term and reversible depression, suggesting that Group III mediated LTD is not a common phenomenon and so far is limited only to the mossy fiber filopodia–interneuron synapse. In addition, although the Group II mGluR agonist DCG-IV is known to trigger a chemical form of LTD at the mossy fiber-CA3 pyramidal cell synapse, this form of LTD is not accompanied by a persistent depression of the presynaptic Ca^{2+} transient (Pelkey et al., 2006). These data indicate that synaptic depression at mossy fiber–CP-AMPAR interneuron synapses arises by an mGluR7-dependent persistent reduction in P/Q calcium channel function.

Of particular interest, mGluR7 not only plays a role as a conventional autoreceptor at these synapses but also acts as a metaplastic switch at MF-SLIN synapses. An "occlusion" experiment designed to prove that chemical- and high-frequency-induced LTD share one and the same mechanism, instead revealed that following L-AP4 application, the same HFS induction protocol that resulted in a depression at naïve synapses, now strengthened or de-depressed synaptic transmission (Pelkey et al., 2006) (Fig. 6). Interestingly, HFS induction restored the synaptic strength close to the original "setting" observed during the control epoch. This series of experiments allowed us to determine that mGluR7 activation and cell surface expression governs the direction of plasticity at mossy fiber–interneuron synapses. In naïve slices, mGluR7 activation during high-frequency stimulation generates MF-SLIN LTD by depressing presynaptic transmitter release through a PKC-dependent mechanism that involves the persistent downregulation of the Ca^{2+} transients arising via P/Q VGCC activation (Fig. 5). However, like many G-protein coupled receptors following agonist exposure, mGluR7 undergoes receptor internalization. Two avenues of investigation allowed us to reach this conclusion. First, using overexpression of a myc-tagged mGluR7 in cultured hippocampal neurons, we demonstrated that the same duration application of L-AP4 resulted in a robust internalization of both isoforms of mGluR7a and b (Pelkey et al., 2005). This observation was then extended to immunoelectron microscopy of mossy fiber filopodial terminals. At the level of the electron microscope, application of L-AP4 was observed to result in both a loss of mGluR7b from mossy fiber filopodia presynaptic grid and its accumulation deep within the

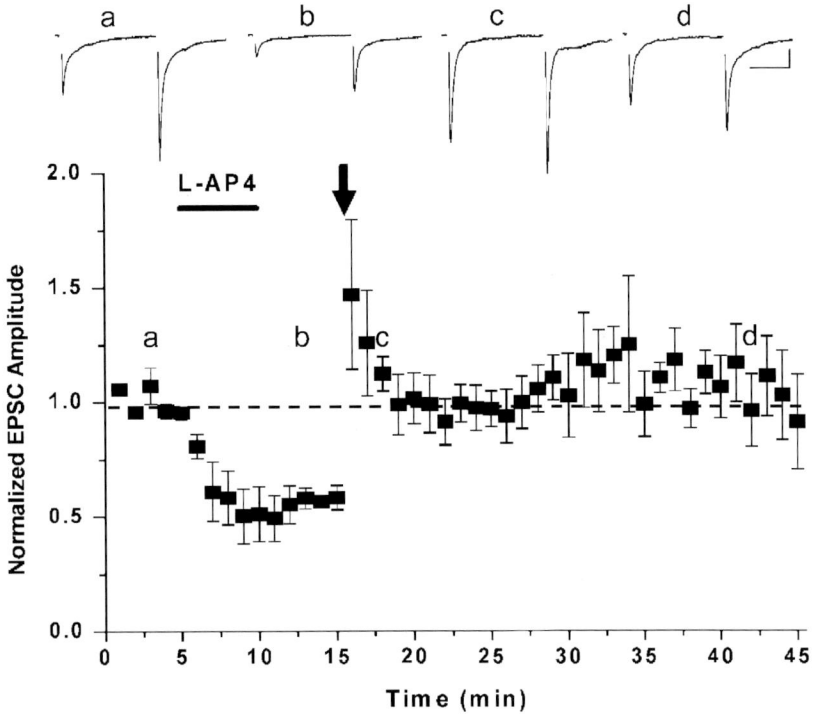

Fig. 6. Activation of mGluR7 unmasks presynaptic MF–interneuron LTP. Application of the Group III mGluR agonist, L-AP4, triggers a chemical form of presynaptic mGluR7-dependent long-lasting depression. Subsequent delivery of a non-associative high-frequency stimulation (indicated by arrow) which in naïve tissue would induce LTD, now induces an LTP of MF–interneuron transmission. Synaptic events across the top of the dot plot are averaged paired pulse data taken from the time points indicated by the letters. Figure has been reproduced from Pelkey et al. (2005).

presynaptic terminal. This internalization unmasks the ability of MF-SLIN synapses to undergo presynaptic potentiation or de-depression in response to the same HFS that induced LTD in naïve slices (Fig. 6). Thus, the selective mGluR7 accumulation at MF terminals contacting SLINs and not PYRs provides cell-target specific plasticity and bi-directional control of feedforward inhibition (Pelkey et al., 2005, 2006).

Unfortunately, at this time the underlying mechanism(s) for this novel form of mossy fiber–interneuron LTP is unknown. However, it is possible that it may simply proceed as the molecular inverse of the mGluR7-mediated LTD or alternatively may arise via an entirely distinct sequence of events. Importantly, it is currently also unclear whether the emergence of the de-depression or potentiation of mossy fiber–interneuron transmission provides the signal necessary to allow reinsertion of mGluR7 back onto the presynaptic terminal. An important issue that remains is, what is the duration of the temporal window controlled by the surface expression and internalization of mGluR in vivo? These are important questions that future research will likely uncover the answers. However, it is highly likely that the presence or absence of mGluR7 on the presynaptic mossy fiber–interneuron terminal represents a key trigger for the bidirectional control of this important feedforward inhibitory pathway.

Implications for the mossy fiber-CA3 circuit

In most systems studied so far, it has been difficult to gauge how long-lasting plasticity at one synapse could influence activity in either the feedforward or feedback inhibitory circuit. However, our data reveal that a common induction paradigm strengthens transmission at the mossy fiber inputs onto CA3 pyramidal cells, while simultaneously *weakening* transmission onto stratum lucidum interneurons. These mossy fiber–interneuron synapses represent the primary feedforward inhibitory drive onto the CA3 pyramidal cells. Consideration of the number of excitatory inputs onto stratum lucidum interneurons, their high initial release probability of synaptic transmission, as well as the large diverging output of these cells onto their CA3 pyramidal cell targets, suggests that plasticity within this circuit could be critically important for the regulation of network excitability. Specifically, a mechanism that simultaneously strengthens excitatory drive of the mossy fiber-CA3 pyramidal cell transmission and weakens the mossy fiber driven di-synaptic feedforward inhibitory input onto the same cells may act to broaden the temporal window for mossy fiber-CA3 pyramidal cell integration.

Typically, in the naïve state, low-frequency mossy fiber synaptic transmission onto CA3 pyramidal cells occurs via a multitude of release sites, each with a low initial release probability. In contrast, mossy fiber inputs onto stratum lucidum interneurons occurs via presynaptic terminals that have a one order of magnitude higher release probability when driven at the same activation frequency. Under these conditions, point to point mossy fiber-pyramidal cell transmission will only sparsely activate the CA3 network. Importantly, this mossy fiber driven activation of CA3 pyramidal cell will arise only in a brief temporal window because the feedforward inhibitory drive to CA3 pyramidal cells will largely constrain the ability of principal cells to fire an action potential in response to its monosynaptic mossy fiber input (Pouille and Scanziani, 2001; Lawrence et al., 2004). Weakening mossy fiber transmission onto inhibitory interneurons will reduce their likelihood of firing an action potential in response to a single mossy fiber input (Toth et al., 2000; Lawrence and McBain, 2003; Lawrence et al., 2004), ultimately contributing to an erosion of the feedforward inhibitory input and the consequent opening of the temporal window for CA3 pyramidal cell action potential initiation. Couple this with the increased probability of transmitter release at potentiated mossy fiber-CA3 pyramidal cell synapses suggests that such a mechanism could exist to increase the net excitation within the CA3 auto-associative network.

Repeated high frequency activation of the mossy fiber pathway will however, likely exacerbate this bifurcation of synaptic drive in the CA3 network to further strengthen transmission onto its pyramidal cell while weakening transmission onto inhibitory interneuron targets. In the absence of a mechanism to re-set or re-establish the balance of synaptic transmission at both mossy fiber targets, the system would be rendered inherently unstable, ultimately resulting in a loss of inhibitory control and the potential for pathological levels of CA3 pyramidal cell network activation. Expression of the Group III mGluR7b at mossy fiber–interneuron synapses may be a mechanism that allows this divergence of signal processing to exist only in a narrow temporal window. Repetitive activation of the mossy fiber–interneuron synapse will trigger mGluR7 activation and its consequent internalization. This internalization is an essential feature that triggers re-strengthening of mossy fiber–interneuron synaptic transmission back to naïve conditions and may exist as a mechanism to reset the inhibitory synaptic tone in the CA3 network.

Clearly, many issues remain to be resolved. However, it is apparent that the mossy fiber-CA3 circuit has designed both an anatomical and physiological template that permits the differential control of basic synaptic transmission as well as endowing this unique synapse with opposing forms of plasticity that depends on the nature of the postsynaptic target. The presence of mGluR7b at the mossy fiber–interneuron filopodial synapse is a key protein in determining the strength and direction of the plasticity induced by a brief non-associative induction paradigm. Typically, metabotropic receptors are considered "autoreceptors" whose function is to decrease presynaptic release probability. At the mossy fiber–interneuron synapse mGluR7 can indeed play this role, however when synaptic activation persists during the epoch of mGluR7 activation this receptor can also trigger a long-lasting form of synaptic depression. mGluR7 is then able to undergo internalization on binding agonist and render the synapse competent for long-lasting potentiation in response to an identical non-associative induction paradigm. It remains to be formally tested as to whether this is a general property of all synapses bearing mGluR7 on their presynaptic terminals, or whether this is a mechanism peculiar only to the mossy fiber filopodial synapse.

Acknowledgment

This work was supported by an NICHD-NIH intramural award to Chris J. McBain.

References

Acsády, L., Kamondi, A., Sik, A., Freund, T. and Buzsaki, G. (1998) GABAergic cells are the major postsynaptic targets of mossy fibers in the rat hippocampus. J. Neurosci., 18(9): 3386–3403.

Amaral, D.G. (1979) Synaptic extensions from the mossy fibers of the fascia dentate. Anat. Embryol., 155: 241–251.

Amaral, D.G. and Dent, J.A. (1981) Development of the mossy fibers of the dentate gyrus: I. A light and electron microscopic study of the mossy fibers and their expansions. J. Comp. Neurol., 195(1): 51–86.

Bischofberger, J., Engel, D., Frotscher, M. and Jonas, P. (2006) Timing and efficacy of transmitter release at mossy fiber synapses in the hippocampal network. Eur. J. Physiol., 453: 361–372.

Bischofberger, J. and Jonas, P. (2002) TwoB or not twoB: differential transmission at glutamatergic mossy fiber–interneuron synapses in the hippocampus. Trends Neurosci., 25: 600–603.

Bowie, D. and Mayer, M.L. (1995) Inward rectification of both AMPA and kainate subtype glutamate receptors generated by polyamine-mediated ion channel block. Neuron, 15: 453–462.

Bragin, A., Jando, G., Nadasday, Z., Hetke, J., Wise, K. and Buzsaki, G. (1995a) Gamma (40–100 Hz) oscillation in the hippocampus of the behaving rat. J. Neurosci., 15: 47–60.

Bragin, A., Jando, G., Nadasday, Z., van Landeghem, M. and Buzsaki, G. (1995b) Dentate EEG spikes and associated interneuronal population bursts in the hippocampal hilar region of the rat. J. Neurophysiol., 73: 1691–1705.

Castillo, P.E., Janz, R., Sudhof, T.C., Tzounopoulos, T., Malenka, R.C. and Nicoll, R.A. (1997) Rab3A is essential for mossy fibre long-term potentiation in the hippocampus. Nature, 388(6642): 590–593.

Castillo, P.E., Schoch, S., Schmitz, F., Sudhof, T.C. and Malenka, R.C. (2002) RIM1alpha is required for presynaptic long-term potentiation. Nature, 415(6869): 327–330.

Claiborne, B.J., Amaral, D.G. and Cowan, W.M. (1986) A light and electron microscopic analysis of the mossy fibers of the rat dentate gyrus. J. Comp. Neurol., 246(4): 435–458.

Frotscher, M. (1985) Mossy fibres form synapses with identified pyramidal basket cells in the CA3 region of the guinea-pig hippocampus: a combined Golgi-electron microscope study. J. Neurocytol., 14: 245–259.

Frotscher, M., Seress, L., Schwerdtfeger, W.K. and Buhl, E. (1991) The mossy cells of the fascia dentata: a comparative study of their fine structure and synaptic connections in rodents and primates. J. Comp. Neurol., 312: 145–163.

Henze, D.A., Urban, N.N. and Barrionuevo, G. (2000) The multifarious hippocampal mossy fiber pathway: a review. Neuroscience, 98(3): 407–427.

Isaac, J.T.R., Ashby, M. and McBain, C.J. (2007) The role of the GluR2 subunit in AMPA receptor function and synaptic plasticity. Neuron, 54: 859–871.

Jonas, P., Bischofberger, J., Fricker, D. and Miles, R. (2004) Interneuron diversity series: fast in, fast out — temporal and spatial signal processing in hippocampal interneurons. Trends Neurosci., 27(1): 30–40.

Jonas, P., Major, G. and Sakmann, B. (1993) Quantal components of unitary EPSCs at the mossy fibre synapse on CA3 pyramidal cells of rat hippocampus. J. Physiol., 472: 615–663.

Jung, M.W. and McNaughton, B. (1993) Spatial selectivity of unit activity in the hippocampal granule granular layer. Hippocampus, 3: 165–182.

Kamboj, S.K., Swanson, G.T. and Cull-Candy, S.G. (1995) Intracellular spermine confers rectification on rat calcium-permeable AMPA and kainate receptors. J. Physiol., 486 (Pt 2): 297–303.

Koh, D.S., Burnashev, N. and Jonas, P. (1995) Block of native Ca^{2+}-permeable AMPA receptors in rat brain by intracellular polyamines generates double rectification. J. Physiol., 486 (Pt 2): 305–312.

Lawrence, J.J., Grinspan, Z.M. and McBain, C.J. (2004) Quantal transmission at mossy fibre targets in the CA3 region of the hippocampus. J. Physiol., 554: 175–193.

Lawrence, J.J. and McBain, C.J. (2003) Interneuron diversity series: containing the detonation — feedforward inhibition in the CA3 hippocampus. Trends Neurosci., 26(11): 631–640.

Lei, S. and McBain, C.J. (2002) Distinct NMDA receptors provide differential modes of transmission at mossy fiber–interneuron synapses. Neuron, 33(6): 921–933.

Lei, S. and McBain, C.J. (2004) Two loci of expression for long-term depression at hippocampal mossy fiber–interneuron synapses. J. Neurosci., 24(9): 2112–2121.

Liu, G., Choi, S. and Tsien, R.W. (1999) Variability of neurotransmitter concentration and nonsaturation of postsynaptic AMPA receptors at synapses in hippocampal cultures and slices. Neuron, 22: 395–409.

Maccaferri, G. and McBain, C.J. (1995) Passive propagation of LTD to stratum oriens-alveus inhibitory neurons modulates the temporoammonic input to the hippocampal CA1 region. Neuron, 15(1): 137–145.

Maccaferri, G. and McBain, C.J. (1996) Long-term potentiation in distinct subtypes of hippocampal nonpyramidal neurons. J. Neurosci., 16: 5334–5343.

Maccaferri, G., Toth, K. and McBain, C.J. (1998) Target-specific expression of presynaptic mossy fiber plasticity. Science, 279: 1368–1370.

McBain, C.J. and Maccaferri, G. (1997) Synaptic plasticity in hippocampal interneurons? A commentary. Can. J. Physiol. Pharmacol., 75: 488–494.

Mori, M., Abegg, M.H., Gahwiler, B.H. and Gerber, U. (2004) A frequency-dependent switch from inhibition to excitation in a hippocampal unitary circuit. Nature, 431: 453–456.

Nicoll, R.A. and Schmitz, D. (2005) Synaptic plasticity at hippocampal mossy fibre synapses. Nat. Rev. Neurosci., 6(11): 863–876.

Pelkey, K.A., Lavezzari, G., Racca, C., Roche, K.W. and McBain, C.J. (2005) mGluR7 is a metaplastic switch controlling bidirectional plasticity of feedforward inhibition. Neuron, 46(1): 89–102.

Pelkey, K.A. and McBain, C.J. (2007) Differential regulation at functionally divergent release sites along a common axon. Curr. Opin. Neurobiol., 17: 366–373.

Pelkey, K.A., Topolnik, L., Lacaille, J.-C. and McBain, C.J. (2006) Compartmentalized Ca^{2+} channel regulation at divergent mossy fiber release sites underlies target-cell dependent plasticity. Neuron, 52: 497–510.

Penttonen, M., Kamondi, A., Sik, A., Acsady, L. and Buzsaki, G. (1997) Feed-forward and feed-back activation of the dentate gyrus in vivo during dentate spikes and sharp wave bursts. Hippocampus, 7: 437–450.

Pouille, F. and Scanziani, M. (2001) Enforcement of temporal fidelity in pyramidal cells by somatic feedforward inhibition. Science, 293: 1159–1163.

Ramon y Cajal, S. (1904, 1995) Histology of the Nervous System, Vol. 2. Translated by Swanson, N., Swanson, L. Oxford University Press, New York.

Rozov, A. and Burnashev, N. (1999) Polyamine-dependent facilitation of postsynaptic AMPA receptors counteracts paired-pulse depression. Nature, 401(6753): 594–598.

Rozov, A., Zilberter, Y., Wollmuth, L.P. and Burnashev, N. (1998) Facilitation of currents through rat Ca^{2+}-permeable AMPA receptor channels by activity-dependent relief from polyamine block. J. Physiol., 511(Pt 2): 361–377.

Salin, P.A., Scanziani, M., Malenka, R.C. and Nicoll, R.A. (1996) Distinct short-term plasticity at two excitatory synapses in the hippocampus. Proc. Natl. Acad. Sci. U.S.A., 93: 13304–13309.

Shigemoto, R., Kinoshita, A., Wada, E. et al. (1997) Differential presynaptic localization of metabotropic glutamate receptor subtypes in the rat hippocampus. J. Neurosci., 17(19): 7503–7522.

Spruston, N. and McBain, C.J. (2006) Structural and functional properties of hippocampal neurons. In: Andersen P., Morris R., Amaral D., Bliss T. and O'Keefe J. (Eds.), The Hippocampus Book. Oxford University Press.

Toth, K. and McBain, C.J. (1998) Afferent-specific innervation of two distinct AMPA receptor subtypes on single hippocampal interneurons. Nat. Neurosci., 1(7): 572–578.

Toth, K., Suares, G., Lawrence, J.J., Philips-Tansey, E. and McBain, C.J. (2000) Differential mechanisms of transmission at three types of mossy fiber synapse. J. Neurosci., 20(22): 8279–8289.

Wadiche, J.I. and Jahr, C.E. (2001) Multivesicular release at climbing fiber–Purkinje cell synapses. Neuron, 32: 301–313.

Walker, H.C., Lawrence, J.J. and McBain, C.J. (2002) Activation of kinetically distinct synaptic conductances on inhibitory interneurons by electrotonically overlapping afferents. Neuron, 35: 161–171.

Yeckel, M.F., Kapur, A. and Johnston, D. (1999) Multiple forms of LTP in hippocampal CA3 neurons use a common postsynaptic mechanism. Nat. Neurosci., 2(7): 625–633.

CHAPTER 14

Long-term synaptic plasticity in hippocampal feedback inhibitory networks

Joe Guillaume Pelletier and Jean-Claude Lacaille*

Département de Physiologie, GRSNC, Université de Montréal, C.P. 6128, Succ. Centre-ville, Montréal, QC H3C 3J7, Canada

Abstract: Recent studies clearly indicate that long-term synaptic plasticity in hippocampal networks not only takes place at excitatory synapses of hippocampal granule and pyramidal cells, but also at excitatory synapses onto inhibitory interneurons. Various forms of long-term potentiation (LTP) and depression (LTD) have now been reported at glutamatergic synapses of interneurons in dentate gyrus (DG), CA3, and CA1 regions of the hippocampus. Importantly, the presence and type of these changes in synaptic efficacy appear to depend on the interneuron subtype, including its specific role within the hippocampal network. The data reviewed here suggest the existence of cell-type specific rules for synaptic plasticity in hippocampal feed-forward and feedback inhibitory networks. This specialized tuning of inhibition is likely important for global hippocampal function.

Keywords: interneuron; LTP; feed-forward; recurrent; hippocampus; facilitation; oriens; alveus

Long-term potentization (LTP) was originally discovered as a persistent increase in synaptic efficacy at glutamatergic synapses on hippocampal granule cells (Bliss and Lomø, 1973). Intensive research yielded comprehensive models of LTP induction and maintenance. Although the mechanisms of LTP at excitatory synapses onto hippocampal principal cells are beyond the scope of this chapter, a key element in induction is a postsynaptic elevation in intracellular Ca^{2+} level (Lynch et al., 1983; Malenka et al., 1988). This finding highlights the necessity of synchronous pre- and postsynaptic activation of hippocampal cells in LTP induction. This form of Hebbian pairing is thought to be the physiological basis for the lasting associations between stimuli that underlie memory (Kandel, 2001; but see also Stevens, 1998; Martin et al., 2000). Interestingly, glutamatergic synapses onto inhibitory interneurons can also display LTP following a Hebbian pairing protocol (Alle et al., 2001; Perez et al., 2001; Lamsa et al., 2005). Furthermore, this form of interneuron LTP also critically relies on postsynaptic Ca^{2+} entry.

However, research in the interneuron LTP field has been exceedingly slow and controversial (McBain et al., 1999; Perez et al., 2001; Lamsa et al., 2005, 2007). Part of this problem derives from the large diversity of hippocampal interneurons (see Freund and Buzsáki, 1996; McBain and Fisahn, 2001; Maccaferri and Lacaille, 2003; Somogyi and Klausberger, 2005). Hippocampal interneurons have highly diverse morphology, as well as heterogeneous expression of calcium

---

*Corresponding author. Tel.: +1 (514) 343-5794; Fax: +1 (514) 343-2111; E-mail: jean-claude.lacaille@umontreal.ca

binding proteins, peptides, receptors, and channels. Not surprisingly, the rules of transmission at glutamatergic synapses onto interneurons are also heterogeneous. The exact reasons for such diversity remain to be clarified.

One possible solution is to consider interneurons in terms of their local circuit synaptic interactions within the hippocampal network and to characterize them as a function of their participation in feed-forward or feedback inhibitory networks (Lacaille et al., 1987; Thompson and Gahwiler, 1989; Pouille and Scanziani, 2001, 2004). Although simplistic, this dichotomy is nonetheless useful. Particularly appealing is the possibility that each pathway contains a collection of interneurons that possess comparable modes of synaptic transmission and plasticity, which may therefore analogously affect global hippocampal function. Thus, the aim of the present chapter is to consider the rules and consequences of short- and long-term plasticity at interneuron excitatory synapses in light of their role in local circuit synaptic networks.

Different local circuit inhibitory networks in the hippocampus

Hippocampal circuitry has been studied extensively by the early anatomists (Cajal, 1911). Most interneurons in the hippocampus have their dendrites and axons confined within their particular sub-region and are thus considered local circuit cells. Inhibitory networks form a continuum between feedback and feed-forward inhibition (Fig. 1; Andersen et al., 1969; Schwartzkroin, 1975; Knowles and Schwartzkroin, 1981). At one extreme, in feedback inhibition, an interneuron is innervated by local recurrent excitatory inputs from projection cells (P-cells). This interneuron consequently inhibits P-cells that excite it, hence creating a feedback inhibitory loop. Feed-forward interneurons are at the other extreme and receive excitatory inputs exclusively from afferent fibers of extrinsic origin. These interneurons in turn inhibit local P-cells, thus providing feed-forward inhibition.

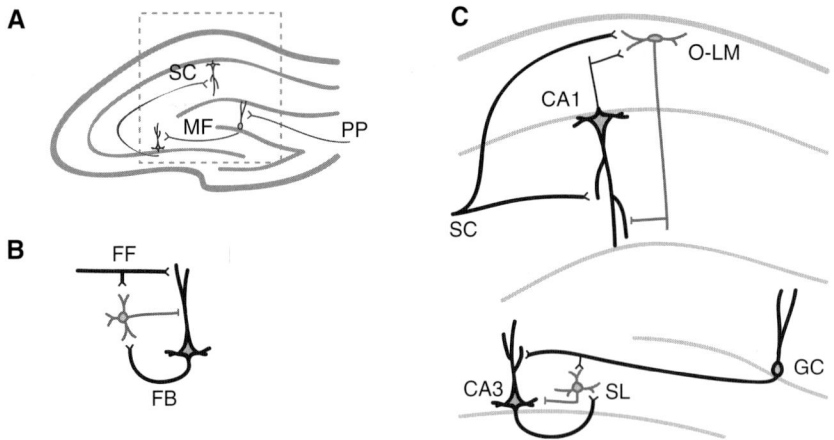

Fig. 1. Plasticity in feed-forward and feedback inhibitory circuits. (A) Schematic representation of the trisynaptic excitatory circuit: the perforant path (PP) connecting the entorhinal cortex and DG, the mossy fiber inputs (MF) from DG to CA3, and the Schaffer collateral (SC) projection from CA3 to CA1. (B) Diagram illustrating feed-forward (FF) and feedback (FB) activation of an inhibitory interneuron (light gray cell). (C) Enlargement of outlined region in (A) showing local connectivity of a CA1 O-LM interneuron (O-LM) and a CA3 stratum lucidum (SL) interneuron. Dentate GCs synapse onto both P-cells and SL interneurons in the CA3 region. Feed-forward inhibition results from the simultaneous excitation of both cell types. Feedback inhibition results from activation of CA3 P-cell recurrent collaterals and stimulation of SL interneurons. LTD of GC to SL interneuron synapses permits more reliable transmission between GCs and CA3 P-cells. In CA1, feedback inhibition also occurs in P-cells by activation of O-LM cells. O-LM interneurons also receive a minor feed-forward input from CA3 P-cells (SC). LTP at feedback excitatory synapses of O-LM interneurons increases the gain of dendritic inhibition of CA1 P-cells.

Many types of interneurons have been identified, possessing dendritic arborizations in the terminal field of perforant path, mossy fiber, or Schaffer collateral axons, and thus participating in feed-forward inhibition (Fig. 1A; Traub and Miles, 1991; Freund and Buzsáki, 1996). The entorhinal cortex constitutes another major extrinsic excitatory afferent to hippocampal interneurons. Interneurons receiving such extra-hippocampal input can also be considered as feed-forward interneurons. For example, interneurons found in the dentate gyrus (DG) molecular layer and CA1 stratum lacunosum-moleculare receive inputs from entorhinal layers II and III, respectively (Freund and Buzsáki, 1996). However, the DG mossy fiber projections to CA3 and the CA3 Schaffer collateral projections to CA1 constitute the major afferents to CA3 and CA1 feed-forward interneurons, respectively (Ishizuka et al., 1990; Acsady et al., 1998). As a result, CA3 stratum lucidum and CA1 stratum radiatum interneurons partake in feed-forward inhibition (Lawrence and McBain, 2003; Mori et al., 2004; Lamsa et al., 2005). Other feed-forward interneurons have been described, but because of a lack of physiological data, will not be discussed further here (Freund and Buzsáki, 1996; Pouille and Scanziani, 2001; Somogyi and Klausberger, 2005).

Feedback inhibitory circuits, on the other hand, are excited by intra-regional recurrent collaterals. For example, interneurons found in CA1 stratum oriens/alveus, especially the so-called stratum oriens-lacunosum moleculare (O-LM) cells, as well as DG hilar interneurons are examples of feedback inhibitory interneurons because they are principally innervated by, and in turn project back to, their respective P-cells. Bistratified and trilaminar cells in oriens/alveus, as well as basket cells, are also considered feedback interneurons. This review mainly focuses on interneurons in stratum oriens/alveus (OA-INs) because LTP has been extensively studied in these cells.

Unfortunately, the picture is not always so clear and many interneurons receive mixed inputs, being innervated to varying degrees by both feed-forward and recurrent afferents (Freund and Buzsáki, 1996; Somogyi and Klausberger, 2005). Moreover, two adjacent neurons within a given region can possess diverse sets of inputs and outputs (Freund and Buzsáki, 1996; Somogyi and Klausberger, 2005). This entails that knowing the location of the cell body relative to hippocampal subfield and layer may not be sufficient to identify interneuron afferent sources. For example, OA-INs, in addition to receiving local recurrent excitatory inputs from CA1 P-cells, are moderately innervated by P-cells located in CA3 (Ishizuka et al., 1990). Similarly, CA3 stratum lucidum interneurons receive feed-forward afferents from DG cells (Acsady et al., 1998) and recurrent inputs from CA3 P-cell collaterals (Ishizuka et al., 1990).

Afferent specific mechanisms of transmission

The fact that interneurons receive afferents from different sources begs the question: can interneurons differentiate between both types of afferent activation? Indeed, interneurons can keep track of these inputs by using distinct mechanisms of transmission. Such afferent-specific dichotomy in mechanisms of transmission was first described for CA3 stratum lucidum interneurons. It is well known that mossy fiber synapses to these interneurons are composed of CP-AMPARs, whereas CA3 recurrent collateral synapses consist of CI-AMPARs on the same cell (Toth and McBain, 1998). A similar situation occurs in CA1, where recurrent afferents from CA1 P-cells form synapses composed of Ca^{2+}-permeable (CP) AMPA receptors on OA-INs. However, feed-forward input from CA3 P-cells on these same interneurons makes synaptic contacts with Ca^{2+}-impermeable (CI) AMPA receptors (Croce and Lacaille, 2005). Moreover, activation of presynaptic group II mGluRs produces depression of CA1 P-cell synapses on OA-INs, but does not affect CA3 P-cell synapses on these cells (Lapointe et al., 2004; Croce and Lacaille, 2005). Thus, interneurons can distinguish between both types of inputs despite the extensive overlap of afferent distribution throughout the somato-dendritic axis (Walker et al., 2002). These results suggest that individual interneurons can participate in multiple inhibitory networks and that distinct synaptic mechanisms ensure a functional dichotomy between feedback and feed-forward activation. However, it is important to note that the mechanisms of transmission (e.g., Ca^{2+} permeability of AMPARs) at synapses from the different types of afferent are the reverse in stratum lucidum compared to OA-INs. Thus, afferent-specific mechanisms of transmission are common principles in interneurons, but the distinct mechanisms associated with each afferent are interneuron subtype-specific.

Afferent-specific short-term synaptic plasticity

The distinction between feedback and feed-forward activation of interneurons is also valid for their properties of short-term plasticity. CA1 OA-INs demonstrate a mixture of facilitation and depression of synaptic responses following repetitive stimulation of their afferents (Losonczy et al., 2002). This variability in short-term plasticity may reflect a heterogeneous sampling of cell types (Losonczy et al., 2002; Pouille and Scanziani, 2004). For example, repetitive activation of excitatory synapses on O-LM cells results in facilitation, whereas it produces depression at synapses on other interneurons such as basket and axo-axonic cells (Pouille and Scanziani, 2004). However, short-term plasticity tends to be highly variable even within a single histochemically confirmed interneuron subtype (Losonczy et al., 2002).

However, it was demonstrated that a single interneuron can display either depression or facilitation of excitatory responses, depending on the type of afferent synapse that is tested. For example, in CA3 stratum lucidum interneurons, repetitive stimulation produces depression of synapses with CI-AMPARs (feedback input from CA3 recurrent collaterals) and facilitation of synapses with CP-AMPARs (feed-forward input from DG granule cell mossy fibers) (Maccaferri et al., 1998; Toth et al., 2000). Similarly in CA1 OA-INs, short-term facilitation or depression is reliably obtained, depending on the type of the presynaptic cell stimulated during paired whole cell recordings. EPSCs recorded in OA-INs and induced by repetitive activation of CA1 P-cells (feedback input) show facilitation, whereas those evoked by repetitive CA3 P-cell stimulation (feed-forward input) display depression (Croce and Lacaille, 2005). These results raise the possibility that the variability in short-term plasticity previously reported for a given interneuron cell type (Losonczy et al., 2002) could be attributed to variable recruitment of feedback and feed-forward afferents by the stimulation paradigm. The disparity of short-term plasticity as a function of specific afferents has not been systematically studied in other hippocampal interneurons. Nonetheless, the available data indicate that afferent-specific properties of short-term plasticity is also a common principle among different interneurons; however, again the distinct properties associated with each afferent appear interneuron subtype-specific.

Long-term plasticity

In recent years, accumulating data have shown that many types of hippocampal interneurons show activity-dependent long-term changes in synaptic efficacy (Maccaferri and McBain, 1996; Laezza et al., 1999; Perez et al., 2001; Alle et al., 2001; Doherty et al., 2004; Lamsa et al., 2005). This issue has been extensively covered previously in feed-forward inhibitory networks (Maccaferri and McBain, 1996; Lawrence and McBain, 2003). However, long-term plasticity in feedback inhibitory networks has been the object of considerable debate (McBain et al., 1999), so the following section will start by focusing on LTP in feedback inhibitory networks.

Interneurons that participate in feedback inhibition develop LTP following high-frequency stimulation of their excitatory inputs. For example, CA1 OA-INs (Perez et al., 2001) and DG basket cells (Alle et al., 2001) show reliable increases in EPSC amplitude following paired synaptic stimulation and postsynaptic depolarization. This Hebbian form of plasticity is not due to disynaptic spread of LTP from neighboring P-cells (McBain et al., 1999). In support of this assertion, LTP of glutamatergic synapses onto OA-INs can be generated following continued bath application of the NMDA receptor blocker AP-5 (Perez et al.,

2001). Under such conditions, NMDA-dependent LTP is not induced in upstream P-cells (Morris et al., 1986). Furthermore, LTP in OA-INs and basket cells does not develop if increases in postsynaptic Ca^{2+} rises are selectively prevented, suggesting that local postsynaptic processes are necessary for induction (Alle et al., 2001; Lapointe et al., 2004; Lamsa et al., 2007).

The molecular mechanisms responsible for LTP induction in feedback neurons are currently under investigation. OA-INs constitute the most extensively studied case of LTP in hippocampal recurrent inhibition. The induction of LTP of excitatory inputs onto OA-INs is critically dependent on postsynaptic group I mGluRs (mGluR1 and 5) (Fig. 2). Studies performed in mGluR1$(-/-)$ knock-out mice, as well as using a specific mGluR1α antagonist, have demonstrated that LTP in OA-INs does not develop under such conditions and is thus mGluR1 dependent (Perez et al., 2001; Lapointe et al., 2004; Topolnik et al., 2005). mGluR1α activation leads to elevations in postsynaptic $[Ca^{2+}]_i$ that are dependent mostly on the activation of a Src/ERK cascade, opening of transient receptor potential channels, as well as liberation from intracellular stores (Topolnik et al., 2006). Important Ca^{2+} elevations also occur as a result of liberation from ryanodine-sensitive stores following mGluR5 activation (Topolnik et al., 2006). Interestingly, recent experiments using the mGluR1/5 agonist DHPG in combination with selective mGluR1 or mGluR5 antagonists suggest that activation of either mGluR1 or mGluR5 are sufficient to induce LTP in OA-INs (Le Vasseur et al., 2007). Therefore, LTP in these interneurons can be induced by multiple independent mechanisms. The end product of mGluR1/5 receptor activation is likely an increase in $[Ca^{2+}]_i$ that is due to both release from intracellular stores as well as entry through transient receptor potential channels (Topolnik et al., 2006). The downstream effects of $[Ca^{2+}]_i$ elevations that are responsible for LTP induction in OA-INs remain to be investigated. In DG basket cells,

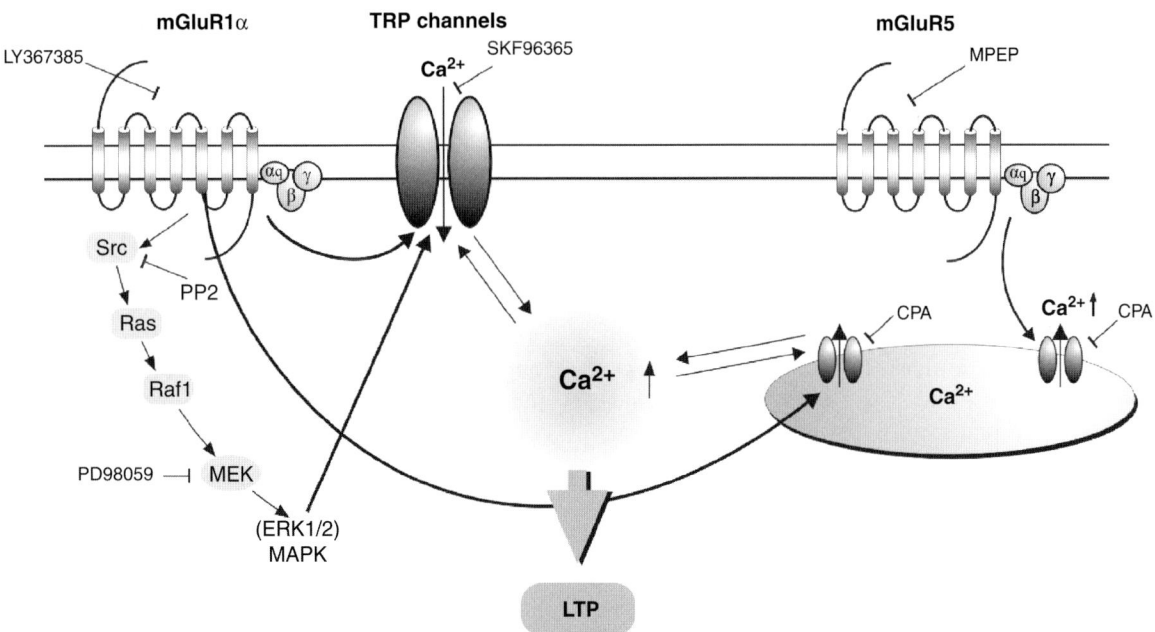

Fig. 2. Schematic representation of mGluR1 and 5 specific Ca^{2+} signaling mechanisms in OA-INs and role in induction of long-term potentiation. Activation of mGluR1 produces a Src and ERK-dependent Ca^{2+} influx via TRP channels and Ca^{2+} release from intracellular stores, leading to LTP induction (Topolnik et al., 2007). In contrast, mGluR5 activation produces solely Ca^{2+} release from intracellular stores. Reproduced with permission from Topolnik et al. (2006).

these mechanisms appear to involve protein kinase C (Alle et al., 2001). Considering that LTP maintenance is expressed as an increase in presynaptic release probability in both of these cell types (Alle et al., 2001; Lapointe et al., 2004), interneuron dendrites may eventually signal via a retrograde messenger.

Bearing in mind that a single interneuron may receive excitatory inputs from multiple sources with functionally distinct synaptic mechanisms (see above), a likely prediction is that the rules for plasticity are also different. At mossy fiber synapses onto CA3 stratum lucidum interneurons for example, two distinct forms of transmission have been identified on the basis of AMPAR subtype activation. Synapses with CP-AMPARs develop a presynaptic form of long-term depression (LTD), whereas at CI-AMPAR synapses, the expression of LTD is postsynaptic (Toth et al., 2000; Lei and McBain, 2004). Moreover feed-forward mossy fiber inputs onto stratum lucidum interneurons reliably develop LTD following high-frequency stimulation, but feedback collateral inputs from CA3 pyramids do not. These results indicate that activity-dependent long-term plasticity can be regulated discriminatively at different excitatory inputs of interneurons (Maccaferri et al., 1998). Interestingly, interneurons in stratum radiatum show both activity-dependent LTD (McMahon and Kauer, 1997) and LTP (Lamsa et al., 2005), although the specific synapses and their underlying mechanisms were not comprehensively investigated. CA1 stratum radiatum interneurons are mostly activated in a feed-forward manner, but their afferents may originate from many different sources (Freund and Buzsáki, 1996), making it difficult to selectively isolate and identify individual inputs. It cannot be excluded, therefore, that in these studies different afferent-specific synaptic mechanisms may be responsible for the distinct type of plasticity observed.

Studies in CA1 OA-INs also point to an afferent-specific plasticity in these cells. The feedback inputs from CA1 P-cells onto OA-INs, identified by their distinctive CP-AMPARs, short-term facilitation, and sensitivity to DCG-IV, develop LTP following theta-burst stimulation coupled to postsynaptic depolarization. However, the feed-forward excitatory inputs from CA3 P-cells to these same interneurons, identified by their CI-AMPARs, short-term depression, and insensitivity to DCG-IV, remain unchanged after the pairing protocol (Croce and Lacaille, 2005; Pelletier et al., 2007). Ca^{2+} signaling coupled to mGluR1, and not Ca^{2+} entry through CP-AMPARs, appears necessary to induce this plasticity (Fig. 2; Topolnik et al., 2007). However, a recent study suggests that the latter may be necessary for an anti-Hebbian type of plasticity at OA-IN synapses (Lamsa et al., 2007). It will be interesting to test if this anti-Hebbian form of LTP involves the same mechanisms as those in mGluR5-mediated LTP (Le Vasseur et al., 2007).

A single interneuron therefore possesses the capacity to distinguish between feed-forward and feedback inputs and to selectively adjust the relative strength of each according to patterns of activity. Interestingly, the specific rules of long-term plasticity depend on the type of interneuron. In CA1 OA-INs, CP-AMPARs at feedback excitatory synapses display LTP (Croce and Lacaille, 2005), whereas in CA3 stratum lucidum interneurons, CP-AMPARs at feed-forward excitatory synapses show LTD (Toth et al., 2000; Lei and McBain, 2004). This indicates that the rules for synaptic plasticity are a function of both interneuron cell type and afferent source. The functional significance for hippocampal processing of such cell type- and afferent-specific rules of long-term plasticity remains to be clearly defined, especially in the context of hippocampal-dependent memory. Nevertheless, these specific mechanisms of plasticity are likely to result in differential long-term synaptic modulation of feed-forward and feedback inhibitory networks, and thus contribute to hippocampal functioning in distinct ways.

Consequences of plasticity in interneurons

Activation of hippocampal interneurons leads to inhibition of P-cells (Bertrand and Lacaille, 2001). Simplistically then, changing the likelihood of interneuron firing will alter P-cell firing probability in the opposite direction. Not surprisingly, interneuron LTP and LTD are thought to respectively increase and decrease P-cell inhibition. It has been shown that LTP at synapses onto CA1 OA-INs (Lapointe et al., 2004) and stratum radiatum interneurons (Lamsa et al., 2005) increases P-cells inhibition. Conversely, it is likely that LTD of MF to CA3 stratum lucidum interneurons decreases disynaptic IPSCs in CA3 P-cells (Mori et al., 2004), although this issue has never been tested directly (Lawrence and McBain, 2003). Considering that the rules for synaptic plasticity are cell-type specific and can differentially affect feedback and feed-forward inhibition, it could be expected that these different mechanisms play distinctive roles in the control of P-cell behavior.

However, feedback and feed-forward interneurons have been shown to affect P-cells in other very discrete ways. For example, in CA1 P-cells feed-forward inhibition is greater in the soma than in dendrites, creating a short time-window (<2 ms) within which excitatory inputs can summate at the soma (Pouille and Scanziani, 2001). Interestingly, LTP at excitatory synapses onto CA1 feed-forward inhibitory interneurons is required to preserve this brief time-window of P-cell activation in the event of P-cell LTP (Lamsa et al., 2005). On the other hand, CA1 recurrent inhibitory OA-INs ramp up dendritic inhibition of P-cells as activity levels increase (Pouille and Scanziani, 2004). Thus, as P-cell firing frequency increases, recruitment switches from somatic-projecting to dendrite-projecting interneurons (Pouille and Scanziani, 2004). Since dendrite-projecting OA-INs undergo mGluR1-dependent Hebbian LTP (Perez et al., 2001), such frequency-dependent shift in dendritic inhibition is likely preserved after LTP but may be associated with an increase in the gain of inhibition to compensate for enhanced P-cell activity (Pelletier et al., 2007).

Another important function of interneurons is their key role in the regulation of hippocampal network oscillations. Membrane oscillations and associated rhythmic population discharges result from intrinsic membrane conductances, in addition to the coordinated interplay between excitatory and inhibitory inputs onto groups of P-cells (Buzsáki et al., 1983; Steriade et al., 1991). It is known, for example, that activity of interneurons is necessary for synchronized gamma oscillations (Mann et al., 2005; Bartos et al., 2007), which dominate hippocampal P-cell activity during memory encoding and retrieval (Bragin et al., 1995; Lisman and Idiart, 1995; Hasselmo et al., 1996). Recently, it was demonstrated that different types of interneurons fire in distinct phase-relationship with P-cells (Klausberger et al., 2003, 2004). These interesting findings suggest that different types of interneurons can participate in hippocampal oscillations in distinct ways. In a network model where LTP onto inhibitory interneurons is not allowed to occur, distant populations of integrate-and-fire P-cells (separated by >8 ms delays) never synchronize (Bibbig et al., 2002). In this model, LTP at interneuron excitatory synapses could reliably reset the phase of both beta and gamma oscillations between distant projection neurons, suggesting that it is necessary for grand-scale coordinated activity. Unfortunately, the model did not distinguish between feed-forward and feedback inhibitory networks. Another model tested the isolated effect of recurrent inhibition on network oscillations and found that preventing LTP at excitatory synapses of feedback interneurons led to poor transmission of oscillatory activity in a multi-layered network following P-cell LTP (Grunze et al., 1996).

In the CA3 region of the hippocampus, feed-forward inhibition is so reliable that it effectively prevents the propagation of excitatory inputs to neighboring P-cells. However, high-frequency activity can diminish the impact of feed-forward inhibition, permitting P-cells to develop highly synchronized activity (Mori et al., 2004). This state-dependent shift in inhibition serves to gate hippocampal inputs. However, a tight cellular regulation of this mechanism is required for proper processing. It will therefore be interesting to examine the consequences of interneuron long-term plasticity on P-cell oscillatory behavior and its possible role in hippocampal function.

Conclusions

Both feed-forward and feedback interneurons are capable of activity-dependent long-term changes in synaptic efficacy of their excitatory inputs. The specific rules of transmission and plasticity are kept distinct by separate pre- and postsynaptic receptors and signaling mechanisms at each type of synapse. This results in a division of labor, whereby feed-forward interneurons appear specialized for input selection, whereas feedback interneurons control a switch between coincidence detection and dendritic temporal integration. Furthermore, the rules for long-term synaptic plasticity appear to be cell-type specific, leading to a rich repertoire of interneuron plasticity and modulation of hippocampal function. Interneuron long-term synaptic plasticity may therefore exert a powerful control over global hippocampal functioning and memory formation.

Abbreviations

AMPA	α-amino-3-hydroxy-5-methylisoxazole-4-propionic acid
CA	cornu amonis
CI	calcium impermeable
CP	calcium permeable
DG	dentate gyrus
EPSC	excitatory postsynaptic current
LTD	long-term depression
LTP	long-term potentiation
mGluR	metabotropic glutamate receptor
NMDA	N-methyl-D-aspartate
OA-IN	oriens-alveus interneuron
O-LM	oriens-lacunosum moleculare
P-cell	principal cell
sl	stratum lucidum

Acknowledgments

J-C.L. is the recipient of a Canada Research Chair in Cellular and Molecular Neurophysiology. This work was also supported by an operating grant from the Canadian Institute of Health Research to J-C.L. and by the Fonds de la Recherche en Santé du Québec (FRSQ) to the GRSNC. J.G.P. is supported by an FRSQ fellowship.

References

Acsady, L., Kamondi, A., Sik, A., Freund, T. and Buzsáki, G. (1998) GABAergic cells are the major postsynaptic targets of mossy fibers in the rat hippocampus. J. Neurosci., 18: 3386–3403.

Alle, H., Jonas, P. and Geiger, J.R. (2001) PTP and LTP at a hippocampal mossy fiber-interneuron synapse. Proc. Natl. Acad. Sci. U.S.A., 98: 14708–14713.

Andersen, P., Gross, G.N., Lomo, T. and Sveen, O. (1969) Participation of inhibitory and excitatory interneurons in the control of hippocampal cortical output. In: Braizier M.A.B. (Ed.), The Interneuron. University of California Press, Los Angeles, CA, pp. 415–465.

Bartos, M., Vida, I. and Jonas, P. (2007) Synaptic mechanisms of synchronized gamma oscillations in inhibitory interneuron networks. Nat. Rev. Neurosci., 8: 45–56.

Bertrand, S. and Lacaille, J.C. (2001) Unitary synaptic currents between lacunosum-moleculare interneurones and pyramidal cells in rat hippocampus. J. Physiol., 532: 369–384.

Bibbig, A., Traub, R.D. and Whittington, M.A. (2002) Long-range synchronization of gamma and beta oscillations and the plasticity of excitatory and inhibitory synapses: a network model. J. Neurophysiol., 88: 1634–1654.

Bliss, T.V. and Lomø, T. (1973) Long-lasting potentiation of synaptic transmission in the dentate area of the anaesthetized rabbit following stimulation of the perforant path. J. Physiol., 232: 331–356.

Bragin, A., Jandó, G., Nadasdy, Z., Hetke, J., Wise, K. and Buzsáki, G. (1995) Gamma (40–100 Hz) oscillation in the hippocampus of the behaving rat. J. Neurosci., 15: 47–60.

Buzsáki, G., Leung, L.W. and Vanderwolf, C.H. (1983) Cellular bases of hippocampal EEG in the behaving rat. Brain Res., 287: 139–171.

Cajal, S.R. (1911) Histologie du système nerveux de l'homme et des vertebras. Maloine, Paris.

Croce, A. and Lacaille, J.C. (2005) Differential mechanisms of transmission and plasticity at feedforward and feedback excitatory synapses of hippocampal oriens alveus interneurons. Program No. 965.6. 2005 Abstract Viewer/Itinerary Planner, Society for Neuroscience, Washington, DC (Online).

Doherty, J.J., Alagarsamy, S., Bough, K.J., Conn, P.J., Dingledine, R. and Mott, D.D. (2004) Metabotropic glutamate receptors modulate feedback inhibition in a developmentally regulated manner in rat dentate gyrus. J. Physiol., 561: 395–401.

Freund, T.F. and Buzsáki, G. (1996) Interneurons of the hippocampus. Hippocampus, 6: 347–370.

Grunze, H.C., Rainnie, D.G., Hasselmo, M.E., Barkai, E., Hearn, E.F., McCarley, R.W. and Greene, R.W. (1996) NMDA-dependent modulation of CA1 local circuit inhibition. J. Neurosci., 16: 2034–2043.

Hasselmo, M.E., Wyble, B.P. and Wallenstein, G.V. (1996) Encoding and retrieval of episodic memories: role of cholinergic and GABAergic modulation in the hippocampus. Hippocampus, 6: 693–770.

Ishizuka, N., Weber, J. and Amaral, D.G. (1990) Organization of intrahippocampal projections originating from CA3 pyramidal cells in the rat. J. Comp. Neurol., 295: 580–623.

Kandel, E.R. (2001) The molecular biology of memory storage: a dialogue between genes and synapses. Science, 294: 1030–1038.

Klausberger, T., Magill, P.J., Marton, L.F., Roberts, J.D., Cobden, P.M., Buzsaki, G. and Somogyi, P. (2003) Brain-state- and cell-type-specific firing of hippocampal interneurons in vivo. Nature, 421: 844–848.

Klausberger, T., Marton, L.F., Baude, A., Roberts, J.D., Magill, P.J. and Somogyi, P. (2004) Spike timing of dendrite-targeting bistratified cells during hippocampal network oscillations in vivo. Nat. Neurosci., 7: 41–47.

Knowles, W.D. and Schwartzkroin, P.A. (1981) Local circuit synaptic interactions in hippocampal brain slices. J. Neurosci., 1: 318–322.

Lacaille, J.C., Mueller, A.L., Kunkel, D.D. and Schwartzkroin, P.A. (1987) Local circuit interactions between oriens/alveus interneurons and CA1 pyramidal cells in hippocampal slices: electrophysiology and morphology. J. Neurosci., 7: 1979–1993.

Laezza, F., Doherty, J.J. and Dingledine, R. (1999) Long-term depression in hippocampal interneurons: joint requirement for pre- and postsynaptic events. Science, 285: 1411–1414.

Lamsa, K., Heeroma, J.H. and Kullmann, D.M. (2005) Hebbian LTP in feed-forward inhibitory interneurons and the temporal fidelity of input discrimination. Nat. Neurosci., 8: 916–924.

Lamsa, K.P., Heeroma, J.H., Somogyi, P., Rusakov, D.A. and Kullmann, D.M. (2007) Anti-Hebbian long-term potentiation in the hippocampal feedback inhibitory circuit. Science, 315: 1262–1266.

Lapointe, V., Morin, F., Ratte, S., Croce, A., Conquet, F. and Lacaille, J.C. (2004) Synapse-specific mGluR1-dependent long-term potentiation in interneurones regulates mouse hippocampal inhibition. J. Physiol., 555: 125–135.

Lawrence, J.J. and McBain, C.J. (2003) Interneuron diversity series: Containing the detonation — feedforward inhibition in the CA3 hippocampus. Trends Neurosci., 26: 631–640.

Lei, S. and McBain, C.J. (2004) Two loci of expression for long-term depression at hippocampal mossy fiber–interneuron synapses. J. Neurosci., 24: 2112–2121.

Le Vasseur, M., Ran, I. and Lacaille, J.C. (2007) Selective induction of metabotropic glutamate receptor 1- and metabotropic glutamate receptor 5-dependent chemical long-term potentiation at oriens/alveus interneuron synapses of mouse hippocampus. Neurosci. doi: 10.1016/j.neuroscience.2007.09.071.

Lisman, J.E. and Idiart, M.A. (1995) Storage of 7 +/− 2 short-term memories in oscillatory subcycles. Science, 267: 1512–1515.

Losonczy, A., Zhang, L., Shigemoto, R., Somogyi, P. and Nusser, Z. (2002) Cell type dependence and variability in the short-term plasticity of EPSCs in identified mouse hippocampal interneurones. J. Physiol., 542: 193–210.

Lynch, G., Larson, J., Kelso, S., Barrionuevo, G. and Schottler, F. (1983) Intracellular injections of EGTA block induction of hippocampal long-term potentiation. Nature, 305: 719–721.

Maccaferri, G. and Lacaille, J.C. (2003) Interneuron diversity series: hippocampal interneuron classifications — making things as simple as possible, not simpler. Trends Neurosci., 26: 564–571.

Maccaferri, G. and McBain, C.J. (1996) Long-term potentiation in distinct subtypes of hippocampal nonpyramidal neurons. J. Neurosci., 16: 5334–5343.

Maccaferri, G., Toth, K. and McBain, C.J. (1998) Target-specific expression of presynaptic mossy fiber plasticity. Science, 279: 1368–1370.

Malenka, R.C., Kauer, J.A., Zucker, R.S. and Nicoll, R.A. (1988) Postsynaptic calcium is sufficient for potentiation of hippocampal synaptic transmission. Science, 242: 81–84.

Mann, E.O., Radcliffe, C.A. and Paulsen, O. (2005) Hippocampal gamma-frequency oscillations: from interneurones to pyramidal cells, and back. J. Physiol., 562: 55–63.

Martin, S.J., Grimwood, P.D. and Morris, R.G. (2000) Synaptic plasticity and memory: an evaluation of the hypothesis. Annu. Rev. Neurosci., 23: 649–711.

McBain, C.J. and Fisahn, A. (2001) Interneurons unbound. Nat. Rev. Neurosci., 2: 11–23.

McBain, C.J., Freund, T.F. and Mody, I. (1999) Glutamatergic synapses onto hippocampal interneurons: precision timing without lasting plasticity. Trends Neurosci., 22: 228–235.

McMahon, L.L. and Kauer, J.A. (1997) Hippocampal interneurons express a novel form of synaptic plasticity. Neuron, 18: 295–305.

Mori, M., Abegg, M.H., Gahwiler, B.H. and Gerber, U. (2004) A frequency-dependent switch from inhibition to excitation in a hippocampal unitary circuit. Nature, 431: 453–456.

Morris, R.G., Anderson, E., Lynch, G.S. and Baudry, M. (1986) Selective impairment of learning and blockade of long-term potentiation by an N-methyl-D-aspartate receptor antagonist, AP5. Nature, 319: 774–776.

Pelletier, J.G., Croce, A. and Lacaille, J.-C. (2007) Intact afferent-specific spike train dynamics and increased gain of input-output function in recurrent inhibition by LTP at oriens/alveus interneuron excitatory synapses. Program No. 48.9. 2007 Neuroscience Meeting Planner, Society for Neuroscience, San Diego, CA (Online).

Perez, Y., Morin, F. and Lacaille, J.C. (2001) A Hebbian form of long-term potentiation dependent on mGluR1a in hippocampal inhibitory interneurons. Proc. Natl. Acad. Sci. U.S.A., 98: 9401–9406.

Pouille, F. and Scanziani, M. (2001) Enforcement of temporal fidelity in pyramidal cells by somatic feed-forward inhibition. Science, 293: 1159–1163.

Pouille, F. and Scanziani, M. (2004) Routing of spike series by dynamic circuits in the hippocampus. Nature, 429: 717–723.

Schwartzkroin, P.A. (1975) Characteristics of CA1 neurons recorded intracellularly in the hippocampal in vitro slice preparation. Brain Res., 85: 423–436.

Somogyi, P. and Klausberger, T. (2005) Defined types of cortical interneurone structure space and spike timing in the hippocampus. J. Physiol., 562: 9–26.

Steriade, M., Dossi, R.C. and Nunez, A. (1991) Network modulation of a slow intrinsic oscillation of cat thalamocortical neurons implicated in sleep delta waves: cortically induced synchronization and brainstem cholinergic suppression. J. Neurosci., 11: 3200–3217.

Stevens, C.F. (1998) A million dollar question: does LTP=memory? Neuron, 20: 1–2.

Thompson, S.M. and Gahwiler, B.H. (1989) Activity-dependent disinhibition. I. Repetitive stimulation reduces IPSP driving force and conductance in the hippocampus in vitro. J. Neurophysiol., 61501–61511.

Topolnik, L., Azzi, M., Morin, F., Kougioumoutzakis, A. and Lacaille, J.C. (2006) mGluR1/5 subtype-specific calcium signalling and induction of long-term potentiation in rat hippocampal oriens/alveus interneurones. J. Physiol., 575: 115–131.

Topolnik, L., Congar, P. and Lacaille, J.C. (2005) Differential regulation of metabotropic glutamate receptor- and AMPA receptor-mediated dendritic Ca^{2+} signals by presynaptic and postsynaptic activity in hippocampal interneurons. J. Neurosci., 25: 990–1001.

Topolnik, L., Pelletier, J.G. and Lacaille, J.C. (2007) mGluR5-induced long-lasting potentiation of L-type calcium channels in dendrites of hippocampal CA1 oriens/alveus interneurons. Program No. 783.10. 2007 Neuroscience Meeting Planner, Society for Neuroscience, San Diego, CA (Online).

Toth, K. and McBain, C.J. (1998) Afferent-specific innervation of two distinct AMPA receptor subtypes on single hippocampal interneurons. Nat. Neurosci., 1: 572–578.

Toth, K., Suares, G., Lawrence, J.J., Philips-Tansey, E. and McBain, C.J. (2000) Differential mechanisms of transmission at three types of mossy fiber synapse. J. Neurosci., 20: 8279–8289.

Traub, R.D. and Miles, R. (1991) Neuronal networks of the hippocampus. Cambridge University Press, New York.

Walker, H.C., Lawrence, J.J. and McBain, C.J. (2002) Activation of kinetically distinct synaptic conductances on inhibitory interneurons by electrotonically overlapping afferents. Neuron, 35: 161–171.

CHAPTER 15

Persistent neural activity in the prefrontal cortex: a mechanism by which BDNF regulates working memory?

Evan M. Galloway[1,3], Newton H. Woo[1] and Bai Lu[1,2,*]

[1]Section on Neural Development and Plasticity, NICHD, National Institutes of Health, Bethesda, MD, USA
[2]Genes, Cognition and Psychosis Program, NIMH, Bethesda, MD, USA
[3]Department of Physiology, Anatomy and Genetics, University of Oxford, Oxford, UK

Abstract: Working memory is the ability to maintain representations of task-relevant information for short periods of time to guide subsequent actions or make decisions. Neurons of the prefrontal cortex exhibit persistent firing during the delay period of working memory tasks. Despite extensive studies, the mechanisms underlying this persistent neural activity remain largely obscure. The neurotransmitter systems of dopamine, NMDA, and GABA have been implicated, but further investigations are necessary to establish their precise roles and relationships. Recent research has suggested a new component: brain-derived neurotrophic factor (BDNF) and its high-affinity receptor, TrkB. We review the research on persistent activity and suggest that BDNF/TrkB signaling in a distinct class of interneurons plays an important role in organizing persistent neural activity at the single-neuron and network levels.

Keywords: TrkB; neurotrophins; plateau potentials; parvalbumin; interneurons; dopamine; ACh; GABA

Introduction: working memory and persistent neuronal firing

Working memory refers to the ability to hold transient representations of task-relevant information to guide subsequent behaviors. Rule learning, planning, decision-making, and many executive functions are likely dependent on this ability. Early lesion studies identified the prefrontal cortex (PFC) as a crucial brain area involved in working memory (Butters and Pandya, 1969). Phylogenetically, the PFC is one of the most recently evolved areas of the brain. The size of the PFC relative to the whole brain increases dramatically in mammals, particularly primates, correlating well with the increasing capacity for more sophisticated executive behaviors. In humans, the PFC is disproportionately large and exceedingly well connected to cortical and subcortical brain regions (for review, see Fuster, 2002). Functional magnetic resonance imaging reveals that the PFC is activated during working memory tasks (McCarthy et al., 1994). The importance of the PFC is further manifested in the severe dysfunction in PFC activation and working memory capacity seen in schizophrenia patients (Weinberger and Berman, 1996).

Classic laboratory tests for working memory are the delayed matching (or non-matching) to

*Corresponding author. Tel.: +1 (301) 435 2970;
Fax: +1 (301) 496 1777; E-mail: bailu@mail.nih.gov

sample, delayed-alternation, and delayed-response tasks. Ultimately, all these tasks consist of three basic parts: (1) the acquisition of a piece of information (e.g. the location of food); (2) a short delay period during which the subject needs to retain the acquired information while no salient information is present (e.g. the location of the food is obscured); (3) a cue indicating that the subject should respond based on the previously acquired information (e.g. point to the location of the hidden food).

Single unit recordings from neurons in the primate PFC demonstrate that neurons persistently fire during the delay period of a delayed response task (Fuster and Alexander, 1971; Kubota and Niki, 1971), suggesting that these neurons are somehow involved in maintaining the "task-relevant information" for a short period of time. This is further clarified by work showing that neurons in the primate dorsolateral prefrontal cortex (DLPFC) persistently fire during the delay period of a spatial delayed response task in response to stimuli in their preferred spatial field (Funahashi et al., 1989). Given that the persistent firing during the "delay" period does not require stimuli to be sustained, this activity is believed to represent the maintenance and encoding of working memory.

Three mechanisms may contribute to working memory-related persistent activity:

(1) intrinsic properties of PFC neurons — individual PFC neurons may have some unique features that allow them to initiate, maintain, and terminate sustained, non-adapting firing;
(2) local synaptic network within the PFC — the short term facilitation of excitatory synapses, in addition to the inhibitory control wielded by GABAergic interneurons forms a strong cellular basis for "reverberant" neuronal firing;
(3) afferents from subcortical areas — the extensive dopaminergic projections and other inputs to the PFC are likely important for tuning the state of a network and initiating its activity.

In this review, we will discuss research on the mechanisms underlying persistent neuronal firing at each of the three levels highlighted above. Given the emerging evidence that brain-derived neurotrophic factor (BDNF) is a key regulator of neuronal excitability and synaptic plasticity (for review, see Poo, 2001; Lu et al., 2005), we will also discuss a potential role for BDNF in regulating the persistent neuronal activity in the PFC.

Persistent activity and its regulation

Intrinsic properties of PFC neurons

A unique combination of ion channels defines the essential properties of PFC neurons and contributes to their ability to persistently fire. In most parts of the brain, a train of action potentials in pyramidal neurons will display rapid adaptation or accommodation in response to a sustained depolarization — an undesirable feature for persistent firing of many seconds in duration. Thus, pyramidal neurons in the PFC must be somewhat unique in their intrinsic ion channel properties to allow them to fire persistently.

While few studies have directly examined the contribution of intrinsic properties to persistent firing in the PFC, studies of carbachol-dependent plateau potentials provide strong evidence for the contribution of intrinsic factors to persistent firing (Egorov et al., 2002). Plateau potentials were originally identified in the entorhinal cortex (EC) (Egorov et al., 2002) — another area where delay period persistent activity is observed (Young et al., 1997) — but a similar phenomenon has been observed in the PFC as well (Andrade, 1991; Haj-Dahmane and Andrade, 1998; unpublished results). In this preparation, persistent activity is initiated by a short (500 ms) depolarizing pulse at threshold in the presence of tonic carbachol ($10 \mu M$), resulting in a burst of firing and a subsequent afterdepolarization. The afterdepolarization results in persistent non-accommodating firing of the neuron, often for an indefinite period of time. Although plateau potentials in EC layer III neurons are resistant to termination by hyperpolarization, a large depolarizing pulse can induce cessation of firing (Tahvildari et al., 2007). In layer V of EC, neurons respond to stimuli in a graded manner, encoding stimulus duration or intensity in the plateau potential firing rate (Egorov et al., 2002). Clearly, this model recapitulates some of the salient features of delay period activity including robust sustained firing and rapid transitions in response to salient input.

Calcium channels and calcium-activated cation or potassium channels may play a role in generating plateau potentials. For instance, blocking L-type calcium channels greatly abbreviates the duration of each plateau potential event. Specifically, during plateau potentials in the EC, continuous application of nifedipine, an L-type calcium channel blocker, greatly decreases the duration of membrane depolarization and curtails all persistent firing (Egorov et al., 2002). In postpubertal rats, D1 type dopamine receptor (D1R) and N-methyl-D-aspartate (NMDA) co-activation is sufficient to induce "up states" in PFC slices, a phenomenon related to synchronized persistent activity (Tseng and O'Donnell, 2005). Here too, application of nifedipine greatly reduced the duration and frequency of spontaneous activity.

Perhaps overshadowing the activity of L-type calcium channels, calcium-dependent non-specific cation (CAN) currents and calcium-activated potassium currents ($K_{Ca^{2+}}$) have proven integral to the generation of plateau potentials in both slice preparations and computational models. Simply blocking CAN currents with flufenamic acid completely abolishes plateau potentials in the EC (Egorov et al., 2002). Furthermore, it has been suggested that the CAN and $K_{Ca^{2+}}$ currents underlie a dynamic attractor for a neuron's firing rate (Fransen et al., 2006). That is, a balance between the two currents and the presence of two activation states for CAN currents create a system where different firing rates can be stably manifested in response to different stimuli (Fransen et al., 2006). A self-terminating and slightly less robust effect has been similarly reported in PFC layer V pyramidal neurons (Andrade, 1991). It too is significantly dependent on CAN currents (Haj-Dahmane and Andrade, 1998).

Potassium channels are generally strong candidates for permitting persistent activity. As noted earlier, in many neurons sustained firing will result in adaptation and eventual cessation. As mediators of the downward phase of the action potential, faster activation of potassium currents can reduce spike width, perhaps allowing for a non-adapting firing pattern (for review, see Bean, 2007). The inactivation and deactivation of potassium channels may also contribute to the frequency and duration of persistent firing.

While many intrinsic elements are likely to play a role in generating sustained activity, research on the contribution of intrinsic properties in PFC neurons, specifically, to persistent neuronal firing remains in its infancy. More studies must now be done to identify the specific elements pertinent to PFC neurons in manifesting persistent activity. In addition, it is important to determine how these intrinsic properties are regulated by intracortical mechanisms or subcortical afferent inputs.

Intracortical synaptic interactions: GABA, glutamate, and synaptic facilitation

Recurrent models of persistent activity suggest that a balance of excitatory and inhibitory inputs onto pyramidal neurons supports the generation and accuracy of persistent activity (Shu et al., 2003a, b). While the story is probably more complicated (Lau and Bi, 2005; Milojkovic et al., 2005), we now know that the synaptic inputs and network architecture of the PFC are integral to persistent activity, both in vivo and in vitro.

GABAergic control

GABAergic interneurons are the primary provider of inhibition in the PFC. Consequently, delay period persistent activity is critically dependent on gamma-aminobutyric acid (GABA) synaptic inputs. Application of a GABA antagonist to the PFC disrupts the working memory activity (Sawaguchi et al., 1989). Recordings of persistently firing pyramidal neurons and GABAergic interneurons reveal that the two cell types have similar receptive fields, and that their responses are phase locked (Wilson et al., 1994). Further work shows that blocking GABA transmission results in a loss of spatial tuning in previously tuned neurons (Rao et al., 2000).

The exact circuit mediating this effect is still unclear, but parvalbumin (PV)-expressing interneurons are likely candidates. This group includes both the chandelier and basket neurons, which innervate pyramidal neurons at the axon initial segment (AIS) and proximal to the soma, respectively. The chandelier neurons are perfectly positioned for gating the output of pyramidal neurons to ensure appropriate spike timing. These neurons exert powerful inhibitory controls over a large number of cortical pyramidal neurons by continuously firing at high frequency (fast spiking). It is conceivable that the transient relief of this inhibition would trigger synchronized firing among pyramidal neurons, initiating sustained firing among this population during the delay period of a working memory task. Similarly, basket neuron synapses at or near the soma could help control the gain and timing of pyramidal neuron responses to dendritic and somatic inputs (for review, see Markram et al., 2004). From a broader perspective, both of these could ensure correct encoding of initial stimuli and provide the correct level of inhibition to balance recurrent excitation during persistent firing.

Glutamatergic control

Using the model of up states — spontaneous sustained depolarizations of the membrane potential — excitatory synaptic transmission via alpha-amino-3-hydroxy-5-methyl-4-isoxazole propionic acid (AMPA) and NMDA receptors has been strongly linked to the generation of persistent activity. Both receptors are important for up state activity, but are most likely responsible for different components of the prolonged depolarization associated with up states. A recent work suggests that an AMPA-dependent component is important for initiation of the up state, while the strong inward current present throughout the up state is NMDA dependent (Seamans et al., 2003).

In mice and rats, spontaneous up states can be induced in acute PFC slices via co-activation of dopamine and NMDA receptors (Tseng and O'Donnell, 2005). NMDA by itself causes the membrane potential to maintain a steady depolarization to about −60 mV. However, NMDA combined with dopamine induces up states in which the potential becomes transiently more depolarized. The importance of NMDA-mediated synaptic inputs and dopaminergic activation in persistent firing was also highlighted in a recent paper by Durstewitz and Gabriel (2007). They suggest that non-linear NMDA conductances induce a potentially meaningful spiking irregularity.

Synaptic facilitation and augmentation

Reverberant network dynamics among assemblies of PFC neurons may yield the sustained synaptic excitation that is thought to underlie persistent activity. Supporting this is the observation that while other cortical areas (such as visual cortex) have synaptic circuits that often exhibit synaptic fatigue or depression, the PFC displays a heterogeneous array of synaptic facilitation, augmentation, and post-tetanic potentiation for a majority of the synapses (Hempel et al., 2000; Wang et al., 2006). These forms of short-term facilitation may help initiate or sustain activity during the delay period of a working memory task. Indeed, in a neurocomputational model of a simple network, synaptic augmentation enables weak recurrent connections to support persistent activity following a transient stimulus (Hempel et al., 2000).

Subcortical afferent inputs

Since most executive functions, including working memory, require the integration of distributed sensory inputs and memory, it is unsurprising that the PFC receives a large number of afferent inputs. Of these, the most prominent are those that release dopamine, norepinephrine, acetylcholine, and serotonin. We will discuss the first three below and simply note that while serotonin has been shown to effect delay period activity and spatial tuning in single neurons (Williams et al., 2002), behavioral data do not indicate a significant effect on delay-dependent working memory (for review, see Ellis and Nathan, 2001).

Dopaminergic afferents

Early studies revealed an apparently simple relationship in which dopamine agonists, especially D1R agonists, increased delay period activity, while antagonists decreased delay period activity (Sawaguchi et al., 1990a, b). However, this picture was soon complicated by behavioral and electrophysiological evidence indicating that optimal dopamine or D1R activation followed an inverted U-shaped tuning curve, in which high or low, but still physiological, levels of dopamine impair working memory (Williams and Goldman-Rakic, 1995; Cai and Arnsten, 1997; Zahrt et al., 1997). A recent study has demonstrated this effect at the single-neuron level after local application of varying concentrations of D1R agonists onto neurons in the PFC during a working memory task. Assuming that weakly spatially tuned neurons have low levels of endogenous dopamine stimulation and strongly tuned neurons have optimal levels, Vijayraghavan et al. (2007) observed that application of a D1R agonist to the former enhanced spatial tuning, while application to the latter caused either a much smaller increase or actually eroded tuning.

Interestingly, stimulation of the VTA — a primary dopaminergic input to the PFC — can cause D1R-mediated up state transitions in the PFC (Lewis and O'Donnell, 2000). Another work showed that up state depolarizations are synchronized between these two regions, and that inactivation of the VTA leaves PFC up states intact, but destroys organized networks of this activity in the PFC (Peters et al., 2004).

The role of dopamine tuning in PFC function is further bolstered by genetic evidence involving a human single nucleotide polymorphism (SNP) in the gene for catechol O-methyltransferase (COMT) that results in a substitution of methionine (met) for valine (val) at codon 158 of the protein (for review, see Savitz et al., 2006). COMT is known to facilitate dopamine catabolism and therefore inactivate dopamine after its release. The val form has higher COMT enzymatic activity compared to the met form, resulting in lower dopamine concentration in the cortex (Egan et al., 2001). Furthermore, COMT allelic background determined subject performance in response to amphetamine (a dopamimetic) during a working memory task. Specifically, those with a met/met genotype, who normally perform better in working memory tasks than val/val individuals, exhibited a decrease in working memory performance when treated with amphetamine, while val/val individuals manifested improved working memory performance after amphetamine treatment (Mattay et al., 2003).

Cholinergic afferents

The role of cholinergic inputs to the PFC in persistent activity and working memory is still unclear. Generally, cholinergic agonism appears to enhance prefrontal-dependent memory task performance, while antagonism disrupts it, especially at muscarinic receptors (Ellis and Nathan, 2001; Dalley et al., 2004). However this interaction is confounded by factors such as smoking history and variances in background agonist levels. Perhaps the strongest physiological evidence that acetylcholine may play role in persistent firing again comes from the plateau potential model (Egorov et al., 2002) as this activity is highly dependent on tonic levels of muscarinic receptor activation. However, the exact contribution of acetylcholine to the generation of plateau potentials remains unclear and warrants further investigation.

Norepinephrinergic afferents

A role for norepinephrine in working memory function is evident from studies in non-human primates using a spatial delayed response task. Activation of the α1-adrenoreceptor in the primate DLPFC both impairs spatial working memory performance in primates and attenuates persistent firing in individual neurons for their preferred spatial stimulus (Mao et al., 1999; Birnbaum et al., 2004). Conversely, activation of the α2A-adrenoreceptors by the agonist guanfacine in the DLPFC enhances spatial working memory performance and persistent firing (Arnsten et al., 1988; Wang et al., 2007). Recently, it was demonstrated that application of an α2A agonist on individual neurons on the DLPFC enhanced their spatial tuning, especially if it was initially weak. Furthermore, the same study indicated that α2A enhances tuning by inhibiting cAMP production, consequently causing the closure of hyperpolarization-activated cyclic nucleotide-gated channels, possibly providing a mechanism for selectively activating networks of spatially tuned neurons (Wang et al., 2007).

Possible roles of BDNF in the PFC: from working memory to persistent neural activity

Among members of the neurotrophin family, BDNF stands out for its high levels of expression in the brain and multi-dimensional regulation by neuronal activity. Initially recognized for its role in neural development and survival, it is now well established that the main function of BDNF in the brain is to regulate synaptic plasticity (Poo, 2001; Lu, 2003a, b). The diverse action of BDNF in the brain results from the numerous signaling cascades downstream of BDNF as well as the activity-dependent transcription of BDNF via promoter III (for review, see Lu, 2003a, b; Lu et al., 2005). This coupling of neuronal activity to changes in synaptic structure and function implicates BDNF as an essential regulator of cellular processes that underlie behavior and cognition. Converging lines of evidence are now suggesting that BDNF and its preferred receptor, neurotrophic receptor tyrosine kinase (TrkB), play an integral role in mediating persistent network activity and maintaining normal PFC function (for review, see Lewis et al., 2005; Savitz et al., 2006; Woo and Lu, 2006).

BDNF cell biology

Transcription of the BDNF gene is mediated by multiple (at least four) promoters, each driving a short 5′ exon that is alternatively spliced onto a common 3′ exon (exon V) that contains the coding region (Timmusk et al., 1993). Thus, there are at least four BDNF mRNAs that give rise to exactly the same BDNF protein (West et al., 2001; for review, see Lu, 2003a, b). BDNF is synthesized in the endoplasmic reticulum as a precursor, proBDNF, and then folded and packaged in the Golgi apparatus (Lu, 2003a, b). After this, it is sorted into either a constitutive or activity-dependent secretion pathway (Mowla et al., 1999). In neurons, the BDNF-containing secretory vesicles are transported along axons or dendrites and some of the vesicles are localized to presynaptic terminals or postsynaptic spines (Lu, 2003a, b). ProBDNF may be cleaved either intracellularly or extracellularly yielding mature BDNF (Seidah et al., 1996). Consequent functions of BDNF are mediated by two receptor systems: TrkB and $p75^{NTR}$. Mature BDNF is a strong ligand for TrkB. Activation of this receptor triggers a number of downstream pathways, including the phosphatidylinositol-3-kinase (PI3K), phospholipase C-gamma (PLC-γ), and MEK-MAPK pathways (Huang and Reichardt, 2003). Conversely, if BDNF remains uncleaved, proBDNF binds preferentially to $p75^{NTR}$, leading to activation of signaling cascades including the nuclear factor-kappa B (NF-κB), c-jun kinase, and sphigomyelin hydrolysis pathways (Chao, 2003). Furthermore, it is now believed that pro- and mature BDNF elicit very different and often opposing biological functions (Lu et al., 2005). Clearly, having multiple promoters, a cleavage-dependent protein, two secretory pathways, and two receptor systems with multiple transduction cascades indicates that BDNF acts in a complex manner and highlights its potential for regulating a number of neuronal processes.

Expression of BDNF and TrkB in the PFC

Expression of BDNF, along with its cognate receptor, TrkB, is widespread in the frontal cortex of mammals, including monkeys and humans (Huntley et al., 1992; Hayashi et al., 2000; Weickert et al., 2003, 2005), as well as rodents (Hofer et al., 1990; Phillips et al., 1990; Connor et al., 1997; Yan et al., 1997a, b; Lipska et al., 2001). In the human dorsolateral PFC, BDNF mRNA is detected in all cortical layers (II–VI). Its expression is developmentally regulated, starting relatively low during infancy, peaking during young adulthood, and then remaining relatively stable throughout later adulthood (Webster et al., 2002). Importantly, BDNF mRNA is expressed in excitatory neurons and not in GABAergic interneurons. The importance of BDNF expression in the PFC is now evident in many CNS disorders where expression levels are significantly altered, including depression, stress-related disorders, and schizophrenia (for review, see Duman and Monteggia, 2006; Castren et al., 2007). As working memory dysfunction is central to the pathology of schizophrenia, we specifically note that expression of BDNF and TrkB is significantly reduced in the dorsolateral PFC of schizophrenic patients (Weickert et al., 2003, 2005; Hashimoto et al., 2005).

In monkeys, BDNF mRNA is highly expressed in large pyramidal neurons of layers III and IV throughout the PFC of fetal and adult monkeys (Huntley et al., 1992). Correspondingly, full-length TrkB receptor immunoreactivity is also expressed in pyramidal cells of layers II, III, V, and VI, and is developmentally regulated (Hayashi et al., 2000). Interestingly, the strongest TrkB immunoreactivity in non-human primates is at six months postnatal, which corresponds to a period of high synapse overproduction in the PFC of macaque monkeys (Bourgeois et al., 1994), suggesting that TrkB and BDNF may play a functional role in synapse formation during development. TrkB mRNA and protein are detected in both excitatory pyramidal neurons and inhibitory interneurons, suggesting that both populations are responsive to BDNF.

An acceptable PFC analogue in rodents has only recently been identified. As a result, much less is known regarding the precise localization of BDNF and TrkB mRNA in this region. However, several recent studies are in agreement that BDNF protein and mRNA are expressed at high levels in the neocortex, particularly the prelimbic and infralimbic cortex (Connor et al., 1997; Yan et al., 1997a, b; Lipska et al., 2001; Blurton-Jones et al., 2004). The situation is similar for TrkB expression, with the anterior cingulate area also showing high expression (Saarelainen et al., 2003; Hashimoto et al., 2005).

While the neurotrophin NT-4 is also a ligand for TrkB and shows relatively similar expression patterns as BDNF, it is expressed at far lower levels (Mori et al., 2004) and its function is less clear.

BDNF/TrkB and PFC-mediated working memory

The discovery of a common SNP resulting in a val to met switch at codon 66 in the pro-domain of the gene encoding for BDNF has greatly facilitated studies on the role of BDNF in cognitive function (Egan et al., 2003; for review, see Savitz et al., 2006). The met-BDNF allele is associated with decreased hippocampal function in humans, impairments in regulated BDNF secretion and trafficking, and deficits in episodic memory (Egan et al., 2003; Hariri et al., 2003). Carriers of the met-BDNF allele show a reduction in PFC gray matter volume (Pezawas et al., 2004), but generally do not show significant impairments in working memory (Egan et al., 2003; Hansell et al., 2007). However, patients with bipolar disorder who carry the met allele have reduced working memory as reflected by the N-back test, and behavior perseveration as reflected by the Wisconsin Card Sorting Test, as compared to patients homozygous for the val allele (Rybakowski et al., 2003, 2006).

In rodents, extensive studies indicate that BDNF is required for hippocampal-dependent learning and memory, but evidence for a role of BDNF in working memory remains sparse. Intracerebroventricular infusion of an antisense BDNF oligonucleotide impaired both reference and working memory as measured by the radial arm maze test (Mizuno et al., 2000). Furthermore, levels of BDNF in the frontal cortex correlate negatively with the number of working memory errors in aged rats (Bimonte et al., 2003), and in Ts65D mice, an animal model of Down syndrome (Bimonte-Nelson et al., 2003). Finally, mice with a forebrain-specific deletion of the TrkB gene manifest schizophrenic-like cognitive deficits, hyperlocomotion, and stereotyped behaviors (Zorner et al., 2003).

BDNF regulation of GABAergic interneurons

Of particular interest in the PFC are the aforementioned GABAergic interneurons that express

PV, which include the basket and chandelier interneurons. They represent a unique class of fast-spiking interneurons, which is important for a number of physiological processes in the mammalian brain, and is regulated by BDNF and TrkB (Woo and Lu, 2006). By innervating the soma and AIS of a large number of pyramidal cells, PV-interneurons are in a powerful position to control persistent and synchronous firing in the PFC (for review, see Markram et al., 2004).

Substantial evidence indicates that BDNF regulates the development of PV-interneurons. In BDNF knockout mice, the number of PV-interneurons is significantly reduced as compared to wild-type counterparts (Jones et al., 1994; Altar et al., 1997). In TrkB hypomorphic mice in which the endogenous TrkB locus is replaced with a floxed TrkB cDNA (Xu et al., 2000), the levels of PV and glutamic acid decarboxylase 67 (GAD67), a GABA synthesizing enzyme, mRNAs were significantly lower in the PFC (Hashimoto et al., 2005). Intriguingly, mice with hypomorphic BDNF expression show no change in GAD67 or PV expression, suggesting that TrkB is more prevalent than BDNF in controlling the GABAergic phenotypes (Monteggia et al., 2004; Hashimoto et al., 2005). This idea is further supported by the observation that PV-interneurons preferentially express TrkB, among other cortical interneurons (Cellerino et al., 1996). Although PV-interneurons do not express BDNF, it is expressed by pyramidal neurons that appear to secrete BDNF only onto local inhibitory neurons, suggesting a tight feedback mechanism (Kohara et al., 2007).

Conversely, in transgenic mice over-expressing BDNF, the maturation of PV-interneurons is significantly accelerated, which is accompanied by an increase in spontaneous network activity (Huang et al., 1999; Aguado et al., 2003). Similarly, application of BDNF to homogenous cell cultures of fast-spiking PV-interneurons resulted in the accelerated formation of reciprocal connections between these cells (Berghuis et al., 2004). It was also found that exogenous and endogenous TrkB ligands accelerate PV expression via PI3 kinase (Patz et al., 2004).

Evidence for a functional consequence of this interaction between BDNF and PV-interneuron development comes from studies of patients with schizophrenia, who usually have severe working memory deficits. Postmortem studies have indicated that schizophrenia is associated with a reduction in BDNF and TrkB expression (Weickert et al., 2003, 2005; Hashimoto et al., 2005). Similarly, the schizophrenic PFC shows decreased expression of GAD67 in a subset of interneurons that also express PV (Hashimoto et al., 2003). Importantly, the number of PV-interneurons is not decreased; only the amount of GAD67 expressed in them is changed. It remains to be established whether the decreased GAD67 expression in GABAergic neurons is due to a reduction in cortical BDNF. More specifically, chandelier neurons, which form cartridge-like synapses at the AIS of pyramidal neurons, have fewer cartridges with significant GABA membrane transporter 1 (GAT1) expression in the PFC of schizophrenic patients (Pierri et al., 1999). Postsynaptically to these cartridges, an increase in $GABA_A\alpha2$ receptors is observed, raising the possibility of compensatory changes to GABAergic neurotransmission (Volk et al., 2002).

Studies of the hippocampus indicate that PV-interneurons are functionally integral to organizing network activity and spontaneous firing (for review, see Bartos et al., 2007). Oscillations can be observed both in vitro and in vivo (for review, see Buzsaki and Draguhn, 2004), and PV-interneurons, especially the basket type, show extensive phase locking with hippocampal oscillations (Penttonen et al., 1998; Klausberger et al., 2003; Hajos et al., 2004). Deletion of glutamate receptor subunits in PV expressing interneurons decreased AMPA currents, leading to diminished gamma oscillation power in field recordings and imprecise spike timing in these interneurons. This was correlated with a deficiency in hippocampal dependent behavior seen in these knockout mice (Fuchs et al., 2007). However, it is unknown whether PV-interneurons in the PFC play a similar role in persistent activity.

Acute BDNF regulation of persistent neuronal activity

Although BDNF is involved in normal PFC and PV-interneuron function, it is unclear whether its action is developmental, tonic, or acute, as BDNF can be secreted both constitutively and in an activity-dependent manner. Certainly, in the case of interneuron function, the influence seems largely developmental or tonic, as BDNF/TrkB signaling levels seem to determine the structure of the inhibitory network. However, considering the effect of the val/met substitution on activity-dependent secretion of BDNF, it also seems likely that BDNF also plays an acute role in regulating excitability and plasticity. That is, individual or successive episodes of persistent firing could be modulated by BDNF signaling. Conceptually, BDNF could regulate this at three different levels.

Regulation of intrinsic excitability

An emerging concept is the homeostatic regulation of neuronal plasticity. By stabilizing neuronal activity, homeostatic plasticity allows neurons to remain responsive to their inputs during periods of change in the strength or number of synaptic inputs. This is accomplished by scaling synaptic strength or intrinsic excitability up or down as a function of activity (LeMasson et al., 1993; Turrigiano et al., 1998; Desai et al., 1999a, b). For example, when tetrodotoxin is chronically applied to cultured cortical neurons, activity is blocked, but, subsequently, their intrinsic excitability is significantly increased (Desai et al., 1999b). Activity-deprived neurons fire in response to smaller current injections and at a higher rate by increasing sodium currents and decreasing persistent potassium currents (Desai et al., 1999b). Interestingly, application of BDNF prevents this increase in excitability, while blocking BDNF/TrkB signaling caused a similar increase in excitability (Desai et al., 1999a). These observations suggest that BDNF-induced changes in excitability as a result of neuronal activity are mediated by alterations in voltage-gated currents, namely sodium and persistent potassium currents. Such regulation by BDNF could serve to fine-tune the intrinsic properties of PFC neurons, altering their ability to fire persistently.

Control of network excitability: a role in synaptic facilitation

Persistent neuronal activity in the PFC is thought to arise from sustained excitation generated by reverberant activity in an assembly of neurons. Recent work indicates that the PFC is unique among other cortical areas in having a heterogeneous array of synaptic dynamics, including facilitation, augmentation, and post-tetanic potentiation that encourage sustained neuronal firing (Hempel et al., 2000; Wang et al., 2006). In principle, these forms of short-term plasticity may allow PFC neurons to more easily reach a threshold for recurrent excitation and reverberant activity. Therefore, synaptic facilitation may be involved in persistent activity in reverberant networks.

In the hippocampus, BDNF has been shown to facilitate several forms of short-term plasticity including paired pulse facilitation and post-tetanic potentiation (Pozzo-Miller et al., 1999). Particularly relevant is the ability of BDNF to facilitate synaptic responses to high-frequency stimulation (HFS). This is achieved by promoting the mobilization/docking of synaptic vesicles to the active zones, leading to an increase in the readily releasable pool of vesicles. For instance, HFS-induced synaptic fatigue, which is thought to result from vesicle depletion of a readily releasable vesicle pool, is attenuated by a short-term treatment with BDNF (Figurov et al., 1996; Gottschalk et al., 1998, 1999). Moreover, optical imaging techniques using FM dyes show that BDNF enhances release from the immediate releasable pool of synaptic vesicles, evident by a significant increase in the rate of destaining (Tyler et al., 2006). Conversely, scavenging endogenous BDNF with TrkB-IgG or deletion of the BDNF gene causes pronounced synaptic fatigue. Quantitative electron microscopy shows that BDNF knockout mice, which have pronounced synaptic fatigue, have fewer docked vesicles at the active zones of hippocampal CA1 synapses (Pozzo-Miller et al., 1999). Similarly, TrkB knockout animals also have fewer docked vesicles at the active zones (Martinez et al., 1998). Taken together, it is clear that BDNF controls the availability of synaptic responses during successive stimuli by regulating the readily releasable pool and probability of release.

Given the profound importance of BDNF in synaptic dynamics across a broad time course — from vesicle release in milliseconds to potentiation for hours — further studies are required to determine whether BDNF also critically regulates the release probabilities and short-term plasticity of PFC neurons. Ultimately, BDNF-induced inhibition of synaptic depression or an enhancement of synaptic augmentation represents a potential mechanism for temporarily boosting the efficacy of recurrent synapses and subsequently enhancing persistent states of neuronal firing.

Regulation of ascending inputs

The dopaminergic inputs to the PFC, especially from the VTA, are important for the generation of meaningful persistent activity. Unsurprisingly, BDNF plays an important role in modulating the plasticity of dopaminergic neurons in this area. More interestingly, BDNF may play a role in modulating feedback between the VTA and the PFC. For instance, weak presynaptic stimulation in the rostral VTA, where PFC inputs may be received, is normally not enough to evoke postsynaptic potentiation in VTA dopaminergic neurons. However, potentiation of these synapses occurs in the presence of BDNF (Pu et al., 2006). Furthermore, BDNF and TrkB are expressed in the rat VTA (Seroogy et al., 1994; Numan and Seroogy, 1999).

BDNF is also expressed and transported to fibers and terminals in noradrenergic neurons (Fawcett et al., 1998). Furthermore, pharmacological activation of noradrenergic neurons results in postsynaptic TrkB activation (Aloyz et al., 1999). Work on noradrenergic system antidepressants has suggested that their effect on the anterior cingulate cortex, a rodent analogue of the PFC, is mediated by TrkB activation (Rantamaki et al., 2007).

Similarly tenuous but promising connections exist for the acetylcholine system. As expected, BDNF promotes the survival of cholinergic neurons of the basal forebrain (Nonomura et al., 1995). At the level of plasticity, BDNF enhances the depolarization-evoked release of acetylcholine in cortical-projecting neurons (Sala et al., 1998).

It is not surprising that BDNF has effects on these different ascending systems. However, as more attention is devoted to the role of these ascending systems in working memory, it will be useful to concomitantly examine the role of powerful neuromodulators like BDNF on these systems. At the very least this may provide greater insight into the etiology of psychiatric disorders of a more global nature.

Conclusions

Historically, mechanistic studies of cognitive functions have benefited greatly from the use of cellular models. For example, the synaptic circuit for hippocampal-dependent spatial memory was defined using long-term potentiation as a cellular model, offering tremendous opportunities for in-depth investigations of the cellular, molecular, and genetic mechanisms underlying learning and memory. In contrast, cellular and molecular studies of PFC-mediated working memory remain at the infancy stage. However, the synaptic network subserving working memory is beginning to be unraveled and persistent neural activity has proven to be a promising cellular correlate for working memory.

Examining this correlate, the evidence is compelling that BDNF may play an important role in persistent neural activity and concomitant network activity in the PFC. However, the nature of this role is still unclear. Evidence from patients with working memory dysfunction suggests that changes to GABAergic synaptic transmission may underlie the etiology of such disorders. Moreover, the changes appear to be targeted at PV-interneurons, which are heavily implicated in organizing network activity and spontaneous firing. Intriguingly, BDNF's high affinity receptor, TrkB, is preferentially expressed in these interneurons and decreased levels of TrkB are associated with GABAergic and working memory dysfunction.

Consequently, we suggest that TrkB plays a role in the modulation of synaptic inputs, by mediating excitability in pyramidal neurons and affecting GABAergic transmission in PV-interneurons. Specifically, we predict that decreases in TrkB signaling in these interneurons may impair GABAergic control of the cortical network, reducing the reliability of firing and fracturing the ensembles of neurons encoding the stimulus.

BDNF may also support the synaptic dynamics required for persistent activity to be sustained in a reverberant network. Modeling suggests that synaptic facilitation may be necessary to initiate persistent firing. Intriguingly, neurons in the PFC are capable of such synaptic facilitation. BDNF is ideally positioned as a modulator of synaptic facilitation and potentiation to play an important role in this mechanism.

We hypothesize that BDNF and TrkB contribute to the etiology of schizophrenia and working memory dysfunction. But, it is not yet clear if they are the keystone of the disease or the result of some upstream modulation. Consequently, much more effort should be directed towards elucidating this process. Modulations of BDNF signaling, especially in PV-expressing interneurons, along with simultaneous single-neuron and network recording, will help identify the role of TrkB in individual neurons, and the role of those neurons in network activity. Furthermore, the relationship between BDNF, high-frequency firing, synaptic augmentation, and persistent firing needs to be more clearly understood. Similarly, BDNF's role as a potent modulator of ascending systems should not be overlooked. Finally, sophisticated behavioral testing of transgenic mice will further clarify the role of these genetic components in working memory and persistent activity.

Abbreviations

AIS	axon initial segment
AMPA	alpha-amino-3-hydroxy-5-methyl-4-isoxazole propionic acid
BDNF	brain-derived neurotrophic factor
CAN	non-specific cation current
COMT	catechol O-methyltransferase
D1	dopamine receptor type 1
D1R	D1 receptor
DLPFC	dorsolateral prefrontal cortex
EC	entorhinal cortex
GABA	gamma-aminobutyric acid
GAD67	glutamic acid decarboxylase 67
GAT1	GABA transporter 1
HFS	high-frequency stimulation
$K_{Ca^{2+}}$	calcium-dependent potassium channel
met	methionine
NF-κB	nuclear factor-kappa B
NMDA	N-methyl-D-aspartate
p75^{NTR}	high-affinity proneurotrophin receptor
PFC	prefrontal cortex
PI3K	phosphatidylinositol-3-kinase
PLC-γ	phospholipase C-gamma
PV	parvalbumin
SNP	single nucleotide polymorphism
TrkB	neurotrophic receptor tyrosine kinase
val	valine
VTA	ventral tegmental area

Acknowledgment

The authors wish to thank the National Institutes of Health — specifically National Institute of Child Health and Human Development and National Institute of Mental Health — for funding their work. E.M.G. wishes to thank Elizabeth Zander for helpful discussions.

References

Aguado, F., Carmona, M.A., Pozas, E., Aguilo, A., Martinez-Guijarro, F.J., Alcantara, S., Borrell, V., Yuste, R., Ibanez, C.F. and Soriano, E. (2003) BDNF regulates spontaneous correlated activity at early developmental stages by increasing synaptogenesis and expression of the K^+/Cl^- co-transporter KCC2. Development, 130: 1267–1280.

Aloyz, R., Fawcett, J.P., Kaplan, D.R., Murphy, R.A. and Miller, F.D. (1999) Activity-dependent activation of TrkB neurotrophin receptors in the adult CNS. Learn. Mem., 6: 216–231.

Altar, C.A., Cai, N., Bliven, T., Juhasz, M., Conner, J.M., Acheson, A.L., Lindsay, R.M. and Wiegand, S.J. (1997) Anterograde transport of brain-derived neurotrophic factor and its role in the brain. Nature, 389: 856–860.

Andrade, R. (1991) Cell excitation enhances muscarinic cholinergic responses in rat association cortex. Brain Res., 548: 81–93.

Arnsten, A.F., Cai, J.X. and Goldman-Rakic, P.S. (1988) The alpha-2 adrenergic agonist guanfacine improves memory in aged monkeys without sedative or hypotensive side effects: evidence for alpha-2 receptor subtypes. J. Neurosci., 8: 4287–4298.

Bartos, M., Vida, I. and Jonas, P. (2007) Synaptic mechanisms of synchronized gamma oscillations in inhibitory interneuron networks. Nat. Rev. Neurosci., 8: 45–56.

Bean, B.P. (2007) The action potential in mammalian central neurons. Nat. Rev. Neurosci., 8: 451–465.

Berghuis, P., Dobszay, M.B., Sousa, K.M., Schulte, G., Mager, P.P., Hartig, W., Gorcs, T.J., Zilberter, Y., Ernfors, P. and Harkany, T. (2004) Brain-derived neurotrophic factor controls functional differentiation and microcircuit formation of selectively isolated fast-spiking GABAergic interneurons. Eur. J. Neurosci., 20: 1290–1306.

Bimonte-Nelson, H.A., Hunter, C.L., Nelson, M.E. and Granholm, A.C. (2003) Frontal cortex BDNF levels correlate with working memory in an animal model of Down syndrome. Behav. Brain Res., 139: 47–57.

Bimonte, H.A., Nelson, M.E. and Granholm, A.C. (2003) Age-related deficits as working memory load increases: relationships with growth factors. Neurobiol. Aging, 24: 37–48.

Birnbaum, S.G., Yuan, P.X., Wang, M., Vijayraghavan, S., Bloom, A.K., Davis, D.J., Gobeske, K.T., Sweatt, J.D., Manji, H.K. and Arnsten, A.F. (2004) Protein kinase C overactivity impairs prefrontal cortical regulation of working memory. Science, 306: 882–884.

Blurton-Jones, M., Kuan, P.N. and Tuszynski, M.H. (2004) Anatomical evidence for transsynaptic influences of estrogen on brain-derived neurotrophic factor expression. J. Comp. Neurol., 468: 347–360.

Bourgeois, J.P., Goldman-Rakic, P.S. and Rakic, P. (1994) Synaptogenesis in the prefrontal cortex of rhesus monkeys. Cereb. Cortex, 4: 78–96.

Butters, N. and Pandya, D. (1969) Retention of delayed-alternation: effect of selective lesions of sulcus principalis. Science, 165: 1271–1273.

Buzsaki, G. and Draguhn, A. (2004) Neuronal oscillations in cortical networks. Science, 304: 1926–1929.

Cai, J.X. and Arnsten, A.F. (1997) Dose-dependent effects of the dopamine D1 receptor agonists A77636 or SKF81297 on spatial working memory in aged monkeys. J. Pharmacol. Exp. Ther., 283: 183–189.

Castren, E., Voikar, V. and Rantamaki, T. (2007) Role of neurotrophic factors in depression. Curr. Opin. Pharmacol., 7: 18–21.

Cellerino, A., Maffei, L. and Domenici, L. (1996) The distribution of brain-derived neurotrophic factor and its receptor trkB in parvalbumin-containing neurons of the rat visual cortex. Eur. J. Neurosci., 8: 1190–1197.

Chao, M.V. (2003) Neurotrophins and their receptors: a convergence point for many signalling pathways. Nat. Rev. Neurosci., 4: 299–309.

Connor, B., Young, D., Yan, Q., Faull, R.L., Synek, B. and Dragunow, M. (1997) Brain-derived neurotrophic factor is reduced in Alzheimer's disease. Brain Res. Mol. Brain Res., 49: 71–81.

Dalley, J.W., Cardinal, R.N. and Robbins, T.W. (2004) Prefrontal executive and cognitive functions in rodents: neural and neurochemical substrates. Neurosci. Biobehav. Rev., 28: 771–784.

Desai, N.S., Rutherford, L.C. and Turrigiano, G.G. (1999a) BDNF regulates the intrinsic excitability of cortical neurons. Learn. Mem., 6: 284–291.

Desai, N.S., Rutherford, L.C. and Turrigiano, G.G. (1999b) Plasticity in the intrinsic excitability of cortical pyramidal neurons. Nat. Neurosci., 2: 515–520.

Duman, R.S. and Monteggia, L.M. (2006) A neurotrophic model for stress-related mood disorders. Biol. Psychiatry, 59: 1116–1127.

Durstewitz, D. and Gabriel, T. (2007) Dynamical basis of irregular spiking in NMDA-driven prefrontal cortex neurons. Cereb. Cortex, 17: 894–908.

Egan, M.F., Goldberg, T.E., Kolachana, B.S., Callicott, J.H., Mazzanti, C.M., Straub, R.E., Goldman, D. and Weinberger, D.R. (2001) Effect of COMT Val108/158 Met genotype on frontal lobe function and risk for schizophrenia. Proc. Natl. Acad. Sci. U.S.A., 98: 6917–6922.

Egan, M.F., Kojima, M., Callicott, J.H., Goldberg, T.E., Kolachana, B.S., Bertolino, A., Zaitsev, E., Gold, B., Goldman, D., Dean, M., Lu, B. and Weinberger, D.R. (2003) The BDNF val66met polymorphism affects activity-dependent secretion of BDNF and human memory and hippocampal function. Cell, 112: 257–269.

Egorov, A.V., Hamam, B.N., Fransen, E., Hasselmo, M.E. and Alonso, A.A. (2002) Graded persistent activity in entorhinal cortex neurons. Nature, 420: 173–178.

Ellis, K.A. and Nathan, P.J. (2001) The pharmacology of human working memory. Int. J. Neuropsychopharmacol., 4: 299–313.

Fawcett, J.P., Bamji, S.X., Causing, C.G., Aloyz, R., Ase, A.R., Reader, T.A., McLean, J.H. and Miller, F.D. (1998) Functional evidence that BDNF is an anterograde neuronal trophic factor in the CNS. J. Neurosci., 18: 2808–2821.

Figurov, A., Pozzo-Miller, L.D., Olafsson, P., Wang, T. and Lu, B. (1996) Regulation of synaptic responses to high-frequency stimulation and LTP by neurotrophins in the hippocampus. Nature, 381: 706–709.

Fransen, E., Tahvildari, B., Egorov, A.V., Hasselmo, M.E. and Alonso, A.A. (2006) Mechanism of graded persistent cellular activity of entorhinal cortex layer v neurons. Neuron, 49: 735–746.

Fuchs, E.C., Zivkovic, A.R., Cunningham, M.O., Middleton, S., Lebeau, F.E., Bannerman, D.M., Rozov, A., Whittington, M.A., Traub, R.D., Rawlins, J.N. and Monyer, H. (2007) Recruitment of parvalbumin-positive interneurons determines

hippocampal function and associated behavior. Neuron, 53: 591–604.
Funahashi, S., Bruce, C.J. and Goldman-Rakic, P.S. (1989) Mnemonic coding of visual space in the monkey's dorsolateral prefrontal cortex. J. Neurophysiol., 61: 331–349.
Fuster, J.M. (2002) Frontal lobe and cognitive development. J. Neurocytol., 31: 373–385.
Fuster, J.M. and Alexander, G.E. (1971) Neuron activity related to short-term memory. Science, 173: 652–654.
Gottschalk, W.A., Jiang, H., Tartaglia, N., Feng, L., Figurov, A. and Lu, B. (1999) Signaling mechanisms mediating BDNF modulation of synaptic plasticity in the hippocampus. Learn. Mem., 6: 243–256.
Gottschalk, W., Pozzo-Miller, L.D., Figurov, A. and Lu, B. (1998) Presynaptic modulation of synaptic transmission and plasticity by brain-derived neurotrophic factor in the developing hippocampus. J. Neurosci., 18: 6830–6839.
Haj-Dahmane, S. and Andrade, R. (1998) Ionic mechanism of the slow afterdepolarization induced by muscarinic receptor activation in rat prefrontal cortex. J. Neurophysiol., 80: 1197–1210.
Hajos, N., Palhalmi, J., Mann, E.O., Nemeth, B., Paulsen, O. and Freund, T.F. (2004) Spike timing of distinct types of GABAergic interneuron during hippocampal gamma oscillations in vitro. J. Neurosci., 24: 9127–9137.
Hansell, N.K., James, M.R., Duffy, D.L., Birley, A.J., Luciano, M., Geffen, G.M., Wright, M.J., Montgomery, G.W. and Martin, N.G. (2007) Effect of the BDNF V166M polymorphism on working memory in healthy adolescents. Genes Brain Behav., 6: 260–268.
Hariri, A.R., Goldberg, T.E., Mattay, V.S., Kolachana, B.S., Callicott, J.H., Egan, M.F. and Weinberger, D.R. (2003) Brain-derived neurotrophic factor val66met polymorphism affects human memory-related hippocampal activity and predicts memory performance. J. Neurosci., 23: 6690–6694.
Hashimoto, T., Bergen, S.E., Nguyen, Q.L., Xu, B., Monteggia, L.M., Pierri, J.N., Sun, Z., Sampson, A.R. and Lewis, D.A. (2005) Relationship of brain-derived neurotrophic factor and its receptor TrkB to altered inhibitory prefrontal circuitry in schizophrenia. J. Neurosci., 25: 372–383.
Hashimoto, T., Volk, D.W., Eggan, S.M., Mirnics, K., Pierri, J.N., Sun, Z., Sampson, A.R. and Lewis, D.A. (2003) Gene expression deficits in a subclass of GABA neurons in the prefrontal cortex of subjects with schizophrenia. J. Neurosci., 23: 6315–6326.
Hayashi, M., Mitsunaga, F., Itoh, M., Shimizu, K. and Yamashita, A. (2000) Development of full-length Trk B-immunoreactive structures in the prefrontal and visual cortices of the macaque monkey. Anat. Embryol. (Berl.), 201: 139–147.
Hempel, C.M., Hartman, K.H., Wang, X.J., Turrigiano, G.G. and Nelson, S.B. (2000) Multiple forms of short-term plasticity at excitatory synapses in rat medial prefrontal cortex. J. Neurophysiol., 83: 3031–3041.
Hofer, M., Pagliusi, S.R., Hohn, A., Leibrock, J. and Barde, Y.A. (1990) Regional distribution of brain-derived neurotrophic factor mRNA in the adult mouse brain. EMBO J., 9: 2459–2464.
Huang, E.J. and Reichardt, L.F. (2003) Trk receptors: roles in neuronal signal transduction. Annu. Rev. Biochem., 72: 609–642.
Huang, Z.J., Kirkwood, A., Pizzorusso, T., Porciatti, V., Morales, B., Bear, M.F., Maffei, L. and Tonegawa, S. (1999) BDNF regulates the maturation of inhibition and the critical period of plasticity in mouse visual cortex. Cell, 98: 739–755.
Huntley, G.W., Benson, D.L., Jones, E.G. and Isackson, P.J. (1992) Developmental expression of brain derived neurotrophic factor mRNA by neurons of fetal and adult monkey prefrontal cortex. Brain Res. Dev. Brain Res., 70: 53–63.
Jones, K.R., Farinas, I., Backus, C. and Reichardt, L.F. (1994) Targeted disruption of the BDNF gene perturbs brain and sensory neuron development but not motor neuron development. Cell, 76: 989–999.
Klausberger, T., Magill, P.J., Marton, L.F., Roberts, J.D., Cobden, P.M., Buzsaki, G. and Somogyi, P. (2003) Brain-state- and cell-type-specific firing of hippocampal interneurons in vivo. Nature, 421: 844–848.
Kohara, K., Yasuda, H., Huang, Y., Adachi, N., Sohya, K. and Tsumoto, T. (2007) A local reduction in cortical GABAergic synapses after a loss of endogenous brain-derived neurotrophic factor, as revealed by single-cell gene knock-out method. J. Neurosci., 27: 7234–7244.
Kubota, K. and Niki, H. (1971) Prefrontal cortical unit activity and delayed alternation performance in monkeys. J. Neurophysiol., 34: 337–347.
Lau, P.M. and Bi, G.Q. (2005) Synaptic mechanisms of persistent reverberatory activity in neuronal networks. Proc. Natl. Acad. Sci. U.S.A., 102: 10333–10338.
LeMasson, G., Marder, E. and Abbott, L.F. (1993) Activity-dependent regulation of conductances in model neurons. Science, 259: 1915–1917.
Lewis, B.L. and O'Donnell, P. (2000) Ventral tegmental area afferents to the prefrontal cortex maintain membrane potential 'up' states in pyramidal neurons via D(1) dopamine receptors. Cereb. Cortex, 10: 1168–1175.
Lewis, D.A., Hashimoto, T. and Volk, D.W. (2005) Cortical inhibitory neurons and schizophrenia. Nat. Rev. Neurosci., 6: 312–324.
Lipska, B.K., Khaing, Z.Z., Weickert, C.S. and Weinberger, D.R. (2001) BDNF mRNA expression in rat hippocampus and prefrontal cortex: effects of neonatal ventral hippocampal damage and antipsychotic drugs. Eur. J. Neurosci., 14: 135–144.
Lu, B. (2003a) BDNF and activity-dependent synaptic modulation. Learn. Mem., 10: 86–98.
Lu, B. (2003b) Pro-region of neurotrophins: role in synaptic modulation. Neuron, 39: 735–738.
Lu, B., Pang, P.T. and Woo, N.H. (2005) The yin and yang of neurotrophin action. Nat. Rev. Neurosci., 6: 603–614.
Mao, Z.M., Arnsten, A.F. and Li, B.M. (1999) Local infusion of an alpha-1 adrenergic agonist into the prefrontal cortex

impairs spatial working memory performance in monkeys. Biol. Psychiatry, 46: 1259–1265.

Markram, H., Toledo-Rodriguez, M., Wang, Y., Gupta, A., Silberberg, G. and Wu, C. (2004) Interneurons of the neocortical inhibitory system. Nat. Rev. Neurosci., 5: 793–807.

Martinez, A., Alcantara, S., Borrell, V., Del Rio, J.A., Blasi, J., Otal, R., Campos, N., Boronat, A., Barbacid, M., Silos-Santiago, I. and Soriano, E. (1998) TrkB and TrkC signaling are required for maturation and synaptogenesis of hippocampal connections. J. Neurosci., 18: 7336–7350.

Mattay, V.S., Goldberg, T.E., Fera, F., Hariri, A.R., Tessitore, A., Egan, M.F., Kolachana, B., Callicott, J.H. and Weinberger, D.R. (2003) Catechol O-methyltransferase val158-met genotype and individual variation in the brain response to amphetamine. Proc. Natl. Acad. Sci. U.S.A., 100: 6186–6191.

McCarthy, G., Blamire, A.M., Puce, A., Nobre, A.C., Bloch, G., Hyder, F., Goldman-Rakic, P. and Shulman, R.G. (1994) Functional magnetic resonance imaging of human prefrontal cortex activation during a spatial working memory task. Proc. Natl. Acad. Sci. U.S.A., 91: 8690–8694.

Milojkovic, B.A., Radojicic, M.S. and Antic, S.D. (2005) A strict correlation between dendritic and somatic plateau depolarizations in the rat prefrontal cortex pyramidal neurons. J. Neurosci., 25: 3940–3951.

Mizuno, M., Yamada, K., Olariu, A., Nawa, H. and Nabeshima, T. (2000) Involvement of brain-derived neurotrophic factor in spatial memory formation and maintenance in a radial arm maze test in rats. J. Neurosci., 20: 7116–7121.

Monteggia, L.M., Barrot, M., Powell, C.M., Berton, O., Galanis, V., Gemelli, T., Meuth, S., Nagy, A., Greene, R.W. and Nestler, E.J. (2004) Essential role of brain-derived neurotrophic factor in adult hippocampal function. Proc. Natl. Acad. Sci. U.S.A., 101: 10827–10832.

Mori, T., Shimizu, K. and Hayashi, M. (2004) Differential expression patterns of TrkB ligands in the macaque monkey brain. Neuroreport, 15: 2507–2511.

Mowla, S.J., Pareek, S., Farhadi, H.F., Petrecca, K., Fawcett, J.P., Seidah, N.G., Morris, S.J., Sossin, W.S. and Murphy, R.A. (1999) Differential sorting of nerve growth factor and brain-derived neurotrophic factor in hippocampal neurons. J. Neurosci., 19: 2069–2080.

Nonomura, T., Nishio, C., Lindsay, R.M. and Hatanaka, H. (1995) Cultured basal forebrain cholinergic neurons from postnatal rats show both overlapping and non-overlapping responses to the neurotrophins. Brain Res., 683: 129–139.

Numan, S. and Seroogy, K.B. (1999) Expression of trkB and trkC mRNAs by adult midbrain dopamine neurons: a double-label in situ hybridization study. J. Comp. Neurol., 403: 295–308.

Patz, S., Grabert, J., Gorba, T., Wirth, M.J. and Wahle, P. (2004) Parvalbumin expression in visual cortical interneurons depends on neuronal activity and TrkB ligands during an early period of postnatal development. Cereb. Cortex, 14: 342–351.

Penttonen, M., Kamondi, A., Acsady, L. and Buzsaki, G. (1998) Gamma frequency oscillation in the hippocampus of the rat: intracellular analysis in vivo. Eur. J. Neurosci., 10: 718–728.

Peters, Y., Barnhardt, N.E. and O'Donnell, P. (2004) Prefrontal cortical up states are synchronized with ventral tegmental area activity. Synapse, 52: 143–152.

Pezawas, L., Verchinski, B.A., Mattay, V.S., Callicott, J.H., Kolachana, B.S., Straub, R.E., Egan, M.F., Meyer-Lindenberg, A. and Weinberger, D.R. (2004) The brain-derived neurotrophic factor val66met polymorphism and variation in human cortical morphology. J. Neurosci., 24: 10099–10102.

Phillips, H.S., Hains, J.M., Laramee, G.R., Rosenthal, A. and Winslow, J.W. (1990) Widespread expression of BDNF but not NT3 by target areas of basal forebrain cholinergic neurons. Science, 250: 290–294.

Pierri, J.N., Chaudry, A.S., Woo, T.U. and Lewis, D.A. (1999) Alterations in chandelier neuron axon terminals in the prefrontal cortex of schizophrenic subjects. Am. J. Psychiatry, 156: 1709–1719.

Poo, M.M. (2001) Neurotrophins as synaptic modulators. Nat. Rev. Neurosci., 2: 24–32.

Pozzo-Miller, L.D., Gottschalk, W., Zhang, L., McDermott, K., Du, J., Gopalakrishnan, R., Oho, C., Sheng, Z.H. and Lu, B. (1999) Impairments in high-frequency transmission, synaptic vesicle docking, and synaptic protein distribution in the hippocampus of BDNF knockout mice. J. Neurosci., 19: 4972–4983.

Pu, L., Liu, Q.S. and Poo, M.M. (2006) BDNF-dependent synaptic sensitization in midbrain dopamine neurons after cocaine withdrawal. Nat. Neurosci., 9: 605–607.

Rantamaki, T., Hendolin, P., Kankaanpaa, A., Mijatovic, J., Piepponen, P., Domenici, E., Chao, M.V., Mannisto, P.T. and Castren, E. (2007) Pharmacologically diverse antidepressants rapidly activate brain-derived neurotrophic factor receptor TrkB and induce phospholipase-Cgamma signaling pathways in mouse brain. Neuropsychopharmacology, 32: 2152–2162.

Rao, S.G., Williams, G.V. and Goldman-Rakic, P.S. (2000) Destruction and creation of spatial tuning by disinhibition: GABA(A) blockade of prefrontal cortical neurons engaged by working memory. J. Neurosci., 20: 485–494.

Rybakowski, J.K., Borkowska, A., Czerski, P.M., Skibinska, M. and Hauser, J. (2003) Polymorphism of the brain-derived neurotrophic factor gene and performance on a cognitive prefrontal test in bipolar patients. Bipolar Disord., 5: 468–472.

Rybakowski, J.K., Borkowska, A., Skibinska, M. and Hauser, J. (2006) Illness-specific association of val66met BDNF polymorphism with performance on Wisconsin card sorting test in bipolar mood disorder. Mol. Psychiatry, 11: 122–124.

Saarelainen, T., Hendolin, P., Lucas, G., Koponen, E., Sairanen, M., MacDonald, E., Agerman, K., Haapasalo, A., Nawa, H., Aloyz, R., Ernfors, P. and Castren, E. (2003) Activation of the TrkB neurotrophin receptor is induced by antidepressant drugs and is required for antidepressant-induced behavioral effects. J. Neurosci., 23: 349–357.

Sala, R., Viegi, A., Rossi, F.M., Pizzorusso, T., Bonanno, G., Raiteri, M. and Maffei, L. (1998) Nerve growth factor and brain-derived neurotrophic factor increase neurotransmitter release in the rat visual cortex. Eur. J. Neurosci., 10: 2185–2191.

Savitz, J., Solms, M. and Ramesar, R. (2006) The molecular genetics of cognition: dopamine, COMT and BDNF. Genes Brain Behav., 5: 311–328.

Sawaguchi, T., Matsumura, M. and Kubota, K. (1989) Delayed response deficits produced by local injection of bicuculline into the dorsolateral prefrontal cortex in Japanese macaque monkeys. Exp. Brain Res., 75: 457–469.

Sawaguchi, T., Matsumura, M. and Kubota, K. (1990a) Catecholaminergic effects on neuronal activity related to a delayed response task in monkey prefrontal cortex. J. Neurophysiol., 63: 1385–1400.

Sawaguchi, T., Matsumura, M. and Kubota, K. (1990b) Effects of dopamine antagonists on neuronal activity related to a delayed response task in monkey prefrontal cortex. J. Neurophysiol., 63: 1401–1412.

Seamans, J.K., Nogueira, L. and Lavin, A. (2003) Synaptic basis of persistent activity in prefrontal cortex in vivo and in organotypic cultures. Cereb. Cortex, 13: 1242–1250.

Seidah, N.G., Benjannet, S., Pareek, S., Chretien, M. and Murphy, R.A. (1996) Cellular processing of the neurotrophin precursors of NT3 and BDNF by the mammalian proprotein convertases. FEBS Lett., 379: 247–250.

Seroogy, K.B., Lundgren, K.H., Tran, T.M., Guthrie, K.M., Isackson, P.J. and Gall, C.M. (1994) Dopaminergic neurons in rat ventral midbrain express brain-derived neurotrophic factor and neurotrophin-3 mRNAs. J. Comp. Neurol., 342: 321–334.

Shu, Y., Hasenstaub, A., Badoual, M., Bal, T. and McCormick, D.A. (2003a) Barrages of synaptic activity control the gain and sensitivity of cortical neurons. J. Neurosci., 23: 10388–10401.

Shu, Y., Hasenstaub, A. and McCormick, D.A. (2003b) Turning on and off recurrent balanced cortical activity. Nature, 423: 288–293.

Tahvildari, B., Fransen, E., Alonso, A.A. and Hasselmo, M.E. (2007) Switching between "On" and "Off" states of persistent activity in lateral entorhinal layer III neurons. Hippocampus, 17: 257–263.

Timmusk, T., Palm, K., Metsis, M., Reintam, T., Paalme, V., Saarma, M. and Persson, H. (1993) Multiple promoters direct tissue-specific expression of the rat BDNF gene. Neuron, 10: 475–489.

Tseng, K.Y. and O'Donnell, P. (2005) Post-pubertal emergence of prefrontal cortical up states induced by D1-NMDA co-activation. Cereb. Cortex, 15: 49–57.

Turrigiano, G.G., Leslie, K.R., Desai, N.S., Rutherford, L.C. and Nelson, S.B. (1998) Activity-dependent scaling of quantal amplitude in neocortical neurons. Nature, 391: 892–896.

Tyler, W.J., Zhang, X.L., Hartman, K., Winterer, J., Muller, W., Stanton, P.K. and Pozzo-Miller, L. (2006) BDNF increases release probability and the size of a rapidly recycling vesicle pool within rat hippocampal excitatory synapses. J. Physiol., 574: 787–803.

Vijayraghavan, S., Wang, M., Birnbaum, S.G., Williams, G.V. and Arnsten, A.F. (2007) Inverted-U dopamine D1 receptor actions on prefrontal neurons engaged in working memory. Nat. Neurosci., 10: 376–384.

Volk, D.W., Pierri, J.N., Fritschy, J.M., Auh, S., Sampson, A.R. and Lewis, D.A. (2002) Reciprocal alterations in pre- and postsynaptic inhibitory markers at chandelier cell inputs to pyramidal neurons in schizophrenia. Cereb. Cortex, 12: 1063–1070.

Wang, M., Ramos, B.P., Paspalas, C.D., Shu, Y., Simen, A., Duque, A., Vijayraghavan, S., Brennan, A., Dudley, A., Nou, E., Mazer, J.A., McCormick, D.A. and Arnsten, A.F. (2007) Alpha2A-adrenoceptors strengthen working memory networks by inhibiting cAMP-HCN channel signaling in prefrontal cortex. Cell, 129: 397–410.

Wang, Y., Markram, H., Goodman, P.H., Berger, T.K., Ma, J. and Goldman-Rakic, P.S. (2006) Heterogeneity in the pyramidal network of the medial prefrontal cortex. Nat. Neurosci., 9: 534–542.

Webster, M.J., Weickert, C.S., Herman, M.M. and Kleinman, J.E. (2002) BDNF mRNA expression during postnatal development, maturation and aging of the human prefrontal cortex. Brain Res. Dev. Brain Res., 139: 139–150.

Weickert, C.S., Hyde, T.M., Lipska, B.K., Herman, M.M., Weinberger, D.R. and Kleinman, J.E. (2003) Reduced brain-derived neurotrophic factor in prefrontal cortex of patients with schizophrenia. Mol. Psychiatry, 8: 592–610.

Weickert, C.S., Ligons, D.L., Romanczyk, T., Ungaro, G., Hyde, T.M., Herman, M.M., Weinberger, D.R. and Kleinman, J.E. (2005) Reductions in neurotrophin receptor mRNAs in the prefrontal cortex of patients with schizophrenia. Mol. Psychiatry, 10: 637–650.

Weinberger, D.R. and Berman, K.F. (1996) Prefrontal function in schizophrenia: confounds and controversies. Philos. Trans. R. Soc. Lond. B Biol. Sci., 351: 1495–1503.

West, A.E., Chen, W.G., Dalva, M.B., Dolmetsch, R.E., Kornhauser, J.M., Shaywitz, A.J., Takasu, M.A., Tao, X. and Greenberg, M.E. (2001) Calcium regulation of neuronal gene expression. Proc. Natl. Acad. Sci. U.S.A., 98: 11024–11031.

Williams, G.V. and Goldman-Rakic, P.S. (1995) Modulation of memory fields by dopamine D1 receptors in prefrontal cortex. Nature, 376: 572–575.

Williams, G.V., Rao, S.G. and Goldman-Rakic, P.S. (2002) The physiological role of 5-HT2A receptors in working memory. J. Neurosci., 22: 2843–2854.

Wilson, F.A., O'Scalaidhe, S.P. and Goldman-Rakic, P.S. (1994) Functional synergism between putative gamma-aminobutyrate-containing neurons and pyramidal neurons in prefrontal cortex. Proc. Natl. Acad. Sci. U.S.A., 91: 4009–4013.

Woo, N.H. and Lu, B. (2006) Regulation of cortical interneurons by neurotrophins: from development to cognitive disorders. Neuroscientist, 12: 43–56.

Xu, B., Gottschalk, W., Chow, A., Wilson, R.I., Schnell, E., Zang, K., Wang, D., Nicoll, R.A., Lu, B. and Reichardt, L.F. (2000) The role of brain-derived neurotrophic factor receptors in the mature hippocampus: modulation of long-term potentiation through a presynaptic mechanism involving TrkB. J. Neurosci., 20: 6888–6897.

Yan, Q., Radeke, M.J., Matheson, C.R., Talvenheimo, J., Welcher, A.A. and Feinstein, S.C. (1997a) Immunocytochemical localization of TrkB in the central nervous system of the adult rat. J. Comp. Neurol., 378: 135–157.

Yan, Q., Rosenfeld, R.D., Matheson, C.R., Hawkins, N., Lopez, O.T., Bennett, L. and Welcher, A.A. (1997b) Expression of brain-derived neurotrophic factor protein in the adult rat central nervous system. Neuroscience, 78: 431–448.

Young, B.J., Otto, T., Fox, G.D. and Eichenbaum, H. (1997) Memory representation within the parahippocampal region. J. Neurosci., 17: 5183–5195.

Zahrt, J., Taylor, J.R., Mathew, R.G. and Arnsten, A.F. (1997) Supranormal stimulation of D1 dopamine receptors in the rodent prefrontal cortex impairs spatial working memory performance. J. Neurosci., 17: 8528–8535.

Zorner, B., Wolfer, D.P., Brandis, D., Kretz, O., Zacher, C., Madani, R., Grunwald, I., Lipp, H.P., Klein, R., Henn, F.A. and Gass, P. (2003) Forebrain-specific trkB-receptor knock-out mice: behaviorally more hyperactive than "depressive". Biol. Psychiatry, 54: 972–982.

SECTION III

Animal Approaches to the Essence of Memory

CHAPTER 16

Animal models and behaviour: their importance for the study of memory

Vincent F. Castellucci*

Department of Physiology, Université de Montréal, Montreal, QC H3C 3J7, Canada

Abstract: In our overview, we will attempt to justify the use of animal models and suggest that it is the only way to make the successive transitions between changes occurring at the molecular and cellular levels and changes at the level of behaviour in the intact organism. We will also stress the importance of criteria that have to be fulfilled in order to unravel the cellular mechanisms of memory: detectability, mimicry, anterograde alteration and retrograde alteration. We will also propose that a large number of animal models should be used to explore the great variety of potential mechanisms that may exist to explain behaviours and their modifications and in particular memory. Finally using the experimental model of *Aplysia* as example we will insist that to explain the total reflex in an intact animal, all the neurons — sensory neurons and different layers of excitatory and inhibitory interneurons — have to be investigated.

Keywords: *Aplysia*; animal models; learning; memory; inhibitory and excitatory interneurons; 5-HT receptors

Why use animal models to begin with?

Even if at times, to talk about the necessity and the usefulness of animal models to study major biological questions seems like trying to sell freezers to the Inuits I think it is useful to revisit the issue. Before considering some examples that have been used to study behaviour and particularly to investigate the mechanisms of memory and learning we should reflect about the postulates that are put forward to justify them. The main goal is to understand the cellular and molecular mechanisms of a variety of behaviours and in our case the various types of memory. The ultimate idea is to understand what is going on in our own brain, but the link between cellular processes and the various manifestations of our behaviour is very difficult to establish. In addition, the understanding of the complex mechanisms that occur at a single synapse during short-term modifications and more permanent ones is only a portion of the story. We should remember that this synapse is only one portion of the neuron which itself is embedded in a network which in turn is related to other ensembles of neurons; all of which contribute to the behaviour of the intact organism. At the end, this is exactly what we want to explain, how the intact subject behaves and modifies his behaviour.

This quest is very difficult to achieve in humans despite the recent great advances in brain imaging that allow the visualization of the working living brain before, during and after various types of

---

*Corresponding author. Tel.: +1 514 343 6300; Fax: +1 514 343 5751; E-mail: vincent.castellucci@umontreal.ca

activities or treatments (see Chapter 21). We therefore turn to animal models, vertebrate or invertebrate, in an attempt to find a more direct link between behaviour and some cellular and molecular mechanisms. The big challenge is to bridge the gaps between genes, proteins, neurons, neural circuits and behaviour.

We have to realize that even in the nematode *C. elegans*, in snails like *Aplysia* or in insects like *Drosophila* where the link is potentially more direct, behavioural learning is not been fully explained (Krasne and Glanzman, 1995; Waddell and Quinn, 2001). In addition, we should keep in mind that animal models can explain many types of behaviour, but many higher functions found in primates or humans are still missing. Whatever the model being used (vertebrate or invertebrate) the link between the molecular and cellular mechanisms, the link between networks of neurons and a given behaviour have to be made.

In this regard, a critical evaluation of the hypothesis that changing the strength of connection betweens neurons is the mechanism by which memory traces are encoded and stored in the central nervous system has to be made as Martin et al. (2000) already have suggested. The postulate is that "activity-dependent synaptic plasticity is induced at appropriate synapses during memory formation and is both necessary and sufficient for the information storage underlying the type of memory mediated by the brain area in which that plasticity is observed". As we can observe in many chapters of this book (see Chapters 2, 6, 8, 10, 13) long-term potentiation and long-term depression are very good examples of activity-dependent synaptic plasticity that may be the cellular substratum of memory traces. These cellular alterations are not the only ones that can be important; various heterosynaptic mechanisms, non-synaptic mechanisms such as changes in the excitability threshold of neurons, the contribution of glial cell activities are some other candidate mechanisms that should be kept in mind as well.

For whatever model system we use and whatever type of synaptic plasticity we observed, we should consider four criteria that should be verified in order to link the cellular mechanism of interest and the behaviour of interest. These are:

(1) detectability, (2) mimicry, (3) anterograde alteration and (4) retrograde alteration. The criterion of detectability means that if an animal displays memory of some previous experience a change in synaptic efficacy should be detectable somewhere in its nervous system. Mimicry means that if it were possible to induce the same spatial pattern of synaptic weight changes artificially the animal should display "apparent" memory for some past experience which did not occur in practice. The criterion of anterograde alteration postulates that intervention that prevents the induction of synaptic weight changes during learning experience should impair the animal's memory of that experience. On the other hand, the criterion of retrograde alteration postulates that intervention that alters the spatial distribution of synaptic weight induced by a prior learning experience should alter the animal's memory of that experience. There are several illustrations of systems (see Chapters 1–15) that have satisfied three of the criteria (detectability, anterograde and retrograde alteration). But the mimicry criterion has not been verified yet, and considering the number of synapses involved for a specific behaviour, the complexity of the interactions between excitatory and inhibitory drives and the great number of neuronal networks involved, this criterion seems to be very difficult, if not impossible, to verify in the immediate future.

Which animal models?

There is a great variety of animal models (invertebrate or vertebrate) that could be used or have been used to tackle the difficult problem of relating cellular and molecular mechanisms and behavioural modifications. We can mention in the invertebrate tradition recent overviews by Crow and Tian (2006) in *Hermissenda*, by Kemenes et al. (2006) in *Lymnaea*, by Frost et al. (2006) in *Tritonia*, by Hawkins et al. (2006) in *Aplysia* and by Bono and Maricq (2005) in *C. elegans*. In this symposium, four excellent experimental models with strategies to deal with these criteria and with a special effort to deal with the whole organism are the work of David Glanzman in the marine mollusc *Aplysia californica* (Chapter 17), the studies of Berry et al. with *Drosophila melanogaster* (Chapter 18), and the studies of Wendy Suzuki in primates (Chapter 19). The common thread of these studies is the emphasis in examining ensembles of neurons in a particular protocol in order to explain behaviour in the intact organisms.

The fact remains that animal models are the preferred way to relate various molecular and cellular mechanisms and behaviour. Moreover, we could state that it is essential that a great variety of animal models should be used to explore the great variety of potential mechanisms that may exist to explain behaviours and their modifications. It is even conceivable that different mechanisms could explain similar behaviours just as different eye structures have been developed during the evolution of various phyla. This is one of the reasons Bullock (1993, 2000) has proposed revisiting the concept of identifiable neurons and an agenda for comparative neurobiology. Some of the questions he was asking were whether more complex brains operate on the same principles as simpler brains, merely employing more of the same; or has evolution produced new principles? These questions are very pertinent to the studies of cellular and molecular mechanisms of memory.

Future challenges for animal models: the case of the *Aplysia* experimental model

Whatever experimental model we choose, the bridge between molecular and cellular mechanisms and the total behaviour in the intact animal is the ultimate goal. In this regard, even if in the *Aplysia* experimental model many fascinating mechanisms have been unravelled at the level of single synapses (see Sossin, Chapter 1) there are still several issues that have to be examined. In this section we will outline briefly what was accomplished and then review some types of future challenges.

In the marine mollusk *A. californica* and in many other invertebrates it is possible to identify neurons and work out the neural circuitry underlying specific behaviours; this is a crucial advantage for relating specific physiological changes to behaviour in general and to learning and memory in particular (Bulloch and Syed, 1992; Hawkins et al., 1993; Krasne and Glanzman, 1995; Lechner and Byrne, 1998). In *Aplysia*, one favourite approach has been the study of withdrawal reflexes of the gill, the siphon or the tail that can undergo simple forms of behavioural changes: habituation (depression of the reflex), sensitization (facilitation), classical conditioning and differential conditioning (Pinsker et al., 1970; Hawkins et al., 1993, 1998a, b). It is possible to study these reflexes in a variety of experimental preparations and to carry out electrophysiological, pharmacological, biochemical and molecular studies from the intact animal down to the level of dissociated cell cultures and single identified neurons. These withdrawal reflexes have two major components: (1) a monosynaptic component or the junctions between sensory neurons and motor neurons and (2) a polysynaptic component which includes excitatory and inhibitory interneurons. Most of the studies so far have been on investigating the modulation in synaptic strength of the sensory neuron underlying short-term, intermediate-term and long-term changes.

During short-term sensitization or dishabituation of the withdrawal reflex, the monosynaptic connections between sensory neurons and motor neurons are enhanced by heterosynaptic facilitation (Castellucci et al., 1970). The transmitter at the sensory-to-motor junction is most likely glutamate (Dale and Kandel, 1993; Levenson et al., 2000). The sensitizing or facilitating stimulation activates several types of facilitator interneurons, some of which have been identified (Hawkins et al., 1981a, b; Hawkins and Schacher, 1989; Trudeau and Castellucci, 1993b). These neurons use at least four different transmitters: 5-hydroxytryptamine (serotonin or 5-HT), the small cardiopeptides SCP_A and SCP_B and a fourth one presently unidentified (Abrams et al., 1984; Glanzman et al., 1989). Most of the work on synaptic facilitation has focused on the action of 5-HT (see Chapter 1). Serotonin induces changes in ionic conductances leading to spike broadening and enhancement of excitability in the sensory neurons and in some motor neurons. Serotonin leads to an increase in synaptic release in two ways, one that is dependent on spike broadening and the other that is independent of spike broadening. These two facilitatory actions of 5-HT are mediated by at least two second messenger-activated protein kinase systems, protein kinase A (PKA) and protein kinase C (PKC). The two biochemical cascades overlap in their contributions to synaptic facilitation; their contributions are not simply synergistic but are state- and time-dependent (Klein, 1995; Byrne and Kandel, 1996; Lechner and Byrne, 1998). Recent observations by Glanzman (2006) (see Chapter 17) have indicated that facilitation of the sensori-motor synapses during and after 5-HT exposure depends also on a rise in postsynaptic intracellular calcium and release of calcium from postsynaptic stores. These studies show the importance of the postsynaptic component in the synaptic modulation at the synapse between the sensory and motor neurons. Finally this synapse can undergo homosynaptic depression (Castellucci et al., 1970; Armitage and Siegelbaum, 1998), homosynaptic facilitation (Lin and Glanzman, 1994; Murphy and Glanzman, 1997) and heterosynaptic depression involving FMRFamide and dopamine (Montarolo et al., 1988).

While 5-HT is a facilitating neurotransmitter at most sensori-motor junctions like the LE sensory cells cluster in the abdominal ganglion or ventrocaudal sensory cluster in the pleural ganglion (VC pleural cells), this is not the case for the RF sensory cluster. Instead of being facilitated, this synaptic junction is depressed by exposure to 5-HT (Storozhuk and Castellucci, 1999). One obvious question is the role of this depression in the intact animal and under which conditions this depression is called upon. This observation also suggests that different types of receptors are present in sensory neurons and it would be important to explore the distribution and the variety of these receptors in the central nervous system of *Aplysia* and determine under what conditions they come into play during behaviour in the intact animal. Several 5-HT receptors have been cloned already and the regional distribution of some of them have been described (Barbas et al., 2005). To summarize, in the abdominal ganglion, the rLE, RE and RF clusters are positive for 5-HT_{ap1} and 5-HT_{ap2} receptors but the LE are not for either while a few of the VC cells are positive for only 5-HT_{ap1}. These observations when combined with earlier results (Dubuc and Castellucci, 1991) indicate a greater heterogeneity among the sensory clusters of the withdrawal reflex. The main challenge will be to examine closely the variety of interactions of these receptors and the conditions they come into play not only at the sensori-motor junctions but also at the level of the interneurons of the withdrawal reflex (see below). For example, it would be important to investigate how various 5-HT receptors on a single cell can contribute to heterosynaptic facilitation and depression and their importance for intermediate- and long-term synaptic modulation.

While examining various sites of plasticity in the neuronal network, we can observe a differential efficacy in the synaptic modulation of different branches of a single sensory neuron (Dubuc and Castellucci, 1991; Trudeau and Castellucci, 1993b). Moreover the same neuromodulator can have different action in a network suggesting again a great diversity of 5-HT receptors (Stark et al., 1996; Barbas et al., 2003). For example, when recording simultaneously from a single sensory neuron of the LE cluster, an excitatory interneuron and a motor neuron, we observe that both sensory-to-motor junction and the sensory-to-interneuron junction are facilitated by the activation of modulator neurons by stimulation of the left pleuroabdominal connective (nerve) which simulates the sensitization stimulation in the intact animal (185 and 93% EPSP increases, respectively) as well as by serotonin (5-HT) (191 and 84%) (Trudeau and Castellucci, 1993b). The difference in the magnitude of facilitation at these two sites is an indication of a branch-specific, differential efficacy in the modulation of different central synapses made by a single neuron. One interesting question would be to examine under what conditions in the behaving animal this differential action is manifested and what its critical role is. One could extend this question to other known modulator of the reflex such as the peptides SCP_b and FMRFamide and dopamine.

The role of dopamine in withdrawal reflex needs to be further examined. While screening for various 5-HT receptors we found a first homologue of members of the vertebrate D1 receptor subfamily isolated from the *Aplysia* CNS (Barbas et al., 2006). Following heterologous expression of the Ap_{dop1} receptor in HEK293 cells, its activation by dopamine (EC_{50} of 35 nM) or serotonin (EC_{50} of 36 μM) leads to cyclic adenosine monophosphate (cAMP) production. Constitutive activity was also observed and blocked by flupentixol, an inverse agonist. The Ap_{dop1} receptor sensitivity to dopamine and also to serotonin makes it a potential candidate to induce a synergic response following serotonin exposure used in many previous protocols. Previous experiments have indicated that dopamine facilitates the neuromuscular junction in the gill (Ruben and Lukowiak, 1983). Brunelli et al. (1976) have compared the effect of 5-HT and dopamine at the LE to gill motor neuron L7 and found that only 5-HT facilitated the junction while dopamine was without effect. A systematic survey of the dopamine action to the other sensory junctions (RE, rLE and RF sensory clusters) needs to be done.

To understand more fully the withdrawal reflex and examine the cellular mechanisms involved during its modulation, it is necessary to pay attention to all the synaptic junctions of its neuronal network. We have estimated the relative importance of the direct pathway, the monosynaptic connection between the sensory neurons and the motor neurons, and the indirect pathway (or polysynaptic) involving one or more layers of excitatory and inhibitory interneurons before reaching the motor neurons. Our results indicated that interneurons are responsible for more than 75% of the afferent input to the motor neurons (Trudeau and Castellucci, 1992). In another set of experiments, we found that functional uncoupling of inhibitory interneurons plays an important role in short-term sensitization of the reflex (Trudeau and Castellucci, 1993a). What is extremely interesting is that the focus of most of the studies has been at the level of one set of monosynaptic connection, the sensory-to-motor synapse which represents around 25% of the total reflex. On the other hand, little attention has been devoted to the other 75% component of the withdrawal reflex involving, in particular, the inhibitory loops on motor neurons. In other words, a great part of the facilitation of the reflex is due to a massive disinhibition of the inhibitory connections. The involvement and predominance of inhibitory transmission in the neuronal network of *Aplysia* was also studied by Frost et al. (1988), Frost and Kandel (1995), Fischer et al. (1997) and Bristol et al. (2001). Thus it would be important to extend these studies by examining the role of interneurons, in particular the inhibitory interneurons during long-term alteration of the reflex.

Once the contribution of the ensemble of the interneurons are assessed, the logical step is to focus on some key interneurons to explore the cellular and molecular mechanisms involved as it was done for the monosynaptic connection. It is very probable that different signalling pathways could be involved. Some of the candidates could be the inhibitory connections of L30 interneurons to L29 interneurons. These L30 inhibitory connections are dramatically reduced during short-term facilitation of the reflex (Frost et al., 1988; Frost and Kandel, 1995). But no investigation has examined these connections during long-term changes of the reflex. Another key interneuron that could be considered is interneuron L16 which is directly excited by the sensory neurons and which inhibits not only several excitatory interneurons and motor neurons but also the sensory neurons themselves. Hawkins et al. (1981a, b) have suggested a possible inhibitory and modulatory role of L16 in the withdrawal reflex and Wright and Carew (1995) provided further evidence indicating the functional importance of L16: but again no studies have examined what would happen to these identified interneurons during long-term effect. Finally among the excitatory interneurons two excellent candidates could be neurons R18 (Byrne, 1980, 1981) and L28 (Hawkins and Schacher, 1989) which have particular strong action on major gill and ink motor neurons.

Conclusion

Animal models are still an essential tool to understand the cellular and molecular mechanisms underlying learning and memory. It is also important to explore a variety of systems for the various solutions that have been used by different animals to sustain learning and memory.

A particular attention must be devoted also to the entire network involved in a given reflex. To explain the total reflex in an intact animal, all the neurons — sensory neurons, different layers of excitatory and inhibitory interneurons — have to be investigated.

Abbreviations

cAMP	cyclic adenosine mono-phosphate
5-HT	5-hydroxy tryptamine
PKA	protein kinase A
PKC	protein kinase C
SCP	small cardiopeptide

Acknowledgement

This work was supported by Canadian Institute of Health Research CIHR grants MOP-14142 and group grant MGC-57079.

References

Abrams, T.W., Castellucci, V.F., Camardo, J.S., Kandel, E.R. and LLoyd, P.E. (1984) Two endogenous neuropeptides modulate the gill- and siphon-withdrawal reflex in *Aplysia* by presynaptic facilitation involving cAMP-dependent closure of a serotonin-sensitive potassium channel. Proc. Natl. Acad. Sci. U.S.A., 81: 7956–7960.

Armitage, B.A. and Siegelbaum, S.A. (1998) Presynaptic induction and expression of homosynaptic depression at *Aplysia* sensorimotor neuron synapses. J. Neurosci., 18: 8770–8779.

Barbas, D., Campbell, A., Castellucci, V.F. and Desgroseillers, L. (2005) Comparative localization of two serotonin receptors and sensorin in the central nervous system of *Aplysia californica*. J. Comp. Neurol., 490: 295–304.

Barbas, D., DesGroseillers, L., Castellucci, V.F., Carew, T.J. and Marinesco, S. (2003) Multiple serotonergic mechanisms contributing to sensitization in *Aplysia*: evidence for diverse serotonin receptor subtypes. Learn. Behav., 10: 373–386.

Barbas, D., Zappula, J.P., Angers, S., Bouvier, M., Byrne, J.H., Castellucci, V.F. and DesGroseillers, L. (2006) An *Aplysia* dopamine1-like receptor: molecular and functional characterization. J. Neurochem., 96: 414–427.

Bono, M. and Maricq, A.V. (2005) Neuronal substrates of complex behaviors in *C. elegans*. Ann. Rev. Neurosci., 28: 451–501.

Bristol, A.S., Fischer, T.M. and Carew, T.J. (2001) Combined effects of intrinsic facilitation and modulatory inhibition of identified interneurons in the siphon withdrawal circuitry of *Aplysia*. J. Neurosci., 21: 8990–9000.

Brunelli, M., Castellucci, V. and Kandel, E.R. (1976) Synaptic facilitation and behavioral sensitization in *Aplysia*: possible role of serotonin and cyclic AMP. Science, 194: 1178–1181.

Bulloch, A.G.M. and Syed, N.I. (1992) Reconstruction of neuronal networks in culture. Trends Neurosci., 15: 422–427.

Bullock, T.H. (1993) How are more complex brains different? One view and an agenda for comparative neurobiology. Brain Behav. Evol., 41: 88–96.

Bullock, T.H. (2000) Revisiting the concept of identifiable neurons. Brain Behav. Evol., 55: 236–240.

Byrne, J.H. (1980) Neural circuit for inking behavior in *Aplysia californica*. J. Neurophysiol., 43: 896–911.

Byrne, J.H. (1981) Comparative aspects of neural circuits for inking behavior and gill withdrawal in *Aplysia californica*. J. Neurophysiol., 45: 98–106.

Byrne, J.H. and Kandel, E.R. (1996) Presynaptic facilitation revisited: state and time dependence. J. Neurosci., 16: 425–435.

Castellucci, V.F., Pinsker, H., Kupferman, I. and Kandel, E.R. (1970) Neuronal mechanisms of habituation and dishabituation of the gill-withdrawal reflex in *Aplysia*. Science, 167: 1745–1748.

Crow, T. and Tian, L.M. (2006) Pavlovian conditioning in *Hermissenda*: a circuit analysis. Biol. Bull., 210: 289–297.

Dale, N. and Kandel, E.R. (1993) L-glutamate may be the fast excitatory transmitter of *Aplysia*. Proc. Natl. Acad. Sci. U.S.A., 90: 7163–7167.

Dubuc, B. and Castellucci, V.F. (1991) Receptive fields and properties of a new cluster of mechanoreceptor neurons innervating the mantle region and the branchial cavity of the marine mollusk *Aplysia californica*. J. Exp. Biol., 156: 315–334.

Fischer, T.M., Blazis, D.E.J., Priver, N.A. and Carew, T.J. (1997) Metaplasticity at identified inhibitory synapses in *Aplysia*. Nature, 389: 860–865.

Frost, W.N., Brandon, C.L. and Van Zyl, C. (2006) Long-term habituation in the marine mollusc *Tritonia diomedea*. Biol. Bull., 210: 230–237.

Frost, W.N., Clark, G.A. and Kandel, E.R. (1988) Parallel processing of short-term memory for sensitization in *Aplysia*. J. Neurobiol., 19: 297–334.

Frost, W.N. and Kandel, E.R. (1995) Structure of the network mediating siphon-elicited siphon withdrawal in *Aplysia*. J. Neurosci., 73: 2413–2427.

Glanzman, D.L. (2006) The cellular mechanisms of learning in *Aplysia*: of blind men and elephants. Biol. Bull., 210: 271–279.

Glanzman, D.L., Mackey, S.L., Hawkins, R.D., Dyke, A.M., Lloyd, P.E. and Kandel, E.R. (1989) Depletion of serotonin in the nervous system of *Aplysia* reduces the behavioral enhancement of gill withdrawal as well as the heterosynaptic facilitation produced by tail shock. J. Neurosci., 9: 4200–4213.

Hawkins, R.D., Castellucci, V.F. and Kandel, E.R. (1981a) Interneurons involved in mediation and modulation of gill-withdrawal reflex in *Aplysia*. I. Identification and characterization. J. Neurophysiol., 45: 304–314.

Hawkins, R.D., Castellucci, V.F. and Kandel, E.R. (1981b) Interneurons involved in mediation and modulation of the gill-withdrawal reflex in *Aplysia*. II. Identified neurons produce heterosynaptic facilitation contributing to behavioral sensitization. J. Neurophysiol., 45: 315–326.

Hawkins, R.D., Cohen, T.E., Greene, W. and Kandel, E.R. (1998a) Relationships between dishabituation, sensitization, and inhibition of the gill- and siphon-withdrawal reflex in *Aplysia californica* — effects of response measure, test time, and training stimulus. Behav. Neurosci., 112: 24–38.

Hawkins, R.D., Greene, W. and Kandel, E.R. (1998b) Classical conditioning, differential conditioning, and second-order conditioning of the *Aplysia* gill-withdrawal reflex in a simplified mantle organ preparation. Behav. Neurosci., 112: 636–645.

Hawkins, R.D., Kandel, E.R. and Bailey, C.H. (2006) Molecular mechanisms of memory storage in *Aplysia*. Biol. Bull., 210: 174–191.

Hawkins, R.D., Kandel, E.R. and Siegelbaum, S.A. (1993) Learning to modulate transmitter release: themes and variations in synaptic plasticity. Ann. Rev. Neurosci., 16: 625–665.

Hawkins, R.D. and Schacher, S. (1989) Identified facilitator neurons L29 and L28 are excited by cutaneous stimuli used in dishabituation, sensitization, and classical conditioning of *Aplysia*. J. Neurosci., 9: 4236–4245.

Kemenes, I., Straub, V.A., Nikitin, E.S., Staras, K., O'Shea, M., Kemenes, G. and Benjamin, P.R. (2006) Role of delayed nonsynaptic neuronal plasticity in long-term associative memory. Curr. Biol., 16(13): 1269–1279.

Klein, M. (1995) Modulation of ion currents and regulation of transmitter release in short-term synaptic plasticity: the rise and fall of the action potential. Invert. Neurosci., 1: 15–24.

Krasne, F.B. and Glanzman, D.L. (1995) What we can learn from invertebrate learning. Ann. Rev. Psychol., 46: 585–624.

Lechner, H.A. and Byrne, J.H. (1998) New perspectives on classical conditioning: a synthesis of Hebbian and non-Hebbian mechanisms. Neuron, 20: 355–358.

Levenson, J., Sherry, D.M., Dryer, L., Chin, J., Byrne, J.H. and Eskin, A. (2000) Localization of glutamate and glutamate transporters in the sensory neurons of *Aplysia*. J. Comp. Neurol., 423: 121–131.

Lin, X.Y. and Glanzman, D.L. (1994) Long-term potentiation of *Aplysia* sensorimotor synapses in cell culture: regulation by postsynaptic voltage. Proc. R. Soc. Lond. B Biol. Sci., 255: 113–118.

Martin, S.J., Grimwood, P.D. and Morris, R.G.M. (2000) Synaptic plasticity and memory: an evaluation of the hypothesis. Ann. Rev. Neurosci., 23: 649–711.

Montarolo, P.G., Kandel, E.R. and Schacher, S. (1988) Long-term heterosynaptic inhibition in *Aplysia*. Nature, 333: 171–174.

Murphy, G.G. and Glanzman, D.L. (1997) Mediation of classical conditioning in *Aplysia californica* by long-term potentiation of sensorimotor synapses. Science, 278: 467–471.

Pinsker, H., Kupfermann, I., Castellucci, V.F. and Kandel, E.R. (1970) Habituation and dishabituation of the gill-withdrawal reflex in *Aplysia*. Science, 167: 1740–1742.

Ruben, P. and Lukowiak, K. (1983) Modulation of the *Aplysia* gill withdrawal reflex by dopamine. J. Neurobiol., 14: 271–284.

Stark, L.L., Mercer, A.R., Emptage, N.J. and Carew, T.J. (1996) Pharmacological and kinetic characterization of two functional classes of serotonergic modulation in *Aplysia* sensory neurons. J. Neurophysiol., 75: 855–866.

Storozhuk, M.V. and Castellucci, V.F. (1999) The synaptic junctions of LE and RF cluster sensory neurones of *Aplysia californica* are differentially modulated by serotonin. J. Exp. Biol., 202: 115–120.

Trudeau, L.-E. and Castellucci, V.F. (1992) Contribution of polysynaptic pathways in the mediation and plasticity of *Aplysia* gill and siphon withdrawal reflex: evidence for differential modulation. J. Neurosci., 12: 3838–3848.

Trudeau, L.-E. and Castellucci, V.F. (1993a) Functional uncoupling of inhibitory interneurons plays an important role in short-term sensitization of *Aplysia* gill and siphon withdrawal reflex. J. Neurosci., 13: 2126–2135.

Trudeau, L.-E. and Castellucci, V.F. (1993b) Sensitization of the gill and siphon withdrawal reflex of *Aplysia*: mulitple sites of changes in the neuronal network. J. Neurophysiol., 70: 1210–1220.

Waddell, S. and Quinn, W.G. (2001) Flies, genes, and learning. Annu. Rev. Neurosci., 24: 1283–1309.

Wright, W.G. and Carew, T.J. (1995) A single identified interneuron gates tail-shock induced inhibition in the siphon withdrawal reflex of *Aplysia*. J. Neurosci., 15: 790–797.

CHAPTER 17

New tricks for an old slug: the critical role of postsynaptic mechanisms in learning and memory in *Aplysia*

David L. Glanzman[1,2,3,]*

[1]*Department of Physiological Science, UCLA College, Los Angeles, CA 90095-1606, USA*
[2]*Department of Neurobiology, David Geffen School of Medicine, UCLA, Los Angeles, CA 90095-1761, USA*
[3]*Brain Research Institute, David Geffen School of Medicine, UCLA, Los Angeles, CA 90095-1761, USA*

Abstract: The marine snail *Aplysia* has served for more than four decades as an important model system for neurobiological analyses of learning and memory. Until recently, it has been believed that learning and memory in *Aplysia* were due predominately, if not exclusively, to presynaptic mechanisms. For example, two nonassociative forms of learning exhibited by *Aplysia*, sensitization and dishabituation of its defensive withdrawal reflex, have been previously ascribed to presynaptic facilitation of the connections between sensory and motor neurons that mediate the reflex. Recent evidence, however, indicates that postsynaptic mechanisms play a far more important role in learning and memory in *Aplysia* than formerly appreciated. In particular, dishabituation and sensitization depend on a rise in intracellular Ca^{2+} in the postsynaptic motor neuron, postsynaptic exocytosis, and modulation of the functional expression of postsynaptic AMPA-type glutamate receptors. In addition, the expression of the persistent presynaptic changes that occur during intermediate- and long-term dishabituation and sensitization appears to require retrograde signals that are triggered by elevated postsynaptic Ca^{2+}. The model for learning-related synaptic plasticity proposed here for *Aplysia* is similar to current mammalian models. This similarity suggests that the cellular mechanisms of learning and memory have been highly conserved during evolution.

Keywords: siphon-withdrawal reflex; sensitization; long-term memory; intermediate-term memory; AMPA receptor trafficking; retrograde messenger

Introduction

When Eric R. Kandel was awarded the Nobel Prize in Physiology or Medicine in 2000, the field of *Aplysia* learning and memory appeared to have reached its apogee. In his Nobel Prize lecture,

*Corresponding author. Tel.: +1 310 206-9972;
Fax: +1 310 267-0306; E-mail: dglanzman@physci.ucla.edu

DOI: 10.1016/S0079-6123(07)00017-9

published as an essay in *Science*, Kandel (2001) masterfully summarized the general insights gained from over 40 years of work by his laboratory and others on the cellular mechanisms of memory in the marine snail, *Aplysia*. As illustrated in Fig. 1, taken from the *Science* essay, Kandel's scheme for learning-related synaptic changes in *Aplysia* focused almost entirely on presynaptic mechanisms of plasticity. Specifically, Kandel proposed that during behavioral sensitization, the form of learning that

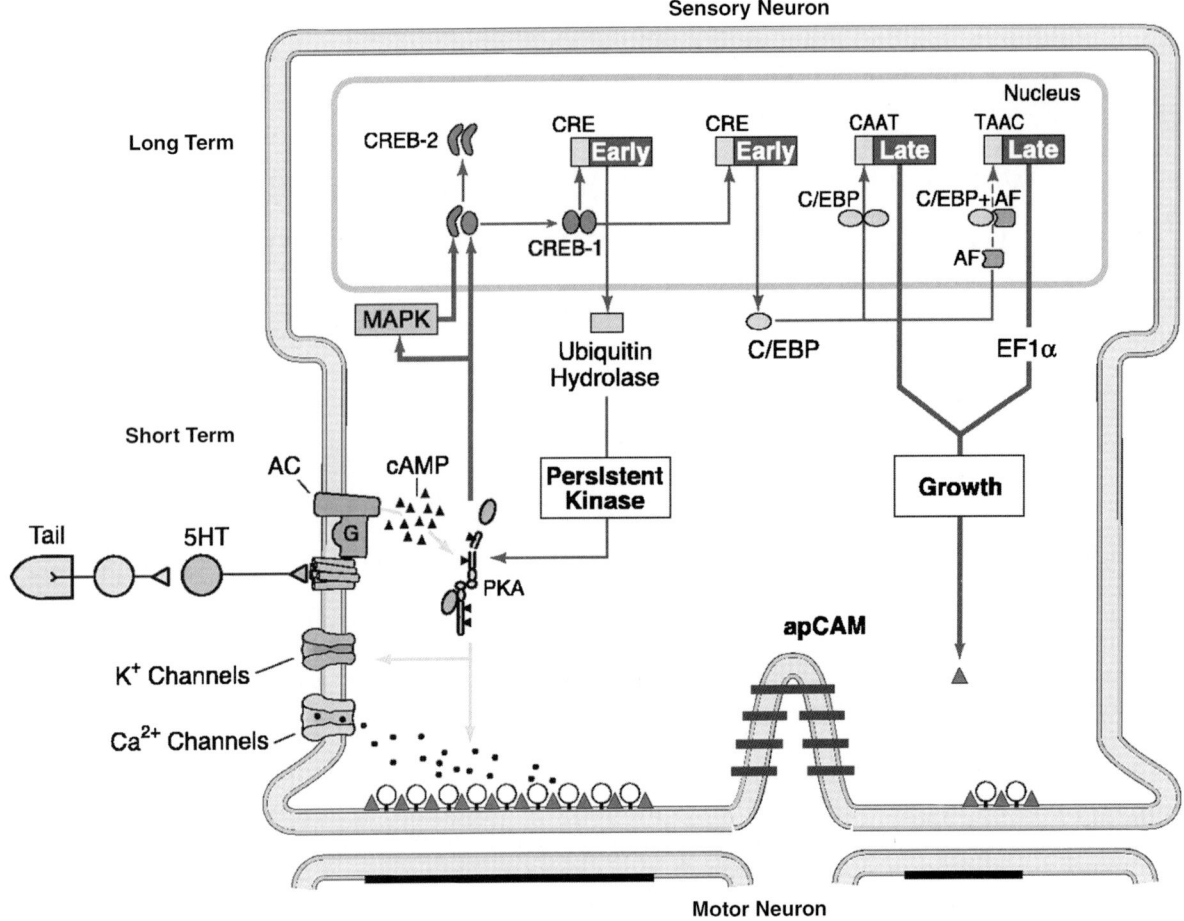

Fig. 1. Presynaptic cellular model for short- and long-term sensitization in *Aplysia*. See the text for details. Adapted with permission from Kandel (2001). (See Color Plate 17.1 in color plate section.)

has been the most studied in *Aplysia*, a monoaminergic transmitter, serotonin (5-HT), is released within the central nervous system (CNS) of *Aplysia*; 5-HT then binds to receptors on the sensory neurons that innervate the siphon and gill of *Aplysia*; and, as a result, a cascade of intracellular changes is initiated in the sensory neurons. These changes involve activation of protein kinases, and, in some cases, activation of genes and synthesis of new proteins. The cellular consequence of these changes is an enhancement of the release of transmitter from the terminals of the sensory neurons. According to the Kandelian model, it is this enhancement of neurotransmitter release, referred to as presynaptic facilitation, that is responsible for the behavioral enhancement during sensitization.

Some version of the diagram in Fig. 1 can be found in practically every introductory textbook in neuroscience. Indeed, the cellular models that have been generated for simple forms of learning and memory, including habituation, sensitization, and classical conditioning, in *Aplysia* are widely regarded as one of the triumphs of modern neuroscience. Several of the mechanisms that have been proposed to explain learning in *Aplysia* — perhaps most prominently, gene transcription triggered by the transcription factor cyclic cAMP response element-binding protein (CREB) (Dash et al., 1990; Kaang et al., 1993) — are not unique to this mollusk, but have been shown to be important for learning in several other important model organisms in neuroscience, including *Drosophila* (Yin et al., 1994, 1995), mice (Kogan et al., 1997; Kida et al., 2002; Bozon et al., 2003; Alarcon et al., 2004), and rats (Josselyn et al., 2001; Brightwell et al., 2004).

Despite the impressive advances toward an understanding of the mechanisms of learning and memory made in *Aplysia*, it has become increasingly evident that the scheme depicted in Fig. 1 is deficient. In particular, postsynaptic mechanisms, which have until relatively recently have been either completely ignored or given short shrift in cellular models of dishabituation and sensitization in *Aplysia*, are indispensable to these forms of learning. As described below, new work reveals an unexpected and striking similarity between the postsynaptic mechanisms that underlie dishabituation/sensitization in *Aplysia* and those that underlie prominent forms of synaptic plasticity and learning in mammals. This similarity suggests that there has been remarkably little change in the biological machinery of memory during the past 555 million years, the approximate period since the last common protostome–deuterostome ancestor (Erwin and Davidson, 2002). Here I review the recent evidence that necessitates a revision of the presynaptic model of dishabituation/sensitization in *Aplysia*, and propose a new model that integrates pre- and postsynaptic mechanisms.

Different phases of memory in *Aplysia*

Behavioral results

To reduce stylistic awkwardness, I will use the term sensitization below to refer to both dishabituation and sensitization, unless otherwise specified. These two cognate forms of nonassociative learning in *Aplysia* are mediated by similar, albeit not entirely identical, mechanisms (see Hochner et al., 1986; Mackey et al., 1987; Marcus et al., 1988; Rankin and Carew, 1988; Glanzman et al., 1989b; Wright et al., 1991; Cohen et al., 1997; Antonov et al., 1999, 2005).

Aplysia exhibits a defensive withdrawal reflex of its gill and siphon that can be sensitized by noxious stimuli, such as electrical shocks applied to the animal's skin (Pinsker et al., 1970; Carew et al., 1971). Repeated application of noxious stimulation can produce sensitization of the withdrawal reflex that persists for days-to-weeks (Pinsker et al., 1973; Frost et al., 1985). Classically, memory for sensitization of the withdrawal reflex has been divided into two phases, short-term memory (STM) and long-term memory (LTM). LTM has been distinguished from the STM by its requirement for protein synthesis and gene transcription (Goelet et al., 1986; Castellucci et al., 1989). In addition, LTM, unlike STM, involves the growth of new synaptic connections between the sensory and motor neurons that mediate the withdrawal reflex (Bailey and Chen, 1983, 1988a, b, 1989; O'Leary et al., 1995; Bailey et al., 1996; Bailey, 1999; Wainwright et al., 2002). Indeed, it seems likely that the protein synthesis and gene transcription that accompany LTM are involved, in part, in the construction of new sensorimotor synapses (see Bailey et al., 1992; Mayford et al., 1992).

Within the last decade a third phase of memory for sensitization has become recognized. This phase, termed intermediate-phase memory (ITM), can be differentiated both temporally and mechanistically from the other two phases of memory. ITM, like LTM, can be induced by multiple tail shocks. Sutton et al. (2001, 2002) reported that a single shock produces sensitization of the siphon-withdrawal reflex that decays within ~ 30 min. By contrast, five individual shocks at 15 min intervals produce sensitization that declines to zero within 3 h after training, but then reappears by 24 h after training. Therefore, training with multiple shocks produces a biphasic pattern of sensitization. The first phase contains two components, an early component (STM) and a later component (ITM); the second phase consists of a single component, LTM. Mechanistically, ITM can be distinguished from STM by its dependence on protein synthesis (but see below); ITM can be distinguished from LTM in not depending on gene transcription (Sutton et al., 2001). In addition, ITM induced by multiple tail shocks requires the persistent activation of protein kinase A (PKA) for its maintenance

(Sutton et al., 2001). More recently, a second form of ITM has been identified. This form, rather than being induced by multiple shocks, is revealed by testing the site on the tail to which a single shock has been delivered (Sutton et al., 2004). This type of learning is referred to as site-specific sensitization; it is believed to result from the conjunction of the firing of sensory neurons that innervate the shocked site on the tail and a neuromodulatory transmitter released from heterosynaptic facilitatory neurons activated by the tail shock (see Walters, 1987; Eliot et al., 1994; Bao et al., 1998). Site-specific ITM differs from ITM induced by multiple shocks (repeated-trial ITM) in a lack of dependence on protein synthesis for its induction (Sutton et al., 2004). Site-specific ITM can be further distinguished from repeated-trial ITM in its dependence on persistent activation of protein kinase C (PKC), rather than PKA. Interestingly, results from experiments using inhibitors differentially selective for different isoforms of PKC indicate that maintenance of site-specific ITM is selectively dependent on the activity of a PKM-like isoform (Sutton et al., 2004). In vertebrates PKM can be generated either through the proteolytic cleavage of atypical PKC, or through de novo synthesis (Hernandez et al., 2003). When generated by proteolytic cleavage, PKM remains persistently active because it lacks autoinhibition from the PKC-regulatory domain. However, the activity of one isoform of PKM, PKMζ, appears to be maintained via ongoing protein synthesis (Hernandez et al., 2003). Recent evidence indicates that PKMζ plays a key role in the persistence of learning-related synaptic plasticity in mammals, as well as in memory maintenance (Sacktor et al., 1993; Serrano et al., 2005; Pastalkova et al., 2006; Shema et al., 2007). A PKM isoform has recently been identified in *Aplysia*, although unlike PKMζ, it appears to be formed by calpain-dependent proteolytic cleavage from atypical PKC (Bougie et al., 2006, 2007).

Synaptic correlates of different memory phases

Each of the three memory phases for sensitization — STM, ITM, and LTM — has a synaptic correlate. These are forms of facilitation of the synaptic connections between the central sensory and motor neurons that participate in the withdrawal reflex (Byrne et al., 1974, 1978a, b; Walters et al., 1983a; Frost et al., 1988; Dubuc and Castellucci, 1991). In addition to producing enhancement of the withdrawal reflex, sensitizing stimuli such as tail shock facilitate the sensorimotor synapse (Castellucci and Kandel, 1976; Walters et al., 1983b; Antonov et al., 1999). Significant evidence indicates that this facilitation is mediated, at least in part, by serotonin (5-HT). Serotonin (5-HT) is an endogenous monoamine (Kistler et al., 1985; Hawkins, 1989) that is released within the CNS of *Aplysia* after sensitizing stimulation (Marinesco and Carew 2002; see also Zhang et al., 2003). Application of 5-HT facilitates sensorimotor connections (Brunelli et al., 1976; Rayport and Schacher, 1986), as does stimulation of an identified serotonergic interneuron, CB1 (Mackey et al., 1989). Depletion of 5-HT within the nervous system impairs sensitization (Glanzman et al., 1989b). Therefore, 5-HT is likely to play a central role in sensitization in *Aplysia*, although other facilitatory transmitters may also participate (Hawkins et al., 1981; Abrams et al., 1984; Ocorr and Byrne, 1985).

A brief (<5 min) application of 5-HT produces short-term facilitation (STF) of the sensorimotor synapse, i.e., facilitation that lasts <30 min (Brunelli et al., 1976; Rayport and Schacher, 1986; Mauelshagen et al., 1996), as does a single tail shock (Walters et al., 1983b; Antonov et al., 1999). Five spaced pulses (typically, a 5–6 min long pulse is used) of 5-HT yields synaptic facilitation that can persist for ≥24 h, or long-term facilitation (LTF) (Montarolo et al., 1986; Mauelshagen et al., 1996). Similarly, multiple spaced shocks to the animal's tail or body wall can also produce LTF of the sensorimotor synapse (Frost et al., 1985; Cleary et al., 1998). Just as training with multiple, spaced sensitizing stimuli recruits not only LTM but also ITM (above), multiple, spaced applications of 5-HT recruit intermediate-term facilitation (ITF), as well as LTF, of the sensorimotor synapse (Ghirardi et al., 1995; Mauelshagen et al., 1996). ITF can also be induced by pairing a brief bout of activity in the sensory neuron with a single pulse of 5-HT (Bao et al., 1998; Sutton and Carew, 2000).

This activity-dependent ITF parallels activity-dependent ITM (above).

Mechanisms of learning-related synaptic facilitation in *Aplysia*

Presynaptic mechanisms

Short-term facilitation

The mechanisms of STF have been extensively reviewed (Byrne and Kandel, 1996; Kandel, 2001). STF appears to be mediated by changes in the presynaptic sensory neurons (but see below). Briefly, 5-HT activates a presynaptic adenylyl cyclase, leading to the synthesis of cyclic AMP (cAMP) and activation of PKA. Activation of PKA closes presynaptic K^+ channels, which results in a longer action potential and greater presynaptic influx of Ca^{2+} during an action potential. In addition to modulation of action potential duration, 5-HT can recruit a second facilitatory mechanism; this mechanism is particularly potent when the sensorimotor synapse is depressed (Gingrich and Byrne, 1985; Hochner et al., 1986). At depressed synapses 5-HT causes a spike duration-independent facilitation of release via a process that is poorly understood but that appears to involve mobilization of presynaptic vesicles. This second process depends predominately on PKC activity (Ghirardi et al., 1992; Manseau et al., 2001).

Intermediate-term facilitation

As outlined above, activity-independent ITF, which results from multiple, spaced exposures of the sensorimotor synapse to 5-HT, depends on protein synthesis and persistent activation of PKA. Until recently, it was unclear whether the necessary protein synthesis occurs in the sensory or motor neuron. However, the locus of persistent PKA activity appears to be presynaptic. Muller and Carew (1998) found that five spaced pulses of 5-HT produced persistent phosphorylation of PKA in sensory neurons; the early phase of this persistent change, which decayed within 3 h, depended on protein, but not RNA, synthesis, and therefore corresponds to ITM. Activity-dependent ITF, which does not require protein synthesis, depends on persistent PKC activity (Sutton and Carew, 2000). A recent study by Zhao et al. (2006) has demonstrated that the sensory neuron is one site for the persistent activity of PKC. Zhao et al. specifically expressed fluorescently tagged isoforms of PKC in sensory neurons. They found that sensory neuron firing paired with a pulse of 5-HT caused the Ca^{2+}-dependent isoform of PKC in *Aplysia*, Apl I (Kruger et al., 1991), to translocate from the cytoplasm to the cell membrane, thereby indicating its activation (see also Sossin et al., 1994). Zhao et al. also found that overexpressing a dominant-negative form of Apl I in sensory neurons blocks activity-dependent ITF.

Long-term facilitation

Most of the work on the cellular mechanisms of LTF has been performed using sensorimotor synapses in dissociated cell culture. LTF is induced by repeated, spaced pulses of 5-HT. In the most common protocol, sensorimotor cocultures are treated with five 5-min pulses of 5-HT, with 15 min between the pulses (Montarolo et al., 1986). The results from studies of LTF have been previously reviewed (Bailey et al., 2004; Hawkins et al., 2006, and Fig. 1; Kandel, 2001), so I will summarize only the major mechanisms of LTF here. LTF depends on both protein and RNA synthesis. LTF is commonly believed to result from the prolonged activation of presynaptic PKA due to the repeated application of 5-HT. The prolonged activation of PKA causes it to translocate from the cytoplasm of the sensory neuron to the nucleus (Bacskai et al., 1993). In the nucleus PKA phosphorylates CREB-1 (Dash et al., 1990; Kaang et al., 1993; Bartsch et al., 1998). CREB-1 is a transcription factor, and its phosphorylation stimulates RNA synthesis. (Activation of CREB-1 also involves its relief from repression by the inhibitory form of CREB, CREB-2, see Bartsch et al., 1995.) Downstream from CREB-1 are the immediate response genes, including the CAAT box enhancer binding protein (C/EBP) (Alberini et al., 1994).

Activation of the immediate response genes, in turn, stimulates the transcription of downstream genes that trigger long-term structural changes in the sensory neurons, including the growth of new presynaptic varicosities and new neurites (Glanzman et al., 1990). These presynaptic structural changes contribute to the formation of new synaptic connections between the sensory and motor neurons (Bailey et al., 1992; Mayford et al., 1992). In addition to the growth of new synapses, LTF is mediated by more subtle cellular changes, including the activation of previously "silent" presynaptic terminals, which lack presynaptic vesicles. During LTF these empty terminals are filled with synaptic vesicles and thereby become functional (Kim et al., 2003). (Note that presynaptic activation of silent synapses also contributes to ITF.)

One presynaptic mechanism recently recognized to be crucial to LTF is the release from the sensory neurons of the sensory neuron-specific neuropeptide sensorin (Brunet et al., 1991). In a series of elegant studies Schacher and colleagues have shown that spaced, repeated pulses of 5-HT cause sensorin to be released from presynaptic terminals, and that the release of sensorin is required for LTF (Hu et al., 2004a, b, 2006, 2007). Release of sensorin is triggered by activation of PKA as well as PKC, depending on the stimulation protocol (see Hu et al., 2007). The released sensorin binds to autoreceptors on the sensory neuron, and thereby stimulates the activity of mitogen-activated protein kinase (MAPK), which, in turn, is believed to play a key role in stimulation of CREB-1 activity (Martin et al., 1997b).

Postsynaptic mechanisms

STF requires only presynaptic changes (Jin et al., 2005; Li et al., 2005), although, as I discuss below, postsynaptic changes may contribute to STF.

ITF and modulation of AMPA receptor trafficking

Chitwood et al. (2001) provided evidence that postsynaptic mechanisms make a critical contribution to activity-independent ITF. Their experiments used isolated siphon motor neurons in dissociated cell culture. Chitwood stimulated the motor neurons individually with brief pulses (puffs) of glutamate, the sensory neuron transmitter (Dale and Kandel, 1993; Levenson et al., 2000) (see also Trudeau and Castellucci, 1993). They electrophysiologically recorded the evoked responses to the glutamate puffs (Glu-EPs) with a sharp microelectrode. A 10-min application of 5-HT to the motor neurons produced enhancement of the Glu-EP that persisted for >40 min after washout of the monamine. (In recent experiments we have found that the enhancement can persist for ≥ 2 h after 5-HT washout [G. Villareal and D.L. Glanzman, unpublished data].) The persistent enhancement of the Glu-EP by 5-HT depended on elevated intracellular Ca^{2+}, because it was blocked by prior injection of the rapid Ca^{2+} chelator BAPTA (Adler et al., 1991) into the motor neuron. Tests with glutamate receptor antagonists indicated that the modulatory effect of 5-HT was specific for AMPA-type receptors. This result suggested that 5-HT modulates the functional expression of AMPA receptors in the motor neuron, possibly by causing exocytotic insertion of new AMPA receptors into the plasma membrane. To test this possibility, Chitwood et al. injected an inhibitor of exocytosis, botulinum toxin, into the motor neuron prior to application of 5-HT. The presence of the toxin blocked the enhancement of the glutamate response.

Li et al. (2005) extended the results of Chitwood et al. (2001) to synaptic facilitation and learning. Using sensorimotor cocultures, Li et al. showed that postsynaptic injection of BAPTA blocked ITF produced by a 10-min application of 5-HT. This result indicates that ITF requires elevated postsynaptic Ca^{2+}. In other experiments Li et al. identified release from intracellular stores as the source of the critical rise in postsynaptic Ca^{2+}. Both inositol 1,4,5-trisphosphate (IP3) receptor-mediated and ryanodine receptor-mediated Ca^{2+} stores appear to mediate 5-HT's facilitatory actions. Additional support for the idea that ITF is mediated by modulation of postsynaptic AMPA receptor trafficking was provided by the finding that postsynaptic injection of botulinum toxin disrupted ITF.

Li et al. (2005) also tested whether facilitation of siphon sensorimotor connections due to sensitizing stimulation — in this case, stimulation of the tail nerves — involves enhancement of the functional expression of AMPA receptors. Tail nerve shock induced persistent facilitation of siphon sensorimotor connections in the abdominal ganglion. Importantly, the nerve shocks produced greater facilitation of the AMPA receptor-mediated component of the excitatory postsynaptic potential (EPSP), than of the NMDA receptor-mediated component. This result provides strong support for the notion that learning-related synaptic facilitation in *Aplysia* involves modulation of AMPA receptor trafficking. Furthermore, the result is inconsistent with the idea that the nerve shock-induced synaptic facilitation is mediated exclusively or predominately by presynaptic facilitation. If this were the case, one would expect the AMPA receptor- and NMDA receptor-mediated components of the EPSP to have been equally facilitated. Li et al. also found that injecting either BAPTA or heparin into the siphon motor neuron prior to tail shock blocked the synaptic facilitation. These results indicate that elevated postsynaptic Ca^{2+}, due to release from intracellular stores, is critical for nerve shock-induced synaptic facilitation, as well as for 5-HT-dependent facilitation. Taken together, the results from the experiments in the abdominal ganglion reinforce those from the in vitro experiments. Both sets of experiments point to a central role for modulation of the functional expression of AMPA receptors — possibly through exocytotic insertion of additional receptors — as well as release of Ca^{2+} from postsynaptic intracellular stores, in mediating persistent synaptic enhancement.

To show that these postsynaptic processes play a role in behavior, Li et al. made use of a reduced preparation, consisting of the CNS, together with the tail and siphon as well as the peripheral nerves that connect the tail and siphon to the CNS. They tested the effect of loading siphon motor neurons with botulinum toxin on dishabituation of the siphon-withdrawal reflex due to tail shock. Li et al. showed that injecting botulinum toxin into just two siphon motor neurons prior to the start of testing blocked dishabituation of the reflex. This result demonstrates that postsynaptic exocytosis is critical not only for synaptic facilitation but also for learning in *Aplysia*.

The results of Chitwood et al. (2001) and Li et al. (2005) are reminiscent of the results from recent studies of synaptic plasticity and learning in mammals. In particular, studies of long-term potentiation (LTP) in the hippocampus have shown that modulation of AMPA receptor trafficking plays a critical role in LTP (Malinow, 2003). Furthermore, an elegant study by Rumpel et al. (2005) has demonstrated that this mechanism is also important in learning in rats. These investigators used an acute gene delivery technique to transfect neurons in the lateral amygdala — a structure critical for fear conditioning — with a construct that blocked the synaptic incorporation of AMPA receptors. This molecular manipulation subsequently impaired fear conditioning to a tone and foot shock. Thus, current evidence indicates that, in both the snail and rat, glutamate receptor trafficking is key to learning and memory.

ITF and rapid, local postsynaptic protein synthesis

Although protein synthesis has been shown to be necessary for activity-independent ITF (Ghirardi et al., 1995; Sutton and Carew, 2000), it has not been shown whether the critical site for protein synthesis is pre- or postsynaptic (or both). We have recently discovered that ITF involves local postsynaptic protein synthesis. Using isolated motor neurons in culture, Villareal et al. (2007) tested whether the enhancement of the glutamate response produced by a 10-min application of 5-HT required protein synthesis. In their initial experiments the irreversible protein synthesis inhibitor emetine was applied to the bath prior to the start of testing. Unexpectedly, the presence of emetine eliminated all enhancement of the Glu-EP due to 5-HT. This result implied a rapid requirement for protein synthesis in the enhancement. It was possible, however, that the increase in the Glu-EP required merely that certain small proteins be present prior to the onset of 5-HT. If these proteins had a rapid turnover, prior application of emetine might impair the enhancement of the glutamate response in the absence of a requirement for de novo protein synthesis. To test this idea, we carried out experiments in which emetine and 5-HT were applied coincidentally. In this case there was some facilitation of the glutamate response while 5-HT was present in the bath; but when 5-HT was washed out, the facilitation declined to zero within 3 min. This result indicates that the latest time at which de novo protein synthesis is necessary for facilitation of the Glu-EP is 3 min after washout of 5-HT. We obtained similar results with another protein synthesis inhibitory, cycloheximide (Villareal et al., 2007).

The rapidity of the requirement for protein synthesis suggested that the critical site for the protein synthesis is the neurites of the motor neuron, rather than the cell body. To test this possibility, we performed experiments on isolated neurites of motor neurons in cell culture. The giant gill motor neuron L7 (Koester and Kandel, 1977; Glanzman et al., 1989a) was used in these experiments. An L7 neuron was dissociated from the CNS, and placed into cell culture. Twenty-four hours later, the major neurite of the motor neuron was severed close to the cell body, and the cell body removed. Experiments were performed on the neurite 24–48 h later. The neurite was impaled with a sharp microelectrode, and the neurite was stimulated with puffs of glutamate, as in the experiments on the whole motor neuron. A 10-min pulse of 5-HT produced persistent enhancement of the glutamate response, and this enhancement was blocked when emetine was present in the bath (Villareal et al., 2007). The results for the isolated neurite were like those for the whole motor neuron. Thus, the critical site for the protein synthesis required for the enhancement of the glutamate response is the process of the motor neuron.

We have extended these results to ITF with experiments on sensorimotor synapses in cell culture using the general protocol of Li et al. (2005). Injecting a cell membrane-impermeant protein synthesis inhibitor, gelonin, into the motor neuron prior to the start of testing disrupted ITF (Villareal et al., 2007). Interestingly, the postsynaptic gelonin had no effect on facilitation while 5-HT was present. This result might appear to support the idea that STF does not depend on postsynaptic processes. However, because STF is commonly regarded as lasting < 30 min (Sutton and Carew, 2002), our results for the isolated motor neuron/neurite raise the possibility that rapid, local postsynaptic protein synthesis may contribute to STF, as well as to ITF.

We do not yet know the identity of the postsynaptic proteins whose local synthesis is stimulated by 5-HT. Recent work indicates that application of dopamine triggers dendritic synthesis of AMPA receptors in hippocampal neurons (Smith et al., 2005). The possibility that a 10-min pulse of 5-HT causes AMPA receptors to be locally synthesized in motor neurons is attractive, particularly in light of our earlier results (Chitwood et al., 2001; Li et al., 2005). We also do not know what signaling pathway mediates the local protein synthesis in motor neurons. One likely candidate, however, is PKC (Villareal et al., 2003). An intriguing possibility is that the local protein synthesis supports the ongoing activity at the synapse of a PKMζ-like kinase (Hernandez et al., 2003). But at present we have no evidence for this idea.

The role of postsynaptic Ca^{2+} and postsynaptic protein synthesis in LTF

Despite the exclusively presynaptic nature of the model of LTF presented in Fig. 1, it has long been apparent that LTF must depend, at least to some extent, on postsynaptic mechanisms. Thus, Glanzman et al. (1990) reported that the presynaptic structural changes that accompany LTF do not occur in the absence of the motor neuron. These investigators treated isolated sensory neurons in culture to the standard five spaced applications of 5-HT. They found that the 5-HT treatment did not produce any structural changes in the sensory neurons when they were not in contact with a motor neuron. More recently, Hu et al. (2007) have found that so-called associative LTF, which is induced by pairing a bout of tetanic stimulation of the sensorimotor synapse with a single pulse of 5-HT, depends, like nonassociative

LTF, on a rapid increase in sensorin expression and secretion. The increase in sensorin expression during associative LTF involves local synthesis of the neuropeptide, and this synthesis requires the presence of the postsynaptic motor neuron.

In addition, several studies have found that LTF is accompanied by an increase in the number of functional AMPA receptors at the sensorimotor synapse (Trudeau and Castellucci, 1995; Zhu et al., 1997; Zhao et al., 2003). This suggests that modulation of postsynaptic AMPA receptor trafficking may contribute to LTF (see also Li et al., 2004). But it remains to be proved that the enhancement of the AMPA receptor-mediated synaptic component shown to accompany LTF is actually *necessary* for LTF. Trudeau and Castellucci (1995) found that LTF of sensorimotor connections in the abdominal ganglion was accompanied by a long-term enhancement of the current evoked by a glutamate receptor agonist, homocysteic acid (HCA), in the L7 motor neuron. However, whereas prior injection of the protein synthesis inhibitor gelonin into the postsynaptic L7 neuron blocked the enhancement of the HCA-evoked current in L7, it did not block LTF of the sensorimotor synapse. This result appears to argue against a functional role for modulation of glutamate receptor trafficking in LTF. Nonetheless, a serious methodological problem with the Trudeau and Castellucci (1995) study is that the investigators measured the response to HCA by voltage-clamp recordings in the soma of the L7 neuron. As our experiments using the isolated neurite of L7 (above) show, 5-HT can produce local enhancement of the motor neurite's sensitivity to glutamate. It is unclear whether the electrophysiological assessments of glutamate receptor function performed by Trudeau and Castellucci in the cell body of L7 were sufficiently sensitive to measure changes in the number of AMPA receptors *in the postsynaptic membrane*. Notice that the majority, if not all, of sites of sensorimotor contact occur on the neurites of L7, rather than on its soma (Winlow and Kandel, 1976; Bailey and Chen, 1988a). For this reason the assertion by Trudeau and Castellucci that their voltage-clamp measurements of the response of the somal membrane of L7 to applied HCA reflected the number of functional glutamate receptors at sensorimotor synapses is problematic, particularly given the enormous size of the L7 cell body (see below).

We have recently reexamined the issue of whether sensory neuron autonomous processes are sufficient to support LTF. All of our experiments were performed using sensorimotor cocultures comprising pleural sensory neurons and small siphon (LFS) motor neurons (Lin and Glanzman, 1994). LTF was induced using the original method of Montarolo et al. (1986), with five spaced pulses of 5-HT. We first asked whether LTF depends on postsynaptic Ca^{2+}. Accordingly, in some experiments the Ca^{2+} chelator BAPTA was injected into the motor neuron before testing the synapse on the first day and prior to 5-HT treatment. Cocultures that received the 5-HT treatment, but not the postsynaptic injection of BAPTA, showed significant LTF compared to control cocultures that received neither 5-HT nor the postsynaptic injection of BAPTA. By contrast, synapses that received the 5-HT treatment plus postsynaptic BAPTA did not exhibit LTF (Cai and Glanzman, 2006). Thus, LTF, like ITF, requires elevation of postsynaptic intracellular Ca^{2+}.

Next we asked whether LTF, like ITF, involves postsynaptic protein synthesis. To test this possibility motor neurons in some cocultures received a prior injection of one of two cell membrane-impermeant inhibitors of protein synthesis, gelonin or the cap analog m^{7}GpppG (Huber et al., 2000). Both inhibitors of protein synthesis blocked LTF when injected postsynaptically (Cai and Glanzman, unpublished data). This result is somewhat surprising. There have been two prior reports that postsynaptic blockade of protein synthesis did not affect LTF. One (Trudeau and Castellucci, 1995) was performed in the CNS, whereas the other was performed using sensorimotor cocultures (Martin et al., 1997a). In the former study 5-HT was applied continuously to the abdominal ganglion for 60 min (note that this procedure produces significant LTF of sensorimotor synapses in the ganglion); in the latter study, 5-HT was iontophoresed onto the synapse in culture, using five spaced applications. A third study that tested the

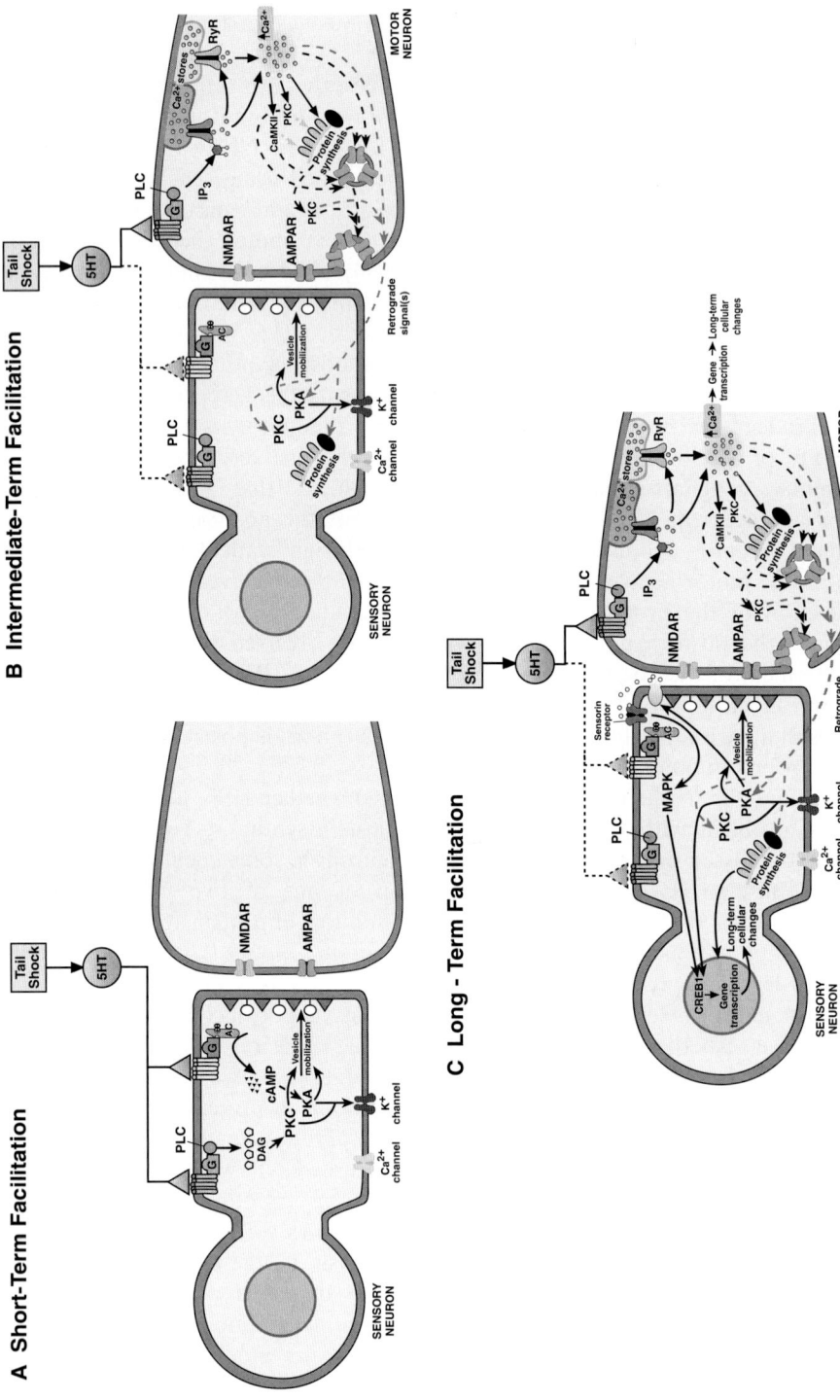

Fig. 2. Revised cellular models for different phases of sensitization-related synaptic facilitation in *Aplysia*. The *dashed lines* depict pathways for which experimental evidence is currently lacking. (A) Short-term facilitation (STF). This phase lasts <30 min. STF can be both induced and expressed presynaptically. (B) Intermediate-term facilitation (ITF). This phase lasts from 20 min to 3 h. According to the model, ITF is induced postsynaptically, by release of Ca^{2+} from intracellular stores in the motor neuron, and expressed both pre- and postsynaptically. The presynaptic expression results from one or more retrograde signals, activated by postsynaptic Ca^{2+}, which stimulate both PKA and PKC within the sensory neuron. (C) Long-term facilitation (LTF). LTF persists for ≥24 h, and involves gene transcription, as well as both pre- and postsynaptic protein synthesis. Like ITF, LTF is induced postsynaptically, and expressed pre- and postsynaptically. LTF is triggered by repeated, spaced applications of 5-HT/sensitizing stimuli. A novel feature of the model in C is that it assumes that prolonged activation of presynaptic PKA, which leads to sensorin release and the translocation of PKA to the presynaptic nucleus, is triggered by elevated postsynaptic Ca^{2+}, via retrograde signaling. In addition, for at least some forms of both LTF and ITF the retrograde signal may activate presynaptic PKC (Jin et al., 2004; Hu et al., 2007) as well as presynaptic PKA. Adapted with permission from Glanzman (2007). (See Color Plate 17.2 in color plate section.)

involvement of postsynaptic protein synthesis in LTF got mixed results. This study, by Sherff and Carew (2004), was performed using the pleural-pedal ganglia. Here, the pleural sensory neurons are physically separated from their postsynaptic targets, which lie in the pedal ganglion. Two separate methods were used to induce LTF. One involved the traditional five spaced applications of 5-HT, which were applied for 25 min to the pleural-pedal ganglia. The second method of 5-HT treatment was the so-called asymmetric method. In this method 5-HT is applied to the pleural ganglion, where the somas of the sensory neurons are located, and to the pedal ganglion, where the sensorimotor synapses (as well as the somas of the motor neurons and other neurons) are located, for the 5-min period corresponding to the end of the 5-HT pulse in the pleural ganglion. This treatment also produces robust LTF. Sherff and Carew found that prior postsynaptic injection of gelonin blocked LTF to the spaced training protocol, but not to the asymmetric protocol.

What can one make of these conflicting results? One possible answer arises from consideration of the postsynaptic target that was used in the Trudeau and Castellucci (1995) and Martin et al. (1997a), the giant motor neuron L7. If, as the results from our studies of ITF (above) indicate, it is local postsynaptic protein synthesis that is critical, then the intrasomal injections used in these two studies may not have delivered a sufficient quantity of gelonin to the critical postsynaptic sites due to the huge volume of the L7 cell. By contrast, because we used the small siphon motor neurons in our experiments, it might have been easier for us to affect local postsynaptic protein synthesis via intrasomal injections of protein synthesis inhibitors. Of course, this explanation does not account for the different results obtained by Sherff and Carew from their postsynaptic gelonin injections. Presumably, the same target neurons were used for the spaced and asymmetric 5-HT treatments. At present, therefore, the question of whether postsynaptic protein synthesis is necessary for LTF must be regarded as unresolved.

The role of retrograde signaling in ITF and LTF

We and others (Jin et al., 2004; Sherff and Carew, 2004; Hu et al., 2007) have shown that both ITF and LTF depend on postsynaptic mechanisms, including elevated postsynaptic Ca^{2+} and postsynaptic protein synthesis (see also Jin and Hawkins, 2003). Furthermore, it is clear that LTF, at least, and possibly ITF as well (Jin et al., 2006), is expressed, in part, through persistent presynaptic changes. How are the pre- and postsynaptic changes coordinated? This coordination would appear to require some form of transsynaptic communication. I propose that the persistent presynaptic changes produced by 5-HT do not result from *direct* actions of this monoamine on the sensory neuron; instead, I suggest that 5-HT's presynaptic effects are *indirect*, and are mediated by a retrograde signal. It is logical to suppose, based on our results, that the retrograde signal is triggered by elevated intracellular Ca^{2+} within the motor neuron (Fig. 2). What might this retrograde signal be? At present, we do not know. But recent evidence from Eric Kandel's laboratory indicates that a transsynaptic interaction between neuroligin and neurexin may subserve retrograde signaling during LTF (Choi et al., 2007). It seems likely, however, that there will be multiple retrograde signals involved in both ITF and LTF. This is an important and challenging area for the field of learning-related synaptic plasticity in *Aplysia*.

Conclusions

As the present review indicates, knowledge of postsynaptic mechanisms is crucial for understanding learning-related synaptic plasticity in *Aplysia*. It is, moreover, intriguing that the postsynaptic mechanisms — such as modulation of AMPA receptor trafficking and rapid, local postsynaptic protein synthesis — that are beginning to be recognized as essential to ITF and LTF have previously been implicated in learning-related synaptic plasticity in vertebrates (see e.g., Huber et al., 2000; Malinow, 2003; Tsokas et al., 2005). This is unlikely to be the result of coincidence. Instead, it is more plausible that the basic cellular and molecular mechanisms of memory arose early in evolution and were maintained.

Although some may find this conclusion surprising, it would not have surprised Charles Darwin. In his final book, *The Formation of Vegetable Mould Through the Action of Worms, with Observations on Their Habits*, published in 1881, Darwin observed that earthworms show "some degree of intelligence." We can be confident that this great biologist was not using the word "intelligence" metaphorically. If as Darwin was convinced, worms and other higher invertebrates do indeed exhibit intelligence — or, as we would say today, cognition — then the complex neuronal machinery of cognition must reside within these relatively humble creatures as well (see Giurfa, 2007; Kristan and Gillette, 2007; Rankin and Dubnau, 2007).

References

Abrams, T.W., Castellucci, V.F., Camardo, J.S., Kandel, E.R. and Lloyd, P.E. (1984) Two endogenous neuropeptides modulate the gill and siphon withdrawal reflex in *Aplysia* by presynaptic facilitation involving cAMP-dependent closure of a serotonin-sensitive potassium channel. Proc. Natl. Acad. Sci. U.S.A., 81: 7956–7960.

Adler, E.M., Augustine, G.J., Duffy, S.N. and Charlton, M.P. (1991) Alien intracellular calcium chelators attenuate neurotransmitter release at the squid giant synapse. J. Neurosci., 11: 1496–1507.

Alarcon, J.M., Malleret, G., Touzani, K., Vronskaya, S., Ishii, S., Kandel, E.R. and Barco, A. (2004) Chromatin acetylation, memory, and LTP are impaired in CBP+/– mice: a model for the cognitive deficit in Rubinstein-Taybi syndrome and its amelioration. Neuron, 42: 947–959.

Alberini, C.M., Ghirardi, M., Metz, R. and Kandel, E.R. (1994) C/EBP is an immediate-early gene required for the consolidation of long-term facilitation in *Aplysia*. Cell, 76: 1099–1114.

Antonov, I., Kandel, E.R. and Hawkins, R.D. (1999) The contribution of facilitation of monosynaptic PSPs to dishabituation and sensitization of the *Aplysia* siphon withdrawal reflex. J. Neurosci., 19: 10438–10450.

Antonov, I., Kandel, E.R. and Hawkins, R.D. (2005) Roles of PKA, PKC, and CaMKII in dishabituation and sensitization of the *Aplysia* siphon-withdrawal reflex. Soc. Neurosci. Abstr., 31: p. 540.6.

Bacskai, B.J., Hochner, B., Mahaut, S.M., Adams, S.R., Kaang, B.K., Kandel, E.R. and Tsien, R.Y. (1993) Spatially resolved dynamics of cAMP and protein kinase A subunits in *Aplysia* sensory neurons. Science, 260: 222–226.

Bailey, C.H. (1999) Structural changes and the storage of long-term memory in *Aplysia*. Can. J. Physiol. Pharmacol., 77: 738–747.

Bailey, C.H., Bartsch, D. and Kandel, E.R. (1996) Toward a molecular definition of long-term memory storage. Proc. Natl. Acad. Sci. U.S.A., 93: 13445–13452.

Bailey, C.H. and Chen, M. (1983) Morphological basis of long-term habituation and sensitization in *Aplysia*. Science, 220: 91–93.

Bailey, C.H. and Chen, M. (1988a) Long-term sensitization in *Aplysia* increases the number of presynaptic contacts onto the identified gill motor neuron L7. Proc. Natl. Acad. Sci. U.S.A., 85: 9356–9359.

Bailey, C.H. and Chen, M. (1988b) Long-term memory in *Aplysia* modulates the total number of varicosities of single identified sensory neurons. Proc. Natl. Acad. Sci. U.S.A., 85: 2373–2377.

Bailey, C.H. and Chen, M. (1989) Time course of structural changes at identified sensory neuron synapses during long-term sensitization in *Aplysia*. J. Neurosci., 9: 1774–1780.

Bailey, C.H., Chen, M., Keller, F. and Kandel, E.R. (1992) Serotonin-mediated endocytosis of apCAM: an early step of learning-related synaptic growth in *Aplysia*. Science, 256: 645–649.

Bailey, C.H., Kandel, E.R. and Si, K. (2004) The persistence of long-term memory: a molecular approach to self-sustaining changes in learning-induced synaptic growth. Neuron, 44: 49–57.

Bao, J.X., Kandel, E.R. and Hawkins, R.D. (1998) Involvement of presynaptic and postsynaptic mechanisms in a cellular analog of classical conditioning at *Aplysia* sensory-motor neuron synapses in isolated cell culture. J. Neurosci., 18: 458–466.

Bartsch, D., Casadio, A., Karl, K.A., Serodio, P. and Kandel, E.R. (1998) CREB1 encodes a nuclear activator, a repressor, and a cytoplasmic modulator that form a regulatory unit critical for long-term facilitation. Cell, 95: 211–223.

Bartsch, D., Ghirardi, M., Skehel, P.A., Karl, K.A., Herder, S.P., Chen, M., Bailey, C.H. and Kandel, E.R. (1995) *Aplysia* CREB2 represses long-term facilitation: relief of repression converts transient facilitation into long-term functional and structural change. Cell, 83: 979–992.

Bougie, J., Lim, T., Ferraro, G., Manjunath, V., Scott, D. and Sossin, W.S. (2006) Cloning and characterization of protein kinase C (PKC) Apl III, a homologue of atypical PKCs in *Aplysia*. Soc. Neurosci. Abstr., 32: p. 669.10.

Bougie, J.K., Lim, T., Manjunath, V., Farah-Abi, C., Nagakura, I. and Sossin, W.S. (2007) The role of atypical protein kinase C (PKC) zeta in synaptic plasticity in *Aplysia*. Soc. Neurosci. Abstr., 33: p. 208.5.

Bozon, B., Kelly, A., Josselyn, S.A., Silva, A.J., Davis, S. and Laroche, S. (2003) MAPK, CREB and zif268 are all required for the consolidation of recognition memory. Philos. Trans. R. Soc. Lond. B Biol. Sci., 358: 805–814.

Brightwell, J.J., Gallagher, M. and Colombo, P.J. (2004) Hippocampal CREB1 but not CREB2 is decreased in aged rats with spatial memory impairments. Neurobiol. Learn. Mem., 81: 19–26.

Brunelli, M., Castellucci, V. and Kandel, E.R. (1976) Synaptic facilitation and behavioral sensitization in *Aplysia*: possible role of serotonin and cyclic AMP. Science, 194: 1178–1181.

Brunet, J.F., Shapiro, E., Foster, S.A., Kandel, E.R. and Iino, Y. (1991) Identification of a peptide specific for *Aplysia* sensory neurons by PCR-based differential screening. Science, 252: 856–859.

Byrne, J., Castellucci, V. and Kandel, E.R. (1974) Receptive fields and response properties of mechanoreceptor neurons innervating siphon skin and mantle shelf in *Aplysia*. J. Neurophysiol., 37: 1041–1064.

Byrne, J.H., Castellucci, V.F., Carew, T.J. and Kandel, E.R. (1978a) Stimulus-response relations and stability of mechanoreceptor and motor neurons mediating defensive gill-withdrawal reflex in *Aplysia*. J. Neurophysiol., 41: 402–417.

Byrne, J.H., Castellucci, V.F. and Kandel, E.R. (1978b) Contribution of individual mechanoreceptor sensory neurons to defensive gill-withdrawal reflex in *Aplysia*. J. Neurophysiol., 41: 418–431.

Byrne, J.H. and Kandel, E.R. (1996) Presynaptic facilitation revisited: state and time dependence. J. Neurosci., 16: 425–435.

Cai, D. and Glanzman, D.L. (2006) Long-term facilitation of *Aplysia* sensorimotor synapses depends on elevated postsynaptic calcium. Soc. Neurosci. Abstr., 32: p. 813.13.

Carew, T.J., Castellucci, V.F. and Kandel, E.R. (1971) An analysis of dishabituation and sensitization of the gill-withdrawal reflex in *Aplysia*. Int. J. Neurosci., 2: 79–98.

Castellucci, V.F., Blumenfeld, H., Goelet, P. and Kandel, E.R. (1989) Inhibitor of protein synthesis blocks long-term behavioral sensitization in the isolated gill-withdrawal reflex of *Aplysia*. J. Neurobiol., 20: 1–9.

Castellucci, V.F. and Kandel, E.R. (1976) Presynaptic facilitation as a mechanism for behavioral sensitization in *Aplysia*. Science, 194: 1176–1178.

Chitwood, R.A., Li, Q. and Glanzman, D.L. (2001) Serotonin facilitates AMPA-type responses in isolated siphon motor neurons of *Aplysia* in culture. J. Physiol., 534: 501–510.

Choi, Y.-B., Kassabov, S.R., Puthenveettil, S.V., Bailey, C.H. and Kandel, E.R. (2007) Role of the autism-linked neurexin-neuroligin trans-synaptic interaction in learning-related synaptic plasticity at the *Aplysia* sensory to motor neuron synapse. Soc. Neurosci. Abstr., 33: p. 131.19.

Cleary, L.J., Lee, W.L. and Byrne, J.H. (1998) Cellular correlates of long-term sensitization in *Aplysia*. J. Neurosci., 18: 5988–5998.

Cohen, T.E., Kaplan, S.W., Kandel, E.R. and Hawkins, R.D. (1997) A simplified preparation for relating cellular events to behavior: mechanisms contributing to habituation, dishabituation, and sensitization of the *Aplysia* gill-withdrawal reflex. J. Neurosci., 17: 2886–2899.

Dale, N. and Kandel, E.R. (1993) L-glutamate may be the fast excitatory transmitter of *Aplysia* sensory neurons. Proc. Natl. Acad. Sci. U.S.A., 90: 7163–7167.

Dash, P.K., Hochner, B. and Kandel, E.R. (1990) Injection of the cAMP-responsive element into the nucleus of *Aplysia* sensory neurons blocks long-term facilitation. Nature, 345: 718–721.

Dubuc, B. and Castellucci, V.F. (1991) Receptive fields and properties of a new cluster of mechanoreceptor neurons innervating the mantle region and the branchial cavity of the marine mollusk *Aplysia californica*. J. Exp. Biol., 156: 315–334.

Eliot, L.S., Hawkins, R.D., Kandel, E.R. and Schacher, S. (1994) Pairing-specific, activity-dependent presynaptic facilitation at *Aplysia* sensory-motor neuron synapses in isolated cell culture. J. Neurosci., 14: 368–383.

Erwin, D.H. and Davidson, E.H. (2002) The last common bilaterian ancestor. Development, 129: 3021–3032.

Frost, W.N., Castellucci, V.F., Hawkins, R.D. and Kandel, E.R. (1985) Monosynaptic connections made by the sensory neurons of the gill- and siphon-withdrawal reflex in *Aplysia* participate in the storage of long-term memory for sensitization. Proc. Natl. Acad. Sci. U.S.A., 82: 8266–8269.

Frost, W.N., Clark, G.A. and Kandel, E.R. (1988) Parallel processing of short-term memory for sensitization in *Aplysia*. J. Neurobiol., 19: 297–334.

Ghirardi, M., Braha, O., Hochner, B., Montarolo, P.G., Kandel, E.R. and Dale, N. (1992) Roles of PKA and PKC in facilitation of evoked and spontaneous transmitter release at depressed and nondepressed synapses in *Aplysia* sensory neurons. Neuron, 9: 479–489.

Ghirardi, M., Montarolo, P.G. and Kandel, E.R. (1995) A novel intermediate stage in the transition between short- and long-term facilitation in the sensory to motor neuron synapse of *Aplysia*. Neuron, 14: 413–420.

Gingrich, K.J. and Byrne, J.H. (1985) Simulation of synaptic depression, posttetanic potentiation, and presynaptic facilitation of synaptic potentials from sensory neurons mediating gill-withdrawal reflex in *Aplysia*. J. Neurophysiol., 53: 652–669.

Giurfa, M. (2007) Invertebrate cognition: nonelemental learning beyond simple conditioning. In: North G. and Greenspan R.J. (Eds.), Invertebrate Neurobiology. Cold Spring Harbor Laboratory Press, New York, pp. 281–308.

Glanzman, D.L. (2007) Simple minds: the neurobiology of invertebrate learning and memory. In: North G. and Greenspan R.J. (Eds.), Invertebrate Neurobiology. Cold Spring Harbor Laboratory Press, New York, pp. 347–380.

Glanzman, D.L., Kandel, E.R. and Schacher, S. (1989a) Identified target motor neuron regulates neurite outgrowth and synapse formation of *Aplysia* sensory neurons in vitro. Neuron, 3: 441–450.

Glanzman, D.L., Kandel, E.R. and Schacher, S. (1990) Target-dependent structural changes accompanying long-term synaptic facilitation in *Aplysia* neurons. Science, 249: 799–802.

Glanzman, D.L., Mackey, S.L., Hawkins, R.D., Dyke, A.M., Lloyd, P.E. and Kandel, E.R. (1989b) Depletion of serotonin in the nervous system of *Aplysia* reduces the behavioral enhancement of gill withdrawal as well as the heterosynaptic facilitation produced by tail shock. J. Neurosci., 9: 4200–4213.

Goelet, P., Castellucci, V.F., Schacher, S. and Kandel, E.R. (1986) The long and the short of long-term memory — a molecular framework. Nature, 322: 419–422.

Hawkins, R.D. (1989) Localization of potential serotonergic facilitator neurons in *Aplysia* by glyoxylic acid histofluorescence combined with retrograde fluorescent labeling. J. Neurosci., 9: 4214–4226.

Hawkins, R.D., Castellucci, V.F. and Kandel, E.R. (1981) Interneurons involved in mediation and modulation of gill-withdrawal reflex in *Aplysia*. II. Identified neurons produce heterosynaptic facilitation contributing to behavioral sensitization. J. Neurophysiol., 45: 315–326.

Hawkins, R.D., Kandel, E.R. and Bailey, C.H. (2006) Molecular mechanisms of memory storage in *Aplysia*. Biol. Bull., 210: 174–191.

Hernandez, A.I., Blace, N., Crary, J.F., Serrano, P.A., Leitges, M., Libien, J.M., Weinstein, G., Tcherapanov, A. and Sacktor, T.C. (2003) Protein kinase Mζ synthesis from a brain mRNA encoding an independent protein kinase Cζ catalytic domain. Implications for the molecular mechanism of memory. J. Biol. Chem., 278: 40305–40316.

Hochner, B., Klein, M., Schacher, S. and Kandel, E.R. (1986) Additional component in the cellular mechanism of presynaptic facilitation contributes to behavioral dishabituation in *Aplysia*. Proc. Natl. Acad. Sci. U.S.A., 83: 8794–8798.

Hu, J.Y., Chen, Y. and Schacher, S. (2007) Protein kinase C regulates local synthesis and secretion of a neuropeptide required for activity-dependent long-term synaptic plasticity. J. Neurosci., 27: 8927–8939.

Hu, J.Y., Glickman, L., Wu, F. and Schacher, S. (2004a) Serotonin regulates the secretion and autocrine action of a neuropeptide to activate MAPK required for long-term facilitation in *Aplysia*. Neuron, 43: 373–385.

Hu, J.Y., Goldman, J., Wu, F. and Schacher, S. (2004b) Target-dependent release of a presynaptic neuropeptide regulates the formation and maturation of specific synapses in *Aplysia*. J. Neurosci., 24: 9933–9943.

Hu, J.Y., Wu, F. and Schacher, S. (2006) Two signaling pathways regulate the expression and secretion of a neuropeptide required for long-term facilitation in *Aplysia*. J. Neurosci., 26: 1026–1035.

Huber, K.M., Kayser, M.S. and Bear, M.F. (2000) Role for rapid dendritic protein synthesis in hippocampal mGluR-dependent long-term depression. Science, 288: 1254–1257.

Jin, I. and Hawkins, R.D. (2003) Presynaptic and postsynaptic mechanisms of a novel form of homosynaptic potentiation at *Aplysia* sensory-motor neuron synapses. J. Neurosci., 23: 7288–7297.

Jin, I., Kandel, E.R. and Hawkins, R.D. (2004) Pre- and postsynaptic mechanisms of facilitation at *Aplysia* sensory-motor synapses: time and state dependence revisited. Soc. Neurosci. Abstr., 30: p. 515.4.

Jin, I., Kandel, E.R. and Hawkins, R.D. (2005) The roles of PKA, PKC, and CamKII in facilitation at *Aplysia* sensory-motor neuron synapses depend on the duration of exposure to 5-HT. Soc. Neurosci. Abstr., 31: p. 540.5.

Jin, I., Kandel, E.R. and Hawkins, R.D. (2006) Presynaptic mechanisms of intermediate-term facilitation. Soc. Neurosci. Abstr., 32: p. 813.2.

Josselyn, S.A., Shi, C., Carlezon, W.A., Jr., Neve, R.L., Nestler, E.J. and Davis, M. (2001) Long-term memory is facilitated by cAMP response element-binding protein overexpression in the amygdala. J. Neurosci., 21: 2404–2412.

Kaang, B.K., Kandel, E.R. and Grant, S.G. (1993) Activation of cAMP-responsive genes by stimuli that produce long-term facilitation in *Aplysia* sensory neurons. Neuron, 10: 427–435.

Kandel, E.R. (2001) The molecular biology of memory storage: a dialogue between genes and synapses. Science, 294: 1030–1038.

Kida, S., Josselyn, S.A., de Ortiz, S.P., Kogan, J.H., Chevere, I., Masushige, S. and Silva, A.J. (2002) CREB required for the stability of new and reactivated fear memories. Nat. Neurosci., 5: 348–355.

Kim, J.H., Udo, H., Li, H.L., Youn, T.Y., Chen, M., Kandel, E.R. and Bailey, C.H. (2003) Presynaptic activation of silent synapses and growth of new synapses contribute to intermediate and long-term facilitation in *Aplysia*. Neuron, 40: 151–165.

Kistler, H.B., Jr., Hawkins, R.D., Koester, J., Steinbusch, H.W.M., Kandel, E.R. and Schwartz, J.H. (1985) Distribution of serotonin-immunoreactive cell bodies and processes in the abdominal ganglion of mature *Aplysia*. J. Neurosci., 5: 72–80.

Koester, J. and Kandel, E.R. (1977) Further identification of neurons in the abdominal ganglion of *Aplysia* using behavioral criteria. Brain Res., 121: 1–20.

Kogan, J.H., Frankland, P.W., Blendy, J.A., Coblentz, J., Marowitz, Z., Schütz, G. and Silva, A.J. (1997) Spaced training induces normal long-term memory in CREB mutant mice. Curr. Biol., 7: 1–11.

Kristan, W. and Gillette, R. (2007) Behavioral choice. In: North G. and Greenspan R.J. (Eds.), Invertebrate Neurobiology. Cold Spring Harbor Laboratory Press, New York, pp. 533–553.

Kruger, K.E., Sossin, W.S., Sacktor, T.C., Bergold, P.J., Beushausen, S. and Schwartz, J.H. (1991) Cloning and characterization of Ca^{2+}-dependent and Ca^{2+}-independent PKCs expressed in *Aplysia* sensory cells. J. Neurosci., 11: 2303–2313.

Levenson, J., Sherry, D.M., Dryer, L., Chin, J., Byrne, J.H. and Eskin, A. (2000) Localization of glutamate and glutamate transporters in the sensory neurons of *Aplysia*. J. Comp. Neurol., 423: 121–131.

Li, H., Huang, B.S.H. and Kandel, E.R. (2004) Innervation-dependent clustering of glutamate receptors during regeneration and plasticity-associated synaptogenesis in *Aplysia*. Soc. Neurosci. Abstr., 30: p. 614.18.

Li, Q., Roberts, A.C. and Glanzman, D.L. (2005) Synaptic facilitation and behavioral dishabituation in *Aplysia*: dependence upon release of Ca^{2+} from postsynaptic intracellular stores, postsynaptic exocytosis and modulation of postsynaptic AMPA receptor efficacy. J. Neurosci., 25: 5623–5637.

Lin, X.Y. and Glanzman, D.L. (1994) Long-term potentiation of *Aplysia* sensorimotor synapses in cell culture: regulation by postsynaptic voltage. Proc. Biol. Sci., 255: 113–118.

Mackey, S.L., Glanzman, D.L., Small, S.A., Dyke, A.M., Kandel, E.R. and Hawkins, R.D. (1987) Tail shock produces inhibition as well as sensitization of the siphon-withdrawal reflex of *Aplysia*: possible behavioral role for presynaptic inhibition mediated by the peptide Phe-Met-Arg-Phe-NH2. Proc. Natl. Acad. Sci. U.S.A., 84: 8730–8734.

Mackey, S.L., Kandel, E.R. and Hawkins, R.D. (1989) Identified serotonergic neurons LCB1 and RCB1 in the cerebral ganglia of *Aplysia* produce presynaptic facilitation of siphon sensory neurons. J. Neurosci., 9: 4227–4235.

Malinow, R. (2003) AMPA receptor trafficking and long-term potentiation. Philos. Trans. R. Soc. Lond. B, 358: 707–714.

Manseau, F., Fan, X., Hueftlein, T., Sossin, W. and Castellucci, V.F. (2001) Ca^{2+}-independent protein kinase C Apl II mediates the serotonin-induced facilitation at depressed *Aplysia* sensorimotor synapses. J. Neurosci., 21: 1247–1256.

Marcus, E.A., Nolen, T.G., Rankin, C.H. and Carew, T.J. (1988) Behavioral dissociation of dishabituation, sensitization, and inhibition in *Aplysia*. Science, 241: 210–213.

Marinesco, S. and Carew, T.J. (2002) Serotonin release evoked by tail nerve stimulation in the CNS of *Aplysia*: characterization and relationship to heterosynaptic plasticity. J. Neurosic., 22: 2299–2312.

Martin, K.C., Casadio, A., Zhu, H., E, Y., Rose, J.C., Chen, M., Bailey, C.H. and Kandel, E.R. (1997a) Synapse-specific, long-term facilitation of *Aplysia* sensory to motor synapses: a function for local protein synthesis in memory storage. Cell, 91: 927–938.

Martin, K.C., Michael, D., Rose, J.C., Barad, M., Casadio, A., Zhu, H. and Kandel, E.R. (1997b) MAP kinase translocates into the nucleus of the presynaptic cell and is required for long-term facilitation in *Aplysia*. Neuron, 18: 899–912.

Mauelshagen, J., Parker, G.R. and Carew, T.J. (1996) Dynamics of induction and expression of long-term synaptic facilitation in *Aplysia*. J. Neurosci., 16: 7099–7108.

Mayford, M., Barzilai, A., Keller, F., Schacher, S. and Kandel, E.R. (1992) Modulation of an NCAM-related adhesion molecule with long-term synaptic plasticity in *Aplysia*. Science, 256: 638–644.

Montarolo, P.G., Goelet, P., Castellucci, V.F., Morgan, J., Kandel, E.R. and Schacher, S. (1986) A critical period for macromolecular synthesis in long-term heterosynaptic facilitation in *Aplysia*. Science, 234: 1249–1254.

Muller, U. and Carew, T.J. (1998) Serotonin induces temporally and mechanistically distinct phases of persistent PKA activity in *Aplysia* sensory neurons. Neuron, 21: 1423–1434.

O'Leary, F.A., Byrne, J.H. and Cleary, L.J. (1995) Long-term structural remodeling in *Aplysia* sensory neurons requires de novo protein synthesis during a critical time period. J. Neurosci., 15: 3519–3525.

Ocorr, K.A. and Byrne, J.H. (1985) Membrane responses and changes in cAMP levels in *Aplysia* sensory neurons produced by serotonin, tryptamine, FMRFamide and small cardioactive peptideB (SCPB). Neurosci. Lett., 55: 113–118.

Pastalkova, E., Serrano, P., Pinkhasova, D., Wallace, E., Fenton, A.A. and Sacktor, T.C. (2006) Storage of spatial information by the maintenance mechanism of LTP. Science, 313: 1141–1144.

Pinsker, H., Kupfermann, I., Castellucci, V. and Kandel, E. (1970) Habituation and dishabituation of the gill-withdrawal reflex in *Aplysia*. Science, 167: 1740–1742.

Pinsker, H.M., Hening, W.A., Carew, T.J. and Kandel, E.R. (1973) Long-term sensitization of a defensive withdrawal reflex in *Aplysia*. Science, 182: 1039–1042.

Rankin, C. and Dubnau, J. (2007) Memories of worms and flies: from gene to behavior. In: North G. and Greenspan R.J. (Eds.), Invertebrate Neurobiology. Cold Spring Harbor Laboratory Press, New York, pp. 309–346.

Rankin, C.H. and Carew, T.J. (1988) Dishabituation and sensitization emerge as separate processes during development in *Aplysia*. J. Neurosci., 8: 197–211.

Rayport, S.G. and Schacher, S. (1986) Synaptic plasticity *in vitro*: cell culture of identified *Aplysia* neuron. J. Neurosci., 6: 759–763.

Rumpel, S., LeDoux, J., Zador, A. and Malinow, R. (2005) Postsynaptic receptor trafficking underlying a form of associative learning. Science, 308: 83–88.

Sacktor, T.C., Osten, P., Valsamis, H., Jiang, X., Naik, M.U. and Sublette, E. (1993) Persistent activation of the ζ isoform of protein kinase C in the maintenance of long-term potentiation. Proc. Natl. Acad. Sci. U.S.A., 90: 8342–8346.

Serrano, P., Yao, Y. and Sacktor, T.C. (2005) Persistent phosphorylation by protein kinase Mζ maintains late-phase long-term potentiation. J. Neurosci., 25: 1979–1984.

Shema, R., Sacktor, T.C. and Dudai, Y. (2007) Rapid erasure of long-term memory associations in the cortex by an inhibitor of PKMζ. Science, 317: 951–953.

Sherff, C.M. and Carew, T.J. (2004) Parallel somatic and synaptic processing in the induction of intermediate-term and long-term synaptic facilitation in *Aplysia*. Proc. Natl. Acad. Sci. U.S.A., 101: 7463–7468.

Smith, W.B., Starck, S.R., Roberts, R.W. and Schuman, E.M. (2005) Dopaminergic stimulation of local protein synthesis enhances surface expression of GluR1 and synaptic transmission in hippocampal neurons. Neuron, 45: 765–779.

Sossin, W.S., Sacktor, T.C. and Schwartz, J.H. (1994) Persistent activation of protein kinase C during the development of long-term facilitation in *Aplysia*. Learn. Mem., 1: 189–202.

Sutton, M.A., Bagnall, M.W., Sharma, S.K., Shobe, J. and Carew, T.J. (2004) Intermediate-term memory for site-specific sensitization in *Aplysia* is maintained by persistent activation of protein kinase C. J. Neurosci., 24: 3600–3609.

Sutton, M.A. and Carew, T.J. (2000) Parallel molecular pathways mediate expression of distinct forms of intermediate-term facilitation at tail sensory-motor synapses in *Aplysia*. Neuron, 26: 219–231.

Sutton, M.A. and Carew, T.J. (2002) Behavioral, cellular, and molecular analysis of memory in *Aplysia* I: intemediate-term memory. Integ. Comp. Biol., 42: 725–735.

Sutton, M.A., Ide, J., Masters, S.E. and Carew, T.J. (2002) Interaction between amount and pattern of training in the induction of intermediate- and long-term memory for sensitization in *Aplysia*. Learn. Mem., 9: 29–40.

Sutton, M.A., Masters, S.E., Bagnall, M.W. and Carew, T.J. (2001) Molecular mechanisms underlying a unique intermediate phase of memory in *Aplysia*. Neuron, 31: 143–154.

Trudeau, L.E. and Castellucci, V.F. (1993) Excitatory amino acid neurotransmission at sensory-motor and interneuronal synapses of *Aplysia californica*. J. Neurophysiol., 70: 1221–1230.

Trudeau, L.E. and Castellucci, V.F. (1995) Postsynaptic modifications in long-term facilitation in *Aplysia*: upregulation of excitatory amino acid receptors. J. Neurosci., 15: 1275–1284.

Tsokas, P., Grace, E.A., Chan, P., Ma, T., Sealfon, S.C., Iyengar, R., Landau, E.M. and Blitzer, R.D. (2005) Local protein synthesis mediates a rapid increase in dendritic elongation factor 1A after induction of late long-term potentiation. J. Neurosci., 25: 5833–5843.

Villareal, G., Li, Q., Cai, D. and Glanzman, D.L. (2007) The role of rapid, local postsynaptic protein synthesis in learning-related synaptic facilitation in *Aplysia* Curr. Biol., 17: 2073–2080.

Wainwright, M.L., Zhang, H., Byrne, J.H. and Cleary, L.J. (2002) Localized neuronal outgrowth induced by long-term sensitization training in *Aplysia*. J. Neurosci., 22: 4132–4141.

Walters, E.T. (1987) Multiple sensory neuronal correlates of site-specific sensitization in *Aplysia*. J. Neurosci., 7: 408–417.

Walters, E.T., Byrne, J.H., Carew, T.J. and Kandel, E.R. (1983a) Mechanoafferent neurons innervating tail of *Aplysia*. I. Response properties and synaptic connections. J. Neurophysiol., 50: 1522–1542.

Walters, E.T., Byrne, J.H., Carew, T.J. and Kandel, E.R. (1983b) Mechanoafferent neurons innervating tail of Aplysia. II. Modulation by sensitizing stimulation. J. Neurophysiol., 50: 1543–1559.

Winlow, W. and Kandel, E.R. (1976) The morphology of identified neurons in the abdominal ganglion of *Aplysia californica*. Brain Res., 112: 221–249.

Wright, W.G., Marcus, E.A. and Carew, T.J. (1991) A cellular analysis of inhibition in the siphon withdrawal reflex of *Aplysia*. J. Neurosci., 11: 2498–2509.

Yin, J.C., Del Vecchio, M., Zhou, H. and Tully, T. (1995) CREB as a memory modulator: induced expression of a dCREB2 activator isoform enhances long-term memory in *Drosophila*. Cell, 81: 107–115.

Yin, J.C.P., Wallach, J.S., Del Vecchio, M., Wilder, E.L., Zhuo, H., Quinn, W.G. and Tully, T. (1994) Induction of a dominant negative CREB transgene specifically blocks long-term memory in *Drosophila*. Cell, 79: 49–58.

Zhang, H., Wainwright, M., Byrne, J.H. and Cleary, L.J. (2003) Quantitation of contacts among sensory, motor, and serotonergic neurons in the pedal ganglion of *Aplysia*. Learn. Mem., 10: 387–393.

Zhao, Y., Hegde, A.N. and Martin, K.C. (2003) The ubiquitin proteasome system functions as an inhibitory constraint on synaptic strengthening. Curr. Biol., 13: 887–898.

Zhao, Y., Leal, K., Abi-Farah, C., Martin, K.C., Sossin, W.S. and Klein, M. (2006) Isoform specificity of PKC translocation in living *Aplysia* sensory neurons and a role for Ca^{2+}-dependent PKC APL I in the induction of intermediate-term facilitation. J. Neurosci., 26: 8847–8856.

Zhu, H., Wu, F. and Schacher, S. (1997) Site-specific and sensory neuron-dependent increases in postsynaptic glutamate sensitivity accompany serotonin-induced long-term facilitation at *Aplysia* sensorimotor synapses. J. Neurosci., 17: 4976–4986.

CHAPTER 18

Olfactory memory traces in *Drosophila*

Jacob Berry[1], William C. Krause[2] and Ronald L. Davis[2,3],*

[1] *Program in Developmental Biology, Baylor College of Medicine, Houston, TX 77030, USA*
[2] *Department of Molecular and Cellular Biology, Baylor College of Medicine, Houston, TX 77030, USA*
[3] *Menninger Department of Psychiatry and Behavioral Sciences, Baylor College of Medicine, Houston, TX 77030, USA*

Abstract: In *Drosophila*, the fruit fly, coincident exposure to an odor and an aversive electric shock can produce robust behavioral memory. This behavioral memory is thought to be regulated by cellular memory traces within the central nervous system of the fly. These molecular, physiological, or structural changes in neurons, induced by pairing odor and shock, regulate behavior by altering the neurons' response to the learned environment. Recently, novel in vivo functional imaging techniques have allowed researchers to observe cellular memory traces in intact animals. These investigations have revealed interesting temporal and spatial dynamics of cellular memory traces. First, a short-term cellular memory trace was discovered that exists in the antennal lobe, an early site of olfactory processing. This trace represents the recruitment of new synaptic activity into the odor representation and forms for only a short period of time just after training. Second, an intermediate-term cellular memory trace was found in the dorsal paired medial neuron, a neuron thought to play a role in stabilizing olfactory memories. Finally, a long-term protein synthesis-dependent cellular memory trace was discovered in the mushroom bodies, a structure long implicated in olfactory learning and memory. Therefore, it appears that aversive olfactory associations are encoded by multiple cellular memory traces that occur in different regions of the brain with different temporal domains.

Keywords: *Drosophila*; mushroom body; olfactory learning; memory trace

Introduction

Two central goals of neuroscience are to understand how and where memories are stored. Over the last century, a variety of experimental approaches have been used to probe these questions in the brains of various mammalian model organisms. The resolution of these studies has improved from early behavioral experiments in which portions of the brain were removed to the development of electrophysiological experiments, which allowed the activity of individual cells to be analyzed. More recently, insects have become appealing model organisms for the study of memory formation. Quinn et al. (1974) showed that *Drosophila melanogaster*, the fruit fly, could form associative memories using an aversive conditioning paradigm involving odor and electrical shock. Fruit flies have the added benefit of being amenable to a growing number of genetic manipulations, and, though the scale of their central nervous system (CNS) is much smaller than mammalian organisms, the anatomy of their olfactory system closely mirrors that of vertebrates

*Corresponding author. Tel.: +1 713-798-6641;
Fax: +1 713-798-8005; E-mail: rdavis@bcm.tmc.edu

(Davis, 2004). Since the initial demonstration of olfactory classical conditioning (Tully and Quinn, 1985), the experimental approaches to studying olfactory memory formation in *Drosophila* have followed a similar trajectory as those used in mammalian studies — from ablation of different brain regions using pharmacological or genetic tools to the use of optical reporters that allow the activity of subsets of cells to be analyzed in response to stimuli in naive and conditioned animals.

The elemental unit of memory formation is the cellular memory trace. It can be defined as the molecular, physiological, or structural changes that occur in neurons to alter their response to the learned environment. The sum of all relevant cellular memory traces is the engram. In *Drosophila*, a variety of experimental approaches have been used in the past to determine which neurons are involved in memory formation. In general, these studies involved mutating a particular gene expressed in a region of the fly brain or using a pharmacological or genetic tool to ablate or inactivate a region of the brain; meanwhile, the effect on memory is observed through behavioral performance. This basic approach utilized in numerous, clever ways has revealed many important features about *Drosophila* memory. For example, the mushroom bodies (MBs), prominent neural structures in the fly, preferentially express a number of proteins that have been found to be important in memory formation (Nighorn et al., 1991; Han et al., 1992; Skoulakis et al., 1993; Skoulakis and Davis, 1996; Grotewiel et al., 1998; Cheng et al., 2001; Folkers et al., 2006); the integrity of the MBs has been found to be critical for memory formation (Heisenberg et al., 1985; de Belle and Heisenberg 1994); and MB synaptic output is required during acquisition, storage, and retrieval of memories (Dubnau et al., 2001; McGuire et al., 2001; Krashes et al., 2007). While these studies have helped to localize the site of memory formation and established a role for at least some of the proteins necessary to form the engram, they lack the spatial and temporal resolution to determine which specific neurons form cellular memory traces, how the activity of these neurons changes to reflect the learned information, and how long these changes persist.

Recently, a new class of tools has been developed that allow experimenters to look at memory trace formation at the level of individual neurons, and even within distinct regions of a neuron. Genetically encoded fluorescent reporters have been developed to monitor a variety of neural physiological processes, such as changes in intracellular calcium concentration, release of synaptic vesicles, changes in cAMP levels, or changes in voltage-sensitive ion channels. In *Drosophila*, the expression of these powerful tools can be controlled using the Gal4–UAS system. In this system, the yeast Gal4 transcriptional activator is expressed in particular cells within the brain using a tissue-specific promoter, while all cells contain a transgene composed of the Gal4 response element, UAS, coupled to the optical reporter. Once expressed, Gal4 seeks out the UAS element and drives expression of the optical reporter in the same tissue defined by the Gal4 promoter. Using this system, optical reporters can be expressed specifically and reproducibly in a particular neuron or set of neurons. Two reporters commonly used for functional imaging are synapto-pHluorin (spH) (Ng et al., 2002) and G-CaMP (Nakai et al., 2001). spH is a pH-sensitive green fluorescent protein (GFP) reporter that is fused to synaptobrevin, which targets it to synaptic vesicles. When these vesicles fuse with the plasma membrane, the reporter is exposed to the more pH-neutral environment of the synaptic cleft and its fluorescence increases. G-CaMP is one of the calcium-sensitive GFP reporters. When the calcium concentration inside the cell increases, the ions are bound by G-CaMP. This binding causes a conformational change that increases the GFP fluorescence. These tools have been utilized in innovative whole animal preparations, where the fly is completely intact except for a tiny window cut out of the head exoskeleton; in this way, flies can be exposed to a variety of stimuli, while the experimenter monitors the reporter activity. This functional in vivo imaging technique has been used to show how various stimuli are represented in the brains of naive flies and how those representations change after conditioning.

The optically recorded cellular memory traces that have been discovered thus far are consistent with the ideas about memory formation that stem from behavioral, genetic, and brain lesioning studies in *Drosophila* and other experimental organisms. Functional imaging has shown that memory traces are dynamic; they form and erode over different time courses. This observation correlates with behavioral conditioning which can generate distinct temporal memory phases often classified as short, intermediate, or long term. These different memory phases can have different molecular requirements, such as the necessity for protein synthesis during formation of the most durable forms of long-term memory; functional imaging experiments have additionally shown that some of these proteins and molecular pathways are necessary to form some cellular memory traces. Finally, the use of optical reporters has revealed that memory traces are distributed across different brain regions.

Olfactory-processing circuit in *Drosophila*

Aversive olfactory conditioning of an animal requires coincident exposure to an odor and an aversive electric shock in order to produce a robust behavioral memory. Therefore, it is critical to have an anatomical understanding of how these two forms of environmental information are received by the CNS. It is important to know which neurons carry this information from the periphery, where they distribute this information in the CNS, and, in particular, in which neurons or neuropil does the odor and shock information intersect. This information provides insights into how associative memory is acquired, stored, and retrieved. Due to the genetic tools available, and the existence of a simple, fairly discrete, and easily identifiable neural architecture, the odor-processing circuitry in the fruit fly has been extensively studied.

The odor information is first received via olfactory receptor neurons (ORNs), which are distributed across the antennae and maxillary palps on the front of the head (for review, see Davis, 2004). For the most part, each of these neurons expresses one and only one olfactory receptor along with a common odorant co-receptor. These receptors bind to volatile odorants in the environment and can lead to excitation or inhibition of the ORN. The ORN's axonal projections bundle together in the antennal nerve (AN), and transfer the odorant information, via excitatory cholinergic output, to the first odor relay station in the fly brain, the antennal lobe (AL) (Fig. 1). The AL is a large structure which is characterized by ∼50 discrete neuropil regions called glomeruli, where ORNs synapse onto either local interneurons or projection neurons (PNs). Even though there is no significant stereotyped arrangement of ORNs on the exterior structures of the head, ORNs with the same olfactory receptor project into the same glomeruli. A particular odor will bind to a particular olfactory receptor, activating its respective ORN, and lead to activation of a discrete set of glomeruli; thus, there is an odor mapping in the AL. The local interneurons are thought to participate in relaying or modulating information between glomeruli. This suggests that the AL may be an early site for modulating or transforming

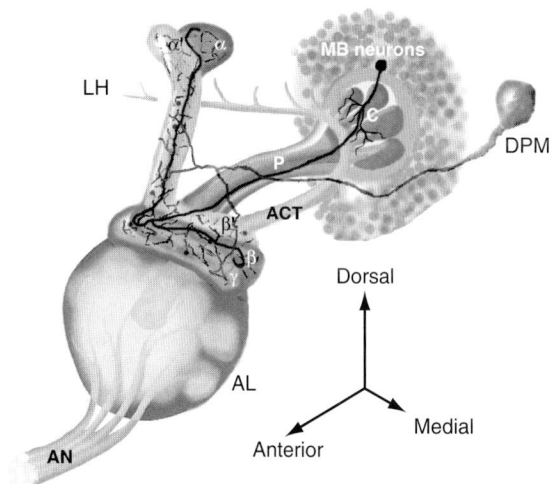

Fig. 1. Olfactory processing circuit in *Drosophila*. A diagram of the important structures in olfactory learning and memory in the fruit fly (one hemisphere of the brain). AN, antennal nerve; ACT, antennal cerebral tract; LH, lateral horn; C, calyx; MB, mushroom body; P, peduncle; DPM, dorsal paired medial neuron. Adapted with permission from Yu et al. (2005). (See Color Plate 18.1 in color plate section.)

olfactory information before sending it further inside the brain. This processed olfactory information exits the AL via PNs, which have dendritic processes innervating specific glomeruli, and transmits olfactory information into two deeper regions in the brain. PNs send their axons through either of the two antennal cerebral tracts (ACTs). Both tracts have a final destination in a region called the lateral horn, LH, a still uncharacterized region of the brain with respect to olfactory learning. However, one PN tract also innervates a region called the calyx on its way to the lateral horn. The calyx is the dendritic neuropil belonging to the Kenyon cells, the cell bodies of prominent structures called mushroom bodies (MBs). Therefore, odor information leaves the AL and is deposited into the MBs and the lateral horn.

The MBs are the most extensively studied structures in the fruit fly and are critical for olfactory learning in insects. Approximately 2500 Kenyon cells in the very posterior of each hemisphere of the brain send out their dendrites into the calyx, where they receive olfactory information from cholinergic PNs. Their axons are bundled into the peduncle that extends anteriorly and bifurcates into vertical and horizontal lobes. These 2500 Kenyon cells are grouped into one of the three classes, γ, α'/β', or α/β, based on the morphology of their bundled axons making up the MB lobes. The MB lobes thus consist of the vertical lobes made of α' and α fibers, and the horizontal lobes consisting of β and β' wrapped diffusely with γ (Fig. 1). Thus, odor information flows into the MB neurons via the calyx; presumably it is processed, and then transmitted to MB extrinsic neurons at axonal synapses in the lobes. Studies have identified various neurons that seem to innervate or surround the MB axonal lobes (Ito et al., 1998). However, the majority of these extrinsic neurons have not been genetically or physiologically characterized and a major unanswered question is where and how the odor information is transmitted from the MB neurons.

Conceptually, aversive olfactory learning would require that information about the negatively reinforcing electric shock must make its way to the learning and memory centers of the fruit fly CNS. How shock information is received at the periphery is less well understood than that of odor information. However, significant work has been performed to understand how shock information is conveyed inside the CNS. It was found that synaptic transmission from dopaminergic (DA) neurons that manufacture the neurotransmitter, dopamine, in the fly is required for *aversive* olfactory conditioning but it is dispensable for *reward* learning (Schwaerzel et al., 2003). The fact that these neurons have a shock-specific role suggests that shock information is conveyed from DA neurons. Early studies have shown that there are more than a hundred TH neurons scattered in clusters throughout the fly brain (Nassel and Elekes, 1992). Importantly, some of the DA neurons seem to be MB extrinsic neurons and thus have extensive innervation of the MB lobes (Riemensperger et al., 2005). Therefore, the MB lobe neuropil appears to represent a nexus in which odor and shock information coincide, and further supports their role in aversive olfactory learning.

One of the few MB extrinsic neurons that have been characterized anatomically and functionally is the dorsal paired medial (DPM) neuron (Fig. 1; Waddell et al., 2000). There is only one very large DPM neuron per hemisphere of the brain. Each of these seemingly sends only one large process out which moves anteriorly until it reaches the MB lobes, where it branches several times to fully innervate all of the MB lobes. In particular, one branch enters the MB lobe neuropil at the tip of the horizontal lobes, one branch enters toward a region near the branch point of the lobes, and the final branch enters at the tip of the vertical lobes. This architecture suggests that there might be functional significance between these branches and the various regions of the MBs they innervate. Based on membrane staining, there are no obvious dendritic processes for the DPM neuron. This type of anatomical evidence suggests the DPM neuron is a unipolar neuron, and thus may play a role in modulating activity in a MB–DPM local circuit.

Short-term memory traces

Experiments in the honeybee that attempted to localize the sites of memory formation found that the AL was involved in the formation of short-term memories. Specifically, it was demonstrated that inactivation of the ALs, using a probe to cool that region of the brain, blocked memory formation (Erber et al., 1980). Fruit flies, like bees, are able to form short-term memories. Prior to the development of functional imaging tools, much of the research on short-term memories in flies was focused on the MBs, where mutations in a number of proteins highly expressed in this structure inhibited memory when it was tested behaviorally 3 min after conditioning (Davis, 2005). The development of optical imaging provided an important tool to investigate the response properties of neurons, including those outside of the MBs. In the AL, anatomical data combined with receptor mapping studies suggested that odors could be encoded by patterns of activated glomeruli (Vosshall et al., 2000; Jefferis et al., 2001). Thus, functional imaging was used in the AL to determine how various stimuli are represented within this brain structure, and to determine whether a cellular memory trace forms within the AL neurons.

Expression of the calcium-sensitive optical reporter, G-CaMP, or the synaptic release reporter, spH, in the PNs of the AL was used initially, in isolated brain preparations, to determine how these neurons responded to odors. It was found that odors, as predicted, elicited stereotypic responses in overlapping subsets of PNs (Ng et al., 2002; Wang et al., 2003). To expand on these findings, spH was used for functional in vivo imaging of the AL neurons to determine how activity was altered by associative olfactory conditioning (Yu et al., 2004).

Using three different Gal4 drivers, spH expression was driven in each of the three neuron types that make up the AL: ORNs, local interneurons, and PNs. The responses of each neuron class were visualized in the synaptically dense glomeruli of the AL; however, because the flies were essentially intact, with only a small window cut in the cuticle covering the brain, only eight of the glomeruli were visible and identifiable. When flies were stimulated with an olfactory cue, all neuron types were found to respond; importantly, these responses were similar to earlier experiments in that each odor tested elicited a distinct, reproducible pattern of response among the visible glomeruli. Because cellular memory traces are thought to form in the neurons where odor and electric shock intersect, each of the AL neuron types was also tested for responses to shock. Unlike the responses to odor, only the PNs were found to respond to electrical shock. Because PNs responded to odor and shock, it was hypothesized that the PNs could encode a memory trace. Accordingly, the response in the PNs to the conditioned stimuli was recorded before and after training. It was found that, after training, the PNs from some of the glomeruli that were inactive in the naive flies were recruited into the representation of the learned odor. The glomeruli containing the newly recruited PNs were odor-specific and reproducible across flies. It was found that this cellular memory trace formed rapidly with recruitment of new PNs observed as quickly as 3 min after conditioning; the memory trace also faded quickly, as the conditioned response was lost by 7 min. Importantly, the cellular memory trace formed only with a delay conditioning protocol in which the odor was paired temporally with the shock, just as for behavioral conditioning (Fig. 2). Other conditioning protocols, such as odor only, shock only, and trace conditioning, in which the shock follows the odor presentation by a significant delay, were ineffective at producing the cellular memory trace and behavioral conditioning.

In summary, functional imaging revealed a short-term memory trace in the PNs of the AL, which correlated with short-term behavioral memory. This memory trace was formed by the recruitment of previously inactive neurons into the representation of the learned odorant. Furthermore, it revealed that, in the *Drosophila* brain, memory traces can be formed in integrative neurons outside of the MBs.

Intermediate-term memory traces

Behavioral experiments have suggested that aversive olfactory memory in fruit flies contains a distinct *intermediate* memory phase, often referred to as middle-term memory. This intermediate

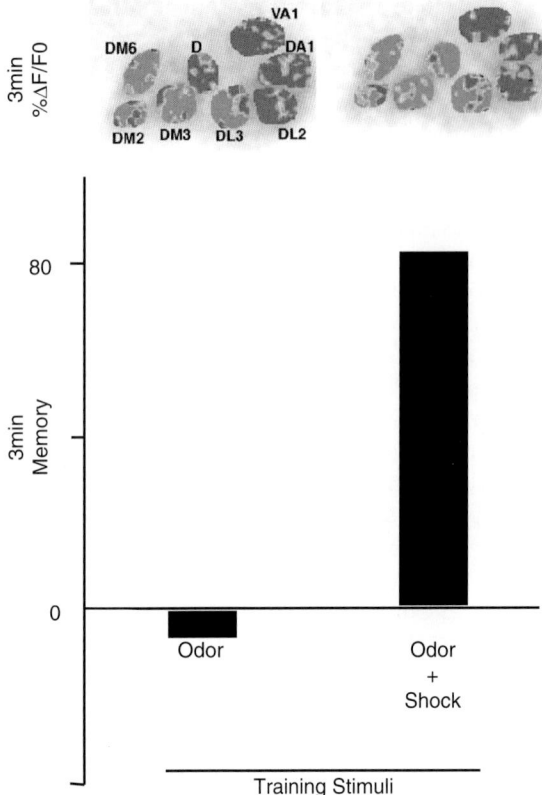

Fig. 2. The AL memory trace involves recruitment of new neurons into the representation of the learned odor and is correlated with behavioral memory. Flies were trained in an olfactory classical conditioning paradigm capable of generating short-term memory. Flies were trained with stimuli consisting of either odor alone, odor paired with shock, or other protocols (not shown). After training, flies were tested, either using a behavioral memory task or using functional imaging, for their response to the trained odor. Three-minute memory scores revealed that paired odor and shock, but not odor alone, was sufficient to form memory and behavioral avoidance of the trained odor (scores indicate the percentage of flies that demonstrated learned behavior). In the flies that were tested using functional imaging, changes in the pattern of optical reporter activity were also observed in response to training with odor and shock; specifically, activity in the PNs of glomerulus D was only elicited when these stimuli were delivered simultaneously. Pseudocolor images indicate the percentage change in fluorescence of the optical reporter during odor stimulation. In this preparation, eight glomeruli were visible and identifiable. These results correlate the cellular memory trace with learned behavior. Adapted with permission from Yu et al. (2004). (See Color Plate 18.2 in color plate section.)

memory was first deduced using flies that had a mutation in the *amnesiac* (*amn*) gene, which is thought to encode a putative neuropeptide (Tully and Quinn, 1985; Moore et al., 1998). These flies had relatively normal memory immediately after training, but this memory decayed more rapidly than normal within the first 30 min to an hour after training. However, if assayed at later time points, the *amn* mutant's memory reaches a level similar to that of wild-type animals. Therefore, *amn* is considered an *intermediate* memory gene due to its preferential role at times between short-term and long-term memories.

An important question is which neurons are involved in intermediate memory. The AMN protein was found to be preferentially expressed in the DPM neuron (Waddell et al., 2000). Importantly, transgenically expressing AMN specifically in the DPM neuron of an *amn* mutant fully rescues the intermediate memory defects. Therefore, this suggests that the DPM neuron plays an important role in intermediate memory. This was further supported by studies utilizing a technique that temporally restricts synaptic transmission in a neuron during defined time points in a behavioral memory assay. This is done by altering the temperature of an animal expressing, within a genetically specified neuron, a temperature-sensitive dominant-negative dynamin protein (*shibire*[ts]) that is critical for synaptic vesicle recycling and thus synaptic transmission. With this, it has been shown that synaptic transmission from the DPM neuron is not required during training or retrieval of the memory, but is required during a window of time that closely correlates with the intermediate memory phase (Keene et al., 2004). Thus, it seems likely that the DPM neuron, via expression of AMN, functions in intermediate memory in *Drosophila*.

Due to its important role in intermediate memory, the DPM neuron is a likely site for an intermediate memory trace. This has been addressed using the optical imaging with the intact animal preparation used in the AL studies above. The DPM neuropil that innervates the MB lobes was found to respond to a large variety of odorants in two important ways (Yu et al., 2005). First, odor exposure evoked a temporary increase in intracellular Ca^{2+} levels, as reported by G-CaMP. This, along with the fact that the DPM neuron responds to many different odors and has no obvious dendritic tree, suggests that the DPM neuron is receiving odor information from the MB neurons and thus it is postsynaptic to the MB lobes. Second, using the spH reporter, odors also evoked synaptic transmission from the DPM neuron. It has been proposed then that this synaptic transmission is releasing neurotransmitters or neuropeptides, such as AMN, back onto the MB axonal fibers. Therefore, it seems likely that the DPM neuron represents a modulatory neuron that is both presynaptic and postsynaptic to the MB neurons.

In addition to odor information, the DPM neuron also responds to electric shock via both increase in Ca^{2+} levels and synaptic transmission. Therefore, since odor and shock information can intersect within the DPM neuron, aversive olfactory memory traces may form in this neuron. To probe this possibility, flies were trained with an odor and electric shock and then their response to this trained odor was assayed at different times after training. Remarkably, it was found that prior coincidence of electric shock with an odor significantly increased the DPM neuron's post-response to this odor but not to an odor that was not paired with shock. Given that this altered response is specific only to the trained odor and to pairing an odor with shock, it is clear that a cellular memory trace forms within the DPM neuron.

The DPM memory trace has intriguing characteristics. First of all, this memory trace is delayed in its formation. The response to the trained odor is not altered at 3 or 15 min after training, but appears at 30 min. Thus, this delayed DPM memory trace interestingly coincides with the temporal requirements for synaptic transmission of this neuron for intermediate memory. A second characteristic is that the DPM memory trace is dependent upon AMN expression within the DPM neuron. All of these observations suggest the intriguing possibility that the DPM memory trace may, in part, regulate intermediate memory. Finally, the DPM memory trace is only observed in the DPM neuropil that innervates the vertical

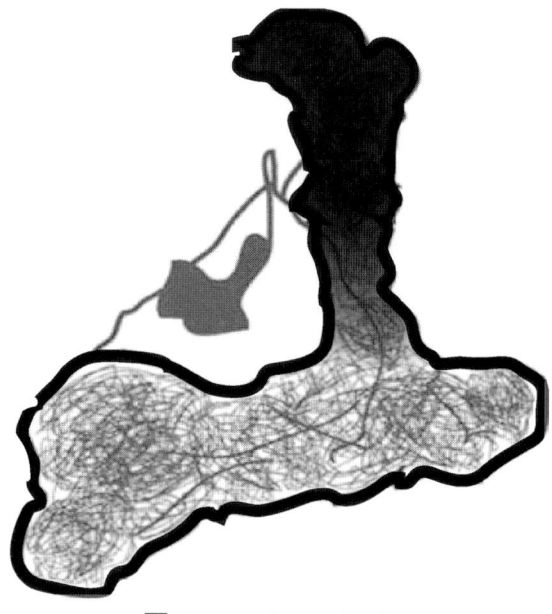

■ Increased synaptic efficacy

Fig. 3. Branch-specific cellular memory trace in the DPM neuron. Diagram of DPM neuron and its innervation of the MB lobes with overlay of branch-specific enhancement of synaptic activity.

branch of the MB lobes (Fig. 3). The role that this branch specificity plays in aversive olfactory memory remains a fascinating unknown.

Another MB extrinsic memory trace

Riemensperger et al. (2005) have studied stimuli-evoked Ca^{2+} responses of neurons that are thought to mediate the transmission of shock information to the CNS. By using a Ca^{2+} fluorescent reporter, it was discovered that DA neurons that innervate the MB lobes respond to odor and electric shock. Therefore, tests were performed to determine whether a memory trace could form in the DA neurons themselves. Unlike in the DPM neuron, the amplitude of the Ca^{2+} responses to an odor did not change in the DA neurons after pairing the odor with shock. However, the calcium responses to the trained odor had a prolonged duration. This indicates that there is a training-induced net increase in Ca^{2+} within the

DA neurons when exposed to a trained odor, thus suggesting the existence of a DA cellular memory trace. Unfortunately, the training protocol used in this study involved multiple presentations of short pairings and was very different from the more classical aversive olfactory paradigm used in the studies of the AL and the DPM neurons. Therefore, the kinetics of the DA neuron cellular memory trace are more obscure. However, the memory trace appears to form by 15 min after the first pairing of odor and shock.

Long-term memory traces

The MBs have long been a focus of cellular and molecular memory research in *Drosophila*. An initial role for this anatomically distinctive collection of neurons came from the realization that a number of proteins involved in cAMP regulation are preferentially expressed in the MB; cAMP signaling has been shown in *Drosophila* and other organisms to be necessary for memory formation (Mayford and Kandel, 1999). Consequently, it was found that ablation of the MBs was sufficient to block memory, while rescue of mutant proteins involved in cAMP regulation by expression of wild-type forms only in the MBs was sufficient to restore memory (Zars et al., 2000; McGuire et al., 2003; Mao et al., 2004). The use of a temperature-sensitive *shibire* to block synaptic output showed that synaptic activity in the MB neurons was required during acquisition, storage, and retrieval of learned information (Dubnau et al., 2001; McGuire et al., 2001; Schwaerzel et al., 2002; Isabel et al., 2004; Krashes et al., 2007). Olfactory conditioning using a spaced training schedule has been shown to form long-term memories in *Drosophila*; these memories specifically require protein synthesis (Tully et al., 1994). Subsequently, genetic studies have suggested that this long-term memory is dependent upon the vertical lobes of the MBs (Pascual and Preat, 2001). The development of optical reporters and their use in functional imaging have allowed researchers to examine the function of MB neurons in olfactory perception and in the formation of cellular memory traces.

As with the AL and DPM neurons, the initial application of functional imaging in the *Drosophila* MBs was focused on determining how these neurons respond to olfactory information. Expression of the G-CaMP reporter was driven in the MBs using the Gal4 system, and the reporter activity was monitored in the calyx and the cell bodies of the MB neurons while olfactory stimulation was delivered (Wang et al., 2004). This technique revealed odor-specific patterns of increasing intracellular calcium concentration in both regions of the MBs. This finding suggested that the spatial code of odor information established in the AL is preserved as that information is transmitted to the MBs.

To determine whether a cellular memory trace was formed in the MB neurons, fluorescence of the G-CaMP reporter was monitored in the horizontal and vertical lobes of the α/β MB neurons (Yu et al., 2006). Changes in the observed G-CaMP fluorescence indicated that both lobes responded to stimulation with odors and electric shock, as would be expected if the neurons were to integrate these stimuli and form a memory trace. To determine whether a memory trace formed, changes in reporter activity were assessed 3, 9, and 24 h after an olfactory conditioning protocol designed to generate either short-term or long-term memory. Only the long-term memory conditioning, consisting of multiple spaced training cycles, was capable of forming a MB cellular memory trace that appeared between 3 and 9 h after training and persisted at 24 h after training. This memory trace took the form of increased amplitude in intracellular calcium levels in response to the learned odor and was, intriguingly mirroring the DPM memory trace, only observed in the α tip of the MB vertical lobe (Fig. 4). This conditioning protocol also generated behavioral memory in trained flies when they were tested at 24 h, which suggests that, under these conditions, the observed MB cellular memory trace underlies the formation of long-term memory. In many organisms, long-term memory has been shown to require protein synthesis as well as a transcription factor, the cyclic AMP response element-binding protein (CREB) (Kandel, 2001). Treatments that interfered with either of these requirements

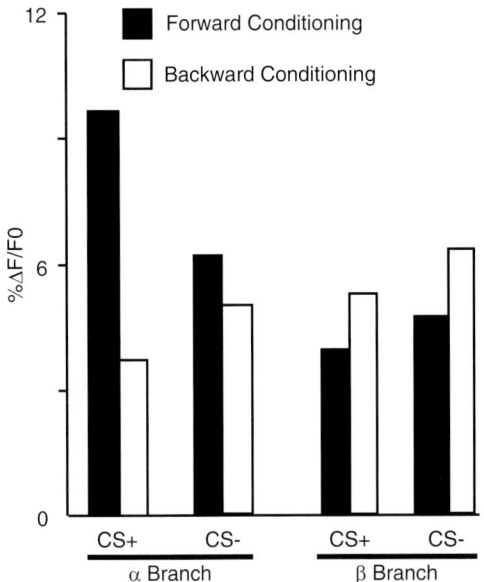

Fig. 4. A cellular memory trace in the MB neurons is localized to the vertical branch. The optical reporter G-CaMP was expressed in both the α (vertical) and the β (horizontal) branches of the MB neurons. Flies were conditioned in a behavioral apparatus such that they received either an odor paired with shock (CS+) or an odor not paired with shock (CS−). Forward conditioning, in which the CS+ and the shock were presented simultaneously, was spread over multiple, spaced trials to generate long-term memory. As a control, backward conditioning, in which the CS+ loses its predictive value because it is delivered at the end of the shock presentation, was also tested. Twenty-four hours post-conditioning, the reporter's response to both the CS+ and the CS− odors was measured and quantified as the percentage change in fluorescence. A significant increase in reporter activity was observed only in response to forward conditioning with the CS+ odor and only in the α branch of the MB neurons. Adapted with permission from Yu et al. (2006).

blocked the formation of both the cellular memory trace and the learned behavior when tested at 24 h.

Functional imaging in the *Drosophila* MBs supports their established role in memory formation. Furthermore, it has demonstrated that a memory trace forms in the vertical lobe of a subset of MB neurons, and only in response to a conditioning schedule that generates long-term behavioral memory. These observations are consistent with an earlier genetic experiment that was performed to focus on the role of specific MB neurons in memory formation (Pascual and Preat, 2001). That study took advantage of a mutant protein, *alpha-lobes-absent*, which can prevent proper development of the MB vertical lobe. In those flies where the mutation ablated only the vertical lobe, no long-term memory was observed 24 h after training; however, other forms of memory were unaffected.

Concluding thoughts and future considerations

As discussed above, associating an odor with an aversive electric shock can create robust behavioral memories in fruit flies that have characteristic temporal phases. In addition, in vivo imaging techniques have revealed that these temporal phases of behavioral memory might be regulated by discrete cellular memory traces, registered as changes in neuronal activity after training. These functional in vivo data offer important insights. First, they strongly suggest that the overall behavioral memory of an animal is regulated by a combination of memory traces that are located in different regions of the brain. So, the memory of an associative event, like a negatively reinforced odor, is encoded in several different regions of the brain. This is evidenced by a short-term memory trace in the AL, an intermediate memory trace in the DPM neurons, and a long-term memory trace in the MBs. Second, the imaging data imply that the spatially segregated memory traces regulate behavioral memory at different times after the association is made. Therefore, in *Drosophila*, it appears that different brain structures form cellular memory traces that regulate the behavior of the animal during dedicated temporal domains (Fig. 5). Since the currently observed memory traces do not cover all temporal domains of behavioral memory, such as times between the AL and DPM memory trace (between 7 and 30 min after training), there are presumably unknown cellular memory traces that exist in undefined locations which would regulate the animal's behavior during these gaps.

There are still many unanswered questions that will be the focus of future research in the field of functional in vivo imaging in *Drosophila*. First, all of the optical memory traces observed have only

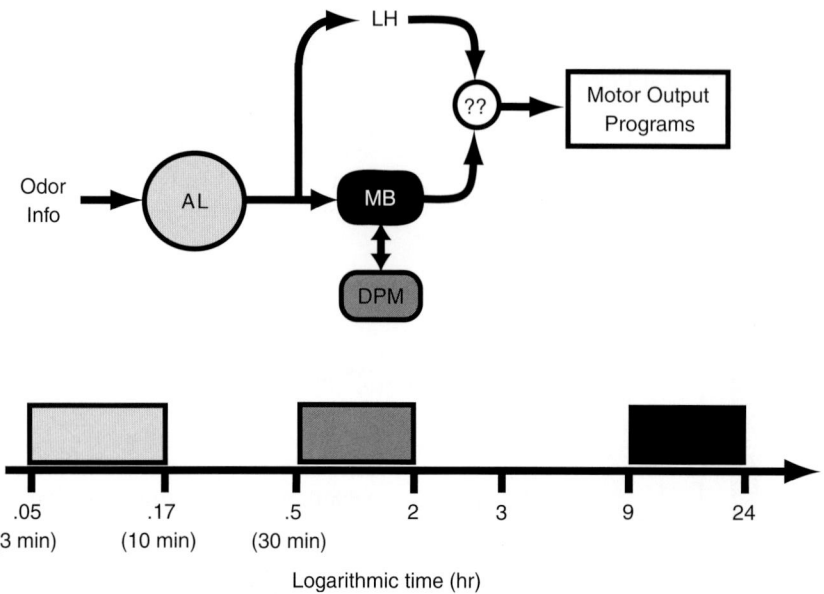

Fig. 5. Spatial and temporal cellular memory traces regulate behavior. A diagram illustrating how odor information flowing through the olfactory circuit can be regulated by distinct brain regions to create a conditioned behavioral response. Behavior is posited to be regulated by multiple regions of the brain, with different regions participating during various time windows. AL, antennal lobe; LH, lateral horn; MB, mushroom bodies; DPM, dorsal paired medial; ??, unknown downstream integrator/effector neurons. The light-gray rectangle indicates the period of activity of the PN memory trace, the medium-gray rectangle the period of activity of the DPM neuron memory trace, and the black rectangle the period of the MB memory trace. Presumably, other memory traces not yet discovered participate during period not covered by these three memory traces.

been shown to exist in animals that display behavioral memory and are absent in animals that do not. This is a correlation that needs to be strengthened. This will require novel whole animal preparations that allow the measurement of conditioned behavior while imaging brain activity. In addition, we will need to understand the molecular mechanisms underlying each of the traces so that we can specifically disrupt the cellular memory trace without a non-specific disruption of all neuronal activity. This will allow us to show that a given memory trace is required for a behavioral output in an animal. One would also like to show sufficiency by artificially creating a memory trace and showing the appearance of a conditioned behavior in an untrained fly. For example, this may be possible by experimentally increasing Ca^{2+} influx or synaptic transmission in the DPM innervation of the vertical MB lobes. Finally, aversive olfactory conditioning may initiate the different cellular memory traces *independently*, but, due to different maturation rates, they have varying windows in which they effectively guide behavior. Alternatively, a subset of the memory traces may be initiated during training, and in addition to guiding early behavior, they are in some way utilized to copy the relevant learned information into a new more stable medium in a different neuronal structure. This hypothesis would suggest that later-occurring memory traces would be *dependent* on earlier-born memory traces. Consequently, further research is required to elucidate the nature and role of spatial and temporally distributed cellular memory traces in the context of the entire olfactory learning and memory circuit.

Abbreviations

ACT	antennal cerebral tract
AL	antennal lobe
AMN	amnesiac
CNS	central nervous system

CREB	cyclic AMP response element-binding protein
CS	conditioned stimulus
DA	dopaminergic
DPM	dorsal paired medial
MB	mushroom body
ORN	olfactory receptor neuron
PN	projection neuron
spH	synapto-pHluorin

Acknowledgments

Research in the laboratory of the authors is supported by grants from the National Institutes of Health, the Mather's Charitable Trust, and the R.P. Doherty-Welch Chair in Science to R.L.D.

References

Cheng, Y., Endo, K., Wu, K., Rodan, A.R., Heberlein, U. and Davis, R.L. (2001) *Drosophila* fasciclin II is required for the formation of odor memories and for normal sensitivity to alcohol. Cell, 105: 757–768.

Davis, R.L. (2004) Olfactory learning. Neuron, 44: 31–48.

Davis, R.L. (2005) Olfactory memory formation in *Drosophila*: from molecular to systems neuroscience. Annu. Rev. Neurosci., 28: 275–302.

de Belle, J.S. and Heisenberg, M. (1994) Associate odor learning in *Drosophila* abolished by chemical ablation of mushroom bodies. Science, 263: 692–695.

Dubnau, J., Grady, L., Kitamoto, T. and Tully, T. (2001) Disruption of neurotransmission in *Drosophila* mushroom body blocks retrieval but not acquisition of memory. Nature, 411: 476–480.

Erber, J., Mashur, T. and Menzel, R. (1980) Localization of short-term memory in the brain of the bee, *Apis mellifera*. Physiol. Entomol., 5: 343–358.

Folkers, E., Waddell, S. and Quinn, W.G. (2006) The *Drosophila* radish gene encodes a protein required for anesthesia-resistant memory. Proc. Natl. Acad. Sci. U.S.A., 103: 17496–17500.

Grotewiel, M.S., Beck, C.D., Wu, K.H., Zhu, X.R. and Davis, R.L. (1998) Integrin-mediated short-term memory in *Drosophila*. Nature, 391: 455–460.

Han, P.L., Levin, L.R., Reed, R.R. and Davis, R.L. (1992) Preferential expression of the *Drosophila* rutabaga gene in mushroom bodies, neural centers for learning in insects. Neuron, 9: 619–627.

Heisenberg, M., Borst, A., Wagner, S. and Byers, D. (1985) *Drosophila* mushroom body mutants are deficient in olfactory learning. J. Neurogenet., 2: 1–30.

Isabel, G., Pascual, A. and Preat, T. (2004) Exclusive consolidated memory phases in *Drosophila*. Science, 304: 1024–1027.

Ito, K., Suzuki, K., Estes, P., Ramaswami, M., Yamamoto, D. and Strausfeld, N.J. (1998) The organization of extrinsic neurons and their implications in the functional roles of the mushroom bodies in *Drosophila melanogaster* Meigen. Learn. Mem., 5: 52–77.

Jefferis, G.S., Marin, E.C., Stocker, R.F. and Luo, L. (2001) Target neuron prespecification in the olfactory map of *Drosophila*. Nature, 414: 204–208.

Kandel, E.R. (2001) The molecular biology of memory storage: a dialogue between genes and synapses. Science, 294: 1030–1038.

Keene, A.C., Stratmann, M., Keller, A., Perrat, P.N., Vosshall, L.B. and Waddell, S. (2004) Diverse odor-conditioned memories require uniquely timed dorsal paired medial neuron output. Neuron, 44: 521–533.

Krashes, M.J., Keene, A.C., Leung, B., Armstrong, J.D. and Waddell, S. (2007) Sequential use of mushroom body neuron subsets during *Drosophila* odor memory processing. Neuron, 53: 103–115.

Mao, Z., Roman, G., Zong, L. and Davis, R.L. (2004) Pharmacogenetic rescue in time and space of the rutabaga memory impairment by using Gene-Switch. Proc. Natl. Acad. Sci. U.S.A., 101: 198–203.

Mayford, M. and Kandel, E.R. (1999) Genetic approaches to memory storage. Trends Genet., 15: 463–470.

McGuire, S.E., Le, P.T. and Davis, R.L. (2001) The role of *Drosophila* mushroom body signaling in olfactory memory. Science, 10: 1126–1129.

McGuire, S.E., Le, P.T., Osborn, A.J., Matsumoto, K. and Davis, R.L. (2003) Spatiotemporal rescue of memory dysfunction in *Drosophila*. Science, 302: 1765–1768.

Moore, M.S., DeZazzo, J., Luk, A.Y., Tully, T., Singh, C.M. and Heberlein, U. (1998) Ethanol intoxication in *Drosophila*: genetic and pharmacological evidence for regulation by the cAMP signaling pathway. Cell, 93: 997–1007.

Nakai, J., Ohkura, M. and Imoto, K. (2001) A high signal-to-noise Ca^{2+} probe composed of a single green fluorescent protein. Nat. Biotechnol., 19: 137–141.

Nassel, D.R. and Elekes, K. (1992) Aminergic neurons in the brain of blowflies and *Drosophila*: dopamine- and tyrosine hydroxylase-immunoreactive neurons and their relationship with putative histaminergic neurons. Cell Tissue Res., 267: 147–167.

Ng, M., Roorda, R.D., Lima, S.Q., Zemelman, B.V., Morcillo, P. and Miesenboeck, G. (2002) Transmission of olfactory information between three populations of neurons in the antennal lobe of the fly. Neuron, 36: 463–474.

Nighorn, A., Healy, M.J. and Davis, R.L. (1991) The cyclic AMP phosphodiesterase encoded by the *Drosophila* dunce gene is concentrated in the mushroom body neuropil. Neuron, 6: 455–467.

Pascual, A. and Preat, T. (2001) Localization of long-term memory within the *Drosophila* mushroom body. Science, 294: 1115–1117.

Quinn, W.G., Harris, W.A. and Benzer, S. (1974) Conditioned behavior in *Drosophila melanogaster*. Proc. Natl. Acad. Sci. U.S.A., 71: 708–712.

Riemensperger, T., Voller, T., Stock, P., Buchner, E. and Fiala, A. (2005) Punishment prediction by dopaminergic neurons in *Drosophila*. Curr. Biol., 15: 1953–1960.

Schwaerzel, M., Heisenberg, H. and Zars, T. (2002) Extinction antagonizes olfactory memory at the subcellular level. Neuron, 35: 951–960.

Schwaerzel, M., Monastirioti, M., Scholz, H., Friggi-Grelin, F., Birman, S. and Heisenberg, M. (2003) Dopamine and octopamine differentiate between aversive and appetitive olfactory memories in *Drosophila*. J. Neurosci., 23: 10495–10502.

Skoulakis, E.M. and Davis, R.L. (1996) Olfactory learning in mutants for *Leonardo*, a *Drosophila* gene encoding a 14-3-3 protein. Neuron, 17: 931–944.

Skoulakis, E.M., Kalderon, D. and Davis, R.L. (1993) Preferential expression in mushroom bodies of the catalytic subunit of protein kinase A and its role in learning and memory. Neuron, 11: 197–208.

Tully, T. and Quinn, W.G. (1985) Classical conditioning and retention in normal and mutant *Drosophila melanogaster*. J. Comp. Physiol., 157: 263–277.

Tully, T., Preat, T., Boynton, S.C. and Del Vecchio, M. (1994) Genetic dissection of consolidated memory in *Drosophila*. Cell, 79: 35–47.

Vosshall, L.B., Wong, A.M. and Axel, R. (2000) An olfactory sensory map in the fly brain. Cell, 102: 147–159.

Waddell, S., Armstrong, J.D., Kitamoto, T., Kaiser, K. and Quinn, W.G. (2000) The *amnesiac* gene product is expressed in two neurons in the *Drosophila* brain that are critical for memory. Cell, 103: 805–813.

Wang, J.W., Wong, A.M., Flores, J., Vosshall, L.B. and Axel, R. (2003) Two-photon calcium imaging reveals an odor-evoked map of activity in the fly brain. Cell, 112: 271–282.

Wang, Y., Guo, H.F., Pologruto, T.A., Hannan, F., Hakker, I., Svoboda, K. and Zhong, Y. (2004) Stereotyped odor-evoked activity in the mushroom body of *Drosophila* revealed by green fluorescent protein-based Ca^{2+} imaging. J. Neurosci., 24: 6507–6514.

Yu, D., Alkalal, D.B. and Davis, R.L. (2006) Drosophila alpha/beta mushroom body neurons form a branch-specific, long-term cellular memory trace after spaced olfactory conditioning. Neuron, 52: 845–855.

Yu, D., Keene, A.C., Srivatsan, A., Waddell, S. and Davis, R.L. (2005) *Drosophila* DPM neurons form a delayed and branch-specific memory trace after olfactory classical conditioning. Cell, 123: 945–957.

Yu, D., Ponomarev, A. and Davis, R.L. (2004) Altered representation of the spatial code for odors after olfactory classical conditioning. Neuron, 42: 437–449.

Zars, T., Fischer, M., Schulz, R. and Heisenberg, M. (2000) Localization of a short-term memory in *Drosophila*. Science, 288: 672–676.

CHAPTER 19

Associative learning signals in the brain

Wendy A. Suzuki[*]

Center for Neural Science, New York University, New York, NY 10003, USA

Abstract: Associative memory is defined as memory for the relationship between two initially unrelated items, like a name and an unfamiliar face. Associative memory is not only one of the most common forms of memory used in everyday situations, but is highly dependent on the structures of the medial temporal lobe (MTL). The goal of this chapter is to review the patterns of neural activity shown to underlie the formation of new associative memories in the MTL, as well as to examine how other extra-MTL areas participate in the learning process. Other areas implicated in various aspects of associative learning include the motor-related areas of the frontal lobe, prefrontal cortex, and striatum. The question of how the MTL and the other cortical and subcortical structures may interact during associative learning will be discussed.

Keywords: hippocampus; striatum; prefrontal cortex; SEF; FEF; premotor cortex

Introduction

In 1957, the description of the now well-known amnesic patient H.M. provided some of the first clues to the mnemonic functions of the medial temporal lobe (MTL) (Scoville and Milner, 1957). The impairment observed in patient H.M. was profound and initially described as global amnesia, that is, extending to all forms of learning and memory. Subsequent and detailed study of patient H.M. together with other human amnesic patients with MTL damage revealed that the memory impairment was not global, but instead was more limited to particular forms of learning and memory including learning and memory for facts (semantic memory) and events (episodic memory) collectively referred to as declarative (Squire et al., 2004) or relational memory (Eichenbaum and Cohen, 2001). One form of declarative/relational memory that has been the focus of extensive experimental research is associative memory, defined as memory for the relationship between initially unrelated items. Findings from both the experimental and clinical literature show that damage to the MTL impairs long-term associative memory for a variety of different kinds of information (Murray et al., 1993, 1998; Vargha-Khadem et al., 1997; Bayley and Squire, 2002; Stark et al., 2002; Stark and Squire, 2003; Liu et al., 2004), and neurophysiological studies have demonstrated a role of the MTL, in particular the perirhinal cortex in the long-term storage of associative information (Sakai and Miyashita, 1991; Murray et al., 1993; Sobotka and Ringo, 1993; Naya et al., 1996, 2003; Booth and Rolls, 1998). In addition to a role in long-term memory for new associations, findings from lesion studies also suggest an important role of the MTL in the

[*]Corresponding author. Tel.: +1 212 998 3734; Fax: +1 212 995 4011; E-mail: wendy@cns.nyu.edu

initial formation of new associative memories (i.e., associative learning; Murray et al., 1993, 1998).

The goal of this chapter is twofold. The first goal is to review the associative learning signals that have been reported in the MTL across species including rabbits, rats, and primates. This review will show that similar patterns of associative learning signals have been reported across species, though the most thorough description to date has been done in the non-human primate model systems. The second related question concerns the other brain areas beyond the MTL that may also contribute to associative learning. While many studies have shown clear impairments in associative learning following MTL damage, the impairment is typically not complete suggesting that other brain areas may also be contributing to associative learning functions. Indeed, in tasks of conditional motor association learning where monkeys are required to associate a particular visual stimulus with a particular motor response (i.e., touch right or touch left), strong associative learning signals have been reported not only in the hippocampus but also throughout other motor-related areas in the frontal lobe and striatum. The second part of this chapter will compare and contrast the associative learning signals across these extra-MTL brain areas. The question of how the MTL and these extra-MTL brain areas might cooperate, compete, and generally interact during new associative learning will also be discussed.

Associative learning in the medial temporal lobe

One of the earliest and most dramatic demonstrations of dynamic learning-related neural signals in the brain came in the 1970s when Berger and colleagues (Berger et al., 1976; Berger and Thompson, 1978) recorded multi-unit activity in the hippocampus in rabbits during a delay eyeblink conditioning task where the air-puff unconditioned stimulus (US) co-terminated with the end of the conditioned stimulus (CS) presentation (a tone). They showed that compared to the responses in unpaired control animals, hippocampal neurons in conditioned animals developed enhanced responses, first to the air-puff US and subsequently to the tone CS, such that the enhanced response to the US appeared to shift forward gradually in time towards the CS presentation with learning. Although it was later shown that delay conditioning (where there is no temporal gap between the CS and US presentation) is not dependent on intact hippocampal function, similar dynamic changes in neural activity were subsequently reported in trace conditioning paradigms where there is a temporal gap between the CS and US presentation. In contrast to delay conditioning, trace conditioning is highly dependent on the integrity of the hippocampus (Solomon et al., 1986; Moyer et al., 1990; Kim et al., 1995; McEchron et al., 2000).

In one key study, hippocampal learning-related activity was examined as animal learned a trace conditioning task where a tone was paired with an air-puff unconditioned response (McEchron and Disterhoft, 1997). This group showed that compared to the unpaired control group, hippocampal cells in the paired trace conditioning group exhibited enhanced responses to the US a full day before the first day animals expressed learning of the CS–US pairing. While the behavioral conditioned responses remained asymptotic on the 2 days following learning, the enhanced neural responses to the CS and US declined back to control levels in the 2 days after learning. The authors suggested that this pattern of neural activity may reflect the relatively transient role of the hippocampus in the consolidation of the CS–US association, though the relationship between the decline and hippocampal activity and consolidation of the CS–US association was not examined directly.

In contrast to studies of associative learning in the rabbit hippocampus which have focused on classical conditioning paradigms, the vast majority of studies in the rodent hippocampus have focused on the neurophysiological correlates of spatial navigation and spatial memory. We have known since the 1970s with the seminal description of O'Keefe and Dostrovsky (1971) that cells in the rat hippocampus signal the relative position of the rat in the environment (place cells). While early theories suggested that place cells represent a spatial cognitive map of the environment

(O'Keefe and Nadel, 1978), more recent theories have suggested that spatial information is one particularly striking example of a more general category of relational information that is highly dependent on the hippocampus (Eichenbaum and Cohen, 2001). Early studies done by Wilson and Mcnaughton (1993) showed that the spatially selective activity could develop very quickly as a rat entered a novel environment. However, fewer studies have attempted to examine these dynamic changes in hippocampal activity in a situation where behavioral learning could be monitored.

One study recorded from hippocampal place cells as rats were exposed to either a familiar set of arms in a modified T-maze or during the first 3 days of exposure to a novel set of arms (Frank et al., 2004). On the first exposure to a novel arm, after an initial period of inactivity, strong and selective place-cell activity developed very quickly. Overall, the most dramatic change in place-field activity in the novel arm occurred on days 1 and 2 and stabilized by day 3. To understand the relationship between the rapid development of place-cell activity and learning, the authors compared various measures of place-cell activity to a behavioral measure of familiarity defined by running speed. This analysis was based on the observation that animals typically run more slowly in order to explore novel environments and speed up in familiar environments. The largest changes in place-field activity occurred in the novel arm on days 1 and 2 and corresponded to the most striking increase in running speed which also occurred between days 1 and 2. These findings suggest that hippocampal place cells play a role in the rapid signaling of novel spatial/relational information. However, even after the place cells on the novel arms had stabilized on day 3, there continued to be a difference between the slower running speed on the novel arm and the faster running speed on familiar arms. These latter findings suggest that other brain areas continue to distinguish between novel and familiar environments even after hippocampal place cells appeared to stabilize.

Rapidly changing place-cell activity was also observed in another spatial learning task in which rats swam in an annular watermaze (Fyhn et al., 2002). Each day, rats were given a "swim only" session followed by another swim session, during which a novel hidden platform was introduced in the maze. This group reported that many hippocampal pyramidal cells fired vigorously the first time the rat encountered the novel platform location and the activity decreased as the animal gained more experience with that platform location. The decreased activity paralleled a decrease in swim time to find the platform, indicative of learning. This transient increase followed by a decrease with learning could play a role in signaling novel spatial information.

While the findings in rabbits and rats suggest that striking changes in neural activity can accompany various forms of associative or spatial/relational learning, less is known about the precise timing of these neural changes relative to clearly defined behavioral learning. To address this question, two groups used conditional motor associative learning tasks to examine learning-related patterns of hippocampal activity in monkeys (Cahusac et al., 1993; Wirth et al., 2003). This particular associative learning paradigm was chosen because previous studies showed that post-training lesions to the MTL in monkeys impair the ability to learn novel conditional motor associations, while well-learned associations remain unaffected (Rupniak and Gaffan, 1987; Murray and Wise, 1996; Wise and Murray, 1999; Murray et al., 2000; Brasted et al., 2002, 2003).

In one study (Wirth et al., 2003), animals were first shown four identical target stimuli superimposed on a large complex visual scene (Fig. 1). Following a delay interval, during which the scene disappeared but the targets remained on the screen, the animal was cued to make a single eye movement to one of the peripheral targets on the screen (Fig. 1). For each visual scene, only one of the four targets was associated with reward. Each day, the animals learned two to four new scenes by trial and error. These new scenes were also randomly intermixed with well-learned "reference" scenes that the animals had seen for many months before the recording experiments began. Responses to the reference scenes were used to control for motor-related activity in the hippocampal cells. A similar task was used by Cahusac et al., (1993), except only two possible response

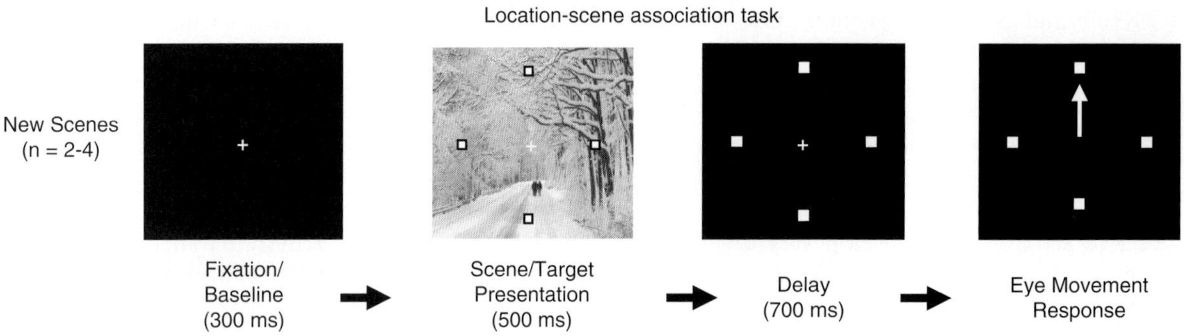

Location-scene association task

Fig. 1. Schematic illustration of the location-scene association task. Adapted with permission from Wirth et al. (2003). Animals initiated each trial by fixating a point on the computer screen. Then four identical targets superimposed on a complex visual scene were presented for 500 ms followed by a 700 ms delay interval in which the scene disappeared but the targets remained on the screen. The trials ended with the fixation point disappearing, which was the monkeys cue to make an eye movement response to one of the targets. Animals typically learned two to four new scenes randomly intermixed with two to four highly familiar "reference" scenes. Each of the four possible reference scenes was associated with a different rewarded target location.

choices instead of four were given and animals made arm-movement rather than eye-movement responses.

Wirth et al. (2003) found that a large proportion of the isolated hippocampal cells (61%) responded differentially (i.e., selectively) to the different stimuli during the scene period, the delay period, or both periods of the task. To identify those selectively responding hippocampal cells with learning-related activity, they compared a moving average of the raw trial-by-trial neural activity with a moving average of the raw behavioral performance during behavioral learning. They found that 28% of the selectively responding cells showed a significant positive or negative correlation with learning. These cells were termed "changing cells". Two categories of changing cells were observed. Sustained changing cells (54% of the population of changing cells) signaled learning with a change in neural activity that was maintained for the duration of the recording session (Fig. 2A). A second category (45% of the population of changing cells) was termed baseline sustained changing cells and these cells started out with a scene-selective response during either the scene or delay period of the task even before the animal learned the association and signaled learning by returning to baseline activity (Fig. 2B). This return to baseline activity was anti-correlated with the animal's learning curve for that particular scene. Neither the sustained or baseline sustained changing cells responded similarly to a highly familiar "reference" scene with the same rewarded target location as they did during learning of a new association with the same rewarded target, suggesting that this change in activity is not associated with a pure motor response.

Consistent with the findings of Wirth et al. (2003), Cahusac et al. (1993) also described sustained-like changing cells in the monkey hippocampus, but they did not observe baseline sustained-type cells. Instead, they described another population of hippocampal learning-related cells that only showed differential activity to the two visual stimuli transiently, near the time of learning before returning to baseline levels of response (transient cells). This category of cell is reminiscent of the transient signal seen in the rat hippocampus during novel spatial/relational learning (Fyhn et al., 2002).

Previous studies have shown that neurons in both the perirhinal cortex and area TE signal long-term associations between visual stimuli by responding similarly to the two items that had been paired in memory (Sakai and Miyashita, 1991; Naya et al., 2003). These findings suggest that the learning of the paired associates may have "tuned" or "shaped" the sensory responses of these cells towards a similar response to the two stimuli paired in memory and raised the possibility that the striking changes in neural activity observed during

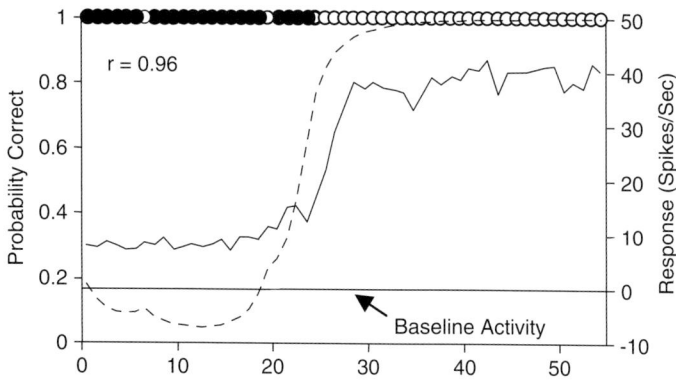

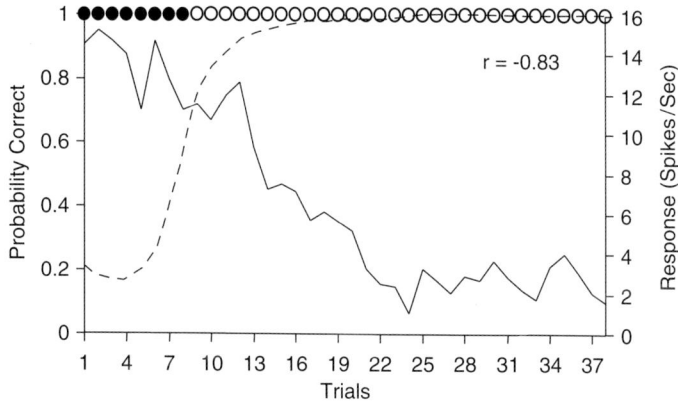

Fig. 2. Panel A shows an example of a sustained changing cell, and panel B shows an example of a baseline sustained changing cell. In both panels, neural activity is shown on the right Y-axis while probability correct is shown on the left Y-axis. Black and white circles at the top of the graph indicate incorrect and correct trials, respectively. The r-values refer to the correlation between the behavioral learning curve and the neural activity across trials. Adapted with permission from Wirth et al. (2003).

the location-scene association task may represent a change in the cell's stimulus-selective response properties with learning. To address this possibility, Wirth et al. (2003) examined the average response of a single changing cell to all new scenes and reference scenes over the course of learning (Fig. 3A). This analysis suggested that sustained changing cells become more highly tuned to a particular scene after learning compared to before learning. A population analysis confirmed this hypothesis, showing that sustained changing cells exhibited a significant increase in selectivity (Fig. 3B) with learning (Fig. 3B). In contrast, the population of baseline sustained cells exhibited a significant decrease in selectivity with learning (Fig. 3C). These findings suggest that hippocampal cells signal new associations with a significant change in their stimulus-selective response properties.

Another important question is that of causality. That is, do these hippocampal changing cells drive associative learning or are they downstream to other brain areas that drive learning? To address this question, Wirth et al. (2003) examined the precise timing of the changes in neural activity (changing cells) on the one hand, and behavioral learning on the other. For each new learning condition for which neural activity changed, comparisons were made between the estimated trial

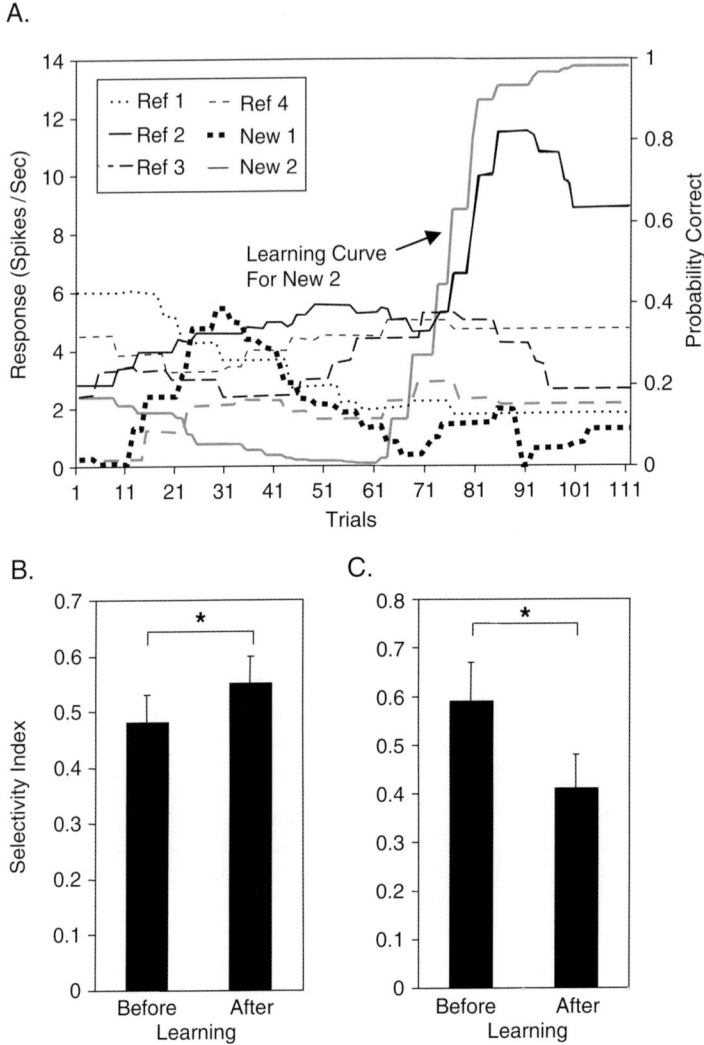

Fig. 3. (A) Average response to four reference scenes and two new scenes over the course of the recording session for a sustained changing cell. The learning curve for New Scene 2 is illustrated in as the thick gray line. (B) Graph showing the significant increase in selectivity index for sustained changing cells after learning compared to before learning. (C) In contrast, sustained changing cells decreased their selectivity after learning compared to before learning.

number of neural change and the estimated trial number of learning. This comparison showed that hippocampal cells can signal learning before ($n=18$), at the same time ($n=1$), as well as after ($n=18$) learning. Hippocampal cells signaled learning staring from as much as 13 trials before learning to 15 trials after learning (Fig. 4). Similar to the Wirth et al. (2003) study, Cahusac et al. (1993) reported that the learning-related signals could occur within a wide range of lag or lead times relative to behavioral learning ranging mainly between 30 trials before learning to 40 trials after learning. Taken together, these findings suggest that hippocampal neurons participate in all aspects of the associative learning process from several trials before learning when changes in neural activity can drive the changes in behavior, to several trials after learning when changes in neural activity can be used to strengthen the newly learned associations.

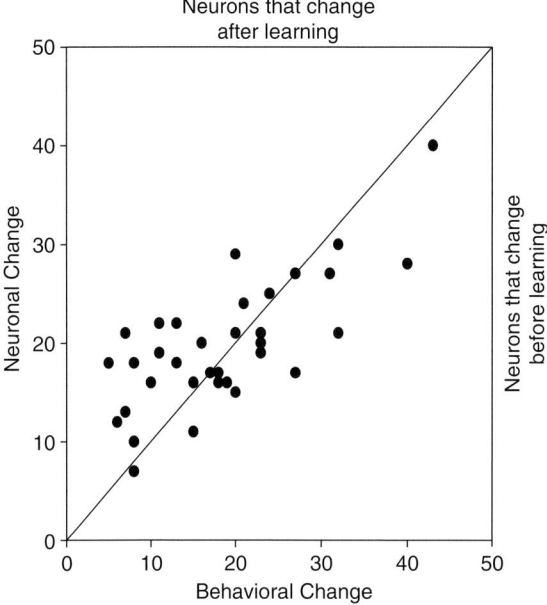

Fig. 4. The estimated relationship between neural activity and learning for the population of sustained and baseline sustained cells. On the X-axis is shown the trial number of learning and the Y-axis shows the trial number when the neuron changed activity as estimated using a change point test. This graph shows that a similar proportion of hippocampal cells change before and after learning. Adapted with permission from Wirth et al. (2003).

In addition to these associative learning signals described in the monkey hippocampus, other studies have also described associative learning signals in the adjacent perirhinal cortex using slightly different associative learning tasks. Messinger et al. (2001) recorded in the perirhinal cortex and the adjacent visual area TE as animals learned novel associations each day during a visual–visual paired-associate task. In this task, animals were first shown a single "sample" object and after a delay interval, two unique visual objects were shown. Animals learned which of the two choice stimuli was the "paired associate" of the sample stimulus. As animals learned the novel associations, neurons in the perirhinal cortex and area TE developed a more correlated visual response to the stimuli that had been paired in memory. Because the changes in neuronal activity appeared to parallel the learning exhibited by the animals, these findings suggested that like hippocampal neurons, the perirhinal neurons signal learning with changes in a neuron's stimulus-selective response properties.

Erickson and Desimone (1999) also recorded activity in the monkey perirhinal cortex as animals performed a task in which a predictor stimulus was followed by a choice stimulus. The choice stimulus could either signal the animals to release a bar (GO condition) or continue holding a bar (NO-GO condition). In this task, animals were not required to learn the explicit association between the predictor and the choice, but knowledge of this association could allow the animal to respond more quickly when the choice was presented. Indeed, animals responded more quickly to learned predictor stimuli after 1 day of training. Neural responses to predictor and choice stimuli were uncorrelated for novel stimuli used for 1 day, but significant correlations were observed after the animal had several days of experience with the stimuli. Thus, in this study, neural activity changed well after learning was expressed. Both the studies in the monkey perirhinal cortex suggest that learning is reflected as a change in a neuron's correlated response to learned pairs of stimuli. The relative speed of this neural change, however, may be dependent on the nature of the behavioral learning task.

To summarize so far, the neurophysiological studies described above showed that neurons in both the monkey hippocampus and perirhinal cortex signal new associative learning with changes in their firing rate which appears to correspond to changes in their stimulus-selective response properties. A detailed analysis of the timing of these learning-related signals suggest that hippocampal neurons participate at all stages of the learning process from several trials before behavioral learning is expressed, when the observed activity may be involved in driving the learned behavior to several trials after learning, and when the activity may be involved in a strengthening process. The associative learning changes in the perirhinal cortex also appear to parallel learning, though the detailed time course of learning relative to neural change was not studied in detail. Another open question concerns the relative contribution of different MTL structures to new associative learning. Findings from functional magnetic resonance image (fMRI) studies have provided important insight into this question.

Following up on the reports of Cahusac et al. (1993) and Wirth et al. (2003), Law et al. (2005) used (fMRI) to examine the patterns of brain activity present during learning of similar conditional-motor associations throughout the structures of the MTL, as well as across other brain areas. They showed that during learning of new conditional-motor associations, increasing fMRI signal was seen that paralleled changes in memory strength throughout the MTL including the hippocampus bilaterally, the right perirhinal cortex, and the parahippocampal cortex bilaterally (Fig. 5). These results confirm the electrophysiological findings in the monkey hippocampus for this task (Cahusac et al., 1993; Wirth et al., 2003) and further suggest that similar learning signals might be particularly prominent in both the perirhinal and parahippocampal cortices.

This study also described similar increasing patterns of activity in a wide range of other brain areas including superior frontal gyrus, medial frontal gyrus, cingulate gyrus, and left fusiform gyrus. A different pattern of learning-related activity consisting of a significant drop-off in activity once the associations were well learned was described for other brain areas including the middle frontal gyrus, inferior frontal gyrus, and right caudate nucleus.

The findings by Law et al. (2005) are also consistent with findings by Toni et al. (1998, 2001a, b) using both fMRI studies and PET studies. The fMRI study by Toni also suggests differential patterns of activation in the MTL and caudate, though the PET study reported similar increases in activity across both areas. Taken together these functional imaging studies not only confirm the important contribution of the structures of the MTL in new conditional-motor learning, but provide specific predictions concerning which other brain areas may be involved in this task. Indeed several of these areas, including the prefrontal cortex, motor portions of the frontal lobe, and the striatum, have

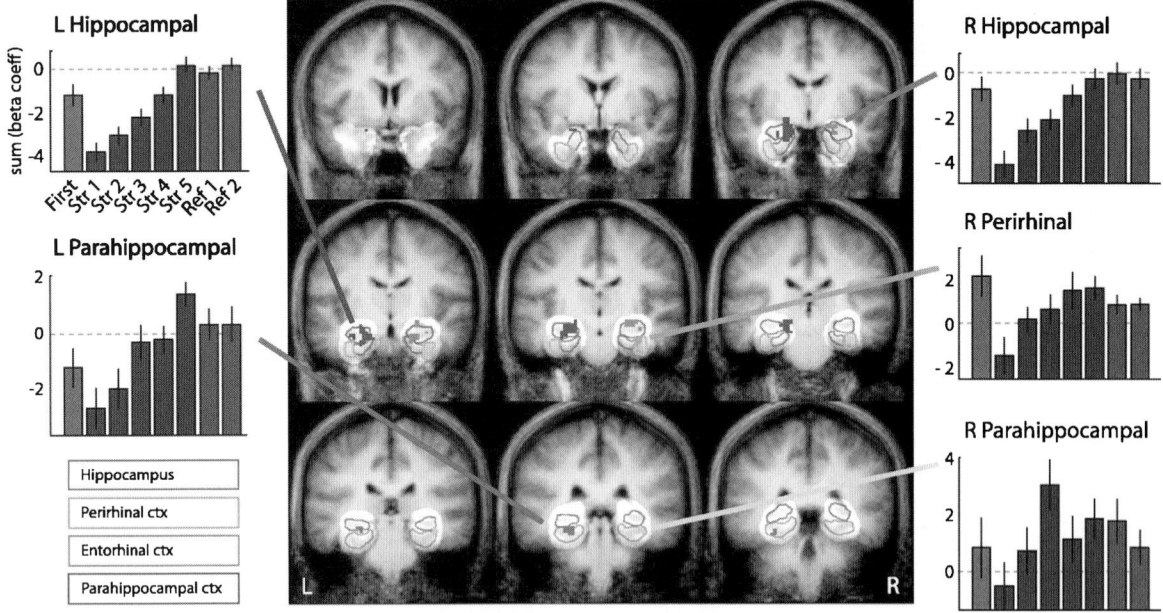

Fig. 5. Results from fMRI studies using an associative learning task similar to the one used in monkeys by Wirth et al. (2003). These findings show an increasing BOLD signal with learning across the left and right hippocampus, right perirhinal cortex and left and right parahippocampal cortex. First, response to the first presentation of the new "scene", Str1–4, memory strength indices 1–5 that correspond to increasingly greater levels of performance with learning. Ref 1–2, responses to the well learned reference images in the first and second half of each run respectively. Adapted with permission from Law et al. (2005).

been examined using single-unit electrophysiological techniques in monkeys. We now turn to these extra-MTL areas to review the patterns of learning-related signals seen in these other brain areas.

Associative learning in motor regions of the frontal lobe

Because conditional motor association learning involves learning to associate a particular visual stimulus with a particular motor response or location, this task has not only been used to study hippocampal associative learning function, but it has also been studied across various motor-related structures of the frontal lobe. For example, Wise and colleagues (Chen and Wise, 1995a, b; Brasted and Wise, 2004) described learning-related activity in the supplementary eye field (SEF) and frontal eye field (FEF) during the performance of a conditional motor task with eye movement responses similar to the task used by Wirth et al. (2003; Fig. 1). These reports describe three major categories of learning-related cells. The largest sub-category of learning-related cells was termed "learning-dependent". These cells exhibited significant changes in their activity during learning of new associations, and these changes were maintained for as long as the neuron was studied (Fig. 6A). Learning-dependent cells were also characterized by having significant task-related activity on familiar trials (analogous to the reference scenes trials in Wirth et al., 2003). Typically, activity during the novel conditions came to resemble activity in the familiar conditions with the same rewarded target location, suggesting a motor-based or direction-based learning signal. Note that motor-based learning signals were not seen in the hippocampus.

A second category of learning-related cells described in the SEF, FEF, and premotor cortex was termed "learning-selective" (Fig. 6B; approximately 24.5% of the learning-related cells in the SEF). Unlike the learning-dependent cells, these cells did not respond to the familiar conditions, but signaled learning for the new conditions with a transient response around the time of learning. A typical pattern of learning-selective activity was an early initial increase in activity, followed by a decrease back down to baseline levels of activity. Control experiments in which the learning-selective cells were examined during a second new learning set showed a similar transient, direction-selective response. Thus, the learning-selective cells in the SEF and FEF signal new learning in a direction-selective frame of reference. This direction-based response in SEF and FEF differentiates these cells from the cells in the hippocampus that did not exhibit a similar response for a second new scene with the same rewarded target location (Wirth et al., 2003).

The third category of learning-related activity described in the SEF and FEF was termed "learning-static" (Fig. 6C; approximately 24.5% of learning-related cells in the SEF). Like the learning-dependent cells, these cells also changed their activity in response to novel conditions and the activity was maintained for as long as the session lasted. In contrast to the learning-dependent cells, when the learning-static cells reached stable performance levels, there was a significant difference between the level of activity in response to the novel condition and the reference condition with the same rewarded target location. In this way, learning-static cells resemble the sustained changing cells observed in the hippocampus (Cahusac et al., 1993; Wirth et al., 2003). Similar to hippocampal changing cells, learning-dependent, learning-selective, as well as learning-static cells were observed during both the visual stimulus presentation and delay intervals in the SEF and FEF. However, unlike hippocampal cells, a relatively large proportion of SEF and FEF cells also signaled learning during the pre- and post-saccadic periods of the task, consistent with their important roles in eye movement responses. Thus, while both hippocampal as well as frontal eye movement regions signal change during stimulus and delay periods, the SEF and FEF appear to play a more prominent role in signaling learning during the motor response periods of the task.

Associative learning in the prefrontal cortex

Asaad et al. (1998) described the activity of cells in the prefrontal cortex during a conditional visual motor task with reversals. In this task, monkeys saw two novel visual stimuli each day and learned to associate those stimuli with either a left or right eye movement response. Once this initial set of two associations was learned, the object-response contingency was reversed. Like cells in SEF and FEF, prefrontal cells described in this study were sensitive to the direction of eye movement. In particular, many prefrontal cells signaled the impending direction of movement (i.e., direction-selective response). However, the appearance of directional selectivity alone did not reflect the learned associations since prefrontal cells continued to reflect the impending movement direction irrespective of whether the response was correct or incorrect. Instead, learning appeared to be most strongly correlated with the decreasing latency of appearance within the trial of direction selectivity. Early in learning, the direction selectivity was observed late in the trial near the time when the response was executed (Fig. 7A). With learning, this direction-selective signal shifted earlier in the trial towards the stimulus presentation period (i.e., cue period in Fig. 7A). These results suggest that the earlier appearance of directional selectivity within prefrontal neurons was related to behavioral learning. However, a quantitative analysis of the precise temporal relationship between the shifts in directional selectivity and behavioral learning was not presented.

Associative learning in the striatum

Three relatively recent studies have described associative learning signals in the striatum during tasks of conditional motor association learning. Brasted and Wise (2004) recorded activity in the caudate and putamen during a conditional motor association task with an arm movement response. They reported that cells in the caudate and putamen exhibit learning-selective, learning-dependent, or learning-static signals similar to their previous reports in the SEF and FEF (Fig. 6; Chen and Wise, 1995a, b). Similar findings have also been reported by Williams and Eskandar (2006) who also recorded in both the caudate and putamen during a similar conditional motor association task. Like the learning-selective cells of Brasted and Wise (2004) and the transient cells described in the hippocampus (Cahusac et al., 1993), they described cells that signaled learning with increases or decreases of learning that were highly dependent on the rate of learning (defined as the slope of the learning curve; Fig. 7C). This latter category of cells signaled learning during the feedback period of the task when animals were informed if they got the trials right or wrong. A second category of cells changed their activity (also mainly during the feedback period of the task) most strongly correlated with the animal's learning curve. These cells resemble the learning-dependent cells of Brasted and Wise (2004; Fig. 7B). Williams and Eskandar also examined the causal link between the learning-related activity in the striatum and learning using microstimulation. They showed that electrical stimulation of the caudate during correct trials but not during error trials could significantly increase the rate of learning. They further suggest that the caudate may be responsible for adjusting the associative weights between sensory cues and motor responses during the learning process.

In contrast, to the reports of Brasted and Wise (2004) and Williams and Eskandar (2006), and Pasupathy and Miller (2005) did not describe cells whose firing rate changed with either learning rate or the learning curve. Instead, like their previous findings in the prefrontal cortex (Asaad et al., 1998), they report that neurons in the caudate reflected learning with the earlier appearance within the trial of directional selectivity as learning progressed (Fig. 7A). The most striking finding reported was that not only was the same shift in direction selectivity seen in caudate neurons, but the speed of the temporal shift with learning was substantially earlier within the learning session compared to the prefrontal cells. Indeed the shift in latency appeared to occur before the relatively slow learning exhibited by the animals, though no direct comparisons were done to determine the precise relationship between the shifts in neural response latency and behavioral learning. The authors argue that their results support the hypothesis that rewarded associations are first identified by the basal ganglia and the output of

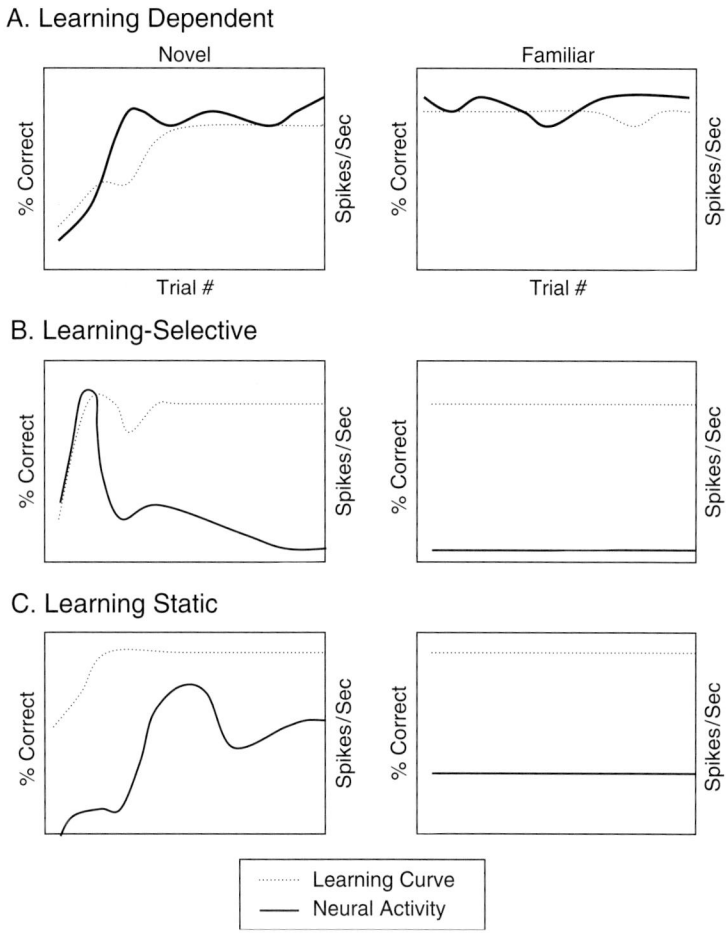

Fig. 6. Schematic representation of (A) learning-dependent, (B) learning-selective, and (C) learning-static responses described in the SEF, FEF, and premotor cortex. Shown in the left-hand column are schematic responses of learning-related cells to novel conditional motor associations, and shown on the right is the corresponding response of the same cell to highly familiar associations with the same motor response. Note that the responses of learning-dependent cells on novel associations come to resemble the responses to familiar associations with the same rewarded target. Learning-selective cells decrease their responses anti-correlated with behavioral learning like the hippocampal baseline sustained cells while the learning-static cells resemble the sustained changing cells in the hippocampus.

this structure may serve to train neurons in the prefrontal cortex. However, an analysis of the error trials showed that caudate neurons did not differentiate between correct and error trials at any time point during the trial (see supplementary Fig. 3 of Pasupathy and Miller, 2005). Thus, it remains unclear whether these early directional signals in the caudate serve as a "teacher" for other brain areas or simply reflect early preparation or anticipation of the motor output. Differences in the precise recording site, differences in behavioral task, as well as possible differences in behavioral strategies used by different monkeys could underlie the differences in the patterns of learning-related activity seen across these studies. Future studies will be needed to sort these differences out.

Discussion

There is now strong evidence from studies in rabbits (McEchron and Disterhoft, 1997), rats (Fyhn et al., 2002; Frank et al., 2004), and monkeys (Cahusac et al., 1993; Erickson and

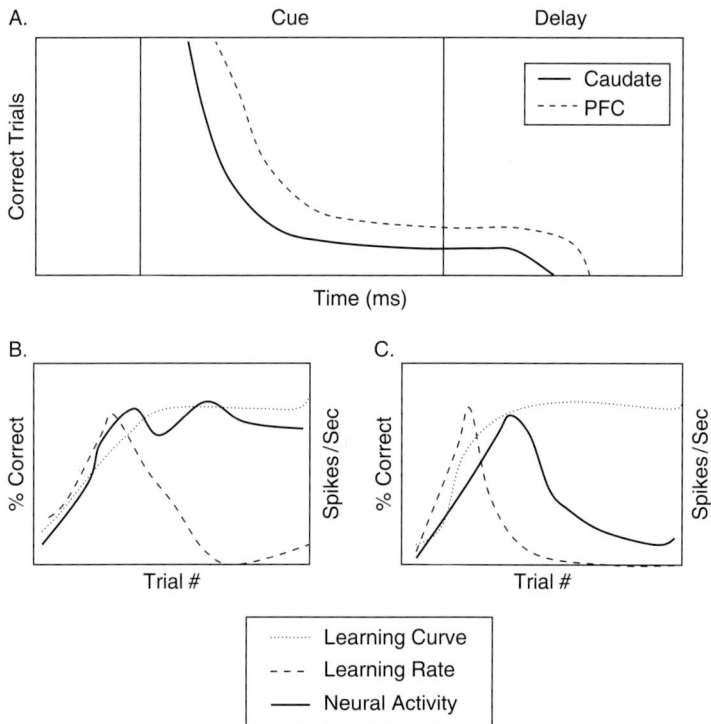

Fig. 7. Illustration of three different patterns of learning-related activity seen in the striatum. (A) Schematic representation of the decreasing latency of the development of direction-selective activity described in the caudate and prefrontal cortex during learning and reversals of a conditional motor task by Pasupathy and Miller (2005). In this graph the rise time denotes the timing within the trial that direction selectivity first developed. Note that the direction-selective response develops more quickly, as well as eventually starts earlier in the trial compared to the direction-selective activity in the prefrontal cortex. (B) Schematic representation of a striatal cell that was highly correlated with the animal's behavioral learning curve described by Williams and Eskandar (2006). (C) Schematic representation of a striatal cell that was highly correlated with learning rate (i.e., the steepest part of the learning curve) described by Williams and Eskandar (2006).

Desimone, 1999; Messinger et al., 2001; Wirth et al., 2003) that neurons throughout the MTL signal new associative learning with either a strong increase or decrease in neural activity that parallels behavioral learning. In some cases, these changes have been linked to changes in a cell's stimulus-selective response properties (Erickson and Desimone, 1999; Messinger et al., 2001; Wirth et al., 2003). In other cases, neurons signal learning by shifting the latency of their response earlier in the trial (McEchron and Disterhoft, 1997).

Consistent with the neurophysiological data, fMRI studies have shown parallel changes in hemodynamic responses during learning across the structures of the MTL (Toni et al., 1998, 2001a, b; Law et al., 2005). These findings make several important points. First, for tasks of new conditional motor learning, the structures of the MTL appear to work together in signaling new learning. These findings are consistent with neurophysiological findings showing that similar kinds of associative learning signals that have been reported in both the monkey hippocampus (Wirth et al., 2003) and the adjacent perirhinal cortex (Erickson and Desimone, 1999; Messinger et al., 2001). Second, the results from these fMRI studies not only emphasized the important role of the MTL in new associative/conditional motor learning, but also revealed the wider range of brain areas engaged during this new learning task. A review of the neurophysiological studies in several of these extra-MTL areas showed both similarities and differences in the learning-related signals observed during tasks of new associative learning.

While many different brain areas reviewed in this chapter signal learning with either increases or decreases in activity that change in parallel with learning, a major difference between brain areas is the relative prominence of direction-based or location-based learning signals. For example, in the hippocampus, there is strong evidence that neither the sustained, baseline sustained, or transient signals signal learning specific for a particular response direction (Cahusac et al., 1993; Wirth et al., 2003). There is also some evidence that the learning-related hippocampal signals are not specific for a particular visual stimuli (Miyashita et al., 1989; Cahusac et al., 1993). Instead, hippocampal signals appear selective for particular the object-place associations being learned. This interpretation is consistent with the relational theory of hippocampal function that stresses the importance of this structure in forming flexible new associations between different stimuli irrespective of modality (Eichenbaum, 2000).

In contrast to hippocampal cells, many cells in the SEF and FEF signal learning in a direction-based or motor-based frame of reference. Cells in these areas signal new learning for a particular target location with either increases (learning-dependent) or decreases (learning-selective) in activity. While some cells signal learning for a particular direction (learning-selective), other cells signal both new learning and previously established associations specific for a particular direction (learning-dependent). These findings are consistent with the idea that these areas are involved in the ability to form arbitrary mappings between objects and particular actions (Murray et al., 2000).

While the associative learning signals in the prefrontal cortex also appeared to be direction-based, the pattern of activity was quite different from the motor/direction-based signals described in SEF and FEF. Prefrontal neurons signaled learning by the earlier appearance within a trial of strong directional activity that signaled the direction/location that the animals would choose at the end of the trial (Asaad et al., 1998).

A question of considerable interest is the relative timing of these prefrontal learning signals relative to behavioral learning. This was not examined in detail in the Asaad et al. (1998) study and will be of particular interest with respect to the timing of hippocampal learning-related activity. Both the hippocampus and prefrontal cortex have been implicated in acquisition of new information, and while interactions between these structures have long been hypothesized (Miller et al., 1995; Tomita et al., 1999), direct evidence for the nature of this interaction is lacking. The studies reviewed above suggest that while both areas contribute to this task, they each convey quite distinct patterns of learning-related activity. A detailed examination of the timing of these learning signals will be important to determine whether these areas signal learning in parallel or one leads the other in the new learning signal.

A comparison of hippocampus and striatal learning signals reveal not only similarities but also striking differences in the patterns of neural signals underlying associative learning. Like the comparison with SEF and FEF, a major distinguishing factor between the learning signals seen in the hippocampus and striatum is the stronger motor-based or direction-based signals observed in the striatum relative to the hippocampus. Brasted and Wise (2004) reported similar learning-selective, learning-dependent, and learning-static signals in the striatum. Williams and Eskandar (2006) also reported clear changes in activity that correlated with either the learning rate or the learning curve. This latter study also emphasized that the strongest correlations were observed during the feedback period of the task right before the reward was given which is different from hippocampal cells that showed strong learning-related signals during the stimulus and delay portions of the task. These comparisons between hippocampal and striatal learning signals are of particular interest with regard to previous studies suggesting that the MTL and striatum are distinct memory systems that may compete for control of behavior. For example, previous studies have suggested that during spatial working memory tasks the hippocampus controls more flexible spatial learning strategies and the caudate controlling more rigid habit-like learning strategies (Packard et al., 1989; Packard and McGaugh, 1996). A similar competitive interaction has been suggested by results from fMRI studies (Poldrack et al., 2001; Foerde et al., 2006). While the neurophysiological studies reviewed above cannot distinguish between a competitive vs. a cooperative interaction between these two brain areas, future studies in which neural activity can be monitored across both brain areas simultaneously together with lesion or inactivation studies will be important to address this question.

Conclusion

Associative learning paradigms offer a unique opportunity to compare and contrast learning-related neural activity across widespread brain areas. These findings suggest that while widespread brain areas participate in new associative learning, they may participate in different aspects of the learning. While the hippocampus appears to specialize in signaling the learning of new associations between arbitrary stimuli, motor-related frontal and prefrontal areas signal learning in a decidedly motor or direction-based frame of reference. Striatal cells also appear to have a strong motor or direction-based learning signal and may also be strongly influenced by information about reward delivery as well. An important challenge in future neuroscience research will be to design experiments that allow us to monitor the activity and possible interactions between these widespread areas during learning to better define how these areas work together to accomplish new associative learning.

Acknowledgments

Support was provided by NIDA grant DA015644 to E.N.B and W.A.S., NIMH grant MH58847 a McKnight Foundation grant and a John Merck Scholars Award to W.A.S.

References

Asaad, W.F., Rainer, G. and Miller, E.K. (1998) Neural activity in the primate prefrontal cortex during associative learning. Neuron, 21: 1399–1407.

Bayley, P.J. and Squire, L.R. (2002) Medial temporal lobe amnesia: gradual acquisition of factual information by nondeclarative memory. J. Neurosci., 22: 5741–5748.

Berger, T.W., Alger, B.E. and Thompson, R.F. (1976) Neuronal substrates of classical conditioning in the hippocampus. Science, 192: 483–485.

Berger, T.W. and Thompson, R.F. (1978) Neuronal plasticity in the limbic system during classical conditioning of the rabbit nictitating membrane response. I. The hippocampus. Brain Res., 145: 323–346.

Booth, M.C. and Rolls, E.T. (1998) View-invariant representations of familiar objects by neurons in the inferior temporal visual cortex. Cereb. Cortex, 8: 510–523.

Brasted, P.J., Bussey, T.J., Murray, E.A. and Wise, S.P. (2002) Fornix transection impairs conditional visuomotor learning in tasks involving nonspatially differentiated responses. J. Neurophysiol., 87: 631–633.

Brasted, P.J., Bussey, T.J., Murray, E.A. and Wise, S.P. (2003) Role of the hippocampal system in associative learning beyond the spatial domain. Brain, 126: 1202–1223.

Brasted, P.J. and Wise, S.P. (2004) Comparison of learning-related neuronal activity in the dorsal premotor cortex and striatum. Eur. J. Neurosci., 19: 721–740.

Cahusac, P.M., Rolls, E.T., Miyashita, Y. and Niki, H. (1993) Modification of the responses of hippocampal neurons in the monkey during the learning of a conditional spatial response task. Hippocampus, 3: 29–42.

Chen, L.L. and Wise, S.P. (1995a) Neuronal activity in the supplementary eye field during acquisition of conditional oculomotor associations. J. Neurophysiol., 73: 1101–1121.

Chen, L.L. and Wise, S.P. (1995b) Supplementary eye field contrasted with the frontal eye field during acquisition of conditional oculomotor associations. J. Neurophysiol., 73: 1122–1134.

Eichenbaum, H. (2000) Cortical-hippocampal networks for declarative memory. Nat. Neurosci. Rev., 1: 41–50.

Eichenbaum, H. and Cohen, N.J. (2001) From Conditioning to Conscious Recollection. Oxford University Press, New York.

Erickson, C.A. and Desimone, R. (1999) Responses of macaque perirhinal neurons during and after visual stimulus association learning. J. Neurosci., 19: 10404–10416.

Foerde, K., Knowlton, B.J. and Poldrack, R.A. (2006) Modulation of competing memory systems by distraction. Proc. Natl. Acad. Sci. U.S.A., 103: 11778–11783.

Frank, L.M., Stanley, G.B. and Brown, E.N. (2004) Hippocampal plasticity across multiple days of exposure to novel environments. J. Neurosci., 24(35): 7681–7689. Ref. type: Magazine article.

Fyhn, M., Molden, S., Hollup, S., Moser, M.B. and Moser, E.I. (2002) Hippocampal neurons responding to first-time dislocation of a target object. Neuron, 35: 555–566.

Kim, J.J., Clark, R.E. and Thompson, R.F. (1995) Hippocampectomy impairs the memory of recently, but not remotely acquired trace eyeblink conditioned responses. Behav. Neurosci., 109: 195–203.

Law, J.R., Flanery, M.A., Wirth, S., Yanike, M., Smith, A.C., Frank, L.M., Suzuki, W.A., Brown, E.N. and Stark, C.E. (2005) Functional magnetic resonance imaging activity during the gradual acquisition and expression of paired-associate memory. J. Neurosci., 25: 5720–5729.

Liu, Z., Richmond, B.J., Murray, E.A., Saunders, R.C., Steenrod, S., Stubblefield, B.K., Montague, D.M. and Ginns, E.I. (2004) DNA targeting of rhinal cortex D2 receptor protein reversibly blocks learning of cues that predict reward. Proc. Natl. Acad. Sci. U.S.A., 101: 12336–12341.

McEchron, M.D. and Disterhoft, J.F. (1997) Sequence of single neuron changes in CA1 hippocampus of rabbits during acquisition of trace eyeblink conditioned responses. J. Neurophysiol., 78: 1030–1044.

McEchron, M.D., Tseng, W. and Disterhoft, J.F. (2000) Neurotoxic lesions of the dorsal hippocampus disrupt auditory-cued trace heart rate (fear) conditioning in rabbits. Hippocampus, 10: 739–751.

Messinger, A., Squire, L.R., Zola, S.M. and Albright, T.D. (2001) Neuronal representations of stimulus associations develop in the temporal lobe during learning. Proc. Natl. Acad. Sci. U.S.A., 98: 12239–12244.

Miller, E.K., Erickson, C.A. and Desimone, R. (1995) Comparison of prefrontal (PF) and inferior temporal (IT) neurons during performance of a memory task. Soc. Neurosci. Abstr. Ref. type: Abstract.

Miyashita, Y., Rolls, E.T., Cahusac, P.M., Niki, H. and Feigenbaum, J.D. (1989) Activity of hippocampal formation neurons in the monkey related to a conditional spatial response task. J. Neurophysiol., 61: 669–678.

Moyer, J.R., Jr., Deyo, R.A. and Disterhoft, J.F. (1990) Hippocampectomy disrupts trace eye-blink conditioning in rabbits. Behav. Neurosci., 104: 243–252.

Murray, E.A., Baxter, M.G. and Gaffan, D. (1998) Monkeys with rhinal cortex damage or neurotoxic hippocampal lesions are impaired on spatial scene learning and object reversals. Behav. Neurosci., 112: 1291–1303.

Murray, E.A., Bussey, T.J. and Wise, S.P. (2000) Role of prefrontal cortex in a network for arbitrary visuomotor mapping. Exp. Brain Res., 133: 114–129.

Murray, E.A., Gaffan, D. and Mishkin, M. (1993) Neural substrates of visual stimulus-stimulus association in rhesus monkeys. J. Neurosci., 13: 4549–4561.

Murray, E.A. and Wise, S.P. (1996) Role of the hippocampus plus subjacent cortex but not amygdala in visuomotor conditional learning in rhesus monkeys. Behav. Neurosci., 110: 1261–1270.

Naya, Y., Sakai, K. and Miyashita, Y. (1996) Activity of primate inferotemporal neurons related to a sought target in pair-association task. Proc. Natl. Acad. Sci. U.S.A., 93: 2664–2669.

Naya, Y., Yoshida, M. and Miyashita, Y. (2003) Forward processing of long-term associative memory in monkey inferotemporal cortex. J. Neurosci., 23: 2861–2871.

O'Keefe, J. and Dostrovsky, J. (1971) The hippocampus as a spatial map. Preliminary evidence from unit activity in the freely-moving rat. Brain Res., 34: 171–175.

O'Keefe, J. and Nadel, L. (1978) The Hippocampus as a Cognitive Map. Oxford University Press, New York.

Packard, M.G., Hirsh, R. and White, N.M. (1989) Differential effects of fornix and caudate nucleus lesions on two radial maze tasks: evidence for multiple memory systems. J. Neurosci., 9: 1465–1472.

Packard, M.G. and McGaugh, J.L. (1996) Inactivation of hippocampus or caudate nucleus with lidocaine differentially affects expression of place and response learning. Neurobiol. Learn. Mem., 65: 65–72.

Pasupathy, A. and Miller, E.K. (2005) Different time courses of learning-related activity in the prefrontal cortex and striatum. Nature, 433: 873–876.

Poldrack, R.A., Clark, J., Pare-Blagoev, E.J., Shohamy, D., Creso, M.J., Myers, C. and Gluck, M.A. (2001) Interactive memory systems in the human brain. Nature, 414: 546–550.

Rupniak, N.M. and Gaffan, D. (1987) Monkey hippocampus and learning about spatially directed movements. J. Neurosci., 7: 2331–2337.

Sakai, K. and Miyashita, Y. (1991) Neural organization for the long-term memory of paired associates. Nature, 354: 152–155.

Scoville, W.B. and Milner, B. (1957) Loss of recent memory after bilateral hippocampal lesions. J. Neurol. Neurosurg. Psychiatry, 20: 11–21.

Sobotka, S. and Ringo, J.L. (1993) Investigation of long-term recognition and association memory in unit responses from inferotemporal cortex. Exp. Brain Res., 96: 28–38.

Solomon, P.R., Vander Schaaf, E.R., Thompson, R.F. and Weisz, D.J. (1986) Hippocampus and trace conditioning of the rabbit's classically conditioned nictitating membrane response. Behav. Neurosci., 100: 729–744.

Squire, L.R., Stark, C.E. and Clark, R.E. (2004) The medial temporal lobe. Annu. Rev. Neurosci., 27: 279–306.

Stark, C.E., Bayley, P.J. and Squire, L.R. (2002) Recognition memory for single items and for associations is similarly impaired following damage to the hippocampal region. Learn. Mem., 9: 238–242.

Stark, C.E. and Squire, L.R. (2003) Hippocampal damage equally impairs memory for single items and memory for conjunctions. Hippocampus, 13: 281–292.

Tomita, H., Ohbayashi, M., Nakahara, K., Hasegawa, I. and Miyashita, Y. (1999) Top-down signal from prefrontal cortex in executive control of memory retrieval. Nature, 401: 699–703.

Toni, I., Krams, M., Turner, R. and Passingham, R.E. (1998) The time course of changes during motor sequence learning: a whole-brain fMRI study. Neuroimage, 8: 50–61.

Toni, I., Ramnani, N., Josephs, O., Ashburner, J. and Passingham, R.E. (2001a) Learning arbitrary visuomotor associations: temporal dynamic of brain activity. Neuroimage, 14: 1048–1057.

Toni, I., Rushworth, M.F. and Passingham, R.E. (2001b) Neural correlates of visuomotor associations. Spatial rules compared with arbitrary rules. Exp. Brain Res., 141: 359–369.

Vargha-Khadem, F., Gadian, D.G., Watkins, K.E., Connelly, A., Van Paesschen, W. and Mishkin, M. (1997) Differential effects of early hippocampal pathology on episodic and semantic memory. Science, 277: 376–380.

Williams, Z.M. and Eskandar, E.N. (2006) Selective enhancement of associative learning by microstimulation of the anterior caudate. Nat. Neurosci., 9: 562–568.

Wilson, M.A. and Mcnaughton, B.L. (1993) Dynamics of the hippocampal ensemble code for space. Science, 261: 1055–1058.

Wirth, S., Yanike, M., Frank, L.M., Smith, A.C., Brown, E.N. and Suzuki, W.A. (2003) Single neurons in the monkey hippocampus and learning of new associations. Science, 300: 1578–1581.

Wise, S.P. and Murray, E.A. (1999) Role of the hippocampal system in conditional motor learning: mapping antecedents to action. Hippocampus, 9: 101–117.

SECTION IV

Human Approaches to the Essence of Memory

CHAPTER 20

What are the differences between long-term, short-term, and working memory?

Nelson Cowan*

Department of Psychological Sciences, University of Missouri, 18 McAlester Hall, Columbia, MO 65211, USA

Abstract: In the recent literature there has been considerable confusion about the three types of memory: long-term, short-term, and working memory. This chapter strives to reduce that confusion and makes up-to-date assessments of these types of memory. Long- and short-term memory could differ in two fundamental ways, with only short-term memory demonstrating (1) temporal decay and (2) chunk capacity limits. Both properties of short-term memory are still controversial but the current literature is rather encouraging regarding the existence of both decay and capacity limits. Working memory has been conceived and defined in three different, slightly discrepant ways: as short-term memory applied to cognitive tasks, as a multi-component system that holds and manipulates information in short-term memory, and as the use of attention to manage short-term memory. Regardless of the definition, there are some measures of memory in the short term that seem routine and do not correlate well with cognitive aptitudes and other measures (those usually identified with the term "working memory") that seem more attention demanding and do correlate well with these aptitudes. The evidence is evaluated and placed within a theoretical framework depicted in Fig. 1.

Keywords: attention; capacity of working memory; control of attention; decay of short-term memory; focus of attention; long-term memory; short-term memory; working memory

Historical roots of a basic scientific question

How many phases of a memory are there? In a naïve view of memory, it could be made all of one cloth. Some people have a good ability to capture facts and events in memory, whereas others have less such ability. Yet, long before there were true psychological laboratories, a more careful observation must have shown that there are separable aspects of memory. An elderly teacher might be seen relating old lessons as vividly as he ever did, and yet it might be evident that his ability to capture the names of new students, or to recall which student made what comment in an ongoing conversation, has diminished over the years.

The scientific study of memory is usually traced back to Hermann Ebbinghaus (1885/1913 translation), who examined his own acquisition and forgetting of new information in the form of series of nonsense syllables tested at various periods upto 31 days. Among many important observations, Ebbinghaus noticed that he often had a "first fleeting grasp ... of the series in moments of special concentration" (p. 33) but that this immediate memory did not ensure that the series had been memorized in a way that would allow its

*Corresponding author. Tel.: +1 573-882-4232;
Fax: +1 573-882-7710; E-mail: CowanN@missouri.edu

recall later on. Stable memorization sometimes required further repetitions of the series. Soon afterward, James (1890) proposed a distinction between primary memory, the small amount of information held as the trailing edge of the conscious present, and secondary memory, the vast body of knowledge stored over a lifetime. The primary memory of James is like the first fleeting grasp of Ebbinghaus.

The Industrial Revolution made some new demands on what James (1890) called primary memory. In the 1850s, telegraph operators had to remember and interpret rapid series of dots and dashes conveyed acoustically. In 1876, the telephone was invented. Three years later, operators in Lowell, Massachusetts started using telephone numbers for more than 200 subscribers so that substitute operators could be more easily trained if the town's four regular operators succumbed to a raging measles epidemic. This use of telephone numbers, complemented by a word prefix, of course spread. (The author's telephone number in 1957 was WHitehall 2-6742; the number is still assigned, albeit as a seven-digit number.) Even before the book by Ebbinghaus, Nipher (1878) reported on the serial position curve obtained among the digits in logarithms that he tried to recall. The nonsense syllables that Ebbinghaus had invented as a tool can be seen to have acquired more ecological validity in an industrial age with expanding information demands, perhaps highlighting the practical importance of primary memory in daily life. Primary memory seems taxed as one is asked to keep in mind aspects of an unfamiliar situation, such as names, places, things, and ideas that one has not encountered before.

Yet, the subjective experience of a difference between primary and secondary memory does not automatically guarantee that these types of memory separately contribute to the science of remembering. Researchers from a different perspective have long hoped that they could write a single equation, or a single set of principles at least, that would capture all of memory, from the very immediate to the very long-term. McGeoch (1932) illustrated that forgetting over time was not simply a matter of an inevitable decay of memory but rather of interference during the retention interval; one could find situations in which memory improved, rather than diminish, over time. From this perspective, one might view what appeared to be forgetting from primary memory as the profound effect of interference from other items on memory for any one item, with interference effects continuing forever but not totally destroying a given memory. This perspective has been maintained and developed over the years by a steady line of researchers believing in the unity of memory, including, among others, Melton (1963), Bjork and Whitten (1974), Wickelgren (1974), Crowder (1982, 1993), Glenberg and Swanson (1986), Brown et al. (2000), Nairne (2002), Neath and Surprenant (2003), and Lewandowsky et al. (2004).

Description of three kinds of memory

In this chapter I will assess the strength of evidence for three types of memory: long-term memory, short-term memory, and working memory. *Long-term memory* is a vast store of knowledge and a record of prior events, and it exists according to all theoretical views; it would be difficult to deny that each normal person has at his or her command a rich, although not flawless or complete, set of long-term memories.

Short-term memory is related to the primary memory of James (1890) and is a term that Broadbent (1958) and Atkinson and Shiffrin (1968) used in slightly different ways. Like Atkinson and Shiffrin, I take it to reflect faculties of the human mind that can hold a limited amount of information in a very accessible state temporarily. One difference between the term "short-term memory" and the term "primary memory" is that the latter might be considered to be more restricted. It is possible that not every temporarily accessible idea is, or even was, in conscious awareness. For example, by this conception, if you are speaking to a person with a foreign accent and inadvertently alter your speech to match the foreign speaker's accent, you are influenced by what was until that point an unconscious (and therefore uncontrollable) aspect of your short-term memory. One might relate short-term memory to a pattern of neural firing that represents a particular idea and one might consider the idea to be in short-term memory only when the firing pattern, or cell assembly, is active (Hebb, 1949). The individual might or might not be aware of the idea during that period of activation.

Working memory is not completely distinct from short-term memory. It is a term that was used by Miller et al. (1960) to refer to memory as it is used to plan and carry out behavior. One relies on working memory to retain the partial results while solving an arithmetic problem without paper, to combine the premises in a lengthy rhetorical argument, or to bake a cake without making the unfortunate mistake of adding the same ingredient twice. (Your working memory would have been more heavily taxed while reading the previous sentence if I had saved the phrase "one relies on working memory" until the end of the sentence, which I did in within my first draft of that sentence; working memory thus affects good writing.) The term "working memory" became much more dominant in the field after Baddeley and Hitch (1974) demonstrated that a single module could not account for all kinds of temporary memory. Their thinking led to an influential model (Baddeley, 1986) in which verbal-phonological and visual-spatial representations were held separately, and were managed and manipulated with the help of attention-related processes, termed the central executive. In the 1974 paper, this central executive possibly had its own memory that crossed domains of representation. By 1986, this general memory had been eliminated from the model, but it was added back again by Baddeley (2000) in the form of an *episodic buffer*. That seemed necessary to explain short-term memory of features that did not match the other stores (particularly semantic information in memory) and to explain cross-domain associations in working memory, such as the retention of links between names and faces. Because of the work of Baddeley et al. (1975), working memory is generally viewed as the combination of multiple components working together. Some even include in that bundle the heavy contribution of long-term memory, which reduces the working memory load by organizing and grouping information in working memory into a smaller number of units (Miller, 1956; Ericsson and Kintsch, 1995). For example, the letter series IRSCIAFBI can be remembered much more easily as a series of acronyms for three federal agencies of the United States of America: the Internal Revenue Service (IRS), the Central Intelligence Agency (CIA), and the Federal Bureau of Investigation (FBI). However, that factor was not emphasized in the well-known model of Baddeley (1986).

What is clear from my definition is that working memory includes short-term memory and other processing mechanisms that help to make use of short-term memory. This definition is different from the one used by some other researchers (e.g., Engle, 2002), who would like to reserve the term working memory to refer only to the attention-related aspects of short-term memory. This, however, is not so much a debate about substance, but rather a slightly confusing discrepancy in the usage of terms.

One reason to pursue the term working memory is that measures of working memory have been found to correlate with intellectual aptitudes (and especially fluid intelligence) better than measures of short-term memory and, in fact, possibly better than measures of any other particular psychological process (e.g., Daneman and Carpenter, 1980; Kyllonen and Christal, 1990; Daneman and Merikle, 1996; Engle et al., 1999; Conway et al., 2005). It has been thought that this reflects the use of measures that incorporate not only storage but also processing, the notion being that both storage and processing have to be engaged concurrently to assess working memory capacity in a way that is related to cognitive aptitude. More recently, Engle et al. (1999) introduced the notion that aptitudes and working memory both depend on the ability to control attention, or to apply the control of attention to the management of both primary and secondary memory (Unsworth and Engle, 2007). However, more research is needed on exactly what we learn from the high correlation between working memory and intellectual aptitudes, and this issue will be discussed further after the more basic issue of the short-term versus the long-term memory distinction is addressed.

Meanwhile, it may be helpful to summarize a theoretical framework (Cowan, 1988, 1995, 1999, 2001, 2005) based on past research. This framework, illustrated in Fig. 1, helps to account for the relation between long-term, short-term, and working memory mechanisms and explains what I see as the relation between them. In this framework, short-term memory is derived from a temporarily activated subset of information in long-term memory. This activated subset may decay as a function of time unless it is refreshed, although the evidence for decay is still tentative at best. A subset of the activated information is the focus of attention, which appears to be limited in chunk capacity (how many separate items can be included at once). New associations between activated elements can form the focus of attention. Now the evidence related to this modeling framework will be discussed.

The short-term memory/long-term memory distinction

If there is a difference between short- and long-term memory stores, there are two possible ways in which these stores may differ: in *duration*, and in *capacity*. A duration difference means that items in short-term storage decay from this sort of storage as a function of time. A capacity difference means that there is a limit in how many items short-term storage can hold. If there is only a limit in capacity, a number of items smaller than the capacity limit could remain in short-term storage until they are replaced by other items. Both types of limit are controversial. Therefore, in order to assess the usefulness of the short-term storage concept, duration and capacity limits will be assessed in turn.

Duration limits

The concept of short-term memory limited by decay over time was present even at the beginning of cognitive psychology, for example in the work of Broadbent (1958). If decay were the only principle affecting performance in an immediate memory experiment, it would perhaps be easy to detect this decay. However, even in Broadbent's work contaminating variables were recognized. To assess decay one must take into account, or

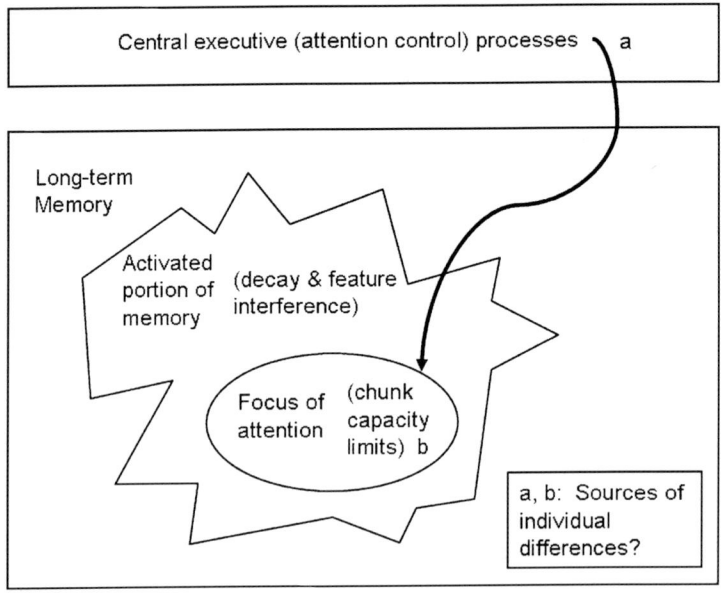

Fig. 1. A depiction of the theoretical modeling framework. Modified from Cowan (1988) and refined in further work by Cowan (1995, 1999, 2005).

overcome, contaminating effects of rehearsal, long-term retrieval, and temporal distinctiveness, which will be discussed one at a time in conjunction with evidence for and against decay.

Overcoming contamination from rehearsal

According to various researchers there is a process whereby one imagines how the words on the list are pronounced without saying them aloud, a process called covert verbal rehearsal. With practice, this process comes to occur with a minimum of attention. Guttentag (1984) used a secondary task to show that rehearsal of a list to be recalled was effortful in young children, but not in adults. If, in a particular experimental procedure, no loss of short-term memory is observed, one can attribute that response pattern to rehearsal. Therefore, steps have been taken to eliminate rehearsal through a process termed articulatory suppression, in which a simple utterance such as the word "the" is repeatedly pronounced by the participant during part or all of the short-term memory task (e.g., Baddeley et al., 1975). There is still the possible objection that whatever utterance is used to suppress rehearsal unfortunately causes interference, which could be the true reason for memory loss over time instead of decay.

That problem of interference would appear moot in light of the findings of Lewandowsky et al. (2004). They presented lists of letters to be recalled and varied how long the participant was supposed to take to recall each item in the list. In some conditions, they added articulatory suppression to prevent rehearsal. Despite that suppression, they observed no difference in performance with the time between items in the response varying between 400 and 1600 ms (or between conditions in which the word "super" was pronounced one, two, or three times between consecutive items in the response). They found no evidence of memory decay.

A limitation of this finding, though, is that covert verbal rehearsal may not be the only type of rehearsal that participants can use. Perhaps there are types that are not prevented by articulatory suppression. In particular, Cowan (1992) suggested that the process of mentally attending to words or searching through the list, an attention-demanding process, could serve to reactivate items to be recalled in a manner similar to covert verbal rehearsal. The key difference is that it would not be expected that articulatory suppression would prevent that type of rehearsal. Instead, to prevent that type of rehearsal an attention-demanding task would have to be used.

Barrouillet et al. (2004, 2007) have results that do seem to suggest that there is another, more attention-demanding type of rehearsal. They have interposed materials between items to be recalled that require choices; they can be numbers to read aloud or multi-choice reaction times. It is found that these interfere with retention to an extent commensurate to the proportion of the inter-item interval used up attending to the distracting items. As the rate of the distracting items goes up, fewer of the to-be-recalled items are recalled. The notion is that when the distracting task does not require attention, the freed-up attention allows an attention-based rehearsal of the items to be recalled. When the interposed task is more automatic and does not require as much attention (e.g., an articulatory suppression task) there is much less effect of the rate of these interposed items.

Based on this logic, one could imagine a version of Lewandowsky's task in which not articulatory suppression but attention-demanding verbal stimuli are placed between items in the response, and in which the duration of this filled time between items in the response varies from trial to trial. The verbal, attention-demanding stimuli should prevent both attention-based rehearsal and articulation-based rehearsal. If there is decay, then performance should decline across serial positions more severely when longer filled intervals are placed between items in the response. Unfortunately, though, such results might be accounted for alternatively as the result of interference from the distracting stimuli, without the need to invoke decay.

What seems to be needed, then, is a procedure to prevent both articulation-based and attention-based rehearsal without introducing interference. Cowan and Aubuchon (in press) tried out one type of procedure that may accomplish this. They presented lists of seven printed digits in which the time between items varied within a list. In addition to some randomly timed filler lists, there were four critical trial types, in which the six inter-digit blank intervals were all short (0.5 s following each item) or all long (2 s following each item), or comprised three short and then three long intervals, or three long and then three short intervals. Moreover, there were two post-list response cues. According to one cue, the participant was to recall the list with the items in the presented order, but at any rate they wished. According to the other response cue, the list was to be recalled using the same timing in which it was presented. The expectation was that the need to remember the timing in the latter response condition would prevent rehearsal of either type. As a consequence, performance should be impaired on trials in which the first three response intervals are long because, on these trials, there is more time for forgetting of most of the list items. Just as predicted, there was a significant interaction between the response cue and the length of the first half of the response intervals. When participants were free to recall items at their own pace, performance was no better with a short first half ($M = .71$) than with a long first half ($M = .74$). The slight benefit of a long first half in that situation could occur because it allowed the list to be rehearsed early on in the response. In contrast, when the timing of recall had to match the timing of the list presentation, performance was better with a short first half ($M = .70$) than with a long first half ($M = .67$). This, then, suggests there could be decay in short-term memory.

Overcoming contamination from long-term retrieval

If there is more than one type of memory storage then there still is the problem of which store provided the information underlying a response. There is no guarantee that, just because a procedure is considered a test of short-term storage, the long-term store will not be used. For example, in a simple digit span task, a series of digits is presented and is to be repeated immediately afterward from memory. If that series turned out to be only slightly different from the participant's telephone number, the participant might be able to memorize the new number quickly and repeat it from long-term memory. The dual-store theories of memory allow this. Although Broadbent (1958) and Atkinson and Shiffrin (1968) drew their models of information processing as a series of boxes representing different memory stores, with long-term memory following short-term memory, these boxes do not imply that memory is exclusively in one box or another; they are better interpreted as the relative times of the first entry of information from a stimulus into one store and then the next. The question remains, then as to how one can determine if a response comes from short-term memory.

Waugh and Norman (1965) developed a mathematical model to accomplish this. The model operated with the assumption that long-term memory occurs for the entire list, including a plateau in the middle of the list. In contrast, by the time of recall, short-term memory is said to remain only at the end of the list. This model assumes that, for any particular serial position within a list, the likelihood of successful short-term storage (S) and long-term storage (L) are independent, so that the likelihood of recalling the item is $S + L - SL$.

A slightly different assumption is that short- and long-term stores are not independent but are used in a complementary fashion. The availability of short-term memory of an item may allow resources needed for long-term memorization to be shifted to elsewhere in the list. The data seem more consistent with that assumption. In several studies, lists to be recalled have been presented to patients with Korsakoff's amnesia and normal control participants (Baddeley and Warrington, 1970; Carlesimo et al., 1995). These studies show that, in immediate recall, performance in amnesic individuals is preserved at the last few serial positions of the list. It is as if the performance in those serial positions is based mostly or entirely on short-term storage, and that there is no decrease in that kind of storage in the amnesic patients. In delayed recall, the amnesic patients show a deficit at all serial positions, as one would expect if short-term memory for the end of the list is lost as a function of a filled delay period (Glanzer and Cunitz, 1966).

Overcoming contamination from temporal distinctiveness

Last, it has been argued that the loss of memory over time is not necessarily the result of decay. Instead, it can be caused by temporal distinctiveness in retrieval. This kind of theory assumes that the temporal context of an item serves as a retrieval cue for that item, even in free recall. An item separated in time from all other items is relatively distinctive and easy to recall, whereas an item that is relatively close to other items is more difficult to recall because it shares their temporal cues to retrieval. Shortly after a list is presented the most recent items are the most distinct temporally (much like the distinctness of a telephone pole you are practically touching compared to poles extending further down the road). Across a retention interval, the relative distinctiveness of the most recent items decreases (much like standing far away from even the last pole in a series).

Although there are data that can be interpreted according to distinctiveness, there also are what look like dissociations between the effects of distinctiveness and a genuine short-term memory effect. One can see this, for example, in the classic procedure of Peterson and Peterson (1959) in which letter trigrams are to be recalled immediately or only after a distracting task, counting backward from a starting number by three, for a period lasting up to 18 s. Peterson and Peterson found severe memory loss for the letter trigram as the filled delay was increased. However, subsequently, sceptics argued that the memory loss occurred because the temporal distinctiveness of the current letter trigram diminished as the filled delay increased. In particular, this delay effect was said to occur because of the increase across test delays in the proactive interference from previous trials. On the first few trials, the delay does not matter (Keppel and Underwood, 1962) and no detrimental effect of delay is observed if delays of 5, 10, 15, and 20 s are tested in separate trial blocks (Turvey et al., 1970; Greene, 1996).

Yet, there may be a true decay effect at shorter test intervals. Baddeley and Scott (1971) set up a trailer in a shopping mall so that they could test a large number of participants for one trial each, so as to avoid proactive interference. They found an effect of the test delay within the first 5 s but not at longer delays. Still, it seems that the concept of decay is not yet on very firm ground and warrants further study. It may be that decay actually reflects not a gradual degradation of the quality of the short-term memory trace, but a sudden collapse at a point that varies from trial to trial. With a control for temporal distinctiveness, Cowan et al. (1997a) found what could be a sudden collapse in the representation of memory for a tone with delays of between 5 and 10 s.

Chunk capacity limits

The concept of capacity limits was raised several times in the history of cognitive psychology. Miller (1956) famously discussed the "magical number seven plus or minus two" as a constant in short-term processing, including list recall, absolute judgment, and numerical estimation experiments. However, his autobiographical essay (Miller, 1989) indicates that he was never very serious about the number seven; it was a rhetorical device that he used to tie together the otherwise unrelated strands of his research for a talk. Although it is true that memory span is approximately seven items in adults, there is no guarantee that each item is a separate entity. Perhaps the most important point of Miller's (1956) article was that multiple items can be combined into a larger, meaningful unit. Later studies suggested that the limit in capacity is more typically only three or four units (Broadbent, 1975; Cowan, 2001). That conclusion was based on an attempt to take into account strategies that often increase the efficiency of use of a limited capacity, or that allow the maintenance of additional information separate from that limited capacity. To understand these methods of discussing capacity limits I will again mention three types of contamination. These come from chunking and the use of long-term memory, from rehearsal, and from non-capacity-limited types of storage.

Overcoming contamination from chunking and the use of long-term memory

A participant's response in an immediate-memory task depends on how the information to be recalled is grouped to form multi-item chunks (Miller, 1956). Because it is not usually clear what chunks have been used in recall, it is not clear how many chunks can be retained and whether the number is truly fixed. Broadbent (1975) proposed some situations in which multi-item chunk formation was not a factor, and suggested on the basis of results from such procedures that the true capacity limit is three items (each serving as a single-item chunk). For example, although memory span is often about seven items, errors are made with seven-item lists and the error-free limit is typically three items. When people must recall items from a category in long-term memory, such as the states of the United States, they do so in spurts of about three items on average. It is as if the bucket of short-term memory is filled from the well of long-term memory and must be emptied before it is refilled. Cowan (2001) noted other such situations in which multi-item chunks cannot be formed. For example, in running memory span, a long list of items is presented with an unpredictable endpoint, making grouping impossible. When the list ends, the participant is to recall a certain number of items from the end of the list. Typically, people can recall three or four items from the end of the list, although the exact number depends on task demands (Bunting et al., 2006). Individuals differ in capacity, which ranges from about two to six items in adults (and fewer in children), and the individual capacity limit is a strong correlate of cognitive aptitude.

Another way to take into account the role of multi-item chunk formation is to set up the task in a manner that allows chunks to be observed. Tulving and Patkau (1962) studied free recall of word lists with various levels of structure, ranging from random words to well-formed English sentences, with several different levels of coherence in between. A chunk was defined as a series of words reproduced by the participant in the same order in which the words had been presented. It was estimated that, in all conditions, participants recalled an average of four to six chunks. Cowan et al. (2004) tried to refine that method by testing serial recall of eight-word lists, which were composed of four pairs of words that previously had been associated with various levels of learning (0, 1, 2, or 4 prior word–word pairings). Each word used in the list was presented an equal number of times (four, except in a non-studied control condition) but what varied was how many of those presentations were as singletons and how many were as a consistent pairing. The number of paired prior exposures was held constant across the four pairs in a list. A mathematical model was used to estimate the proportion of recalled pairs that could be attributed to the learned association (i.e., to a two-word chunk) as opposed to separate recall of the two words in a pair. This model suggested that the capacity limit was about 3.5 chunks in every learning condition, but that the ratio of two-word chunks to one-word chunks increased as a function of the number of prior exposures to the pairs in the list.

Overcoming contamination from rehearsal

The issue of rehearsal is not entirely separate from the issue of chunk formation. In the traditional concept of rehearsal (e.g., Baddeley, 1986), one imagines that the items are covertly articulated in the presented order at an even pace. However, another possibility is that rehearsal involves the use of articulatory processes in order to put the items into groups. In fact, Cowan et al. (2006a) asked participants in a digit span experiment how they carried out the task and by far the most common answer among adults was that they grouped the items; participants rarely mentioned saying the items to themselves. Yet, it is clear that suppressing rehearsal affects performance.

Presumably, the situations in which items cannot be rehearsed are for the most part the same as the situations in which items cannot be grouped. For example, Cowan et al. (2005) relied on a running memory span procedure in which the items were presented at the rapid rate of 4 per second. At that rate, it is impossible to rehearse the items as they are presented. Instead, the task is probably accomplished by retaining a passive store (sensory or phonological memory) and then transferring the last few items from that store into a more attention-related store at the time of recall. In fact, with a fast presentation rate in running span, instructions to rehearse the items is detrimental, not helpful, to performance (Hockey, 1973). Another example is memory for lists that were ignored at the time of their presentation (Cowan et al., 1999). In these cases, the capacity limit is close to the three or four items suggested by Broadbent (1975) and Cowan (2001).

It is still quite possible that there is a speech-based short-term storage mechanism that is by and large independent of the chunk-based mechanism. In terms of the popular model of Baddeley (2000), the former is the phonological loop and the latter, the episodic buffer. In terms of Cowan (1988, 1995, 1999, 2005), the former is part of activated memory, which may have a time limit due to decay, and the latter is the focus of attention, which is assumed to have a chunk capacity limit.

Chen and Cowan (2005) showed that the time limit and chunk capacity limit in short-term memory are separate. They repeated the procedure of Cowan et al. (2004) in which pairs of words sometimes were presented in a training session preceding the list recall test. They combined lists composed of pairs as in that study. Now, however, both free and serial recall tasks were used, and the length of list varied. For long lists and free recall, the chunk capacity limit governed the recall. For example, lists of six well-learned pairs were recalled as well as lists of six unpaired singletons (i.e., were recalled at similar proportions of words correct). For shorter lists and serial recall strictly scored, the time limit instead governed the recall. For example, lists of four well-learned pairs were not recalled nearly as well as lists of four unpaired singletons, but only as well as lists of eight unpaired singletons. For intermediate conditions it appeared as if chunk capacity limits and time limits operate together to govern recall. Perhaps the capacity-limited mechanism holds items and the rehearsal mechanism preserves some serial order memory for those held items. The exact way in which these limits work together is not yet clear.

Overcoming contamination from non-capacity-limited types of storage

It is difficult to demonstrate a true capacity limit that is related to attention if, as I believe, there are other types of short-term memory mechanisms that complicate the results. A general capacity should include chunks of information of all sorts: for example, information derived from both acoustic and visual stimuli, and from both verbal and nonverbal stimuli. If this is the case, there should be cross-interference between one type of memory load and another. However, the literature often has shown that there is much more interference between similar types of memoranda, such as two visual arrays of objects or two acoustically presented word lists, than there is between two dissimilar types, such as one visual array and one verbal list. Cocchini et al. (2002) suggested that there is little or no interference between dissimilar lists. If so, that would appear to provide an argument against the presence of a general, cross-domain, short-term memory store.

Morey and Cowan (2004, 2005) questioned this conclusion. They presented a visual array of colored spots to be compared to a second array that matched the first or differed from it in one spot's color. Before the first array or just after it, participants sometimes heard a list of digits that was then to be recited between the two arrays. In a low-load condition, the list was their own seven-digit telephone number whereas, in a high-load condition, it was a random seven-digit number. Only the latter condition interfered with array-comparison performance, and then only if the list was to be recited aloud between the arrays. This suggests that retrieving seven random digits in a way that also engages rehearsal processes relies upon some type of short-term memory mechanism that also is needed for the visual arrays. That shared mechanism may be the focus of attention, with its capacity limit. Apparently, though, if the list was maintained silently rather than being recited aloud, this silent maintenance occurred without much use of the common, attention-based storage mechanism, so visual array performance was not much affected.

The types of short-term memory whose contribution to recall may obscure the capacity limit

Melton, A.W. (1963) Implications of short-term memory for a general theory of memory. J. Verbal Learn. Verbal Behav., 2: 1–21.

Miller, G.A. (1956) The magical number seven, plus or minus two: some limits on our capacity for processing information. Psychol. Rev., 63: 81–97.

Miller, G.A. (1989) George A. Miller. In: Lindzey G. (Ed.), A History of Psychology in Autobiography, Vol. VIII. Stanford University Press, Stanford, CA, pp. 391–418.

Miller, G.A., Galanter, E. and Pribram, K.H. (1960) Plans and the structure of behavior. Holt, Rinehart and Winston, Inc., New York.

Morey, C.C. and Cowan, N. (2004) When visual and verbal memories compete: evidence of cross-domain limits in working memory. Psychon. Bull. Rev., 11: 296–301.

Morey, C.C. and Cowan, N. (2005) When do visual and verbal memories conflict? The importance of working-memory load and retrieval. J. Exp. Psychol. Learn. Mem. Cogn., 31: 703–713.

Nairne, J.S. (2002) Remembering over the short-term: the case against the standard model. Annu. Rev. Psychol., 53: 53–81.

Naveh-Benjamin, M., Cowan, N., Kilb, A. and Chen, Z. (2007) Age-related differences in immediate serial recall: dissociating chunk formation and capacity. Mem. Cognit., 35: 724–737.

Neath, I. and Surprenant, A. (2003) Human memory (2nd ed.). Wadsworth, Belmont, CA.

Nipher, F.E. (1878) On the distribution of errors in numbers written from memory. Trans. Acad. Sci. St. Louis, 3: ccx–ccxi.

Peterson, L.R. and Peterson, M.J. (1959) Short-term retention of individual verbal items. J. Exp. Psychol., 58: 193–198.

Saults, J.S. and Cowan, N. (2007) A central capacity limit to the simultaneous storage of visual and auditory arrays in working memory. J. Exp. Psychol., 136: 663–684.

Sperling, G. (1960) The information available in brief visual presentations. Psychol. Monogr., 74. (Whole No. 498.)

Stoltzfus, E.R., Hasher, L. and Zacks, R.T. (1996) Working memory and retrieval: an inhibition-resource approach. In: Richardson J.T.E., Engle R.W., Hasher L., Logie R.H., Stoltzfus E.R. and Zacks R.T. (Eds.), Working Memory and Human Cognition. Oxford University Press, New York, pp. 66–88.

Talmi, D., Grady, C.L., Goshen-Gottstein, Y. and Moscovitch, M. (2005) Neuroimaging the serial position curve: a test of single-store versus dual-store models. Psychol. Sci., 16: 716–723.

Tulving, E. and Patkau, J.E. (1962) Concurrent effects of contextual constraint and word frequency on immediate recall and learning of verbal material. Can. J. Psychol., 16: 83–95.

Turvey, M.T., Brick, P. and Osborn, J. (1970) Proactive interference in short-term memory as a function of prior-item retention interval. Q. J. Exp. Psychol., 22: 142–147.

Tzeng, O.J.L. (1973) Positive recency effect in a delayed free recall. J. Verbal Learn. Verbal Behav., 12: 436–439.

Unsworth, N. and Engle, R.W. (2007) The nature of individual differences in working memory capacity: active maintenance in primary memory and controlled search from secondary memory. Psychol. Rev., 114: 104–132.

Vogel, E.K., McCollough, A.W. and Machizawa, M.G. (2005) Neural measures reveal individual differences in controlling access to working memory. Nature, 438: 500–503.

Vogel, E.K., Woodman, G.F. and Luck, S.J. (2006) The time course of consolidation in visual working memory. J. Exp. Psychol. Hum. Percept. Perform., 32: 1436–1451.

Waugh, N.C. and Norman, D.A. (1965) Primary memory. Psychol. Rev., 72: 89–104.

Wickelgren, W.A. (1974) Single-trace fragility theory of memory dynamics. Mem. Cogn., 2: 775–780.

CHAPTER 21

Encoding-retrieval overlap in human episodic memory: a functional neuroimaging perspective

Michael D. Rugg*, Jeffrey D. Johnson, Heekyeong Park and Melina R. Uncapher

Center for the Neurobiology of Learning and Memory, and Department of Neurobiology and Behavior, University of California, Irvine, CA 92697-3800, USA

Abstract: The principle of transfer-appropriate processing and the cortical reinstatement hypothesis are two influential theoretical frameworks, articulated at the psychological and neurobiological levels of explanation, respectively, that each propose that the processes supporting the encoding and retrieval of episodic information are strongly interdependent. Here, we integrate these two frameworks into a single model that generates predictions that can be tested using functional neuroimaging methods in healthy humans, and then go on to describe findings that are in accord with these predictions. Consistent with the transfer-appropriate processing and cortical reinstatement frameworks, the neural correlates of successful encoding vary according to how retrieval is cued, and the neural correlates of retrieval are modulated by how items are encoded. Thus, encoding and retrieval should not be viewed as separate stages of memory that can be investigated in isolation from one another.

Keywords: cortical reinstatement; transfer-appropriate processing; fMRI; recollection; subsequent memory

Introduction

Episodic memory is the collection of processes that together support the recollection of unique events (Tulving, 1983). By 'unique event', we are referring to events that are individuated by their contexts, such as where you parked your car today rather than yesterday. A key feature of episodic memories is the establishment of associations between the different elements of an event; in the example above, remembering where your car is currently parked depends on retrieval of associations between the act of parking the car, the location of the act, and when it occurred. Episodic memory can be contrasted with two other types of explicit (conscious) memory in which contextual associations play little or no role. *Semantic memory* supports general knowledge about the world that is acquired through repeated exposure to the same information in a variety of different contexts, such that memory for the information becomes contextually aspecific (Tulving, 1972). And memory based on a sense of *familiarity* can support simple recognition, recency, or frequency judgments, but provides no access to qualitative information about a prior event, and hence no specific information about where or when the event occurred (for review, see Yonelinas, 2002).

In the present paper, we bring together two frameworks — one rooted in experimental psychology and the other in neurobiology — that have each been highly influential in guiding ideas

*Corresponding author. Tel.: +1 949 824 8861;
Fax: +1 949 824 4807; E-mail: mrugg@uci.edu

about how episodic memory operates. Both frameworks propose that there is an intimate relationship between the processes set in train when an event is experienced, and those that are engaged when it is later remembered. Although they derive from different experimental traditions, and are articulated at different levels of explanation, the two frameworks are highly complementary. Integrating the frameworks leads to an account of episodic memory encoding and retrieval that can be framed at the explanatory level of cognitive neuroscience. This account generates predictions about the neural correlates of episodic encoding and retrieval that can be tested in healthy humans with functional neuroimaging methods, notably, event-related functional magnetic resonance imaging (fMRI).

Below, we briefly describe the two frameworks and our attempt to integrate them into a single model. After noting some important caveats, we then describe recent findings from fMRI studies of episodic encoding and retrieval, and discuss how these findings can be understood in the context of the model.

Transfer-appropriate processing

For more than 40 years, experimental psychologists have investigated the relationship between the encoding and retrieval of episodic information, emphasizing the interdependency of these seemingly distinct mnemonic functions (e.g., Tulving and Thomson, 1973; Morris et al., 1977). One outcome of this line of research has been the principle of *transfer-appropriate processing* (TAP) (Morris et al., 1977). TAP is predicated on the twin assumptions that memories are represented in terms of the cognitive operations engaged by an event as it is initially processed, and that successful memory retrieval occurs when those earlier operations are recapitulated (Kolers, 1973; for review, see Roediger et al., 2002). According to the first of these assumptions, the same event will ostensibly give rise to different memory representations depending on which aspects of the event are emphasized or attended at study. According to the second assumption, the effectiveness of a retrieval cue will depend on the similarity between the processing engaged by the cue and the processing that occurred during encoding: the greater the similarity (or 'study-test overlap'), the greater the likelihood of successful retrieval (Roediger et al., 1989; Roediger and Guynn, 1996; see Nairne, 2002, for caveats). From this perspective, the question of what constitutes the most effective way of encoding information into memory can be fully addressed only if the retrieval conditions are also specified. Similarly, the question of what constitutes an effective retrieval cue has little meaning if the processes engaged during encoding are undefined.

Cortical reinstatement and episodic retrieval

The idea at the heart of TAP principle, namely that episodic retrieval involves the reinstatement of processes that were active at the time of encoding, is also found in several neurobiologically based models of memory retrieval (e.g., Alvarez and Squire, 1994; Rolls, 2000; Shastri, 2002; Norman and O'Reilly, 2003). According to such models, recollection of a recent event occurs when a pattern of cortical activity corresponding to the event is reinstated by activation of a hippocampally stored representation of that pattern. Through this mechanism, anatomically distinct cortical regions that were concurrently active during the online processing of an event will also be co-activated during its retrieval, preserving associations between the features of the event that, together, make it distinct from other similar occurrences. In the model of Norman and O'Reilly (2003), for example, memory encoding occurs through the establishment of a sparsely encoded representation of the pattern of cortical activity engendered by an event in the hippocampal CA3 region. Retrieval occurs through the reactivation of this representation, which in turn leads to the reinstatement of the pattern of cortical activity encoded in the representation. Crucially, reactivation of the hippocampal representation does not depend on the presence of perfect overlap between the originally encoded activity and the activity engendered during a retrieval attempt. Because the hippocampus is highly effective at pattern completion (Marr, 1971; Wallenstein et al., 1998), activity that only partially overlaps the encoded information can be sufficient to cause reactivation of the entire representation, and hence reinstatement of the original cortical pattern. Because of this pattern-completion mechanism, memories can be retrieved in response to retrieval cues that elicit activity only partially resembling the activity elicited by the original event.

The complementary nature of TAP and reinstatement theory

The principles of TAP and cortical reinstatement share several key concepts. These include the idea that memory retrieval involves the recapitulation of processes and representations that were active during encoding, and that the likelihood of successful retrieval is a function of the extent to which the processing engaged by a retrieval cue overlaps with that engaged at encoding.

Figure 1 illustrates one way in which these ideas can be schematized in terms of large-scale patterns of brain activity. The figure attempts to capture the twin ideas that the retrieval of a prior episode involves reinstatement of the pattern of neural activity engaged during the original experience, and that retrieval cues need only elicit a fraction of that original activity in order to trigger the reinstatement of the entire pattern. Predictions about encoding- and retrieval-related neural activity arising from this framework are outlined below. Before doing so, however, some caveats and qualifications are in order.

(i) Encoding–retrieval overlap is represented in Fig. 1 in terms of co-activation of cortical regions on a relatively coarse spatial scale concordant with what can be detected with fMRI. Cortical activity can, however, be differentiated on a considerably more fine-grained spatial scale, and in terms of the temporal dynamics as well as the loci of the activated neural elements. For example, online processing of two different faces would likely be differentiated not by activity in distinct cortical regions, but by differences in the patterning of activity within a common region (or set of regions; e.g., Haxby et al., 2001). Thus, the task of the hippocampus, or any other structure whose function is to capture and later reinstate patterns of activity associated with the processing of an event, is considerably more complex than simply registering which cortical areas were co-activated by the event.

(ii) Figure 1 is far from being a complete specification of the component processes underlying successful retrieval. Indeed, if this is all that there was to retrieval it is unclear how we would be able to distinguish between the perception of an event and our later memory of it! Furthermore, if retrieval occurred every time there was overlap between current and past processing we would be in a state of almost continuous remembering, which would be highly maladaptive. For example, it is not helpful when struggling to park your car in the sole remaining parking spot to be distracted by a vivid recollection of the last time you parked in the same lot. As noted by Tulving (1983), these and related considerations suggest that episodic retrieval is subject to some kind of control mechanism. Tulving proposed that stimulus events are processed as retrieval cues only when the individual adopts a specific cognitive state or 'set', which he termed *retrieval mode*. By this proposal, depending on whether or not retrieval mode is engaged, the same stimulus event will be processed either as an episodic retrieval cue or in terms of its online significance. In addition, it is only when retrieval mode is engaged that information about an experience will be attributed to the past (Wheeler et al., 1997). It is currently unclear how this and related ideas about retrieval processing (e.g., Rugg and Wilding, 2000) should be incorporated into the framework outlined in Fig. 1.

Another important class of retrieval processes operates on the products of a

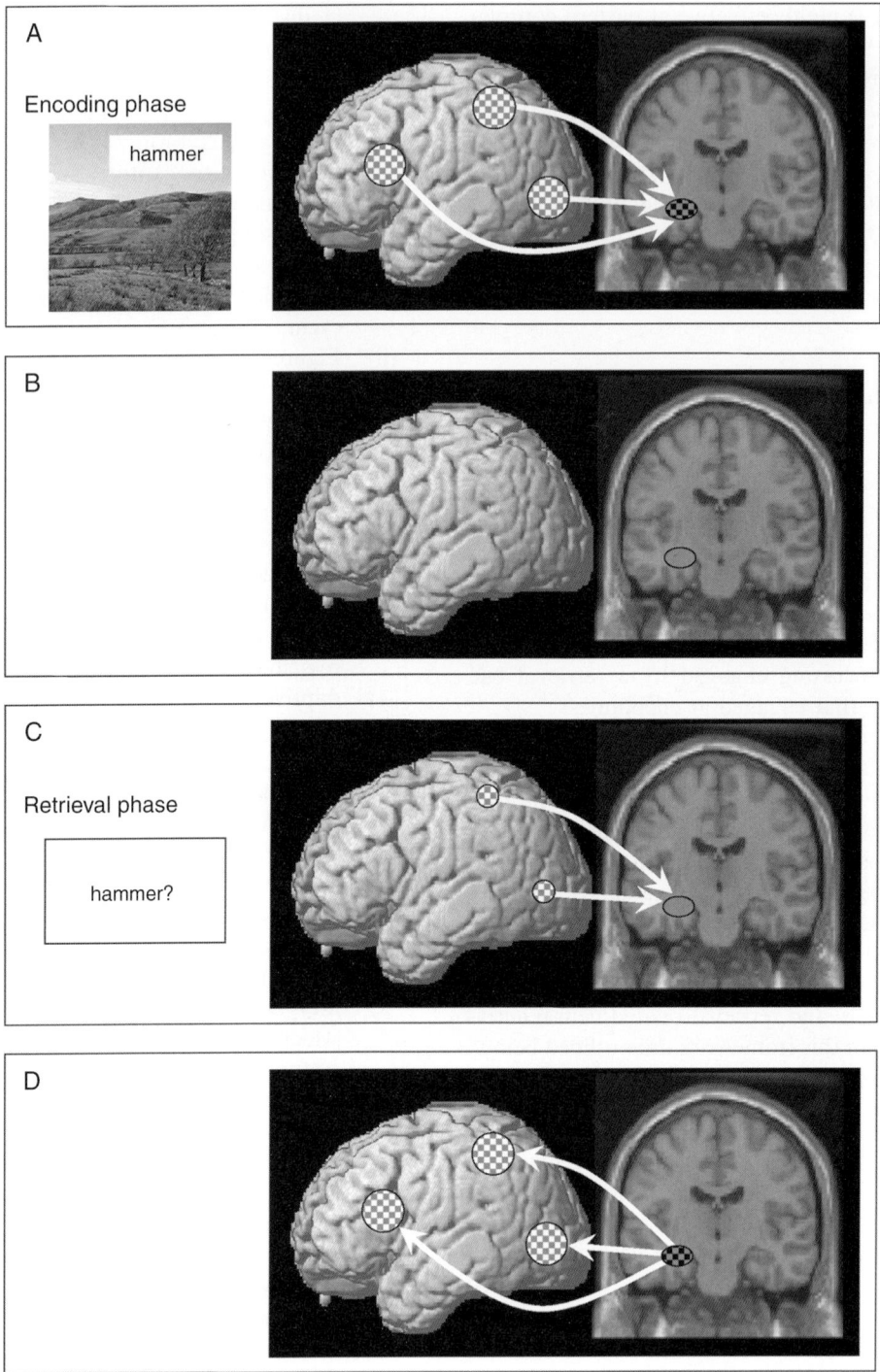

Fig. 1. Schematic depiction of the proposed relationship between encoding- and retrieval-related processing in episodic memory. (A) Presentation of a stimulus event (a word superimposed on a scene) activates a diverse set of cortical regions. The resulting pattern of cortical activity is encoded by the hippocampus. (B) Following the event, a representation of the pattern of activity that it elicited is stored in the hippocampus. (C) Subsequent presentation of part of the event (the *retrieval cue*) leads to partial reinstatement of the original pattern of activity, which feeds forward to the hippocampus. (D) Overlap between the activity elicited by the cue and the stored pattern of activity causes the hippocampal representation to be re-activated, which in turn leads to full reinstatement at the cortical level.

retrieval attempt (for review, see Rugg, 2004). Without these 'post-retrieval' processes, which connect the products of retrieval to current behavioral goals, episodic memory would have little impact on goal-directed behavior. These processes too are absent from Fig. 1.

(iii) The framework outlined in Fig. 1 implies that retrieval consists of little more than the 'replaying' of the processing engaged by the original experience. If this was actually the case, then recollection would be 'all or nothing'; setting aside inefficiency resulting from system noise, either everything that was registered in the brain as an event unfolded would be retrieved, or nothing would be. Moreover, memories would be largely veridical. Clearly, neither of these scenarios is accurate. Memories are a very imperfect mirror of experience; they are invariably partial, and often highly distorted, records of the original event (e.g., Bartlett, 1932; Loftus and Palmer, 1974; Schacter, 2002). Among the many factors contributing to this imperfect relation between an event and our later memory of it, two stand out. First, the different features of an event are not equally likely to be successfully encoded. Other things being equal, those aspects of the event that are attended the most fully are most likely to be later remembered (Moscovitch, 1992). Second, there is a wealth of evidence indicating that episodic retrieval is a constructive process, in which retrieved information is combined with other knowledge about the event and the result then interpreted in light of current expectations and biases (e.g., Bransford and Franks, 1971; Brewer and Treyens, 1981; Brewer, 1987; Schacter et al., 1998). Thus, retrieved episodic information may only partially determine the content of the resulting 'memory'. Together, these two factors will act to reduce the amount of overlap between encoding- and retrieval-related neural activity to well below the 100% illustrated in Fig. 1. Specifically, the modulatory role of attention during encoding implies that cortical activity elicited during an event by attended information is more likely to be incorporated into the resulting hippocampal representation than is the activity elicited by unattended information. And the constructive nature of memory retrieval means that retrieval-related neural activity will reflect not only the reinstatement of activity elicited at encoding but also activation of whatever non-episodic information is recruited into the 'constructed' memory representation.

Empirical findings

The foregoing caveats aside, the framework illustrated in Fig. 1 leads to a number of predictions about encoding- and retrieval-related neural activity that can be tested in functional neuroimaging studies in healthy humans. Below, we outline these predictions and discuss relevant fMRI data, focusing on recent findings from our laboratory.

Before moving on to the data, a brief methodological comment is in order. Several of the studies discussed below investigate the neural correlates of encoding with variants of the 'subsequent memory' procedure (Paller and Wagner, 2002). In this procedure, which was first employed with fMRI by Wagner et al. (1998) and Brewer et al. (1998), the neural activity elicited by a series of study items is segregated according to performance on a later memory test, for example, according to whether items are accurately recognized, or whether recognized items attract accurate versus inaccurate source memory judgments. The procedure therefore permits identification of regions where activity at study 'predicts' later memory performance. Regions that are more active for study items later accorded accurate rather than inaccurate memory judgments are assumed to have played a role in encoding the relevant attribute or attributes of the items.

The loci of cortical regions supporting successful encoding differ according to the online processing engaged by an event

According to the TAP principle, encoding is a 'by-product' of online processing, and therefore there is no single cognitive system subserving memory encoding (Lockhart, 1992). Thus, the loci of encoding-related cortical activity should reflect the cognitive operations engaged by a study task, rather the engagement of a dedicated 'encoding circuit'. fMRI evidence supports this prediction (Otten and Rugg, 2001; Otten et al., 2002; Mitchell et al., 2004; for review, see Rugg et al., 2002). In the study of Otten and Rugg (2001), for example, each of a series of study words was preceded by a cue that instructed the subjects to perform either a semantic (living/non-living) or a phonological (odd/even number of syllables) judgment. For both tasks, items that were recognized with high confidence on a memory test undertaken some 15 min later elicited greater study activity than did items that failed to be recognized. As is illustrated in Fig. 2, the loci of these 'subsequent memory effects' differed according to study task. Whereas effects associated with the semantic task were localized primarily to medial and left inferior frontal cortex, effects for the syllable task predominated in posterior cortical regions (see Otten et al., 2002, for similar findings). Importantly, the regions demonstrating the subsequent memory effects in each case overlapped with regions that were engaged selectively by the two task requirements, as identified by differences in the activity elicited by all of the items subjected to each class of study judgment. Thus, in accordance with the TAP principle, these findings suggest that regions supporting successful memory encoding vary according to the nature of the online processing engaged by study items, and constitute a subset of the regions engaged in service of those processing demands.

Findings from a second study (Uncapher et al., 2006) lead to a similar conclusion. In this experiment, subjects were scanned while they studied words presented in one of four different possible locations and in one of four different colors. A subsequent memory test required the subjects to discriminate studied from unstudied items (now all presented in a single color and at the center of the display) and, for each recognized item, to indicate the color and location in which it had been studied. Using as a baseline the study activity elicited by items that were later recognized but for which neither color nor location information could be retrieved correctly ('recognition only' items), we investigated whether successful retrieval of each of the two contextual features was associated with distinct subsequent memory effects. In addition, we asked whether there were any brain regions where subsequent memory effects were specific for study items for which both features were later retrieved. The key findings are illustrated in Fig. 3. Relative to items attracting recognition only judgments, study items whose location was accurately remembered elicited additional activity in, among other regions, retrosplenial cortex, whereas items whose color was remembered accurately elicited greater study activity in posterior inferior temporal cortex. These regions have been previously implicated in the processing of location information and color knowledge, respectively (Chao and Martin, 1999; Kellenbach et al., 2001; Mayes et al., 2004; Frings et al., 2006). The findings for location and color memory thus provide strong support for the idea that successful encoding of the different attributes of an event is supported by the cortical regions engaged to process those attributes online.

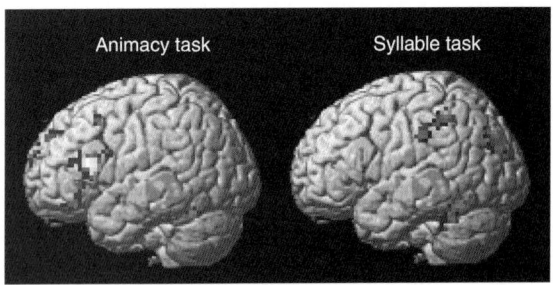

Fig. 2. Data from Otten and Rugg (2001) illustrating the differently localized subsequent memory effects (thresholded at $p < .001$) associated with words subjected to animacy judgments (left) and syllable judgments (right). Results are overlaid on a rendered canonical brain. (See Color Plate 21.2 in color plate section.)

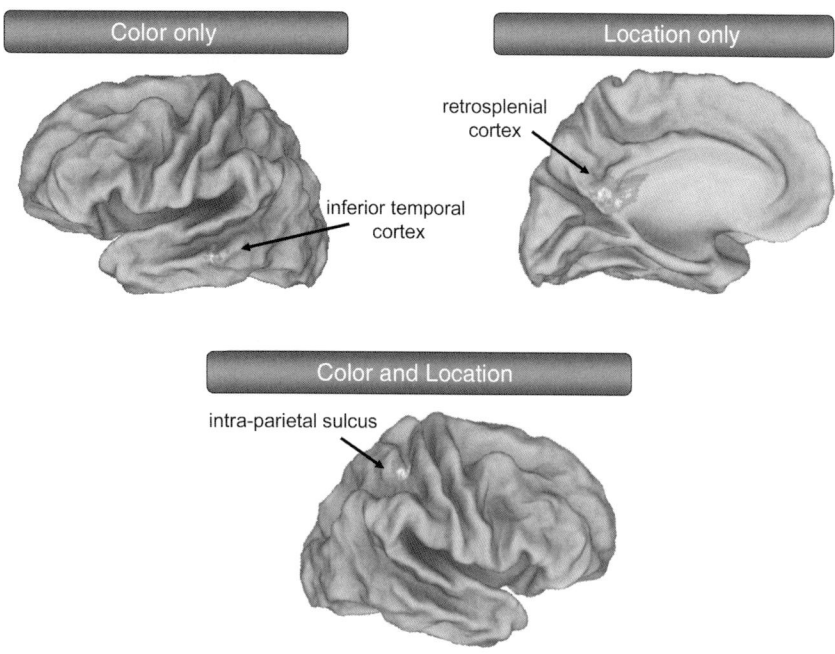

Fig. 3. Data from Uncapher et al. (2006) illustrating feature-specific and multi-featural subsequent memory effects. Top panel: subsequent memory effects (relative to correctly recognized study words for which no contextual feature could be accurately recalled) associated with accurate memory for the location (left) and color (right) of the study word. Bottom panel: Subsequent memory effects in the right intra-parietal sulcus (IPS) uniquely associated with accurate memory for *both* color and location. All effects thresholded at $p<.001$. Results are displayed onto PALS atlas (Van Essen, 2005) with Caret5 Software (Van Essen et al., 2001; http://brainmap.wustl.edu/caret). (See Color Plate 21.3 in color plate section.)

Figure 3 also illustrates a region — right intra-parietal sulcus (IPS) — where subsequent memory effects were uniquely associated with items whose color and location were conjointly remembered on the later test. These effects indicate that conjoint encoding of the two features was not simply the result of the concurrent engagement of encoding operations supporting memory for each feature alone. Rather, it appears that conjoint encoding engaged processing that operated across color and location information, perhaps binding the two features into a single memory representation. The pattern of behavioral performance on the retrieval test is consistent with this proposal. As was also reported in two prior behavioral studies employing a similar task (Meiser and Broder, 2002; Starns and Hicks, 2005), the proportion of items for which the two contextual features were both accurately remembered was greater than that predicted from the proportions of trials where memory was accurate for only one of the features. This finding is consistent with the idea that recollection of both features together benefited from a mechanism separate from the one supporting memory for the individual features. The locus of the subsequent memory effect illustrated in Fig. 3 provides an important clue about the workings of this mechanism. A wealth of evidence points to a key role for the IPS in attentionally mediated perceptual binding (Humphreys, 1998; Shafritz et al., 2002; Cusack, 2005; for review, see Robertson, 2003). In light of this evidence, Uncapher et al. (2006) suggested that the IPS subsequent memory effect in Fig. 3 reflected the mnemonic benefit of allocating attention to the 'object level' rather than to individual features of a stimulus event. This allowed the features to be conjoined in a unitized perceptual representation, and hence to be encoded together in a single memory representation.

Together, the findings reviewed in this section provide strong evidence in support of one of the central tenets of TAP. The distinction between processes subserving memory encoding on the one hand, and online processing on the other, does not appear to be honored at the cortical level.

The loci of cortical regions supporting successful encoding differ according to how memory is tested

According to the TAP principle, a retrieval cue will be effective when the processing it elicits recapitulates or overlaps processing engaged during encoding. Thus, not only should the cortical loci supporting encoding differ according to the nature of the study task (see above), but these loci should also vary depending on how memory is subsequently cued. The question whether subsequent memory effects vary according to retrieval conditions has received only limited attention, but there are a few relevant findings. For example, Uncapher and Rugg (2005) reported a regional double-dissociation between subsequent memory effects associated with successful recollection of visually presented study words (as operationalized by the 'Remember/Know' procedure; Tulving, 1985) as a function of study-test delay (30 min vs. 48 h). Effects in a left lateral fusiform region were found only when memory was tested after the short delay, whereas study items recollected after the longer delay elicited subsequent memory effects in bilateral ventral frontal cortex. These findings were interpreted as evidence for a delay-dependent reliance on study processes emphasizing perceptual versus semantic features of the study words, respectively. A more direct demonstration of the sensitivity of encoding-related activity to a retrieval manipulation is provided by the findings of Otten (2007). Memory for visually presented words was assessed with a Remember/Know recognition test in which the retrieval cues were either spoken words or pictures. Eight regions were identified, distributed widely over the cortical surface, where subsequent memory effects were uniquely associated with successful recollection in response to word (three regions) or picture (five regions) retrieval cues. Otten (2007) suggested that the regions supported the encoding of different attributes of the study episode that were differentially accessible to the two classes of retrieval cue.

In a recent study we investigated whether, as predicted by TAP, retrieval cues are most effective when the processing they engage resembles the processing engaged at the time of encoding (Park and Rugg, in press). We took advantage of the 'cue-congruency' effect, the well-established finding that memory performance is higher when study and test materials are presented in the same format (e.g., word–word) than when formats are crossed (word–picture). Subjects studied a randomly intermixed series of visually presented words and pictures, judging whether the referent of each item would be more likely to be found indoors or outdoors. In a later test phase outside the scanner, memory for the study items was tested using a Remember/Know recognition test. The test items also comprised words and pictures, permitting half of each class of the study items to be congruently cued (word–word and picture–picture) and the other half to be incongruently cued (word–picture and picture–word). As expected, memory performance for congruently cued study items was markedly higher than it was for incongruently cued items. This effect was found only for recognized items that were endorsed as 'Remembered', and not for items recognized on the basis of familiarity alone (i.e., items accorded 'Know' judgments), indicating that congruency selectively modulated the likelihood of recollection (episodic retrieval).

To identify encoding-related activity associated with these behavioral cue-congruency effects we contrasted the activity elicited at study by items that were later successfully recognized in response to congruent versus incongruent cues. We predicted that these neural cue-congruency effects would overlap with the activity elicited by the retrieval cues themselves. Since memory for the study items was tested outside the scanner, this prediction could not be tested directly. Instead, we contrasted the activity elicited by the two classes of study item, identifying regions that were selectively activated for each class of items (words or pictures) during encoding regardless of how memory was later cued. We could then ask whether regions exhibiting material-selective activity overlapped with regions demonstrating material-specific cue-congruency effects. That is, we investigated whether studied words that were later retrieved with congruent cues were processed in a more 'word-like' manner during encoding than words retrieved with incongruent cues, which had less potential for high levels of study-test overlap. Analogously, we also investigated whether the study activity elicited by congruently cued pictures was more 'picture-like' than the activity elicited by pictures later retrieved with incongruent cues. The key findings are illustrated in Fig. 4. Studied words that were later recognized with a congruent cue elicited greater activity than incongruently cued words in a part of the left superior temporal cortex that also showed greater activity in response to words than pictures. By contrast, study activity elicited by congruently cued pictures was greater than the activity elicited by incongruently cued pictures in right parietal and occipito-temporal cortex, regions that overlapped with those where pictures elicited more activity than words. These findings suggest that congruent retrieval cues are more effective than incongruent cues only when study and retrieval processing are aligned. The findings provide strong support both for the general proposal that the neural correlates of successful encoding depend upon how memory is later tested, and the more specific hypothesis that retrieval cues are most effective when their processing overlaps or recapitulates processing engaged during study.

In summary, although only a few studies have specifically addressed the question whether encoding-related activity varies depending on how memory is cued at retrieval, the findings are promising. They provide persuasive support for the idea that the cortical substrates of successful encoding can be defined only in relation to specific retrieval conditions. As was noted by Otten (2007), these findings also have important implications for the interpretation of between-group differences in the localization of encoding-related neural activity, such as those reported in studies of aging (e.g., Morcom et al., 2003; Gutchess et al., 2005). These differences are usually interpreted as evidence for age-related differences in the cognitive/neural processes engaged during encoding. The findings reviewed in this section raise the possibility, however, that differences in the localization of subsequent memory effects may reflect group-wise differences in the processing engaged at retrieval rather than at encoding.

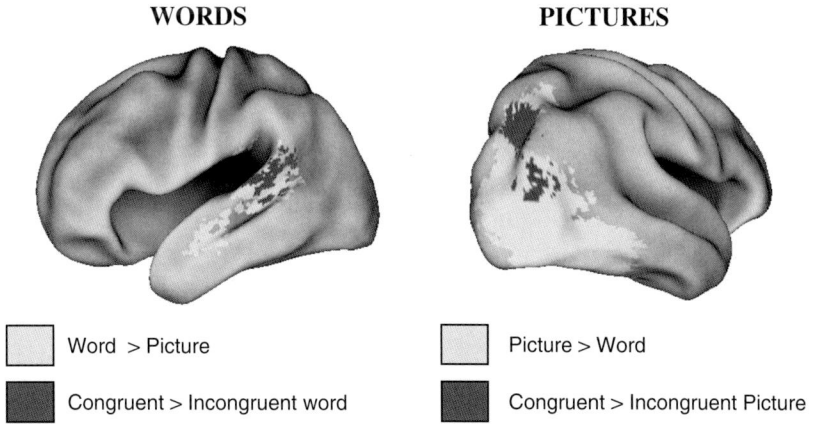

Fig. 4. Data from Park and Rugg (in press) illustrating overlap between cue-congruency effects and the main effect of study material. Left panel: Overlap between word congruency effects and regions where activity was greater for words than for pictures. Right panel: Overlap between picture congruency effects and regions activity was greater for pictures than for words. Results are displayed onto PALS atlas by using Caret5 (Van Essen et al., 2001; http://brainmap.wustl.edu/caret). Cue-congruency effects thresholded at $p < .01$, material effects thresholded at $p < .005$.

Successful retrieval is associated with reinstatement of cortical activity elicited at encoding

The most obvious prediction that can be derived from Fig. 1, central to both TAP and the cortical reinstatement hypothesis, is that the patterns of neural activity elicited during encoding and retrieval should overlap. This prediction has motivated a number of functional neuroimaging studies over the past few years. In several studies employing blocked experimental designs, it was reported that cortical regions where activity differed during the encoding of one stimulus class relative to another were also differentially active during later memory tests for the same stimulus class (Nyberg et al., 2000, 2001; Persson and Nyberg, 2000; Vaidya et al., 2002). Although these findings are consistent with the cortical reinstatement hypothesis, they are largely inconclusive: the blocked designs of the studies leave open the possibility that the material-specific retrieval effects reflect the attempt to retrieve targeted memories, rather than the consequences of successful retrieval.

Findings from more recent event-related fMRI studies also suggest that the neural correlates of recollection overlap with encoding-related activity (Wheeler et al., 2000, 2006; Wheeler and Buckner, 2003, 2004; Gottfried et al., 2004; Kahn et al., 2004; Khader et al., 2005; Woodruff et al., 2005). However, findings from four of these studies (Wheeler et al., 2000, 2006; Wheeler and Buckner, 2003; Khader et al., 2005) were based on the retrieval of items that had been studied on multiple occasions, raising the possibility that the effects reflect a form of learning distinct from that supporting memory for unique events. Furthermore, in three of the studies (Gottfried et al., 2004; Wheeler and Buckner, 2004; Wheeler et al., 2006), evidence for the content sensitivity of the neural correlates of recollection took the form of a single association between cortical activity and retrieved content, rather than the more convincing evidence provided by a double-dissociation in patterns of encoding-retrieval overlap associated with two different classes of content. Two studies employing trial-unique study presentations reported content-specific double-dissociations of recollection-related activity, in both cases showing a crossover pattern in two different cortical regions according to the nature of the recollected information (Kahn et al., 2004; Woodruff et al., 2005). These findings constitute powerful evidence that the neural correlates of recollection are, at least in part, content-specific. However, in neither study was the relation between recollection- and encoding-related activity assessed within-subjects. Thus, the findings provide only indirect support for the reinstatement hypothesis.

In a recent study, we subjected the cortical reinstatement hypothesis to what is arguably its strongest test to date (Johnson and Rugg, 2007). Subjects studied a series of words, which were presented in the context of one of two distinct encoding tasks designed so as to differentially engage multiple cortical regions. In one task, words were superimposed on a picture of a landscape scene, and subjects were required to imagine the object depicted by the word at any location within the scene. In the other task, words were presented on a blank background, with the requirement to covertly generate a sentence that incorporated the word. Memory for the words was later tested with a Remember/Know procedure (Tulving, 1985), permitting identification of items for which recognition was accompanied by the recollection of details surrounding the study episode. fMRI data were obtained at both study and test, allowing the encoding- and recollection-related activity associated with the two classes of study episode to be directly compared.

The principal findings from this study are illustrated in Fig. 5. As shown in the upper panel of the figure, the two encoding tasks elicited robustly different patterns of neural activity in multiple cortical regions. The lower panel of Fig. 5 illustrates regions where recollection-related activity that was specific to words from each encoding condition overlapped with the corresponding encoding-related activity. Four regions were identified. Three of these regions, in left occipital and fusiform cortex, were selectively activated during recollection of words from the 'scene' task. The remaining region, in ventromedial prefrontal cortex, was engaged by recollection of words that had been presented in the 'sentence' task. These

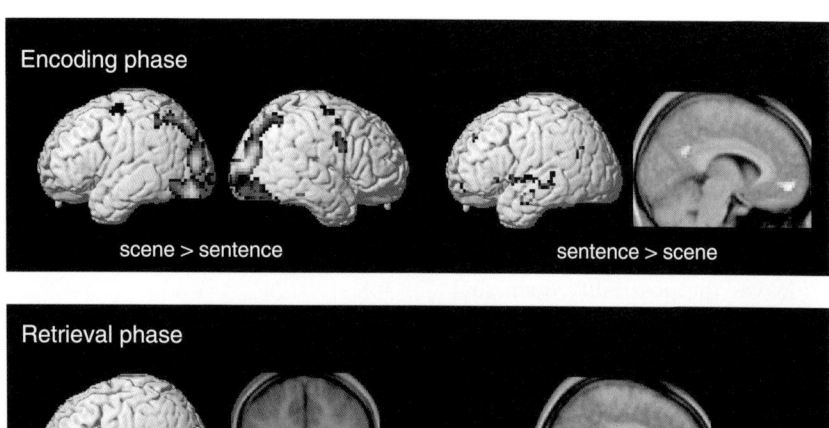
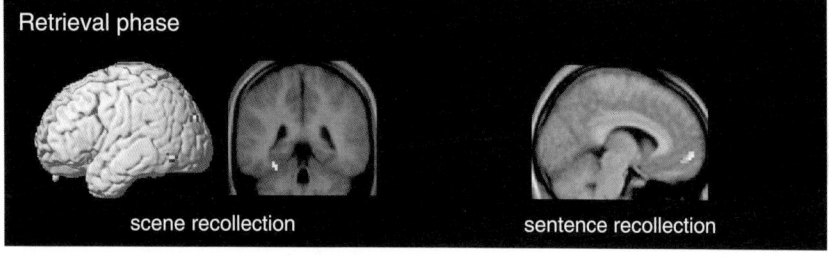

Fig. 5. Data from Johnson and Rugg (2007) illustrating overlap of content-specific brain activity during encoding and recollection. Top panel: Regions where activity differed during the encoding of words in a scene task compared to a sentence-generation task. Bottom panel: Regions where recollection-related activity (remember > know; thresholded at $p < .001$) was greater for words from one encoding condition ($p < .05$) and overlapped with the corresponding contrast showing greater encoding-related activity for those words ($p < .001$). Results are overlaid on a rendered canonical brain (with the cerebellum artificially removed) and sagittal sections of the across-subjects mean of normalized T_1-weighted images. (See Color Plate 21.5 in color plate section.)

findings represent a double-dissociation between cortical regions demonstrating encoding-retrieval overlap for two different classes of study episode.

Together with previous results, the present findings provide strong support for the cortical reinstatement hypothesis. At the same time, they highlight the question of why content-selective recollection-related activity overlaps with such a small fraction of the cortical area activated during encoding. One obvious possibility is that the methods used to detect the neural activity associated with recollection lack power, and that alternative methods of fMRI analysis (e.g., multi-voxel pattern classification; Polyn et al., 2005; for review, see Norman et al., 2006) may reveal more extensive effects. This seems unlikely to be the whole answer, however; even when assessed at a markedly more liberal statistical threshold, overlap between recollection- and encoding-related activities increased by only a small amount. We think that the limited overlap illustrated in Fig. 5 reflects the contribution of two more interesting factors. First, as was noted previously, the 'gating'

influence of attention during encoding means that it is very unlikely that all of the activity elicited by a stimulus event is incorporated into a memory representation. Thus, only an unknown fraction of the activity illustrated in the upper panel of Fig. 5 would have been 'available' for reinstatement at retrieval. Second, we assume that not all of the information encoded about an event is equally likely to be retrieved on any given retrieval attempt. In particular, given that the memory test did not explicitly require the retrieval of highly differentiated information about each word's study context, it is possible that little effort was made to recollect such details.

As already noted, the findings discussed in this section support the widely held view that episodic retrieval involves reinstatement of neural activity elicited during the occurrence of the encoded event, allowing the event to be 're-experienced'. An important caveat remains, however (for similar concerns, see Maratos et al., 2001; Kahn et al., 2004; Woodruff et al., 2005; Johnson and Rugg, 2007). The findings reviewed above all come from studies that employed hemodynamic imaging techniques (mainly fMRI). Although the spatial resolution that can be achieved with these methods is very impressive, their temporal resolution, which is of the order of several hundred milliseconds, is much less impressive. Thus, it is not possible to accurately determine the onset of recollection-related fMRI effects in relation to the timing of the retrieval cue. This leaves the possibility open that fMRI 'reinstatement effects' are a consequence, rather than a cause, of successful recollection. By this argument, recollection depends on processes associated with activity in regions distinct from those exhibiting content-specific effects; these effects, in turn, reflect post-recollective processes — such as the deployment of attention toward particular types of recollected information or the maintenance of that information in working memory — that operate in service of task goals. Resolution of this issue will require employment of methods for measuring retrieval-related neural activity — such as those based on electro- or magneto-encephalography — that have higher temporal resolution than fMRI.

Concluding comments

The findings reviewed in the preceding three sections offer broad support for the framework illustrated in Fig. 1. The framework emphasizes the intimacy of the relationship between episodic encoding and retrieval that exists at both psychological and neural levels of analysis, and serves a useful role in integrating the two levels. As was discussed in "The complementary nature of TAP and reinstatement theory", however, Fig. 1 offers what is at best a simplistic and incomplete account that neglects many subtleties. Also, it has little or nothing to say about the attentional processes that influence what gets encoded, or the control processes responsible for either intentional retrieval processing or the employment of retrieved information in service of behavioral goals. The incorporation of all of these processes into the framework is a major challenge for the cognitive neuroscience of episodic memory.

Acknowledgments

The research described in the chapter was supported by the Wellcome Trust, the Medical Research Council of the United Kingdom, and the National Institute of Mental Health (grant numbers R01-MH072966 and R01-MH074528).

References

Alvarez, P. and Squire, L.R. (1994) Memory consolidation and the medial temporal lobe: a simple network model. Proc. Natl. Acad. Sci. U.S.A., 91: 7041–7045.

Bartlett, F.C. (1932) Remembering: A Study in Experimental and Social Psychology. Cambridge University Press, Cambridge, England.

Bransford, J.D. and Franks, J.J. (1971) The abstraction of linguistic ideas. Cognit. Psychol., 2: 331–350.

Brewer, J.B., Zhao, Z., Desmond, J.E., Glover, G.H. and Gabrieli, J.D.E. (1998) Making memories: brain activity that predicts how well visual experience will be remembered. Science, 281: 1185–1187.

Brewer, W.F. (1987) Schemas versus mental models in human memory. In: Morris P. (Ed.), Modelling Cognition. Wiley, Chichester, pp. 187–197.

Brewer, W.F. and Treyens, J.C. (1981) Role of schemata in memory for places. Cognit. Psychol., 13: 207–230.

Chao, L.L. and Martin, A. (1999) Cortical regions associated with perceiving, naming, and knowing about colors. J. Cogn. Neurosci., 11: 25–35.

Cusack, R. (2005) The intraparietal sulcus and perceptual organization. J. Cogn. Neurosci., 17: 641–651.

Frings, L., Wagner, K., Quiske, A., Schwarzwald, R., Spreer, J., Halsband, U. and Schulze-Bonhage, A. (2006) Precuneus is involved in allocentric spatial location encoding and recognition. Exp. Brain Res., 173: 661–672.

Gottfried, J.A., Smith, A.P.R., Rugg, M.D. and Dolan, R.J. (2004) Remembrance of odors past: human olfactory cortex in cross-modal recognition memory. Neuron, 42: 687–695.

Gutchess, A.H., Welsh, R.C., Hedden, T., Bangert, A., Minear, M., Liu, L. and Park, D.C. (2005) Aging and the neural correlates of successful picture encoding: frontal activations compensate for decreased medial temporal activity. J. Cogn. Neurosci., 17: 84–96.

Haxby, J.V., Gobbini, M.I., Furey, M.L., Ishai, A., Schouten, J.L. and Pietrini, P. (2001) Distributed and overlapping representations of faces and objects in ventral temporal cortex. Science, 293: 2425–2430.

Humphreys, G.W. (1998) Neural representation of objects in space: a dual coding account. Philos. Trans. R. Soc. Lond. B Biol. Sci., 353: 1341–1351.

Johnson, J.D. and Rugg, M.D. (2007) Recollection and the reinstatement of encoding-related cortical activity. Cereb. Cortex., 17: 2507–2515.

Kahn, I., Davachi, L. and Wagner, A.D. (2004) Functional-neuroanatomic correlates of recollection: implications for models of recognition memory. J. Neurosci., 28: 4172–4180.

Kellenbach, M.L., Brett, M. and Patterson, K. (2001) Large, colorful, or noisy? Attribute- and modality-specific activations during retrieval of perceptual attribute knowledge. Cogn. Affect. Behav. Neurosci., 1: 207–221.

Khader, P., Burke, M., Bien, S., Ranganath, C. and Rosler, F. (2005) Content-specific activatoin during associative long-term memory retrieval. Neuroimage, 27: 805–816.

Kolers, P.A. (1973) Remembering operations. Mem. Cognition, 1: 347–355.

Lockhart, R.S. (1992) Levels of processing. In: Squire L.R. (Ed.), Encyclopedia of Learning and Memory. Macmillan, New York, pp. 106–108.

Loftus, E.F. and Palmer, J.C. (1974) Reconstruction of automobile destruction. J. Verbal Learn. Verbal Behav., 13: 585–589.

Maratos, E.J., Dolan, R.J., Morris, J.S., Henson, R.N.A. and Rugg, M.D. (2001) Neural activity associated with episodic memory for emotional context. Neuropsychologia, 39: 910–920.

Marr, D. (1971) Simple memory: a theory for archicortex. Philos. Trans. R. Soc. London Ser. B, 262: 23–81.

Mayes, A.R., Montaldi, D., Spencer, T.J. and Roberts, N. (2004) Recalling spatial information as a component of recently and remotely acquired episodic or semantic memories: an fMRI study. Neuropsychology, 18: 426–441.

Meiser, T. and Broder, A. (2002) Memory for multidimensional source information. J. Exp. Psychol. Learn. Mem. Cogn., 28: 116–137.

Mitchell, J.P., Macrae, C.N. and Banaji, M.R. (2004) Encoding-specific effects of social cognition on the neural correlates of subsequent memory. J. Neurosci., 24: 4912–4917.

Morcom, A.M., Good, C.D., Frackowiak, R.S. and Rugg, M.D. (2003) Age effects on the neural correlates of successful memory encoding. Brain, 126: 213–229.

Morris, C.D., Bransford, J.D. and Franks, J.J. (1977) Levels of processing versus transfer appropriate processing. J. Verbal Learn. Verbal Behav., 16: 519–533.

Moscovitch, M. (1992) Memory and working with memory: a component process model based on modules and central systems. J. Cogn. Neurosci., 4: 257–267.

Nairne, J.S. (2002) The myth of the encoding-retrieval match. Memory, 10: 389–395.

Norman, K.A. and O'Reilly, R.C. (2003) Modeling hippocampal and neocortical contributions to recognition memory: a complementary-learning-systems approach. Psychol. Rev., 110: 611–646.

Norman, K.A., Polyn, S.M., Detre, G.J. and Haxby, J.V. (2006) Beyond mind-reading: multi-voxel pattern analysis of fMRI data. Trends. Cogn. Sci., 10: 424–430.

Nyberg, L., Habib, R., McIntosh, A.R. and Tulving, E. (2000) Reactivation of encoding-related brain activity during memory retrieval. Proc. Natl. Acad. Sci. U.S.A., 97: 11120–11124.

Nyberg, L., Petersson, K.M., Nilsson, L.G., Sandblom, J., Aberg, C. and Ingvar, M. (2001) Reactivation of motor brain areas during explicit memory for actions. Neuroimage, 14: 521–528.

Otten, L.J. (2007) Fragments of a larger while: retrieval cues constrain observed neural correlates of memory encoding. Cereb. Cortex, 17: 2030–2038.

Otten, L.J., Henson, R.N.A. and Rugg, M.D. (2002) State-related and item-related neural correlates of successful memory encoding. Nat. Neurosci., 5: 1339–1344.

Otten, L.J. and Rugg, M.D. (2001) Task-dependency of the neural correlates of episodic encoding as measured by fMRI. Cereb. Cortex, 11: 1150–1160.

Paller, K.A. and Wagner, A.D. (2002) Observing the transformation of experience into memory. Trends Cogn. Sci., 6: 93–102.

Park, H. and Rugg, M.D. (in press) The relationship between study processing and the effects of cue congruency at retrieval: fMRI support for transfer appropriate processing. Cereb. Cortex.

Persson, J. and Nyberg, L. (2000) Conjunction analyses of cortical activations common to encoding and retrieval. Microsc. Res. Tech., 51: 39–44.

Polyn, S.M., Natu, V.S., Cohen, J.D. and Norman, K.A. (2005) Category-specific cortical activity precedes retrieval during memory search. Science, 310: 1963–1966.

Robertson, L. (2003) Binding, spatial attention and perceptual awareness. Nat. Rev. Neurosci., 4: 93–102.

Roediger, H.L., Gallo, D.A. and Geraci, L. (2002) Processing approaches to cognition: the impetus from the levels of processing framework. Memory, 10: 319–332.

Roediger, H.L. and Guynn, M.J. (1996) Retrieval processes. In: Bjork E.L. and Bjork R.A. (Eds.), Memory. Academic Press, San Diego, pp. 197–236.

Roediger, H.L., Weldon, M.S. and Challis, B. (1989) Explaining dissociations between implicit and explicit measures of retention: a processing account. In: Roediger H. and Craik F.I.M. (Eds.), Varieties of Memory and Consciousness: Essays in Honor of Endel Tulving. Erlbaum, Hillsdale, pp. 3–41.

Rolls, E.T. (2000) Memory systems in the brain. Annu. Rev. Psychol., 51: 599–630.

Rugg, M.D. (2004) Retrieval processing in human memory: electrophysiological and fMRI evidence. In: Gazzaniga M.S. (Ed.), The Cognitive Neurosciences (3rd ed.). MIT press, Cambridge, pp. 727–738.

Rugg, M.D., Otten, L.J. and Henson, R.N. (2002) The neural basis of episodic memory: evidence from functional neuroimaging. Philos. Trans. R. Soc. Lond. B Biol. Sci., 357: 1097–1110.

Rugg, M.D. and Wilding, E.L. (2000) Retrieval processing and episodic memory. Trends. Cogn. Sci., 4: 108–115.

Schacter, D.L. (2002) The Seven Sins of Memory: How the Mind Forgets and Remembers. Houghton Mifflin, New York.

Schacter, D.L., Norman, K.A. and Koutstaal, W. (1998) The cognitive neuroscience of constructive memory. Annu. Rev. Psychol., 49: 289–318.

Shafritz, K.M., Gore, J.C. and Marois, R. (2002) The role of the parietal cortex in visual feature binding. Proc. Natl. Acad. Sci. U.S.A., 99: 10917–10922.

Shastri, L. (2002) Epidodic memory and cortico-hippocampal interactions. Trends Cogn. Sci., 6: 162–168.

Starns, J.J. and Hicks, J.L. (2005) Source dimensions are retrieved independently in multidimensional monitoring tasks. J. Exp. Psychol. Learn. Mem. Cogn., 31: 1213–1220.

Tulving, E. (1972) Episodic and semantic memory. In: Tulving E. and Donaldson W. (Eds.), Organization of Memory. Academic Press, New York, pp. 381–403.

Tulving, E. (1983) Elements of Episodic Memory. Oxford University Press, New York.

Tulving, E. (1985) Memory and consciousness. Can. J. Psychol., 26: 1–12.

Tulving, E. and Thomson, D.M. (1973) Encoding specificity and retrieval processes in episodic memory. Psychol. Rev., 80: 352–373.

Uncapher, M.R., Otten, L.J. and Rugg, M.D. (2006) Episodic encoding is more than the sum of its parts: an fMRI investigation of multifeatural contextual encoding. Neuron, 52: 547–556.

Uncapher, M.R. and Rugg, M.D. (2005) Encoding and the durability of episodic memory: a functional magnetic resonance imaging study. J. Neurosci., 25: 7260–7267.

Vaidya, C.J., Zhao, M., Desmond, J.E. and Gabrieli, J.D.E. (2002) Evidence for cortical encoding specificity in episodic memory: memory-induced re-activation of picture processing areas. Neuropsychologia, 40: 2136–2143.

Van Essen, D.C. (2005) A population-average, landmark- and surface-based (PALS) atlas of human cerebral cortex. Neuroimage, 28: 635–662.

Van Essen, D.C., Dickson, J., Harwell, J., Hanlon, D., Anderson, C.H. and Drury, H.A. (2001) An integrated software system for surface-based analysis of cerebral cortex. J. Am. Med. Int. Assoc., 28: 443–459.

Wagner, A.D., Schacter, D.L., Rotte, M., Koutstaal, W., Maril, A., Dale, A.M., Rosen, B. and Buckner, R.L. (1998) Building memories: remembering and forgetting of verbal experiences as predicted by brain activity. Science, 281: 1188–1191.

Wallenstein, G.V., Eichenbaum, H. and Hasselmo, M.E. (1998) The hippocampus as an associator of discontiguous events. Trends Neurosci., 21: 317–323.

Wheeler, M.A., Stuss, D.T. and Tulving, E. (1997) Toward a theory of episodic memory: the frontal lobes and autonoetic consciousness. Psychol. Bull., 121: 331–354.

Wheeler, M.E. and Buckner, R.L. (2003) Functional dissociation among components of remembering: control, perceived oldness, and content. J. Neurosci., 23: 3869–3880.

Wheeler, M.E. and Buckner, R.L. (2004) Functional-anatomic correlates of remembering and knowing. Neuroimage, 21: 1337–1349.

Wheeler, M.E., Petersen, S.E. and Buckner, R.L. (2000) Memory's echo: vivid remembering reactivates sensory-specific cortex. Proc. Natl. Acad. Sci. U.S.A., 97: 11125–11129.

Wheeler, M.E., Shulman, G.L., Buckner, R.L., Miezin, F.M., Velanova, K. and Petersen, S.E. (2006) Evidence for separate perceptual reactivation and search processes during remembering. Cereb. Cortex, 16: 949–959.

Woodruff, C.C., Johnson, J.D., Uncapher, M.R. and Rugg, M.D. (2005) Content-specificity of the neural correlates of recollection. Neuropsychologia, 43: 1022–1032.

Yonelinas, A.P. (2002) The nature of recollection and familiarity: a review of 30 years of research. J. Mem. Lang., 46: 441–517.

CHAPTER 22

Cognitive aging and increased distractibility: costs and potential benefits

M. Karl Healey[1], Karen L. Campbell[1,2] and Lynn Hasher[1,2,*]

[1]*Department of Psychology, University of Toronto, Toronto, ON M5S 3G3, Canada*
[2]*The Rotman Research Institute, Baycrest, 3560 Bathurst Street, Toronto, ON, M6A 2E1, Canada*

Abstract: Older adults show a characteristic pattern of impaired and spared functioning relative to younger adults. Elsewhere we have argued that many age-related changes in cognitive function are rooted in an impaired ability to inhibit irrelevant information and inappropriate responses. In this chapter we review evidence that as a direct result of impaired inhibitory processes, older adults tend to be highly susceptible to distraction. We suggest that because the distinction between relevant and irrelevant is seldom either clear or static, distractibility can manifest as either a cost or a benefit depending on the situation. We review evidence that in situations in which it interferes with the current task, distraction is disproportionately detrimental to older adults compared to university aged adults, but that when previously distracting information becomes relevant, older adults show a benefit whereas younger adults do not.

Keywords: cognitive aging; inhibition; suppression; distractibility

A person's cognitive abilities and intellectual proficiencies are not static across their life but rather follow a distinct developmental trajectory. The most salient phase of this trajectory may be the rapid increase in ability during childhood and adolescence, followed by changes in ability as people age. Relative to young adults (aged 18–30), older adults (typically 65+ years old) do more poorly on laboratory tests measuring a variety of abilities such as attention control (Cohn et al., 1984; Gazzaley et al., 2005; Hasher et al., 2007), working memory (May et al., 1999), and long-term memory (Grady and Craik, 2000; Park et al., 2002).

Despite the fact that older adults experience difficulty with a wide variety of cognitive tasks, aging is not characterized by a generalized decline in cognitive ability. For example, older adults show performance comparable to (or in some cases superior to) younger adults on tests of semantic memory and verbal ability (older adults routinely outperform younger adults on vocabulary tests: Park et al., 2002; Verhaeghen, 2003), and decision making (Kim and Hasher, 2005; Kim et al., 2005; Mather, 2006; Peters et al., 2007). Moreover, as will be seen, factors that contribute to poor performance in some situations can lead to superior performance in other situations. Therefore, aging is characterized by changes in cognitive function that manifest as a distinct pattern of preserved, impaired, and occasionally enhanced performance.

A major goal of cognitive aging research has been to determine what underlies this pattern of impaired and preserved functioning, and to this end, a number of theories have been advanced.

*Corresponding author. Tel.: +1 416 978 1557;
Fax: +1 416 978 4811; E-mail: Hasher@psych.utoronto.ca

For example, it has been suggested that older adults' difficulties stem from a general reduction in processing speed (Salthouse, 1996), or from structural and functional changes in the prefrontal cortex (West, 1996, 2000); still others have argued that age-related changes in memory arise from a deficit in the ability to form associations between the various aspects of an episode (e.g., a fact and the context in which it was learned: Naveh-Benjamin, 2000; Naveh-Benjamin et al., 2004a, b; Oberauer, 2005). Our own view and that of an increasing number of researchers is that reductions in the ability of attention to regulate distraction underlies many age-related deficits (Hasher and Zacks, 1988; Hasher et al., 1999, 2007). In this chapter we discuss some of the unique predictions that derive from this view and review the relevant empirical evidence. We begin with a brief overview of the attentional dysregulation account of cognitive aging.

The environment constantly presents us with massive amounts of information and it is far beyond our ability to actively process all of it and carefully consider all possible responses. However, to accomplish our goals we must process at least some of this information and decide upon an appropriate response. Therefore, one of the primary obstacles to successful information processing and interaction with the environment is winnowing the relevant from the irrelevant, the appropriate from the inappropriate. We argue that this obstacle is overcome by using attention to actively inhibit information that is currently irrelevant, and to suppress prepotent but momentarily inappropriate responses. That is, we view inhibitory processes as narrowing the scope of information processing and the resulting overt responses by excluding information and responses that are situationally inappropriate (for a more detailed exposition of this view, see Hasher and Zacks, 1988; Hasher et al., 1999, 2007).

A direct consequence of this inhibitory deficit is that older adults will actually process more total information than will younger adults, with a greater proportion of that information being irrelevant. To the extent that successful task performance depends on selectively attending to only relevant information, older adults will be disadvantaged relative to younger adults. That is, in situations that demand a narrow focus of attention, older adults are likely to be more distracted than are younger adults. However, in some situations it is not clear which information is relevant and which is irrelevant and it is also often the case that information that was irrelevant at one point in time becomes relevant at a later point. In situations such as these, a tendency to process irrelevant information can actually be beneficial, with older adults ultimately showing better performance than younger adults.

In the following sections we review work, by ourselves and colleagues, demonstrating the various impacts that increased distractibility has on older adults' cognition. First we will consider ways in which distractibility can hamper the performance of older adults. Then we will consider ways in which distractibility (or perhaps more appropriately, a wide scope of attention) can enhance performance.

Disruptive effects of distraction

On processing speed

Older adults are substantially slower than younger adults on simple measures of processing speed such as the rate at which a participant can compare two strings of letters and determine whether they are identical or not. Performance on these tasks accounts for a considerable proportion of age-related variance on memory tasks such as free recall and paired associate learning, prompting the claim that reduced processing speed is a major cause of age-related cognitive impairments (see Salthouse, 1996, for a review). However, even though the basic task in most processing speed measures is quite simple (e.g., compare two letter strings), many such items are actually presented on a single page, producing a cluttered display. If older adults are especially vulnerable to distraction, such clutter could have a negative impact on their processing speed. That is, distractibility and not reduced processing speed per se may cause older adults to be slow on these measures.

To test the idea that distraction partially determines older adults' performance on speed measures, Lustig et al. (2006) created low and high distraction versions of two common speed measures. In the letter comparison task (Salthouse and Babcock, 1991), participants are shown two strings of three, six, or nine letters (e.g., RXL___RXL) and must indicate whether the two strings are the same or different. The standard version of the task consists of two pages each with 21 pairs of strings. For the high distraction version, Lustig et al. presented the strings on a computer with 24 pairs per screen (48 total). For the low distraction condition they presented the same 48 pairs but one at a time (each stayed onscreen until the participant responded).

Unsurprisingly, reaction time increased as the number of letters per string increased, and overall, older adults were slower than younger adults (see Fig. 1). More interestingly, the older but not younger adults were faster in the low distraction condition than in the high distraction condition. That is, the standard presentation of multiple items in a single cluttered array disproportionately slowed older adults' reaction times and thus exaggerated age differences in processing speed. Moreover, older adults' performance on the high distraction version of the computerized task showed higher correlations with their performance on more traditional pen and paper speed tests than did performance on the low distraction version.

Lustig et al. (2006) performed a similar distraction reducing manipulation on another widely used speed task, The Symbol Digit Substitution test (Royer et al., 1981). It consists of a page with 90 unfamiliar symbols and participants must substitute a digit (1–9) for each symbol according to a provided translation key. The speed with which they can carry out this translation provides a measure of processing speed. Lustig et al.'s high distraction version presented 93 symbols on each screen, whereas the low distraction version presented the same symbols but one at a time. As with the letter comparison task, younger adults were equally fast across distraction conditions (far right column of Fig. 1), but older adults, though slower than younger adults overall, were considerably faster in the low distraction condition than in the high distraction condition.

Lustig et al. (2006) showed that older adults are indeed especially susceptible to distraction and that this susceptibility contributes to slowing on tasks once thought to be relatively pure measures of processing speed. While use of an uncluttered display did not completely eliminate age-related slowing, it did reduce it substantially, clearly

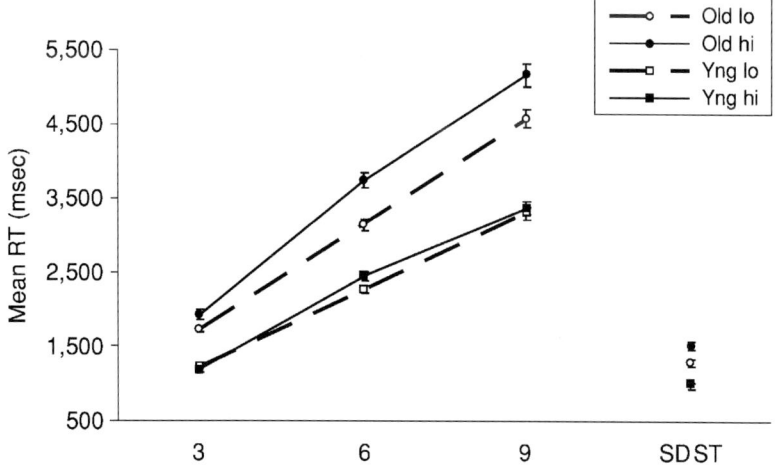

Fig. 1. Mean per item reaction time on the letter comparison task and the symbol digit substitution task (SDST; far right) as a function of age and distraction (hi or lo). Adapted from Lustig et al. (2006). Published by the Psychonomic Society. (Reprinted with permission.)

illustrating the influence distraction can have on older adults.

On reading speed

The speed measures discussed in the previous section assess the rapidity with which one can complete a simple but unfamiliar cognitive task. But even in a familiar, well-learned task the presence of distracting information can have a differentially disruptive impact on the performance of older adults. That is, even if older adults usually have no problem completing a given task and show minimum deficits relative to younger adults, performing the task under conditions of unusually high distraction will cause a precipitous decline in their performance while having a much smaller impact on the performance of younger adults.

One domain that should be particularly familiar for older adults is reading. Most older adults have a lifetime of reading experience and are quite skilled at it. However, sometimes older adults have to read a given piece of text in the midst of other, distracting, information. A common example is reading a newspaper or magazine: the article you are interested in is surrounded by other articles and advertisements that could potentially draw your attention away from the relevant text. To determine whether distracting information does indeed impede older adults' reading, Connelly et al. (1991) used a task in which participants had to read a short story with irrelevant text interspersed with the relevant text (the irrelevant text was distinguished by a different type face; see Fig. 2A). They found that for control stories in which no distracting information was presented, older adults' reading speeds were only slightly slower than younger adults', but that when distracting words and phrases, which were related to the meaning of the relevant text, were included (text-related condition), the older adults were disproportionately slowed (see Fig. 3). Once again, introducing distraction to a task increased age differences in processing speed.

In a second experiment, Connelly et al. (1991) manipulated the nature of the distracting information. In the text-related condition the information was related to the content of the story (as in Experiment 1); in the text-unrelated condition words were also used as distraction but were unrelated to the story; in the x-string condition, strings of X's were interspersed with the text (see Fig. 2B). As seen in Fig. 4, both older and younger adults were reliably slowed relative to the no distraction control by all forms of distraction, but in each case the effect of distraction was greater for older adults. Note that for both groups, distracting words produced more slowing than did x-strings, but that for younger adults it did not matter whether the words were related to the text or not, in contrast, related words produced more slowing

```
         A                              B
              The Dig                         The Dig
    The car lottery ride wheel was    The car xxxxxxx ride xxxxxxx
    getting basket bumpy outside      was getting xxxxxxx bumpy
    now that George video had left    xxxxxxx now that George xxxxxxx
    addition the main trainer         had left xxxxxxx the main
    highway to use notion the dirt    xxxxxxx highway to use xxxxxxx
    silence road. He stamp was out    the dirt xxxxxxx road. He
    river of school, assembly not     xxxxxxx was out xxxxxxx of
    having lottery to study stamp     school, xxxxxxx not having
    during the summer notion break.   xxxxxxx to study xxxxxxx during
                                      the summer xxxxxxx break.
```

Fig. 2. Examples of the reading with distraction task. The text-related condition is shown on the left and the x-string condition is shown on the right.

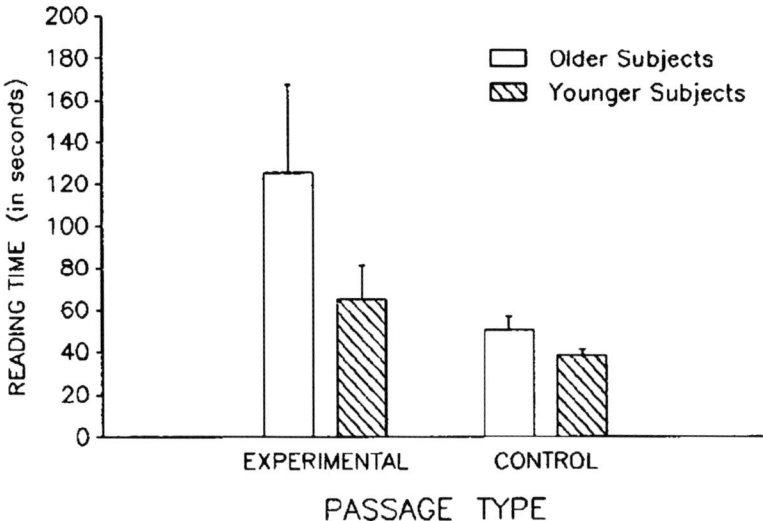

Fig. 3. Reaction times for older and younger adults on the reading with distraction task (experimental) and the control condition (Control) where no distracting information was interspersed with the relevant text. Adapted from Connelly et al. (1991). Published by APA. (Reprinted with permission.)

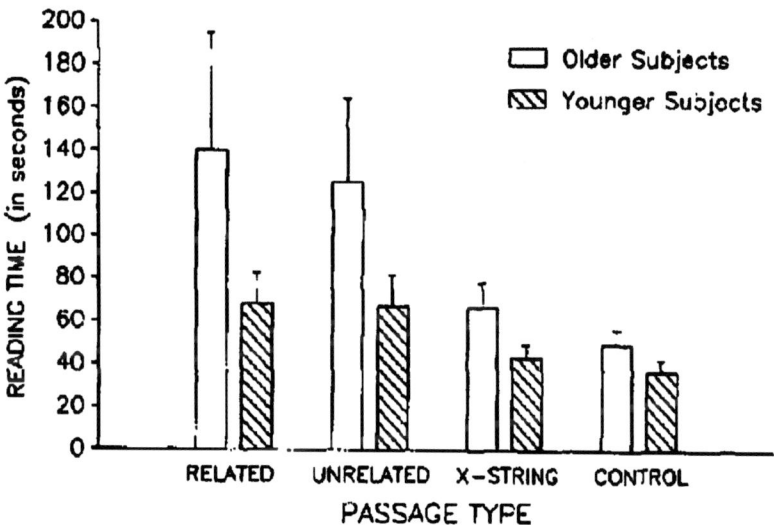

Fig. 4. Reaction times for older and younger adults on text-related, text-unrelated, x-string, and control passages. Adapted from Connelly et al. (1991). Published by APA. (Reprinted with permission.)

than unrelated words for older adults. This differential effect of relatedness suggests that while younger adults were distracted by the words, they were able to ignore any relation to the text material; older adults apparently processed the meaning of the words and incurred an additional cost when they were related to the main text. Overall, distraction was much more disruptive for older adults (a difference of ~90s between the control and related conditions) than for younger adults (~30s). Thus, Connelly et al. showed that even if older adults are highly familiar and skilled with a task, their performance is extremely sensitive to distraction.

On problem solving

The two studies discussed thus far have focused on the speed of cognitive processing. But slowed performance is not the only effect of increased susceptibility to distraction. For instance, May (1999) demonstrated that distraction impacts not only how *fast* older adults process information, but also the way in which the information is processed. Participants performed the remote associates task (RAT: Mednick, 1962) in which they are shown three words that are (distantly) related to a fourth word and they must produce this fourth linking word (e.g., "space" for the three words, Ship, Outer, and Crawl). In the standard task, the word triplets are presented alone, but on some trials May presented ostensibly irrelevant words along with the word triplet (one for each word in the triplet, presented below the relevant word). These distracting words were in fact related to the triplet word to which they were paired and suggested a meaning inconsistent with the relevant meaning of its paired triplet word. For example, inconsistent distractors for "ship, outer, crawl" would be "ocean, inner, floor" which do not suggest the "space" meaning of the triplet words. Participants were told that paying attention to the distractors would always impede solving the RAT problem. The rationale of the design is that if older adults are unable to ignore distraction, they should attend to the distracting words, which would prime an irrelevant meaning of the RAT words and make detection of the link between the words less likely.

There is evidence that the efficiency of inhibitory processes varies in a circadian fashion, and that there are individual and age differences in the time of peak efficiency; generally being in the morning for older adults and the evening for university aged adults (Hasher et al., 1999; Yoon et al., 2007). By testing participants at peak and off-peak times of day, May (1999) capitalized on this circadian variation in distractibility to determine whether susceptibility to distraction influences the impact of irrelevant information on RAT solution rates within an age group.

Younger adults tested at their peak time of day were not influenced by the distracting information. In contrast, older adults tested at their peak time of day did show a significant cost of distraction: they solved approximately 10% fewer RAT problems when inconsistent distractors were presented compared to the no-distraction condition. However, when tested at off-peak times of day, both age groups showed a negative impact of distraction. Younger adults tested in the morning were no longer able to successfully ignore the distracting information, leading them to solve approximately 10% fewer RAT problems when distraction was present. Older adults tested in the evening were even more impaired by the distraction, solving approximately 17% fewer RAT problems when distraction was present. Thus, May (1999) demonstrated that older adults' susceptibility to distraction extends beyond simple slowing; older adults are not only distracted by irrelevant information, they process the information and it influences the products (i.e., RAT solutions) of their cognitive processing. Moreover, susceptibility to distraction varies across the day, and even younger adults can incorporate irrelevant information into their information processing when tested at off-peak times.

Neural correlates of distractibility

The behavioral evidence reviewed above clearly indicates that older adults have difficulty controlling distraction, and there is now some evidence regarding the neural signatures of older adults' increased susceptibility to distraction. For example, Jonides et al. (2000) used the recent negatives task, which requires participants to resolve interference between relevant and irrelevant memory traces. On each trial of the task, participants are shown a set of four letters to remember followed by a probe letter. Participants must indicate if the probe matches any of the letters from the memory set. There are two types of trials that require negative responses: non-recent negatives on which the probe was not a member of any recent memory set, and recent negatives on which the probe is not a member of the current set, but had been a member of the just previous set. Thus, correct responding on recent negative trials requires participants to ignore the familiarity of the probe and to say "no." For younger adults, reaction times on recent negative trials were slower than on non-recent negative trials, and the former were associated with increased activity in the left lateral prefrontal cortex. Older adults showed even greater slowing to recent negatives relative to non-recent negative trials but showed *less* activation in the same left lateral prefrontal area. These results suggest that older adults' inability to ignore currently irrelevant information (in this case the familiarity of previous trial items) is related to decreased activation in the prefrontal cortex areas believed to be responsible for distraction control.

Further evidence of the neural correlates of distractibility comes from a study by Gazzaley et al. (2005). They presented participants with a series of faces and scenes (e.g., a sunset). In one condition participants were told to ignore the faces and remember the scenes, in another condition they were told to remember the faces and ignore the scenes. Gazzaley et al. found that when they were told to remember the scenes and ignore the faces, both older and younger adults showed increased activity in the parahippocampal place area (PPA), a region known to be involved in the processing of scenes, relative to a control condition where participants passively viewed the pictures. In contrast, when told to ignore the scenes, activity in the PPA decreased below baseline, but only for younger adults: older adults did not show decreased activity when scenes were irrelevant. That is, when scenes were relevant, there were no age differences in PPA activity but when scenes were irrelevant distraction, younger adults suppressed PPA activity but older adults did not. Moreover, the extent to which activity in the PPA was suppressed when remembering faces, predicted memory accuracy. Thus, one neural signature of older adults' distractibility appears to be activity in processing areas that fails to discriminate between relevant and irrelevant information.

Taken together, the Jonides et al. (2000) and Gazzaley et al. (2005) studies suggest that older adults are impaired at distraction control abilities mediated by the prefrontal cortex and this impairment results in processing of information regardless of relevance as reflected by indiscriminant activation in processing areas such as the PPA.

Fortuitous effects of distraction

To this point we have discussed the negative consequences associated with being distracted by irrelevant information. However, outside the laboratory the delineation between irrelevant and relevant is often fuzzy and tends to change unpredictably. For example, if you are reading a journal article with the aim of finding evidence to support a claim you want to make in a paper you are writing, any data or arguments not directly relevant to your claim could be considered distraction that should be inhibited. When you begin to write your next paper, however, some of the data that were previously a distraction may now be very relevant. Indeed, the ability to connect disparate, seemingly unrelated (i.e., mutually irrelevant) ideas might be one key aspect of creativity (Peterson and Carson, 2000). In this light the distractibility resulting from older adults' inefficient inhibitory processes can be seen as a wider scope of attention. If this wider scope leads to processing information that interferes with the current task, a cost is incurred, but it is also possible that a less constrained focus of attention may lead to processing information that initially seems irrelevant but later turns out to be quite relevant, either to the current task or to a subsequent task; in such cases distractibility may actually be beneficial. In the following sections we will review several empirical demonstrations of the positive consequences of distractibility.

Benefiting from concurrent distraction

In addition to investigating how distraction can impede RAT performance, May (1999) tested whether distracting information can improve RAT performance. This was accomplished by including a condition in which distractors were consistent rather than inconsistent with the relevant meanings of the RAT triplet words. For example, for "space, outer, crawl" relevant distractors would be "rocket, atmosphere, attic," all of which suggest the "space" meaning of the triplet words. The pattern of results found when these consistent distractors were presented was virtually a mirror image of the inconsistent distractor data. For younger adults tested at peak times, distraction had no impact on the likelihood of reaching a solution. In contrast, older adults tested at peak times were marginally more likely to solve the problems if distraction was present. When participants were tested at off-peak times the effects of distraction were increased: both younger and older adults were more likely to solve the RAT problems if consistent distractors were present. The effect of consistent distractors clearly illustrates that ostensibly irrelevant, distracting information does not always impair cognitive processing. Moreover, it is not simply a case of older adults always attending to distraction but younger adults flexibility attending or ignoring depending on the situation: when their distraction control abilities were at peak efficiency, younger adults effectively ignored the distraction even though it would have been beneficial for them to attend to it.

Benefiting from previously relevant information

Although proactive interference is usually associated with a decrement in performance on conventional cognitive tasks, there may be certain situations in which previously relevant information becomes relevant once again. One such situation was demonstrated by May and Hasher (1998), who showed that older but not younger adults can make use of non-relevant words in a later task. In the first phase of that study, young and older adults generated endings to normatively high-Cloze sentences (e.g., "Before you go to bed turn off the ____. Expected ending: "lights"). On half of these trials, the generated ending (e.g., "lights") was disconfirmed and was replaced by an experimenter-provided target ending (e.g., "stove"), which participants were told to remember for a later memory test. The critical question was whether participants would suppress the disconfirmed endings (e.g., "lights") and remember only the target endings (e.g., "stove").

In order to assess their memory for the items, participants were given a new set of medium-Cloze sentences that could each be completed with several different endings. Importantly, some of these sentences could be completed with the disconfirmed endings (e.g., "The baby was fascinated by the bright ____," for "lights"), some could be completed with the target endings (e.g., "She remodeled the kitchen and replaced the old ____," for "stove"), and some could be completed with previously unseen control words. The main question was whether participants would show priming for the critical items. That is, would they use these endings (e.g., "lights" and "stove") to complete more sentences than participants who had seen a different set of endings at study? Furthermore, would young and older adults show different amounts of priming for the target and disconfirmed items?

As can be seen in Fig. 5, young and older adults showed the same amount of priming for the to-be-remembered target words. However, a very different pattern of results emerged for the disconfirmed items. Older adults demonstrated similar amounts of priming to disconfirmed and target items, suggesting that their failure to inhibit no-longer-relevant words left them as accessible as words they intended to remember. Young adults, on the other hand, demonstrated below-baseline priming for the disconfirmed items, suggesting that they were so effective at suppressing this information that it became even less accessible than usual. Thus, older adults demonstrated greater implicit memory for the disconfirmed items than younger adults. Although we traditionally speak of the disadvantages of failing to inhibit information that is no longer relevant (e.g., Hasher et al., 1999), whenever that information becomes relevant again, older adults may well be at an advantage.

Benefiting from information that was never relevant

At any given moment, we are bombarded with information, some of it relevant, some of it not. However, information that is distraction at one moment may become the focus of attention in the next. In this section, we discuss two studies in which information that served as distraction on one task later became relevant on another task. Both studies point to the same conclusion: older

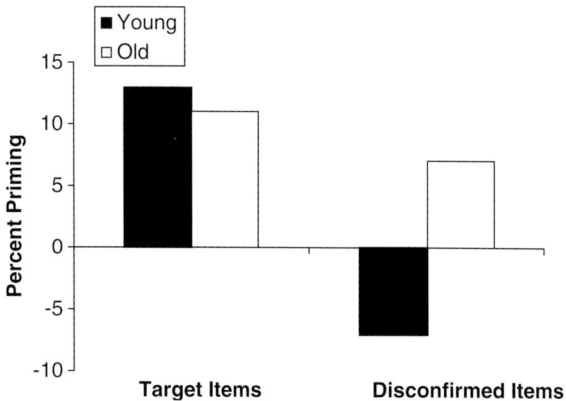

Fig. 5. Priming of disconfirmed and target items for each age group. Adapted from May and Hasher (1998). Published by APA. (Reprinted with permission.)

adults can outperform younger adults when past distraction becomes relevant.

As previously stated, young adults are quite proficient at ignoring distracting information. In fact, they are even capable of ignoring words that are presented at central fixation (Rees et al., 1999). In a seminal fMRI study, participants were shown rapid streams of letter strings (words and nonwords) superimposed on objects and were told to attend to either the letters or the objects. Their task was to press a button whenever an item in the attended stream repeated. In the attend-letters condition, participants' brain activity distinguished between words and nonwords. That is, words activated parts of the brain associated with word-processing, while nonwords did not. Conversely, in the attend-objects condition, brain activity did not distinguish between words and nonwords, suggesting that these young participants were capable of blocking out the irrelevant letter stimuli, even though they were looking right at it.

In a recent behavioral adaptation of this paradigm, Rowe et al., (2006) showed that older adults fail to ignore the superimposed words *and* they can use their implicit memory for these words to aid performance on a future task. Young and older adults were shown overlapping pictures and letter strings (words and nonwords) and were told to press a button whenever the same picture was shown twice in a row. Thus, in order to perform this task efficiently, participants should have always tried to ignore the distracting words/nonwords. After a brief filled interval, memory for the distracting words was tested implicitly with a word fragment completion task, which included a number of fragments that could be solved with distracting words from the previous task. For each participant, priming scores were calculated as the difference between the proportion of target-word fragments they correctly solved and the baseline-completion rates for those fragments.

Older adults demonstrated a substantial amount of priming for the distractor words when tested at their peak time of day (14%) and an even greater amount of priming when tested at their off-peak time of day (33%). Thus, older adults benefited from their attention to the distracting words at both peak and off-peak times, although this benefit was more pronounced in the afternoon when their ability to ignore the words was presumably lower. By contrast, young adults showed no priming for the distractor words when tested at their peak time of day (~0%) and only a small amount of priming when tested at their off-peak time of day (9%). Thus, similar to the findings of Rees et al. (1999), the young adults in this study were capable of ignoring words presented at central fixation, as indexed by their lack of implicit memory for the words. Only at their off-peak time of day, when inhibitory control was poorest, did younger adults show a slight benefit of distraction. Overall, older adults demonstrated greater implicit memory for the distracting words than young adults. These results suggest that

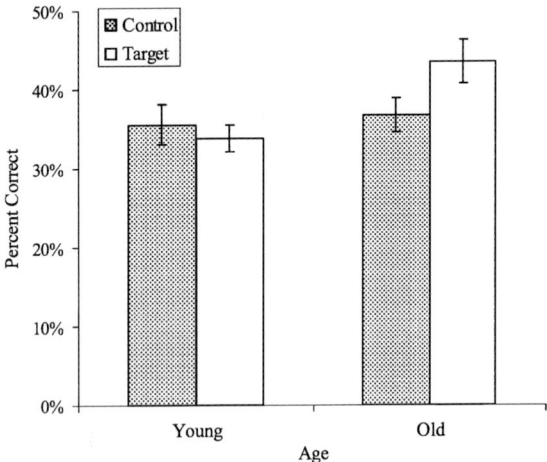

Fig. 6. Mean percent correct scores for control and target RAT problems for each age group. Adapted from Kim et al. (2007). Published by the Psychonomic Society. (Reprinted with permission.)

susceptibility to distraction can sometimes be helpful to both young and older adults, although by far the greatest advantage is afforded to older adults.

In a related study, Kim et al. (2007) demonstrated that older adults can also use their implicit memory for distracting information to aid in their performance on the RAT. Participants first performed a modified version of the reading with distraction task, in which the distracting words were actually solutions for the upcoming RAT problems. The reading task was followed by a 15 min filled interval and finally, participants received 50 RAT problems. A third of these problems could be solved by distracting words seen during the initial reading with distraction task, while another third served as control items (solutions for which were seen by participants in the alternate counterbalance condition). Older participants were expected to solve more of the target problems than young adults, as they were expected to read more of the distracting words and to be unable to inhibit them to baseline levels once activated.

As can be seen in Fig. 6, this is precisely the pattern of results that was found. Young and older adults did not differ in their performance on the control RAT problems. That is, when the solutions had not been previously viewed as distraction, young and older adults solved an equal proportion of problems. However, when the solutions to those problems had previously served as distraction on the reading task, older adults outperformed their younger counterparts on the problem-solving task. These surprising results demonstrate a clear benefit of older adults' greater access to distraction. In fact, the benefit of distraction reported here is even more astounding than that reported by May (1999), as the useful distracting information in this study was *not* shown concurrently with the RAT problems. Older participants capitalized on their access to information that was never relevant in the first place. This 'downstream' effect is quite similar to the 'far transfer' effects widely sought after in younger adults in the training and problem-solving literature (e.g., Barnett and Ceci, 2002). Older adults, with their reduced ability to suppress the past, may be better suited to transferring information from one situation to another in order to solve a problem.

Conclusion

Aging is associated with a decrease in the ability to inhibit irrelevant information. As a result, older adults are less able to regulate their attention and they end up processing more distracting information than younger adults. In this chapter, we have discussed some of the deleterious effects that increased susceptibility to distraction can have on older adults' cognitive performance. Inadvertently attending to irrelevant information can slow down processing on simple cognitive tasks (Lustig et al., 2006), disrupt a skilled activity such as reading (Connelly et al., 1991), and hinder problem solving on the remote associates task (May, 1999). However, we have also reviewed some exciting new work pointing to the potential benefits of increased susceptibility to distraction. For instance, older adults demonstrate greater implicit memory for distracting information (May and Hasher, 1998; Rowe et al., 2006) and when that distracting information is actually pertinent, they can use it to outperform younger adults at problem solving (May, 1999; Kim et al., 2007). These findings highlight the notion that cognitive aging is characterized by both losses and gains, and that whether to consider reduced inhibitory control as a help or a hindrance depends entirely on the situation.

Acknowledgments

Much of the research reviewed here was supported by a grant from the National Institute on Aging (R37 AGO4306). We thank all of the people who contributed to the various projects.

References

Barnett, S.M. and Ceci, S.J. (2002) When and where do we apply what we learn? A taxonomy for far transfer. Psychol. Bull., 128: 612–637.

Cohn, N.B., Dustman, R.E. and Bradford, D.C. (1984) Age-related decrements in stroop color test performance. J. Clin. Psychol., 40: 1244–1250.

Connelly, S.L., Hasher, L. and Zacks, R.T. (1991) Age and reading: the impact of distraction. Psychol. Aging, 6: 533–541.

Gazzaley, A., Cooney, J.W., Rissman, J. and D'Esposito, M. (2005) Top-down suppression deficit underlies working memory impairment in normal aging. Nat. Neurosci., 8: 1298–1300.

Grady, C.L. and Craik, F.I. (2000) Changes in memory processing with age. Curr. Opin. Neurobiol., 10(2): 224–231.

Hasher, L., Lustig, C. and Zacks, R.T. (2007) Inhibitory mechanisms and the control of attention. In: Conway A., Jarrold C., Kane M., Miyake A. and Towse J. (Eds.), Variation in Working Memory. Oxford University Press, New York.

Hasher, L. and Zacks, R.T. (1988) Working memory, comprehension, and aging: a review and a new view. In: Bower G.H. (Ed.), The Psychology of Learning and Motivation, Vol. 22. Academic Press, San Diego, CA, pp. 193–225.

Hasher, L., Zacks, R.T. and May, C.P. (1999) Inhibitory control, circadian arousal, and age. In: Gopher D. and Koriat A. (Eds.), Attention and Performance, XVII. MIT Press, Cambridge, MA, pp. 653–675.

Jonides, J., Marshuetz, C., Smith, E.E., Reuter-Lorenz, P.A., Koeppe, R.A. and Hartley, A. (2000) Age differences in behavior and PET activation reveal differences in interference resolution in verbal working memory. J. Cogn. Neurosci., 12: 188–196.

Kim, S., Goldstein, D., Hasher, L. and Zacks, R.T. (2005) Framing effects in young and older adults. J. Gerontol: Psychol. Sci., 60: P215–P218.

Kim, S. and Hasher, L. (2005) The attraction effect in decision making: superior performance by older adults. Q. J. Exp. Psychol., 58: 120–133.

Kim, S., Hasher, L. and Zacks, R.T. (2007) Aging and a benefit of distractibility. Psychon. Bull. Rev., 14(2): 301–305.

Lustig, C., Hasher, L. and Tonev, S.T. (2006) Distraction as a determinant of processing speed. Psychon. Bull. Rev., 13: 619–625.

Mather, M. (2006) A review of decision-making processes: weighing the risks and benefits of aging. In: Carstensen L.L. and Hartel C.R. (Eds.), When I'm 64. The National Academies Press, Washington, DC, pp. 145–173.

May, C.P. (1999) Synchrony effects in cognition: the costs and a benefit. Psychon. Bull. Rev., 6: 142–147.

May, C.P. and Hasher, L. (1998) Synchrony effects in inhibitory control over thought and action. J. Exp. Psychol. Hum., 24: 263–279.

May, C.P., Hasher, L. and Kane, M.J. (1999) The role of interference in memory span. Mem. Cogn., 27: 759–767.

Mednick, S.A. (1962) The associative basis of the creative process. Psychol. Rev., 69: 220–232.

Naveh-Benjamin, M. (2000) Adult age differences in memory performance: tests of an associative deficit hypothesis. J. Exp. Psychol. Learn., 26(5): 1170–1187.

Naveh-Benjamin, M., Guez, J., Kilb, A. and Reedy, S. (2004a) The associative memory deficit of older adults: further support using face-name associations. Psychol. Aging, 19: 541–546.

Naveh-Benjamin, M., Guez, J. and Shulman, S. (2004b) Older adults' associative deficit in episodic memory: assessing the role of decline in attentional resources. Psychon. Bull. Rev., 11: 1067–1073.

Oberauer, K. (2005) Binding and inhibition in working memory: individual and age differences in short-term recognition. J. Exp. Psychol. Gen., 134: 368–387.

Park, D.C., Lautenschlager, G., Hedden, T., Davidson, N.S., Smith, A.D. and Smith, P.K. (2002) Models of visuospatial and verbal memory across the adult life span. Psychol. Aging, 17: 299–320.

Peters, E., Hess, T.M., Västfjäll, D. and Auman, C. (2007) Adult age differences in dual information processes: implications for the role of affective and deliberative processes in older adults' decision making. Persp. Psychol. Sci., 2: 1–23.

Peterson, J.B. and Carson, S. (2000) Latent inhibition and openness to experience in a high-achieving student population. Person Indiv. Diff., 28: 323–332.

Rees, G., Russell, C., Frith, C.D. and Driver, J. (1999) Inattentional blindness versus inattentional amnesia for fixated but ignored words. Science, 286: 2504–2507.

Rowe, G., Valderrama, S., Hasher, L. and Lenartowicz, A. (2006) Attentional disregulation: a benefit for implicit memory. Psychol. Aging, 21: 826–830.

Royer, F.L., Gilmore, G.C. and Gruhn, J.J. (1981) Normative data for the symbol digit substitution task. J. Clin. Psychol., 37: 608–614.

Salthouse, T.A. (1996) The processing-speed theory of adult age differences in cognition. Psychol. Rev., 103(3): 403–428.

Salthouse, T.A. and Babcock, R.L. (1991) Decomposing adult age differences in working memory. Dev. Psychol., 27: 763–776.

Verhaeghen, P. (2003) Aging and vocabulary scores: a meta-analysis. Psychol. Aging, 18: 332–339.

West, R.L. (1996) An application of prefrontal cortex function theory to cognitive aging. Psycholl. Bull., 120(2): 272–292.

West, R. (2000) In defense of the frontal lobe hypothesis of cognitive aging. J. Int. Neuropsychol. Soc., 6(6): 727–729.

Yoon, C., May, C.P., Goldstein, D. and Hasher, L. (2007) Aging, circadian arousal patterns and cognition. In: Park D. and Schwarz N. (Eds.), Cognitive Aging: A Primer (2nd ed.). Psychology Press, Philadelphia, PA.

CHAPTER 23

Characterizing the memory changes in persons with mild cognitive impairment

Sylvie Belleville*, Stéphanie Sylvain-Roy, Chloé de Boysson and Marie-Claude Ménard

Department of Psychology, Université de Montréal and Research Center, Institut Universitaire de Gériatrie de Montréal, Montreal, QC H3W 1W5, Canada

Abstract: Persons with mild cognitive impairment (MCI) do not meet criteria for Alzheimer's disease (AD) but are at high risk for developing the disease. Presence of a memory deficit is a key component in the characterization of MCI. This chapter presents empirical studies that attempt to describe and understand the nature of the memory deficit in MCI with a focus on episodic memory and working memory. Cross-sectional studies report prominent deficits of episodic memory characterized by impaired encoding of the contextual information that makes up complex events. This results in reduced free and cued recall, impaired recognition, and impaired associative learning. Although semantic encoding is found to be impaired in conditions that rely on explicit and intentional retrieval, preserved semantic processing is found with automatic conditions of testing. Studies indicate the presence of a partial deficit of working memory with the ability to divide attention being most severely impaired. However, there appears to be heterogeneity as to the extent of the working memory impairment. The presence of vascular anomalies on MRI, as well as being in a more advanced stage in the continuum from MCI to AD, are associated with more severe and more pervasive working memory deficits. Finally, longitudinal studies indicate that the combination of episodic and working memory deficits represents a strong predictor of progression from MCI to AD.

Keywords: memory; mild cognitive impairment; working memory; Alzheimer's disease; aging

Age-associated neurodegenerative disorders have made an enormous contribution to the scientific investigation of the nature of memory processes and their underlying brain organization. This is due, in large part, to the fact that those are diseases that strike the brain structures responsible for memory processes. As a result, they cause marked, as well as selective memory deficits. Memory studies have thus investigated the way these diseases dissect memory processes. Indeed, dissociations among memory processes or systems in neurodegenerative diseases have largely contributed to the development of memory models. The study of Alzheimer's disease (AD) has been key in this endeavor. Presence of a memory deficit is indeed central to the diagnosis of AD and represents the main complaint from affected persons. In addition, the medial temporal lobe — typically implicated in episodic memory — is one of the brain structures that is most affected in the earliest phase of AD. A large number of studies have thus attempted to describe and characterize memory processes in AD.

---

*Corresponding author. Tel.: +1 514-340-3540, Ext. 4779; Fax: +1 514-340-3548; E-mail: sylvie.belleville@umontreal.ca

This has led to comprehensive descriptions of impaired and preserved memory processes in AD (for a review, see Becker and Overman, 2002; Morris and Becker, 2004; Ergis et al., 2005). While this area of research has led to significant contributions in recent years, there has been a major shift of interest toward the earliest phase of the disease.

AD is a progressive disorder and studies have shown that the first signs of the disease can actually precede by many years the time at which patients actually meet criteria for diagnosis (Elias et al., 2000). Figure 1 shows a representation of cognitive changes in healthy aging and AD. In recent years, there has been tremendous interest in this very early phase of AD, for which the term mild cognitive impairment (MCI) has been proposed (for a review, see Gauthier et al., 2006). Figure 1 shows how MCI can be conceptualized as a function of the changes that occur during the development of AD. Typically, persons with MCI have a memory complaint and show objective deficits on formal tests of memory, but they do not meet criteria for dementia. They have neither global cognitive impairment, nor functional impact on activities of daily life (Petersen et al., 1999; Petersen, 2003). Yet older persons have a ten-fold increase in their risk of developing AD when they meet criteria for MCI. For this reason, it is now generally agreed that a large proportion of persons with MCI are in the process of developing AD.

The goal of this chapter is to present the current state of knowledge regarding memory functions in MCI. Investigating memory in MCI is justified on numerous grounds. First, persons with MCI have marked atrophy of the hippocampus (Jack et al., 1999). MCI thus represents an ideal model for the study of hippocampal-related memory. Second, the cognitive impairment in AD is pervasive, which is likely to impact profoundly on the performance that those patients are able to maintain on memory tasks. In contrast, persons with MCI do not show a global cognitive impairment, which permits assessment of memory deficits without the detrimental contribution of global cognitive deficits. Finally, it is critical to understand how the memory deficits of persons with MCI compare with those of persons with AD, to determine whether they differ in a qualitative or quantitative manner. If MCI results in an impairment that strikes different memory processes than AD, this would confirm the presence of qualitative differences between the two clinical groups. In turn, if MCI causes impairment to the same processes as is typical of AD, but to a lesser degree, this would confirm quantitative differences. Resolving this issue is highly relevant to clinicians. It is also of importance to those involved in elucidating the natural history of AD along its continuum from MCI to dementia. Third, it is of prime interest for memory researchers. To investigate the pattern of memory breakdown during a degenerative process provides exciting information at the neurobiological level, as some areas become increasingly dysfunctional during the process, while others, undamaged at first, get involved during the evolution of the disease. Finally, the study of memory changes during the MCI phase provides cognitive data that complement those made available by studies of static brain lesions. Indeed, the particular dynamic of impairment with accumulation of lesions can provide a rich set of information for theorizing, as it should constrain current theoretical models of memory organization. It also sheds light on the manner in which the older brain compensates for slowly occurring lesions.

Episodic memory in mild cognitive impairment

Because episodic memory is impaired early in the evolution of AD, it is the cognitive component that has been the most studied in MCI. Episodic

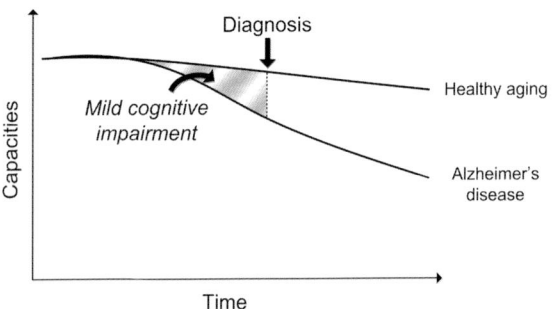

Fig. 1. Theoretical changes in cognitive functions in persons with healthy aging and persons with Alzheimer's disease.

memory refers to encoding and retrieval of events embedded in their spatio-temporal context. Studies of MCI have shown an episodic memory deficit in MCI when tested for lists of words, short texts, as well as visuospatial material (Petersen et al., 1997, 1999; Kawas et al., 2003; Loewenstein et al., 2004; Moulin et al., 2004; Della Sala et al., 2005; Ivanoiu et al., 2005). Typically, the performance of individuals with MCI is about 1.5–2 standard deviations under that of matched controls. The level of performance stands between that of healthy older adults and that of AD patients (Petersen et al., 1999), although some studies have reported that impairment can be as profound as that found in AD on some task components. Recent studies have attempted to identify the mechanisms that contribute to the episodic memory deficit found in MCI. Typically, this is done by manipulating, encoding, and retrieval conditions to identify the impaired mechanism underlying performance decline.

Most studies have reported impairment using both free recall and cued recall conditions (Petersen et al., 1999; Ivanoiu et al., 2005). In one of these studies, impairment was reported with a cued recall task that presented a large number of items in order to increase task sensitivity (Ivanoiu et al., 2005; Adam et al., 2007). The test involved learning 48 words under a condition that oriented encoding toward the categorical properties of items (for example, to encode the visually presented word "palm", the examiner would ask "show me the tree"). This was followed by categorical cued recall (for example, the examiner would ask to recall the word that was a tree). The test showed a good discriminating validity to distinguish healthy older adults from persons with MCI (Adam et al., 2007). One of the studies compared the cued recall test with measures of free recall and of visual recognition memory (Ivanoiu et al., 2005). The cued recall test was as sensitive and specific to AD and MCI as delayed free recall tasks, but cued recall was better at classifying participants as a function of their diagnosis.

Impaired cued recall can be interpreted as arising from greater difficulties in MCI than in healthy aging to benefit from rich cues at encoding and retrieval. Thus, tests promoting deep encoding as well as providing cues at recall may be particularly useful in a perspective of early diagnosis of AD. Similar results have been found in our laboratory (Hudon et al., 2006). We used the cued recall task from the Memoria battery (Belleville et al., 2002). Two conditions of free recall are compared in this task: one that promotes categorical encoding by asking the patients to point and memorize items that belong to particular semantic categories, and one where no orientation is provided during encoding. Persons with MCI were comparable to healthy older adults when tested under the no-orientation condition. In contrast, they were markedly impaired under the condition that orients them toward the semantic properties of words. This was due to the fact that categorical orienting improved performance, probably by promoting a better encoding, in healthy controls but not in persons with MCI. These findings suggest that persons with MCI have a marked deficit in encoding material even when they are provided with strong supportive cues.

Support for an encoding deficit is also provided by studies reporting impaired recognition. Dudas et al. (2005) reported that the face recognition level of persons with MCI was as low as that of persons suffering from AD. Once scores were corrected for the participants' response bias (d'), persons with MCI showed better face recognition than persons with AD, but their scores were still below those of controls. Regarding recognition of verbal material, some studies have reported normal performance (e.g., Hudon et al., 2006), while others report the presence of a deficit (e.g., Perri et al., 2005). Bennett et al. (2006) compared MCI persons with healthy older adults using verbal free recall, Yes/No verbal recognition, and three-alternative forced-choice verbal recognition. Persons with MCI showed deficits on all tasks, but the Yes/No recognition was the best to predict the group to which participants belonged.

The deficits observed in cued recall and in recognition suggest that persons with MCI have difficulty in efficiently encoding information during the learning phase, a pattern that is also characteristic of AD (for a review, see Lekeu and Van der Linden, 2005). Some studies evaluated the processes that take place during the memory encoding process. Recently, Hudon et al. (2006) evaluated persons with MCI to see if they encoded the general or schematic properties of information (the gist) to the same extent as healthy older adults. This was assessed with a text memory task that distinguishes recall of the main ideas from the secondary ideas of a text (Memo-text; Cadilhac et al., 1997). Importantly, the task used here was constructed by relying on a theoretical model of discourse processing (Kintsch and Van Dijk, 1978) and was not overloaded by detail information; on the contrary, there was a balance between general and detail items. Results indicated that both healthy older adults and persons with MCI recall more items related to main ideas (gist) than detail items. Importantly, MCI were impaired for the two types of items to the same degree and the main ideas were thus not selectively preserved. Thus, even if memory for gist was not preserved in MCI, those patients were sensitive to the hierarchical structure of a text since they reported more general ideas than secondary ones, as is typically found in normal participants. Persons with MCI differed quantitatively from AD patients on this task but they did not differ qualitatively. Persons with AD were more impaired than persons with MCI and this was true for all item types. Furthermore, like MCI persons and healthy controls, AD patients recalled more gist than detail items from the story. Thus, both persons with AD and individuals with MCI appear able to process the general semantic or schematic dimensions of material but are unable to use this information to support their episodic memory. This is coherent with our findings that persons with AD and MCI cannot make use of semantic orientation to improve their recall. Similarly, Perri et al. (2005) have reported reduced semantic clustering when patients were asked to recall a list of categorically related words, supporting the finding that MCI persons fail to use semantic knowledge to support encoding and/or retrieval.

In line with this hypothesis, we showed that MCI persons are sensitive to the overall semantic context of a list in the Deese–Roediger–McDermot (DRM) false recognition paradigm. In this paradigm, participants first learn a series of words that are all associated to a non-presented critical lure. In the retrieval phase, participants are shown studied items, the non-studied critical lure, and unrelated non-studied items. Typically, healthy adults falsely recognize the critical lure because the semantic context of the list provides the gist upon which participants base their recognition. When taking into account their tendency for false alarms by subtracting the level of false alarms to unrelated items, AD patients exhibit lower level of false recognition for the critical lure than healthy controls (Hudon et al., 2006). This indicates that AD impairs the ability to process and memorize the general semantic context of the list. In contrast, persons with MCI have a level of false recognition for the lure comparable to that of healthy controls. They are thus influenced by the general semantic context of lists and are able to encode the gist of the list. The combination of our results on text memory and on the DRM paradigms suggests that the nature of the task modulates the capacity of individuals with MCI to encode the general ideas (gist) of events and/or impacts on their influence on performance. Persons with MCI are sensitive to the semantic or schematic properties of the to-be learned material. However, this is insufficient for normalizing their performance in tasks that measure episodic memory.

This is perhaps due to their inability to perform contextual binding between the to-be remembered items and the properties that form the learning episode. Indeed, many studies have reported deficits in associative memory, which is the ability to associate previously unrelated items presented together during the encoding phase, or to associate one item to its spatial or visual context (Collie et al., 2002; Dudas et al., 2005; Nordahl et al., 2005). Similarly, Loewenstein et al. (2004) reported a deficit in source memory, as measured by indicating the list to which the items belonged. Those data are supportive of problems in encoding items with their context, a deficit that was related to hippocampal dysfunctions (Collie et al., 2002).

Likewise, studies evaluating the impact of interference revealed a sensitivity to both proactive (Loewenstein et al., 2004) and retroactive interference (Loewenstein et al., 2004; Della Sala et al., 2005). In the study by Loewenstein et al. (2004), participants first learned 10 common objects. They were then asked to learn 10 new objects that were semantically related to the first objects (to measure proactive interference), followed by a recall of the objects from the first series (to measure retroactive interference). Persons with AD and persons with MCI showed greater proactive and retroactive interference than healthy controls. Whereas proactive interference was greater in AD than in MCI, retroactive interference effects were as great in MCI as they were in AD. This sensitivity to interference suggests a difficulty in efficiently encoding the temporal context in which items were presented. This observation is coherent with an encoding problem that would arise from a difficulty in binding items with their contextual elements.

Delay effects and forgetting rate were examined to evaluate storage capacity in persons with MCI. Most studies have reported impairment in both immediate and delayed recall (Masur et al., 1990; Petersen et al., 1999; Ivanoiu et al., 2005; Perri et al., 2005; Hudon et al., 2006), but some failed to report impairment when testing is immediate (Della Sala et al., 2005). In contrast, delayed recall was systematically impaired (Loewenstein et al., 2004; Della Sala et al., 2005; Ivanoiu et al., 2005) and a few studies obtained greater impairment with delayed recall (e.g., Maruff et al., 2004; Perri et al., 2005). Furthermore, Moulin et al. (2004) observed rapid forgetting in persons with MCI. Those two findings — delayed recall impairment and rapid forgetting — are compatible with a deficit in information storage. Contrary to the study by Moulin and collaborators, Grober and Kawas (1997) did not find an accelerated forgetting in the pre-clinical phase of AD. These divergent results could be related to methodological components of the two studies. First, Grober and Kawas used an epidemiological sample, and participants were retrospectively identified as having been, three years prior, in a pre-clinical phase of AD. Thus, it is unclear whether those patients actually met criteria for MCI when the measure of their forgetting rate was taken. In contrast, Moulin and colleagues recruited their participants from a memory clinic and used the classical criteria for MCI. Participants recruited from a clinical sample may represent a more homogeneous group of MCI or could have a greater impairment (Petersen, 2003).

One last aspect that should be addressed is the effect of material type on the memory capacity of persons with MCI. Typically, AD patients show deficits that cover a wide range of material. Thus, memory for verbal and non-verbal material is impaired in AD. However, the pervasive effect of memory deficits in MCI has been challenged. Indeed, some authors have proposed that non-verbal memory may be more vulnerable in persons with MCI. The findings with recognition procedures are coherent with this view. Recognition was found to be systematically impaired when it was tested with non-verbal material such as faces or visual patterns, whereas the pattern is more incoherent with verbal material. In a longitudinal follow-up of persons with MCI, Ivanoiu et al. (2005) observed that those who would later progress to AD exhibited problems in both verbal and visual episodic memory five years prior to progression, whereas those who did not progress to AD showed deficits in the verbal domain only. Findings of a larger and more frequent impairment of visual recognition could be due to the fact that verbal recognition is most often tested with well-known material that can be supported by semantic encoding, which is not the case for most non-verbal tests. Thus in the case of a slight recognition deficit, semantic support could increase performance with verbal but not with non-verbal material. It could also be related to the fact that encoding non-verbal material is more dependent on the initiation of controlled or demanding attentional strategies. The presence of an executive or controlled impairment in some or many of those patients would thus contribute to increase the recognition deficit when tested with visual material. This latter hypothesis is coherent with the presence of a partial impairment of the attentional component of working memory in persons with MCI, as will be shown below.

Working memory in MCI

Working memory is an attentional control system involved in the online manipulation and maintenance of information. One of the most influent models of working memory, the Baddeley and

Hitch model (1974), conceptualizes it as a tripartite system which includes a phonological loop, involved in the short-term maintenance of verbal material; a visuospatial sketch pad, involved in the maintenance of image-based and spatial information; and an attentional controller, the central executive. Working memory is among the most severely impaired cognitive functions in AD, even in the early stage of the disease (Belleville et al., 1996, 2003). Persons with AD have a reduced span capacity when using words and visuospatial sequences (for a review, see Belleville and Bélanger, 2006). They are also markedly impaired in more complex working memory tasks that involve concurrent retention and processing (Baddeley et al., 1986, 2001; Morris, 1986; Belleville et al., 1996, 2003). Considering that working memory is sensitive to the earliest phase of AD, it is surprising that its direct evaluation in individuals with MCI has only been carried out in a few studies.

Using a psychometric approach, some studies showed results that are compatible with the presence of a working memory impairment in MCI. Difficulties on the Digit Symbol and Block Design subtests of the Wechsler Adult Intelligence Scale (Flicker et al., 1991; Goldman et al., 1999), as well as on the Mental Control subtest of the Wechsler Memory Scale (Tierney et al., 1996) have been reported. A difficulty in maintaining complex tasks subgoals within working memory could account for these impairments. However, these measures are neither pure nor classical indicators of working memory. For instance, many of these tests are timed, and impaired performance could be explained by psychomotor slowing. Indeed, studies have reported impaired performance on psychomotor tasks measuring capacities to perform complex motor tasks, speed tasks, and tasks measuring fine manual dexterity in persons with MCI (Kluger et al., 1997; Levinoff et al., 2005; but see also Goldman et al., 1999).

More recently, we directly assessed working memory in individuals with MCI and compared their performance with that of persons with AD (Belleville et al., 2007). Our goal was to investigate whether different attentional control mechanisms within working memory were selectively impaired in individuals with MCI and, more precisely, whether certain components would be impaired while others remain functional in the MCI phase. Three working memory components were measured. An adapted version of the Brown-Peterson procedure was used to assess dual-tasking. The procedure, taken from the Memoria battery (Belleville et al., 2002), involves asking participants to recall three consonants after short delays (0-10-20-30 s) during which they perform an addition task. Manipulation capacities were measured with the alphabetical recall procedure, where participants were asked to report a series of words in the order of the alphabet (Belleville et al., 1998). Semantic inhibition was measured with the Hayling paradigm, which involves completing sentences with an item that is unrelated to the actual semantic meaning of the sentence (Burgess and Shallice, 1996, 1997). These tasks were selected because they are tasks for which AD persons are known to be profoundly impaired (Belleville et al., 1996, 2003, 2006).

Three groups of participants were involved in the study: individuals with AD, persons with MCI, and healthy older adults. AD patients were impaired on all tasks, confirming previously reported findings of a pervasive working memory deficit. On the contrary, persons with MCI showed selective impairment. First, their performance on the Hayling task was normal, even though this is a paradigm known to be highly sensitive to the early stage of AD. Thus, semantic inhibition was intact in MCI. In contrast, performance on the Brown-Peterson procedure was severely impaired. As was mentioned above, patients are asked in this procedure to maintain three consonants over short delays (0–10–20–30 s) while completing an addition task. Whereas AD patients were severely impaired on all four delays, persons with MCI were only impaired with the longest one (30 s). Of clinical significance was the fact that impairment at the 30-s delay was as severe in persons with MCI as it was in AD. Finally, performance on alphabetical recall was inferior to that of normal elderly persons, but this effect was non-significant, and there was great variability among individuals with MCI.

All MCI persons who participated in the above study were then tested yearly in a longitudinal follow-up to determine those who would later progress to AD. This allowed the researchers to identify retrospectively those individuals with MCI who would later deteriorate and assess whether that subgroup of decliners would exhibit more severe or widespread working memory deficit. It was found that the group of later decliners was significantly impaired on the alphabetical recall. Therefore, impairment in alphabetical recall could predict future progression of symptoms in MCI or reflect a more advanced stage in the MCI/AD continuum. Strikingly, MCI decliners did not have a more impaired performance than non-decliners on the Hayling test. Our failure to find an inhibition deficit on the Hayling test is surprising because it is a task that is severely impaired in early AD. This could be due to a lack of task sensitivity, a hypothesis that we are currently investigating. Furthermore, the Hayling test implicates semantic inhibition. It is possible that other types of inhibition are impaired in MCI. In line with this suggestion, Wylie et al. (2007) reported that persons with MCI have difficulty inhibiting an incongruent response in the Flanker task where participants are asked to indicate the direction of a target arrow flanked by distractor arrows pointing in the opposite direction.

Other studies have provided support for the presence of a working memory deficit in MCI. Alescio-Lautier et al. (2007) reported that persons with MCI were impaired when visual and visuo-spatial short-term recognition was delayed by a short interval (1–30 s) filled with an interference task. Dannhausser et al. (2005) reported deficits in a divided attention paradigm that involved concurrent detection of letters and numbers presented in separate visual (letters) and auditory (digits) streams.

There have also been studies indicating that the presence of vascular anomalies in MCI patients may relate to the presence and extent of their working memory deficits. Nordahl et al. (2005) distinguished two subgroups of individuals with MCI based on their neuroradiological characteristics as shown with MRI. One group was composed of those MCI with marked hippocampal atrophy, whereas the other group was composed of those MCI with significant sub-cortical white matter anomalies. The authors proposed that the former group corresponded to a prodrome of AD, while the latter would be a prodrome for sub-cortical vascular dementia. A set of clinical and cognitive tasks was then administered in order to compare the two groups. Results showed that both MCI groups had episodic memory impairment. However, working memory, as measured by the N-back, a task that reflects updating capacities, was only impaired in those MCI with vascular anomalies. Results from our laboratory are congruent with this finding (Labelle et al., 2007). We compared MCI persons with persons meeting criteria for the sub-cortical vascular form of MCI (Frisoni et al., 2002). The two groups were equally impaired on the Brown-Peterson procedure and none was impaired on alphabetical recall. However, the group of vascular MCI showed more severe impairment than healthy controls and MCI on the stop signal, a task that measures motor inhibition, and on the third plate of the Stroop test, a task that measures verbal inhibition.

Natural history of MCI

Little is known regarding the natural history of cognitive deficits as people progress from MCI to AD. According to some authors, only episodic memory is impaired during the early MCI phase, and this is followed by executive control deficits, and then by language and perception deficits (Perry and Hodges, 1999; Petersen, 2003). However, few studies have actually used longitudinal follow-up to assess the way in which cognitive impairments unravel during the MCI period. One approach has been to use longitudinal follow-up to identify cognitive predictors of later progression to dementia. Indeed, 10–20% of persons meeting criteria for MCI at a particular time will not progress to dementia (Petersen, 2003), and some may even revert to normal levels (Ritchie, 2004). Longitudinal follow-up of individuals with MCI allows identification of cognitive variables that better predict future progression to AD. Such longitudinal studies have reported that impairment of both episodic memory and executive control is good predictors of progression toward dementia during the following years (Tierney et al., 1996; Petersen et al., 1997; Kluger et al., 1999; Chen et al., 2000). This suggests that impairment in those two domains occurs quite early in the progression from MCI to AD.

Another approach has been to assess cognitive functions repeatedly in cohorts of MCI persons. In a longitudinal study following a cohort of persons with MCI and a cohort of individuals with AD, Lambon Ralph et al. (2003) showed that the cognitive changes in MCI were relatively gradual, beginning with memory, and later extending to language and visuoperceptual functions. They also reported that from the moment individuals with MCI convert to AD, their progression is similar to that of individuals with AD, supporting common neuropathological processes. Although distinct profiles could be identified during the MCI phase, a significant homogeneity was observed overall in the progression of patients. Unfortunately, the progression of working memory or executive function was not investigated in this study and, for this reason, this study does not shed light on whether this component is impaired as early as the episodic memory component is.

Bennett et al. (2002) followed 211 individuals with MCI and 587 individuals with AD annually over a 6-year period. On the basis of an extended cognitive assessment, they derived composite scores for episodic memory, semantic memory, working memory, perceptual speed, and visuospatial abilities. They reported impairment in each domain at the first evaluation, followed by a differential decline across domains. Persons with MCI showed a more rapid decline than healthy controls for episodic memory and semantic memory. They also showed a more rapid decline of working memory than healthy controls but this was only marginally significant.

To summarize, little is known regarding the rate and pattern of memory decline during the MCI period. The few studies available suggest a differential rate of decline across general cognitive domains. However, none of these studies have focused on fine components of memory.

Conclusion

Studies of MCI can provide important information to further our understanding of memory processes and of their breakdown following brain damage. Indeed, persons with dementia have a global cognitive impairment, which renders difficult a fine characterization of cognitive functions. The selectivity in deficits expected to occur in persons with MCI has the potential to provide neuropsychological support to dissociate among memory processes.

This is the case in the domain of working memory, where three components, divided attention, inhibition, and manipulation, known to be equally impaired in AD, exhibit selective impairment in MCI. Divided attention is severely impaired, manipulation is only impaired in patients who are close to progression, whereas semantic inhibition is unimpaired. At a theoretical level, this pattern supports the hypothesis according to which those are independent working memory components that rely on neural substrates differently affected in the early phase of the disease. Similarly, persons with MCI are sensitive to the semantic context of a list in the false recognition paradigm, which contrasts with AD patients who show reduced encoding of semantic properties whether tested directly or indirectly.

In addition, studies of MCI allow a better approach to the natural history of the MCI/AD continuum, as well as to the way in which precise parts of memory decline along with the progression of the neurobiological anomalies. Interestingly, memory studies have shown that the changes from MCI to AD reflect both qualitative effects and quantitative ones. Qualitative changes characterize the domain of working memory because some of those domains that are unimpaired during the MCI phase become impaired as patients progress to AD. Episodic memory reflects quantitative changes because impairment appears to gain in severity but the processes impaired as patients progress to AD are also impaired in the MCI phase. Yet, much work remains to be done on the natural history from MCI to AD and on its relationship with the underlying brain modifications.

Studies of memory in MCI also have implications for clinical diagnosis and management. For example, knowledge of impaired and intact capacities in MCI can be exploited through intervention studies, the goal of which is to optimize memory performance. Because memory deficit is selective during that phase, MCI may be envisioned as an optimal period to teach patients how to use encoding strategies. In a recent study, we have shown that persons with MCI can improve their memory capacities to the same extent as healthy controls when they are taught mnemotechnics because they promote a more elaborate encoding of information (Belleville et al., 2006). This illustrates that studies of memory in MCI, in addition to having important theoretical significance, can have profound clinical implications.

Abbreviations

AD	Alzheimer's disease
DRM	Deese–Roediger–McDermot paradigm
MCI	mild cognitive impairment

Acknowledgments

Sylvie Belleville is supported by an FRSQ Chercheur national. Her studies on MCI and AD benefit from a financial support from CIHR, VRQ-CONCOV, HSF-ASC-HSF-Pfizer partnership, REPAR and CAREC. Marie-Claude Ménard bene fits from NSERC and FRSQ Ph.D. scholarships. Stéphanie Sylvain-Roy is supported by an NSERC student research award. We thank Marc-Antoine Labelle for editorial assistance.

References

Adam, S., Van der Linden, M., Ivanoiu, A., Juillerat, A.-C., Bechet, S. and Salmon, E. (2007) Optimization of encoding specificity for the diagnosis of early Alzheimer's disease: the RI-48 task. J. Clin. Exp. Neuropsychol., 29(5): 477–487.

Alescio-Lautier, B., Michel, B.F., Herrera, C., Elahmadi, A., Chambon, C., Touzet, C. and Paban, V. (2007) Visual and visuospatial short-term memory in mild cognitive impairment and Alzheimer disease: role of attention. Neuropsychologia, 45(8): 1948–1960.

Baddeley, A.D., Baddeley, H.A., Bucks, R.S. and Wilcock, G.K. (2001) Attentional control in Alzheimer's disease. Brain, 124(8): 1492–1508.

Baddeley, A.D. and Hitch, G.J. (1974) Working memory. In: Bower G.H. (Ed.), The Psychology of Learning and Motivation, Vol. 8. Academic Press, New York, pp. 47–90.

Baddeley, A.D., Logie, R., Bressi, S., Della Sala, S. and Spinnler, H. (1986) Dementia and working memory. Q. J. Exp. Psychol. A, 38(4): 603–618.

Becker, J.T. and Overman, A.A. (2002) The memory deficit in Alzheimer's disease. In: Baddeley A.D., Kopelman M.D. and Wilson B. (Eds.), The Handbook of Memory Disorders (2nd ed.). Wiley, West Sussex, pp. 569–589.

Belleville, S. and Bélanger, S. (2006) Maladie d'Alzheimer: Mémoire de travail, fonctions exécutives et attention. In: Belin C., Ergis A.M. and Moreaud O. (Eds.), Actualités sur les Démences: Aspects Cliniques et Neuropsychologiques. Solal, Marseille, pp. 35–50.

Belleville, S., Chatelois, J., Fontaine, F. and Peretz, I. (2002) Mémoria: Batterie informatisée d'évaluation de la mémoire pour Mac et PC. Institut Universitaire de Gériatrie de Montréal, Montréal.

Belleville, S., Chertkow, H. and Gauthier, S. (2007) Working memory and control of attention in persons with Alzheimer's disease and mild cognitive impairment. Neuropsychology, 21(4): 458–469.

Belleville, S., Peretz, I. and Malenfant, D. (1996) Examination of the working memory components in normal aging and in dementia of the Alzheimer type. Neuropsychologia, 34(3): 195–207.

Belleville, S., Rouleau, N. and Caza, N. (1998) Effect of normal aging on the manipulation of information in working memory. Mem. Cogn., 26(3): 572–583.

Belleville, S., Rouleau, N. and Van der Linden, M. (2006) Use of the Hayling task to measure inhibition of prepotent responses in normal aging and Alzheimer's disease. Brain Cogn., 62(2): 113–119.

Belleville, S., Rouleau, N., Van der Linden, M. and Collette, F. (2003) Effect of manipulation and irrelevant noise on working memory capacity of patients with Alzheimer's dementia. Neuropsychology, 17(1): 69–81.

Bennett, D.A., Wilson, R.S., Schneider, J.A., Evans, D.A., Beckett, L.A., Aggarwal, N.T., Barnes, L.L., Fox, J.H. and Bach, J. (2002) Natural history of mild cognitive impairment in older persons. Neurology, 59(2): 198–205.

Bennett, I.J., Golob, E.J., Parker, E.S. and Starr, A. (2006) Memory evaluation in mild cognitive impairment using recall and recognition tests. J. Clin. Exp. Neuropsychol., 28(8): 1408–1422.

Burgess, P.W. and Shallice, T. (1996) Response suppression, initiation and strategy use following frontal lobe lesions. Neuropsychologia, 34(4): 263–272.

Burgess, P.W. and Shallice, T. (1997) The Hayling and Brixton tests. Hartcourt Assessment. The Psychological Corporation, San Antonio, TX.

Cadilhac, C., Gély-Nargeot, M.-C., Virbel, J. and Nespoulous, J.L. (1997) De l'essentiel aux détails: le rappel des structures narratives par des sujets âgés normaux et déments. In: Lambert J. and Nespoulous J.-L. (Eds.), Perception auditive et compréhension du langage. Solal, Marseille, pp. 295–317.

Chen, P., Ratcliff, G.D., Belle, S.H., Cauley, J.A., DeKosky, S.T. and Ganguli, M. (2000) Cognitive tests that best discriminate between presymptomatic AD and those who remain nondemented. Neurology, 55(12): 1847–1853.

Collie, A., Myers, C., Schnirman, G., Wood, S. and Maruff, P. (2002) Selectively impaired associative learning in older people with cognitive decline. J. Cogn. Neurosci., 14(3): 484–492.

Dannhausser, T.M., Walkerm, Z., Stevens, T., Lee, L., Seal, M. and Shergill, S.S. (2005) The functional anatomy of divided attention in amnestic mild cognitive impairment. Brain, 128(6): 1418–1427.

Della Sala, S., Cowan, N., Beschin, N. and Perini, M. (2005) Just lying there, remembering: improving recall of prose in amnesic patients with mild cognitive impairment by minimising interference. Memory, 13(3–4): 435–440.

Dudas, R.B., Clague, F., Thompson, S.A., Graham, K.S. and Hodges, J.R. (2005) Episodic and semantic memory in mild cognitive impairment. Neuropsychologia, 43(9): 1266–1276.

Elias, M.F., Beiser, A., Wolf, P.A., Au, R., White, R.F. and D'Agostino, R.B. (2000) The preclinical phase of Alzheimer's disease: a 22-year prospective study of the Framingham Cohort. Arch. Neurol., 57(6): 808–813.

Ergis, A.M., Gély-Nargeot, M.-C. and Van der Linden, M. (2005) Les troubles de la mémoire dans la maladie d'Alzheimer. Solal, Marseille, pp. 1–387.

Flicker, C., Ferris, S.J. and Reisberg, B. (1991) Mild cognitive impairment in the elderly: predictors of dementia. Neurology, 41(7): 1006–1009.

Frisoni, G.B., Galluzzi, S., Bresciani, L., Zanetti, O. and Geroldi, C. (2002) Mild cognitive impairment with subcortical vascular features: clinical characteristics and outcome. J. Neurol., 249(10): 1423–1432.

Gauthier, S., Reisberg, B., Zaudig, M., Petersen, R.C., Richie, K., Broich, K., Belleville, S., Brodaty, H., Bennett, D., Chertkow, H., Cummings, J.L., de Leon, M.J., Feldman, H., Ganguli, M., Hampel, H., Scheltens, P., Tierney, M.C., Whitehouse, P. and Winblad, B. (2006) Mild cognitive impairment. Lancet, 367(9518): 1262–1270.

Goldman, W.P., Baty, J.D., Buckles, V.D., Sahrmann, S. and Morris, J.C. (1999) Motor dysfunction in mildly demented AD individuals without extrapyramidal signs. Neurology, 53(5): 956–962.

Grober, E. and Kawas, C. (1997) Learning and retention in preclinical and early Alzheimer's disease. Psychol. Aging, 12(1): 183–188.

Hudon, C., Belleville, S., Souchay, C., Gély-Nargeot, M.-C., Chertkow, H. and Gauthier, S. (2006) Memory for gist and detail information in Alzheimer's disease and mild cognitive impairment. Neuropsychology, 20(5): 566–577.

Ivanoiu, A., Adam, S., Van der Linden, M., Salmon, E., Juillerat, A.-C., Mulligan, R. and Seron, X. (2005) Memory evaluation with a new cued recall test in patient with mild cognitive impairment and Alzheimer's disease. J. Neurol., 252(1): 47–55.

Jack, C.R., Petersen, R.C., Xu, Y.C., O'Brien, P.C., Smith, G.E., Ivnik, R.J., Boeve, B.F., Waring, S.C., Tangalos, E.G. and Kolmen, E. (1999) Prediction of AD with MRI-based hippocampal volume in mild cognitive impairment. Neurology, 52(7): 1397–1403.

Kawas, C.H., Corrada, M.M., Brookmeyer, R., Morrison, A., Resnick, S.M., Zonderman, A.B. and Arenberg, D. (2003) Visual memory predicts Alzheimer's disease more than a decade before diagnosis. Neurology, 60(7): 1089–1093.

Kintsch, W. and Van Dijk, T.A. (1978) Toward a model of text comprehension and production. Psychol. Rev., 85(5): 363–394.

Kluger, A., Ferris, S.H., Golomb, J., Mittelman, M.S. and Reisberg, B. (1999) Neuropsychological prediction of decline to dementia in nondemented elderly. J. Geriatr. Psychiatry Neurol., 12(4): 168–179.

Kluger, A., Gianutsos, J.G., Golomb, J., Ferris, S.H. and Reisberg, B. (1997) Motor/psychomotor dysfunction in normal aging, mild cognitive decline, and early Alzheimer's disease: diagnostic and differential diagnostic features. Int. Psychogeriatr., 9(suppl. 1): 307–316.

Labelle, M.-A., Belleville, S., Bocti, C., Massoud, F., Mellah, S.,Villeneuve, S., Joncas, S., Gilbert B., Fontaine, F., Enriquez, A. and Gauthier, S. (2007) Caractérisation de la mémoire de travail et des fonctions exécutives dans le trouble cognitif léger d'origine vasculaire. Colloquium for the 25th anniversary of the Centre de recherche Institut universitaire de gériatrie de Montréal, Montréal, Canada.

Lambon Ralph, M.A., Patterson, K., Graham, N., Dawson, K. and Hodges, J.R. (2003) Homogeneity and heterogeneity in mild cognitive impairment and Alzheimer's disease: a cross-sectional and longitudinal study of 55 cases. Brain, 126(11): 2350–2362.

Lekeu, F. and Van der Linden, M. (2005) Le fonctionnement de la mémoire épisodique dans la maladie d'Alzheimer. In: Ergis A.M., Nargeot M.C. and Van der Linden M. (Eds.), Les troubles de la mémoire dans la maladie d'Alzheimer. Solal, Marseille, pp. 73–117.

Levinoff, E., Saumier, D. and Cherkow, H. (2005) Focused attention deficits in patients with Alzheimer's deficits and mild cognitive impairment. Brain Cogn., 57(2): 127–130.

Loewenstein, D.A., Acevedo, A., Luis, C., Crum, T., Barker, W. and Duara, R. (2004) Semantic interference deficits and the detection of mild Alzheimer's disease and mild cognitive impairment without dementia. J. Int. Neuropsychol. Soc., 10(1): 91–100.

Maruff, P., Collie, A., Darby, D., Weaver-Cargin, J., Masters, C. and Currie, J. (2004) Subtle memory decline over 12 months in mild cognitive impairment. Dement. Geriatr. Cogn. Disord., 18(3–4): 342–348.

Masur, D.M., Fuld, P.A., Blau, A.D., Crystal, H. and Aronson, M.K. (1990) Predicting development of dementia in the elderly with the selective reminding test. J. Clin. Exp. Neuropsychol., 12(4): 529–538.

Morris, R.G. (1986) Short-term forgetting in senile dementia of the Alzheimer's type. Cogn. Neuropsychol., 3: 77–97.

Morris, R.G. and Becker, J.T. (2004) A cognitive neuropsychology of Alzheimer's disease. In: Morris R.G. and Becker J.T. (Eds.), Cognitive Neuropsycology of Alzheimer's Disease. Oxford University Press, New York, pp. 3–10.

Moulin, C.J.A., James, N., Freeman, J.E. and Jones, R.W. (2004) Deficient acquisition and consolidation: intertrial free recall performance in Alzheimer's disease and mild cognitive impairment. J. Clin. Exp. Neuropsychol., 26(1): 1–10.

Nordahl, C.W., Ranganath, C., Yonelinas, A.P., DeCarli, C., Reed, B.R. and Jagust, W.J. (2005) Different mechanisms of episodic memory failure in mild cognitive impairment. Neuropsychologia, 43(11): 1688–1697.

Perri, R., Carlesimo, G.A., Serra, L. and Caltagirone, C. and The Early Diagnosis Group of the Italian Interdisciplinary Network on Alzheimer's Disease. (2005) Characterization of memory profile in subjects with amnestic mild cognitive impairment. J. Clin. Exp. Neuropsychol., 27(8): 1033–1055.

Perry, R.J. and Hodges, J.R. (1999) Attention and executive deficits in Alzheimer's disease: a critical review. Brain, 122(3): 383–404.

Petersen, R.C. (2003) Mild cognitive impairment: aging to Alzheimer's disease. University Press, Oxford.

Petersen, R.C., Smith, G.E., Waring, S.C., Ivnik, R.J., Kokmen, E. and Tangelos, E.G. (1997) Aging, memory, and mild cognitive impairment. Int. Psychogeriatr., 9(suppl. 1): 65–69.

Petersen, R.C., Smith, G.E., Waring, S.C., Ivnik, R.J., Tangalos, E.G. and Kokmen, E. (1999) Mild cognitive impairment: clinical characterization and outcome. Arch. Neurol., 56(3): 303–308.

Ritchie, K. (2004) Mild cognitive impairment: an epidemiologic perspective. Dialogues Clin. Neurosci., 6(4): 401–408.

Tierney, M.C., Szalai, J.P., Snow, W.G., Fisher, R.H., Nores, A., Nadon, G., Dunn, E. and George-Hyslop, P.H. (1996) Prediction of probable Alzheimer's disease in memory-impaired patients: a prospective longitudinal study. Neurology, 46(3): 661–665.

Wylie, S.A., Ridderinkhof, K.R., Eckerle, M.K. and Manning, C.A. (2007) Inefficient response inhibition in individuals with mild cognitive impairment. Neuropsychologia, 45(7): 1408–1419.

CHAPTER 24

Aging, metamemory regulation and executive functioning

Michel Isingrini[1,*], Audrey Perrotin[1] and Céline Souchay[2]

[1]*University François-Rabelais de Tours, UMR CNRS 6215 LMDC, 3 rue des Tanneurs, BP 4103, 37041 Tours Cedex 1, France*
[2]*University of Leeds, Leeds, UK*

Abstract: In this chapter we deal with metamemory regulation processes and concentrate mainly on how they are related to learning in episodic memory. In recent years an increasing amount of the literature has emphasized conceptual similarities between metamemory regulation and executive-frontal functioning. Different data have also highlighted that age-related cognitive differences might, in many cases, be explained by the decline of executive-frontal functioning that accompanies aging. Thus, in the present chapter we evaluate the relationship of aging and metamemory regulation among the cognitive decline frontal hypothesis of aging. We focus specifically on two measures of metamemory regulation allowing evaluating monitoring and control processes: feeling-of-knowing (FOK) and capacity to adjust study strategies to task demand, respectively. After having presented evidence supporting the executive-frontal hypothesis of FOK, we present a series of experiments addressing the questions of age-related differences in metamemory monitoring and control, and of possible mediation of this age effect by the age-related decline in executive-frontal functioning. The findings support the ideas that the monitoring process of episodic memory FOK and the control process of adjusting study time to task difficulty are impaired in older adults. Moreover, these declines can be explained by the decline of executive-frontal functioning associated to aging. Finally, types of mechanisms pertaining to FOK monitoring and to adjustment control process on which executive-frontal functioning and aging may have an impact are discussed.

Keywords: aging; memory; metamemory; monitoring; control; executive functioning

Examining metamemory in older people is of theoretical interest in helping to characterize deficit mechanisms about episodic memory which is particularly vulnerable in the aging process. Much of the interest in aging and metamemory has focused on the hypothesis that deficits in metamemory may explain age differences in memory performance.

Metamemory can be simply defined as cognition about memory. A wide variety of concepts have been included under this broad concept, including the so-called metamemory regulation. In this chapter, we deal with the two attributes of metamemory regulation reported by Nelson and Narens (1990): memory monitoring and memory control. Within this topic, we will concentrate mainly on processes relating to learning and retrieval in episodic memory. In recent years an increasing amount of the literature has emphasized

*Corresponding author. Tel.: +33 2 47 36 81 21;
E-mail: isingrini@univ-tours.fr

that, on the one hand, metacognition and metamemory processes are dependent on executive-frontal functioning, and on the other hand that age-related cognitive differences might in many cases be explained by the decline of executive-frontal functioning that accompanies aging. Thus, in the present chapter we also evaluate the outcomes of studies on aging and metamemory regulation in relation to the cognitive decline frontal hypothesis of aging.

Metacognition and metamemory regulation

According to Nelson and Narens (1990), cognitive processes in humans can be divided into two interrelated levels: the metalevel and the cognitive object level. The metalevel contains a cognitive model of the object level that is thought to be continuously updated with information from the cognitive object level. The metalevel in return controls the object level by providing information, and initiating and terminating actions performed by the object level. According to this view, metacognition can be conceptualized as a system that modulates cognitive processes. Many authors have defined metacognition as a broad term that encompasses knowledge and regulation of cognitive activity. Metacognitive knowledge is knowledge people have about their cognitive abilities ("I have a bad memory"), whereas metacognitive regulation refers to processes that coordinate cognition in the course of carrying out a task. Metacognitive regulation has been defined by Nelson and Narens (1990) as consisting of two processes: monitoring and control. Monitoring is the collection of information issuing from the evaluation about one's own knowledge and performance, whereas control is a process of self-regulating one's own behavior.

Metacognition in relation to memory is known as metamemory. Like metacognition, metamemory encompasses metamemory regulation processes such as monitoring and control of the memory process. Metamemory regulation fits well with the meta level-object level of metacognition proposed by Nelson and Narens (1990). For example, the role of the metamemory monitoring process (metalevel) may be to provide information about memory of task difficulty (object level), and the role of the metamemory control process (metalevel) may in turn be to produce information likely to adjust learning and retrieval strategies according to the estimated difficulty of the task. Thus, metamemory regulation is seen as an essential part of the process of adapting control behavior in order to optimize learning and memory retrieval. According to Nelson and Narens (1990), there is a close relationship between the monitoring and the control processes, given that knowledge about the state of the object level constitutes an indispensable source of information for implementing adjustment strategies when carrying out a task. The link between the two variables is commonly referred to as the *"monitoring-affects-control hypothesis."* Monitoring provides evidence about ongoing learning that an individual needs to either maintain or change a particular strategic approach to learning. For example, an individual who perceives that a particular strategy used to learn new information is not proving useful could switch to a new and perhaps more effective strategy.

The ability to monitor and control memory performance has considerable importance in everyday life and is among the metamemory dimensions which have been widely investigated (Dixon et al., 1988). It increasingly seems that understanding metamemory processes is of major importance in understanding the functioning and deterioration of the human memory. Thus, it is accepted that a deficit in the components of metamemory regulation (monitoring, control or monitoring-control relationship) can result in memory dysfunctions to the same degree as those caused by a deficit directly affecting specific memory operations (Metcalfe, 1993).

Evaluation of metamemory regulation processes

A common and useful method to evaluate metamemory monitoring accuracy is to ask individuals to predict how well they will remember items during an upcoming memory test. The implicit logic is that the better one's knowledge about one's memory, the more accurate the predictions are likely to be. Two different types of predictions have been frequently elicited. Global predictions, in which participants judge how many items from an entire study list they will subsequently recall, and item-by-item predictions, in which they predict the likelihood of subsequent recall separately for each item. In the literature on aging, the two most widely assessed types of item-by-item monitoring are judgment of learning (JOL), in which predictions are made about the likelihood of subsequent recollection of recently studied items and feeling-of-knowing (FOK), in which predictions are made about the likelihood of subsequent recognition of non-recalled information (Nelson and Narens, 1990). Thus, JOL involves monitoring at the time the information is being studied, whereas FOK involves monitoring during attempted retrieval. In a typical JOL experiment, participants are instructed to memorize a list of paired associates. Following the study phase, they are asked to make predictions regarding the likelihood of recalling the target word in a subsequent test. In the test phase, participants are presented with each of the cue words and their recall of the corresponding target is tested. The FOK procedure proposed by Hart (1965) requires subjects to estimate the likelihood that they will subsequently recognize a piece of information that they have failed to recall, either from semantic memory (Hart, 1965; Nelson and Narens, 1990), or from recently learned episodic memory information (Schacter, 1983; Souchay et al., 2000).

Nelson (1984) argued that the best index of metamemory accuracy to measure monitoring resolution was the individual Goodman-Kruskal gamma correlation, which is an index of the rank order agreement of metacognitive judgment with memory performance (recall or recognition). This index is based on the difference between concordant and discordant pairs, and is calculated by the equation $(P-Q)/(P+Q)$, where P corresponds to concordant pairs ("I predict that I will recognize the word and I actually recognize it" or "I predict that I will not recognize the word and I actually do not recognize it") and Q corresponds to discordant pairs ("I predict that I will recognize the word but actually I do not recognize it" or "I predict that I will not recognize the word but actually I recognize it"). As with the Pearson correlation, a gamma of 1.0 indicates perfect agreement between prediction and memory performance, a gamma of 0 indicates independence of judgment and probability of recall or recognition, and a gamma of −1.0 indicates a perfect inverse relationship. The gamma correlation index is used for calculating both FOK and JOL accuracy.

To evaluate the metamemory control process, some aging experiments have used the "recall readiness" paradigm. This procedure requires subjects to study verbal stimuli and declare their "recall readiness" for those stimuli. For example, the subject is asked to study a list of words, one word at a time, at his/her own pace. The next word is presented only when the subject indicates the previous word has been adequately studied for future recall (Murphy et al., 1981). Following the recall-readiness trial, the subject is asked to recall the studied words, in order to determine the accuracy of the previous recall-readiness judgments. The metamemory control process is also evaluated by assessing the relationship between monitoring and control (*monitoring-affects-control hypothesis*), generally by using correlations (gamma index correlation) as an individual measure of the relationship between monitoring outcomes and strategic control behaviors such as time allotted to study a list of words (Dunlosky and Connor, 1997).

Metamemory and frontal executive function

Several authors have emphasized conceptual similarities between executive functions and metamemory characteristics (Shimamura, 1995; Fernandez-Duque et al., 2000; Souchay et al., 2000; Pannu and Kaszniak, 2005). The term "executive functions" refers to a set of higher order cognitive processes encompassing a wide variety of mental activities such as dealing with novelty, planning, implementing strategies, monitoring performance, using feedback to adjust future responses, and inhibiting task-irrelevant information (Rabbitt, 1997). It is generally conceptualized as a superordinate- or meta-cognitive activity (Luszcz and Bryan, 1999). The bulk of empirical evidence supports the view that executive functions depend mainly on the integrity of the frontal cortex (Raz et al., 1998). In line with the hypothesis of a link between executive and metamemory processes, some studies have demonstrated that FOK judgment depends on the prefrontal cortex. Janowsky et al. (1989), and more recently Schnyer et al. (2004), showed that patients with frontal lobe lesions were impaired on an FOK task involving recently learned information. Also consistent with this hypothesis, in a recent study, Schnyer et al. (2005), using event-related neuroimaging fMRI study, have shown that the accuracy of FOK judgments is associated with the activation of frontal-temporal network, with a critical role of the ventral medial prefrontal cortex (VMPC) in such judgments.

Overall, these results support the hypothesis that metamemory monitoring is one of the cognitive processes that contributes to the supervisory process of the frontal cortex in memory functioning. However, according to Shimamura (2000), the nature of this contribution is not clear. Investigation of executive functioning has assessed and defined specific components — such as selecting stimulus information, maintaining, updating and inhibiting information in working memory, and global shift of information processing — that may also be involved in the metamemory process. Thus, the linking of metacognition to aspects of executive functions offers opportunities to improve the definition of specific cognitive components of metacognition.

On the other hand, some results suggest that metamemory monitoring may not always be related to frontal lobe or executive functioning. For example, research with frontal lobe patients (Janowsky et al., 1989, exp. 2) and patients suffering from various subcortical diseases such as Huntington's and Parkinson's (see Pannu and Kasmiak, 2005) who often perform poorly on frontal tests, demonstrates normal FOK accuracy about general knowledge. These observations suggest that frontal or executive functioning may be differentially implicated in metamemory measures.

Executive decline hypothesis of memory and metamemory in aging

The hypothesis that executive functions are the first cognitive functions to decline with aging is supported by neurobiological and neuropsychological data. These data have led to the hypothesis that executive deficit associated with normal aging is the cause of age-related effects on certain cognitive functions such as memory and metamemory (Crawford et al., 2000; Raz, 2000; Souchay et al., 2000). This is well supported by data from various sources which show that a number of changes during aging are in agreement with the executive deficit hypothesis.

Studies have revealed that neurological modifications primarily affect the frontal cortex. In a meta-analysis Raz (2000) showed that the mean correlation between age and brain volume varied from one region to another. The region most affected by aging is the prefrontal cortex $(-.47)$, while the temporal $(-.27)$ and parietal $(-.29)$ regions show only a moderate correlation. Moreover, several studies investigating blood flow in response to memory tasks or executive functions have demonstrated an age-related decrease in frontal activation (see Cabeza, 2002).

Executive functioning is traditionally evaluated using neuropsychological tests. The fact that these tests have been shown to be sensitive to frontal lobe dysfunctioning, which is assumed to be the cause of the executive syndrome, is evidence that these tests involve executive functions. Several studies have analyzed the effect of aging on some of these tests as for example the Wisconsin Card Sorting Test (WCST), Stroop's test, and verbal fluency tests. Most of the research carried out in the field of aging has shown a significant decrease in all these tests in elderly subjects (see Salthouse et al., 1996; Isingrini and Vazou, 1997).

Examination of the memory profile of frontal lesion patients and neuroimaging studies have revealed a close link between memory and frontal functioning (Wheeler et al., 1995; Cabeza, 2002). The main hypothesis relating to how executive processes intervene in memory functioning is that executive functions correspond to a regulatory system which contributes to memory operations (Shimamura, 1995). Several studies have shown similar changes in memory profiles between frontal lesion patients and elderly subjects as lower organization index for the material to be memorized (see Stuss et al., 1996; Isingrini and Taconnat, 1997), more marked deficit in free-recall tasks than in cued-recall and recognition tasks (Shimamura, 1995), increase in false-recall or false-recognition errors (see Schacter, 1997; Taconnat et al., 2008), and difficulties in remembering the learning context (see Spencer and Raz, 1995). Studies have shown that these deficits in elderly subjects are correlated with executive functioning, as evaluated using the WCST (Craik et al., 1990; Troyer et al., 1994; Glisky et al., 1995; Fabiani and Friedman, 1997; Bryan et al., 1999; Glisky et al., 2001).

Overall, these observations are in line with the idea that normal aging is associated with an executive deficit likely to affect cognitive functions. Anderson and Craik (2000) suggest a cascade model to account for this hypothesis. The main consequence of aging would be a significant modification of the frontal cortex. These modifications would be at the basis of executive dysfunctioning, which would explain why normal aging is associated with decreased performances on cognitive tasks, which depends on executive resources, such as episodic memory or metamemory monitoring.

Aging, metamemory regulation, and executive functioning

The study of age effects on metamemory regulation processes is of interest for two reasons. Firstly to identify the metamemory tasks that change with age and to understand what causes these changes, and secondly to test the hypothesis that memory modifications are partly the result of a deterioration in the metamemory monitoring and control processes. According to this last idea, age-related memory difficulties could thus be partly explained by the following sequence: (1) normal aging is associated with a decline of the frontal cortex, (2) this decline leads to an executive control deficit, (3) with regard to memory, this deficit results in less efficient metamemory processes in recall and recognition tasks. This model is based on neuropsychological cognitive aging models which assume that the frontal cortex is not directly involved in information storage in the memory, but plays an essential role in implementing control processes allowing the subject to optimize his/her memory performance (Moscovitch and Winocur, 2002). The close relationship between executive functions and episodic FOK (Janowsky et al., 1989) would suggest that this task is of prime interest to test the hypothesis that age-related executive deficit is an essential factor in explaining metamemory aging. It would thus be logical for there to observe an age effect on episodic FOK mediated by the deterioration of executive functions. Another way of studying the link between the three variables — age, executive functions, and metamemory — is to investigate the metamemory control processes used by subjects to select strategies to adjust their behavior according to the task in order to optimize performance.

To provide some answers to these questions, the following section presents a review of work carried out on the relation between age and metamemory regulation. In addition, we report on a series of experiments conducted in our laboratory aimed at exploring the effects of aging on the measure of episodic FOK and metamemory control. The objective of these experiments was also to examine the role of executive functions in explaining these aging effects, using a correlational approach.

Aging and FOK accuracy

In the aging literature, several studies have examined the effect of age on the item-by-item accuracy of memory monitoring in JOL and FOK. A typical finding for JOL accuracy is that younger and older adults do not differ in their ability to predict which items they will be able to remember (Rabinowitz et al., 1982; Lovelace and Marsh, 1985; Shaw and Craik, 1989; Bieman-Copland and Charness, 1994; Connor et al., 1997). A similar pattern of results was observed for semantic FOK accuracy (Lachman et al., 1979; Bäckman and Karlsson, 1985; Butterfield et al., 1988). In a study by Butterfield et al. (1988), individual gamma correlation was used to evaluate accuracy of FOK and they found similar accuracy for younger and older adults. Given these results, according to Hertzog and Hultsch (2000) it appears that monitoring accuracy per se does not seem to be a viable explanatory variable for age differences in memory. Indeed, in JOL studies, metacognitive monitoring ability appears to be spared even when memory performance is impaired by aging. However, while such findings suggest that the accuracy of adults' on-line memory monitoring appears to be unrelated to age, assessed on either the JOL or the FOK procedure, there is still some evidence that older adults have difficulty with memory monitoring on items that have to be learned. Studies have found that older adults make more overall prediction errors when asked to judge how many items out of an entire list they will subsequently recall (Bruce et al., 1982; Rebok and Balcerak, 1989; Devolder et al., 1990). On the other hand, the FOK studies with older adults presented above have only examined prediction performance for well-learned semantic memory information. Since age-related differences in memory performance are typically much greater in episodic than semantic memory tasks (Craik et al., 1995), it can be assumed that age will have a reliable effect on episodic memory FOK accuracy. Furthermore, given Janowsky et al.'s findings (1989) that episodic memory FOK and semantic memory FOK can be dissociated in patients with frontal lobe damage which only affects episodic memory FOK accuracy, and on the basis of individual differences in frontal lobe functioning in elderly subjects, it can be hypothesized that the age effect will specifically alter episodic memory FOK accuracy. Our laboratory has specifically studied this question and below are the results of a series of experiments which aimed to test the hypothesis on three points: (1) that there is an aging effect on the FOK measure only when it involves episodic memory, (2) that there is a relationship between this measure and executive functioning, and (3) that executive deficit associated with aging is a pertinent factor in explaining the effect of age on episodic FOK.

In the first experiment (Souchay et al., 2000), a group of young and a group of elderly subjects were compared using an episodic memory FOK procedure similar to that developed by Schacter (1983). The episodic FOK protocol provided three measures, two relating to memory (cued-recall and recognition) and one to metamemory (FOK accuracy assessed by the individual gamma index correlation). The subjects also carried out two executive function tasks, the WCST and a verbal fluency test (FAS), and two independent measures of memory, story recall. The results revealed a significant age effect on the FOK measure. They also indicated that this measure was significantly correlated with a composite executive index reflecting the performance on the two executive tests but not with the memory measures. Finally, after a regression analysis, they showed that this effect was dependent on the executive deficit observed in elderly subjects. Statistical control of the executive score reduced the age effect on the episodic FOK measure by 86%, taking it to a statistically non-significant level. These data support the executive hypothesis of cognitive aging and the idea that certain monitoring measures (i.e., episodic FOK) are closely linked to executive functioning.

In a second experiment (Perrotin et al., 2006), the aim was to confirm that a specific relationship exists between executive functions and episodic FOK by concurrently analysing the roles played by executive deficit and the age-related processing speed slowing. The executive decline hypothesis (West, 1996; Raz, 2000) and processing speed slowing hypothesis (Salthouse, 1996) are two concurrent hypotheses that dominate current research to explain age-related declines in memory. Executive functioning and processing speed are impaired with normal aging, making them potential mediators of age differences in cognitive abilities. To explain memory aging, many studies have examined the processing speed hypothesis, and speed has consistently emerged as a mediating factor (see Luszcz and Bryan, 1999). This robust effect has also been found in studies that have examined the contribution of speed in memory performance together with a wide range of other cognitive predictors (see Luszcz et al., 1997).

Within the metamemory monitoring field, the slowing hypothesis has never been explored as a potential mediator between age and a specific metamemory FOK accuracy measure. Thus, the first objective of the experiment was to confirm the effect of age on metamemory monitoring whenever episodic memory is involved by comparing younger and older adults' groups on episodic FOK accuracy. The second aim was to examine the respective involvement of executive functioning and processing speed in episodic FOK. The third objective was to identify the major predictor of the age effect on episodic metamemory measure when executive and speed measures were evaluated concurrently. In line with the executive decline view of metamemory aging, we hypothesized that the age effect on metamemory FOK accuracy would be better accounted for as a function of executive dysfunctioning. The results showed a significant relationship between episodic memory FOK and executive functioning but not processing speed. A hierarchical regression analysis showed that when executive functioning and processing speed were evaluated concurrently on FOK, the executive score was the better mediator of age-related variance in FOK. Regarding metamemory processes, this result confirmed the significant age effect on FOK accuracy and confirmed the hypothesis that FOK accuracy is closely linked to executive functioning, suggesting that metamemory monitoring largely depends on the integrity of executive-frontal functioning.

In a most recent study (Perrotin et al., in press), the aim was to carry out a more specific analysis of cognitive mechanisms which allow subjects to make successful FOK judgments associated with episodic memory. More precisely, using a regression analysis, the study investigated the potential cognitive processes underlying FOK accuracy. The aim was to contrast four general cognitive constructs — executive functioning, processing speed, fluid intelligence, and episodic memory — as plausible mediators of FOK accuracy. The findings provide evidence that the main factor involved in accuracy of FOK judgments is executive functioning followed by memory. Analyses show that the most relevant model accounting for FOK accuracy combines two factors, executive functioning and episodic memory, the former having by far the greater role. Once executive and memory indices had been taken into consideration, fluid intelligence and processing speed no longer appeared as significant factors for FOK accuracy.

Episodic FOK and executive-frontal functioning

Overall, the results of these three studies confirm the hypothesis that FOK accuracy is closely linked to executive-frontal functioning. However, in contrast to the findings of the study of Souchay et al. (2000), which also included memory-independent measures, the data of Perrotin et al.'s study (in press) show that, after the executive involvement, memory functioning makes a weak but significant independent contribution to FOK accuracy. This suggests that memory and executive functioning may be two crucial factors entering into the computation of FOK. In recent years, FOK mechanisms have been thoroughly documented (see Metcalfe et al., 1993; Koriat, 1995; Koriat and Levy-Sadot, 2001) and the literature has proposed relevant conceptual frameworks against which findings of neuropsychological studies may now be examined. Koriat and Levy-Sadot (2001) recently put forward an interactive model integrating these different views. It assumes that FOK is determined both by the overall familiarity of the presented cue (Reder, 1987; Metcalfe et al., 1993) and by the accessibility of a variety of pertinent information, such as structural, contextual and semantic attributes, fragments of the target, etc., during the course of the search-and-retrieval process for the target (Koriat, 1993). Subjects may infer their FOK judgment on the basis of these various activated clues and their intensity (e.g., ease of access, vividness, specificity, persistence). Within this FOK model, it could be hypothesized that executive functioning is essentially involved during the accessibility assessment stage of partial information rather than during the preliminary cue familiarity stage, given the rapid and automatic aspect of familiarity judgments (Reder, 1987). In support of this hypothesis, Schnyer et al. (2004) found that patients with frontal lesions displayed an intact ability to monitor the familiarity of the cue, but failed when monitoring was based on retrieved information.

Consistent with this view, in a recent fMRI study, Schnyer et al. (2005) have proposed an accuracy system that engages an interactive retrieval network on which a monitoring function is exerted. The retrieval network encompasses the lateral temporal cortex, the hippocampus, and the inferior frontal cortex (IFC). On the basis of this retrieval network, a directional connection pathway leads from the temporal cortex to the VMPC, and this latter region is postulated to serve a monitoring function during retrieval (Schnyer et al., 2004). With respect to the FOK model proposed by Koriat and Levy-Sadot (2001), this FOK accuracy network is assumed to be activated at the accessibility assessment stage of FOK (Schnyer et al., 2005). Moreover, according to Moscovitch and Winocur (2002), since neuroimaging studies have shown that dorsolateral prefrontal cortex (DLPFC) is mainly involved in initiating and orienting memory retrieval operations, it can be hypothesized that this region should play an important role at the beginning of the search-and-retrieval process for the target associated with the FOK process.

These observations are consistent with the idea that memory is an important FOK component in that it provides the basis on which monitoring exerts its action to produce FOK judgments. In line with this view, a growing metamemory literature on neurological populations support that patients with memory loss in addition to frontal lobe dysfunction, typical of the syndromes that occur in dementia, Korsakoff's syndrome, and frontal patients, perform the most poorly on monitoring metamemory tasks (Pannu and Kaszniak, 2005). Then it can be proposed that episodic FOK accuracy involves memory components of storage and retrieval ensured by temporal cortex areas, whose by-products are controlled by executive functioning subserved by frontal structures.

Aging and frontal model of FOK

Returning to aging, several neuroimaging studies in recent years have shown that the three prefrontal structures (DLPFC, VMPFC, IFC) assumed to be involved in the FOK process are less activated in elderly than young subjects when performing memory or executive tasks (see Cabeza, 2002). These observations are in line with the hypothesis that the deficit in episodic FOK observed in elderly subjects could be explained by prefrontal cortex dysfunctioning. More precisely, in light of the data observed (Souchay et al., 2000; Perrotin et al., 2006; Perrotin et al., in press) regarding the link between aging and the episodic FOK task, and neuroimaging data (Schnyer et al., 2005), it is now possible to hypothesize more precisely that the age effect on episodic FOK could be the consequence of a reduced activation of these three prefrontal structures (DLPFC, VMPFC, IFC). This reduced activation would be associated with the memory accessibility and decision-making processes involved in the FOK judgment. Thus the impact of aging due to executive-frontal underfunctioning would intervene in the main steps of the accessibility process: strategic initiation of the memory search, recollection of partial information from long-term memory to working memory, evaluation of the quality and correctness of the information retrieved, and finally the decision-making stage relating to the plausibility of subsequent recognition of the solicited target.

According to Koriat and Levy-Sadot (2001) model, FOK is also determined by the overall familiarity of the presented cue. The hypothesis that aging affects the mechanism of familiarity-novelty of the presented cue may also account for age effects on episodic memory FOK. However, it should be noted that available data on the aging of memory appear to argue more in favor of the hypothesized episodic information accessibility deficit. For example, while older adults have reduced memory for the features of events, as stated above, some studies, using the remember-know paradigm, have also found them to be less likely than younger adults to report that they consciously recollect contextual details of a remembered event, although their confidence in the event's familiarity is not necessarily impaired (Parkin and Walter, 1992; Clarys et al., 2002).

Nevertheless, the finding that an age-related change may appear in an FOK task whenever episodic memory is involved can appear inconsistent with the findings of initial studies on the relationship between aging and FOK accuracy in which no significant effect of aging on FOK accuracy measures has been observed (Lachman et al., 1979; Bäckman and Karlsson, 1985; Butterfield et al., 1988). However, these experiments differ from ours in several respects, the main difference being the type of memory task the participants performed and on which they based their recognition judgments. In the above studies, a semantic memory task involving recognition predictions about information already stored in memory was used, whereas the studies of Souchay et al. (2000) and Perrotin et al. (2006) were based on an episodic memory task involving recognition predictions about newly learned information. Thus, when compared with the findings of the previous experiments about aging and FOK, our results suggest that only episodic memory FOK accuracy declines during normal aging. This hypothesis requires testing directly by examining age-related effects on episodic and semantic FOK accuracy in the same sample of subjects. In a recent study (Souchay et al., 2007), we aimed to address this question specifically. The main purpose of the experiment was to confirm the hypothesis that FOK accuracy is impaired in older adults when assessed by episodic but not semantic FOK accuracy. This was evaluated by directly comparing age effects on both semantic and episodic FOK within the same individuals, with the same type of materials, and with comparable FOK judgment procedures. The results confirmed that age affects the accuracy of the FOK when predictions are made on an episodic memory task but not on a semantic task. Following on from these results, we conducted a second experiment to examine the quality of the subjective experience accompanying episodic FOK judgments in older adults by using the remember-know distinction paradigm (Tulving, 1985). The main aim was to assess the relationship between FOK accuracy and the availability of contextual cues and partial information available at test as indexed by remember responses. This second experiment showed that those individuals who made more accurate FOK predictions prior to the test phase made more recognition judgments on the basis of remembering. Moreover, the age difference in FOK accuracy was removed when measures of remembering were taken into account. Thus, it can be argued that the aging FOK deficit specific to episodic memory tasks seems based on a lack of episodic richness, contextual information, or retrieval cues that are brought into play when participants are asked to report their recollective experience. Thus, it can be suggested that the cue generation triggering memory in the accessibility process of FOK proposed by Koriat (1995) is mainly based on the ability to generate learning of contextual cues in the case of episodic FOK, while cues are more directly related to the target information in the case of semantic FOK. One main difference between episodic FOK and semantic FOK that may plausibly explain the differential involvement of aging in these two tasks is that only the former involves monitoring information associated with the target during the study phase (*i.e., source memory*). Consistent with this view, many studies have already demonstrated that older adults are less likely to remember various contextual features (for a review, see Craik and Jennings, 1992; Henkel et al., 1998), and that they are less efficient at integrating contextual information into the target memory trace (Chalfonte and Johnson, 1996; Glisky et al., 2001). Thus, the FOK judgments of older adults can be expected to be particularly impaired in the episodic memory condition, in which context retrieval processes are assumed to be highly critical. In this case, the contextual episodic cues generated by elderly subjects would be insufficient in quality or quantity to trigger the episodic memory. Moreover, since generating episodic cues, probably less familiar, may be more processing resource-dependent than generating cues about well-learned semantic information, and because there is evidence that the magnitude of age-related differences in a variety of cognitive tasks rises as the processing demand of the task increases (Salthouse, 1991; Luszcz and Bryan, 1999), it would also be expected that elderly subjects' performance would be impaired in episodic FOK accuracy. These ideas remain in line with the executive-frontal hypothesis of the age effect on episodic FOK insofar as on the one hand several studies have shown that the ability to retrieve contextual cues in episodic memory depends on frontal functioning (Glisky et al., 1995), and on the other hand, it is also accepted that executive-frontal functions are more specifically involved in strongly resource-dependent tasks.

Aging and metacognitive control

Previous research has shown that older adults fail to produce the types of effective strategies that young adults spontaneously adopt (Schmitt et al., 1981; see Hertzog and Hultsch, 2000). As suggested by Murphy et al. (1987), such strategy production deficiencies may be explained by a metamemory deficit. Two approaches have been investigated regarding the relationship between aging and metamemory control: the influence of memory monitoring and of knowledge about memory on the use of strategies in order to optimize learning and memory retrieval.

A few studies have examined the influence of memory monitoring on the use of strategies. An increasing body of evidence has revealed that metacognitive judgments influence subsequent study-time-allocation strategies, a phenomenon termed the *monitoring-affects-control hypothesis* (Mazzoni and Cornoldi, 1993; Dunlosky and Connor, 1997). Correlational evidence in support of this hypothesis has been found in JOLs elicited during study (Mazzoni et al., 1990) and in FOK judgment (Nelson and Leonesio, 1988). In these two studies an inverse relationship between judgments and study time was observed. The self-regulation of behavior also appeared to be based on memory outcomes. Previous research has shown that subjects allocated less study time to items they had recalled than to items they had not (Mazzoni and Cornoldi, 1993; Dunlosky and Connor, 1997).

In a recent study, Dunlosky and Connor (1997) investigated this issue in elderly subjects by examining how aging affects the utilization of monitoring in the allocation of paired-associates study time during multiple study-test trials. During each trial, subjects paced the presentation of individual items and later judged the likelihood of recalling each item on an upcoming test (JOL judgments). Recall took place after all items had been studied and judged. Individual gamma correlation between JOL ratings on one trial and study times on the next trial were negative, which suggests that subjects utilized monitoring to allocate study time. However, the magnitude of these correlations was smaller for older than younger adults. Moreover, the authors found that the magnitude of the correlations between recall performance on one trial and study time on a second trial was also lower for older than younger adults. These results suggest that older adults do not utilize on-line monitoring (i.e., memorability ratings and memory outcome) to allocate study time at the same level as young adults.

To extend these results to FOK judgments, we carried out an experiment (Souchay and Isingrini, 2004a) to investigate whether aging affects the degree to which metacognitive judgments made in the course of the retrieval process are utilized to allocate study time. According to the *monitoring-affects-control hypothesis*, we predicted that there would be an inverse relationship between FOK and the amount of self-paced study time allocated to various items. In other words, items lower in FOK should receive more self-paced study time. In the light of Dunlosky and Connor's (1997) results showing an age effect on the gamma correlation between JOL judgments and study time, we expected the magnitude of the relationship between FOK ratings and study time to be weaker for older than for younger adults. A measure of the allocation of self-paced study time was computed: the individual gamma correlation between FOK ratings on one trial and self-paced study on the next trial. This procedure also enabled comparison of the impact of memorability ratings (FOK) with the memory outcome (recall) on time allocation strategies in aging. To the extent that metamemory control processes mediate subsequent memory (Nelson and Narens, 1990), individual differences found in the allocation of study time were expected to account for individual differences in recall performance. Regardless of why older adults are less fine-tuned at allocating study time, such deficits may contribute to their poorer memory performance during self-regulated learning. Thus, given that age-related differences appear in allocation measures, a secondary goal of the study was to determine whether individual differences in allocation measures would account for age-related differences in recall performance. To examine these different questions, two groups, one of young and one of elderly subjects, were compared while learning paired associates during two study-recall tests. First, participants learned 40 critical cue-target pairs. Next, they had to recall the target word of each pair and make a FOK judgment. Finally, they studied the 40 cue-target pairs on a self-paced study trial and then recalled the items. The results indicated that both groups allocated their self-paced study time in accordance with their FOK judgments, with items higher in FOK receiving less self-paced study time. The magnitude of this relationship was significantly different in young and elderly subjects. Finally, a regression analysis indicated that age-related differences in the allocation of study time accounted for the age-related differences in memory performance.

To extend this research on aging and metacognitive control, a second study was carried out (Souchay and Isingrini, 2004b) which examined the effect of aging on strategy during learning as a consequence of knowledge about memory. The first aim of this study was to evaluate whether aging affects the capacity to regulate study time and item rehearsal at encoding in relation to task difficulty in a readiness-recall task of words. The second aim was to investigate what role executive functions play in this ability. In the literature, few studies have investigated the age effect on metamemory control using a readiness-recall task. Murphy et al. (1981) used this task to show that young subjects adjust their study time to the task difficulty better than elderly adults. This first result was confirmed in another study by Murphy et al. (1987) which also showed that young subjects repeated the words to learn more as the difficulty increased. Overall, these data are in line with the idea that aging is associated with a deficit in strategy manipulation during learning.

Based on Murphy et al.'s study (1987), we compared groups of young and elderly adults on their capacity to regulate study time and item rehearsal at encoding according to task difficulty in a readiness-recall task. As mentioned above, this procedure allows learning strategies to be assessed directly independent of recall performance, by asking participants to determine their own study time and rehearsal. Moreover, to test rehearsal directly, we used the overt rehearsal procedure (Rundus and Atkinson, 1970) which has been used successfully with older adults (Sanders et al., 1980). Furthermore, following Nelson and Naren's (1990) suggestion that metacognitive control processes (i.e., the capacity to adopt and modify strategies) mediate subsequent memory performance, the aim of the study was also to examine whether age-related differences in strategy manipulation mediate age-related differences in memory performance. To determine the metamemory control effect on memory performance, several authors have suggested contrasting the memory effects of self-paced and experimenter-paced procedures (Jacoby, 1973; Mazzoni and Cornoldi, 1993). Thus, two memory tests were carried out: the first was a self-paced task which enabled investigation of spontaneous learning strategies, and the second was an experimenter-paced task. Performance on these tasks were compared in order to determine whether strategy utilization had an impact on memory performance and to evaluate whether the age-related difference in strategies at encoding could explain the age-related differences in memory performance. Finally, to investigate the hypothesis that metacognitive control assessed by subjects' capacity to manipulate strategies at encoding is related to executive functions, participants completed standard neuropsychological tests used to assess executive functioning. In line with the executive decline hypothesis of metamemory, we also attempted to determine whether the assumed metamemory decline in older people is reliably related to deterioration of executive functions.

The task used for metamemory control assessment was a word-recall test. The subjects had to learn and then recall words on lists of varying lengths (7, 9, and 11 words). Two metamemory control measures were chosen corresponding to the proportional increase in study time and number of word repetitions for conditions with 9 and 11 words compared to the condition with 7 words which was taken as the control condition. Higher indexes indicate greater awareness by the subject that a 9- or 11-word list requires longer study time and more repetitions per word than a 7-word list. These metamemory control measure can thus be interpreted as task-difficulty adjustment index according to the difficulty of the memory task. Each subject performed two executive function tests: the WSTC and a FAS. The results highlighted a significant age effect on the task-difficulty adjustment index, showing that the elderly, unlike the young, do not increase their study time or number of repetitions as task difficulty increases. The results also indicate that memory performance is better overall in the *self-paced study procedure* than in the *experimenter-paced procedure*. However, the significant interaction between the age factor and the learning condition revealed that the elderly benefited less than the young from the *self-paced procedure*. In line with the idea that there is a link between memory control and executive control, the data highlighted a significant partial correlation (age, years of education, and verbal ability partialled out) between an executive index and the task-difficulty adjustment index. Finally, the proportion of variance accounted for by age in this adjustment index dropped by 86% when the executive index was controlled for, indicating that executive deficit can be taken as a relevant mediator of the age effect on metamemory control.

Overall, the results of the studies carried out on aging and metacognitive control confirm a significant age effect in metamemory control measures. Younger adults were consistently found to be better in adjusting their strategies at encoding to task demand than older adults. The results of Souchay and Isingrini (2004b) suggest that this deficit may be explained by the executive decline associated to aging. Moreover, the results of Souchay and Isingrini (2004a), in line with those of Mazzoni and Cornoldi (1993), by contrasting self-paced and experimenter-based learning conditions, provide evidence that memory performance depends on metamemory processes. Younger adults recalled significantly more words in the self-paced than in the experimenter-paced test, while this was not the case for older subjects.

Altogether these results suggest also that older subjects' difficulties to adjust their strategies to task difficulty or to by-product information from monitoring judgment processes may in part explain their poorer memory performance. These results are in line with the view of Berry et al. (1989) that the age effect on metamemory may mediate the age-related differences in memory.

This raises the question of why older subjects seem to have more difficulties than younger subjects to adjust their strategies to task difficulty or to information resulting from a monitoring process. The metamemory and aging literature proposes two types of explanation. First, older subjects may lack some key element(s) of metamemory information, pertaining to metacognitive knowledge and to metacognitive monitoring, required to develop an appropriate strategy or to use it efficiently. A second and complementary explanation is that older subjects may have difficulty self-initiating the coordination of meta-memorial information (i.e., metacognitive knowledge and/or on-line monitoring) and control processes spontaneously (Dunlosky and Connor, 1997; Souchay and Isingrini, 2004a). This account refers to Craik's (1986) hypothesis that older adults are less likely to self-initiate processes that would yield maximal memory performance.

Concluding remarks and future research

The fact that metamemory regulation processes are dependent on executive-frontal abilities and that these same abilities are impaired with aging constitute a strong argument supporting the hypothesis that there is an age effect on at least some metamemory monitoring and control measures. With regard to monitoring, no aging effect has been observed from the initial studies reported in the literature investigating measures of either FOK linked to semantic memory or JOL. However, the studies recently carried out in our laboratory (Souchay et al., 2000; Perrotin et al., 2006; Souchay et al., 2007) have clearly demonstrated that FOK associated with episodic memory in elderly subjects is significantly less accurate than in young adults. Following these observations, future research should be directed at understanding the mechanisms that constitute episodic FOK and which are susceptible to modification with age. In line with this and with reference to Koriat and Levy-Sadot's (2001) FOK model, we have put forward the hypothesis that aging mainly affects accessibility to partial information which allows the FOK judgment to be made. This hypothesis of an age-related impairment in the accessibility process opens up several possibilities for future research in particular, to check that, unlike semantic FOK, the reduced accuracy in episodic FOK of elderly subjects is in fact the consequence of difficulties in retrieving features linked to the learning context. At a neuropsychological level, it will involve checking that, during aging, the prefrontal regions linked to the accessibility process (DLPFC, IFC, VMPFC) are activated less when an episodic FOK task is performed. It will be important to investigate which of the events involved in the accessibility process (i.e., initiation of the search, generation of memory triggering cues, feeling of rightness, overt judgment, and decision making) are specifically affected by aging, using an approach enabling measures for each process to be obtained.

The results relating to the age effect on metamemory control have until now provided less contradictory data than those observed for monitoring. All the experiments conducted in this field have confirmed an age effect on the ability to adapt learning strategies to the task demands or to monitoring judgments (JOL and FOK). In this context, strategies linked to memory retrieval process have yet to be investigated. It would also be interesting to try to identify the factors underlying the aging effect on metamemory control. Several contenders have been suggested in this chapter: lack of metacognitive knowledge, metacognitive monitoring difficulty, and difficulty to initiate the coordination of metamemory information and control processes. In recent studies (Souchay and Isingrini, 2004a, b), we have highlighted that the difficulties in metamemory control experienced by elderly subjects could explain their memory difficulties. This question remains open, and future research is required to investigate this hypothesis in depth in connection with the theory that metamemory monitoring and control influence memory performance.

The experiments carried out in our laboratory using a correlational approach associated with neuroimaging observations of frontal lesion patients have confirmed a close link between executive-frontal functions and metamemory regulation. Furthermore, with regard to aging, all our data converge towards the idea that the age-related executive-frontal deficit explains the age effect on certain monitoring tasks (episodic FOK) and on metamemory control. These observations raise the question of identifying the executive processes (selecting, maintaining, updating, shifting, inhibiting) specifically involved in control and monitoring processes and which could also explain the age effect. Another interesting question is to know whether the involvement of executive functions can be generalized to all the metamemory regulation situations. Certain data in the literature seem to indicate that this is not the case. Likewise, it would seem that aging does not affect all metamemory tasks, with JOL and semantic FOK remaining unaltered. These data raise the question as to whether the aging-metamemory link parallels the link between executive-frontal functions and metamemory, as well as to the unicity of metamemory monitoring. By studying the dissociations in metamemory linked to age and to executive-frontal functions, it should be possible to understand better the specific mechanisms involved in each metamemory task and as a result the possible effects of aging on these mechanisms.

Finally, understanding the effects of normal aging on metamemory measures should help us understand how this function changes in age-related pathologies such as Alzheimer's disease. The experimental measures developed for metamemory regulation, in particular the monitoring judgment measures, could be valuable indicators of the anosognosia phenomenon which appears in the early stages of AD. In this context, the observation that executive-frontal functions are affected in the early stages of AD leads to the hypothesis of a premature deficit in metamemory monitoring and control in this disease. This is supported by a number of studies (see Pannu and Kasczniak, 2005). In our laboratory, two studies have investigated this question using an episodic FOK paradigm in AD patients (Souchay et al., 2002) and in subjects with Mild Cognitive Impairment (MCI: term used to define a transitional zone between normal aging and dementia) (Perrotin et al., 2007). A significant reduction in FOK accuracy, measured using individual gamma correlation, was observed in patients with AD and MCI compared to control subjects. However, in contrast to the control groups, the ability of people diagnosed with AD or MCI to predict with accuracy their recognition performance was shown to be related primarily to their memory abilities rather than to their executive functioning. These results suggest that these two populations do not base their FOK judgments on the same mechanisms than normal older adults.

Acknowledgment

We thank Badiâa Bouazzaoui for her helpful comments.

References

Anderson, N.D. and Craik, F.I.M. (2000) Memory in the aging brain. In: Tulving E. and Craik F.I.M. (Eds.), The Oxford Handbook of Memory. Oxford University Press, Oxford, pp. 411–426.

Bäckman, L. and Karlsson, T. (1985) The relation between level of general knowledge and feeling-of-knowing: an adult age study. Scand. J. Psychol., 26: 249–258.

Berry, J., West, R. and Dennehey, D. (1989) Reliability and validity of the memory self-efficacy questionnaire. Dev. Psychol., 25(5): 701–713.

Bieman-Copland, S. and Charness, N. (1994) Memory knowledge and memory monitoring in adulthood. Psychol. Aging, 9: 287–302.

Bruce, P.R., Coyne, A.C. and Botwinick, J. (1982) Adult age difference in metamemory. J. Gerontol., 37(3): 354–357.

Bryan, J., Luszcz, M.A. and Pointer, S. (1999) Executive function and processing resources as predictors of adult age differences in the implementation of encoding strategies. Aging Neuropsychol. Cogn., 6(4): 273–287.

Butterfield, E., Nelson, T. and Peck, V. (1988) Developmental aspects of the feeling of knowing. Dev. Psychol., 24(5): 654–663.

Cabeza, R. (2002) Functional neuroimaging of cognitive aging. In: Cabeza R. and Kingstone A. (Eds.), Handbook of Functional Neuroimaging of Cognition. MIT Press, Cambridge, MA, pp. 331–378.

Chalfonte, B.L. and Johnson, M.K. (1996) Feature memory and binding in young and older adults. Mem. Cogn., 24(4): 403–416.

Clarys, D., Isingrini, M. and Gana, K. (2002) Aging and episodic memory: mediators of age-related differences in remembering and knowing. Acta Psychol., 109: 315–329.

Connor, L., Dunlosky, J. and Hertzog, C. (1997) Age-related differences in absolute but not relative metamemory accuracy. Psychol. Aging, 12(1): 50–71.

Craik, F., Anderson, N., Kerr, S. and Li, K. (1995) Memory changes in normal ageing. In: Baddeley A.D., Wilson B.A. and Watts F.N. (Eds.), Handbook of Memory Disorders. Wiley, Chichester, England, pp. 211–241.

Craik, F.I.M. (1986) A functional account of age difference in memory. In: Flix F. and Hagendorf H. (Eds.), Human Memory and Cognitive Capabilities, Mechanisms and Performances. Elsevier Science, New York, pp. 409–442.

Craik, F.I.M. and Jennings, J.M. (1992) Human memory. In: Craik F.I.M. and Salthouse T.A. (Eds.), The Handbook of Aging and Cognition. Laurence Erlbaum Associates, Hillsdale, NJ, pp. 51–110.

Craik, F.I.M., Morris, L.W., Morris, R.G. and Loewen, E.R. (1990) Relations between source amnesia and frontal lobe functioning in older adults. Psychol. Aging, 5(1): 148–151.

Crawford, J.R., Bryan, J., Luszcz, M.A., Obonsawin, M.C. and Stewart, L. (2000) The executive decline hypothesis of cognitive aging: do executive deficits qualify as differential deficits and do they mediate age-related memory decline. Aging Neuropsychol. Cogn., 7: 9–31.

Devolder, P., Brigham, M. and Pressley, M. (1990) Memory performance awareness in younger and older adults. Psychol. Aging, 5(2): 291–303.

Dixon, A., Hultsch, D. and Hertzog, C. (1988) The metamemory in adulthood (MIA) questionnaire. Psychopharmacol. Bull., 24: 671–688.

Dunlosky, J. and Connor, L. (1997) Age difference in the allocation of study time account for age differences in memory performance. Mem. Cogn., 25(5): 691–700.

Fabiani, M. and Friedman, D. (1997) Dissociations between memory for temporal order and recognition memory in aging. Neuropsychologia, 35: 129–141.

Fernandez-Duque, D., Baird, J. and Posner, M. (2000) Executive attention and metacognitive regulation. Conscious Cogn., 9: 288–307.

Glisky, L., Polster, M.R. and Routhieaux, B.C. (1995) Double dissociation between item and source memory. Neuropsychologia, 9(2): 229–235.

Glisky, E.L., Rubin, S.R. and Davidson, P.S. (2001) Source memory in older adults: an encoding or a retrieval problem? J. Exp. Psychol. Learn. Mem. Cogn., 27: 1131–1146.

Hart, J.T. (1965) Memory and the feeling-of-knowing experiments. J. Educ. Psychol., 56: 208–216.

Henkel, L.A., Johnson, M.K. and De Leonardis, D.M. (1998) Aging and source monitoring: cognitive processes and neuropsychological correlates. J. Exp. Psychol. Gen., 127: 251–268.

Hertzog, C. and Hultsch, D.F. (2000) Metacognition in adulthood and old age. In: Craik F.I.M. and Salthouse T.A. (Eds.), The Handbook of Aging and Cognition (2nd eds.).

Lawrence Erlbaum Associates Publishers, Mahwah, NJ, pp. 417–466.

Isingrini, M. and Taconnat, L. (1997) Aspect du vieillissement normal de la mémoire. Psychol. Française, 42: 319–331.

Isingrini, M. and Vazou, F. (1997) Relation between fluid and frontal lobe functioning in older adults. Int. J. Aging Hum. Dev., 45: 99–109.

Jacoby, L. (1973) Encoding processes, rehearsal, and recall requirements. J. Verbal Learn. Verbal Behav., 12: 302–310.

Janowsky, J.S., Shimamura, A.P. and Squire, L.R. (1989) Memory and metamemory: comparisons between patients with frontal lobe lesions and amnesic patients. Psychobiology, 17(1): 3–11.

Koriat, A. (1993) How do we know that we know? The accessibility model of the feeling of knowing. Psychol. Rev., 100: 609–639.

Koriat, A. (1995) Dissociating knowing and the feeling of knowing: further evidence for the accessibility model. J. Exp. Psychol. Gen., 124(3): 311–333.

Koriat, A. and Levy-Sadot, R. (2001) The combined contributions of the cue-familiarity and accessibility heuristics to feeling-of-knowing. J. Exp. Psychol. Learn. Mem. Cogn., 27: 34–53.

Lachman, J., Lachman, R. and Thronesbery, C. (1979) Metamemory through the adult life span. Dev. Psychol., 15(5): 543–551.

Lovelace, E. and Marsh, G. (1985) Prediction and evaluation of memory performance by young and old adults. J. Gerontol., 40(2): 192–197.

Luszcz, M. and Bryan, J. (1999) Toward understanding age-related memory loss in late adulthood. Gerontology, 45(1): 2–9.

Luszcz, M., Bryan, J. and Kent, P. (1997) Predicting episodic memory performance of very old men and women: contributions from age, depression, activity, cognitive ability, and speed. Psychol. Aging, 12(2): 340–351.

Mazzoni, G. and Cornoldi, C. (1993) Strategies in study item allocation: why is study time sometimes not effective? J. Exp. Psychol. Gen., 122(1): 47–60.

Mazzoni, G., Cornoldi, C. and Marchitelli, G. (1990) Do memorability ratings affect study-time allocation? Mem. Cogn., 18(2): 196–204.

Metcalfe, J. (1993) Novelty monitoring, metacognition, and control in a composite holographic associative recall model: implications for Korsakoff amnesia. Psychol. Rev., 100: 3–22.

Metcalfe, J., Schwartz, B.L. and Joaquim, S.G. (1993) The cue-familiarity heuristic in metacognition. J. Exp. Psychol. Learn. Mem. Cogn., 19: 851–864.

Moscovitch, M. and Winocur, G. (2002) The frontal cortex and working with memory. In: Stuss D.T. and Knight R.T. (Eds.), Principles of Frontal Lobe Function. Oxford University Press, New York, pp. 188–209.

Murphy, M., Sanders, R., Gabriesheski, A. and Schmitt, F. (1981) Metamemory in the aged. J. Gerontol., 36(2): 185–193.

Murphy, M., Schmitt, F., Caruso, M. and Sanders, R. (1987) Metamemory in older adults: the role of monitoring in serial recall. Psychol. Aging, 2(4): 331–339.

Nelson, T.O. (1984) A comparison of current measures of the accuracy of feeling-of-knowing predictions. Psychol. Bull., 95: 109–133.

Nelson, T. and Leonesio, J. (1988) Allocation of self-paced study time and the labor in vain effect. J. Exp. Psychol. Learn. Mem. Cogn., 14(4): 676–686.

Nelson, T.O. and Narens, L. (1990) Metamemory: a theoretical framework and new findings. In: Bower G.H. (Ed.), The Psychology of Learning and Motivation: Advances in Research and Theory, Vol. 26. Academic Press, San Diego, CA, pp. 125–173.

Pannu, J.K. and Kaszniak, A.W. (2005) Metamemory experiments in neurological populations: a review. Neuropsychol. Rev., 15: 105–130.

Parkin, A.J. and Walter, B.M. (1992) Recollective experience, normal aging, and frontal dysfunction. Psychol. Aging, 7: 290–298.

Perrotin, A., Belleville, S. and Isingrini, M. (2007) Metamemory monitoring in mild cognitive impairment: evidence of a less accurate episodic feeling-of-knowing. Neuropsychologia, 45: 2811–2826.

Perrotin, A., Isingrini, M., Souchay, C., Clarys, D. and Taconnat, L. (2006) Episodic feeling-of-knowing accuracy and cued recall in the elderly: evidence for double dissociation involving executive functioning and processing speed. Acta Psychol., 122: 58–73.

Perrotin, A., Tournelle, L. and Isingrini, M. (in press) Executive functioning and memory as potential mediators of the episodic feeling of knowing accuracy. Brain Cogn.

Rabbitt, P. (1997) Introduction: methodologies and models in the study of executive function. In: Rabbitt P. (Ed.), Methodology of Frontal and Executive Function. Psychology Press, Hove, England, pp. 1–38.

Rabinowitz, J., Ackerman, B., Craik, F.I.M. and Hinckley, J. (1982) Aging and metamemory: the role of relatedness and imagery. J. Gerontol., 37: 688–695.

Raz, N. (2000) Aging of the brain and its impact on cognitive performance: integration of structural and functional findings. In: Craik F.I.M. and Salthouse T.A. (Eds.), The Handbook of Aging and Cognition (2nd ed.). Erlbaum, Mahwah, NJ, pp. 1–90.

Raz, N., Gunning-Dixon, F., Acker, J., Head, D. and Dupuis, J.H. (1998) Neuroanatomical correlates of cognitive aging: evidence from structural magnetic resonance in aging. Neuropsychologia, 12: 95–114.

Rebok, G. and Balcerak, L. (1989) Memory self-efficacy and performance differences in young and old adults: the effect of mnemonic training. Dev. Psychol., 25(5): 714–721.

Reder, L.M. (1987) Strategy selection in question answering. Cognit. Psychol., 19: 90–138.

Rundus, D. and Atkinson, C. (1970) Rehearsal processes in free recall: a procedure for direct observation. J. Verbal Learn. Verbal Behav., 9: 99–105.

Salthouse, T. (1991) Theoretical perspectives and cognitive aging. Erlbaum, Hillsdale, NJ.

Salthouse, T.A. (1996) The processing speed theory of adult age differences in cognition. Psychol. Rev., 103: 403–428.

Salthouse, T., Fristoe, N. and Rhee, S. (1996) How localized age-related effects on neuropsychological measures? Neuropsychologia, 10(2): 272–285.

Sanders, E., Murphy, M., Schmitt, F. and Walsh, K. (1980) Age differences in free recall rehearsal strategies. J. Gerontol., 35(4): 550–558.

Schacter, D.L. (1983) Feeling-of-knowing in episodic memory. J. Exp. Psychol. Learn. Mem. Cogn., 9: 39–54.

Schacter, D.L. (1997) False recognition and the brain. Curr. Dir. Psychol. Sci., 3: 65–70.

Schmitt, F.A., Murphy, M.D. and Sanders, R.E. (1981) Training older adult rehearsal strategies. J. Gerontol., 36: 329–337.

Schnyer, D.M., Nicholls, L. and Verfaellie, M. (2005) The role of VMPC in metamemorial judgments of content retrievability. J. Cogn. Neurosci., 17: 832–846.

Schnyer, D.M., Verfaellie, M., Alexander, M.P., LaFleche, G., Nicholls, L. and Kaszniak, A.W. (2004) A role for right medial prefrontal cortex in accurate feeling-of-knowing judgments: evidence from patients with lesions to frontal cortex. Neuropsychologia, 42: 957–966.

Shaw, R. and Craik, F.I.M. (1989) Age differences in predictions and performance on a cued recall task. Psychol Aging, 14(2): 131–135.

Shimamura, A.P. (1995) Memory and frontal lobe function. In: Gazzaniga M.S. (Ed.), The Cognitive Neurosciences. MIT Press, Cambridge, MA, pp. 803–813.

Shimamura, A.P. (2000) Toward a cognitive neuroscience of metacognition. Conscious Cogn., 9: 313–323.

Souchay, C., Isingrini, M. and Espagnet, L. (2000) Relations between feeling-of-knowing and frontal lobe functioning in older adults. Neuropsychologia, 14(2): 299–309.

Souchay, C. and Isingrini, M. (2004a) Age-related differences in the relation between monitoring and control learning. Exp. Aging Res., 30: 179–194.

Souchay, C. and Isingrini, M. (2004b) Age related differences in metacognitive control: role of executive functioning. Brain Cogn., 56: 89–99.

Souchay, C., Isingrini, M. and Gil, R. (2002) Alzheimer's disease and feeling-of-knowing in episodic memory. Neuropsychologia, 40: 2386–2396.

Souchay, C., Moulin, C.J.A., Clarys, D., Taconnat, L. and Isingrini, M. (2007) Diminished episodic memory awareness in older adults: evidence from feeling-of-knowing and recollection. Conscious Cogn., 16: 769–784.

Spencer, W.D. and Raz, N. (1995) Differential effects of aging on memory for content and context: a meta-analysis. Psychol. Aging, 10(4): 527–539.

Stuss, D.T., Craik, F.I.M., Sayer, L., Franchi, D. and Alexander, M.P. (1996) Comparison of older elderly subjects to patients with frontal lesions: evidence from word list learning. Psychol. Aging, 11: 387–395.

Taconnat, L., Froger, C., Sacher, M. and Isingrini, M. (2008) Generation and associative encoding in young and old adults. Exp. Psychol., 55: 23–30.

Troyer, A.K., Graves, R.E. and Cullum, C.M. (1994) Executive functioning as a mediator of the relationship between age and episodic memory in healthy aging. Aging Neuropsychol. Cogn., 1(1): 45–53.

Tulving, E. (1985) Memory and consciousness. Can. Psychol., 26(1): 1–12.

West, R. (1996) An application of prefrontal cortex function theory to cognitive aging. Psychol. Bull., 120(2): 272–292.

Wheeler, M.A., Stuss, D.T. and Tulving, E. (1995) Frontal lobe damage produces episodic memory impairment. J. Int. Neuropsychol. Soc., 1: 525–536.

CHAPTER 25

Cognitive neuroscience studies of semantic memory in Alzheimer's disease

Howard Chertkow[1,2,3,4,*], Christine Whatmough[1,2,4], Daniel Saumier[2,4,5] and Anh Duong[1,5]

[1]*Bloomfield Centre for Research in Aging, Sir Mortimer B. Davis–Jewish General Hospital, McGill University, Montréal, QC, Canada*
[2]*Department of Neurology and Neurosurgery, McGill University, Montréal, QC, Canada*
[3]*Division of Geriatric Medicine, Department of Medicine, Sir Mortimer B. Davis–Jewish General Hospital, McGill University, Montréal, QC, Canada*
[4]*Centre de Recherche, Institut Universitaire de Gériatrie de Montréal, Université de Montréal, Montréal, QC, Canada*
[5]*Neurochem Inc., Laval, QC, Canada*

Abstract: Semantic memory is the component of long-term memory that stores our concepts about the world. The disruption of semantic memory as a result of brain damage may have profound negative consequences on an individual's ability to name objects and process concepts. This can be disrupted as a result of many forms of brain damage, particularly Alzheimer's disease (AD). The current paper reviews research demonstrating that semantics deteriorates early in AD, particularly on effortful semantic tasks. There is a "category effect", meaning that AD preferentially affects concepts dealing with living things and abstract concepts compared to non-living objects and verbs/actions. While this pattern of deterioration, specific for AD, may reflect a breakdown within a distributed semantic system (where living things are distinguished by a high rate of inter-correlations between concepts or by a particular mode of being learned), it is equally possible that there is a regional distribution of semantic knowledge, with living things preferentially involving left temporal regions which become damaged early on in AD. Evidence from patients with strokes and semantic dementia, as well as activation studies in normal individuals, implicates the left posterior temporal region in semantic processing for pictures, abstract words, and concrete words. AD individuals, who are impaired in a variety of semantic tasks, show functional deficits in this area, and fail to activate it normally.

Keywords: memory; Alzheimer's disease; semantic memory

Introduction

Semantic memory is the sub-component of long-term memory that is responsible for the acquisition, representation, and processing of conceptual information. This critically important system of brain function is implicated in a wide range of cognitive functions, including the ability to assign meaningful interpretations to words and sentences, recognize objects, recall specific information from previously learned concepts, and acquire new information from reasoning and

*Corresponding author. Tel.: +1 (514) 340 8222, Ext. 5116; Fax: +1 (514) 340 8295; E-mail: howard.chertkow@mcgill.ca

DOI: 10.1016/S0079-6123(07)00025-8

perceptual experience. One of the basic goals of cognitive neuroscience is to understand how the brain represents and processes semantic knowledge, which regions of the brain are involved in the retrieval and use of semantic knowledge, and how the dissolution of semantic memory can occur as a result of brain damage.

There are several types of neurological syndromes in which the integrity of semantic memory may be compromised. Well-documented examples often include left temporal lobe vascular and neoplastic lesions producing aphasia, and herpes simplex encephalitis. In addition, semantic memory has been now recognized as a critical cognitive domain affected in Alzheimer's disease (AD) and semantic dementia. Evaluating semantic memory loss in individual cases through mental status and neuropsychological testing is not a straightforward matter, since the pattern of semantic impairment may vary in subtle ways either within or across these different patient types. Semantic memory was historically considered as diffuse or poorly localized. More recent evidence suggests that there are multiple focal brain regions implicated in semantic processing, but the different brain areas may contribute to similar semantic impairments.

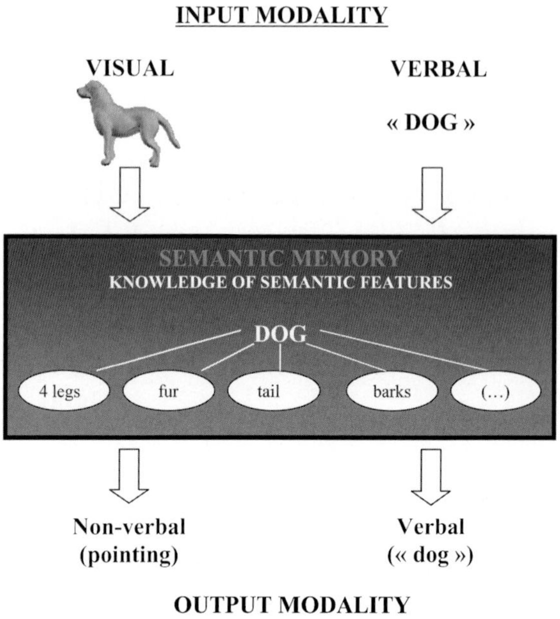

Assessing semantic loss

Assessment of the integrity of the semantic system is complicated by the fact that semantic memory performance may be affected by the disruption of different component stages of semantic processing, including damage to non-semantic systems that interact with semantic memory. The process of naming a picture (say, of a dog) can be disrupted by purely visual processing problems (pre-semantic), or difficulty accessing the sound form of words (post-semantic processing), in addition to disruption of the concept memory for <dog> and its features in semantic memory (Saumier and Chertkow, 2002) (Fig. 1).

When evaluating a patient with signs of semantic disturbance, such as anomia, for instance, the clinician must initially determine whether the naming problem may stem from difficulties at perceiving the object correctly. If the

Fig. 1. A standard model of semantic memory and modality-specific input and output mechanisms. The central "semantic memory" box includes an example features that may be associated with the semantic representation of a dog.

patient perceives the objects adequately, then the problem may reside in the patient's ability to gain access to the object's semantic information. Alternatively, an anomic individual may be able to perceive the form of an object normally and map it onto appropriate semantic information, but be unable to retrieve the correct phonological representation for naming. Thus, evaluating semantic memory in brain-damaged individuals requires the use of a constellation of tasks that recruit low-level perceptual processes, access to semantic knowledge across different sensory modalities, and output responses such as speech production. Examples of semantic memory tests are listed in Table 1, along with an illustration of a semantic probe task shown in Fig. 2.

In the context of an information processing view, it is generally held that there is a common semantic knowledge store that may be accessed and used by way of modality-specific input and output mechanisms within the brain. Thus, semantic knowledge networks may be activated not only when objects are visually perceived, but

Table 1. Examples of tests that are used to measure non-semantic and semantic performance when assessing semantic memory deficits

Tests that are often used as a preliminary assessment of semantic skills
    Naming pictures of objects
    Providing names to semantic descriptions

Tests that are used to assess the integrity of visual input to semantic memory
    Matching two pictures of an object that are depicted from different viewpoints
    Copying simple geometric figures
    Discriminate between real and unreal objects

Tests are used to assess the integrity of word representations
    Deciding whether or not a spoken utterance or a written letter sequence is a word

Tests that emphasize central semantic processing
    Semantic probes — see Fig. 2
    Sorting pictures of objects according to specific semantic properties
    Naming fluency — name as many animals as possible within a 1 min period; name as many words beginning with the letter 'T'
    Picture-word matching — see Fig. 2

Source: Adapted with permission from Saumier and Chertkow (2002), Vol. 2, Philadelphia, Current Medicine Group, LLC, 2002.

also when they are heard, smelled, or touched, or when the name of the object is read or heard. Semantic knowledge may also be retrieved when writing or speaking about concepts. One of the major goals of cognitive neuropsychological research on semantic memory is to understand how semantic features in the brain are represented and how such features become linked together to form coherent concepts.

Effortful semantic memory deteriorates early on in Alzheimer's disease

Lexico-semantic impairments are one of the earliest manifestations of language breakdown in AD. AD patients typically have difficulty naming pictures, generating words according to semantic or formal criteria (Appell et al., 1982; Bayles et al., 1987), and answering detailed questions about an object's characteristics (Chertkow et al., 1989; Chertkow and Bub, 1990). Because these tasks are characterized by conscious and even effortful processing of semantic information, some have argued that this pattern of impairment reflects a disorder of intentional access to semantic memory (Nebes, 1989; Ober and Shenaut, 1995).

In contrast, there are tasks in which access to semantic information, and semantic processing, appear to occur in a more automatic manner. Lexical (word/non-word) decision tasks, for example, are not overtly semantic although a certain degree of semantic checking may occur without conscious effort (Neely, 1990).

There is a family of tasks which demonstrate priming effects, conceptualized as a form of implicit memory. Many of these include demonstration of priming effects related to previous exposure to a subset of test items, extending over many minutes, on tasks such as the word stem completion task. It has been shown that priming effects in such paradigms are preserved in AD for more "identification" (vs. "generation") tasks, and more "perceptual" (vs. conceptual) priming tasks (Gabrieli et al., 1994). A separate set of tasks examine priming effects over a second or less, based on "spreading activation" within lexical-semantic fields (Collins and Loftus, 1975). Automatic access to semantic knowledge is best captured through such short-term semantic priming tasks, and it is such a task which we will administer in the current study. It has been demonstrated that subjects responded faster and more accurately to a target item when it followed a semantically related item than when it followed a semantically unrelated item (Meyer and Schvaneveldt, 1971). Priming is most often explained according to the network model of semantic memory (Collins and Loftus, 1975) in which concepts are represented in semantic memory through a network of interconnected nodes. Activation of one node will automatically spread to neighboring nodes producing a decrease in response time on a variety of tasks. Although a number of studies have reported normal (Balota and Duchek, 1991) or decreased priming (Ober and Shenaut, 1988), the majority of studies have found increased priming, often termed hyperpriming, in patients with mild-to-moderate AD (Chertkow et al., 1989, 1994; Giffard et al., 2001, 2002). Abnormal priming has been suggested to reflect disorders in automatic access to semantic memory.

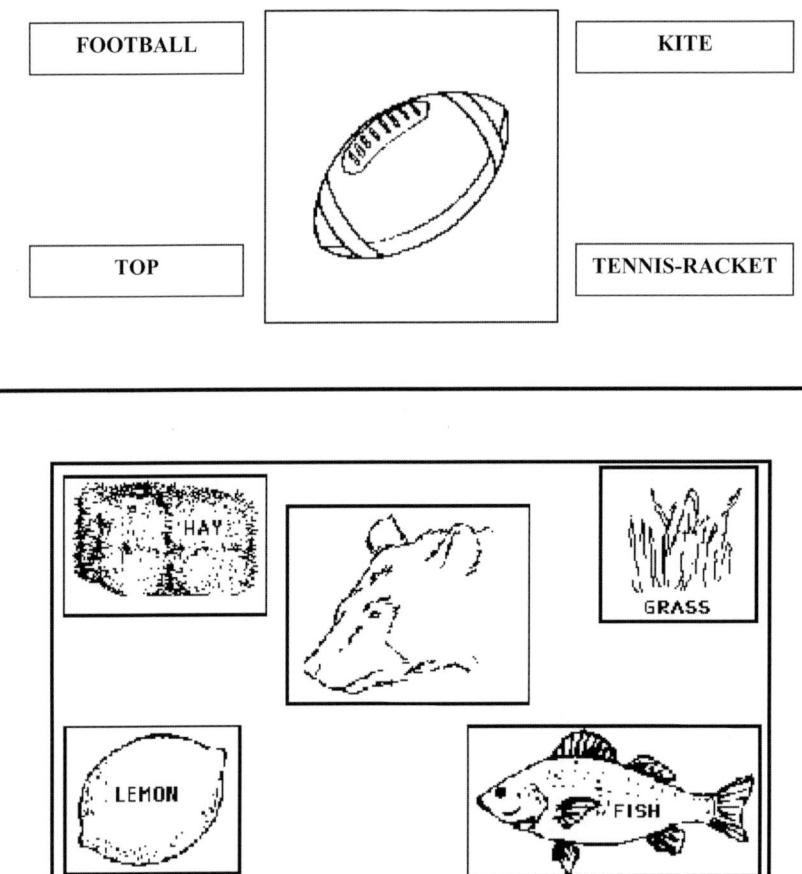

Fig. 2. Examples of test items that may be found in a semantic probes (top) and word-picture matching (bottom) tasks. (Reprinted with permission from Saumier and Chertkow (2002), Vol. 2, Philadelphia, Current Medicine Group, LLC, 2002.)

Impairments on language tests and abnormal priming results suggest that both intentional and automatic semantic processes may be altered in patients with mild-to-moderate AD. However, the order in which these processes are impaired during the course of the disease is unclear. We sought to answer this question by documenting the pattern of lexico-semantic impairments in patients with mild cognitive impairment (MCI), a transitional stage considered by many to be a pre-AD stage (Collie and Maruff, 2000; Petersen et al., 2001; Chertkow, 2002). While the "amnestic" form of MCI is characterized by prominent episodic memory impairments, subtle changes in semantic memory and naming can also often be detected. By carrying out comparisons between tasks that tap into intentional and automatic processes and between MCI and AD patients, patterns of impairment in these two groups were established.

We examined lexico-semantic impairments in 61 patients with MCI and compared them to those of 39 AD patients and 60 normal elderly (NE) participants (Duong et al., 2006). All subjects were tested on tasks that required intentional (picture naming and semantic probes), and automatic (lexical decision and lexical semantic priming) access to semantic memory in addition to tasks that assessed frontal executive functions (Stroop and Stroop-Picture naming). Results indicated that the MCI group performed at a level between that of the NE and AD group on tasks of intentional access but did not differ from the NE group on tasks of automatic access. On frontal executive tests, the MCI group performed at a similar level than the NE group on the Stroop but was between that of the NE and AD groups on the Stroop-Picture naming task. In other words, we found that, relative to NE participants, MCI patients were impaired on intentional tasks but not on automatic tasks, whereas AD patients would show impairments on both types of tasks.

Given that most MCI patients go on to develop AD, this pattern of results suggests that intentional access to semantic memory is impaired earlier in the course of the disease than automatic access. In addition, lexico-semantic impairments in MCI may be related in part not to inhibition deficits per se, but to inhibition deficits during semantic search.

Localization of semantic memory may explain its breakdown in Alzheimer's disease

Numerous functional neuroimaging and brain lesion studies have been conducted in an attempt to elucidate the neurological basis of semantic memory. Several of these have sought to determine whether there is a localizable neuroanatomical semantic system that is common to all forms of semantic processing. Others studies have focused on the issue of category-specificity, seeking to determine whether there are dedicated neural regions for the representation and processing of specific categories of objects, such as animals, fruits and vegetables, and artefacts.

Focal brain damage resulting in aphasia is one form of brain lesion pathology that has been investigated to elucidate brain regions that subserve semantic memory. Many aphasic individuals have difficulties in understanding words, either heard or seen. Their verbal impairment may be associated with poor semantic knowledge (Goodglass and Baker, 1976; Caramazza et al., 1982; Goodglass et al., 1997). Our laboratory has argued (Chertkow et al., 1997) using results from patients with focal left hemispheric cerebral infarctions that a lesion of a common region of brain damage situated within the lateral surface of the left posterior temporal lobe that included the superior, middle, and inferior temporal regions, affects verbal as well as non-verbal semantic memory processing. From these results, we proposed that the left lateral temporal region is part of an amodal neural network that is responsible for general semantic processing. This proposal is largely in agreement with other investigations of semantic knowledge (Hart and Gordon, 1990).

Both PET and fMRI brain imaging techniques have been used to detect brain areas that are active during tasks designed to elicit activity in a common semantic system. One of most frequently localized region of activation found in these studies has been the left inferior frontal cortex (Posner et al., 1988). This finding was originally interpreted as implying that the left inferior frontal region subserved the storage of semantic knowledge. Such a claim runs counter to clinical experience with focal left frontal lesions, which appear to spare picture naming and other semantic functions (Goodglass and Baker, 1976). Researchers currently view this frontal activation region as being responsible for searching, selecting, and effortful retrieval of information from semantic memory (Demb et al., 1995).

A number of imaging investigations point to the temporal lobe as the actual locus of semantic memory Noppeney and Price (2002), for instance, conducted a PET study which contrasted brain activity regions where activity was common to the semantic processing of words and semantic decisions (e.g., is the sound described by the stimulus word "usually loud"?) relative to semantic decisions alone. They found a functional dissociation between the left inferior temporal cortex subserving the activation of semantic representations and the left inferior frontal cortex extending to the left anterior temporal pole associated with the effortful retrieval of semantic information. The interpretation of these findings is consistent with other studies pointing to the involvement of the left temporal lobes during semantic memory tasks (Fig. 3).

Whatmough et al. (2002) used an approach of increasingly difficult semantic tasks in a PET study in which a group of participants were required to name pictures of objects with high or low levels of familiarity. They found that picture naming difficulty, as reflected by reaction times to carry out the task, correlated with rCBF changes in the left posterior middle temporal gyrus and the prefrontal cortex. The left posterior temporal activation was interpreted as reflecting the access to semantic representations of the pictorial stimuli and the prefrontal activation was ascribed to general semantic mechanisms. Consistent with this formulation, partial blocking of such focal temporal lobe activation using transcranial magnetic stimulation (TMS) slows the picture naming responses (Fung et al., 2002) (Fig. 4).

If the inferior region of the temporal cortex is an area where semantic representations of objects are stored, then one would expect that deterioration of this area would result in problems in semantic knowledge, as well as abnormal cortical activity in this region when performing semantic tasks. This prediction was tested by Chertkow et al. (2000) in a PET activation study involving AD patients performing the same task that was used by Whatmough et al. (2002) with elderly healthy adults. AD patients tend to suffer from diffuse damage to brain regions associated with the processing of semantic information, particularly the temporal lobe and frontal lobe association cortices (Braak and Braak, 1991; Hyman, 1997). In line with these observations, Chertkow et al. (2000) found that AD patients failed to show the same left posterior middle frontal cortical activity that had been obtained with increase picture naming difficulty in healthy elderly adults, despite having activated many of the other cortical regions associated with object recognition in the controls (e.g., visual cortex, left frontal cortex). Together, these results suggest that the posterior inferior temporal lobes are critical neural areas for storing and processing semantic knowledge (see also Grossman et al., 1997).

Category-specific semantic memory loss in Alzheimer's disease

Numerous clinical studies of patients with semantic memory impairments have shown that the severity of semantic deficits is often asymmetric

a)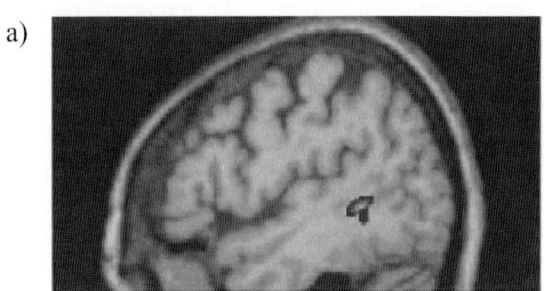

b)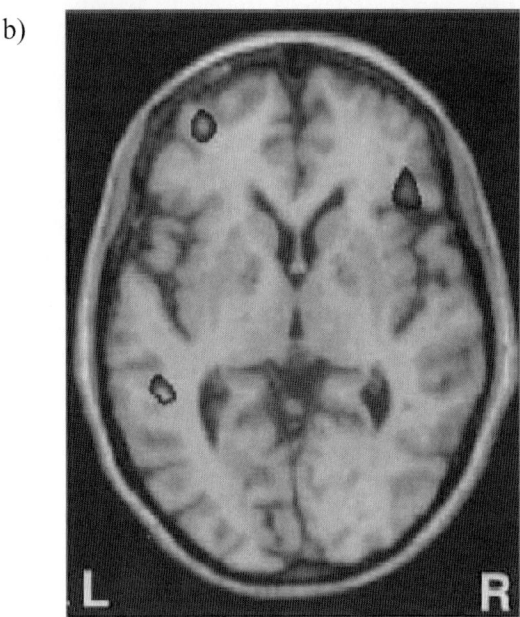

Fig. 3. Cortical regions showing semantic effects in horizontal (a) and coronal (b) slices of the brain. (Reprinted with permission from Whatmough et al., 2002, © 2002, with permission from Elsevier.)

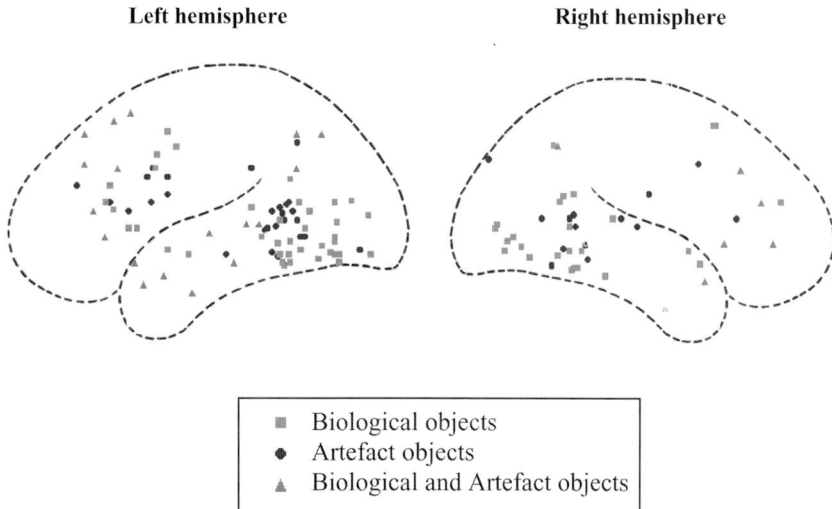

Fig. 4. A summary of results obtained from neuroimaging studies that have examined semantic tasks involving biological and artefact categories. (Reprinted with permission from Saumier and Chertkow (2002), Vol. 2, Philadelphia, Current Medicine Group, LLC, 2002.)

and does not equally affect all categories of concepts. This is termed "category-specificity". Biological categories, such as animals, and fruits and vegetables tend to be more impaired than artefacts, such as tools and furniture. Warrington and Shallice (1984), for instance, reported four case studies of patients recovering from Herpes simplex encephalitis demonstrating selective impairments in either defining names or recognizing pictures of biological objects (e.g., animals, plants, foods), compared to preserved abilities regarding common manufactured objects. The same pattern is generally also encountered in AD and semantic dementia (Hodges and Patterson, 1995). Rarely artefacts are selectively impaired, with biological items remaining relatively intact (Sacchett and Humphreys, 1992).

Only a few possible categorical divisions have been proposed and actually tested for their validity in AD. Those who claim that the categories have a neurological basis usually attempt to show that AD patients are less accurate when tested for their knowledge about certain categories than they are for others (e.g., animals vs. tools). Dramatic examples of this phenomenon have been found in people who have suffered acute brain damage (Warrington and Shallice, 1984; Riddoch and Humphreys, 1987; Sartori and Job, 1988; Silveri and Gainotti, 1988). In the case of AD patients the effect is never as sizable as in these cases, instead AD patients are found to perform poorly across most categories, but worse in others. For this reason the phenomenon is called a "category effect", and not "category specificity", in AD. Because of the smaller effect it has been important to be sure that a seeming category effect does not result from poor matching across categories for other more common semantic factors known to affect patients' performance. Some of the factors that need to be controlled for are familiarity, visual complexity (in the case of picture naming), and lexical frequency.

The most frequently contrasted categories are those of biological and non-biological items, and the finding is that knowledge for biological items such as animals, fruits and vegetables is more severely compromised in AD than knowledge for man-made objects such as tools, clothing, or vehicles. Other category divisions that have been compared include objects vs. actions, and abstract concepts vs. concrete objects.

Our group sought more definitive proof of a semantic category effect in AD patients (Fung et al., 2001). One weakness of previous studies they wished to avoid was that often the elderly control group performed most of the tasks almost perfectly. When a group demonstrates this ceiling effect, it can mask smaller effects of level of difficulty that only become apparent in the impaired. Patients then may have been poorer on one category not because of a category-specific semantic disorder, but because the category was more difficult even for the healthy. Another shortcoming of previous studies that Fung et al. wished to avoid was the use of familiarity ratings provided mostly by young college students. For our study, we developed lists of objects that were matched on lexical frequency from published sources, but for familiarity and visual complexity ratings from a group of 60 healthy elderly individuals. Furthermore, the lists were selected such that the accuracy of the healthy elderly was not at ceiling but at 80% for each category.

One particular category that had been previously understudied was that of action words. Some had claimed that ADs were poorer at naming actions than objects (Robinson et al., 1996). Fung et al. pointed out, however, that in those studies subjects were not shown actions but instead static images of a part of an action, requiring some degree of inference on the part of the participant. Furthermore, in some cases, the actions were intimately linked to knowing the name of an object in the scene. For example, knowing that someone wearing 'skates' was skating and that a 'sail'-boat was sailing. In their study, therefore, Fung et al. also compared naming of static images with naming actions carried out by animated figurines that did not have any objects. Animated actions depicted involved tool use (e.g., typing) and body movement, whole and part (e.g., diving and clapping).

The results of their study showed that the ADs were poorer at naming biological items (31%) than non-biological items (41%). As we had predicted, ADs were better at naming animations than line drawings of actions. In fact they were better at naming animations than all other categories. The elderly controls although not at ceiling, demonstrated the same level of performance across all categories (80–81%) (see Table 2).

Although picture naming is often used to evaluate semantic memory impairments, the task

Table 2. Performance on categories in picture naming and synonym judgment in Alzheimer disease

	Accuracy (%)				$t(56)$	P
	Control		DAT			
	M	SD	M	SD		
Naming performance by control and DAT subjects						
Biological objects	81	16	31	16	10.9	<0.001
Non-biological objects	81	15	41	15	9.4	<0.001
Line-drawings of actions	80	15	36	19	9.2	<0.001
Animations	80	10	59	15	6.2	<0.001

Note: DAT, dementia of the Alzheimer's type. See text for details of within group comparisons.
Source: Adapted with permission from Fung et al. (2001), © 2001, with permission from Elsevier.

requires more than semantic processing, such as visual analysis and lexical-phonological retrieval. A demonstration of a category effect in other semantic tasks is therefore important to support claims that the category effect is indeed a semantic effect. Fung et al. tested the same patients for their knowledge of semantic word associations. Participants were presented with a printed target word (e.g., lamb) and below, side-by-side, two possible semantic associates (e.g., goat, sheep) and were asked to indicate which of the two possible associates was closest in meaning to the target word. Several categories were tested. Once again, when elderly normals were tested they performed at less than perfect but they showed no effect of category. The AD patients, however, were poorer in responding to biological items (animals, fruits and vegetables) and abstract concepts (e.g., chaos) than they were to non-biological items (tools, clothing, and furniture) or action words (e.g., sailing). This study therefore showed a category effect in a semantic task that could not be attributed to difficulty in verbal production. Furthermore, there was a high correlation between AD scores on the two tasks.

Further studies of category effects

Having found fairly good evidence of a category effect, the question is what is the neurological basis for the effect: are different categories stored in different areas of the brain, with one area being more susceptible than the other to AD type pathology? Or are all categories in a common area and there is something in the instantiation of certain categories that makes them more vulnerable to neuronal degeneration.

Many researchers (McClelland, 1994; Caramazza and Shelton, 1998) maintain that "items" in semantic memory are in fact different patterns of distributed activation across more primitive features. There is divergence, however, as to what the effect of gradually eroding the features will cause on knowledge for items for different categories. One proposal, made by Gonnerman et al. (1997), was that the relative importance of intercorrelated features in different categories was key to predicting how one category would deteriorate relative to another. They predicted that because in the biological category, features are highly intercorrelated (e.g., beak, wings, and feathers frequently co-occur) initial damage to the semantic system would maintain the integrity of the category members until a critical loss of features at which point there would be a dramatic decline in knowledge for that category. They provided modest support for their prediction in two AD patients. We (Whatmough et al., 2003) examined this proposal using data from a large group of dementia patients ($n = 72$) ranging from mildly to moderately impaired. We used the Chertkow lab "Category picture naming task" (Chertkow et al., submitted), a task that has been normed on elderly subjects for familiarity, visual complexity, and accuracy and in which biological and non-biological items are matched for these factors. We did not find an early preservation of the biological category. Instead we found that even mildly impaired individuals were poorer at naming biological items and that this category effect increased with increasing anomia (see Fig. 5). The spread between accuracy for biological and non-biological items increased from 10% in the mildly impaired to 21% in the moderately impaired patients.

All of the patients in this large group had been diagnosed with either MCI or probable AD with no additional diagnoses. We also looked at 15 patients who either had AD plus an additional diagnosis (e.g., depression) or were diagnosed with

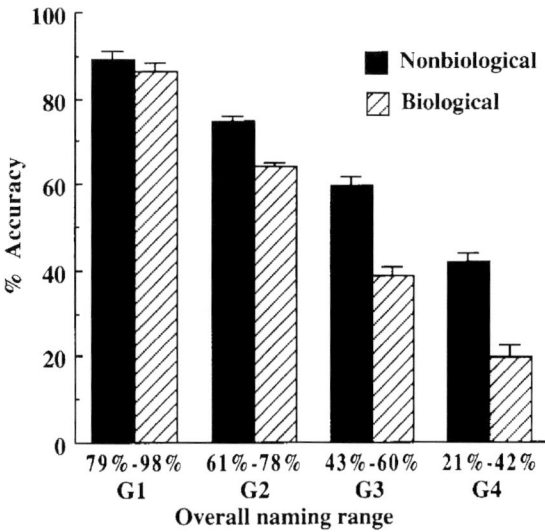

Fig. 5. Mean naming accuracy of DAT/MCI participants on non-biological and biological items at successive ranges of overall naming impairment. Error bars indicate the standard error of the mean. (Reprinted with permission from Whatmough et al., 2003, © 2003, with permission from Elsevier.)

another type of dementia. Our findings here were interesting. In half of these "other dementia" patients, the biological/non-biological difference was either non-existent or in the opposite direction (better performance on the biological). This indicates, first of all, that the stimuli from one category were not inherently more difficult. More importantly, those who differed the most from the AD pattern were patients with other dementias: fronto-temporal dementia, vascular cognitive impairment, and mixed dementia. On the other hand, one severely impaired patient with a diagnosis of "primary progressive aphasia vs. AD" completely conformed to the AD profile. These findings suggest that the category effect is linked to the pattern of neural degeneration that occurs in AD, suggesting that the temporal association cortex, an area of early degeneration, is more critical to the category effects than is the frontal cortex.

Brain imaging to elucidate category effects in semantic memory

We carried out brain imaging studies in healthy elderly in order to evaluate whether particular brain areas are important for semantic processing and category effects. One prominent theory is that category effects arise because categories are heavily reliant on the lower processing areas by which they are experienced (Martin, 1998; Chao et al., 1999). So for instance, tools implicate movement and they will be found close to areas associated with movement. Animals must be distinguished by visual features, therefore, their category will be represented closer to visual areas etc. We were unconvinced by the justification for this theory, since one could just as easily argued that animals are associated with movement and tools are inanimate. Further, the proposal has not found wide support from lesion studies. Instead lesion studies tend to indicate that the posterior lateral and inferior temporal cortex are areas common for all semantic processing.

We carried out two studies of category effects using the $H_2^{15}O$ methodology of positron emission tomography. In the first (Whatmough et al., 2002) healthy elderly subjects named pictures of animals and tools for which there was an easy and a harder-to-name block of stimuli for each category. We found that category differences were apparent in the ventral surface of the brain. Naming pictures of tools vs. picture of animals produced robust differences in the ventral tempero-occipital area of the brain, the fusiform gyrus. The fusiform is an area that also distinguishes between visioning faces or buildings. In general it suggests that the different categories rely on different visual processing areas. Distinguishing between animals (wolf vs. fox) requires a finer level of visual memory than a shovel and a fork. The area that is most associated with semantic deficits per se is the lateral temporal cortex in the posterior middle gyrus. When we contrasted easy vs. hard blocks we found that more intense semantic processing (hard blocks) was associated with increased activation in left posterior middle temporal cortex, irrespective of category.

The second PET study (Whatmough et al., 2004) was designed in a similar fashion except that here the stimuli were words, and subjects were presented with two word pairs (e.g., error-mistake, error-fright) and were asked to read aloud the pair of words that was closer in meaning (error-mistake). There were four blocks of trials two of concrete concepts and two of abstract concepts with a hard and an easy block of each category of stimuli. Our findings were similar to those of the preceding experiment. Category effects were found in the ventral surface of the brain in the temporo-occipital cortex, but difficulty effects were again found in the lateral and inferior temporal cortices for both categories. There were differences, however, in the specifics of these areas. The concrete stimuli activated the left fusiform gyrus whereas the abstract stimuli activated the right fusiform gyrus. This is intriguing in that people with semantic dementia suffer severe atrophy of the left inferior temporal cortex and temporal pole; in some cases such individuals are also more impaired on concrete concepts than on abstract ones (Garrard and Hodges, 1999; Graham, 1999; Bird et al., 2000). The difficulty effect area on the left lateral temporal cortex was anterior to the "difficulty" area for the picture-naming task. When we put our findings in the context of other studies which had found semantics in the lateral temporal cortex we found a certain pattern. As tasks required greater and greater association between words, from single items to word phrases up to the level of narratives there was a demarcation between posterior temporal activation and mid-temporal lobe activation. Together then the brain imaging studies indicate that some categories are activated in different areas on the ventral surface of the brain but that there are common areas of activation to all categories in the lateral temporal cortex. Furthermore there appears to be a posterior–anterior progression in activation, as semantic tasks require greater and greater lexical complexity or associations.

What causes category effects?

McCarthy and Warrington (1988) interpreted category-specific semantic disorders as reflecting pathological involvement of distinct semantic features for different categories of knowledge. In particular, biological objects are distinguished primarily by their sensory (essentially visual) features, while artefacts are mainly distinguished by their functional features. While perceptual and functional features are not assumed do divide category distinctions precisely (many artefacts may be identified according to their perceptual attributes, some biological objects may be defined according to their function), damage to perceptual feature knowledge might result in a disproportionate impairment of biological objects, while damage to functional features would give rise to impairments that preferentially affect artefacts. The notion that sensory and functional features may be differentially disturbed by brain damage, hence accounting for category-specific semantic impairments has also been developed in computational models of semantic processing (Farah and McClelland, 1991).

There are alternative theoretical explanations for category-specificity. Not all cases of category-specific semantic deficits show selective impairments for specific types of features. Rare cases reveal deficits that are restricted to either animals (fruits and vegetables, or artefacts, but without the disproportionate impairment of either perceptual or functional features observed in other patients with semantic deficits (e.g., Caramazza and Shelton, 1998). It has been hypothesized that these deficits reflect damage to abstract representations within independent brain regions that have evolved through evolutionary pressures for storing and processing the categories of animals, fruits and vegetables, or artefacts (Caramazza and Shelton, 1998). Caramazza's conclusion that category-specific impairments are both categorical and abstract is interpreted as being consistent with the observation that early in their development children acquire distinctions between biological and non-biological categories that cannot be accounted for by perceptual similarities (e.g., Carey, 1985; Gelman, 1988; Keil, 1989).

A third class of theory suggests that dissociations between object categories may arise because of the way features of semantic knowledge are linked together in the brain. Evidence for this position stems partly from research involving patients affected by the early stages of AD who, like the Herpes encephalitis patients, tend to show greater difficulties at recognizing biological than artefact objects. It has been argued (Gonnerman et al., 1997) that these deficits cannot be accounted for by specific deficits to perceptual features alone since the widespread and patchy brain atrophy resulting from AD is likely to affect both perceptual and functional features equally (but see also Garrard et al., 1998). There is also evidence from cases of herpes encephalitis (e.g., Moss et al., 1998) and semantic dementia (e.g., Moss et al., 1995) patients who show biological object deficits, but show neither poorer performance on perceptual than functional features nor impairments that are strictly limited to biological objects (some artefact concepts are impaired as well). These semantic memory deficits have been interpreted as reflecting the fact that biological objects tend to have greater overlapping and intercorrelated sensory and functional features than artefacts. Since the perceptual and functional features are more widely shared among biological objects than artefacts, disruption of these features will tend to interfere more with representations of biological concepts than artefacts. Importantly, however, artefact categories with many shared and intercorrelated features (e.g., musical instruments, makes of cars) would also show a similar vulnerability as a result of brain damage. This is consistent with the observation that such categories are frequently impaired in cases of category-specific deficits that predominantly affect biological objects (Dixon et al., 1998).

Most cases of individuals with encephalitis, AD or semantic dementia which demonstrate the co-occurrence of semantic loss with preferential involvement of biological categories, have brain damage affecting various structures of the left temporal lobe. It has proven difficult to use such cases to specify whether specific categories are functionally represented in distinct brain regions. Regarding the localization of category-specific brain damage affecting artefact categories, the current literature of cases amounts to 10 subjects with either cerebral infarction or encephalitis, and 10 subjects with progressive degenerative disorders, either semantic dementia, or AD (see reviews in Laiacona et al., 2000). With one exception (Warrington and McCarthy, 1987) all of these cases also point to a temporal lobe involvement.

Unfortunately, localization of the functional impairment in these cases is problematic. First, naturally occurring lesions may vary from one case to another and according to vascularization. For instance, most cases of cerebral infarction or encephalitis involve large left middle cerebral artery territory infarction, the effects of which invariably extends to frontal, parietal, as well as temporal lobe cortical regions, as documented on CT or MRI scanning (Grossman et al., 2002). This problem is further compounded by the fact that patients do not always undergo functional neuroimaging (e.g., none of the 10 patients showing preferential artefact impairments had PET scanning). Thus, the overall conclusion from patient studies is that while semantic impairments for either biological or artefact categories generally involve the left temporal lobe as the major neuroanatomic substrate, it remains unclear whether or not different semantic categories are represented via different neuroanatomic substrates within this region. A wealth of neuroimaging studies in normals have attempted to address this question, but the discussion is beyond the scope of this paper (Cappa et al., 1998; Chao and Martin, 1999; Devlin et al., 2002; Gerlach et al., 1999; Gorno-Tempini et al., 2000; Kraut et al., 2002; Martin et al., 1996; Moore and Price, 1999; Mummery et al., 1996, 1998; Perani et al., 1995, 1999; Thompson-Schill et al., 1999).

Regarding the situation in AD, it has been quite clearly shown that a category effect is reliably seen, even if the theoretical explanation for this phenomenon appears difficult to pin down.

Summary and conclusions

The main goal of the current paper was to provide an overview of recent literature regarding the neural basis of semantic memory and it's impairment in AD. To this aim, we have noted that different types of neurological disorders may give rise to complex and varying forms of semantic memory impairments. An amalgam of brain imaging and neuropsychological tests is essential in detecting and distinguishing subtle forms of semantic impairments that may occur in patients with brain damage. We reviewed clinical and neuroimaging studies that have investigated the neuroanatomical regions of the brain that are involved in storing and processing general semantic knowledge. These studies point towards the posterior region of left temporal lobe as the major neuroanatomic substrate of semantic memory, which appears to interact with semantic search mechanisms situated in the frontal lobes.

We also surveyed several theories regarding category distinctions within semantic memory, with a particular emphasis on the frequently investigated category-specific semantic disorders associated with living and artefact category knowledge. Theories based on sensory/functional knowledge, general category membership, and feature correlations have been proposed to account for different documented patterns of category-specific deficits. While lesion-based data is limited regarding the neural localization of category-specific effects, functional imaging studies involving neurologically intact subjects have been used to provide more precise information regarding the localization of category distinctions. The results of these studies considered together reveal that both biological and artefact objects tend to activate a broad and overlapping range of cortical regions that implicate executive, semantic, and name retrieval processes, with biological objects preferentially activating the posterior visual areas of the brain. These findings suggest that semantic memory invloves multiple levels of knowledge that are common to different categories, with regions involved in perceptual processing that may be recruited to disambiguate visually similar categories of biological and artefact objects. It is suggested that these findings are relevant for both the diagnosis and assessment of various forms of semantic memory impairments.

Acknowledgment

This study was supported in part by operating grants from the CIHR and the Fonds de la recherche en santé du Québec (FRSQ) awarded to HC. HC is a "chercheur national" of the FRSQ (Fonds de la Recherche en Santé du Québec).

References

Appell, J., Kertesz, A. and Fisman, M. (1982) A study of language functioning in Alzheimer patients. Brain Lang., 17: 73–91.

Balota, D.A. and Duchek, J.M. (1991) Semantic priming effects, lexical repetition effects, and contextual disambiguation effects in healthy aged individuals and individuals with senile dementia of the Alzheimer type. Brain Lang., 40(2): 181–201.

Bayles, K.A., Kaszniak, A.W. and Tomoeda, C.K. (1987) Communication and Cognition in Normal Aging and Dementia. College-Hill Press/Little, Brown & Co., Boston, MA.

Bird, H., Lambon Ralph, M.A., Patterson, K. and Hodges, J.R. (2000) The rise and fall of frequency and imageability: noun and verb production in semantic dementia. Brain Lang., 73(1): 17–49.

Braak, H. and Braak, E. (1991) Neuropathological staging of Alzheimer-related changes. Acta Neuropathol., 82(4): 239–259.

Cappa, S.F., Perani, D., Schnur, T., Tettamanti, M. and Fazio, F. (1998) The effects of semantic category and knowledge type on lexical-semantic access: a PET study. Neuroimage, 8(4): 350–359.

Caramazza, A., Berndt, R.S. and Brownell, H.H. (1982) The semantic deficit hypothesis: perceptual parsing and object classification by aphasic patients. Brain Lang., 15: 161–189.

Caramazza, A. and Shelton, J. (1998) Domain-specific knowledge systems in the brain: the animate–inanimate distinction. J. Cogn. Neurosci., 10(1): 1–34.

Carey, S. (1985) Conceptual Change in Childhood. MIT Press, Cambridge, MA.

Chao, L.L., Haxby, J.V. and Martin, A. (1999) Attribute-based neural substrates in temporal cortex for perceiving and knowing about objects. Nat. Neurosci., 2(10): 913–919.

Chao, L.L. and Martin, A. (1999) Cortical regions associated with perceiving, naming, and knowing about colors. J. Cogn. Neurosci., 11(1): 25–35.

Chertkow, H. (2002) Mild cognitive impairment. Curr. Opin. Neurol., 15(1999): 401–407.

Chertkow, H. and Bub, D. (1990) Semantic memory loss in Alzheimer-type dementia. In: Schwartz M.F. (Ed.), Modular Deficits in Alzheimer-Type Dementia. Issues in the Biology of Language and Cognition. MIT Press, Cambridge, MA, pp. 207–244.

Chertkow, H., Bub, D., Bergman, H., Bruemmer, A., Merling, A. and Rothfleisch, J. (1994) Increased semantic priming in patients with dementia of the Alzheimer's type. J. Clin. Exp. Neuropsychol., 16(4): 608–622.

Chertkow, H., Bub, D., Deaudon, C. and Whitehead, V. (1997) On the status of object concepts in aphasia. Brain Lang., 58(2): 203–232.

Chertkow, H., Bub, D.N. and Seidenberg, M. (1989) Priming and semantic memory loss in Alzheimer's disease. Brain Lang., 36(3): 420–446.

Chertkow, H., Murtha, S., Whatmough, C., McKelvey, R., et al. (2000) Altered activation of cerebral cortex in Alzheimer's disease during picture naming: a positron emission tomographic study. Paper presented at the Cognitive Neuroscience Society, San Francisco.

Chertkow, H., Whatmough, C., Whitehead, V. and Frederickson, K. (submitted) Categorical picture naming and semantic probes: deficits in early dementia.

Collie, A. and Maruff, P. (2000) The neuropsychology of preclinical Alzhemier's disease and mild cognitive impairment. Neurosci. Biobehav. Rev., 24(3): 365–374.

Collins, A.M. and Loftus, E.F. (1975) A spreading-activation theory of semantic processing. Psychol. Rev., 82: 407–428.

Demb, J.B., Desmond, J.E., Wagner, A.D., Vaidya, C.J., Glover, G.H. and Gabrieli, J.D. (1995) Semantic encoding and retrieval in the left inferior prefrontal cortex: a functional MRI study of task difficulty and process specificity. J. Neurosci., 15(9): 5870–5878.

Devlin, J.T., Russell, R.P., Davis, M.H., Price, C.J., Moss, E.H., Fadili, M.J., et al. (2002) Is there an anatomical basis for category-specificity? Semantic memory study in PET and fMRI. Neuropsychologia, 40: 54–75.

Dixon, M.J., Bub, D.N. and Arguin, M. (1998) Semantic and visual determinants of face recognition in a prosopagnosic patient. J. Cogn. Neurosci., 10(3): 362–376.

Duong, A., Whitehead, V., Hanratty, K. and Chertkow, H. (2006) The nature of lexico-semantic processing deficits in mild cognitive impairment. Neuropsychologia, 44(10): 1928–1935.

Farah, M.J. and McClelland, J.L. (1991) A computational model of semantic memory impairment: modality specificity and emergent category specificity. J. Exp. Psychol. Gen., 120: 339–357.

Fung, T.D., Chertkow, H., Murtha, S., Whatmough, C., Péloquin, L., Whitehead, V., et al. (2001) The spectrum of category effects in object and action knowledge in Alzheimer's disease. Neuropsychology, 15: 371–379.

Fung, T.D., Chertkow, H., Paus, T. and Whatmough, C. (2002) TMS of the left inferior temporal cortex slow picture naming. J. Int. Neuropsychol. Soc., 8(Suppl. 2): p. 193.

Gabrieli, J.D.E., Keane, M.M., Stanger, B.Z., Kjelgaard, M.M., et al. (1994) dissociations among structural-perceptual, lexical-semantic, and event-fact memory systems in Alzheimer, amnesic, and normal subjects. Cortex, 30(1): 75–103.

Garrard, P. and Hodges, J.R. (1999) Semantic dementia: implications for the neural basis of language and meaning. Aphaisology, 13(8): 609–623.

Garrard, P., Patterson, K., Watson, P.C. and Hodges, J.R. (1998) Category specific semantic loss in dementia of Alzheimer's type. Brain, 121(Pt. 4): 633–646.

Gelman, S. (1988) Development of induction within natural kinds and artificial categories. Cogn. Psychol., 20: 65–95.

Gerlach, C., Law, I., Gade, A. and Paulson, O.B. (1999) Perceptual differentiation and category effects in normal object recognition: a PET study. Brain, 122(Pt. 11): 2159–2170.

Giffard, B., Desgranges, B., Nore-Mary, F., Lalevee, C., Beaunieux, H., de la Sayette, V., et al. (2002) The dynamic time course of semantic memory impairment in Alzheimer's disease: clues from hyperpriming and hypopriming effects. Brain, 125(9): 2044–2057.

Giffard, B., Desgranges, B., Nore-Mary, F., Lalevee, C., de la Sayette, V., Pasquier, F., et al. (2001) The nature of semantic memory deficits in Alzheimer's disease: new insights from hyperpriming effects. Brain, 124(Pt. 8): 1522–1532.

Gonnerman, L., Andersen, E., Devlin, J., Kempler, D. and Seidenberg, M. (1997) Double dissociation of semantic categories in Alzheimer's disease. Brain Lang., 57(2): 254–279.

Goodglass, H. and Baker, E. (1976) Semantic field, naming, and auditory comprehension in aphasia. Brain Lang., 3: 359–374.

Goodglass, H., Wingfield, A. and Ward, S.E. (1997) Judgments of concept similarity by normal and aphasic subjects: relation to naming and comprehension. Brain Lang., 56(1): 138–158.

Gorno-Tempini, M.L., Cipolotti, L. and Price, C.J. (2000) Category differences in brain activation studies: where do they come from? Proc. Biol. Sci., 267(1449): 1253–1258.

Graham, K.S. (1999) Semantic dementia: a challenge to the multiple-trace theory? Trends Cogn. Sci., 3(3): 85–87.

Grossman, M., Koenig, P., DeVita, C., Glosser, G., Alsop, D., Detre, J., et al. (2002) The neural basis for category-specific knowledge: an fMRI study. Neuroimage, 15: 936–948.

Grossman, M., Payer, F., Onishi, K., White-Devine, T., Morrison, D., D'Esposito, M., et al. (1997) Constraints on the cerebral basis for semantic processing from neuroimaging studies of Alzheimer's disease. J. Neurol. Neurosurg. Psychiatry, 63(2): 152–158.

Hart, J., Jr. and Gordon, B. (1990) Delineation of single-word semantic comprehension deficits in aphasia, with anatomical correlation. Ann. Neurol., 27(3): 226–231.

Hodges, J.R. and Patterson, K. (1995) Is semantic memory consistently impaired early in the course of Alzheimer's disease? Neuroanatomical and diagnostic implications. Neuropsychologia, 33(4): 441–459.

Hyman, B.T. (1997) The neuropathological diagnosis of Alzheimer's disease: clinical-pathological studies. Neurobiol. Aging, 18(S4): S27–S32.

Keil, F.C. (1989) Concepts, Kinds, and Cognitive Development. MIT Press, Cambridge, MA.

Kraut, A.M., Moo, L.R., Segal, J.B. and Hart, J.J. (2002) Neural activation during an explicit categorization task: category- or feature-specific effects? Cogn. Brain Res., 13: 213–220.

Laiacona, M., Capitani, E. and Barbarotto, R. (2000) Do living and nonliving categories need further fractionation? A study of picture naming in a pathological sample. Brain Cogn., 43(1–3): 291–296.

Martin, A. (1998) Organization of semantic knowledge and the origin of words in the brain. In: Jablonski N.G. and Aiello L.C. (Eds.), The Origin and Diversification of Language. Memoirs of the California Academy of Sciences California, LA, pp. 69–87.

Martin, A., Wiggs, C.L., Ungerleider, L.G. and Haxby, J.V. (1996) Neural correlates of category-specific knowledge. Nature, 379(6566): 649–652.

McCarthy, R.A. and Warrington, E.K. (1988) Evidence for modality-specific meaning systems in the brain. Nature, 334: 428–430.

McClelland, J.L. (1994) The organization of memory. A parallel distributed processing perspective. Rev. Neurol., 150(8–9): 570–579.

Meyer, D.M. and Schvaneveldt, R.W. (1971) Facilitation in recognizing pairs of words: evidence of a dependence between retrieval operations. J. Exp. Psychol., 90: 227–234.

Moore, C.J. and Price, C.J. (1999) A functional neuroimaging study of the variables that generate category-specific object processing differences. Brain, 122(Pt. 5): 943–962.

Moss, D.E., Tyler, L.K., Durrant-Peatfield, M. and Bunn, E.M. (1998) Two-eyes of a see-through: impaired and intact semantic knowledge in a case of selective deficit for living things. Neurocase, 10: 362–376.

Moss, H., Tyler, L., Patterson, K. and Hodges, J. (1995) Exploring the loss of semantic memory in semantic dementia: evidence from a primed monitoring study. Neuropsychology, 9(1): 16–26.

Mummery, C.J., Patterson, K., Hodges, J.R. and Price, C.J. (1998) Functional neuroanatomy of the semantic system: divisible by what? J. Cogn. Neurosci., 10(6): 766–777.

Mummery, C.J., Patterson, K., Hodges, J.R. and Wise, R.J. (1996) Generating 'tiger' as an animal name or a word beginning with T: differences in brain activation [published erratum appears in Proc. R. Soc. Lond. B Biol. Sci. 1996 Dec. 22; 263(1377):1755–1756]. Proc. R. Soc. Lond. B Biol. Sci., 263(1373): 989–995.

Nebes, R.D. (1989) Semantic memory in Alzheimer's disease. Psychol. Bull., 106(3): 377–394.

Neely, J.H. (1990) Semantic priming effects in visual word recognition: a selective review of current findings and theories. In: Besner D. and Humphreys G. (Eds.), Basic Processes in Reading: Visual Word Recognition. Lawrence Erlbaum Associates, Hillsdale, NJ.

Noppeney, U.P. and Price, C.J. (2002) A PET study of stimulus- and task-induced semantic processing. Neuroimage, 15: 927–935.

Ober, B.A. and Shenaut, G.K. (1995) Theoretical and practical issues in semantic priming research with Alzheimer's disease subjects. In: Dollinger S.M.C. and DiLalla L.F. (Eds.), Assessment of Biological Mechanisms Across the Life Span, Vol. 203. Lawrence Erlbaum Associates, Inc., Hillsdale, NJ, pp. 121–150.

Ober, B.A. and Shenaut, G.K. (1988) Lexical decision and priming in Alzheimer's disease. Neuropsychologia, 26(2): 273–286.

Perani, D., Cappa, S.F., Bettinardi, V., Bressi, S., Gorno-Tempini, M., Matarrese, M., et al. (1995) Different neural systems for the recognition of animals and man-made tools. Neuroreport, 6(12): 1637–1641.

Perani, D., Schnur, T., Tettamanti, C., Gorno-Tempini, M., Cappa, S.F. and Fazio, F. (1999) Word and picture matching: a PET study of semantic category effects. Neuropsychologia, 37(3): 293–306.

Petersen, R.C., Doody, R., Kurz, A., Mohs, R.C., Morris, J.C., Rabins, P.V., et al. (2001) Current concepts in mild cognitive impairment. Arch. Neurol., 58(12): 1985–1992.

Posner, M.I., Petersen, S.E. and Raichle, M.E. (1988) Localization of cognitive operations in the human brain. Science, 240: 1627–1631.

Riddoch, M.J. and Humphreys, G.W. (1987) Picture naming. In: Humphreys G.W. and Riddoch M.J. (Eds.), Visual Object Processing A Cognitive Neuropsychological Approach. Lawrence Erlbaum Associates, Hillsdale, NJ.

Robinson, K., Grossman, M., White-Devine, T. and D'Esposito, M. (1996) Category-specific difficulty naming with verbs in Alzheimer's disease. Neurology, 47(1): 178–182.

Sacchett, C. and Humphreys, G. (1992) Calling a squirrel a squirrel but a canoe a wigwam: a category-specific deficit for artefactual objects and body parts. Cogn. Neuropsychol., 9(1): 73–86.

Sartori, G. and Job, R. (1988) The oyster with four legs: a neuropsychological study on the interaction of visual and semantic information. Cogn. Neuropsychol., 5(1): 105–132.

Saumier, D. and Chertkow, H. (2002) Semantic memory. Curr. Neurol. Neurosci. Rep., 2(6): 516–522.

Silveri, M.C. and Gainotti, G. (1988) Interaction between vision and language in category-specific semantic impairment. Cogn. Neuropsychol., 5(6): 677–709.

Thompson-Schill, S.L., Aguirre, G.K., D'Esposito, M. and Farah, M.J. (1999) A neural basis for category and modality specificity of semantic knowledge. Neuropsychologia, 37(6): 671–676.

Warrington, E.K. and McCarthy, R.A. (1987) Categories of knowledge: further fractionations and an attempted integration. Brain, 110: 1273–1296.

Warrington, E.K. and Shallice, T. (1984) Category specific semantic impairments. Brain, 107: 829–854.

Whatmough, C., Chertkow, H., Murtha, S. and Hanratty, K. (2002) Dissociable brain regions are involved in processing an object's meaning and structure during picture naming. Neuropsychologia, 40: 174–186.

Whatmough, C., Chertkow, H., Murtha, S., Templeman, D., Babins, L. and Kelner, N. (2003) The semantic category effect increases with worsening anomia in Alzheimer's type dementia. Brain Lang., 84(1): 134–147.

Whatmough, C., Verret, L., Fung, D. and Chertkow, H. (2004) Common and contrasting areas of activation for abstract and concrete concepts: an H2 15O PET study. J. Cogn. Neurosci., 16(7): 1211–1226.

CHAPTER 26

The effects of surgery and anesthesia on memory and cognition

Nicole Caza[1,2,*], Rame Taha[3], Yanqin Qi[3] and Gilbert Blaise[3]

[1]*Centre de Recherche, Institut Universitaire de Gériatrie de Montréal, 4565 chemin Queen-Mary, Montréal, QC H3W 1W5, Canada*
[2]*Psychology Department, Université de Montréal, C.P. 6128, succursale Centre-Ville, Montréal, QC H3C 3J7, Canada*
[3]*CHUM, Hôpital Notre-Dame, 1560 Sherbrooke Est, Pavillon Deschamps-FS-1136, Montréal, QC H2L 4M1, Canada*

Abstract: This chapter describes current findings from the research into postoperative cognitive dysfunction (POCD) following cardiac and non-cardiac surgery in older adults. The evidence suggests that a significant proportion of patients show POCD in the early weeks following surgery and anesthesia. Specific domains of cognition are affected, especially memory. Much less evidence supports the presence of POCD several months or years after surgery, suggesting that POCD may be transient. However, several methodological issues make it difficult to compare findings across studies. Increasing age is among the most consistently reported patient-related risk factor. Other factors more directly related to the surgery and anesthesia are likely to contribute to the pathogenesis of POCD, including inflammatory processes triggered by the surgical procedure. Animal studies have provided valuable findings otherwise not possible in human studies; these include a correlation between the inflammatory response in the hippocampus and the development of POCD in rodents.

Keywords: anesthesia; post-operative cognitive dysfunction; inflammatory cells; aging; neuropsychology

Introduction

A few decades ago, having a surgical procedure under anesthesia was a very dangerous event as the mortality and morbidity following surgery and anesthesia were very high. It was common to say that the patient did not tolerate anesthesia if perioperative mortality happened. Surgical procedures at that time had to be very short in order to survive anesthesia. Nowadays, the safety of anesthesia is so high that perioperative mortality due to anesthesia has almost disappeared, and when it happens, it makes front-page headlines. Though we no longer die from anesthesia, can anesthesia and the surgical procedure have short and long-term effects on cognitive function?

Postoperative cognitive dysfunction (POCD) was first described after cardiac surgery using cardiopulmonary bypass (CPB) technique. It is defined as a subtle dysfunction in one or more cognitive domains, of which memory is typically affected. Such cognitive changes should be distinguished from postoperative delirium defined as an acute deficit of attention and cognition with fluctuating levels of consciousness as well as disturbed sleep–wake cycles (for a review, see

---

*Corresponding author. Tel.: +1 514 340 3540, Ext. 3362; Fax: +1 514 340 3548; E-mail: nicole.caza@umontreal.ca

Dyer et al., 1995; Parikh and Chung, 1995). Contrary to POCD, postoperative delirium tends to occur mainly on the first days following surgery and is generally considered to be transient.

On a subjective level, patients, and especially older adults, complain of forgetfulness and difficulty to concentrate after surgery and anesthesia. These can be illustrated by problems with remembering names of familiar people or what one has read in the morning paper. However, many older adults not having undergone surgery and anesthesia have similar complaints, adding to the difficulty of establishing the presence of POCD. This underlines the importance of the neuropsychological evaluation in detecting subtle cognitive changes after surgery and anesthesia. Family members also report that their relative is not the same as before the operation. Although such complaints are usually anecdotal, and studies assessing their incidence with more objective measures such as questionnaires are few (Moller et al., 1993; Dijkstra et al., 1999), they are compatible with current evidence suggesting the existence of POCD at a time which exceeds that for pharmacological action of anesthetic drugs on the brain.

How do we define POCD? Is it transient or permanent? Who are the people at risk? What are the causes of POCD? Can animal models help us understand the pathogenesis of POCD? We will try to answer these questions based on the current literature. Our focus will be on older adults given that much of the evidence suggesting POCD has been gathered in this population, and most surgeries occur later in life.

Incidence of POCD

A large part of the evidence for POCD comes from older patients undergoing cardiac surgery (for a recent review, see Selnes and McKhann, 2005). In such cases, the incidence of POCD in the first 2 weeks post surgery ranges from 24 to 79%, and from 10 to 60%, 12 weeks postoperatively (Savageau et al., 1982b; Shaw et al., 1987; McKhann et al., 1997; Newman et al., 2001; Van Dijk et al., 2002). Factors specific to cardiac surgery, such as the use of CPB or heart–lung machine have been suggested as contributing heavily to such high incidences (e.g., Lee et al., 2003). CPB is associated with embolization, reduction of cerebral perfusion, and loss of the pulsatile component of flow which may contribute to cerebral injury (Moody et al., 1990; Anderson et al., 1999).

Although the type of surgery appears as a determinant factor, other factors are likely involved in POCD given that it is observed in older patients undergoing non-cardiac surgery (for a recent review, see Rasmussen, 2006; Newman et al., 2007). The incidence of POCD, independent of the cognitive domain, ranges from 16 to 59%, 7 days postoperatively, and from 10 to 34%, 12 weeks after surgery (Williams-Russo et al., 1995; Moller et al., 1998; Ancelin et al., 2001; Rasmussen et al, 2003; Monk and Phillips-Bute, 2004). In one of the largest multicenter studies to date (1218 patients), the International Study of Post-Operative Cognitive Dysfunction (ISPOCD1; Moller et al., 1998), cognitive status of older patients was examined the day before surgery, as well as 1 and 12 weeks after a major non-cardiac surgery (abdominal and orthopedic surgeries). Performance was compared to that of a control group ($n=176$) not admitted to the hospital but assessed at similar time intervals as patients. By comparing changes in performance from baseline (preoperative assessment) to 1 week after surgery in control participants, the investigators were able to obtain an estimate for the learning effects on each test. Similar change scores were calculated for patients but the average learning effect obtained from the control group was subtracted from this score, which was then converted into a z-score for each test. Results showed that 25.8% of patients had POCD after 1 week relative to 3.4% in control participants. At 12 weeks after surgery, the investigators found POCD in 9.9% of patients compared with 2.8% in the control group. They also found a significant correlation between a decline in the activities of daily living and POCD at 12 weeks. However, no correlation was found between subjective complaints and POCD. Some evidence suggests that self-assessed measures of POCD tend to overestimate its incidence (Rödig et al., 1999; Johnson et al., 2002). Subjective feelings of cognitive dysfunction in some patients may reflect sudden awareness of age-related cognitive changes.

A key issue in the incidence of POCD in older adults is whether patients have pre-existing cognitive impairment. Such patients were excluded from the ISPOCD1 (Moller et al., 1998), resulting in a possible underestimation of POCD in the general population. An exception to this rule is the study by Ancelin et al. (2001). They examined the incidence of POCD after orthopedic surgery in patients over the age of 64 years, and deliberately chose not to use formal exclusion criteria so that a more representative sample of the general population would be assessed rather than a subpopulation of very healthy older adults. Pre-existing cognitive deficits in patients was assessed with an informant questionnaire (Détérioration Cognitive Observée or DéCO) that measures changes in cognitive performance over the past year (Ritchie and Fuhrer, 1995). From the 140 patients who completed baseline assessment, 15% were identified as being at risk of having early-stage dementia, a percentage that is similar to what is typically observed in the general elderly population (Canadian Study on Health and Aging or CSHA Working Group, 1994). The investigators also looked at which domains of cognition were more likely to be affected by surgery and anesthesia. Six cognitive domains were investigated preoperatively, as well as 1 and 12 weeks after surgery, using a comprehensive computerized cognitive battery (Ritchie et al., 1993). These included attention, short-term memory, episodic memory, implicit memory, visuo-spatial ability, and language. Decline in each domain was calculated by subtracting scores at 1 and 12 weeks from scores obtained preoperatively. A positive score indicated cognitive deterioration, and a negative score indicated improved performance. Results showed differences in susceptibility to POCD across cognitive domains. Depending on which aspect of cognition was measured, the incidence of POCD ranged from 0 to 25.4% after 1 week, and from 0 to 19.1% after 12 weeks. Cognitive domains most affected by surgery and anesthesia were episodic and implicit memory, visuo-spatial ability (measured by a reasoning task), and language (measured by verbal fluency). The same domains tended to be systematically affected at 1 and 12 weeks. Interestingly, no patient met criteria for dysfunction in attention tasks, and a significant deterioration in short-term memory tasks was found in less than 5% of patients after 12 weeks. Importantly, among the patients identified as being at risk of dementia, half of them withdrew before completing the neuropsychological evaluation at 12 weeks; withdrawal from study may be indicative of POCD, leading to underestimation of incidence in the general population. However, there was no control group so that POCD may have been underestimated by excluding patients that showed learning effects, albeit reduced. On the other hand, Ancelin and colleagues used a less stringent criterion than the one used in the ISPOCD1 (Moller et al., 1998). POCD was defined as a deterioration of more than 1 SD on at least one cognitive measure.

The issue of late POCD

Another important question regarding POCD is whether it is transient or permanent. Only a few studies using a longitudinal design have assessed the incidence of late POCD, that is, several months or years after surgery and anesthesia. Most of these studies concern cardiac surgery patients. The incidence of cognitive dysfunction in studies that extended follow-up to 6–12 months, ranges from 24 to 57% (Savageau et al., 1982a; Shaw et al., 1987; McKhann et al., 1997; Van Dijk et al., 2002;). Newman et al. (2001) found that 42% of their patients showed decline when assessed 5 years after surgery. Selnes et al. (2001) found significant declines in visuoconstruction and psychomotor speed tests 5 years after cardiac surgery. Findings from these studies suggest a pattern defined by early decline, followed by improvement, and then later decline at 5 years. However, as with many studies, there was no control group allowing measurement of confounding factors such as aging and history of cardiovascular disease. In a more recent study, Selnes et al. (2003) compared cognitive changes in coronary artery bypass graft patients and a non-surgical control group with similar risk factor for coronary artery disease.

They found no evidence for POCD at 12 months after cardiac surgery (nor at 3 months postoperatively, similar to Jones et al., 1990, in non-cardiac patients). The investigators concluded that decline in the early postoperative period may be transient, and later cognitive changes may be associated with progressive cerebrovascular disease. However, in a retrospective study that included a total of 9170 patients, Lee et al. (2005) compared the emergence of Alzheimer's disease following CPB surgery ($n = 5216$) and non-surgical percutaneous transluminal coronary angioplasty ($n = 3954$). After adjusting for age, length of hospitalization, comorbidity and number of procedures, results showed that CPB patients had an increased risk for emergence of Alzheimer's disease compared to the angioplasty group.

In non-cardiac surgery patients, Williams-Russo et al. (1995) found that only 5% of patients showed a late (6 months postoperatively) deterioration in cognitive function. However, the study lacked a non-surgical control group that takes into account learning effects from repeated testing. Goldstein and Fogel (1993) found that immediate postoperative scores on the Mini Mental State Examination (MMSE; Folstein et al., 1975) predicted 10-month postoperative MMSE scores. Using a subgroup of older patients ($n = 336$) from the ISPOCD1 (Moller et al., 1998), Abildstrom et al. (2000) found an incidence of decline of 10.4% after 1 or 2 years postsurgery. A similar incidence of cognitive decline (10.6%) was observed in their smaller group of controls ($n = 47$), suggesting that early POCD is transient. However, Monk and Phillips-Bute (2004) reported an incidence of decline of 42%, 2 years after surgery, but they did not use a control group. Clearly, more studies are needed to determine whether surgery and anesthesia lead to more permanent cognitive changes. However, the evidence so far seems to indicate that POCD is transient.

Methodological issues in the study of POCD

A striking feature is the important variability in the incidences of POCD reported across studies. Many of the earlier studies were designed to compare factors such as surgical methods or type of anesthesia (general vs. regional) rather than establishing the incidence of POCD. This partly explains why some of these studies only used global scales to assess cognitive changes, such as the MMSE. The MMSE is a commonly used screening test for dementia and may be used to quantify dementia severity. However, studies have shown that the MMSE is insensitive to patients in the earlier stages of dementia, and ceiling effects have been reported in patients with mild cognitive impairment (MCI) (Tombaugh and Melaryne, 1992; Wind et al., 1997; Nasreddine et al., 2005). Similarly, the use of comprehensive batteries, such as the Wechsler Adult Intelligence Scale (WAIS; Wechsler, 1997), has been found to be insensitive to the detection of subtle cognitive changes in cardiac surgery patients (Newman, 1995). Neuropsychological tests should be selected on the basis of their sensitivity to detect even small modifications in cognitive functioning (for a review, see Rasmussen et al., 2001). Test–retest validity is a crucial aspect to consider in the selection of cognitive tests, given that patients are assessed repeatedly over time. The use of validated parallel versions appears essential to minimize learning effects, although such effects cannot be completely eliminated; participants become familiar with the task structure and some of them may develop novel strategies to perform the task. Thus, studies assessing POCD should include a control group evaluated at a similar timetable as patients in order to measure learning effects. The absence or reduction of learning effects on repeated testing may be a strong indicator of subtle cognitive dysfunction.

The time at which cognitive changes are measured both before and after the operation varies greatly across studies. Many studies have assessed POCD within 1 week after the surgery while patients are still in the hospital. However, given the greater likelihood that potential confounds would also occur at that time, such as hypoactive state of delirium or severe pain, we contend that a minimum of 2 weeks might constitute better clinical practice for assessment of short-term POCD. Similarly, assessment of patients before surgery often occurs 1–2 days before surgery; factors associated with impending surgery such as stress and anxiety may negatively affect test performance. In turn, postoperative improvement in scores may be due to stress reduction rather than recovery from POCD. Ideally, preoperative assessment should be performed a few weeks before surgery and a placebo group not receiving surgery and anesthesia, but hospitalized, should be included to allow measurement of effects associated with stress or pain, which are known to affect memory performance.

There is very little consensus in the literature on which criteria to use for defining POCD. How many cognitive tests should be used? What level of change constitutes a significant decline? In the earlier studies, cut-off scores on screening tests were often used to define POCD. The neuropsychological assessment should include different domains of cognition. However, for practical reason, the assessment should be limited in time and not go beyond 2 h. In more recent studies, individual change scores comparing performance on different cognitive measures before and after surgery are typically used. Some investigators used a more lenient criterion such as deterioration in performance equal to or greater than 1 SD on at least one neuropsychological measure while others used more stringent criteria. In the ISPOCD1 (Moller et al., 1998), the investigators used a change score that was superior to 2 z-scores on at least two distinct cognitive measures or on a combined z-scores of all tests. Such criteria have the advantage of taking into consideration a decline in specific tests as well as a general deterioration in all tests.

Patient-related factors

POCD is likely to be multi determined and some of the factors contributing to its occurrence may be related to the patients themselves. Among patient-related risk factors, increasing age is consistently reported in the literature although most of these studies focused on older adults and did not encompass the entire age spectrum (Smith et al., 1986; Shaw et al., 1987; Moller et al., 1998; Ancelin et al., 2001; Johnson et al., 2002; Newman et al., 1994). Basic functions of organs are generally not compromised by normal aging, however, the aged brain is different from the younger brain in several important aspects, including size, distribution of neurotransmitters, and metabolic function. In turn, this leads to a diminished functional reserve which may reduce capacity to compensate for physiological stress such as surgery and anesthesia. In Ancelin et al.'s (2001) study, patients aged 75 years and older showed reduced learning effects on several cognitive tests relative to patients aged between 65 and 75 years old. Johnson et al. (2002) measured POCD in a large cohort of patients ($n = 508$) between the ages of 40 and 59 years. The investigators also included an age-matched control group not hospitalized ($n = 185$). The incidence of POCD in patients undergoing general surgery and anesthesia was 19.2%, 1 week after surgery, and dropped at 6.2%, 3 months postoperatively. In the control group, the incidence was 4.0 and 4.1%, respectively for the same time period. Thus, only the incidence 1 week postoperatively was statistically different between patients and controls, suggesting that structural and physiological changes that occur in the aging brain may be modulating the response to surgery and anesthesia. Education may also act as a protective factor against cognitive decline in older adults by increasing cognitive reserve in more educated individuals. Several studies have found that poor education constituted a significant risk factor for POCD (Moller et al., 1998; Ancelin et al., 2001).

Some studies have found a correlation between POCD and depression and/or anxiety (Ancelin et al., 2001; Newman et al., 2001). The patient's operation could trigger symptoms of depression/anxiety which in turn may affect performance on neuropsychological test, and especially those assessing memory. It is thus important to measure psychological status of patients before surgery. Ancelin et al. (2001) found that 21% of participants meet criteria for major depressive episode before surgery and anesthesia compared with 14% in the general elderly population (Fuhrer et al., 1992). These patients showed decreased performance in episodic memory. However, the investigators found that depression status was not modified by the surgery and anesthesia.

Pre-existing cognitive impairment might constitute an important risk factor for POCD (Dijkstra et al., 1999; Rankin et al., 2003). Some researchers have suggested that POCD reflects a preoperative dementia unmasked by the surgery. However, as we have already mentioned, most studies have used screening tests for dementia that are not sensitive enough to identify people with pre-existing MCI or early dementia. MCI is seen as a transitional state between the cognition of normal aging and mild dementia. In such cases, there is a memory complaint, and performance on objective memory test is at least 1.5 SD below that of an age-matched control group. However, the person remains functional and does not meet formal criteria for dementia according to the Diagnostic and Statistical Manual of Mental Disorders (American Psychiatric Association, 1994). Importantly, people with MCI have an increased risk of developing a dementia, although a sizeable portion remain stable or even improve their performance on subsequent testing (Palmer et al., 2002). Because of limited cognitive reserve to cope with the physiological stress of surgery and anesthesia, individuals with MCI may be those more susceptible to POCD; they should be identified preoperatively using a comprehensive evaluation that includes more sensitive screening tests such as the Montreal Cognitive Assesment (Nasreddine et al., 2005).

The role of surgery and anesthesia in POCD

There are several risk factors more directly related to surgery and anesthesia that may be involved in the pathogenesis of POCD (Rolfson et al., 1999). Surgery is associated with the stress response, with increasing secretion of cortisol and catecholamines. Persistently high levels of stress may inhibit memory and interfere with hippocampal function. Surgery alone also activates specific homeostatic responses, triggering immune mechanisms and the inflammatory cascade through the release of various inflammatory mediators (Westaby et al., 2001). Intraoperative hypotension, hypoxia, embolization, medications, and postoperative infections have also been described as risk factors for POCD. Because the incidence of POCD does not seem to be influenced by the type of anesthesia (i.e., general vs. regional), attention has begun to focus on the role of the surgical intervention itself in the genesis of this condition. Postoperative pain is a possible etiological factor in POCD mechanisms. Epidural analgesia with local anesthetics and/or opioids has been found to be probably better than parenteral opioids for the control of postoperative pain and the prevention of early POCD (Rasmussen and Moller, 2000; Bekker and Weeks, 2003; Cohendy et al., 2005). Furthermore, those who received postoperative analgesia orally were at least risk of experiencing POCD compared with parenteral analgesia (Bekker and Weeks, 2003; Wang et al., 2007).

Depression of central nervous system function is a part of anesthesia. This condition is expected to be perfectly reversible and transient, but several complications may arise, some of them causing serious disability. General anesthesia affects brain function at all levels, including neuronal membranes, receptors, ion channels, neurotransmitters, cerebral blood flow, and metabolism. Impairment of cognitive and psychomotor performance in the days following general anesthesia is common and typically attributed to incomplete drug clearance.

The possibility that general anesthesia contributes to cognitive deterioration has not been tested directly; partly because clinical studies have not controlled for the anesthetics used and cannot differentiate between the effects of illness, hospitalization, surgery, and anesthesia. In that respect, animal models may provide some insights into potential mechanisms by which anesthesia may lead to POCD as will be discussed below.

Inflammation and POCD pathogenesis

We and other investigators believe that inflammatory processes play a key role in the pathogenesis of POCD (Gao et al., 2005; Mathew et al., 2007). The cells involved in brain inflammation are mixtures of structural and inflammatory cells. In the brain, supporting cells of the glial family, known as microglial cells, act as scavengers, in much the same fashion as macrophages (Von Bernhardi and Ramirez, 2001; Guo et al., 2007). They engulf and eliminate dead neurons that have been damaged by injury or illness. The presence of activated microglial cells is an indicator of chronic inflammation. Astrocytes and/or microglia secrete most cytokines in the brain, such as interleukin(IL)-1β, IL-6, and tumor necrosis factor-α (TNF-α), which circulate in the blood and communicate with neurons. As it is now well recognized, inflammatory responses develop within the brain under a variety of pathological conditions. Cytokines are hardly detectable in the central nervous system under physiological conditions, but they become rapidly upregulated by pathological events, like ischemia (Minamikawa et al., 1992; Wang and McCubrey, 1997), excitotoxicity (Minami et al., 1991), lipopolysaccharide (LPS) injection (Gabellec et al., 1995), or viral infection (Marquette et al., 1996). Glial activation and the consequent release of proinflammatory cytokines within the hippocampus interfere with cognitive function as evidenced by abnormal memory and learning in the intact organism and/or inability to develop long-term potentiation in hippocampal slice preparation (Von Bernhardi and Ramirez, 2001). There is evidence that under certain conditions, neurons can also produce cytokines (Guo et al., 2007). Cerebral endothelial cells are actively engaged in processes of microvascular stasis as well as leukocyte infiltration by evoking a plethora of bioactive inflammatory cytokines and chemokines (Matsumura and Kobayashi, 2004).

It has been suggested that proinflammatory cytokines could play a role in the development of cognitive decline that may acutely follow surgery. The peripheral IL-6 response to surgical injury has extended our understanding of the role of proinflammatory cytokine response to surgery. Within 2 h of skin incision, IL-6 levels increase, peak between 4 and 12 h, and remain elevated for up to 3 days postoperatively (Biffl et al., 1996). Extent of surgical trauma is an important factor; there was a greater degree of IL-6 elevation after abdominal aortic and colorectal procedures than after hip replacement despite similar surgical procedure times. Further, four separate studies illustrate that the increment in IL-6 is less after laparoscopic versus open cholecystectomy, despite similar surgical times. The IL-6 response is not modified by anesthetic regimen but is decreased by use of antiinflammatory analgesic agents. Cytokines originating from the periphery, such as IL-1β, can act on the central nervous system in a wide variety of ways, including increases in slow-wave sleep, modulation of long-term potentiation, changes in monoamine release and turnover, and more global effects on mood and cognition (Von Bernhardi and Ramirez, 2001; Gao et al., 2005; Mathew et al., 2007). Cytokines exert effects within the central nervous system through both direct and indirect means. Both IL-1β and TNF-α have been shown to gain direct entry into the central nervous system through the relatively permeable blood–brain barrier in the periventricular regions. Further, IL-1β can also directly bind to its cognate receptors on the endothelial cells within the brain microvasculature where these can elaborate a central inflammatory response. Indirectly, cytokines can induce changes within the central nervous system through vagal afferent nerves. In many studies, either LPS or IL-1β, when administered peripherally, induced brainstem and hippocampal IL-1β production and central nervous system expression of IL-1β, IL-6, and TNF-α, all of which could be abolished by vagotomy (Hansen et al., 2000; Chauvet et al., 2001).

Several studies also suggest that the marked and sustained expression of inflammation-related enzymes such as cyclooxygenase-2 (COX-2) plays an important role in secondary events that amplify cerebral injury after ischemia. The contribution of COX-2 to peripheral inflammation is well documented, but little is known about its involvement in brain inflammation (Minghetti, 2004). It has been reported that COX-2 is significantly induced in astrocyte and microglial cultures by radiation injury (Kyrkanides et al., 2002). It has been shown that COX-2 inhibitor protects the brain against amyloid beta-induced memory disturbances in mice (Giovannini et al., 2003; Cakala et al., 2007).

Matrix metalloproteinases (MMPs) have been implicated in early breakdown of the blood–brain barrier in neuroinflammatory disease. MMPs comprise a group of proteolytic enzymes that act as mediators of brain injury in a wide variety of disease processes, including multiple sclerosis, Alzheimer's disease (AD), stroke, tumor invasion, and other inflammatory brain disorders (Anthony et al., 1997; Kolb et al., 1998; Romanic et al., 1998; Yong et al., 1998). Interruption of the MMPs proteolytic cascade may be a possible therapeutic approach to preventing the secondary progression of damage after brain injury.

Considerable evidence gained over the past decade supports the conclusion that neuroinflammation is associated with AD pathology (Cacquevel et al., 2004; Yao et al., 2004; Tuppo and Arias, 2005). Inflammatory brain cells, such as microglia and astrocytes, as well as cytokines, including IL-6, TNF-α, and transforming growth factor-beta, have clearly been implicated in this inflammatory process (Meeuwsen et al., 2003). Many varieties of chemokines, such as IL-8, monocyte chemoattractant protein, and RANTES, are also expressed in brain tissue from humans in conjunction with dementia. It has been shown that IL-8 is a key mediator of neuroinflammation in severe traumatic brain injuries, and is constitutively expressed in the brain (Kushi et al., 2003).

These inflammatory mediators should be widely investigated and considered as targets in the inflammatory process associated with POCD, and especially after CPB, given that the incidence of AD has been linked to cardiac surgery. Sparks et al. (2000) found evidence of AD-like lesions in the brains of non-demented individuals with mitral valve prolapse. They and others suggested that cognitive dysfunction occurring after CPB with coronary artery grafting or valve repair/replacement is a functional sequel of AD-like neuropathology. In cardiac surgery, non-pulsatile flow generated by the CPB machine could lower shear stress on endothelial cells, reduce nitric oxide (NO) release, and induce non-homogenous blood flow distribution in ischemic areas presenting reperfusion injury after weaning from CPB. Deficient NO production affects the inflammatory cascade, allowing the vascular adhesion of inflammatory cells primed by contact with the extracorporeal circuit. This initial deficit in endothelial NO synthase is particularly marked in older patients (Wan et al., 1997; Maffei et al., 2004).

As suggested by some researchers, genetics may play a role in the pathogenesis of POCD. This is indicated by the presence of the apolipoprotein ε4 (APO-ε4) allele in a subgroup of POCD patients. Previous data have confirmed the association between AD and APO-ε4, and support the hypothesis that the APO-ε4 allele either confers genetic susceptibility to AD or may be in linkage disequilibrium with another susceptibility locus (Abildstrom et al., 2004; Hsiung et al., 2004). Ethnic variability in the allelic frequency of APO-ε4 in the elderly warrants further investigation.

Animal models and POCD

In the past decade, although the mechanisms for POCD remain unclear, progress has been made with the help of animal models. The role of cytokine-mediated inflammation within the central nervous system in the development of cognitive dysfunction has been studied in rodents. To elucidate the effect of surgery and anesthesia on learning and memory, Wan et al. (2007) used the Y-maze test to evaluate spatial learning and memory in rodents. They found a correlation between the inflammatory response in the hippocampus and the development of POCD. They also found that after surgery (splenectomy), rats displayed impaired memory that was associated with glial activation and proinflammatory cytokine expression in the hippocampus.

Neurons in the hippocampus of splenectomized rats increased expression of both B-cell CLL/lymphoma 2 (Bcl-2) and Bcl-2-associated X protein (Bax). However, the ratio of Bcl-2:Bax was decreased after surgery, suggesting that neurons in the hippocampus are inclined toward apoptosis. So the cognitive dysfunction is not necessarily associated with the presence of actual cell death in hippocampus; rather, there may be abnormal interaction between neurons and glial cells in the hippocampus (Jarrard, 1995; Tanaka et al., 2006).

IL-1β has consistently been detected in central nervous system after injury to the brain or peripheral immune activation. For example, IL-1β bioactivity and immunoreactivity and mRNA have all been found to be present in brain after peripheral administration of LPS in the rat (Buttini and Boddeke, 1995). It has been demonstrated that IL-1β may lead to some "sickness behavior" which reduces food and water intake, decreases exploration, decreases social interactions, and somnogenesis (Dantzer et al., 1998), and may produce alterations in cognitive processes, including learning and memory. Nathan et al. (2005) have shown that systematic LPS or IL-1β may affect performance in various learning tasks in mice. Specifically, LPS stimulates Toll-like receptors and induces expression of proinflammatory cytokines IL-1β, IL-6, and TNFα, primarily from macrophages (Gibertini et al., 1995; Dantzer et al., 1998; Dantzer, 2001). Although these cytokines may exert neurobiological effects, the strongest case has been made for IL-1β (Gibertini et al., 1995; Gibertini, 1996; Bianchi et al., 1998; Dantzer et al., 1998). Furthermore, many of the physiological and behavioral effects associated with LPS administration can be removed with the administration of IL-1 receptor antagonist. In our laboratory we are currently injecting LPS intraperitoneally to induce systemic inflammation in rats, and then testing their spatial memory using the Morris water maze, a hippocampus-dependent task to examine brain inflammation by molecular biology technique to test for the level of proinflammatory cytokines.

Oitz et al. (1993) compared the influences of two proinflammatory cytokines on Morris water maze learning by infusing IL-1β or IL-6 intracerebroventricularly either 1 h before or immediately prior to the first of 2 days of testing. Results showed that animals treated with IL-1β 1 h prior to testing showed significantly longer latencies and distances on the first trial of days compared to vehicle controls and animals receiving IL-1β immediately before testing. IL-6-treated animals did not differ from controls.

In addition, peripheral inflammatory responses to immune activating agents, as well as brain cytokine responses to stimulation are altered with normal aging. For example, Saito et al. (2003) reported that the levels of IL-6 in blood produced by cecal ligation and puncture, as well as by LPS, were elevated more in aged than in young mice. Within the brain, IL-1β and TNF-α responses to peripheral LPS administration appear to increase with aging (Terao et al., 2002; Xie et al., 2003). Barrientos et al. (2006) report that a peripheral injection of *Escherichia coli* produces both anterograde and retrograde amnesia in 24 month old, but not 3 month old rats for memories that depend on the hippocampus, which is memory for context, contextual fear, and place learning. And it produced a masked increased in IL-1β levels in the hippocampus, but not in parietal cortex or serum. These findings manifested that age is a vulnerability factor that increases the likelihood that an immune challenge will produce a cognitive impairment. It is possible that this cognitive vulnerability is mediated by age-related changes in the glial environment that results in an exaggerated brain pro-inflammatory response to infection.

Some studies have previously demonstrated lasting impairment on a spatial memory task in rats after a single-2 h isoflurane (ISO)-nitrous oxide (N_2O) anesthetic, where rats were trained on a radial arm maze for 2 months before anesthesia and then tested for 8 weeks beginning 24 h after anesthesia. The ability of aged rats to improve their maze performance was worse than that of identically treated, unanesthetized control rats (Culley et al., 2003, 2004b). These results showed that general anesthesia affects performance for longer than would be expected based on the pharmacology of the drugs and suggested that aged rats might be more susceptible to such disruption. This radial arm maze task tests spatial working memory, assesses the integrity of the frontal cortex, entorhinal cortex, and hippocampus (Decker and Gallagher, 1987; Baxter et al., 1997), and can detect subtle differences in learning caused by aging, sedatives and anesthetics (Decker and Gallagher, 1987; Luine and Rodriguez, 1994; Borde et al., 1998; Culley et al., 2003, 2004b). Culley et al. (2004a) also demonstrated that general anesthesia produces long-lasting impairment in the ability of rats to acquire and perform a spatial memory task and the aged rats' performance on a spatial memory task remains impaired for at least 2 weeks after general anesthesia. These results cannot be explained by the pharmacokinetics of the drugs involved.

Conclusion

With increasing longevity of the general population, more and more surgical interventions will be performed on older adults. As older patients must be informed about the risks associated with their surgery, understanding the effects of surgery and anesthesia on memory and cognition has become a key issue. There is evidence for short-term POCD affecting the memory and cognition of older adults following cardiac surgery and anesthesia, and to a lesser extent that of patients having undergone non-cardiac surgery. More research is needed to determine whether POCD is permanent but the evidence so far suggests that cognitive dysfunction is reversible. Importantly, however, this research area suffers from a large number of methodological difficulties. These include the differences in surgery and participants, the diversity, number, and range of neuropsychological tests used with varying sensitivity to change and learning, and the variety of definitions used to classify individuals as having POCD. These differences make it difficult to compare results across studies. To overcome some of the methodological issues, it would be useful to recognize the arbitrariness of any definition of POCD and to consider whether it is timely to establish a consensus that specifies a limited number of tests to be used in all studies and the value of pooling data across studies to increase power in secondary analyses.

Patient-related risk factors, such as increased age, have been identified in several studies as well as factors more directly related to surgery and anesthesia. Inflammatory processes appear to play a key role in the pathogenesis of POCD. Biochemical markers and new sensitive methods are needed to detect early POCD after cardiac and non-cardiac surgery. Stable NO products represent a potential biochemical predictor of POCD. It has been demonstrated that preoperative and postoperative plasma concentrations of stable NO products (nitrate/nitrite) are associated with the early detection of POCD after cardiac surgery. Animal models provide a way to test the role of cytokine-mediated inflammation within the central nervous system in POCD. Future research should also focus on prevention of POCD. It is conceivable that therapeutic interventions aimed at attenuating the inflammatory response might result in better outcomes. A prevention program that prepares patients before surgery and anesthesia might also help alleviate complaints regarding memory and cognition and reduce stress. POCD diminishes the quality of the patient's life and adds costs to hospitalization and out-of-hospital care. Research must be geared towards understanding POCD in older adults and identifying those at risk. Clearly, a multidisciplinary approach appears as the most fruitful means for study and management of POCD.

Abbreviations

AD	Alzeimer's disease
APO-ε4	Apolipoprotein ε4
Bax	Bcl2-associated X protein
Bcl-2	B-cell CLL/lymphoma 2
COX-2	cyclooxygenase-2
CPB	cardiopulmonary bypass
CSHA	Canadian Study on Health and Aging
IL-1β	interleukin-1β
IL-6	interleukin-6
IL-8	interleukin-8
ISO	isoflurane
ISPOCD1	International Study of Post-Operative Cognitive Dysfunction
LPS	lipopolysaccharide
MCI	mild cognitive impairment
MMPs	matrix metalloproteinases
MMSE	Mini Mental State Examination
mRNA	messenger ribonucleic acid
NO	nitric oxide
N_2O	nitrous oxide
POCD	postoperative cognitive dysfunction
RANTES	regulated upon activation, normal T-cell expressed and secreted
SD	standard deviation
TNF-α	tumor necrosis factor-α
WAIS	Wechsler Adult Intelligence Scale

Acknowledgments

This work was supported by a CIHR grant MOP-68890 and a FRSQ grant awarded to N.C. The authors would like to thank Michèle Moreau for her assistance in editing the chapter.

References

Abildstrom, H., Christiansen, M. and Feldman, H. (2004) ISPOCD2 Investigators: apolipoprotein E genotype and cognitive dysfunction after noncardiac surgery. Anesthesiology, 101(4): 855–861.

Abildstrom, H., Rasmussen, L.S., Rentowl, P., Hanning, C.D., Rasmussen, H., Kristensen, P.A., et al. (2000) Cognitive dysfunction 1-2 years after non-cardiac surgery in the elderly. ISPOCD group. International study of post-operative cognitive dysfunction. Acta Anaesthesiol. Scand., 44(10): 1246–1251.

American Psychiatric Association. (1994) Diagnostic and Statistical Manual of Mental Disorders (Fourth Edition). American Psychiatric Association, Washington, DC.

Ancelin, M.-L., De Roquefeuil, G., Ledesert, B., Bonnel, F., Cheminal, J.-C. and Ritchie, K. (2001) Exposure to anaesthetic agents, cognitive functioning and depressive symptomatology in the elderly. Br. J. Psychiatry, 178: 360–366.

Anderson, R.E., Li, T.Q., Jindmarsh, T., Settergren, G. and Vaage, J. (1999) Increased extracellullar brain water after coronary artery bypass grafting is avoided by off pump surgery. J. Cardiothorac Vasc. Anesth., 13: 698–702.

Anthony, D.C., Ferguson, B., Matyzak, M.K., Miller, K.M., Esiri, M.M. and Perry, V.H. (1997) Differential matrix metalloproteinase expression in cases of multiple sclerosis and stroke. Neuropathol. Appl. Neurobiol., 23: 406–415.

Barrientos, R.M., Higgins, E.A., Biedenkapp, J.C., Sprunger, D.B., Wright-Hardesty, K.J., Watkins, L.R., et al. (2006) Peripheral infection and aging interact to impair hippocampal memory consolidation. Neurobiol. Aging, 27(5): 723–732.

Baxter, M.G., Holland, P.C. and Gallagher, M. (1997) Disruption of decrements in conditioned stimulus processing by selective removal of hippocampal cholinergic input. J. Neurosci., 17(13): 5230–5236.

Bekker, A.Y. and Weeks, E.J. (2003) Cognitive function after anaesthesia in the elderly. Best Pract. Res. Clin. Anaesthesiol., 17(2): 259–272.

Bianchi, M., Sacerdote, P. and Panerai, A.E. (1998) Cytokines and cognitive function in mice. Biol. Signals Recept., 7(1): 45–54.

Biffl, W.L., Moore, E.E., Moore, F.A. and Peterson, V.M. (1996) Interleukin-6 in the injured patient: marker of injury or mediator of inflammation? Ann. Surg., 224(5): 647–664.

Borde, N., Jaffard, R. and Beracochea, D. (1998) Effects of chronic alcohol consumption or Diazepam administration on item recognition and temporal ordering in a spatial working memory task in mice. Eur. J. Neurosci., 10(7): p. 2380.

Buttini, M. and Boddeke, H. (1995) Peripheral lipopolysaccharide stimulation induces interleukin-1 beta messenger RNA in rat brain microglial cells. Neuroscience, 65(2): 523–530.

Cacquevel, M., Lebeurrier, N., Cheenne, S. and Vivien, D. (2004) Cytokines in neuroinflammation and Alzheimer's disease. Curr. Drug Targets, 5(6): 529–534.

Cakala, M., Malik, A.R. and Strosznajder, J.B. (2007) Inhibitor of cyclooxygenase-2 protects against amyloid beta peptide-evoked memory impairment in mice. Pharmacol. Rep., 59(2): 164–172.

Chauvet, N., Palin, K., Verrier, D., Poole, S., Dantzer, R. and Lestage, J. (2001) Rat microglial cells secrete predominantly the precursor of interleukin-1beta in response to lipopolysaccharide. Eur. J. Neurosci., 14(4): 609–617.

Cohendy, R., Brougere, A. and Cuvillon, P. (2005) Anaesthesia in the older patient. Curr. Opin. Clin. Nutr. Metab. Care, 8(1): 17–21.

CSHA Working Group. (1994) Canadian study of health and aging: study methods and prevalence of dementia. J. Can. Med. Assoc., 150: 899–913.

Culley, D.J., Baxter, M., Yukhananov, R. and Crosby, G. (2003) The memory effects of general anesthesia persist for weeks in young and aged rats. Anesth. Analg., 96(3): 1004–1009.

Culley, D.J., Baxter, M.G., Crosby, C.A., Yukhananov, R. and Crosby, G. (2004a) Impaired acquisition of spatial memory 2 weeks after isoflurane and isoflurane-nitrous oxide anesthesia in aged rats. Anesth. Analg., 99(5): 1393–1397.

Culley, D.J., Baxter, M.G., Yukhananov, R. and Crosby, G. (2004b) Long-term impairment of acquisition of a spatial memory task following isoflurane-nitrous oxide anesthesia in rats. Anesthesiology, 100(2): 309–314.

Dantzer, R. (2001) Cytokine-induced sickness behavior: mechanisms and implications. Ann. N.Y. Acad. Sci., 933: 222–234.

Dantzer, R., Bluthe, R.M., Laye, S., Bret-Dibat, J.L., Parnet, P. and Kelley, K.W. (1998) Cytokines and sickness behavior. Ann. N.Y. Acad. Sci., 840: 586–590.

Decker, M.W. and Gallagher, M. (1987) Scopolamine-disruption of radial arm maze performance: modification by noradrenergic depletion. Brain Res., 417(1): 59–69.

Dijkstra, J.B., Houx, P.J. and Jolles, J. (1999) Cognition after major surgery in the elderly: test performance and complaints. Br. J. Anaesth., 82(6): 867–874.

Dyer, C.B., Ashton, C.M. and Teasdale, T.A. (1995) Postoperative delirium. A review of 80 primary data-collection studies. Arch. Intern. Med., 155(5): 461–465.

Folstein, M.F., Folstein, S.E. and McHugh, P.R. (1975) Mini-mental state: a practical method for grading the cognitive state of patients for the clinician. J. Psychiatr. Res., 12(3): 189–198.

Fuhrer, R., Antonucci, T.C., Gagnon, M., Dartigues, J.-F., Barberger-Gateau, P. and Alperovitch, A. (1992) Depressive symptomatology and cognitive functioning: an epidemiological survey in an elderly community sample in France. Psychol. Med., 22(1): 159–172.

Gabellec, M.M., Grifais, R., Fillion, G. and Haour, F. (1995) Expression of interleukin 1 alpha, interleukin 1 beta and interleukin 1 receptor antagonist mRNA in mouse brain: regulation by bacterial lipopolysaccharide (LPS) treatment. Mol. Brain Res., 31: 122–130.

Gao, L., Taha, R., Gauvin, D., Othmen, L., Wang, Y. and Blaise, G. (2005) Postoperative cognitive dysfunction after cardiac surgery. Chest, 128(5): 3664–3670.

Gibertini, M. (1996) IL1 beta impairs relational but not procedural rodent learning in a water maze task. Adv. Exp. Med. Biol., 402: 207–217.

Gibertini, M., Newton, C., Klein, T.W. and Friedman, H. (1995) *Legionella pneumophila*-induced visual learning impairment reversed by anti-interleukin-1 beta. Proc. Soc. Exp. Biol. Med., 210(1): 7–11.

Giovannini, M.G., Scali, C., Prosperi, C., Belluci, A., Pepeu, G. and Casamenti, F. (2003) Experimental brain inflammation and neurodegeneration as model of Alzheimer's disease: protective effects of selective COX-2 inhibitors. Int. J. Immunopathol. Pharmacol., 16(2): 31–40.

Goldstein, M.Z. and Fogel, B.S. (1993) Cognitive change after elective surgery in nondemented older adults. Am. J. Geriatr. Psychiatry, 1(2): 118–125.

Guo, W., Wang, H., Watanabe, M., Shimizu, K., Zou, S., LaGraize, S.C., et al. (2007) Glial-cytokine-neuronal interactions underlying the mechanisms of persistent pain. J. Neurosci., 27(22): 6006–6018.

Hansen, M.K., Nguyen, K.T., Goehler, L.E., Gaykema, R.P., Fleshner, M., Maier, S.F., et al. (2000) Effects of vagotomy on lipopolysaccharide-induced brain interleukin-1beta protein in rats. Auton. Neurosci., 85(1–3): 119–126.

Hsiung, G.Y., Sadovnick, A.D. and Feldman, H. (2004) Apolipoprotein E epsilon4 genotype as a risk factor for cognitive decline and dementia: data from the Canadian Study of Health and Aging. CMAJ, 171(8): 863–867.

Jarrard, L.E. (1995) What does the hippocampus really do? Behav. Brain Res., 71(1–2): 1–10.

Johnson, T., Monk, T., Rasmussen, L.S., Abildstrom, H., Houx, P., Korttila, K., et al. (2002) Postoperative cognitive dysfunction in middle-aged patients. Anesthesiology, 96(6): 1351–1357.

Jones, M.J.T., Piggott, S.E., Vaughan, R.S., Bayer, A.J., Newcombe, R.G., Twining, T.C., et al. (1990) Cognitive and functional competence after anaesthesia in patients aged over 60: controlled trial of general anaesthesia for elective hip or knee replacement. Br. Med. J., 300(6741): 1683–1687.

Kolb, S.A., Lahrtz, F., Paul, R., Leppert, D., Nadal, D., Pfister, H.W., et al. (1998) Matrix metalloproteinases and tissue inhibitors of metalloproteinases in viral meningitis: upregulation of MMP-9 and TIMP-1 in cerebrospinal fluid. J. Neuroimmunol., 84(2): 143–150.

Kushi, H., Saito, T., Makino, K., et al. (2003) IL-8 is a key mediator of neuroinflammation in severe traumatic brain injuries. Acta Neurochir., 86: p. 347.

Kyrkanides, S., Moore, A.H., Olschowka, J.A., Daeschner, J.C., Williams, J.P., Hansen, J.T., et al. (2002) Cyclooxygenase-2 modulates brain inflammation-related gene expression in central nervous system radiation injury. Brain Res. Mol. Brain Res., 104(2): 159–169.

Lee, J.D., Lee, S.J., Tsushima, W.T., Yamauchi, H., Lau, W.T., Popper, J., et al. (2003) Benefits of off-pump bypass on neurologic and clinical morbidity: a prospective randomized trial. Ann. Thorac. Surg., 76: 18–26.

Lee, T.A., Wolozin, B., Weiss, K.B. and Bednard, M.M. (2005) Assessment of the emergence of Alzheimer's disease following coronary artery bypass graft surgery or percutaneous transluminal coronary angioplasty. J. Alzheimers Dis., 7(4): 319–324.

Luine, V. and Rodriguez, M. (1994) Effects of estradiol on radial arm maze performance of young and aged rats. Behav. Neural. Biol., 62(3): 230–236.

Maffei, C.M., Mirels, L.F., Sobel, R.A., Clemons, K.V. and Stevens, D.A. (2004) Cytokine and inducible nitric oxide synthase mRNA expression during experimental murine cryptococcal meningoencephalitis. Infect. Immunol., 72(4): 2338–2349.

Marquette, C., Van Dam, A.M., Ceccaldi, P.E., Weber, P., Haour, F. and Tsiang, H. (1996) Induction of immunoreactive interleukin-1 beta and tumor necrosis factor-alpha in the brains of rabies virus infected rats. J. Neuroimmunol., 68(1–2): 45–51.

Mathew, J.P., Podgoreanu, M.V., Grocott, H.P., White, W.D., Morris, R.W., Stafford-Smith, M., et al. (2007) Genetic variants in P-selectin and C-reactive protein influence susceptibility to cognitive decline after cardiac surgery. J. Am. Coll. Cardiol., 49(19): 1934–1942.

Matsumura, K. and Kobayashi, S. (2004) Signaling the brain in inflammation: the role of endothelial cells. Front Biosci., 9: 2819–2826.

McKhann, G.M., Goldsborough, M.A., Borowicz, Jr., L.M., Selnes, O.A., Mellits, E.D., Enger, C., et al. (1997) Cognitive outcome after coronary artery bypass: a one-year prospective study. Ann. Thorac. Surg., 63(2): 510–515.

Meeuwsen, S., Persoon-Deen, C., Bsibsi, M., Ravid, R. and Van Noort, J.M. (2003) Cytokine, chemokine and growth factor gene profiling of cultured human astrocytes after exposure to proinflammatory stimuli. Glia, 43(3): 243–253.

Minami, M., Kuraishi, Y. and Satoh, M. (1991) Effects of kainic acid on messenger RNA levels of IL-1 beta, IL-6, TNF alpha and LIF in the rat brain. Biochem. Biophys. Res. Commun., 176(2): 593–598.

Minamikawa, K., Wada, H., Ohiwa, M., Kaneko, T., Tsukada, T., Kageyama, S., et al. (1992) Plasma interleukin-6 in patients with disseminated intravascular coagulation. Rinsho Ketsueki, 33(12): 1797–1801.

Minghetti, L. (2004) Cyclooxygenase-2 (COX-2) in inflammatory and degenerative brain diseases. Neuropathol. Exp. Neurol., 63(9): 901–910.

Moller, J.T., Cluitmans, P., Rasmussen, L.S., Houx, P., Rasmussen, H., Canet, J., et al. (1998) Long-term postoperative cognitive dysfunction in the elderly ISPOCD1 study. ISPOCD investigators. International Study of Post-Operative Cognitive Dysfunction. Lancet, 351(9106): 857–861.

Moller, J.T., Svennild, I., Johannessen, N.W., Jensen, P.F., Espersen, K., Gravenstein, J.S., et al. (1993) Perioperative monitoring with pulse oximetry and late postoperative cognitive dysfunction. Br. J. Anaesth., 71(3): 340–347.

Monk, T.G. and Phillips-Bute, B.G. (2004) Longitudinal assessment of neurocognitive function in elderly patients after major, noncardiac surgery [abstract A-62]. Paper presented at the ASA Annual Scientific Session, Las Vegas, NV.

Moody, D.M., Bell, M.A., Challa, V.R., Johnston, W.E. and Prough, D.S. (1990) Brain microemboli during cardiac surgery or aortography. Ann. Neurol., 28: 477–486.

Nasreddine, Z.S., Phillips, N.A., Bédirian, V., Charbonneau, S., Whitehead, V., Collin, I., et al. (2005) The Montreal cognitive assessment, MoCA: a brief screening tool for mild cognitive impairment. JAGS, 53: 695–699.

Nathan, L.S., Lucie, A.M., William, S.C. and Gray, W.B. (2005) Effects of intraperitoneal lipopolysaccharide on Morris maze performance in year-old and 2-month-old female C57BL/6J mice. Behav. Brain Res., 159: 145–151.

Newman, M.F., Croughwell, N.D., Blumenthal, J.A., White, W.D., Lewis, J.B., Smith, L.R., et al. (1994) Effect of aging on cerebral autoregulation during cardiopulmonary bypass. Circulation, 90: 243–249.

Newman, M.F., Grocott, H.P., Mathew, J.P., White, W.D., Landolfo, K., Reves, J.G., et al. (2001) Report of the substudy assessing the impact of neurocognitive function on quality of life 5 years after cardiac surgery. Stroke, 32: 2874–2881.

Newman, S., Stygall, J., Hirani, S., Shaefi, S. and Maze, M. (2007) Postoperative cognitive dysfunction after non-cardiac surgery: a systematic review. Anesthesiology, 106(3): 572–590.

Newman, S.P. (1995) Analysis and interpretation of neuro-psychologic tests in cardiac surgery. Ann. Thorac. Surg., 59: 1351–1355.

Oitz, M.S., Van Oers, H., Schobitz, B. and De Kloet, E.R. (1993) Interleukin-1 beta, but not interleukin-6, impairs spatial navigation learning. Brain Res., 613(1): 160–163.

Palmer, K., Wang, H.X., Bäckman, L., Winblad, B. and Fratiglioni, L. (2002) Differential evolution of cognitive impairment in nondemented older persons: results from the Kungsholmen project. Am. J. Psychiatry, 159: 436–442.

Parikh, S. and Chung, C. (1995) Postoperative delirium in the elderly. Anesth. Analg., 80(6): 1223–1232.

Rankin, K.P., Kochamba, G.S., Boone, K.B., Petitti, D.B. and Buckwalter, J.G. (2003) Presurgical cognitive deficits in patients receiving coronary artery bypass graft surgery. J. Int. Neuropsychol. Soc., 9(6): 913–924.

Rasmussen, L.S. (2006) Postoperative cognitive dysfunction: incidence and prevention. Best Pract. Res., 20(2): 315–330.

Rasmussen, L.S., Johnson, T., Kuipers, H.M., Kristensen, D., Siersma, V.D., Vila, P., et al. (2003) Does anaesthesia cause postoperative cognitive dysfunction? A randomised study of regional versus general anaesthesia in 438 elderly patients. Acta Anaesthesiol. Scand., 47: 260–266.

Rasmussen, L.S., Larsen, K., Houx, P., Skovgaard, L.T., Hanning, C.D. and Moller, J.T. (2001) The assessment of postoperative cognitive function. Acta Anaesthesiol. Scand., 45(3): 275–289.

Rasmussen, L.S. and Moller, J.T. (2000) Central nervous system dysfunction after anesthesia in the geriatric patient. Anesthesiol. Clin. North America, 18(1): 59–70.

Ritchie, K., Allard, M., Huppert, F.A., Nargeot, C., Pinek, B. and Ledesert, B. (1993) Computerized cognitive examination of the elderly (ECO): the development of a neuropsychological examination for clinic and population use. Int. J. Geriatr. Psychiatry, 8: 899–914.

Ritchie, K. and Fuhrer, R. (1995) The validation of an informant screening test for irreversible cognitive decline in the elderly: performance characteristics within a general population sample. Int. J. Geriatr. Psychiatry, 2: 149–156.

Rödig, G., Rak, A., Kasprzak, P. and Hobbhahn, J. (1999) Evaluation of self-reported failures in cognitive function after cardiac and noncardiac surgery. Anaesthesia, 54: 826–830.

Rolfson, D.B., McElhaney, J.E., Rockwood, K., Finnegan, B.A., Entwistle, L.M., Wang, J.F., et al. (1999) Incidence and risk factors for delirium and other adverse outcomes in older adults after coronary artery bypass graft surgery. Can. J. Cardiol., 15: 771–776.

Romanic, A.M., White, R.F., Arleth, A.J., Ohlstein, E.H. and Barone, F.C. (1998) Matrix metalloproteinase expression increases after cerebral focal ischemia in rats: inhibition of matrix metalloproteinase-9 reduces infarct size. Stroke, 29: 1020–1030.

Saito, H., Seherwood, E.R., Varma, T.K. and Evers, B.M. (2003) Effects of aging on mortality, hypothermia, and cytokine induction in mice with endotoxemia or sepsis. Mech. Ageing Dev., 124(10–12): 1047–1058.

Savageau, J.A., Stanton, B.A., Jenkins, C.D. and Frater, R.W. (1982a) Neuropsychological dysfunction following elective cardiac operation. II. A six-month reassessment. J. Thorac. Cardiovasc. Surg., 84(4): 595–600.

Savageau, J.A., Stanton, B.A., Jenkins, C.D. and Klein, M.D. (1982b) Neuropsychological dysfunction following elective cardiac operation. I. Early assessment. J. Thorac. Cardiovasc. Surg., 84(4): 585–594.

Selnes, O.A., Grega, M.A., Borowicz, Jr., L.M., Royall, R.M., McKhann, G.M. and Baumgartner, W.A. (2003) Cognitive changes with coronary artery disease: a prospective study of coronary artery bypass graft patients and nonsurgical controls. Ann. Thorac. Surg., 75(5): 1377–1384.

Selnes, O.A. and McKhann, G.M. (2005) Neurocognitive complications after coronary artery bypass surgery. Ann. Neurol., 57: 615–621.

Selnes, O.A., Royall, R.M., Grega, M.A., Borowicz, Jr., L.M., Quaskey, S. and McKhann, G. (2001) Cognitive changes 5 years after coronary artery bypass grafting. Arch. Neurol., 58(4): 598–604.

Shaw, P.J., Bates, D., Cartlidge, N.E., French, J.M., Heaviside, D., Julian, D.G., et al. (1987) Long-term intellectual dysfunction following coronary artery bypass graft surgery: a six month follow-up study. Q. J. Med., 62(239): 259–268.

Smith, R.J., Roberts, N.M., Rodgers, R.J. and Bennett, S. (1986) Adverse cognitive effects of general anaesthesia in young and elderly patients. Int. Clin. Psychopharmacol., 1(3): 253–259.

Sparks, D.L., Gross, D.R. and Hunsaker, J.C. (2000) Neuropathology of mitral valve prolapse in man and cardiopulmonary bypass (CPB) surgery in adolescent Yorkshire pigs. Neurobiol. Aging, 21(2): 363–372.

Tanaka, S., Ide, M., Shibutani, T., Ohtaki, H., Numazawa, S., Shioda, S., et al. (2006) Lipopolysaccharide-induced microglial activation induces learning and memory deficits without neuronal cell death in rats. J. Neurosci., 83(4): 557–566.

Terao, A., Apte-Deshpande, A., Dousman, L., Morairty, S., Eynon, B.P., Kilduff, T.S., et al. (2002) Immune response gene expression increases in the aging murine hippocampus. J. Neuroimmunol., 132(1–2): 99–112.

Tombaugh, T.N. and Melaryne, N.J. (1992) The mini-mental state examination: a comprehensive review. J. Am. Geriatr. Soc., 40: 922–935.

Tuppo, E.E. and Arias, H.R. (2005) The role of inflammation in Alzheimer's disease. Int. J. Biochem. Cell Biol., 37(2): 289–305.

Van Dijk, D., Jansen, E.W., Hijman, R., Nierich, A.P., Diephuis, J.C., Moons, K.G., et al. (2002) Cognitive outcome after off-pump and on-pump coronary artery bypass graft surgery: a randomized trial. JAMA, 287(11): 1405–1412.

Von Bernhardi, R. and Ramirez, G. (2001) Microglia-astrocyte interaction in Alzheimer's disease: friends or foes for the nervous system? Biol. Res., 34(2): 123–128.

Wan, S., LeClerc, J.L. and Vincent, J.L. (1997) Cytokine responses to cardiopulmonary bypass: lessons learned from cardiac transplantation. Ann. Thorac. Surg., 63(1): 269–276.

Wan, Y., Xu, J., Ma, D., Zeng, Y., Cibelli, M. and Maze, M. (2007) Postoperative impairment of cognitive function in rats: a possible role for cytokine-mediated inflammation in the hippocampus. Anesthesiology, 106(3): 436–443.

Wang, X.Y. and McCubrey, J.A. (1997) Differential affects of retroviral long terminal repeat on interleukin-3 gene expression and autocrine transformation. Leukemia, 11(10): 1711–1725.

Wang, Y., Sands, L.P., Vaurio, L., Mullen, E.A. and Leung, J.M. (2007) The effects of postoperative pain and its management on postoperative cognitive dysfunction. Am. J. Geriatr. Psychiatry, 15(1): 50–59.

Wechsler, D. (1997) Wechsler Adult Intelligence Scale (3rd ed.). Psychological Corporation, San Antonio, TX.

Westaby, S., Saatvedt, K., White, S., Katsumata, T., van Oeveren, W. and Halligan, P.W. (2001) Is there a relationship between cognitive dysfunction and systemic inflammatory response after cardiopulmonary bypass? Ann. Thorac. Surg., 71(2): 667–672.

Williams-Russo, P., Sharrock, N., Mattis, S., Szatrowski, T. and Charlson, M. (1995) Cognitive effects after epidural vs. general anesthesia in older adults. A randomized trial. JAMA, 274(1): 44–50.

Wind, A.W., Schellevis, F.G., Van Scaveren, G., Scholten, R.P., Jonker, C. and Van Eijk, J.T. (1997) Limitations of the mini-mental state examination in diagnosing dementia in general practice. Int. J. Geriatr. Psychiatry, 12: 101–108.

Xie, Z., Morgan, T.E., Rozovsky, I. and Finch, C.E. (2003) Aging and glial responses to lipopolysaccharide in vitro: greater induction of IL-1 and IL-6, but smaller induction of neurotoxicity. Exp. Neurol., 182(1): 135–141.

Yao, Y., Chinnici, C., Tang, H., Trojanowski, J.Q., Lee, V.M. and Praticò, D. (2004) Brain inflammation and oxidative stress in a transgenic mouse model of Alzheimer-like brain amyloidosis. J. Neuroinflammation, 1(1): p. 21.

Yong, V.W., Krekoski, C.A., Forsyth, P.A., Bell, R. and Edwards, D.R. (1998) Matrix metalloproteinases and diseases of the CNS. Trends Neurosci., 21: 75–80.

Subject Index

aging 377–390
Alzheimers disease 365–372, 393–404
amygdyla 73, 102–103
aphasia 397
aplysia 4–7, 179–194, 277–287

behavioral test
    Morris water maze 72–73, 88, 100, 104, 106, 417
    radial arm maze 100, 257, 417
    Stroop's test 371, 380, 397
    Wisconsin card sorting test (WCST) 257, 380–382
brain-derived neurotrophic factor (BDNF) 15, 18, 47, 65, 67, 251–260

cerebellum 73, 106, 226
cerebral blood flow (CBF) 414
cognition 409–418
cognitive impairment
    mild cognitive impairment (MCI) 365–372, 390, 396–397

Drosophila 35, 62, 99, 293–302

entorhinal cortex 120, 243, 252–253
executive function 254, 377–390

frontal lobe 313, 380
fusiform gyrus 312, 402

glutamate receptor
    AMPA receptor 31–33, 104–106, 159–175, 228–231
    GluR1 (GluRA) 10, 15–17, 104–106, 159–175
    GluR2 (GluRB) 32–33, 148, 228–231
    NMDA receptor 104, 168–169, 228–231, 254
    NMDAR1 170

hippocampus
    CA1 5–7, 15, 123–124, 127–129, 133–134, 136–137, 243–244, 246–247
    CA3 6–7, 15, 225–228, 237–239, 243–244
    dentate gyrus 106, 133–135, 169, 243
    mossy fiber 106, 225–239
    Schaffer collateral 64, 72–73, 87, 104, 108, 120, 243
homeostasis 18, 72, 212–213, 217, 221

lexical 395, 399–400
long-term potentiation (LTP) 27–37, 199–205
    early LTP (E-LTP) 15–17, 52, 67–68, 69–70, 87–89, 106–107
    late LTP (L-LTP) 15–18, 33, 52, 67–70, 86–89, 92, 106–107
long-term facilitation (LTF) 186–189, 193–194, 281–287

memory
    associative memory 36, 305–318
    episodic memory 339–350, 366–369, 382–385
    executive memory 254, 377–390
    fear conditioning 63, 88, 92, 101–102
    intermediate memory 297–299
    long-term memory 14, 28–29, 35–36, 73, 86–89, 102–103, 121, 179–195
    memory trace 3–20, 293–302
    meta-memory 377–390
    semantic memory 339, 393–404
    short-term memory 323–336
    spatial memory 35–36, 100–101, 151–154, 160–175
    working memory 160–175, 251–261, 323–336, 369–372
messenger ribonucleic acid (mRNA)
    mRNA binding protein 41–53

mouse
  knock out mouse (KO) 69–74
  transgenic mouse 100–101, 258

neuronal imaging
  functional magenetic resonance imaging (fMRI) 138–139, 311–313, 339–350
  magnetic resonance imaging (MRI) 138–139, 311–313, 339–350
  positron emission tomography (PET) 312, 397–398, 402
neuronal circuits/neuronal networks 89, 136–137, 182, 193, 213, 272–273
neurotransmitters
  acetylcholine 255, 260
  dopamine 123–124, 131–134, 138–139, 252–255, 260, 272–273, 284
  GABA 137, 253–254, 258
  glutamate 123–124, 146–147, 150, 152–153, 159–175
  noradrenaline 131, 134, 260
  serotonin (5-HT) 9–12, 92, 183, 187–194, 271–273, 278, 280–287

prefrontal cortex 251–261
priming 173–175, 360–361, 395–396
protein kinases
  calcium-calmodulin dependent protein kinase (CAMK) 12, 43, 129, 200
  cAMP-dependent protein kinase (PKA) 99
  extracellular signal regulated kinase; mitogen activated protein kinase (ERK; MAP kinase) 52, 62, 67–69

protein kinase C (PKC) 6, 9–12, 28–29, 31, 67, 272, 280, 281
protein kinase M ζ (PKMζ) 11–13, 27–37, 280
S6 kinase 5

rat 358–360, 362

stress 145–154
structural changes
  active zone 180, 183–186, 188–189, 194
  post-synaptic density 12–13, 51, 180–182, 203
  spine 199–205
  varicosity 183–190, 194
synaptic plasticity 145–154, 211–221, 241–248
synaptic tag 18–19, 34, 103, 127, 129, 133, 135

temporal lobe 306–313, 397–398, 403–404
transcranial magnetic stimulation (TMS) 398
transcription factor
  cAMP-response element binding protein (CREB) 13–14, 89–90, 102, 106–107, 281–282
translation factor
  eukaryotic initiation factor 2α (eIF2α) 83–87, 89–92
  eukaryotic initiation factor 4E (eIF4E) 60–63, 65–69, 70–72, 81, 83
  eIF4E binding protein (4EBP) 60, 65–70, 72–74
  target of rapamycin (TOR) (mTOR) 31, 65–69, 75

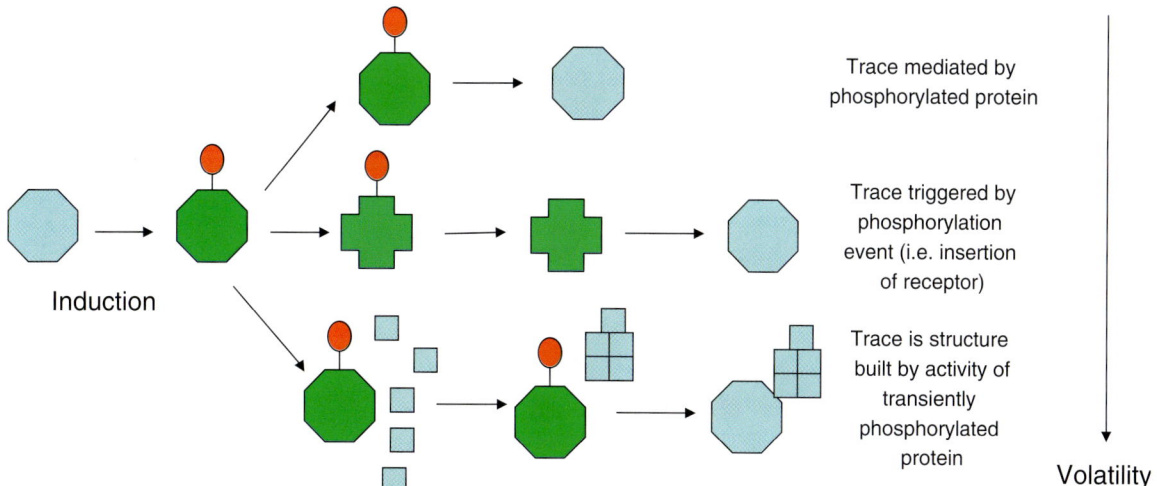

Plate 1.1. Different molecular traces induced by phosphorylation have different volatility. (A) The simplest memory trace involves activation of kinase that phosphorylates a substrate. Phosphorylation of the substrate itself leads to an increase in synaptic strength through either increasing transmitter release (e.g. SNAP-25) or transmitter reception (e.g. CAMKII phosphorylation of AMPA receptors). Once the kinase is no longer active (degradation of second messengers), phosphatases remove the phosphate and the trace is erased. (B) Phosphorylation can induce a change in the protein (shift from hexagon to cross) that lasts longer than the phosphorylation itself. For example, phosphorylation of AMPA receptors may lead to insertion of the receptor that can persist after the phosphorylation has been removed. Some additional event then has to occur to erase the memory trace. (C) Phosphorylation can activate a protein (i.e. a regulator of actin polymerization, cofilin) that leads to a structural change (assembly of blocks into a house-like structure) that persists after the phosphorylation is removed and requires an active process to disassemble. The three changes are graded in their volatility. (For B/W version, see page 8 in this volume.)

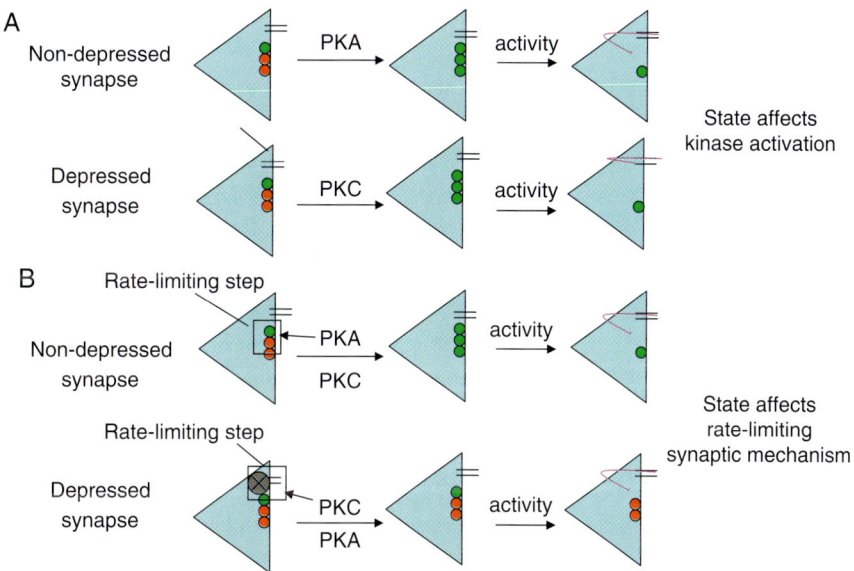

Plate 1.2. The state of the synapse can determine the requirements for increases in synaptic strength. (A) This figure describes the possibility that the kinase activated can be determined by the state of the synapse. In the non-depressed synapse PKA is activated, increases the readiness of the synaptic vesicles to be released (shift from red-green), and thus more transmitters are released after an action potential (activity). At the depressed synapse, PKC is activated instead of PKA, but the change at the synapse is the same. (B) This figure describes the possibility that the state of the synapse determines which kinase is required. At the non-depressed synapse, the rate-limiting step is the readiness of the synaptic vesicles to be released. While both PKA and PKC are activated by 5-HT, only PKA phosphorylations that increase the readiness of the synaptic vesicles to be released are affected. At the depressed synapse, calcium-secretion coupling becomes the rate-limiting step and activation of PKC, but not PKA can remove this inhibition. The bars represent calcium channels and the arrow represents the influx of calcium leading to the release of synaptic vesicles. (For B/W version, see page 9 in this volume.)

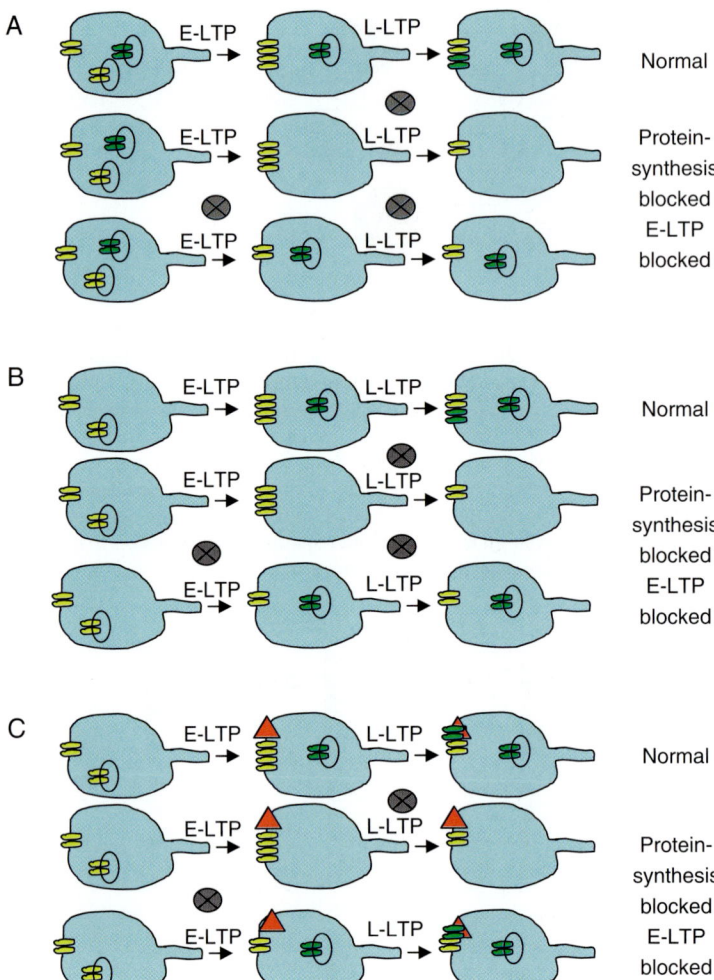

Plate 1.3. Three models for the role of new protein synthesis in memory. (A) A model where only basal protein synthesis is required. Old AMPA receptors 'yellow' are inserted into the membrane to give E-LTP. These receptors are replaced with more newly synthesized AMPA receptors (green) to give L-LTP. In the presence of a protein synthesis inhibitor, the newer AMPA receptors were never synthesized and thus there are no AMPA receptors to replace the old ones and L-LTP is blocked. Blocking the initial insertion of AMPA receptors blocks L-LTP as the new AMPA receptors are only serving to replace the old ones. (B) A model where regulated protein synthesis is required to consolidate the synaptic change. The same as in (A) but the newly synthesized AMPA receptors (green) are newly synthesized by the E-LTP stimulation. (C) A parallel model. In this case the stimulus for E-LTP both inserts AMPA receptors and sets a tag (red diamond) such that newly synthesized receptors can insert into the membrane. Protein synthesis blocks the production of these new receptors and thus blocks L-LTP. In contrast if E-LTP is blocked but not protein synthesis, or tag production, L-LTP will exist in the absence of E-LTP. (For B/W version, see page 16 in this volume.)

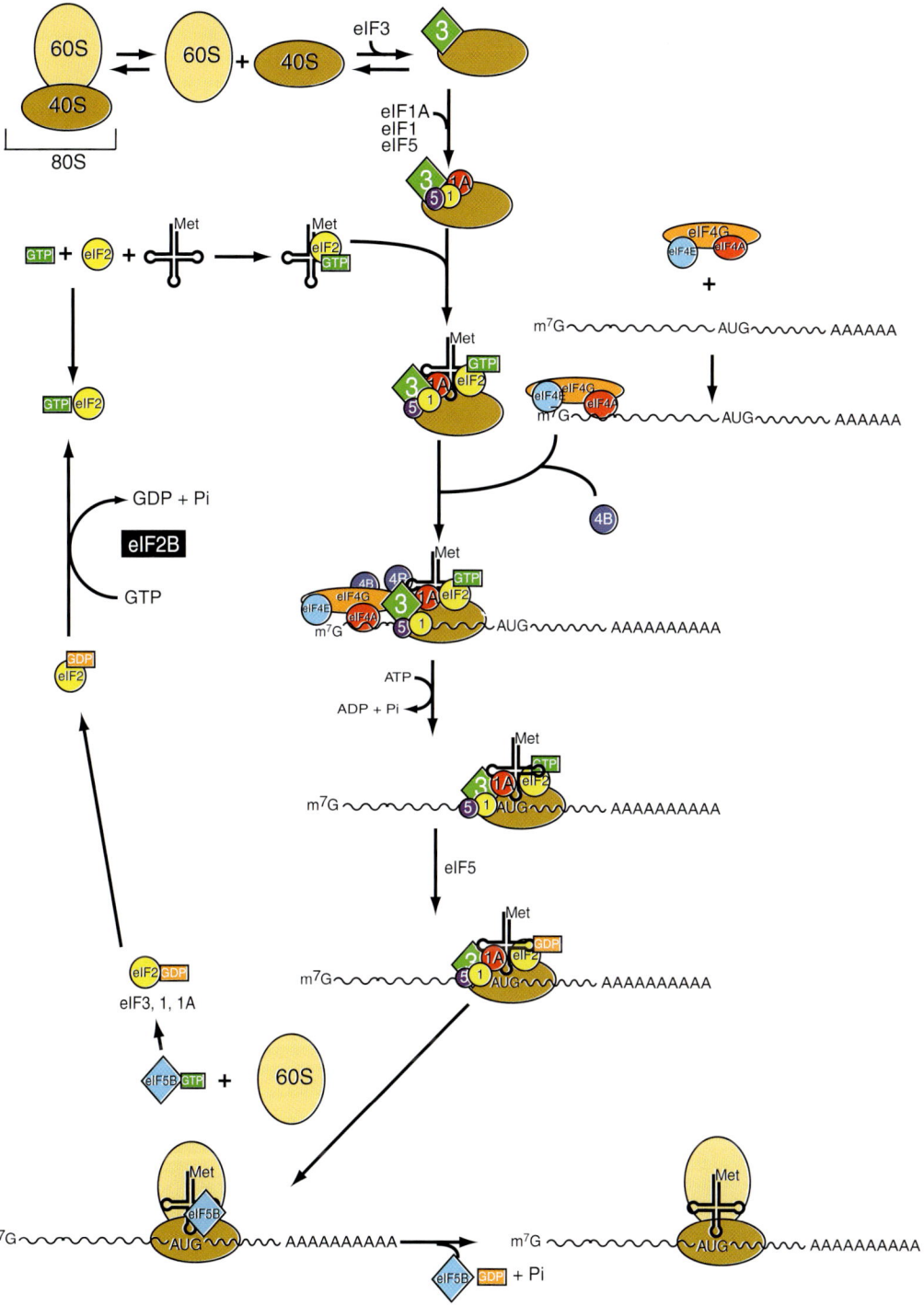

Plate 5.1. Schematic representation of the translation initiation pathway in eukaryotes. Eukaryotic translation initiation factors are indicated as color-coded circles. (For B/W version, see page 82 in this volume.)

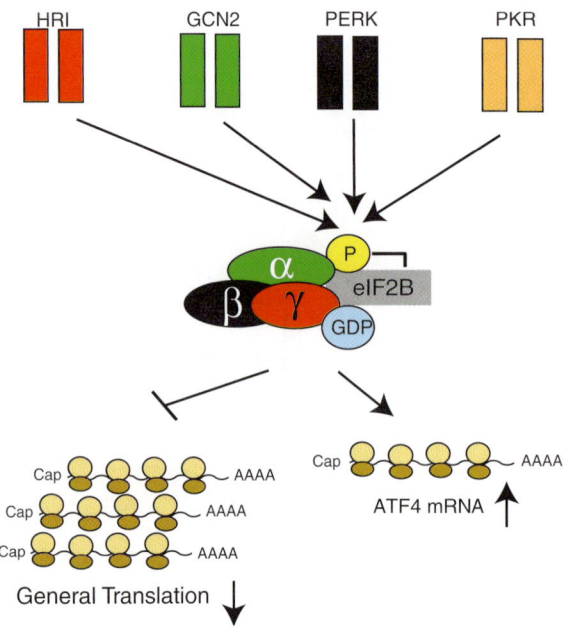

Plate 5.2. Schematic representation of the eIF2α signaling pathway. The four eIF2α kinases (GCN2, PERK, PKR, and HRI), whose activity is regulated by different stress signals, phosphorylate Ser51 on the α subunit of eIF2. Phosphoryation of eIF2α leads to inhibition of general translation but it stimulates translation of ATF4 mRNA. (For B/W version, see page 84 in this volume.)

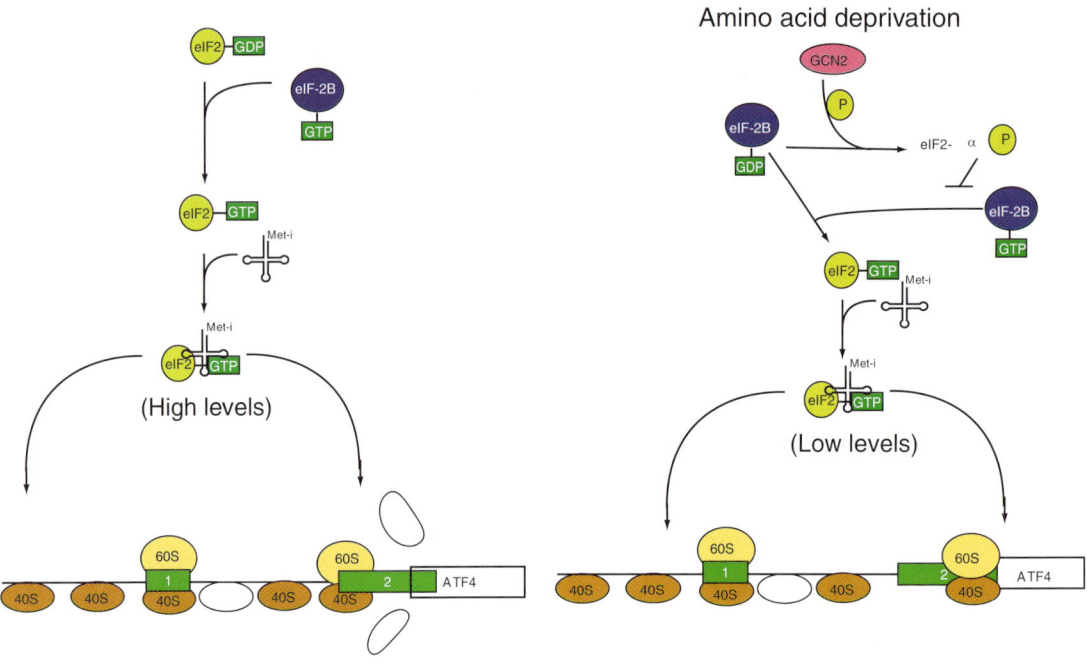

Plate 5.3. A model for ATF4 mRNA translation. Schematic diagram of the 5′ untranslated region of ATF4 mRNA. The open reading frames (ORFs) are shown as green boxes, and the ATF4 mRNA authentic ORF as an open rectangle. Under normal growing conditions (left panel) the 40S ribosome initiates at ORF1 and reinitiates at ORF2. Under amino acid deprivation conditions (right panel), due to a low concentration of ternary complex, 40S ribosomes conditions failed to reinitiate at ORF2 but reinitiated instead at the authentic ORF. (For B/W version, see page 85 in this volume.)

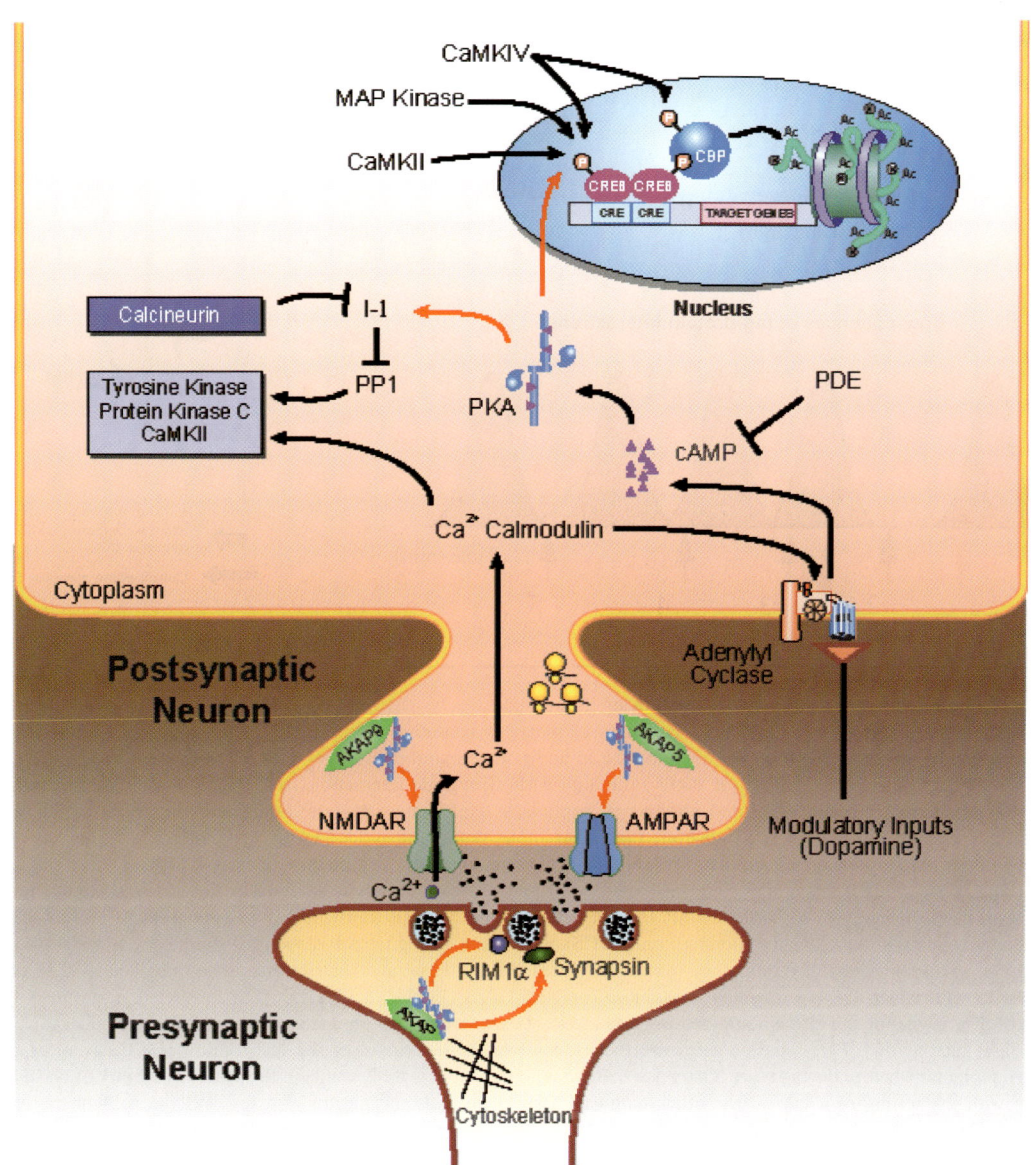

Plate 6.1. The cAMP/PKA signaling pathway critically regulates molecular components underlying long-term potentiation and long-term memory. In the postsynaptic neuron, PKA targets include NMDA and AMPA receptors, inhibitor-1 (I-1) and CREB. In the presynaptic terminal, PKA targets include RIM1α and synapsin. (For B/W version, see page 98 in this volume.)

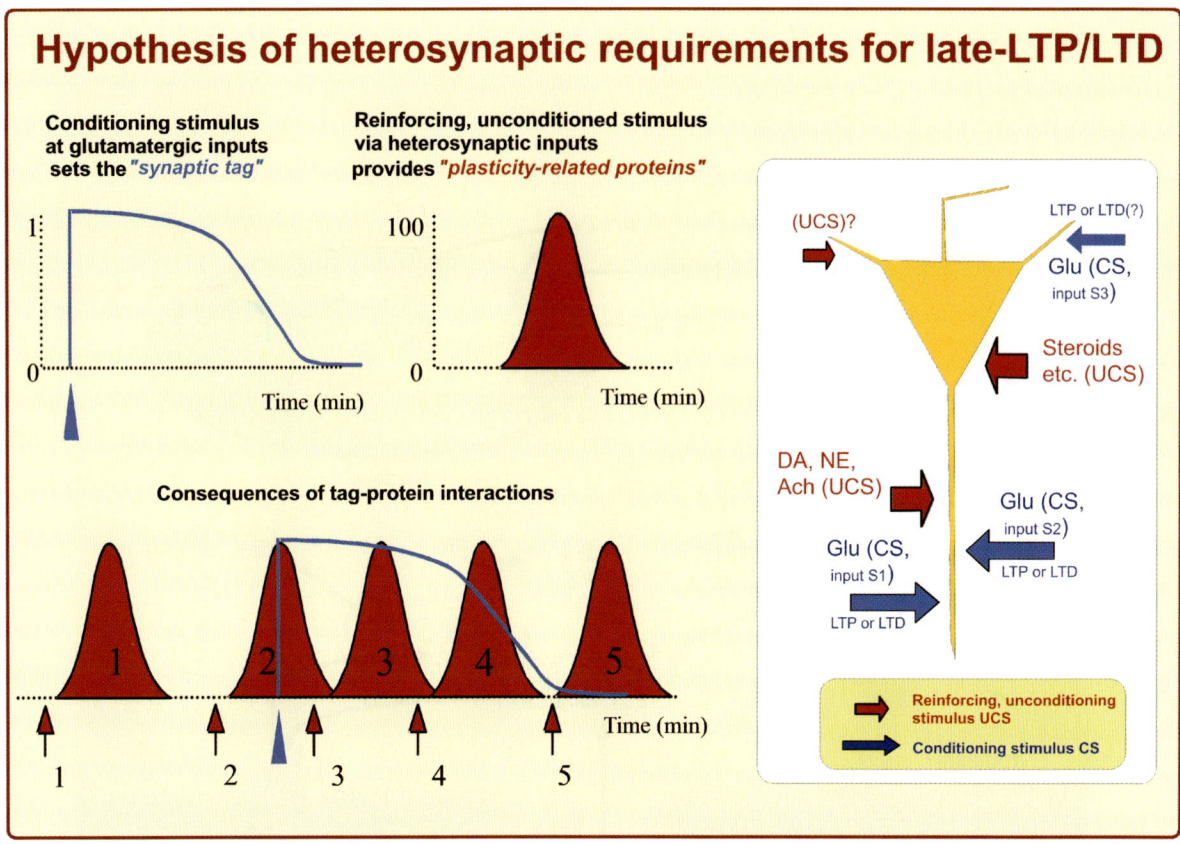

Plate 7.2. Hypothesis of a heterosynaptic induction of long-lasting plastic changes. The left part shows specific properties of the interaction of setting a synaptic tag and the availability of plasticity-related proteins (PRPs). Our studies revealed that most likely glutamatergic inputs are required to set the synaptic tag in addition to maintaining early-LTP/LTD (conditioning stimulus, CS) and modulatory inputs regulate the synthesis of PRPs (although with synergistic interactions with NMDA-receptor function). We have identified dopaminergic, noradrenergic and cholinergic inputs as reinforcing, unconditioning stimuli (UCS) which seems to be substructure-dependent. We hypothesize, however, that other molecules can also act as UCS (e.g., steroids during stress, etc.). Our current hypothesis of synaptic tagging required for late-LTP/LTD is as follows: The induction of early-LTP/LTD on a glutamatergic CS pathway (blue arrow symbol) sets a tag with a probability of 1 and this happens immediately. The tag lasts a period of time and then decays in a probabilistic manner. We could identify that the activation of the tag is transient and decays within 30–60 min. Stimulation of a non-glutamatergic reinforcing, unconditioning input (UCS) either via a strong tetanus/strong low-frequency stimulation or via direct stimulation of extrahippocampal, modulatory systems, induces the synthesis and distribution to appropriate postsynaptic sites of putative PRPs with which the tag interacts to stabilize (or permissively enable) LTP/LTD. The lower left part shows the possible dynamics of heterosynaptic tag–protein interactions. Induction of early-LTP/LTD on a CS pathway is shown at the centre of the time diagram (light blue arrow). Five separate cases are then considered. Case 1 involves the distribution and decay of macromolecule availability a long time before the tag is set. There can then be no interaction. Case 2 results in the peak availability of plasticity-proteins coinciding with the setting of a tag on the other pathway. Case 3 involves the setting of a tag and the subsequent synthesis and distribution of plasticity-proteins. In case 4, the interval is longer and only a limited interaction can occur, resulting in a smaller proportion of synapses becoming stabilized. Finally, in case 5, the interval is too long for any effective interaction to occur. These dynamics may represent the more native possibilities of interstructural interactions important for the induction of long-lasting plastic neuronal changes, schematically shown on the right part of the figure (DA: dopamine, NE: norepinephrine, Glu: glutamate, ACh: acetylcholine). Modified from Frey and Morris (1998b). Our later work suggests that glutamatergic inputs can transiently store sensory information under distinct circumstances. If the system then decides that this information is system-relevant, it can influence the duration of the normally transient storage of that sensory information by reinforcing heterosynaptic associative interactions within a time-window of about 30 min, i.e. the transient event can be "reinforced" into a long-term memory trace. The neuromodulatory input thereby regulates specifically the synthesis of PRPs. The right schema represents a CA1 pyramidal neuron with inputs representing the sensory information on glutamatergic synapses (conditioning stimulus, CS) and adequate neuromodulatory inputs (reinforcing, unconditioning stimulus, UCS) regulating and providing the availability of PRPs required for the transformation of the originally transient trace stored at CS-inputs into long-lasting memory traces. (For B/W version, see page 126 in this volume.)

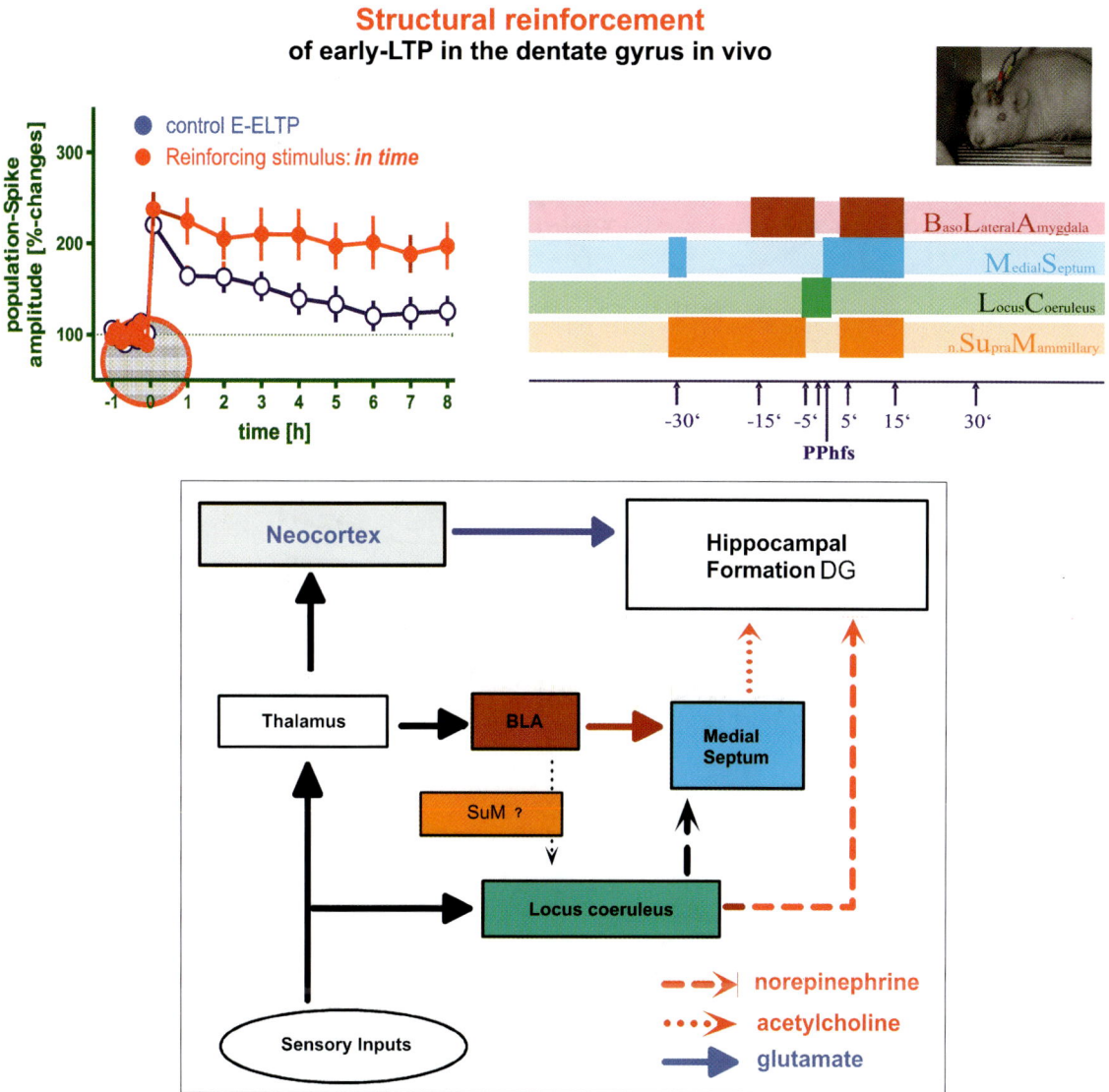

Plate 7.4. Structural reinforcement of LTP in vivo. Hippocampal early-LTP (blue curve in the upper left graphs, which sets a synaptic tag) can be reinforced into late-LTP (red curve) by the associative heterosynaptic stimulation of modulatory brain structures in a hippocampus-substructure-specific manner (here LTP-reinforcement of the dentate gyrus is presented). The specific neuromodulatory inputs must, however, interact with LTP in the DG within characteristic time-windows for the reinforcement to occur ("in time" stimulation). The lower schema represents identified modulatory brain structures taking part in dentate gyrus LTP-reinforcement, i.e. it represents the putative neural system and neurotransmitters involved (Bergado et al., 2007). Similar results, however with different modulatory brain structures involved, can be obtained from the study of CA1-LTP-reinforcement (Reymann and Frey, 2007). The upper right schema represents our current state regarding the determination of the specific time-points when the different modulatory brain structures have to be stimulated to be capable of transforming a protein synthesis-independent early-LTP into a protein synthesis-dependent late-LTP in the rat DG (Frey et al., 2001; Straube and Frey, 2003; Bergado et al., 2007). The darker-coloured boxes indicate the effective time-window for the corresponding structure. So far, it has been shown that the basal lateral amygdala, the medial septum, the locus coeruleus and the nucleus supramammillaris (SuM) interfere specifically with DG-early-LTP. Depending on specific time-windows within 30 min before or after DG-tetanization and the form of stimulation (e.g., high- or low-frequency stimulation) the activation of these structures resulted in a reinforcement of early- into late-LTP. (For B/W version, see page 132 in this volume.)

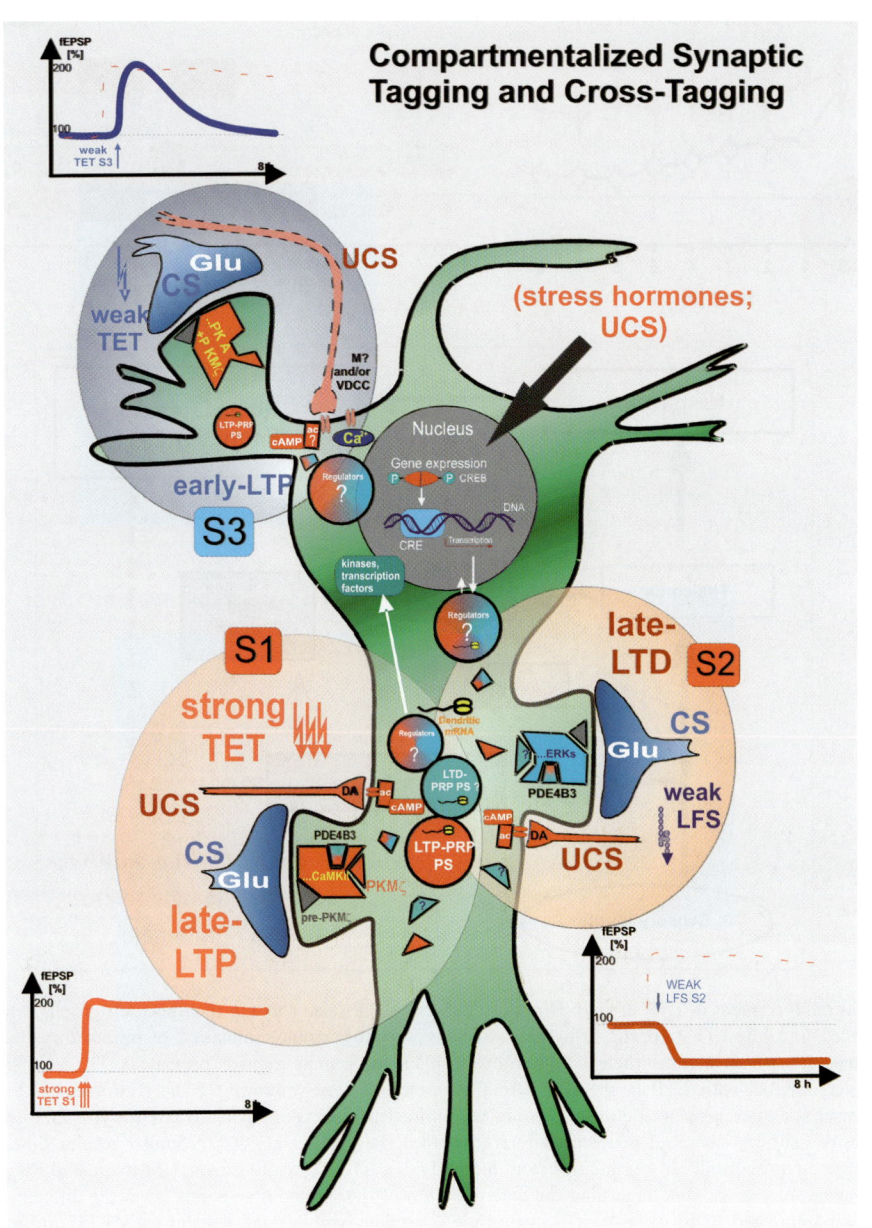

Plate 7.3. Compartmentalized functional plasticity versus uncompartmentalization in a pyramidal CA1-neuron. The schematical pyramidal CA1-neuron represents our cellular hypothesis of compartmentalized synaptic tagging in the hippocampal CA1 region during cognitive information-processing (in contrast to uncompartmentalization by stress — see text). The induction of late-LTP (but also similarly late-LTD) in the apical synaptic input S1 activates postsynaptically, via a heterosynaptic synergistic interaction of glutamatergic (conditioning stimulus, CS) as well as dopaminergic receptors (modulatory input; unconditioning stimulus, UCS), a transient, protein synthesis-independent synaptic tag, which is made specific for LTP by CaMKII-dependent processes (dark red symbol at the postsynaptic site in S1) and the synthesis of a pool of PRPs (triangles, trapeziums). PRPs consist of LTP-specific proteins such as PKMζ (red triangles) and regulatory proteins such as PDE4B4 (red-blue symbols). Both the tag and the PRPs have a distinct half-life during that time-window an interaction of the two can occur, thus maintaining LTP in S1 for at least 8 h. If within this time-window of available PRPs within this dendritic compartment a second, independent synaptic input S2 in the apical dendritic compartment is stimulated to set a synaptic tag (e.g., by induction of early-LTD via a weak low-frequency stimulation (weak LFS)), the latter can benefit from the compartment-restricted availability of PRPs, thus transforming the normally transient plastic form into an enduring one, in the example shown, from early-LTD into late-LTD in S2 by processes of synaptic tagging, cross-tagging and capturing of the PRPs provided through input S1. LTD-tags in the apical CA1-dendrites however, are not dependent on CaMKII but rather on MAPK-activity. If within the same time-window at a basal dendritic compartment S3 an early-LTP is induced even with a synaptic tag set (here early-LTP by a weak tetanization in S3 (weak TET)), these tags cannot benefit from the compartmentalized availability of PRPs in the apical dendrites, thus resulting in a transient expression of LTP, in early-LTP. Synaptic tagging or better cross-compartmental tagging/capturing does not occur. Interestingly, the basal compartment is also characterized by different key-players responsible for setting the local tag-here, either PKA and/or PKMζ is required to mediate the tag. It remains unclear, whether basal plastic events require an activation of modulatory inputs and which molecules are the PRPs in basal dendritic compartments. For a prolonged maintenance of a plastic event beyond 8 h the expression of genes has been shown to be an obvious requirement, which is also schematically represented in the figure. Furthermore, under high stress or life-threatening situations (grey arrow on gene expression) an immediate gene expression can also occur, which results in an uncompartmentalization of the plasticity events, i.e. the transformation of early- into late-LTP in the transcompartments, as between S1 and S3 (see text and Sajikumar et al., 2007b). The graphs at the synaptic inputs S1, S2 and S3 schematically illustrate compartmentalized "synaptic tagging" in CA1 (with the exception of uncompartmentalization by stress hormones (see text). Induction of late-LTP by a strong tetanus (strong TET) in a synaptic input S1 in the apical dendrites leads to late-LTP (red curve). If a weak low-frequency stimulation (weak LFS) was applied within a distinct time-window to a separate synaptic input S2 in the apical dendrite the normally early-LTD was transformed into late-LTD by processes of synaptic cross-tagging (red curve in right lower graph). Weak TET to a synaptic input S3 in the basal dendrites — even if applied within the normally effective time-window of tagging within a dendritic compartment — revealed, however, that strong TET in apical S1 was ineffective in influencing early-LTP in S3 at basal dendrites (blue curve in the upper graph) during cognitive information-processing. Thus tagging is restricted to distinct dendritic compartments (stress however, is able to break the compartmentalization — see text). Modified from Frey and Morris (1998a); Reymann and Frey (2007); Sajikumar et al. (2007a). (For B/W version, see page 128 in this volume.)

Behavioural reinforcement
of early-LTP in the dentate gyrus in vivo

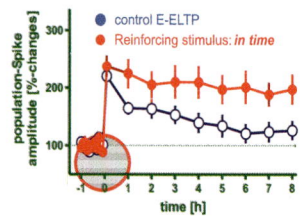

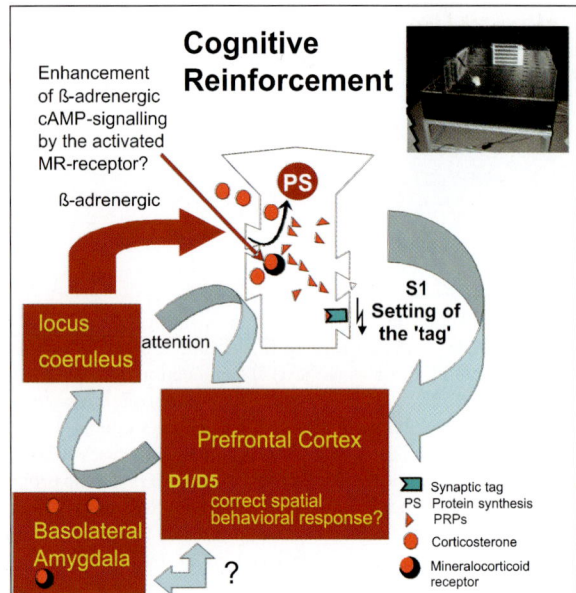

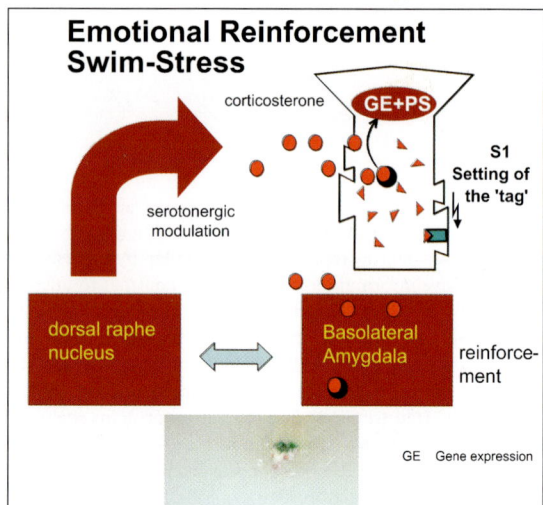

Plate 7.5. Behavioural LTP-reinforcement. Hippocampal early-LTP in the DG (blue curve in the upper right graph, which sets a synaptic tag) can be reinforced into late-LTP (red curve) by the associative heterosynaptic stimulation of structure-specific, modulatory brain structures (provided in the lower schemes) during behavioural tasks. Like structural reinforcement distinct behavioural paradigms with different content can activate these modulatory structures to reinforce early-LTP into late-LTP. The modulatory brain structures regulate the synthesis of PRPs which then interact with the synaptic tags, thus expressing late-LTP. Emotional, high stress conditions (lower right scheme) revealed that a 2-min-swim in a water tank 15 min after a weak tetanus also prolongs early-LTP up to 24 h in naïve animals. The prolongation is dependent on gene expression, and protein synthesis but not on β-adrenergic activation. Through such a high stress paradigm, LTP is maintained by corticosterone-induced activation of mineralocorticoid receptors (MR). Because the activated MR-complex acts as a transcription factor (circles) and modulates the activity of other factors, resulting translational products may include PRPs (triangles) being involved in the maintenance of LTP. Indeed, this form of LTP-reinforcement requires immediate gene expression in addition to protein synthesis. LTP can also be reinforced by the mastering of a spatial task in a holeboard (lower left panel). In contrast to emotional reinforcement this cognitive reinforcement appears only after repetitive training in the same environment and is related to the formation of a long-lasting spatial reference memory, suggesting that proteins involved in memory consolidation also contribute to the maintenance of hippocampal LTP. This type of LTP-reinforcement does not require gene expression but involves protein synthesis. Different cellular signalling is involved in emotional and cognitive reinforcement of LTP as indicated by the boxes in each scheme. Adapted from Korz and Frey (2004). (For B/W version, see page 135 in this volume.)

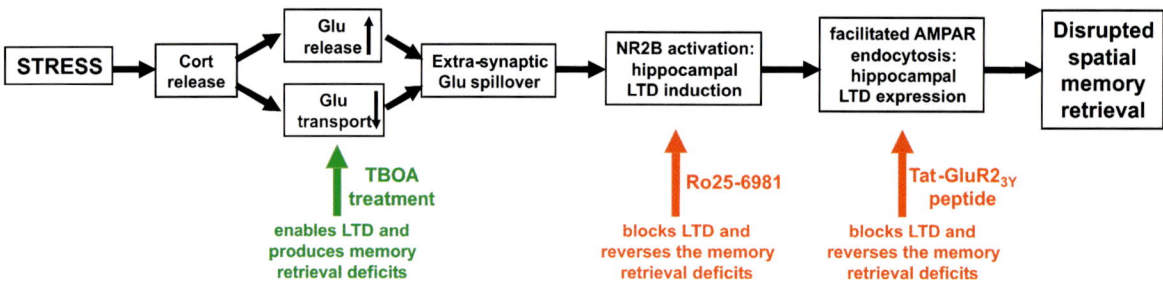

Plate 8.1. Schematic describing the hypothetical mechanisms by which acute stress enables induction of long-term depression (LTD) in the hippocampus, thereby causing impaired spatial memory retrieval and the steps at which experimental treatments performed in Wong et al. (2007) interfere with this process. Acute stress causes the release of corticosterone (Cort) which then increases glutamate concentration in the synaptic cleft either through increased glutamate (Glu) release and/or decreased glutamate transport in the hippocampus. The increased glutamate concentration enables the induction of LTD via a spill-over activation of extra-synaptically localized NR2B-containing NMDARs, and hence the expression of LTD via facilitating the endocytosis of post-synaptic AMPARs, thereby leading to the disrupted spatial memory retrieval. The treatments used to experimentally induce and inhibit LTD are indicated in green and red, respectively. Figure used with permission, copyright (2007), National Academy of Sciences, U.S.A. (Wong et al., 2007). (For B/W version, see page 153 in this volume.)

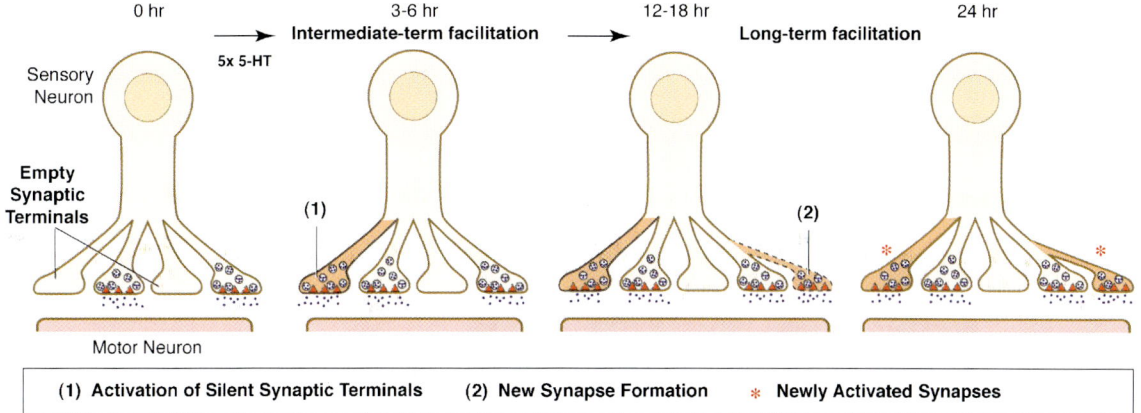

Plate 10.2. Time course and functional contribution of two distinct presynaptic structural changes associated with intermediate- and long-term facilitation in *Aplysia*. Repeated pulses of 5-HT in sensory to motor neuron co-cultures trigger two distinct classes of presynaptic structural changes: (1) the rapid clustering of synaptic vesicles to pre-existing silent sensory neuron varicosities (3–6 h) and (2) the slower generation of new sensory neuron synaptic varicosities (12–18 h). The resultant newly filled and newly formed varicosities are functionally competent (capable of evoked transmitter release) and contribute to the synaptic enhancement that underlies LTF. The rapid filling and activation of silent presynaptic terminals at 3 h suggests that, in addition to its role in LTF, this modification of pre-existing varicosities may also contribute to the intermediate phase of synaptic plasticity. Red triangles represent transmitter release sites (active zones). (Modified from Kim et al., 2003.) (For B/W version, see page 188 in this volume.)

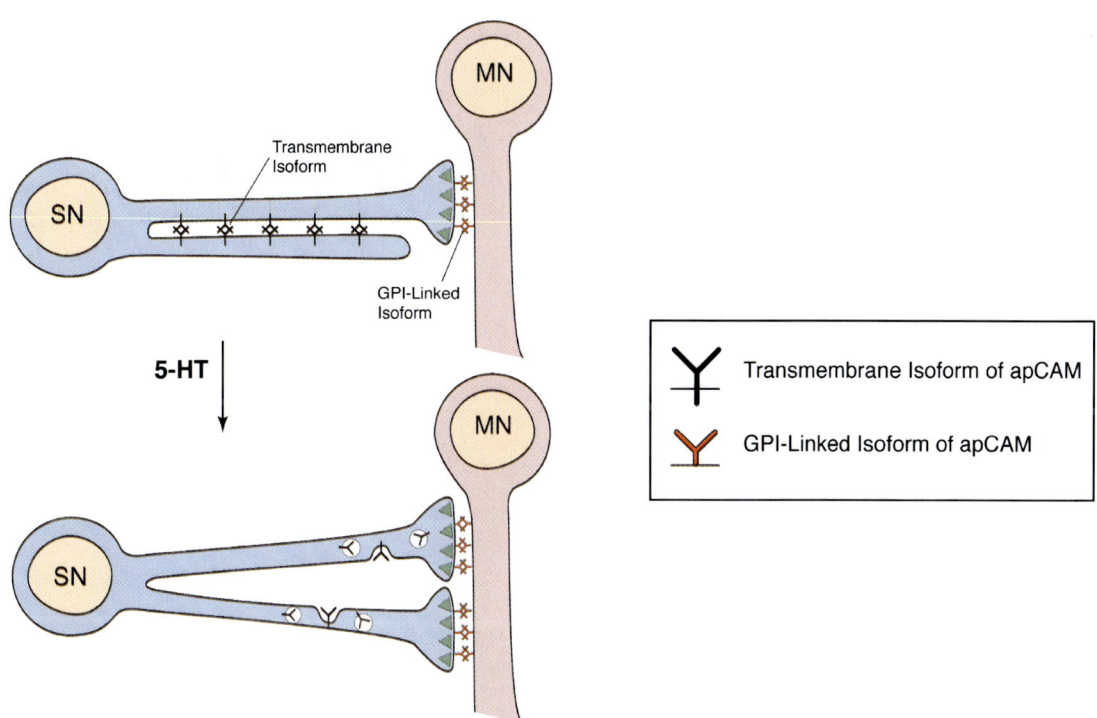

Plate 10.3. Regional specific down-regulation of the transmembrane isoform of apCAM. This model is based on the assumption that the relative concentration of the GPI-linked versus transmembrane isoforms of apCAM is highest at points of synaptic contact between the sensory neuron and motor neuron and reflects the results of studies done in dissociated cell culture. Thus, previously established connections might remain intact following exposure to 5-HT since they would be held in place by the adhesive, homophilic interactions of the GPI-linked isoforms and the process of outgrowth from sensory neuron axons would be initiated by down-regulation of the transmembrane form at extrasynaptic sites of membrane apposition. In the intact ganglion, the axons of sensory neurons are likely to fasciculate not only with other sensory neurons but also with the processes of other neurons and perhaps even glia. One of the attractive features of this model is that the mechanism for down-regulation is intrinsic to the sensory neurons. Thus, even if some of the sensory neuron axonal contacts in the intact ganglion were heterophilic in nature, i.e., with other neurons or glia, we would still expect the selective internalization of apCAM at the sensory neuron surface membrane at these sites of heterophilic apposition to destabilize adhesive contacts and to facilitate disassembly. (From Bailey et al., 1997.) (For B/W version, see page 192 in this volume.)

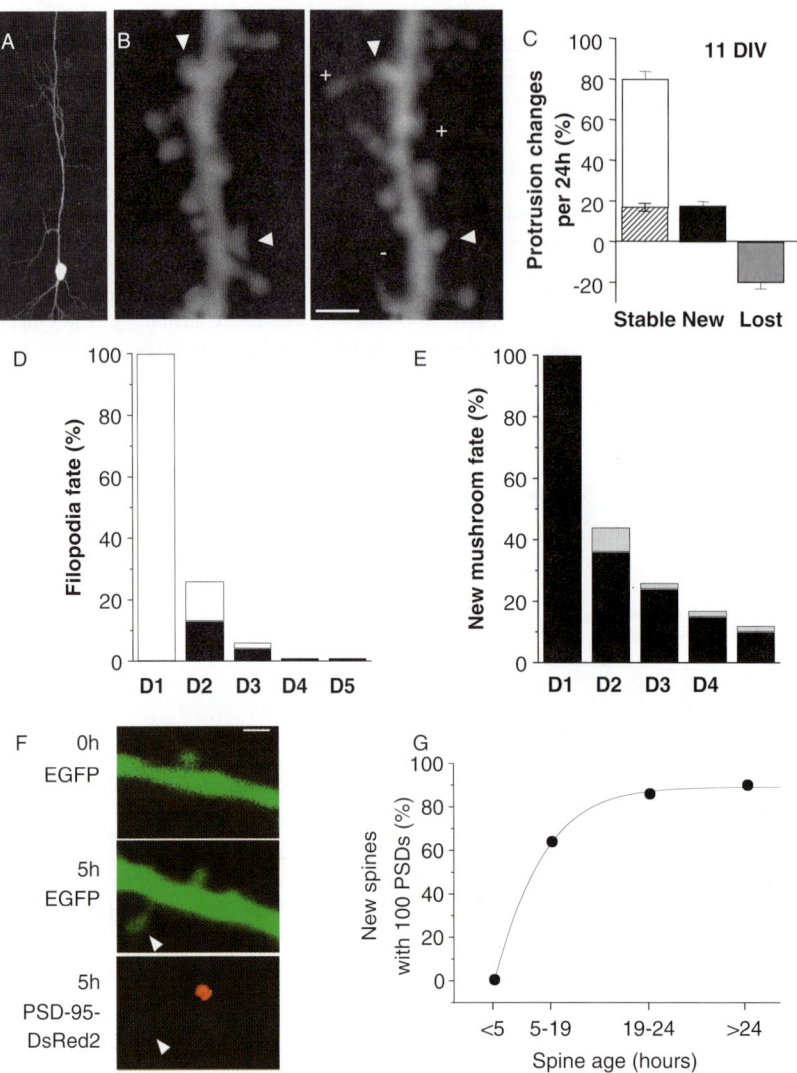

Plate 11.1. Spine dynamics in hippocampal organotypic slice cultures. (A) EGFP transfected CA1 pyramidal neuron. (B) Repetitive imaging of a dendritic segment at 24 h interval reveals the occurrence of new and lost protrusions. (C) Summary of the proportion of stable spines (open column), which include spines exhibiting changes in morphology (dashed column), of newly formed (black column) and disappearing (gray column) spines. (D) Stability over 5 days of newly formed filopodia. Note that most of them disappear within 1–2 days and only exceptionally lead to the formation of a stable spine. (E) Stability of newly formed spines. (F) Illustration of a newly formed spine (age < 5 h; arrow head, middle panel) which do not express PSD-95-DsRed2 (arrow head, lower panel). (G) Time course of PSD-95-DsRed2 expression in newly formed spines. (For B/W version, see page 204 in this volume.)

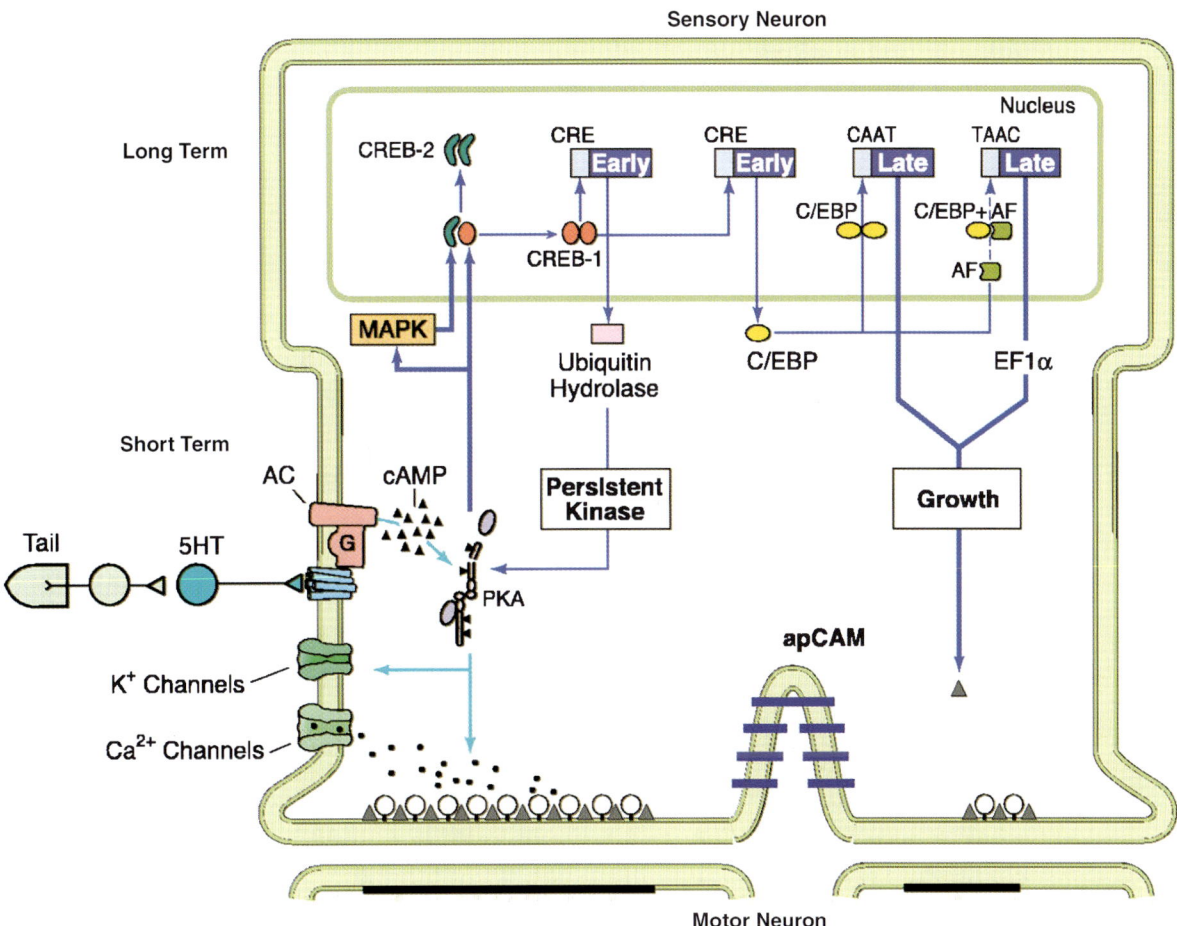

Plate 17.1. Presynaptic cellular model for short- and long-term sensitization in *Aplysia*. See the text for details. Adapted with permission from Kandel (2001). (For B/W version, see page 278 in this volume.)

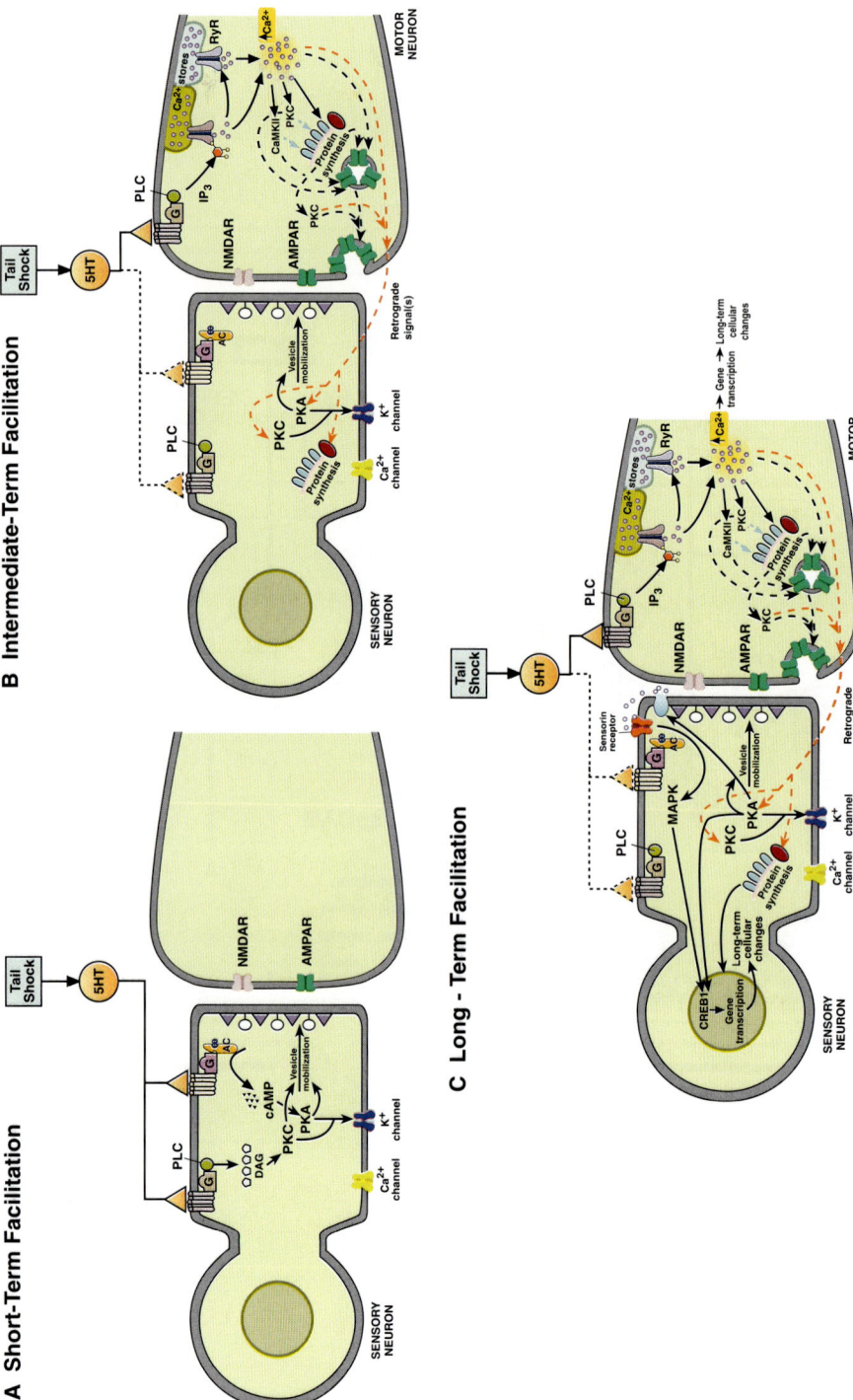

Plate 17.2. Revised cellular models for different phases of sensitization-related synaptic facilitation in *Aplysia*. The *dashed lines* depict pathways for which experimental evidence is currently lacking. (A) Short-term facilitation (STF). This phase lasts <30 min. STF can be both induced and expressed presynaptically. (B) Intermediate-term facilitation (ITF). This phase lasts from 90 min to 3 h. According to the model, ITF is induced postsynaptically, by release of Ca^{2+} from intracellular stores in the motor neuron, and expressed both pre- and postsynaptically. The presynaptic expression results from one or more retrograde signals, activated by postsynaptic Ca^{2+}, which stimulate both PKA and PKC within the sensory neuron. (C) Long-term facilitation (LTF). LTF persists for ≥24 h, and involves gene transcription, as well as both pre- and postsynaptic protein synthesis. Like ITF, LTF is induced postsynaptically, and expressed pre- and postsynaptically. LTF is triggered by repeated, spaced applications of 5-HT/sensitizing stimuli. A novel feature of the model in C is that it assumes that prolonged activation of presynaptic PKA, which leads to sensorin release and the translocation of PKA to the presynaptic nucleus, is triggered by elevated postsynaptic Ca^{2+}, via retrograde signaling. In addition, for at least some forms of both LTF and ITF the retrograde signal may activate presynaptic PKC (Jin et al., 2004; Hu et al., 2007) as well as presynaptic PKA. Adapted with permission from Glanzman (2007). (For B/W version, see page 286 in this volume.)

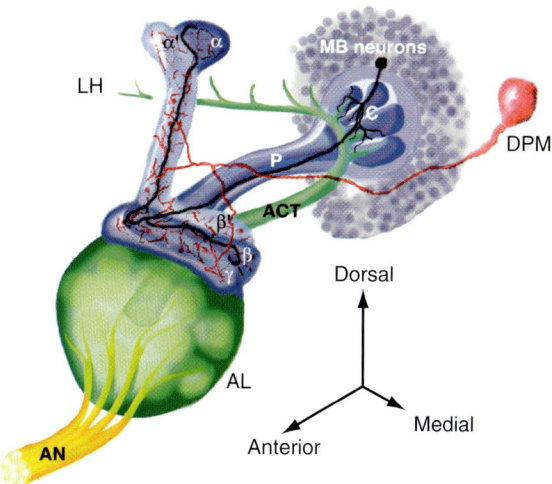

Plate 18.1. Olfactory processing circuit in *Drosophila*. A diagram of the important structures in olfactory learning and memory in the fruit fly (one hemisphere of the brain). AN, antennal nerve; ACT, antennal cerebral tract; LH, lateral horn; C, calyx; MB, mushroom body; P, peduncle; DPM, dorsal paired medial neuron. Adapted with permission from Yu et al. (2005). (For B/W version, see page 295 in this volume.)

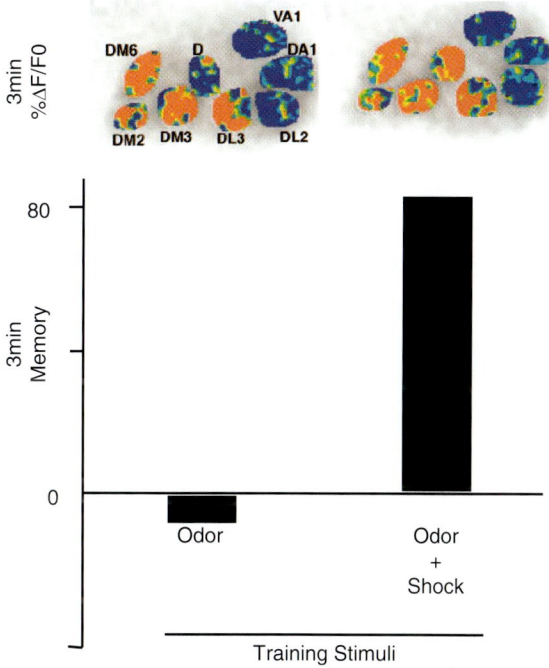

Plate 18.2. The AL memory trace involves recruitment of new neurons into the representation of the learned odor and is correlated with behavioral memory. Flies were trained in an olfactory classical conditioning paradigm capable of generating short-term memory. Flies were trained with stimuli consisting of either odor alone, odor paired with shock, or other protocols (not shown). After training, flies were tested, either using a behavioral memory task or using functional imaging, for their response to the trained odor. Three-minute memory scores revealed that paired odor and shock, but not odor alone, was sufficient to form memory and behavioral avoidance of the trained odor (scores indicate the percentage of flies that demonstrated learned behavior). In the flies that were tested using functional imaging, changes in the pattern of optical reporter activity were also observed in response to training with odor and shock; specifically, activity in the PNs of glomerulus D was only elicited when these stimuli were delivered simultaneously. Pseudocolor images indicate the percentage change in fluorescence of the optical reporter during odor stimulation. In this preparation, eight glomeruli were visible and identifiable. These results correlate the cellular memory trace with learned behavior. Adapted with permission from Yu et al. (2004). (For B/W version, see page 298 in this volume.)

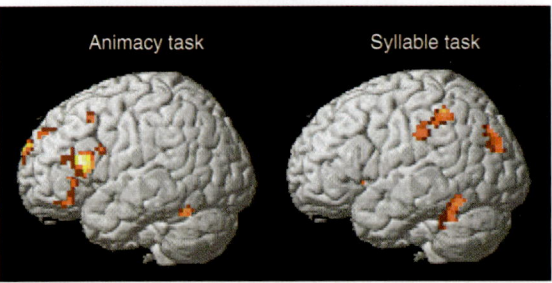

Plate 21.2. Data from Otten and Rugg (2001) illustrating the differently localized subsequent memory effects (thresholded at $p < .001$) associated with words subjected to animacy judgments (left) and syllable judgments (right). Results are overlaid on a rendered canonical brain. (For B/W version, see page 344 in this volume.)

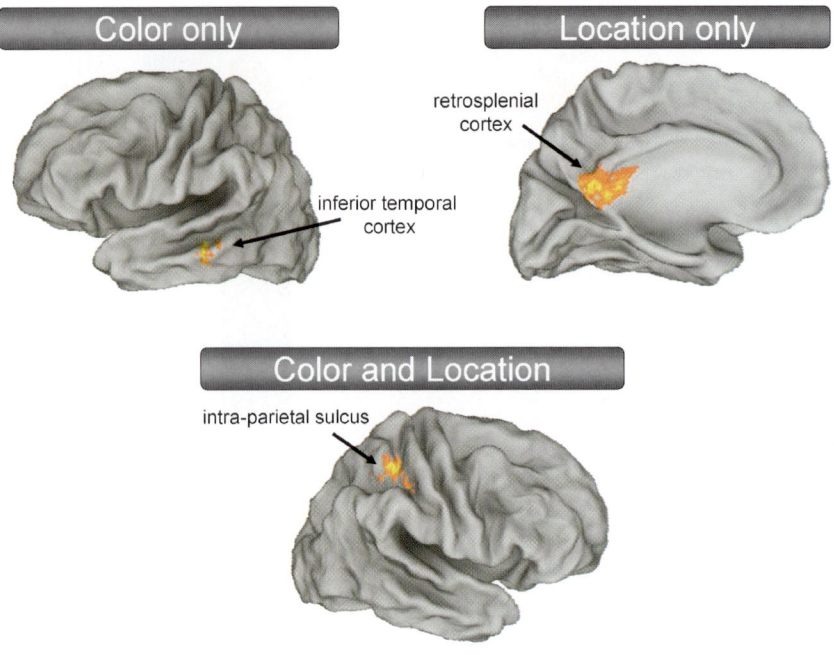

Plate 21.3. Data from Uncapher et al. (2006) illustrating feature-specific and multi-featural subsequent memory effects. Top panel: subsequent memory effects (relative to correctly recognized study words for which no contextual feature could be accurately recalled) associated with accurate memory for the location (left) and color (right) of the study word. Bottom panel: Subsequent memory effects in the right intra-parietal sulcus (IPS) uniquely associated with accurate memory for *both* color and location. All effects thresholded at $p < .001$. Results are displayed onto PALS atlas (Van Essen, 2005) with Caret5 Software (Van Essen et al., 2001; http://brainmap.wustl.edu/caret). (For B/W version, see page 345 in this volume.)

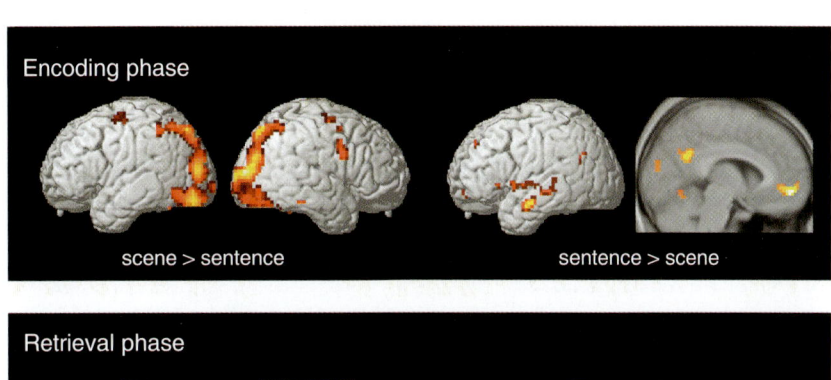
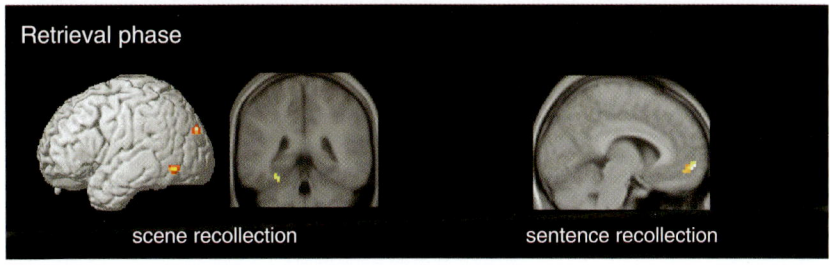

Plate 21.5. Data from Johnson and Rugg (in press) illustrating overlap of content-specific brain activity during encoding and recollection. Top panel: Regions where activity differed during the encoding of words in a scene task compared to a sentence-generation task. Bottom panel: Regions where recollection-related activity (remember > know; thresholded at $p < .001$) was greater for words from one encoding condition ($p < .05$) and overlapped with the corresponding contrast showing greater encoding-related activity for those words ($p < .001$). Results are overlaid on a rendered canonical brain (with the cerebellum artificially removed) and sagittal sections of the across-subjects mean of normalized T_1-weighted images. (For B/W version, see page 349 in this volume.)